DIGITAL FUNDAMENTALS
A SYSTEMS APPROACH

THOMAS L. FLOYD

PEARSON

Boston Columbus Indianapolis New York San Francisco Upper Saddle River
Amsterdam Cape Town Dubai London Madrid Milan Munich Paris Montreal Toronto
Delhi Mexico City São Paulo Sydney Hong Kong Seoul Singapore Taipei Tokyo

Editorial Director: Vernon R. Anthony
Senior Acquisitions Editor: Lindsey Prudhomme
Development Editor: Dan Trudden
Editorial Assistant: Yvette Schlarman
Director of Marketing: David Gesell
Marketing Manager: Harper Coles
Senior Marketing Coordinator: Alicia Wozniak
Senior Marketing Assistant: Les Roberts
Senior Managing Editor: JoEllen Gohr
Senior Project Manager: Rex Davidson
Senior Operations Supervisor: Pat Tonneman
Creative Director: Andrea Nix
Art Director: Diane Y. Ernsberger
Text and Cover Designer: Candace Rowley
Cover Image: Kayros Studio "Be Happy!"/Shutterstock.com
Media Project Manager: Karen Bretz
Full-Service Project Management: Penny Walker/Aptara®, Inc.
Composition: Aptara®, Inc.
Printer/Binder: Courier Kendallville
Cover Printer: Lehigh/Phoenix Color Hagerstown
Text Font: Times Roman

Credits and acknowledgments for materials borrowed from other sources and reproduced, with permission, in this textbook appear on the appropriate page within text.

Library of Congress Cataloging-in-Publication Data

Floyd, Thomas L.
  Digital fundamentals: a systems approach / Thomas L. Floyd.
    p. cm.
  ISBN-13: 978-0-13-293395-7
  ISBN-10: 0-13-293395-0
  1. Digital electronics. I. Title.
  TK7868.D5F534 2013
  621.39'5—dc23
                          2012020323

10 9 8 7 6 5 4 3

PEARSON

ISBN 10:   0-13-293395-0
ISBN 13: 978-0-13-293395-7

CONTENTS

PREFACE

This first edition of *Digital Fundamentals: A Systems Approach* provides a unique coverage of digital technology with a system emphasis. This textbook provides a fundamental grounding in the basic concepts of digital technology and systems reinforced by an abundance of illustrations, examples, applications, and exercises. There are system examples and system notes throughout many chapters in addition to traditional worked examples. Many chapters have a system section that presents a certain type of system and discusses its operation as related to topics covered in that chapter and to its implementation in programmable logic. Most chapters include a troubleshooting section that emphasizes the system approach. Additionally, system level chapters cover digital data transmission; data processing and control; and buses, networks, and interfacing.

Core fundamentals and basic logic functions are presented using a practical approach with emphasis on operation and application rather than on analysis and design. Mathematical topics are limited to only essential coverage that a technician or technologist will need to understand the basic concepts. Programmable logic is emphasized whereas fixed-function logic is introduced on a limited basis.

Features

- Core fundamentals are presented without being intermingled with more advanced topics.
- Many chapters feature an entire section devoted to a specific type of system.
- System examples are used to illustrate how basic concepts and logic elements are applied in a system application.
- System notes present interesting facts and information about system-related issues.
- Multisim is used in selected examples, figures, and problems to provide practice in simulating logic circuits and systems and in troubleshooting.
- Worked examples illustrate core fundamentals and logic functions and require some basic analytical thought.
- Related problems in each worked example relate to the coverage of the example.
- Hands-on-tips (HOT) provide useful and practical information.
- Many chapters include a section on hardware description languages (VHDL and Verilog), which are used to show how logic functions and systems can be described and implemented in a programmable logic device (PLD).
- Many chapters have a troubleshooting section that relates to topics covered in the chapter and emphasizes troubleshooting techniques, and the use of instrumentation, and circuit simulation (Multisim).
- Each chapter begins with a list of sections (Outline), chapter objectives, introduction, key terms list, and website reference.
- Each section within a chapter begins with an introduction and objectives.
- Each section concludes with Checkup exercises that emphasize the main concepts presented in the section.
- Each chapter ends with a summary, key term glossary, true/false quiz, self-test, and sectionalized problem set.
- Answers to related problems, section checkups, true/false quiz, and self-test are at the end of each chapter.
- An end-of-book glossary contains all bold and color terms in the text.

- Answers to odd-numbered problems are at the end of the book
- Website (www.pearsonhighered.com) includes files related to the text such as tutorials and Multisim files.

Student Resources

- *Experiments in Digital Fundamentals: A Systems Approach* (ISBN 0132989840) by David Buchla and Doug Joksch. Lab exercises are coordinated with the text.
- *Multisim Experiments for the DC/AC, Digital, and Devices Courses* (ISBN 0132113880) by Gary Snyder and David Buchla. Students take data, analyze results, and write a conclusion to simulate an actual laboratory experience.
- **Multisim Files Available on the Website**   Circuit files coordinated with this text in Versions 11 and 12 of Multisim are available for download from www.pearsonhighered.com/floyd. Circuit files with prefix F are figure circuits and files with prefix P are problem circuits. Also, a few files are prefixed with E or T, represesting examples or tables.

In order to use the Multisim circuit files, you must have Multisim software installed on your computer. Multisim software is available at **www.ni.com/Multisim**. Although the Multisim circuit files are intended to complement classroom, textbook, and laboratory study, these files are not essential to successfully using this text.

Instructor Resources

Instructor resources are available from Pearson's **Instructor's Resource Center**.

- PowerPoint® slides (ISBN 013298962x) support the topics in each chapter.
- Instructor's Resource Manual (ISBN 0132989832) contains the solutions to the text problems and the solutions to the lab manual.
- TestGen (ISBN 0132988615) This electronic bank of test questions can be used to develop customized quizzes, tests, and/or exams.

To access supplementary materials online, instructors need to request an instructor access code. Go to **www.pearsonhighered.com/irc**, where you can register for an instructor access code. Within 48 hours after registering, you will receive a confirming e-mail, including an instructor access code. Once you have received your code, go to the site and log on for full instructions on downloading the materials you wish to use.

Illustrations of Textbook Features

Chapter Opener A typical chapter opener is shown in Figure P–1.

Worked Example and Related Problem Worked-out examples illustrate basic concepts or specific procedures. A Related Problem reinforces or expands on the content of the example. A typical worked-out example with a Related Problem is shown in Figure P–2.

Section Opener Each section in a chapter begins with a brief introduction that includes a general overview and section objectives, as shown in Figure P–2.

Section Checkup A typical Section Checkup is shown in Figure P–2. (Answers to the Section Checkups are at the end of the chapter.)

Hands On Tip A typical Hands On Tip is shown in Figure P–2.

System Section Typical pages from a System Section are shown in Figure P–3.

FIGURE P–1   Chapter opener.

CHAPTER 9

PROGRAMMABLE LOGIC

OUTLINE

9–1   Simple Programmable Logic Devices (SPLDs)
9–2   Complex Programmable Logic Devices (CPLDs)
9–3   Macrocell Modes
9–4   Field-Programmable Gate Arrays (FPGAs)
9–5   Programmable Logic Software
9–6   Boundary Scan Logic
9–7   Troubleshooting

OBJECTIVES

- Discuss the types of programmable logic, SPLDs and CPLDs, and explain their basic structure
- Describe the basic architecture of two types of SPLDs—the PAL and the GAL
- Explain the basic structure of a programmable logic array (PLA)
- Discuss the operation of macrocells
- Distinguish between CPLDs and FPGAs
- Explain the basic operation of a look-up table (LUT)
- Define *intellectual property* and *platform FPGA*
- Discuss embedded functions
- Show a basic software design flow for a programmable device
- Explain the design flow elements of design entry, functional simulation, synthesis, implementation, timing simulation, and downloading
- Discuss several methods of testing a programmable logic device, including boundary scan logic

KEY TERMS

PAL	Design flow
GAL	Target device
Macrocell	Schematic entry
Registered	Text entry
CPLD	Functional simulation
LAB	Compiler
LUT	Timing simulation
FPGA	Downloading
CLB	Break point
Intellectual property	Boundary scan

INTRODUCTION

The distinction between hardware and software is hazy. Today, new digital circuits are programmed into hardware using languages like VHDL and Verilog. The density (number of equivalent gates on a single chip) has increased dramatically over the past few years. The maximum number of gates in an FPGA (a type of PLD known as a field-programmable gate array) is over 500,000 and doubling every 18 months, according to Moore's law. At the same time, the price for a PLD is decreasing.

PLDs, such as the FPGA, can be used in conjunction with processors and software in an embedded system, or the FPGA can be the sole component with all the

VISIT THE WEBSITE
Study aids for this chapter are available at
http://pearsonhighered.com/floyd

Worked Example with Related Problem

Hands On Tip

Section Checkup

Section Opener

FIGURE P–2   Worked example with Related Problem, Hands On Tip, Section Checkup, and Section Opener.

FIGURE P–3   Partial System Section.

System Example A typical System Example is shown in Figure P–4.

FIGURE P–4   System Example.

System Notes A typical System Note is shown in Figure P–5.

Programmable Logic Coverage The hardware description languages VHDL and Verilog are used in programmable logic applications. Figure P–6 shows two typical pages.

FIGURE P–5   System Note

System Note

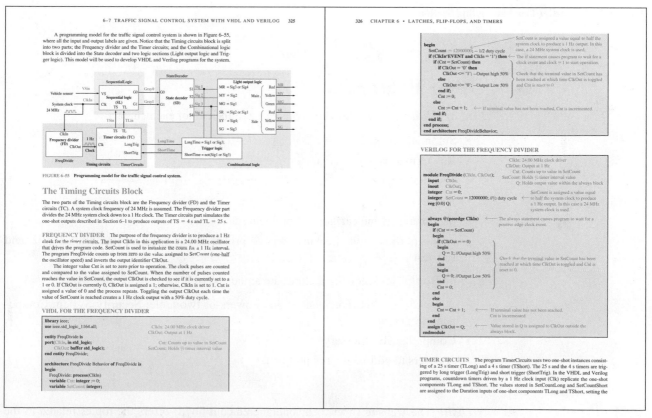

FIGURE P–6   Example pages of system implementation with programmable logic.

Troubleshooting Section A portion of a typical troubleshooting section is shown in Figure P–7.

FIGURE P–7   Partial Troubleshooting Section.

Other Features

End of Chapter The following features are at the end of each chapter.

- Summary
- Key term glossary
- True/False quiz
- Self-test
- Sectionalized and categorized problem set
- Answers to section checkups, related problems for examples, true/false quiz, and self-test

End of Book The following features are at the end of book.

- Appendices: Code conversions and Powers-of-two table; Security System Component programs
- Comprehensive glossary
- Answers to odd-numbered problems
- Index

Website The website (www.pearsonhighered.com/floyd) offers the topics related to the textbook for reference or advanced informations, as shown in Figure P–8.

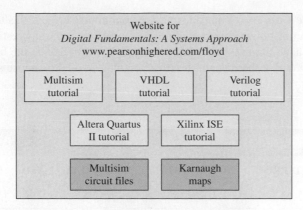

FIGURE P–8   Website for *Digital Fundamentals: A Systems Approach.*

To the Student

Today, it seems that digital technology pervades most everything and is continuously changing. This makes it essential that you obtain a thorough grounding in the fundamentals because even though technology continues to change, the fundamentals remain intact. By fully understanding fundamental concepts, you can adapt to changing conditions.

A digital system is a combination of many logic functions that operate together to produce a desired result. This book not only covers the fundamentals of digital technology but presents several basic systems and shows how fundamental concepts and individual logic devices are used in them. Those working in electronics technology should have a basic grasp of the system concept and a practical knowledge of how to apply the fundamentals.

Today, logic devices are integrated circuits that can be programmed to perform a desired function. Although some fixed-function devices remain, the trend is mainly toward the use and application of programmable logic devices (PLDs). Therefore, two major programming languages used for logic programming are introduced in this text (VHDL and Verilog). The coverage in this book is not intended to make you an expert on the programming languages, but only to familiarize you with some basic concepts. Tutorials are available at the website, and more advanced coverage can be found from many sources.

Anyone working in the field needs to be able to troubleshoot systems. The troubleshooting methods and examples provided in this book will help you to get started as a troubleshooter.

To the Instructor

Time limitations and/or program emphasis generally are major factors in the amount of material and the topics covered in a course. It is not uncommon to omit or condense selected topics or to alter the sequence of certain topics in order to accommodate the requirements of a course. To this end, topics have been organized with a "modular" approach so that certain topics are not integrated or intermingled with the more basic core topics. Topics such as programmable logic, PLD programming, and troubleshooting are contained in separate chapters, dedicated sections within a chapter, or on the website to permit more flexible treatment.

A fairly strong emphasis is placed on programmable logic devices because they are so prevalent in the implementation of today's systems. PLD and programming coverage using VHDL and Verilog is introduced at a fundamental level to provide a basic foundation and is not intended to be a comprehensive treatment. You may choose to cover either VHDL or Verilog or both. Tutorials for PLD programming (VHDL, Verilog, Altera Quartus II, and Xilinx ISE) are provided on the Internet to assist students in their study. Of course, much more extensive coverage of these topics can be found at many Internet sites if that is deemed necessary. Fixed-function logic devices (7400 series for example) are introduced but given a very light treatment due to their declining availability and use.

Customizing the Contents You can structure your course around this text from a minimal coverage to a full-blown coverage. Table P–1 provides some suggestions with Option 1 being minimal and Option 9 being full coverage. You can decide the best approach for your course using one of these suggestions or you may decide to choose some other combination of topics.

Table P–1 • Some suggested combinations of topics. Many others are possible.

OPTIONS	CORE + SYSTEMS	PLDS	PLD PROG	TROUBLESHOOTING	CH 11	CH12	CH13	CH14
1	Yes	No	No	No	No	No	No	No
2	Yes	No	No	Yes	Yes	No	No	No
3	Yes	No	No	Yes	Yes	Yes	No	No
4	Yes	No	No	No	Yes	Yes	Yes	Yes
5	Yes	No	No	Yes	Yes	Yes	Yes	Yes
6	Yes	Yes	No	Yes	No	No	No	No
7	Yes	Yes	Yes	No	No	No	No	No
8	Yes	Yes	Yes	Yes	Yes	No	No	Yes
9	Yes	Yes	Yes	Yes	Yes	Yes	Yes	Yes

Acknowledgments

The concept of this series of systems-oriented textbooks is credited to suggestions and discussions with senior instructional staff at ITT Schools and Vern Anthony at Pearson Education. The staff and others at Pearson Education, by their hard work and dedication, have helped make the textbook a reality. Lois Porter, who did the manuscript editing, has done a great job and she has helped me to turn my original rough manuscript into a top quality marketable product. Rex Davidson skillfully guided the work through its many detailed phases of production to create the end product that you are now looking at. Lindsey Prudhomme, acquisitions editor, and Dan Trudden, development editor, have provided effective overall guidance for this project.

Many thanks also to Gary Snyder and Doug Joksch for their extensive contributions to this book. Doug developed the VHDL and Verilog programs and much of the supporting text in the area of programmable logic. Gary provided the Multisim circuit files and wrote much of the chapter on data processing and control.

TOM FLOYD

CHAPTER 1

INTRODUCTION TO DIGITAL SYSTEMS

OUTLINE

OBJECTIVES

- Explain the basic differences between digital and analog quantities
- Show how voltage levels are used to represent digital quantities
- Describe various parameters of a pulse waveform such as rise time, fall time, pulse width, frequency, period, and duty cycle
- Explain the logic operations of NOT, AND, and OR
- Describe several types of logic functions
- Describe programmable logic, discuss the various types, and describe how PLDs are programmed using VHDL and Verilog with system software
- Describe the basics of a microcontroller
- Identify fixed-function digital integrated circuits according to their technology and the type of packaging
- Discuss how various logic functions are used in a digital system
- Recognize various instruments and understand how they are used in measurement and trouble-shooting digital devices and systems

KEY TERMS

Key terms are in order of appearance in the chapter.

Analog	NOT
Digital	Inverter
Digital system	AND
Binary	OR
Bit	Programmable logic device
Pulse	
Duty cycle	SPLD
Clock	CPLD
Timing diagram	FPGA
Data	Compiler
Serial	Microcontroller
Parallel	Embedded system
Logic	Integrated circuit (IC)
Input	Fixed-function logic
Output	Troubleshooting
Gate	

VISIT THE WEBSITE
Study aids for this chapter are available at
http://pearsonhighered.com/floyd

INTRODUCTION

The term *digital* is derived from the way operations are performed, by counting digits. For many years, applications of digital electronics were confined to computer systems. Today, digital technology is applied in a wide range of systems in addition to computers. Such applications as television, communications systems, radar, navigation and guidance systems, military systems, medical instrumentation, industrial process control, and consumer electronics use digital techniques. Over the years digital technology has progressed from vacuum-tube circuits to fixed-function integrated circuits to programmable logic and embedded microcontrollers.

This chapter introduces you to digital electronics and provides a broad overview of many important concepts, applications, and methods.

1–1 DIGITAL AND ANALOG SIGNALS AND SYSTEMS

Electronic systems can be divided into two broad categories, digital and analog. Digital electronics involves quantities with discrete values, and analog electronics involves quantities with continuous values. Although you will be studying digital fundamentals in this book, you should also know something about analog because many applications require both; and interfacing between analog and digital is important.

After completing this section, you should be able to

- Define *analog*
- Define *digital*
- Explain the difference between digital and analog signals
- State the advantages of digital over analog
- Discuss modulation methods
- Describe two types of digital systems

An **analog**[*] quantity is one having continuous values. A **digital** quantity is one having a discrete set of values. Most things that can be measured quantitatively occur in nature in analog form. For example, the air temperature changes over a continuous range of values. During a given day, the temperature does not go from, say, 70° to 71° instantaneously; it takes on all the infinite values in between. If you graphed the temperature on a typical summer day, you would have a smooth, continuous curve similar to the curve in Figure 1–1. Other examples of analog quantities are time, pressure, distance, and sound.

Rather than graphing the temperature on a continuous basis, suppose you just take a temperature reading every hour. Now you have sampled values representing the temperature at discrete points in time (every hour) over a 24-hour period, as indicated in Figure 1–2. You have effectively converted an analog quantity to a form that can now be digitized by representing each sampled value by a digital code. It is important to realize that Figure 1–2 itself is not the digital representation of the analog quantity.

---

[*]All bold terms are important and are defined in the end-of-book glossary. The bold terms in color are key terms and are included in a Key Term glossary at the end of each chapter.

FIGURE 1–1   Graph of an analog quantity (temperature versus time).

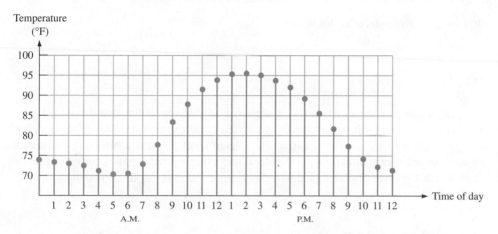

FIGURE 1–2   Sampled-value representation (quantization) of the analog quantity in Figure 1–1. Each value represented by a dot can be digitized by representing it as a digital code that consists of a series of 1s and 0s.

THE DIGITAL ADVANTAGE   Digital representation has certain advantages over analog representation in electronics applications. For one thing, digital data can be processed and transmitted more efficiently and reliably than analog data. Also, digital data has a great advantage when storage is necessary. For example, music when converted to digital form can be stored more compactly and reproduced with greater accuracy and clarity than is possible when it is in analog form. Noise (unwanted voltage fluctuations) does not affect digital data nearly as much as it does analog signals.

Analog Signals

An analog quantity, such as voltage, that is repetitive or varies in a certain manner is an analog signal. An analog signal can be a repetitive waveform, such as the sine wave in Figure 1–3(a), or a continuously varying audio signal that carries information (music, the spoken word, or other sounds), as shown in part (b). Other examples of analog signals are amplitude-modulated signals (AM) and frequency-modulated signals (FM), as illustrated in parts (c) and (d). In AM, a lower-frequency information signal, such as voice, varies the amplitude of a high-frequency sine wave. In FM, the information signal varies the frequency of the sine wave.

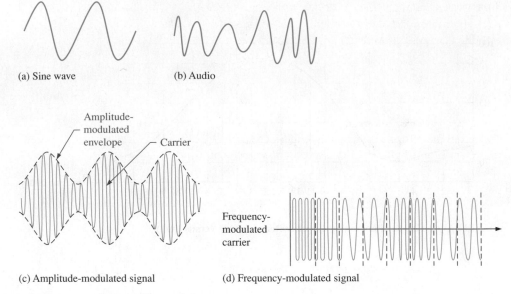

(a) Sine wave   (b) Audio

(c) Amplitude-modulated signal   (d) Frequency-modulated signal

FIGURE 1–3   **Examples of analog signals.**

Digital Signals

A digital signal is a representation of a sequence of discrete values that are coded into a stream of 1s and 0s. A bit stream appears as a train of pulses or voltage levels where a high voltage level conveys a binary 1 and a low voltage level conveys a binary 0. Bit streams are used in telecommunications, computers, and other system applications. Figure 1–4 illustrates one type of digital signal. The duration of each bit (bit time) is indicated by the hash marks.

```
1   1   1   0   0   1   1   1   0   0   0   1   0
```

FIGURE 1–4   **Example of a digital waveform.**

```
1   1   1   0   0   1   1   1   1   0   0   0   1   0
```

FIGURE 1–5   **Example of a digital-modulated signal.**

DIGITAL MODULATION   In some applications, analog and digital signals are combined with a sine wave, called a *carrier*, by amplitude modulating the sine wave with the digital waveform. A common example is a modem that turns digital data from a computer into modulated signals in the voice frequency range for transmission over telephone lines. A digital-modulated signal is shown in Figure 1–5 where the digital signal (bit stream) in Figure 1–4 modulates the sine wave. Dashed lines mark the bit times. The frequency of the sine wave is shown arbitrarily low in relation to the digital-modulating signal for illustration.

PULSE-CODE MODULATION (PCM)   A PCM signal represents sampled analog signals with a sequence of digital codes. It is used in computers for digital audio, in Blu-ray, compact disc and DVD formats, and in digital telephone systems. The sampling process results in a "stair-step" voltage as shown in Figure 1–6. The analog signal is sampled at each step, and each sampled value is converted (quantized) to a digital code. The digital signal would be the time sequence of the digital codes where the binary numbers shown for each step appear in sequence beginning at the left. The more steps there are the more accurate is the digital representation. The length of the code depends on the number of steps.

Digital Systems

A **digital system** is an arrangement of the individual logic functions connected to perform a specified operation or produce a defined output. An example of a digital system is a computer, as shown in Figure 1–7 in basic block diagram form. A computer processes, transfers, and stores data in digital form (1s and 0s). To make a complete system, the computer is interfaced with peripheral devices such as a modem, a mouse, a keyboard, and a monitor.

FIGURE 1–6   **Illustration of pulse-code modulation.**

FIGURE 1–7   **Basic block diagram of a computer.**

Figure 1–8, another example of a digital system, shows the traffic light controller that you will study in Chapter 6. All of the digital signals that the system uses to properly sequence the traffic light are internally generated, making the controller a type of finite state machine.

FIGURE 1–8   **A digital traffic light controller.**

Analog Systems

An **analog system** is one that processes data in analog form only. One example is a public address system, used to amplify sound so that it can be heard by a large audience. The basic diagram in Figure 1–9 illustrates that sound waves, which are analog in nature, are picked up by a microphone and converted to a small analog voltage called the audio signal. This voltage varies continuously as the volume and frequency of the sound changes and is applied to the input of a linear amplifier. The output of the amplifier, which is an increased reproduction of input voltage, goes to the speaker(s). The speaker changes the amplified audio signal back to sound waves that have a much greater volume than the original sound waves picked up by the microphone.

FIGURE 1–9   **A basic audio public address system.**

Another example of an analog system is the FM receiver. The system processes the incoming frequency-modulated carrier signal, extracts the audio signal for amplification, and produces audible sound waves. A block diagram is shown in Figure 1–10 with a representative signal shown at each point in the system.

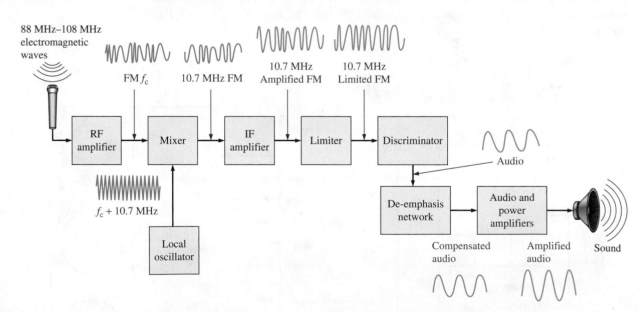

FIGURE 1–10   **Block diagram of superheterodyne FM receiver.**

A Combination Digital and Analog System

The compact disk (CD) player is an example of a system in which both digital and analog elements are used. The simplified block diagram in Figure 1–11 illustrates the basic system. Music in digital form is stored on the compact disk. A laser diode optical system picks up

FIGURE 1–11 Simplified diagram of a compact disk player.

the digital data from the rotating disk and transfers it to the **digital-to-analog converter (DAC).** The DAC changes the digital data into an analog signal that is an electrical reproduction of the original music. This signal is amplified and sent to the speaker for you to enjoy. When the music was originally recorded on the CD, a process, essentially the reverse of the one described here, using an **analog-to-digital converter (ADC)** was used.

SECTION 1–1 CHECKUP*

1. Define *analog.*

2. Define *digital.*

3. Explain the difference between a digital quantity and an analog quantity.

4. Give an example of a system that is analog and one that is a combination of both digital and analog. Name a system that is entirely digital.

*Answers are at the end of the chapter.

1–2 BINARY DIGITS, LOGIC LEVELS, AND DIGITAL WAVEFORMS

Digital systems involve operations in which there are only two possible states. These states are represented by two different voltage levels: A HIGH and a LOW. The two states can also be represented by current levels or pits and lands on a CD or DVD. In digital systems such as computers, combinations of the two states, called *codes,* are used to represent numbers, symbols, alphabetic characters, and other types of information. The two-state number system is called *binary,* and its two digits are 0 and 1. A binary digit is called a *bit.*

After completing this section, you should be able to

- **Define *binary***
- **Define *bit***
- **Name the bits in a binary system**
- **Explain how voltage levels are used to represent bits**
- **Explain how voltage levels are interpreted by a digital circuit**
- **Describe the general characteristics of a pulse**
- **Determine the amplitude, rise time, fall time, and width of a pulse**
- **Identify and describe the characteristics of a digital waveform**
- **Determine the amplitude, period, frequency, and duty cycle of a digital waveform**
- **Explain what a timing diagram is and state its purpose**
- **Explain serial and parallel data transfer and state the advantage and disadvantage of each**

Binary Digits

Each of the two digits in the **binary** system, 1 and 0, is called a **bit**, which is a contraction of the words *binary digit.* In digital circuits, two different voltage levels are used to represent the two bits. Generally, 1 is represented by the higher voltage, which we will refer to as a HIGH, and a 0 is represented by the lower voltage level, which we will refer to as a LOW. This is called **positive logic** and will be used throughout the book.

$$\text{HIGH} = 1 \quad \text{and} \quad \text{LOW} = 0$$

Another system in which a 1 is represented by a LOW and a 0 is represented by a HIGH is called *negative logic.*

Groups of bits (combinations of 1s and 0s), called *codes,* are used to represent numbers, letters, symbols, instructions, and anything else required in a given application.

The concept of a digital computer can be traced back to Charles Babbage, who developed a crude mechanical computation device in the 1830s. John Atanasoff was the first to apply electronic processing to digital computing in 1939. In 1946, an electronic digital computer called ENIAC was implemented with vacuum-tube circuits. Even though it took up an entire room, ENIAC didn't have the computing power of your handheld calculator.

SYSTEM NOTE

Logic Levels

The voltages used to represent a 1 and a 0 are called *logic levels.* Ideally, one voltage level represents a HIGH and another voltage level represents a LOW. In a practical digital circuit, however, a HIGH can be any voltage between a specified minimum value and a specified maximum value. Likewise, a LOW can be any voltage between a specified minimum and a specified maximum. There can be no overlap between the accepted range of HIGH levels and the accepted range of LOW levels.

Figure 1–12 illustrates the general range of LOWs and HIGHs for a digital circuit. The variable $V_{H(max)}$ represents the maximum HIGH voltage value, and $V_{H(min)}$ represents the minimum HIGH voltage value. The maximum LOW voltage value is represented by $V_{L(max)}$, and the minimum LOW voltage value is represented by $V_{L(min)}$. The voltage values between $V_{L(max)}$ and $V_{H(min)}$ are unacceptable for proper operation. A voltage in the unacceptable range can appear as either a HIGH or a LOW to a given circuit. For example, the HIGH input values for a certain type of digital circuit technology called CMOS may range from 2 V to 3.3 V and the LOW input values may range from 0 V to 0.8 V. If a voltage of 2.5 V is applied, the circuit will accept it as a HIGH or binary 1. If a voltage of 0.5 V is applied, the circuit will accept it as a LOW or binary 0. For this type of circuit, voltages between 0.8 V and 2 V are unacceptable.

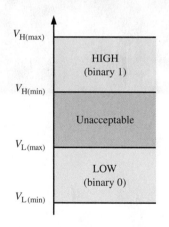

FIGURE 1–12 Logic level ranges of voltage for a digital circuit.

Digital Waveforms

Digital waveforms consist of voltage levels that are changing back and forth between the HIGH and LOW levels or states. Figure 1–13(a) shows that a single positive-going **pulse** is generated when the voltage (or current) goes from its normally LOW level to its HIGH level and then back to its LOW level. The negative-going pulse in Figure 1–13(b) is generated when the voltage goes from its normally HIGH level to its LOW level and back to its HIGH level. A digital waveform is made up of a series of pulses.

THE PULSE   As indicated in Figure 1–13, a pulse has two edges: a **leading edge** that occurs first at time t_0 and a **trailing edge** that occurs last at time t_1. For a positive-going pulse, the leading edge is a rising edge, and the trailing edge is a falling edge. The pulses

(a) Positive–going pulse   (b) Negative–going pulse

FIGURE 1–13   **Ideal pulses.**

in Figure 1–13 are ideal because the rising and falling edges are assumed to change in zero time (instantaneously). In practice, these transitions never occur instantaneously, although for most digital work you can assume ideal pulses.

Figure 1–14 shows a nonideal pulse. In reality, all pulses exhibit some or all of these characteristics. The overshoot and ringing are sometimes produced by stray inductive and capacitive effects. The droop can be caused by stray capacitance and circuit resistance, forming an *RC* circuit with a low time constant.

The time required for a pulse to go from its LOW level to its HIGH level is called the **rise time** (t_r), and the time required for the transition from the HIGH level to the LOW level is called the **fall time** (t_f). In practice, it is common to measure rise time from 10% of the pulse **amplitude** (height from baseline) to 90% of the pulse amplitude and to measure the fall time from 90% to 10% of the pulse amplitude, as indicated in Figure 1–14. The bottom 10% and the top 10% of the pulse are not included in the rise and fall times because of the nonlinearities in the waveform in these areas. The **pulse width** (t_{PW}) is a measure of the duration of the pulse and is often defined as the time interval between the 50% points on the rising and falling edges, as indicated in Figure 1–14.

FIGURE 1–14   **Nonideal pulse characteristics.**

WAVEFORM CHARACTERISTICS   Most waveforms encountered in digital systems are composed of series of pulses, sometimes called *pulse trains,* and can be classified as either periodic or nonperiodic. A **periodic** pulse waveform is one that repeats itself at a fixed interval, called a **period** (**T**). The **frequency** (**f**) is the rate at which it repeats itself and is measured in hertz (Hz). A nonperiodic pulse waveform, of course, does not repeat itself at fixed intervals and may be composed of pulses of randomly differing pulse widths and/or randomly differing time intervals between the pulses. An example of each type is shown in Figure 1–15.

Period $= T_1 = T_2 = T_3 = \ldots = T_n$

Frequency $= \frac{1}{T}$

(a) Periodic (square wave)   (b) Nonperiodic

FIGURE 1–15   **Examples of digital waveforms.**

The frequency (f) of a pulse (digital) waveform is the reciprocal of the period. The relationship between frequency and period is expressed as follows:

$$f = \frac{1}{T} \tag{1-1}$$

$$T = \frac{1}{f} \tag{1-2}$$

An important characteristic of a periodic digital waveform is its **duty cycle,** which is the ratio of the pulse width (t_{PW}) to the period (T). It can be expressed as a percentage.

$$\text{Duty cycle} = \left(\frac{t_{PW}}{T}\right)100\% \tag{1-3}$$

EXAMPLE 1–1

A portion of a periodic digital waveform is shown in Figure 1–16. The measurements are in milliseconds. Determine the following:

(a) period   **(b)** frequency   **(c)** duty cycle

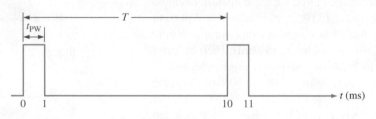

FIGURE 1–16

SOLUTION

(a) The period is measured from the edge of one pulse to the corresponding edge of the next pulse. In this case T is measured from leading edge to leading edge, as indicated. T equals **10 ms.**

(b) $f = \dfrac{1}{T} = \dfrac{1}{10 \text{ ms}} =$ **100 Hz**

(c) Duty cycle $= \left(\dfrac{t_{PW}}{T}\right)100\% = \left(\dfrac{1 \text{ ms}}{10 \text{ ms}}\right)100\% =$ **10%**

RELATED PROBLEM*

A periodic digital waveform has a pulse width of 25 μs and a period of 150 μs. Determine the frequency and the duty cycle.

*Answers are at the end of the chapter.

A Digital Waveform Carries Binary Information

Binary information that is handled by digital systems appears as waveforms that represent sequences of bits. When the waveform is HIGH, a binary 1 is present; when the waveform is LOW, a binary 0 is present. Each bit in a sequence occupies a defined time interval called a **bit time.**

The speed at which a computer can operate depends on the type of microprocessor used in the system. The speed specification, for example 3.5 GHz, of a computer is the maximum clock frequency at which the microprocessor can run.

SYSTEM NOTE

THE CLOCK   In digital systems, all waveforms are synchronized with a basic timing waveform called the **clock**. The clock is a periodic waveform in which each interval between pulses (the period) equals the time for one bit.

An example of a clock waveform is shown in Figure 1–17. Notice that, in this case, each change in level of waveform *A* occurs at the leading edge of the clock waveform. In other cases, level changes occur at the trailing edge of the clock. During each bit time of the clock, waveform *A* is either HIGH or LOW. These HIGHs and LOWs represent a sequence of bits as indicated. A group of several bits can be used as a piece of binary information, such as a number or a letter. The clock waveform itself does not carry information.

FIGURE 1–17   Example of a clock waveform synchronized with a waveform representation of a sequence of bits.

TIMING DIAGRAMS   A **timing diagram** is a graph of digital waveforms showing the actual time relationship of two or more waveforms and how each waveform changes in relation to the others. By looking at a timing diagram, you can determine the states (HIGH or LOW) of all the waveforms at any specified point in time and the exact time that a waveform changes state relative to the other waveforms. Figure 1–18 is an example of a timing diagram made up of four waveforms. From this timing diagram you can see, for example, that the three waveforms *A*, *B*, and *C* are HIGH only during bit time 7 (shaded area) and they all change back LOW at the end of bit time 7.

FIGURE 1–18   Example of a timing diagram.

Data Transfer

Data refers to groups of bits that convey some type of information. Binary data, which are represented by digital waveforms, must be transferred from one circuit to another within a digital system or from one system to another in order to accomplish a given purpose. For example, numbers stored in binary form in the memory of a computer must be transferred to the computer's central processing unit in order to be added. The sum of the addition must then be transferred to a monitor for display and/or transferred back to the memory. In computer systems, as illustrated in Figure 1–19, binary data are transferred in two ways: serial and parallel.

(a) Serial transfer of 8 bits of binary data from computer to modem. Interval t_0 to t_1 is first.

(b) Parallel transfer of 8 bits of binary data from computer to printer. The beginning time is t_0.

FIGURE 1–19 **Illustration of serial and parallel transfer of binary data. Only the data lines are shown.**

When bits are transferred in **serial** form from one point to another, they are sent one bit at a time along a single line, as illustrated in Figure 1–19(a) for the case of a computer-to-modem transfer. During the time interval from t_0 to t_1, the first bit is transferred. During the time interval from t_1 to t_2, the second bit is transferred, and so on. To transfer eight bits in series, it takes eight time intervals.

Universal Serial Bus (USB) is a serial bus standard for device interfacing. It was originally developed for the personal computer but has become widely used on many types of handheld and mobile devices. USB is expected to replace other serial and parallel ports. USB operated at 12 Mbps (million bits per second) when first introduced in 1995, but it now operates at up to 5 Gbps.

SYSTEM NOTE

When bits are transferred in **parallel** form, all the bits in a group are sent out on separate lines at the same time. There is one line for each bit, as shown in Figure 1–19(b) for the example of eight bits being transferred from a computer to a printer or other device. To transfer eight bits in parallel, it takes one time interval compared to eight time intervals for the serial transfer.

To summarize, an advantage of serial transfer of binary data is that a minimum of only one line is required. In parallel transfer, a number of lines equal to the number of bits to be transferred at one time is required. A disadvantage of serial transfer is that it can take longer to transfer a given number of bits than with parallel transfer at the same clock frequency. For example, if one bit can be transferred in 1 μs, then it takes 8 μs to serially transfer eight bits but only 1 μs to parallel transfer eight bits. A disadvantage of parallel transfer is that it takes more lines than serial transfer.

EXAMPLE 1–2

(a) Determine the total time required to serially transfer the eight bits contained in waveform *A* of Figure 1–20, and indicate the sequence of bits. The left-most bit is the first to be transferred. The 1 MHz clock is used as reference.

(b) What is the total time to transfer the same eight bits in parallel?

Clock

A

FIGURE 1–20

SOLUTION

(a) Since the frequency of the clock is 1 MHz, the period is

$$T = \frac{1}{f} = \frac{1}{1 \text{ MHz}} = 1 \ \mu s$$

It takes 1 μs to transfer each bit in the waveform. The total transfer time for 8 bits is

$$8 \times 1 \ \mu s = \mathbf{8 \ \mu s}$$

To determine the sequence of bits, examine the waveform in Figure 1–20 during each bit time. If waveform *A* is HIGH during the bit time, a 1 is transferred. If waveform *A* is LOW during the bit time, a 0 is transferred. The bit sequence is illustrated in Figure 1–21. The left-most bit is the first to be transferred.

| 1 | 1 | 0 | 1 | 0 | 1 | 0 | 0 |

FIGURE 1–21

(b) A parallel transfer would take **1 μs** for all eight bits.

RELATED PROBLEM

If binary data are transferred on a USB at the rate of 480 million bits per second (480 Mbps), how long will it take to serially transfer 16 bits?

SECTION 1–2 CHECKUP

1. Define *binary*.

2. What does *bit* mean?

3. What are the bits in a binary system?

4. How are the rise time and fall time of a pulse measured?

5. Knowing the period of a waveform, how do you find the frequency?

6. Explain what a clock waveform is.

7. What is the purpose of a timing diagram?

8. What is the main advantage of parallel transfer over serial transfer of binary data?

1–3 LOGIC OPERATIONS

In its basic form, logic is the realm of human reasoning that tells you a certain proposition (declarative statement) is true if certain conditions are true. Propositions can be classified as true or false. Many situations and processes that you encounter in your daily life can be expressed in the form of propositional, or logic, functions. Since such functions are true/false or yes/no statements, digital circuits with their two-state characteristics are applicable.

After completing this section, you should be able to

- List three basic logic operations
- Define the NOT operation
- Define the AND operation
- Define the OR operation

Several propositions, when combined, form propositional, or logic, functions. For example, the propositional statement "The light is on" will be true if "The bulb is not burned out" is true and if "The switch is on" is true. Therefore, this logical statement can be made: *The light is on only if the bulb is not burned out and the switch is on.* In this example the first statement is true only if the last two statements are true. The first statement ("The light is on") is then the basic proposition, and the other two statements are the conditions on which the proposition depends.

In the 1850s, the Irish logician and mathematician George Boole developed a mathematical system for formulating logic statements with symbols so that problems can be written and solved in a manner similar to ordinary algebra. Boolean algebra, as it is known today, is applied in the design and analysis of digital systems and will be covered in detail in Chapter 3.

The term **logic** is applied to digital circuits used to implement logic functions. Several kinds of digital logic **circuits** are the basic elements that form the building blocks for such complex digital systems as the computer. We will now look at these elements and discuss their functions in a very general way. Later chapters will cover these circuits in detail.

Three basic logic operations (NOT, AND, and OR) are indicated by standard distinctive shape symbols in Figure 1–22. An alternate standard symbol for each of these logic operations will be introduced in Chapter 3. The lines connected to each symbol are the **inputs** and **outputs.** The inputs are on the left of each symbol and the output is on the right. A circuit that performs a specified logic operation (AND, OR) is called a logic **gate.** AND and OR gates can have any number of inputs, as indicated by the dashes in the figure.

NOT   AND   OR

FIGURE 1–22 The basic logic operations and symbols.

In logic operations, the true/false conditions mentioned earlier are represented by a HIGH (true) and a LOW (false). Each of the three basic logic operations produces a unique response to a given set of conditions.

NOT

The **NOT** operation changes one logic level to the opposite logic level, as indicated in Figure 1–23. When the input is HIGH (1), the output is LOW (0). When the input is LOW,

HIGH (1) ———▷o— LOW (0)   LOW (0) ———▷o— HIGH (1)

FIGURE 1–23 The NOT operation.

the output is HIGH. In either case, the output is *not* the same as the input. The NOT operation is implemented by a logic circuit known as an **inverter.**

AND

The **AND** operation produces a HIGH output only when all the inputs are HIGH, as indicated in Figure 1–24 for the case of two inputs. When one input is HIGH *and* the other input is HIGH, the output is HIGH. When any or all inputs are LOW, the output is LOW. The AND operation is implemented by a logic circuit known as an *AND gate*.

FIGURE 1–24   The AND operation.

OR

The **OR** operation produces a HIGH output when one or more inputs are HIGH, as indicated in Figure 1–25 for the case of two inputs. When one input is HIGH *or* the other input is HIGH *or* both inputs are HIGH, the output is HIGH. When both inputs are LOW, the output is LOW. The OR operation is implemented by a logic circuit known as an *OR gate*.

FIGURE 1–25   The OR operation.

SECTION 1–3 CHECKUP

1. When does the NOT operation produce a HIGH output?

2. When does the AND operation produce a HIGH output?

3. When does the OR operation produce a HIGH output?

4. What is an inverter?

5. What is a logic gate?

1-4 COMBINATIONAL AND SEQUENTIAL LOGIC FUNCTIONS

The three basic logic elements AND, OR, and NOT can be combined to form various types of logic functions: comparison, arithmetic, code conversion, encoding, decoding, data selection, counting, and storage. This section provides an overview of important logic functions and illustrates how they can be used in a specific system.

After completing this section, you should be able to

- List several types of logic functions
- Describe comparison and list the four arithmetic functions
- Describe code conversion, encoding, and decoding
- Describe multiplexing and demultiplexing
- Describe the counting function
- Describe the storage function

The Comparison Function

Magnitude comparison is performed by a logic circuit called a **comparator**. A comparator compares two quantities and indicates whether or not they are equal. For example, suppose you have two numbers and wish to know if they are equal or not equal and, if not equal, which is greater. The comparison function is represented in Figure 1–26. One number in binary form (represented by logic levels) is applied to input A, and the other number in binary form (represented by logic levels) is applied to input B. The outputs indicate the relationship of the two numbers by producing a HIGH level on the proper output line. Suppose that a binary representation of the number 2 is applied to input A and a binary representation of the number 5 is applied to input B. (The binary representation of numbers and symbols is discussed in Chapter 2.) A HIGH level will appear on the $A < B$ (A is less than B) output, indicating the relationship between the two numbers (2 is less than 5). The wide arrows represent a group of parallel lines on which the bits are transferred.

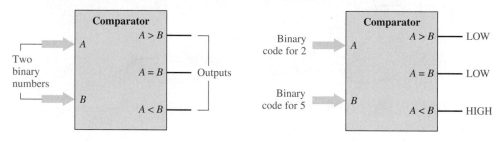

(a) Basic magnitude comparator

(b) Example: A is less than B ($2 < 5$) as indicated by the HIGH output ($A < B$)

FIGURE 1–26   The comparison function.

The Arithmetic Functions

ADDITION   Addition is performed by a logic circuit called an **adder.** An adder adds two binary numbers (on inputs A and B) with a carry input C_{in} and generates a sum (Σ) and a carry output (C_{out}), as shown in Figure 1–27(a). Figure 1–27(b) illustrates the addition of 3 and 9. You know that the sum is 12; the adder indicates this result by producing the code for 2 on the sum output and 1 on the carry output. Assume that the carry input in this example is 0.

SUBTRACTION   Subtraction is also performed by a logic circuit. A **subtracter** requires three inputs: the two numbers that are to be subtracted and a borrow input. The

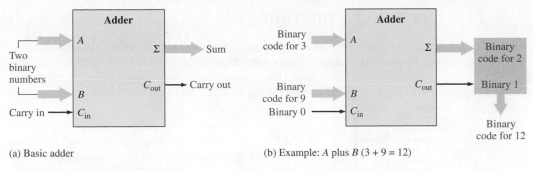

(a) Basic adder (b) Example: *A* plus *B* (3 + 9 = 12)

FIGURE 1–27   The addition function.

two outputs are the difference and the borrow output. When, for instance, 5 is subtracted from 8 with no borrow input, the difference is 3 with no borrow output. You will see in Chapter 2 how subtraction can actually be performed by an adder because subtraction is simply a special case of addition.

MULTIPLICATION   Multiplication is performed by a logic circuit called a *multiplier*. Numbers are always multiplied two at a time, so two inputs are required. The output of the multiplier is the product. Because multiplication is simply a series of additions with shifts in the positions of the partial products, it can be performed by using an adder in conjunction with other circuits.

In a microprocessor, the arithmetic logic unit (ALU) performs the operations of add, subtract, multiply, and divide as well as the logic operations on digital data as directed by a series of instructions. A typical ALU is constructed of many thousands of logic gates.

SYSTEM NOTE

DIVISION   Division can be performed with a series of subtractions, comparisons, and shifts, and thus it can also be done using an adder in conjunction with other circuits. Two inputs to the divider are required, and the outputs generated are the quotient and the remainder.

The Code Conversion Function

A **code** is a set of bits arranged in a unique pattern and used to represent specified information. A code converter changes one form of coded information into another coded form. Examples are conversion between binary and other codes such as the binary coded decimal (BCD) and the Gray code.

The Encoding Function

The encoding function is performed by a logic circuit called an **encoder.** The encoder converts information, such as a decimal number or an alphabetic character, into some coded form. For example, one certain type of encoder converts each of the decimal digits, 0 through 9, to a binary code. A HIGH level on the input corresponding to a specific decimal digit produces logic levels that represent the proper binary code on the output lines.

Figure 1–28 is a simple illustration of an encoder used to convert (encode) a calculator keystroke into a binary code that can be processed by the calculator circuits.

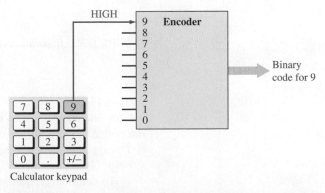

FIGURE 1–28   A calculator keystroke encoded into a binary code for processing by the calculator system.

The Decoding Function

The decoding function is performed by a logic circuit called a **decoder.** The decoder converts coded information, such as a binary number, into a noncoded form, such as a decimal

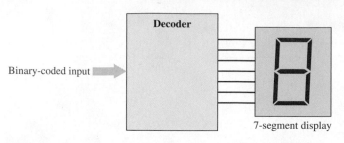

form. For example, one particular type of decoder converts a 4-bit binary code into the appropriate decimal digit.

Figure 1–29 is a simple illustration of one type of decoder that is used to activate a 7-segment display. Each of the seven segments of the display is connected to an output line from the decoder. When a particular binary code appears on the decoder inputs, the appropriate output lines are activated and light the proper segments to display the decimal digit corresponding to the binary code.

FIGURE 1–29   A decoder used to convert a special binary code into a 7-segment decimal readout.

The Data Selection Function

Two types of circuits that select data are the multiplexer and the demultiplexer. The **multiplexer,** or mux for short, is a logic circuit that switches digital data from several input lines onto a single output line in a specified time sequence. Functionally, a multiplexer can be represented by an electronic switch operation that sequentially connects each of the input lines to the output line. The **demultiplexer** (demux) is a logic circuit that switches digital data from one input line to several output lines in a specified time sequence. Essentially, the demux is a mux in reverse.

Multiplexing and demultiplexing are used when data from several sources are to be transmitted over one line to a distant location and redistributed to several destinations. Figure 1–30 illustrates this type of application where digital data from three sources are sent out along a single line to three terminals at another location.

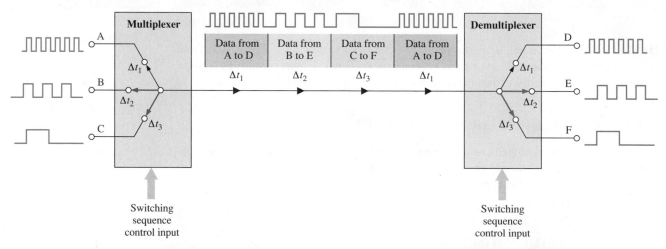

FIGURE 1–30   Illustration of a basic multiplexing/demultiplexing application.

In Figure 1–30, data from input A are connected to the output line during time interval Δt_1 and transmitted to the demultiplexer that connects them to output D. Then, during interval Δt_2, the multiplexer switches to input B and the demultiplexer switches to output E. During interval Δt_3, the multiplexer switches to input C and the demultiplexer switches to output F.

To summarize, during the first time interval, input A data go to output D. During the second time interval, input B data go to output E. During the third time interval, input C data go to output F. After this, the sequence repeats. Because the time is divided up among several sources and destinations where each has its turn to send and receive data, this process is called *time division multiplexing* (TDM).

The Memory and Storage Functions

Memory and **storage** are functions that are required in most digital systems, and their purpose is to retain binary data for a period of time. Generally, *memory* refers to relatively short-term data retention, and *storage* refers to long-term data retention. A storage device can "memorize" a bit or a group of bits and retain the information as long as necessary. Memories include flip-flops, registers, and semiconductor memory. Storage includes magnetic disks (hard drives), optical disks (CDs), and magnetic tape.

FLIP-FLOPS   A **flip-flop** is a bistable (two stable states) logic circuit that can store only one bit at a time, either a 1 or a 0. The output of a flip-flop indicates which bit it is storing. A HIGH output indicates that a 1 is stored and a LOW output indicates that a 0 is stored. Flip-flops are implemented with logic gates and are covered in Chapter 6.

REGISTERS   A **register** is formed by combining several flip-flops so that groups of bits can be stored. For example, an 8-bit register is constructed from eight flip-flops. In addition to storing bits, registers can be used to shift the bits from one position to another within the register or out of the register to another circuit; therefore, these devices are known as *shift registers.*

   The two basic types of shift registers are serial and parallel. The bits are stored in a serial shift register one at a time. A good analogy to the serial shift register is loading passengers onto a bus single file through the door. They also exit the bus single file. The bits are stored in a parallel register simultaneously from parallel lines. For this case, a good analogy is loading and unloading passengers on a roller coaster where they enter all of the cars in parallel and exit in parallel.

SEMICONDUCTOR MEMORIES   Semiconductor memories are devices typically used for storing large numbers of bits. In one type of memory, called the *read-only memory* or ROM, the binary data are permanently or semipermanently stored and cannot be readily changed. In the *random-access memory* or RAM, the binary data are temporarily stored and can be easily changed. Memories are covered in Chapter 10.

MAGNETIC MEMORIES   Magnetic disk memories are used for mass storage of binary data. An example is the computer's internal hard disk. Magneto-optical disks use laser beams to store and retrieve data. Magnetic tape is still used in memory applications and for backing up data from other storage devices.

> The internal computer memories, RAM and ROM, as well as the smaller caches are semiconductor memories. The registers in a microprocessor are constructed of semiconductor flip-flops. Opto-magnetic disk memories are used in the internal hard drive and for the CD-ROM.
>
> SYSTEM NOTE

The Counting Function

A **counter** is a sequential device and is a type of state machine because it has a unique internal sequence of states. The counting function is important in digital systems. There are many types of digital counters, but their basic purpose is to count events or to generate sequences represented by changing levels or pulses. To count, the counter must "remember" the present number so that it can go to the next proper number in sequence. Therefore, storage capability is an important characteristic of all counters, and flip-flops are generally used to implement them. Figure 1–31 illustrates the basic idea of counter operation.

Parallel
output lines

| | 1 | 2 | 3 | 4 | 5 |
| | | | | | |

Input pulses → Counter → | Binary code for 1 | Binary code for 2 | Binary code for 3 | Binary code for 4 | Binary code for 5 |

Sequence of binary codes that represent
the number of input pulses counted.

FIGURE 1–31   **Illustration of basic counter operation.**

SECTION 1–4 CHECKUP

1. What does a comparator do?

2. What are the four basic arithmetic operations?

3. Describe encoding and give an example.

4. Describe decoding and give an example.

5. Explain the basic purpose of multiplexing and demultiplexing.

6. Name four types of memory and storage devices.

7. What does a counter do?

1–5 PROGRAMMABLE LOGIC

A **programmable logic device (PLD)** is a type of integrated circuit (IC) that starts as a "blank slate" and into which a logic design is programmed. Programmable logic requires both hardware and software. PLDs can be programmed to perform specified logic functions by the manufacturer or by the user. One advantage of programmable logic over fixed-function logic is that the devices use much less board space for an equivalent amount of logic. Another advantage is that, with programmable logic, designs can be readily changed without rewiring or replacing components. Also, a logic design can generally be implemented faster and with less cost with programmable logic than with fixed-function ICs.

After completing this section, you should be able to

* State the major types of programmable logic and discuss the differences
* Discuss methods of programming
* List the major programming languages used for programmable logic
* Discuss the programmable logic design process

Types of Programmable Logic Devices (PLDs)

Many types of programmable logic devices are available, ranging from small devices that can replace a few fixed-function devices to complex high-density devices that can replace thousands of fixed-function devices. Two major categories of user-programmable logic are **PLD** (programmable logic device) and **FPGA** (field-programmable gate array), as indicated in Figure 1–32. PLDs are either SPLDs (simple PLDs) or CPLDs (complex PLDs).

SIMPLE PROGRAMMABLE LOGIC DEVICE (SPLD)   The SPLD was the original PLD and is still available for small-scale applications. Generally, an **SPLD** can replace up to ten fixed-function ICs and their interconnections, depending on the type of functions and the specific SPLD. Most SPLDs are in one of two categories: PAL and GAL. A **PAL** (programmable array logic) is a device that can be programmed one time. It consists of a programmable array of AND gates and a fixed array of OR gates, as shown in Figure 1–33(a). A **GAL** (generic array logic) is a device that is basically a PAL that can be

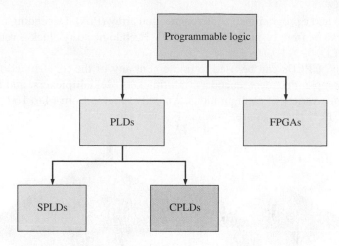

FIGURE 1–32   **Programmable logic hierarchy.**

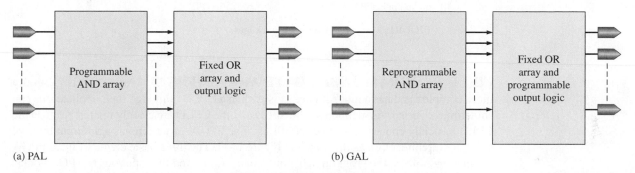

(a) PAL

(b) GAL

FIGURE 1–33   **Block diagrams of simple programmable logic devices (SPLDs).**

reprogrammed many times. It consists of a reprogrammable array of AND gates and a fixed array of OR gates with programmable ouputs, as shown in Figure 1–33(b). A typical SPLD package is shown in Figure 1–34 and generally has from 24 to 28 pins.

FIGURE 1–34   **A typical SPLD package.**

COMPLEX PROGRAMMABLE LOGIC DEVICE (CPLD)   As technology progressed and the amount of circuitry that could be put on a chip (chip density) increased, manufacturers were able to put more than one SPLD on a single chip and the CPLD was born. Essentially, the **CPLD** is a device containing multiple SPLDs and can replace many fixed-function ICs. Figure 1–35 shows a basic CPLD block diagram with four logic array

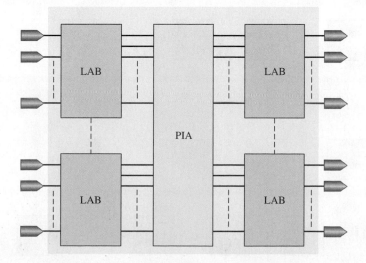

FIGURE 1–35   **General block diagram of a CPLD.**

blocks (LABs) and a programmable interconnection array (PIA). Depending on the specific CPLD, there can be from two to sixty-four LABs. Each logic array block is roughly equivalent to one SPLD.

Generally, CPLDs can be used to implement any of the logic functions discussed earlier, for example, decoders, encoders, multiplexers, demultiplexers, and adders. They are available in a variety of configurations, typically ranging from 44 to 160 pin packages. Examples of CPLD packages are shown in Figure 1–36.

(a) 80-pin PQFP   (b) 128-pin PQFP

FIGURE 1–36   **Typical CPLD packages.**

FIELD-PROGRAMMABLE GATE ARRAY (FPGA)   An **FPGA** is generally more complex and has a much higher density than a CPLD, although their applications can sometimes overlap. As mentioned, the SPLD and the CPLD are closely related because the CPLD basically contains a number of SPLDs. The FPGA, however, has a different internal structure (architecture), as illustrated in Figure 1–37. The three basic elements in an FPGA are the logic block, the programmable interconnections, and the input/output (I/O) blocks.

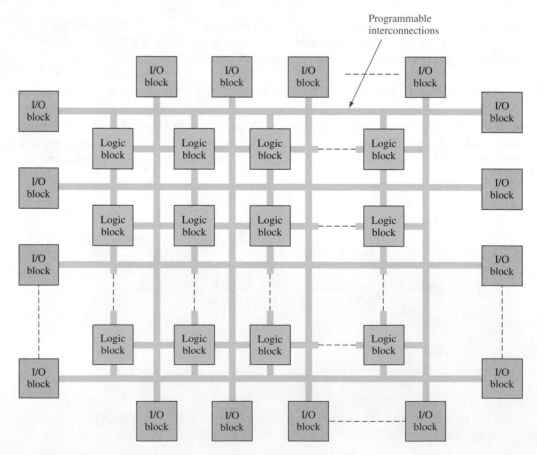

FIGURE 1–37   **Basic structure of an FPGA.**

The logic blocks in an FPGA are not as complex as the logic array blocks (LABs) in a CPLD, but generally there are many more of them. When the logic blocks are relatively simple, the FPGA architecture is called *fine-grained.* When the logic blocks are larger and more complex, the architecture is called *coarse-grained.* The I/O blocks are on the outer edges of the structure and provide individually selectable input, output, or bidirectional access to the outside world. The distributed programmable interconnection matrix provides for interconnection of the logic blocks and connection to inputs and outputs. Large FPGAs can have tens of thousands of logic blocks in addition to memory and other resources. A typical FPGA ball-grid array package is shown in Figure 1–38. These types of packages can have over 1000 input and output pins.

(a) Top view   (b) Bottom view

FIGURE 1–38   **A typical ball-grid array (BGA) package.**

The Programming Process

An SPLD, CPLD, or FPGA can be thought of as a "blank slate" on which you implement a specified system design using a certain process. This process requires a software development package installed on a computer to implement a circuit design in the programmable chip. The computer must be interfaced with a development board or programming fixture containing the device, as illustrated in Figure 1–39.

PLD development board

Programmable logic device

FIGURE 1–39   **Basic setup for programming a PLD or FPGA.**   (Photo courtesy of Digilent, Inc.)

Several steps, called the *design flow,* are involved in the process of implementing a digital logic design in a programmable logic device. A block diagram of a typical programming process is shown in Figure 1–40. As indicated, the design flow has access to a design library.

FIGURE 1–40 **Basic programming flow block diagram.**

DESIGN ENTRY This is the first programming step. The circuit or system design must be entered into the design application software using text-based entry, graphic entry (schematic capture), or state diagram description. Design entry is device independent. Text-based entry is accomplished with a hardware description language (HDL) such as VHDL, Verilog, or AHDL. Graphic (schematic) entry allows prestored logic functions from a library to be selected, placed on the screen, and then interconnected to create a logic design. State-diagram entry requires specification of both the states through which a sequential logic circuit progresses and the conditions that produce each state change.

Once a design has been entered, it is compiled. A **compiler** is a program that controls the design flow process and translates source code into object code in a format that can be logically tested or downloaded to a target device. The source code is created during design entry, and the object code is the final code that actually causes the design to be implemented in the programmable device.

FUNCTIONAL SIMULATION The entered and compiled design is simulated by software to confirm that the logic circuit functions as expected. The functional simulation will verify that correct outputs are produced for a specified set of inputs. A device-independent software tool for doing this is generally called a *waveform editor*. Any flaws demonstrated by the simulation would be corrected by going back to design entry and making appropriate changes.

SYNTHESIS **Synthesis** is where the design is translated into a netlist, which has a standard form and is device independent.

IMPLEMENTATION **Implementation** is where the logic structures described by the netlist are mapped into the actual structure of the specific device being programmed. The implementation process is called *fitting* or *place and route* and results in an output called a bitstream, which is device dependent.

TIMING SIMULATION This step comes after the design is mapped into the specific device. The timing simulation is basically used to confirm that there are no design flaws or timing problems due to propagation delays.

DOWNLOAD Once a bitstream has been generated for a specific programmable device, it has to be downloaded to the device to implement the software design in hardware. Some programmable devices have to be installed in a special piece of equipment called a *device programmer* or on a development board. Other types of devices can be programmed while in a system—called in-system programming (ISP)—using a standard JTAG (Joint Test Action Group) interface. Some devices are volatile, which means they lose their contents when reset or when power is turned off. In this case, the bitstream data must be stored in a memory and reloaded into the device after each reset or power-off. Also, the contents of an ISP device can be manipulated or upgraded while it is operating in a system. This is called "on-the-fly" reconfiguration.

The Microcontroller

A microcontroller is different than a PLD. The internal circuits of a microcontroller are fixed, and a program (series of instructions) directs the microcontroller operation in order to achieve a specific outcome. The internal circuitry of a PLD is programmed into it, and once programmed, the circuitry performs required operations. Thus, a program determines microcontroller operation, but in a PLD a program determines the logic function. Microcontrollers are generally programmed with either the C language or the BASIC language.

A **microcontroller** is basically a special-purpose small computer. Microcontrollers are generally used for embedded system applications. An **embedded system** is one that is designed to perform one or a few dedicated functions. By contrast, a general-purpose computer, such as a laptop, is designed to perform a wide range of functions and applications.

Embedded microcontrollers are used in many common applications. The embedded microcontroller is part of a complete system, which may include additional electronics and mechanical parts. For example, a microcontroller in a television set displays the input from the remote unit on the screen and controls the channel selection, audio, and various menu adjustments like brightness and contrast. In an automobile a microcontroller takes engine sensor inputs and controls spark timing and fuel mixture. Other applications include home appliances, thermostats, cell phones, and toys.

SECTION 1–5 CHECKUP

1. List three major categories of programmable logic devices and specify their acronyms.

2. How does a CPLD differ from an SPLD?

3. Name the steps in the programming process.

4. Briefly explain each step named in question 3.

5. What are the two main functional characteristics of a microcontroller?

1–6 FIXED-FUNCTION LOGIC DEVICES

All the logic elements and functions that have been discussed are generally available in integrated circuit (IC) form. A fixed-function device is one that cannot be programmed like a PLD. Digital systems have incorporated ICs for many years because of their small size, high reliability, low cost, and low power consumption. It is important to be able to recognize the IC packages and to know how the pin connections are numbered, as well as to be familiar with the way in which circuit complexities and circuit technologies determine the various IC classifications.

After completing this section, you should be able to

- **Recognize the difference between through-hole devices and surface-mount fixed-function devices**
- **Identify dual in-line packages (DIP)**
- **Identify small-outline integrated circuit packages (SOIC)**
- **Identify plastic leaded chip carrier packages (PLCC)**
- **Identify leadless ceramic chip carrier packages (LCC)**
- **Determine pin numbers on various types of IC packages**
- **Explain the complexity classifications for fixed-function ICs**

A monolithic **integrated circuit (IC)** is an electronic circuit that is constructed entirely on a single small chip of silicon. All the components that make up the circuit—transistors, diodes, resistors, and capacitors—are an integral part of that single chip. Fixed-function logic and programmable logic are two broad categories of digital ICs. In **fixed-function logic**, the logic functions are set by the manufacturer and cannot be altered.

Figure 1–41 shows a cutaway view of one type of fixed-function IC package with the circuit chip shown within the package. Points on the chip are connected to the package pins to allow input and output connections to the outside world.

FIGURE 1–41   Cutaway view of one type of fixed-function surface-mount IC package, showing the chip mounted inside and connections to input and output pins.

(a) Dual in-line package (DIP)   (b) Small-outline IC (SOIC)

FIGURE 1–42   Examples of through-hole (DIP) and surface-mounted devices. The DIP is larger than the SOIC with the same number of leads.

IC Packages

Integrated circuit (IC) packages are classified according to the way they are mounted on printed circuit (PC) boards as either through-hole mounted or surface mounted. The through-hole type packages have pins (leads) that are inserted through holes in the PC board and can be soldered to conductors on the opposite side. The most common type of through-hole package is the dual in-line package (**DIP**) shown in Figure 1–42(a). The DIP is useful in lab work because it plugs in easily to a protoboard.

Most IC packages use surface-mount technology (**SMT**). Surface mounting is a space-saving alternative to through-hole mounting. The holes through the PC board are unnecessary for SMT. The pins of surface-mounted packages are soldered directly to conductors on one side of the board, leaving the other side free for additional circuits. Also, for a circuit with the same number of pins, a surface-mounted package is much smaller than a dual in-line package because the pins are placed closer together. An example of a surface-mounted package is the small-outline integrated circuit (**SOIC**) shown in Figure 1–42(b).

Various types of SMT packages are available in a range of sizes, depending on the number of leads (more leads are required for more complex circuits and lead configurations). Examples of several types are shown in Figure 1–43. As you can see, the leads of the **SSOP** (shrink small-outline package) are formed into a "gull-wing" shape. The leads of the **PLCC** (plastic-leaded chip carrier) are turned under the package in a J-type shape. Instead of leads, the **LCC** (leadless ceramic chip) has metal contacts molded into its ceramic body. The LQFP also has gull-wing leads. Both the CSP (chip scale package) and the FBGA (fine-pitch ball grid array) have contacts embedded in the bottom of the package.

(a) SSOP (153 × 193 mils)   (b) PLCC (350 × 350 mils)   (c) LCC (350 × 350 mils)

(d) LQFP (7 × 7 mm)   (e) Laminate CSP (3.5 × 3.5 mm)   (f) FBGA (4 × 4 mm)

FIGURE 1–43   Typical SMT package configurations. Parts (e) and (f) show bottom views.

Pin Numbering

All IC packages have a standard format for numbering the pins (leads). The dual in-line packages (DIPs) and the shrink small-outline packages (SSOP) have the numbering arrangement illustrated in Figure 1–44(a) for a 16-pin package. Looking at the top of the package, pin 1 is indicated by an identifier that can be either a small dot, a notch, or a beveled edge. The dot is always next to pin 1. Also, with the notch oriented upward, pin 1 is always the top left pin, as indicated. Starting with pin 1, the pin numbers increase as you go down, then across and up. The highest pin number is always to the right of the notch or opposite the dot.

The PLCC and LCC packages have leads arranged on all four sides. Pin 1 is indicated by a dot or other index mark and is located at the center of one set of leads. The pin numbers increase going counterclockwise as viewed from the top of the package. The highest pin number is always to the right of pin 1. Figure 1–44(b) illustrates this format for a 20-pin PLCC package.

(a) DIP or SSOP

(b) PLCC or LCC

FIGURE 1–44 **Pin numbering for standard types of IC packages. Top views are shown.**

Complexity Classifications for Fixed-Function ICs

Fixed-function digital ICs are classified according to their complexity. They are listed here from the least complex to the most complex. The complexity figures stated here for SSI, MSI, LSI, VLSI, and ULSI are generally accepted, but definitions may vary from one source to another.

- **Small-scale integration (SSI)** describes fixed-function ICs that have up to ten equivalent gate circuits on a single chip, and they include basic gates and flip-flops.

- **Medium-scale integration (MSI)** describes integrated circuits that have from 10 to 100 equivalent gates on a chip. They include logic functions such as encoders, decoders, counters, registers, multiplexers, arithmetic circuits, small memories, and others.

- **Large-scale integration (LSI)** is a classification of ICs with complexities of from more than 100 to 10,000 equivalent gates per chip, including memories.

- **Very large-scale integration (VLSI)** describes integrated circuits with complexities of from more than 10,000 to 100,000 equivalent gates per chip.

- **Ultra large-scale integration (ULSI)** describes very large memories, larger **microprocessors,** and larger single-chip computers. Complexities of more than 100,000 equivalent gates per chip are classified as ULSI.

Integrated Circuit Technologies

The types of transistors with which all integrated circuits are implemented are either MOSFETs (metal-oxide semiconductor field-effect transistors) or bipolar junction transistors. A circuit technology that uses MOSFETs is **CMOS** (complementary MOS). **Bipolar** is a type of fixed-function digital circuit technology that uses bipolar junction transistors and is sometimes called **TTL** (transistor-transistor logic). **BiCMOS** uses a combination of both CMOS and bipolar. All the types of logic gates and logic functions that have been discussed are generally available as ICs.

All gates and other functions can be implemented with either type of circuit technology. SSI and MSI circuits are generally available in both CMOS and bipolar in the 74XX series, but CMOS is the most common.

SECTION 1–6 CHECKUP

1. What is an integrated circuit?

2. Define the terms DIP, SMT, SOIC, SSI, MSI, LSI, and VLSI.

3. Generally, in what classification does a fixed-function IC with the following number of equivalent gates fall?

(a) 10   (b) 75   (c) 500   (d) 15,000   (e) 200,000

A tablet-bottling system illustrates how the logic functions covered in this chapter can be used in a system environment. The functions used in this system are the encoder, decoder, code converter, adder, multiplexer, demultiplexer, register, and counter. This system could be implemented in three ways: with a PLD, with a microcontroller, or with fixed-function ICs. The first two are how all digital systems are currently implemented.

After completing this section, you should be able to

- Understand basic system operation and how certain components work together
- Explain the purpose of each logic function in the total system
- Describe the transfer of digital data throughout the system

A Process Control System

A system for bottling vitamin tablets is shown in the block diagram of Figure 1–45. To begin, the tablets are fed into a large funnel-type hopper. The narrow neck of the hopper creates a serial flow of tablets into a bottle on the conveyor belt below. Only one tablet at a time passes the sensor, so the tablets can be counted.

The system controls the number of tablets into each bottle and displays a continually updated readout of the total number of tablets bottled. This system utilizes all of the basic logic functions that have been introduced and illustrates how these functions can be connected to work together to produce a specified result. This system is purely for instructional purposes and is not intended to necessarily represent the most efficient or best way to implement the operation.

GENERAL OPERATION   The maximum number of tablets per bottle is entered from the keypad, changed to a code by the *Encoder,* and stored in *Register A. Decoder A* changes the code stored in the register to a form appropriate for turning on the display. *Code converter A* changes the code to a binary number and applies it to the *A* input of the *Comparator* (Comp).

An optical sensor in the neck of the hopper detects each tablet that passes and produces a pulse. This pulse goes to the *Counter* and advances it by one count; thus, any time during the filling of a bottle, the binary state of the counter represents the number of tablets in the bottle. The binary count is transferred from the counter to the *B* input of the comparator (Comp). The *A* input of the comparator is the binary number for the maximum tablets per bottle. Now, let's say that the present number of tablets per bottle is 50. When the binary number in the counter reaches 50, the $A = B$ output of the comparator goes HIGH, indicating that the bottle is full.

The HIGH output of the comparator causes the valve in the neck of the hopper to close and stop the flow of tablets. At the same time, the HIGH output of the comparator activates the conveyor, which moves the next empty bottle into place under the hopper. When the bottle is in place, the conveyor control issues a pulse that resets the counter to zero. As a result, the output of the comparator goes back LOW and causes the hopper valve to restart the flow of tablets.

For each bottle filled, the maximum binary number in the counter is transferred to the *A* input of the *Adder.* The *B* input of the adder comes from *Register B* that stores the total number of tablets bottled up through the last bottle filled. The adder produces a new cumulative sum that is then stored in register B, replacing the previous sum. This keeps a running total of the tablets bottled during a given run.

The cumulative sum stored in register B goes to *Decoder B*, which detects when register B has reached its maximum capacity and enables the *MUX.* The binary sum in register B is converted from parallel to serial form by the MUX and transmitted over the single line to the remote *Demultiplexer* (DEMUX), which changes the number back to parallel form for storage in a remote computer for keeping track of the total tablets bottled in a specified time period.

FIGURE 1–45   Block diagram of a tablet-bottling system.

SECTION 1–7 CHECKUP

1. How is the number of tablets per bottle entered into the system?

2. How does the system determine when a bottle is full?

3. When is the counter reset?

Troubleshooting is the process of systematically isolating, identifying, and correcting a fault in a circuit or system. A variety of instruments are available for use in troubleshooting and testing. Some common types of instruments are introduced and discussed in this section.

After completing this section, you should be able to

- Distinguish between an analog and a digital oscilloscope
- Recognize common oscilloscope controls
- Determine amplitude, period, frequency, and duty cycle of a pulse waveform with an oscilloscope
- Discuss the logic analyzer and some common formats
- Describe the purpose of the data pattern generator, the digital multimeter (DMM), the dc power supply, the logic probe, and the logic pulser

The Oscilloscope

The oscilloscope (scope for short) is one of the most widely used instruments for general testing and troubleshooting. The scope is basically a graph-displaying device that traces the graph of a measured electrical signal on its screen. In most applications, the graph shows how signals change over time. The vertical axis of the display screen represents voltage, and the horizontal axis represents time. Amplitude, period, and frequency of a signal can be measured using the oscilloscope. Also, the pulse width, duty cycle, rise time, and fall time of a pulse waveform can be determined. Most scopes can display at least two, and many can display four signals on the screen at one time, enabling their time relationship to be observed. A typical 4-channel digital oscilloscope is shown in Figure 1–46.

Two basic types of oscilloscopes, analog and digital, can be used to view digital waveforms. An analog scope works by applying the measured waveform directly to control the up and down motion of the electron beam in the cathode-ray tube (CRT) as it sweeps across the display screen. As a result, the beam traces out the waveform pattern on the screen. A digital scope converts the measured waveform to digital information by a sampling process in an analog-to-digital converter (ADC). The digital information is then used to reconstruct the waveform on the screen.

FIGURE 1–46   A digital oscilloscope.   Used with permission from Tektronix, Inc.

The digital scope is more widely used than the analog scope. However, either type can be used in many applications; each has characteristics that make it more suitable for certain situations. An analog scope displays waveforms as they occur in "real time." Digital scopes are useful for measuring transient pulses that may occur randomly or only once. Also, because information about the measured waveform can be stored in a digital scope, it may be viewed at some later time, printed out, or thoroughly analyzed by a computer or other means.

BASIC OPERATION OF ANALOG OSCILLOSCOPES   To measure a voltage, a **probe** must be connected from the scope to the point in a circuit at which the voltage is present. Generally, a ×10 probe is used that reduces (attenuates) the signal amplitude by ten. The signal goes through the probe into the vertical circuits where it is either further attenuated or amplified, depending on the actual amplitude and on where you set the vertical control of the scope. The vertical circuits then drive the vertical deflection plates of the CRT. Also, the signal goes to the trigger circuits that trigger the horizontal circuits to initiate repetitive horizontal sweeps of the electron beam across the screen using a sawtooth waveform. There are many sweeps per second so that the beam appears to form a solid line across the screen in the shape of the waveform. This basic operation is illustrated in Figure 1–47.

FIGURE 1–47   **Block diagram of an analog oscilloscope.**   (Photo courtesy of Digilent, Inc.)

BASIC OPERATION OF DIGITAL OSCILLOSCOPES   Some parts of a digital scope are similar to the analog scope. However, the digital scope is more complex than an analog scope and typically has an LCD screen rather than a CRT. Rather than displaying a waveform as it occurs, the digital scope first acquires the measured analog waveform and converts it to a digital format using an analog-to-digital converter (ADC). The digital data is stored and processed. The data then goes to the reconstruction and display circuits for display in its original analog form. Figure 1–48 shows a basic block diagram for a digital oscilloscope.

FIGURE 1–48   **Block diagram of a digital oscilloscope.**   (Photo courtesy of Digilent, Inc.)

OSCILLOSCOPE CONTROLS   A front panel view of a typical digital oscilloscope is shown in Figure 1–49. Instruments vary depending on model and manufacturer, but most have certain common features. For example, the four vertical sections contain a Position control, a channel menu button, and a volts/div control. The horizontal section contains a sec/div control.

Some of the main oscilloscope controls are now discussed. Refer to the user manual for complete details of your particular scope.

VERTICAL CONTROLS   In the vertical section of the scope in Figure 1–49, there are identical controls for each of the four channels (1, 2, 3, and 4). The Position control lets you move a displayed waveform up or down vertically on the screen. The buttons on the right side of the screen provide for the selection of several items that appear on the screen, such as the coupling modes (ac, dc, or ground), coarse or fine adjustment for the volts/div, signal inversion, and other parameters. The volts/div control adjusts the number of volts represented by each vertical division on the screen. The volts/div setting for each channel is displayed on the bottom of the screen.

FIGURE 1–49   A digital oscilloscope front panel.   Used with permission from Tektronix, Inc.

HORIZONTAL CONTROLS   In the horizontal section, the controls apply to all channels. The Position control lets you move a displayed waveform left or right horizontally on the screen. The sec/div control adjusts the time represented by each horizontal division or main time base. The sec/div setting is displayed at the bottom of the screen.

TRIGGER CONTROLS   In the Trigger control section, the Level control determines the point on the triggering waveform where triggering occurs to initiate the sweep to display input waveforms. The Menu button provides for the selection of several items that appear on the screen, including edge or slope triggering, trigger source, trigger mode, and other parameters. There is also an input for an external trigger signal.

Triggering stabilizes a waveform on the screen or properly triggers on a pulse that occurs only one time or randomly. Also, it allows you to observe time delays between two waveforms. Figure 1–50 compares a triggered to an untriggered signal. The untriggered signal tends to drift across the screen, producing what appears to be multiple waveforms.

(a) Untriggered waveform display

(b) Triggered waveform display

FIGURE 1–50   Comparison of an untriggered and a triggered waveform on an oscilloscope.

COUPLING A SIGNAL INTO THE SCOPE   Coupling is the method used to connect a signal voltage to be measured into the oscilloscope. DC and AC coupling are usually selected from the Vertical menu on a scope. DC coupling allows a waveform including its dc component to be displayed. AC coupling blocks the dc component of a signal so that you see the waveform centered at 0 V. The Ground mode allows you to connect the channel input to ground to see where the 0 V reference is on the screen. Figure 1–51 illustrates the result of DC and AC coupling using a pulse waveform that has a dc component.

(a) DC coupled waveform   (b) AC coupled waveform

FIGURE 1–51   Displays of the same waveform having a dc component.

The voltage probe, shown in Figure 1–46, is essential for connecting a signal to the scope. Since all instruments tend to affect the circuit being measured due to loading, most scope probes provide a high series resistance to minimize loading effects. Probes that have a series resistance ten times larger than the input resistance of the scope are called ×10 probes. Probes with no series resistance are called ×1 probes. The oscilloscope adjusts its calibration for the attenuation of the type of probe being used. For most measurements, the ×10 probe should be used. However, if you are measuring very small signals, a ×1 may be the best choice.

The probe has an adjustment that allows you to compensate for the input capacitance of the scope. Most scopes have a probe compensation output that provides a calibrated square wave for probe compensation. Before making a measurement, you should make sure that the probe is properly compensated to eliminate any distortion introduced. Typically, there is a screw or other means of adjusting compensation on a probe. Figure 1–52 shows scope waveforms for three probe conditions: properly compensated, undercompensated, and overcompensated. If the waveform appears either over- or undercompensated, adjust the probe until the properly compensated square wave is achieved.

Properly compensated   Undercompensated   Overcompensated

FIGURE 1–52   Probe compensation conditions.

EXAMPLE 1–3

Based on the readouts, determine the amplitude and the period of the pulse waveform on the screen of a digital oscilloscope as shown in Figure 1–53. Also, calculate the frequency.

FIGURE 1–53

SOLUTION

The volts/div setting is 1 V. The pulses are three divisions high. Since each division represents 1 V, the pulse amplitude is

$$\text{Amplitude} = (3 \text{ div})(1 \text{ V/div}) = \mathbf{3\ V}$$

The sec/div setting is 10 μs. A full cycle of the waveform (from beginning of one pulse to the beginning of the next) covers four divisions; therefore, the period is

$$\text{Period} = (4 \text{ div})(10 \text{ μs/div}) = \mathbf{40\ \mu s}$$

The frequency is calculated as

$$f = \frac{1}{T} = \frac{1}{40\ \mu s} = \mathbf{25\ kHz}$$

RELATED PROBLEM

For a volts/div setting of 4 V and sec/div setting of 2 ms, determine the amplitude and period of the pulse shown on the screen in Figure 1–53.

The Logic Analyzer

Logic analyzers are used for measurements of multiple digital signals and measurement situations with difficult trigger requirements. Basically, the logic analyzer came about as a result of microprocessors in which troubleshooting or debugging required many more inputs than an oscilloscope offered. Many oscilloscopes have two input channels and some are available with four. Logic analyzers are available with from 34 to 136 input channels. Generally, an oscilloscope is used either when amplitude, frequency, and other timing parameters of a few signals at a time or when parameters such an rise and fall times, overshoot, and delay times need to be measured. The logic analyzer is used when the logic levels of a large number of signals need to be determined and for the correlation of simultaneous signals based on their timing relationships. A typical logic analyzer is shown in Figure 1–54, and a simplified block diagram is in Figure 1–55.

FIGURE 1–54 Typical logic analyzer. Used with permission from Tektronix, Inc.

DATA ACQUISITION   The large number of signals that can be acquired at one time is a major factor that distinguishes a logic analyzer from an oscilloscope. Generally, the two types of data acquisition in a logic analyzer are the timing acquisition and the state acquisition. Timing acquisition is used primarily when the timing relationships among the various signals need to be determined. State acquisition is used when you need to view the sequence of states as they appear in a system under test.

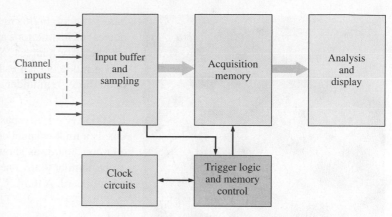

FIGURE 1–55   **Simplified block diagram of a logic analyzer.**

It is often helpful to have correlated timing and state data, and most logic analyzers can simultaneously acquire that data. For example, a problem may initially be detected as an invalid state. However, the invalid condition may be caused by a timing violation in the system under test. Without both types of information available at the same time, isolating the problem could be very difficult.

CHANNEL COUNT AND MEMORY DEPTH   Logic analyzers contain a real-time acquisition memory in which sampled data from all the channels are stored as they occur. Two features that are of primary importance are the channel count and the memory depth. The acquisition memory can be thought of as having a width equal to the number of channels and a depth that is the number of bits that can be captured by each channel during a certain time interval.

Channel count determines the number of signals that can be acquired simultaneously. In certain types of systems, a large number of signals are present, such as on the data bus in a microprocessor-based system. The depth of the acquisition memory determines the amount of data from a given channel that you can view at any given time.

ANALYSIS AND DISPLAY   Once data has been sampled and stored in the acquisition memory, it can typically be used in several different display and analysis modes. The waveform display is much like the display on an oscilloscope where you can view the time relationship of multiple signals. The listing display indicates the state of the system under test by showing the values of the input waveforms (1s and 0s) at various points in time (sample points). Typically, this data can be displayed in hexadecimal or other formats. Figure 1–56 shows simplified versions of these two display modes. The listing display samples correspond to the sampled points shown in red on the waveform display. You will study binary and hexadecimal (hex) numbers in the next chapter.

Sample	Binary	Hex	Time
1	1111	F	1 ns
2	1110	E	10 ns
3	1101	D	20 ns
4	1100	C	30 ns
5	1011	B	40 ns
6	1010	A	50 ns
7	1001	9	60 ns
8	1000	8	70 ns

1 2 3 4 5 6 7 8

(a) Waveform display   (b) Listing display

FIGURE 1–56   **Two logic analyzer display modes.**

Two more modes that are useful in computer and microprocessor-based system testing are the instruction trace and the source code debug. The instruction trace determines and displays instructions that occur, for example, on the data bus in a microprocessor-based system. In this mode the op-codes and the mnemonics (English-like names) of instructions are generally displayed as well as their corresponding memory address. Many logic analyzers also include a source code debug mode, which essentially allows you to see what is actually going on in the system under test when a program instruction is executed.

PROBES   Three basic types of probes are used with logic analyzers. One is a multichannel compression probe that can be attached to points on a circuit board, as shown in Figure 1–57. Another type of multichannel probe, similar to the one shown, plugs into dedicated sockets mounted on a circuit board. A third type is a single-channel clip-on probe.

FIGURE 1–57   A typical multichannel logic analyzer probe.   Used with permission from Tektronix, Inc.

Signal Generators

LOGIC SIGNAL SOURCE   These instruments are also known as pulse generators and data pattern generators. They are specifically designed to generate digital signals with precise edge placement and amplitudes and to produce the streams of 1s and 0s needed to test computer buses, microprocessors, and other digital systems.

WAVEFORM AND DATA PATTERN GENERATORS   The arbitrary waveform generator can be used to generate standard signals like sine waves, triangular waves, and pulses as well as signals with various shapes and characteristics. Waveforms can be defined by mathematical or graphical input. A typical arbitrary waveform generator is shown in Figure 1–58(a).

The data pattern generator, shown in Figure 1–58(b), provides digital waveforms with programmable bit patterns. The function generator, shown in part (c), provides pulse, sine, and triangular waveforms, often with programmable capability. Signal generators have logic-compatible outputs to provide the proper level and drive for inputs to digital circuits.

(a) Arbitrary waveform generator   (b) Data pattern generator   (c) Function generator

FIGURE 1–58   Typical signal generators.   Used with permission from Tektronix, Inc.

The Digital Multimeter (DMM)

The digital multimeter (DMM) is a versatile instrument found on virtually all workbenches. All DMMs can make basic ac and dc voltage, current, and resistance measurements. Voltage and resistance measurements are the principal quantities measured with DMMs. For current measurements, the leads are switched to a separate set of jacks and placed in series with the current path. In this mode, the meter acts like a short circuit, so serious problems can occur if the meter is incorrectly placed in parallel.

In addition to the basic measurements, most DMMs can also test diodes and capacitors and frequently will have other capabilities such as frequency measurements. Most new DMMs have an autoranging feature, meaning that the user is not required to select a range for making a measurement. If the range is not set automatically, the user needs to set the range switch for voltage measurements *higher* than the expected reading to avoid damage to the meter.

In digital systems, DMMs are the preferred instrument for setting dc power supply voltages or checking the supply voltage on various points in the circuit. Because digital signals are nonsinusoidal, the DMM is generally *not* used for measurements of digital signals (although the average or rms value can be determined in some cases). For signal measurements, the oscilloscope is the preferred instrument.

In addition, DMMs are used in digital systems for testing continuity between points in a circuit and checking resistors with the ohmmeter function. For checking a circuit path or looking for a short, DMMs are the instrument of choice. Many DMMs sound a beep or tone when there is continuity between the leads, making it handy to trace paths without having to look at the display. If the DMM is not equipped with a continuity test, the ohmmeter function can be used instead. Measurements of continuity or resistance are never done in "live" circuits, as any circuit voltage will disrupt the readings and can be dangerous.

Typical test bench and handheld DMMs are shown in Figure 1–59.

FIGURE 1–59   Typical DMMs.   Courtesy of B&K Precision Corporation.

The DC Power Supply

This instrument is an indispensable instrument on any test bench. The power supply converts ac power from the standard wall outlet into regulated dc voltage. All digital circuits require dc voltage. Most logic circuits require from 1.2 V to 5 V to operate. The power supply is used to power circuits during design, development, and troubleshooting when in-system power is not available. Typical test bench dc power supplies are shown in Figure 1–60.

FIGURE 1–60   Typical dc power supplies.   Courtesy of B&K Precision Corporation.

The Logic Probe and Logic Pulser

The logic probe is a convenient, inexpensive handheld tool that provides a means of troubleshooting a digital circuit by sensing various conditions at a point in a circuit. A probe can detect high-level voltage, low-level voltage, single pulses, repetitive pulses, and opens on a PC board. A probe lamp indicates the condition that exists at a certain point.

The logic pulser produces a repetitive pulse waveform that can be applied to any point in a circuit. You can apply pulses at one point in a circuit with the pulser and check another point for resulting pulses with a logic probe.

SECTION 1–8 CHECKUP

1. What is the main difference between a digital and an analog oscilloscope?

2. Name two main differences between a logic analyzer and an oscilloscope?

3. What does the volts/div control on an oscilloscope do?

4. What does the sec/div control on an oscilloscope do?

5. What is the purpose of a function generator?

SUMMARY

- An analog quantity has continuous values.
- A digital quantity has a discrete set of values.
- A binary digit is called a bit.
- A pulse is characterized by rise time, fall time, pulse width, and amplitude.
- The frequency of a periodic waveform is the reciprocal of the period. The formulas relating frequency and period are

$$f = \frac{1}{T} \quad \text{and} \quad T = \frac{1}{f}$$

- The duty cycle of a pulse waveform is the ratio of the pulse width to the period, expressed by the following formula as a percentage:

$$\text{Duty cycle} = \left(\frac{t_{\text{PW}}}{T} \right) 100\%$$

- A timing diagram is an arrangement of two or more waveforms showing their relationship with respect to time.
- Three basic logic operations are NOT, AND, and OR. The standard symbols for these are given in Figure 1–61.

NOT   AND   OR

FIGURE 1–61

- The basic logic functions are comparison, arithmetic, code conversion, decoding, encoding, data selection, storage, and counting.
- PLDs are integrated circuits into which logic designs can be programmed.
- Two types of SPLDs (simple programmable logic devices) are PAL (programmable array logic) and GAL (generic array logic).
- The CPLD (complex programmable logic device) contains multiple SPLDs with programmable interconnections.
- The FPGA (field-programmable gate array) has a different internal structure than the CPLD and is generally used for more complex circuits and systems.
- Fixed-function logic consists of integrated circuits that are manufactured with logic elements that cannot be changed.
- The two broad physical categories of IC packages are through-hole mounted and surface mounted.
- Three families of fixed-function integrated circuits are CMOS, bipolar, and BiCMOS.
- Bipolar is also known as TTL (transistor-transistor logic).
- The categories of ICs in terms of circuit complexity are SSI (small-scale integration), MSI (medium-scale integration), LSI, VLSI, and ULSI (large-scale, very large-scale, and ultra large-scale integration).
- Common instruments used in testing and troubleshooting digital circuits are the oscilloscope, logic analyzer, arbitrary waveform generator, data pattern generator, function generator, dc power supply, digital multimeter, logic probe, and logic pulser.

KEY TERMS

Key terms and other bold terms in the chapter are defined in the end-of-book glossary.

Analog   Being continuous or having continuous values.

AND   A basic logic operation in which a true (HIGH) output occurs only when all the input conditions are true (HIGH).

Binary   Having two values or states; describes a number system that has a base of two and utilizes 1 and 0 as its digits.

Bit   A binary digit, which can be either a 1 or a 0.

Clock   The basic timing signal in a digital system; a periodic waveform used to synchronize operation.

Compiler   A program that controls the design flow process and translates source code into object code in a format that can be logically tested or downloaded to a target device.

CPLD   A complex programmable logic device that consists basically of multiple SPLD arrays with programmable interconnections.

Data   Information in numeric, alphabetic, or other form.

Digital   Related to digits or discrete quantities; having a set of discrete values.

Digital system   An arrangement of the individual logic functions connected to perform a specified operation or produce a defined output.

Duty cycle   The ratio of the pulse width to the period of a digital waveform, expressed as a percentage.

Embedded system   Generally, a single-purpose system, such as a processor, built into a larger system for the purpose of controlling the system.

Fixed-function logic   A category of digital integrated circuits having functions that cannot be altered.

FPGA   Field-programmable gate array.

Gate   A logic circuit that performs a basic logic operation such as AND or OR.

Input   The signal or line going into a circuit.

Integrated circuit (IC)   A type of circuit in which all of the components are integrated on a single chip of semiconductive material of extremely small size.

Inverter   A NOT circuit; a circuit that changes a HIGH to a LOW or vice versa.

Logic   In digital electronics, the decision-making capability of gate circuits, in which a HIGH represents a true statement and a LOW represents a false one.

Microcontroller   An integrated circuit consisting of a complete computer on a single chip and used for specified control functions.

NOT   A basic logic operation that performs inversions.

OR   A basic logic operation in which a true (HIGH) output occurs when one or more of the input conditions are true (HIGH).

Output   The signal or line coming out of a circuit.

Parallel   In digital systems, data occurring simultaneously on several lines; the transfer or processing of several bits simultaneously.

Programmable logic device   A type of integrated circuit (IC) that starts as a "blank state" and into which a logic design is programmed.

Pulse   A sudden change from one level to another, followed after a time, called the pulse width, by a sudden change back to the original level.

Serial   Having one element following another, as in a serial transfer of bits; occurring in sequence rather than simultaneously.

SPLD   Simple programmable logic device.

Timing diagram   A graph of digital waveforms showing the time relationship of two or more waveforms.

Troubleshooting   The technique or process of systematically identifying, isolating, and correcting a fault in a circuit or system.

TRUE/FALSE QUIZ

Answers are at the end of the chapter.

1. An analog signal is one having continuous values.

2. A digital quantity has ten discrete values.

3. There are two digits in the binary system.

4. The term *bit* is short for binary digit.

5. In positive logic, a LOW level represents a binary 1.

6. If the period of a pulse waveform increases, the frequency also increases.

7. A timing diagram shows the timing relationship of two or more digital waveforms.

8. The basic logic operations are AND, OR, and MAYBE.

9. If the input to an inverter is a 1, the output is a 0.

10. CPLD stands for *complex programmable logic device.*

11. FPGA stands for *functionally programmed gate array.*

12. An oscilloscope is used to observe, measure, and analyze waveforms.

SELF-TEST

Answers are at the end of the chapter.

1. A quantity having continuous values is
   (a) a digital quantity   (b) an analog quantity
   (c) a binary quantity   (d) a natural quantity

2. The term *bit* means
   (a) a small amount of data   (b) a 1 or a 0
   (c) binary digit   (d) both answers (b) and (c)

3. The time interval on the leading edge of a pulse between 10% and 90% of the amplitude is the
   (a) rise time   (b) fall time   (c) pulse width   (d) period

4. A pulse in a certain waveform occurs every 10 ms. The frequency is
   (a) 1 kHz   (b) 1 Hz   (c) 100 Hz   (d) 10 Hz

5. In a certain digital waveform, the period is twice the pulse width. The duty cycle is
   (a) 100%   (b) 200%   (c) 50%

6. An inverter
   (a) performs the NOT operation   (b) changes a HIGH to a LOW
   (c) changes a LOW to a HIGH   (d) does all of the above

7. The output of an AND gate is HIGH when
   (a) any input is HIGH   (b) all inputs are HIGH
   (c) no inputs are HIGH   (d) both answers (a) and (b)

8. The output of an OR gate is HIGH when
   (a) any input is HIGH   (b) all inputs are HIGH
   (c) no inputs are HIGH   (d) both answers (a) and (b)

9. The device used to convert a binary number to a 7-segment display format is the
   (a) multiplexer   (b) encoder   (c) decoder   (d) register

10. An example of a data storage device is
    (a) the logic gate   (b) the flip-flop   (c) the comparator
    (d) the register   (e) both answers (b) and (d)

11. A fixed-function IC package containing four AND gates is an example of
    (a) MSI   (b) SMT   (c) SOIC   (d) SSI

12. An LSI device has a circuit complexity of from
    (a) 10 to 100 equivalent gates
    (b) more than 100 to 10,000 equivalent gates
    (c) 2000 to 5000 equivalent gates
    (d) more than 10,000 to 100,000 equivalent gates

13. VHDL is a
    (a) logic device   (b) PLD programming language
    (c) computer language   (d) very high density logic

14. A CPLD is a
    (a) controlled program logic device   (b) complex programmable logic driver
    (c) complex programmable logic device   (d) central processing logic device

15. An FPGA is a
    (a) field-programmable gate array   (b) fast programmable gate array
    (c) field-programmable generic array   (d) flash process gate application

PROBLEMS

Answers to odd-numbered problems are at the end of the book.

SECTION 1–1   Digital and Analog Signals and Systems

1. Name two advantages of digital data as compared to analog data.

2. Name an analog quantity other than temperature and sound.

3. List three common products that can have either a digital or analog output.

SECTION 1–2   Binary Digits, Logic Levels, and Digital Waveforms

4. Explain the difference between positive and negative logic.

5. Define the sequence of bits (1s and 0s) represented by each of the following sequences of levels:
    (a) HIGH, HIGH, LOW, HIGH, LOW, LOW, LOW, HIGH
    (b) LOW, LOW, LOW, HIGH, LOW, HIGH, LOW, HIGH, LOW

6. List the sequence of levels (HIGH and LOW) that represent each of the following bit sequences:
    (a) 1 0 1 1 1 0 1   (b) 1 1 1 0 1 0 0 1

7. For the pulse shown in Figure 1–62, graphically determine the following:
    (a) rise time   (b) fall time   (c) pulse width   (d) amplitude

8. Determine the period of the digital waveform in Figure 1–63.

9. What is the frequency of the waveform in Figure 1–63?

10. Is the pulse waveform in Figure 1–63 periodic or nonperiodic?

11. Determine the duty cycle of the waveform in Figure 1–63.

FIGURE 1–62

FIGURE 1–63

12. Determine the bit sequence represented by the waveform in Figure 1–64. A bit time is 1 μs in this case.

13. What is the total serial transfer time for the eight bits in Figure 1–64? What is the total parallel transfer time?

14. What is the period if the clock frequency is 3.5 GHz?

FIGURE 1–64

SECTION 1–3   Logic Operations

15. Form a single logical statement from the following information:
    (a) The light is ON if SW1 is closed.
    (b) The light is ON if SW2 is closed.
    (c) The light is OFF if both SW1 and SW2 are open.

16. A logic circuit requires HIGHs on all its inputs to make the output HIGH. What type of logic circuit is it?

17. A basic 2-input logic circuit has a HIGH on one input and a LOW on the other input, and the output is LOW. Identify the circuit.

18. A basic 2-input logic circuit has a HIGH on one input and a LOW on the other input, and the output is HIGH. What type of logic circuit is it?

SECTION 1–4   Combinational and Sequential Logic Functions

19. Name the logic function of each block in Figure 1–65 based on your observation of the inputs and outputs.

(a)   (b)   (c)   (d)

FIGURE 1–65

20. A pulse waveform with a frequency of 10 kHz is applied to the input of a counter. During 100 ms, how many pulses are counted?

21. Consider a register that can store eight bits. Assume that it has been reset so that it contains zeros in all positions. If you transfer four alternating bits (0101) serially into the register, beginning with a 1 and shifting to the right, what will the total content of the register be as soon as the fourth bit is stored?

SECTION 1–5   Programmable Logic

22. Which of the following acronyms do not describe a type of programmable logic?

    PAL, GAL, SPLD, VHDL, CPLD, AHDL, FPGA

23. What do each of the following stand for?
    (a) SPLD   (b) CPLD   (c) HDL   (d) FPGA   (e) GAL

24. Define each of the following PLD programming terms:
    (a) design entry   (b) simulation   (c) compilation   (d) download

25. Describe the process of place-and-route.

SECTION 1–6   Fixed-Function Logic Devices

26. A fixed-function digital IC chip has a complexity of 200 equivalent gates. How is it classified?

27. Explain the main difference between the DIP and SMT packages.

28. Label the pin numbers on the packages in Figure 1–66. Top views are shown.

(a)   (b)

FIGURE 1–66

SECTION 1–7   A System

29. List three ways that the tablet-bottling system in Figure 1–45 can be implemented.

30. Which function(s) determines when each bottle is full?

31. Identify the purpose of each register in Figure 1–45.

SECTION 1–8   Measuring Instruments

32. A pulse is displayed on the screen of an oscilloscope, and you measure the base line as 1 V and the top of the pulse as 8 V. What is the amplitude?

33. A waveform is measured on the oscilloscope and its amplitude covers three vertical divisions. If the vertical control is set at 2 V/div, what is the total amplitude of the waveform?

34. The period of a pulse waveform measures four horizontal divisions on an oscilloscope. If the time base is set at 2 ms/div, what is the frequency of the waveform?

ANSWERS TO SECTION CHECKUPS

SECTION 1–1   Digital and Analog Signals and Systems

1. *Analog* means continuous.

2. *Digital* means discrete.

3. A digital quantity has a discrete set of values and an analog quantity has continuous values.

4. A public address system is analog. A CD player is analog and digital. A computer is all digital.

SECTION 1–2   Binary Digits, Logic Levels, and Digital Waveforms

1. Binary means having two states or values.

2. A bit is a binary digit.

3. The bits are 1 and 0.

4. Rise time: from 10% to 90% of amplitude. Fall time: from 90% to 10% of amplitude.

5. Frequency is the reciprocal of the period.

6. A clock waveform is a basic timing waveform from which other waveforms are derived.

7. A timing diagram shows the time relationship of two or more waveforms.

8. Parallel transfer is faster than serial transfer.

SECTION 1–3   Logic Operations

1. When the input is LOW

2. When all inputs are HIGH

3. When any or all inputs are HIGH

4. An inverter is a NOT circuit.

5. A logic gate is a circuit that performs a logic operation (AND, OR).

SECTION 1–4   Combinational and Sequential Logic Functions

1. A comparator compares the magnitudes of two input numbers.

2. Add, subtract, multiply, and divide

3. Encoding is changing a familiar form such as decimal to a coded form such as binary.

4. Decoding is changing a code to a familiar form such as binary to decimal.

5. Multiplexing puts data from many sources onto one line. Demultiplexing takes data from one line and distributes it to many destinations.

6. Flip-flops, registers, semiconductor memories, magnetic disks

7. A counter counts events with a sequence of binary states.

SECTION 1–5   Programmable Logic

1. Simple programmable logic device (SPLD), complex programmable logic device (CPLD), and field-programmable gate array (FPGA)

2. A CPLD is made up of multiple SPLDs.

3. Design entry, functional simulation, synthesis, implementation, timing simulation, and download

4. *Design entry:* The logic design is entered using development software. *Functional simulation:* The design is software simulated to make sure it works logically. *Synthesis:* The design is translated into a netlist. *Implementation:* The logic developed by the netlist is mapped into the programmable device. *Timing simulation:* The design is software simulated to confirm that there are no timing problems. *Download:* The design is placed into the programmable device.

5. The microcontroller has fixed internal circuits, and its operation is directed by a program.

SECTION 1–6   Fixed-Function Logic Devices

1. An IC is an electronic circuit with all components integrated on a single silicon chip.

2. DIP—dual in-line package; SMT—surface-mount technology; SOIC—small-outline integrated circuit; SSI—small-scale integration; MSI—medium-scale integration; LSI—large-scale integration; VLSI—very large-scale integration.

3. **(a)** SSI   **(b)** MSI   **(c)** LSI   **(d)** VLSI   **(e)** ULSI

SECTION 1–7   A System

1. Number of tablets is entered via keypad.

2. The system counts tablets and compares to preset number to determine when a bottle is full.

3. Counter is reset after maximum count and when next bottle is in place.

SECTION 1–8   Measuring Instruments

1. The analog scope applies the measured waveform directly to the display circuits. The digital scope first converts the measured signal to digital form.

2. The logic analyzer has more channels than the oscillosope and has more than one data display format.

3. The volts/div control sets the voltage for each division on the screen.

4. The sec/div control sets the time for each division on the screen.

5. The function generator produces various types of waveforms.

ANSWERS TO RELATED PROBLEMS FOR EXAMPLES

1–1 $f = 6.67$ kHz; Duty cycle $= 16.7\%$

1–2 Serial transfer: 3.33 ns

1–3 Amplitude $= 12$ V; $T = 8$ ms

ANSWERS TO TRUE/FALSE QUIZ

1. T   **2.** F   **3.** T   **4.** T   **5.** F   **6.** F
7. T   **8.** F   **9.** T   **10.** T   **11.** F   **12.** T

ANSWERS TO SELF-TEST

1. (b)   **2.** (d)   **3.** (a)   **4.** (c)   **5.** (c)   **6.** (d)   **7.** (b)   **8.** (d)
9. (c)   **10.** (e)   **11.** (d)   **12.** (d)   **13.** (b)   **14.** (c)   **15.** (a)

CHAPTER 2

NUMBER SYSTEMS, OPERATIONS, AND CODES

OUTLINE

OBJECTIVES

- Review the decimal number system
- Count in the binary number system
- Convert from decimal to binary and from binary to decimal
- Apply arithmetic operations to binary numbers
- Determine the 1's and 2's complements of a binary number
- Express signed binary numbers in sign-magnitude, 1's complement, 2's complement, and floating-point format
- Carry out arithmetic operations with signed binary numbers
- Convert between the binary and hexadecimal number systems
- Add numbers in hexadecimal form
- Convert between the binary and octal number systems
- Express decimal numbers in binary coded decimal (BCD) form
- Add BCD numbers
- Convert between the binary system and the Gray code
- Interpret the American Standard Code for Information Interchange (ASCII)
- Explain how to detect code errors
- Discuss the cyclic redundancy check (CRC)

KEY TERMS

LSB	BCD
MSB	Alphanumeric
Byte	ASCII
Floating-point number	Parity
Hexadecimal	Cyclic redundancy check (CRC)
Octal	

INTRODUCTION

The binary number system and digital codes are fundamental to computers and to digital electronics in general. In this chapter, the binary number system and its relationship to

VISIT THE WEBSITE
Study aids for this chapter are available at
http://pearsonhighered.com/floyd

other number systems such as decimal, hexadecimal, and octal is presented. Arithmetic operations with binary numbers are covered to provide a basis for understanding how computers and many other types of digital systems work. Also, digital codes such as binary coded decimal (BCD), the Gray code, and the ASCII are covered. The parity method for detecting errors in codes is introduced. The tutorials on the use of the calculator in certain operations are based on the TI-36X calculator. The procedures shown may vary on other types.

2–1 THE DECIMAL NUMBER SYSTEM

You are familiar with the decimal number system because you use decimal numbers every day. Although decimal numbers are commonplace, their weighted structure is often not understood. In this section, the structure of decimal numbers is reviewed. This review will help you more easily understand the structure of the binary number system, which is important in computers and digital electronics.

After completing this section, you should be able to

- **Explain why the decimal number system is a weighted system**
- **Explain how powers of ten are used in the decimal system**
- **Determine the weight of each digit in a decimal number**

The decimal number system has ten digits.

In the **decimal** number system each of the ten digits, 0 through 9, represents a certain quantity. As you know, the ten symbols (**digits**) do not limit you to expressing only ten different quantities because you use the various digits in appropriate positions within a number to indicate the magnitude of the quantity. You can express quantities up through nine before running out of digits; if you wish to express a quantity greater than nine, you use two or more digits, and the position of each digit within the number tells you the magnitude it represents. If, for example, you wish to express the quantity twenty-three, you use (by their respective positions in the number) the digit 2 to represent the quantity twenty and the digit 3 to represent the quantity three, as illustrated below.

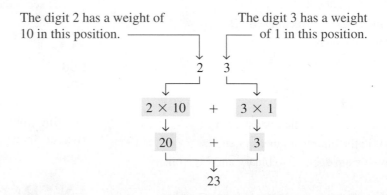

The decimal number system has a base of 10.

The position of each digit in a decimal number indicates the magnitude of the quantity represented and can be assigned a **weight.** The weights for whole numbers are positive powers of ten that increase from right to left, beginning with $10^0 = 1$.

$$\ldots 10^5 \; 10^4 \; 10^3 \; 10^2 \; 10^1 \; 10^0$$

For fractional numbers, the weights are negative powers of ten that decrease from left to right beginning with 10^{-1}.

$$10^2 \; 10^1 \; 10^0.10^{-1} \; 10^{-2} \; 10^{-3} \ldots$$

$\uparrow$— Decimal point

The value of a decimal number is the sum of the digits after each digit has been multiplied by its weight, as Examples 2–1 and 2–2 illustrate.

The value of a digit is determined by its position in the number.

EXAMPLE 2–1

Express the decimal number 47 as a sum of the values of each digit.

SOLUTION

The digit 4 has a weight of 10, which is 10^1, as indicated by its position. The digit 7 has a weight of 1, which is 10^0, as indicated by its position.

$$47 = (4 \times 10^1) + (7 \times 10^0)$$
$$= (4 \times 10) + (7 \times 1) = \mathbf{40 + 7}$$

RELATED PROBLEM*

Determine the value of each digit in 939.

*Answers are at the end of the chapter.

EXAMPLE 2–2

Express the decimal number 568.23 as a sum of the values of each digit.

SOLUTION

The whole number digit 5 has a weight of 100, which is 10^2, the digit 6 has a weight of 10, which is 10^1, the digit 8 has a weight of 1, which is 10^0, the fractional digit 2 has a weight of 0.1, which is 10^{-1}, and the fractional digit 3 has a weight of 0.01, which is 10^{-2}.

$$568.23 = (5 \times 10^2) + (6 \times 10^1) + (8 \times 10^0) + (2 \times 10^{-1}) + (3 \times 10^{-2})$$
$$= (5 \times 100) + (6 \times 10) + (8 \times 1) + (2 \times 0.1) + (3 \times 0.01)$$
$$= \mathbf{500} + \mathbf{60} + \mathbf{8} + \mathbf{0.2} + \mathbf{0.03}$$

RELATED PROBLEM

Determine the value of each digit in 67.924.

CALCULATOR TUTORIAL

Powers of Ten

EXAMPLE

Find the value of 10^3.

TI-36X   Step 1:   [1] [0] [yˣ]

Step 2:   [3] [=] | 1000 |

SECTION 2–1 CHECKUP*

1. What weight does the digit 7 have in each of the following numbers?

   (a) 1370   (b) 6725   (c) 7051   (d) 58.72

2. Express each of the following decimal numbers as a sum of the products obtained by multiplying each digit by its appropriate weight:

   (a) 51   (b) 137   (c) 1492   (d) 106.58

*Answers are at the end of the chapter.

The binary number system is another way to represent quantities. It is less complicated than the decimal system because it has only two digits. The decimal system with its ten digits is a base-ten system; the binary system with its two digits is a base-two system. The two binary digits (bits) are 1 and 0. The position of a 1 or 0 in a binary number indicates its weight, or value within the number, just as the position of a decimal digit determines the value of that digit. The weights in a binary number are based on powers of two.

After completing this section, you should be able to

- Count in binary
- Determine the largest decimal number that can be represented by a given number of bits
- Convert a binary number to a decimal number

Counting in Binary

The binary number system has two digits (bits).

Learning to count in binary will help you to basically understand how digital circuits can be used to count events. This can be anything from counting items on an assembly line to counting operations in a computer. To learn to count in the binary system, first look at how you count in the decimal system. You start at zero and count up to nine before you run out of digits. You then start another digit position (to the left) and continue counting 10 through 99. At this point you have exhausted all two-digit combinations, so a third digit position is needed to count from 100 through 999.

The binary number system has a base of 2.

A comparable situation occurs when you count in binary, except that you have only two digits, called *bits*. Begin counting: 0, 1. At this point you have used both digits, so include another digit position and continue: 10, 11. You have now exhausted all combinations of two digits, so a third position is required. With three digit positions you can continue to count: 100, 101, 110, and 111. Now you need a fourth digit position to continue, and so on. A binary count of zero through fifteen is shown in Table 2–1. Notice the patterns with which the 1s and 0s alternate in each column.

The value of a bit is determined by its position in the number.

As you can see in Table 2–1, four bits are required to count from zero to 15. In general, with n bits you can count up to a number equal to $2^n - 1$.

$$\text{Largest decimal number} = 2^n - 1$$

For example, with five bits ($n = 5$) you can count from zero to thirty-one.

$$2^5 - 1 = 32 - 1 = 31$$

With six bits ($n = 6$) you can count from zero to sixty-three.

$$2^6 - 1 = 64 - 1 = 63$$

A table of powers of two is given in Appendix A.

In computer operations, there are many cases where adding or subtracting 1 to a number stored in a counter is necessary. Computers have special instructions that use less time and generate less machine code than the ADD or SUB instructions. For the Intel processors, the INC (increment) instruction adds 1 to a number. For subtraction, the corresponding instruction is DEC (decrement), which subtracts 1 from a number.

SYSTEM NOTE

TABLE 2–1				
DECIMAL NUMBER	**BINARY NUMBER**			
0	0	0	0	0
1	0	0	0	1
2	0	0	1	0
3	0	0	1	1
4	0	1	0	0
5	0	1	0	1
6	0	1	1	0
7	0	1	1	1
8	1	0	0	0
9	1	0	0	1
10	1	0	1	0
11	1	0	1	1
12	1	1	0	0
13	1	1	0	1
14	1	1	1	0
15	1	1	1	1

CALCULATOR TUTORIAL

Powers of Two

EXAMPLE

Find the value of 2^5.

TI-36X   Step 1: [2] [y^x]

Step 2: [5] [=]   32

SYSTEM EXAMPLE 2–1

BALL-COUNTING SYSTEM

Figure 2–1 illustrates the counting of tennis balls that go into a box from a conveyor belt. Nine balls go into each box. The counter counts the pulses from a sensor that detects the passing of a ball and produces a sequence of logic levels (digital waveforms) on each of its four parallel outputs. Each set of logic levels represents a 4-bit binary number (HIGH = 1 and LOW = 0), as indicated. As the decoder receives these waveforms, it decodes each set of four bits and converts it to the corresponding decimal number in the 7-segment display. When the counter gets to the binary state of 1001, it has counted nine tennis balls, the display shows decimal 9, and a new box is moved under the conveyor. Then the counter goes back to its zero state (0000), and the process starts over. (The number 9 was used only in the interest of single-digit simplicity.)

FIGURE 2–1   Illustration of a simple binary counting application.

The Weighting Structure of Binary Numbers

The weight or value of a bit increases from right to left in a binary number.

A binary number is a weighted number. The right-most bit is the **LSB** (least significant bit) in a binary whole number and has a weight of $2^0 = 1$. The weights increase from right to left by a power of two for each bit. The left-most bit is the **MSB** (most significant bit); its weight depends on the size of the binary number.

Fractional numbers can also be represented in binary by placing bits to the right of the binary point, just as fractional decimal digits are placed to the right of the decimal point. The left-most bit is the MSB in a binary fractional number and has a weight of $2^{-1} = 0.5$. The fractional weights decrease from left to right by a negative power of two for each bit.

The weight structure of a binary number is

$$2^{n-1} \ldots 2^3 \, 2^2 \, 2^1 \, 2^0 . \, 2^{-1} \, 2^{-2} \ldots 2^{-n}$$

$$\underset{\text{Binary point}}{\uparrow}$$

where n is the number of bits from the binary point. Thus, all the bits to the left of the binary point have weights that are positive powers of two, as previously discussed for whole numbers. All bits to the right of the binary point have weights that are negative powers of two, or fractional weights.

Computers use binary numbers to select memory locations. Each location is assigned a unique number called an *address*. Some microprocessors, for example, have 32 address lines which can select 2^{32} (4,294,967,296) unique locations.

SYSTEM NOTE

The powers of two and their equivalent decimal weights for an 8-bit binary whole number and a 6-bit binary fractional number are shown in Table 2–2. Notice that the weight doubles for each positive power of two and that the weight is halved for each negative power of two. You can easily extend the table by doubling the weight of the most significant positive power of two and halving the weight of the least significant negative power of two; for example, $2^9 = 512$ and $2^{-7} = 0.0078125$.

TABLE 2–2 • Binary weights.														
POSITIVE POWERS OF TWO (WHOLE NUMBERS)									NEGATIVE POWERS OF TWO (FRACTIONAL NUMBER)					
2^8	2^7	2^6	2^5	2^4	2^3	2^2	2^1	2^0	2^{-1}	2^{-2}	2^{-3}	2^{-4}	2^{-5}	2^{-6}
256	128	64	32	16	8	4	2	1	1/2 0.5	1/4 0.25	1/8 0.125	1/16 0.625	1/32 0.03125	1/64 0.015625

Binary-to-Decimal Conversion

The decimal value of any binary number can be found by adding the weights of all bits that are 1 and discarding the weights of all bits that are 0.

Add the weights of all 1s in a binary number to get the decimal value.

EXAMPLE 2–3

Convert the binary whole number 1101101 to decimal.

SOLUTION

Determine the weight of each bit that is a 1, and then find the sum of the weights to get the decimal number.

$$\text{Weight:}\quad 2^6\,2^5\,2^4\,2^3\,2^2\,2^1\,2^0$$
$$\text{Binary number:}\quad 1\ 1\ 0\ 1\ 1\ 0\ 1$$
$$1101101 = 2^6 + 2^5 + 2^3 + 2^2 + 2^0$$
$$= 64 + 32 + 8 + 4 + 1 = \mathbf{109}$$

RELATED PROBLEM

Convert the binary number 10010001 to decimal.

EXAMPLE 2–4

Convert the fractional binary number 0.1011 to decimal.

SOLUTION

Determine the weight of each bit that is a 1, and then sum the weights to get the decimal fraction.

$$\text{Weight:}\quad 2^{-1}\ 2^{-2}\ 2^{-3}\ 2^{-4}$$
$$\text{Binary number:}\quad 0.1\ \ 0\ \ 1\ \ 1$$
$$0.1011 = 2^{-1} + 2^{-3} + 2^{-4}$$
$$= 0.5 + 0.125 + 0.0625 = \mathbf{0.6875}$$

RELATED PROBLEM

Convert the binary number 10.111 to decimal.

SECTION 2–2 CHECKUP

1. What is the largest decimal number that can be represented in binary with eight bits?
2. Determine the weight of the 1 in the binary number 10000.
3. Convert the binary number 10111101.011 to decimal.

In Section 2–2 you learned how to convert a binary number to the equivalent decimal number. Now you will learn two ways of converting from a decimal number to a binary number.

After completing this section, you should be able to

- Convert a decimal number to binary using the sum-of-weights method
- Convert a decimal whole number to binary using the repeated division-by-2 method
- Convert a decimal fraction to binary using the repeated multiplication-by-2 method

Sum-of-Weights Method

To get the binary number for a given decimal number, find the binary weights that add up to the decimal number.

One way to find the binary number that is equivalent to a given decimal number is to determine the set of binary weights whose sum is equal to the decimal number. An easy way to remember binary weights is that the lowest is 1, which is 2^0, and that by doubling any weight, you get the next higher weight; thus, a list of seven binary weights would be 64, 32, 16, 8, 4, 2, 1 as you learned in the last section. The decimal number 9, for example, can be expressed as the sum of binary weights as follows:

$$9 = 8 + 1 \text{ or } 9 = 2^3 + 2^0$$

Placing 1s in the appropriate weight positions, 2^3 and 2^0, and 0s in the 2^2 and 2^1 positions determines the binary number for decimal 9.

$$
\begin{array}{cccc}
2^3 & 2^2 & 2^1 & 2^0 \\
1 & 0 & 0 & 1
\end{array} \quad \text{Binary number for decimal 9}
$$

EXAMPLE 2–5

Convert the following decimal numbers to binary:

(a) 12   (b) 25   (c) 58   (d) 82

SOLUTION

(a) $12 = 8 + 4 = 2^3 + 2^2$ $\longrightarrow$ **1100**

(b) $25 = 16 + 8 + 1 = 2^4 + 2^3 + 2^0$ $\longrightarrow$ **11001**

(c) $58 = 32 + 16 + 8 + 2 = 2^5 + 2^4 + 2^3 + 2^1$ $\longrightarrow$ **111010**

(d) $82 = 64 + 16 + 2 = 2^6 + 2^4 + 2^1$ $\longrightarrow$ **1010010**

RELATED PROBLEM

Convert the decimal number 125 to binary.

Repeated Division-by-2 Method

To get the binary number for a given decimal number, divide the decimal number by 2 until the quotient is 0. Remainders form the binary number.

A systematic method of converting whole numbers from decimal to binary is the *repeated division-by-2* process. For example, to convert the decimal number 12 to binary, begin by dividing 12 by 2. Then divide each resulting quotient by 2 until there is a 0 whole-number quotient. The **remainders** generated by each division form the binary number. The first remainder to be produced is the LSB (least significant bit) in the binary number, and the last remainder to be produced is the MSB (most significant bit). This procedure is illustrated as follows for converting the decimal number 12 to binary.

Remainder

$$\frac{12}{2} = 6 \qquad 0$$

$$\frac{6}{2} = 3 \qquad 0$$

$$\frac{3}{2} = 1 \qquad 1$$

$$\frac{1}{2} = 0 \qquad 1$$

Stop when the whole-number quotient is 0.

1 1 0 0

MSB ⤴   ⤴ LSB

EXAMPLE 2–6

Convert the following decimal numbers to binary:

(a) 19   **(b)** 45

SOLUTION

(a) Remainder

$$\frac{19}{2} = 9 \qquad 1$$

$$\frac{9}{2} = 4 \qquad 1$$

$$\frac{4}{2} = 2 \qquad 0$$

$$\frac{2}{2} = 1 \qquad 0$$

$$\frac{1}{2} = 0 \qquad 1$$

1 0 0 1 1

MSB ⤴   ⤴ LSB

(b) Remainder

$$\frac{45}{2} = 22 \qquad 1$$

$$\frac{22}{2} = 11 \qquad 0$$

$$\frac{11}{2} = 5 \qquad 1$$

$$\frac{5}{2} = 2 \qquad 1$$

$$\frac{2}{2} = 1 \qquad 0$$

$$\frac{1}{2} = 0 \qquad 1$$

1 0 1 1 0 1

MSB ⤴   ⤴ LSB

RELATED PROBLEM

Convert decimal number 39 to binary.

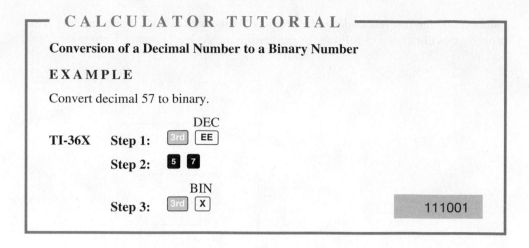

Converting Decimal Fractions to Binary

Examples 2–5 and 2–6 demonstrated whole-number conversions. Now let's look at fractional conversions. An easy way to remember fractional binary weights is that the most significant weight is 0.5, which is 2^{-1}, and that by halving any weight, you get the next lower weight; thus a list of four fractional binary weights would be 0.5, 0.25, 0.125, 0.0625.

SUM-OF-WEIGHTS   The sum-of-weights method can be applied to fractional decimal numbers, as shown in the following example:

$$0.625 = 0.5 + 0.125 = 2^{-1} + 2^{-3} = 0.101$$

There is a 1 in the 2^{-1} position, a 0 in the 2^{-2} position, and a 1 in the 2^{-3} position.

REPEATED MULTIPLICATION BY 2   As you have seen, decimal whole numbers can be converted to binary by repeated division by 2. Decimal fractions can be converted to binary by repeated multiplication by 2. For example, to convert the decimal fraction 0.3125 to binary, begin by multiplying 0.3125 by 2 and then multiplying each resulting fractional part of the product by 2 until the fractional product is zero or until the desired number of decimal places is reached. The carry digits, or **carries,** generated by the multiplications produce the binary number. The first carry produced is the MSB, and the last carry is the LSB. This procedure is illustrated as follows:

Continue to the desired number of decimal places ———
or stop when the fractional part is all zeros.

SECTION 2–3 CHECKUP

1. Convert each decimal number to binary by using the sum-of-weights method:

(a) 23   (b) 57   (c) 45.5

2. Convert each decimal number to binary by using the repeated division-by-2 method (repeated multiplication-by-2 for fractions):

(a) 14   (b) 21   (c) 0.375

2–4 BINARY ARITHMETIC

Binary arithmetic is essential in all digital computers and in many other types of digital systems. To understand digital systems, you must know the basics of binary addition, subtraction, multiplication, and division. This section provides an introduction that will be expanded in later sections.

After completing this section, you should be able to

- **Add binary numbers**
- **Subtract binary numbers**
- **Multiply binary numbers**
- **Divide binary numbers**

Binary Addition

The four basic rules for adding binary digits (bits) are as follows:

In binary $1 + 1 = 10$, not the decimal digit 2.

$$0 + 0 = 0 \qquad \text{Sum of 0 with a carry of 0}$$
$$0 + 1 = 1 \qquad \text{Sum of 1 with a carry of 0}$$
$$1 + 0 = 1 \qquad \text{Sum of 1 with a carry of 0}$$
$$1 + 1 = 10 \qquad \text{Sum of 0 with a carry of 1}$$

Notice that the first three rules result in a single bit and in the fourth rule the addition of two 1s yields a binary two (10). When binary numbers are added, the last condition creates a sum of 0 in a given column and a carry of 1 over to the next column to the left, as illustrated in the following addition of $11 + 1$:

```
      Carry  Carry
       1 ←    1 ←
       0      1     1
     + 0      0     1
     ─────────────────
       1     └0    └0
```

In the right column, $1 + 1 = 0$ with a carry of 1 to the next column to the left. In the middle column, $1 + 1 + 0 = 0$ with a carry of 1 to the next column to the left. In the left column, $1 + 0 + 0 = 1$.

When there is a carry of 1, you have a situation in which three bits are being added (a bit in each of the two numbers and a carry bit). This situation is illustrated as follows:

Carry bits ⟶

$$1 + 0 + 0 = 01 \qquad \text{Sum of 1 with a carry of 0}$$
$$1 + 1 + 0 = 10 \qquad \text{Sum of 0 with a carry of 1}$$
$$1 + 0 + 1 = 10 \qquad \text{Sum of 0 with a carry of 1}$$
$$1 + 1 + 1 = 11 \qquad \text{Sum of 1 with a carry of 1}$$

┌─ **E X A M P L E 2 – 7** ─────────────────────────────────┐

Add the following binary numbers:

(a) 11 + 11   **(b)** 100 + 10   **(c)** 111 + 11   **(d)** 110 + 100

SOLUTION

The equivalent decimal addition is also shown for reference.

(a)	11	3	**(b)**	100	4	**(c)**	111	7	**(d)**	110	6
	+11	+3		+10	+2		+ 11	+3		+100	+4
	110	6		**110**	6		**1010**	10		**1010**	10

RELATED PROBLEM

Add 1111 and 1100.

└──┘

Binary Subtraction

In binary $10 - 1 = 1$, not the decimal digit 9.

The four basic rules for subtracting bits are as follows:

$$0 - 0 = 0$$
$$1 - 1 = 0$$
$$1 - 0 = 1$$
$$10 - 1 = 1 \qquad 0 - 1 \text{ with a borrow of } 1$$

When subtracting numbers, you sometimes have to borrow from the next column to the left. A borrow is required in binary only when you try to subtract a 1 from a 0. In this case, when a 1 is borrowed from the next column to the left, a 10 is created in the column being subtracted, and the last of the four basic rules just listed must be applied. Examples 2–8 and 2–9 illustrate binary subtraction; the equivalent decimal subtractions are also shown.

┌─ **E X A M P L E 2 – 8** ─────────────────────────────────┐

Perform the following binary subtractions:

(a) 11 − 01   **(b)** 11 − 10

SOLUTION

(a)	11	3	**(b)**	11	3
	− 01	− 1		− 10	− 2
	10	2		**01**	1

No borrows were required in this example. The binary number 01 is the same as 1.

RELATED PROBLEM

Subtract 100 from 111.

└──┘

┌─ **E X A M P L E 2 – 9** ─────────────────────────────────┐

Subtract 011 from 101.

SOLUTION

101	5
− 011	− 3
010	2

└──┘

Let's examine exactly what was done to subtract the two binary numbers since a borrow is required. Begin with the right column.

Left column:
When a 1 is borrowed,
a 0 is left, so $0 - 0 = 0$.

Middle column:
Borrow 1 from next column
to the left, making a 10 in
this column, then $10 - 1 = 1$.

$$\begin{array}{r} \overset{0}{\cancel{1}}01 \\ -0\,11 \\ \hline 0\,10 \end{array}$$

Right column:
$1 - 1 = 0$

RELATED PROBLEM

Subtract 101 from 110.

Binary Multiplication

The four basic rules for multiplying bits are as follows:

$$0 \times 0 = 0$$
$$0 \times 1 = 0$$
$$1 \times 0 = 0$$
$$1 \times 1 = 1$$

Binary multiplication of two bits is the same as multiplication of the decimal digits 0 and 1.

Multiplication is performed with binary numbers in the same manner as with decimal numbers. It involves forming partial products, shifting each successive partial product left one place, and then adding all the partial products. Example 2–10 illustrates the procedure; the equivalent decimal multiplications are shown for reference.

EXAMPLE 2–10

Perform the following binary multiplications:

(a) 11×11 **(b)** 101×111

SOLUTION

(a)

$$\begin{array}{r} 11 \\ \times 11 \\ \hline 11 \\ +11 \\ \hline \mathbf{1001} \end{array} \qquad \begin{array}{r} 3 \\ \times 3 \\ \hline 9 \end{array}$$

Partial products

(b)

$$\begin{array}{r} 111 \\ \times 101 \\ \hline 111 \\ 000 \\ +111 \\ \hline \mathbf{100011} \end{array} \qquad \begin{array}{r} 7 \\ \times 5 \\ \hline 35 \end{array}$$

Partial products

RELATED PROBLEM

Multiply 1101×1010.

Binary Division

Division in binary follows the same procedure as division in decimal, as Example 2–11 illustrates. The equivalent decimal divisions are also given.

EXAMPLE 2–11

Perform the following binary divisions:

(a) $110 \div 11$ **(b)** $110 \div 10$

SOLUTION

```
           10    2                    11    3
(a)    11)110  3)6          (b)   10)110  2)6
           11    6                    10    6
          000    0                    10    0
                                      10
                                      10
                                      00
```

RELATED PROBLEM

Divide 1100 by 100.

SECTION 2–4 CHECKUP

1. Perform the following binary additions:

    (a) $1101 + 1010$ **(b)** $10111 + 01101$

2. Perform the following binary subtractions:

    (a) $1101 - 0100$ **(b)** $1001 - 0111$

3. Perform the indicated binary operations:

    (a) 110×111 **(b)** $1100 \div 011$

2–5 1'S AND 2'S COMPLEMENTS OF BINARY NUMBERS

The 1's complement and the 2's complement of a binary number are important because they permit the representation of negative numbers. The method of 2's complement arithmetic is commonly used in computers to handle negative numbers.

After completing this section, you should be able to

- **Convert a binary number to its 1's complement**
- **Convert a binary number to its 2's complement using either of two methods**

Finding the 1's Complement

Change each bit in a number to get the 1's complement.

The 1's **complement** of a binary number is found by changing all 1s to 0s and all 0s to 1s, as illustrated below:

$$1\,0\,1\,1\,0\,0\,1\,0 \qquad \text{Binary number}$$
$$\downarrow\downarrow\downarrow\downarrow\downarrow\downarrow\downarrow\downarrow$$
$$0\,1\,0\,0\,1\,1\,0\,1 \qquad \text{1's complement}$$

The simplest way to obtain the 1's complement of a binary number with a digital circuit is to use parallel inverters (NOT circuits), as shown in Figure 2–2 for an 8-bit binary number.

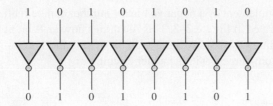

FIGURE 2–2   **Example of inverters used to obtain the 1's complement of a binary number.**

Finding the 2's Complement

The 2's complement of a binary number is found by adding 1 to the LSB of the 1's complement.

$$2\text{'s complement} = (1\text{'s complement}) + 1$$

Add 1 to the 1's complement to get the 2's complement.

E X A M P L E   2 – 1 2

Find the 2's complement of 10110010.

S O L U T I O N

10110010	Binary number
01001101	1's complement
+ 1	Add 1
01001110	2's complement

R E L A T E D   P R O B L E M

Determine the 2's complement of 11001011.

An alternative method of finding the 2's complement of a binary number is as follows:

1. Start at the right with the LSB and write the bits as they are up to and including the first 1.
2. Take the 1's complements of the remaining bits.

Change all bits to the left of the least significant 1 to get 2's complement.

E X A M P L E   2 – 1 3

Find the 2's complement of 10111000 using the alternative method.

S O L U T I O N

                        10111000   Binary number
1's complements   **01001000**   2's complement
of original bits ───────┘ └─These bits stay the same.

R E L A T E D   P R O B L E M

Find the 2's complement of 11000000.

The 2's complement of a negative binary number can be realized using inverters and an adder, as indicated in Figure 2–3. This illustrates how an 8-bit number can be converted to its 2's complement by first inverting each bit (taking the 1's complement) and then adding 1 to the 1's complement with an adder circuit.

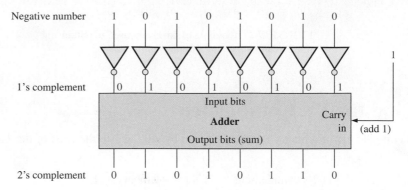

FIGURE 2–3   **Example of obtaining the 2's complement of a negative binary number.**

To convert from a 1's or 2's complement back to the true (uncomplemented) binary form, use the same two procedures described previously. To go from the 1's complement back to true binary, reverse all the bits. To go from the 2's complement form back to true binary, take the 1's complement of the 2's complement number and add 1 to the least significant bit.

SECTION 2–5 CHECKUP

1. Determine the 1's complement of each binary number:

  (a) 00011010   (b) 11110111   (c) 10001101

2. Determine the 2's complement of each binary number:

  (a) 00010110   (b) 11111100   (c) 10010001

2–6 SIGNED NUMBERS

Digital systems, such as the computer, must be able to handle both positive and negative numbers. A signed binary number consists of both sign and magnitude information. The sign indicates whether a number is positive or negative, and the magnitude is the value of the number. There are three forms in which signed integer (whole) numbers can be represented in binary: sign-magnitude, 1's complement, and 2's complement. Of these, the 2's complement is the most important and the sign-magnitude is the least used. Noninteger and very large or small numbers can be expressed in floating-point format.

After completing this section, you should be able to

  • **Express positive and negative numbers in sign-magnitude**
  • **Express positive and negative numbers in 1's complement**
  • **Express positive and negative numbers in 2's complement**
  • **Determine the decimal value of signed binary numbers**
  • **Express a binary number in floating-point format**

The Sign Bit

The left-most bit in a signed binary number is the **sign bit,** which tells you whether the number is positive or negative.

A 0 sign bit indicates a positive number, and a 1 sign bit indicates a negative number.

Sign-Magnitude Form

When a signed binary number is represented in sign-magnitude, the left-most bit is the sign bit and the remaining bits are the magnitude bits. The magnitude bits are in true (uncomplemented) binary for both positive and negative numbers. For example, the decimal number +25 is expressed as an 8-bit signed binary number using the sign-magnitude form as

$$00011001$$

Sign bit ⌐↑   ↑⌐ Magnitude bits

The decimal number −25 is expressed as

$$10011001$$

Notice that the only difference between +25 and −25 is the sign bit because the magnitude bits are in true binary for both positive and negative numbers.

In the sign-magnitude form, a negative number has the same magnitude bits as the corresponding positive number but the sign bit is a 1 rather than a zero.

1's Complement Form

Positive numbers in 1's complement form are represented the same way as the positive sign-magnitude numbers. Negative numbers, however, are the 1's complements of the corresponding positive numbers. For example, using eight bits, the decimal number −25 is expressed as the 1's complement of +25 (00011001) as

$$11100110$$

In the 1's complement form, a negative number is the 1's complement of the corresponding positive number.

2's Complement Form

Positive numbers in 2's complement form are represented the same way as in the sign-magnitude and 1's complement forms. Negative numbers are the 2's complements of the corresponding positive numbers. Again, using eight bits, let's take decimal number −25 and express it as the 2's complement of +25 (00011001). Inverting each bit and adding 1, you get

$$-25 = 11100111$$

In the 2's complement form, a negative number is the 2's complement of the corresponding positive number.

Computers use the 2's complement for negative integer numbers in arithmetic operations. The reason is that subtraction of a number is the same as adding the 2's complement of the number. Computers form the 2's complement by inverting the bits and adding 1, using special instructions that produce the same result as the adder in Figure 2–3.

SYSTEM NOTE

EXAMPLE 2–14

Express the decimal number −39 as an 8-bit number in the sign-magnitude, 1's complement, and 2's complement forms.

SOLUTION

First, write the 8-bit number for +39.

$$00100111$$

In the *sign-magnitude form,* −39 is produced by changing the sign bit to a 1 and leaving the magnitude bits as they are. The number is

10100111

In the *1's complement form,* −39 is produced by taking the 1's complement of +39 (00100111).

11011000

In the *2's complement form,* −39 is produced by taking the 2's complement of +39 (00100111) as follows:

$$
\begin{array}{ll}
11011000 & \text{1's complement} \\
\underline{+\qquad 1} & \\
\mathbf{11011001} & \text{2's complement}
\end{array}
$$

RELATED PROBLEM

Express +19 and −19 as 8-bit numbers in sign-magnitude, 1's complement, and 2's complement.

The Decimal Value of Signed Numbers

SIGN-MAGNITUDE   Decimal values of positive and negative numbers in the sign-magnitude form are determined by summing the weights in all the magnitude bit positions where there are 1s and ignoring those positions where there are zeros. The sign is determined by examination of the sign bit.

EXAMPLE 2–15

Determine the decimal value of this signed binary number expressed in sign-magnitude: 10010101.

SOLUTION

The seven magnitude bits and their powers-of-two weights are as follows:

$$
\begin{array}{ccccccc}
2^6 & 2^5 & 2^4 & 2^3 & 2^2 & 2^1 & 2^0 \\
0 & 0 & 1 & 0 & 1 & 0 & 1
\end{array}
$$

Summing the weights where there are 1s,

$$16 + 4 + 1 = 21$$

The sign bit is 1; therefore, the decimal number is **−21.**

RELATED PROBLEM

Determine the decimal value of the sign-magnitude number 01110111.

1'S COMPLEMENT Decimal values of positive numbers in the 1's complement form are determined by summing the weights in all bit positions where there are 1s and ignoring those positions where there are zeros. Decimal values of negative numbers are determined by assigning a negative value to the weight of the sign bit, summing all the weights where there are 1s, and adding 1 to the result.

EXAMPLE 2–16

Determine the decimal values of the signed binary numbers expressed in 1's complement:

(a) 00010111   **(b)** 11101000

SOLUTION

(a) The bits and their powers-of-two weights for the positive number are as follows:

$$-2^7 \quad 2^6 \quad 2^5 \quad 2^4 \quad 2^3 \quad 2^2 \quad 2^1 \quad 2^0$$
$$0 \quad\; 0 \quad\; 0 \quad\; 1 \quad\; 0 \quad\; 1 \quad\; 1 \quad\; 1$$

Summing the weights where there are 1s,

$$16 + 4 + 2 + 1 = \mathbf{+23}$$

(b) The bits and their powers-of-two weights for the negative number are as follows. Notice that the negative sign bit has a weight of -2^7 or -128.

$$-2^7 \quad 2^6 \quad 2^5 \quad 2^4 \quad 2^3 \quad 2^2 \quad 2^1 \quad 2^0$$
$$1 \quad\; 1 \quad\; 1 \quad\; 0 \quad\; 1 \quad\; 0 \quad\; 0 \quad\; 0$$

Summing the weights where there are 1s,

$$-128 + 64 + 32 + 8 = -24$$

Adding 1 to the result, the final decimal number is

$$-24 + 1 = \mathbf{-23}$$

RELATED PROBLEM

Determine the decimal value of the 1's complement number 11101011.

2'S COMPLEMENT Decimal values of positive and negative numbers in the 2's complement form are determined by summing the weights in all bit positions where there are 1s and ignoring those positions where there are zeros. The weight of the sign bit in a negative number is given a negative value.

EXAMPLE 2–17

Determine the decimal values of the signed binary numbers expressed in 2's complement:

(a) 01010110   **(b)** 10101010

SOLUTION

(a) The bits and their powers-of-two weights for the positive number are as follows:

$$-2^7 \quad 2^6 \quad 2^5 \quad 2^4 \quad 2^3 \quad 2^2 \quad 2^1 \quad 2^0$$
$$0 \quad\; 1 \quad\; 0 \quad\; 1 \quad\; 0 \quad\; 1 \quad\; 1 \quad\; 0$$

Summing the weights where there are 1s,

$$64 + 16 + 4 + 2 = \textbf{+86}$$

(b) The bits and their powers-of-two weights for the negative number are as follows. Notice that the negative sign bit has a weight of $-2^7 = -128$.

-2^7	2^6	2^5	2^4	2^3	2^2	2^1	2^0
1	0	1	0	1	0	1	0

Summing the weights where there are 1s,

$$-128 + 32 + 8 + 2 = \textbf{-86}$$

RELATED PROBLEM

Determine the decimal value of the 2's complement number 11010111.

From these examples, you can see why the 2's complement form is preferred for representing signed integer numbers: To convert to decimal, it simply requires a summation of weights regardless of whether the number is positive or negative. The 1's complement system requires adding 1 to the summation of weights for negative numbers but not for positive numbers. Also, the 1's complement form is generally not used because two representations of zero (00000000 or 11111111) are possible.

Range of Signed Integer Numbers

The range of magnitude of a binary number depends on the number of bits (n).

We have used 8-bit numbers for illustration because the 8-bit grouping is common in most computers and has been given the special name **byte**. With one byte or eight bits, you can represent 256 different numbers. With two bytes or sixteen bits, you can represent 65,536 different numbers. With four bytes or 32 bits, you can represent 4.295×10^9 different numbers. The formula for finding the number of different combinations of n bits is

$$\text{Total combinations} = 2^n$$

For 2's complement signed numbers, the range of values for n-bit numbers is

$$\text{Range} = -(2^{n-1}) \text{ to } +(2^{n-1} - 1)$$

where in each case there is one sign bit and $n - 1$ magnitude bits. For example, with four bits you can represent numbers in 2's complement ranging from $-(2^3) = -8$ to $2^3 - 1 = +7$. Similarly, with eight bits you can go from -128 to $+127$, with sixteen bits you can go from $-32,768$ to $+32,767$, and so on. There is one less positive number than there are negative numbers because zero is represented as a positive number (all zeros).

Floating-Point Numbers

To represent very large **integer** (whole) numbers, many bits are required. There is also a problem when numbers with both integer and fractional parts, such as 23.5618, need to be represented. The floating-point number system, based on scientific notation, is capable of representing very large and very small numbers without an increase in the number of bits and also for representing numbers that have both integer and fractional components.

A **floating-point number** (also known as a *real number*) consists of two parts plus a sign. The **mantissa** is the part of a floating-point number that represents the magnitude of the number and is between 0 and 1. The **exponent** is the part of a floating-point number that represents the number of places that the decimal point (or binary point) is to be moved.

A decimal example will be helpful in understanding the basic concept of floating-point numbers. Let's consider a decimal number which, in integer form, is 241,506,800. The mantissa is .2415068 and the exponent is 9. When the integer is expressed as a floating-point number, it is normalized by moving the decimal point to the left of all the digits so that the mantissa is a fractional number and the exponent is the power of ten. The floating-point number is written as

$$0.2415068 \times 10^9$$

For binary floating-point numbers, the format is defined by ANSI/IEEE Standard 754-1985 in three forms: *single-precision, double-precision,* and *extended-precision.* These all have the same basic formats except for the number of bits. Single-precision floating-point numbers have 32 bits, double-precision numbers have 64 bits, and extended-precision numbers have 80 bits. We will restrict our discussion to the single-precision floating-point format.

In addition to the CPU (central processing unit), computers use *coprocessors* to perform complicated mathematical calculations using floating-point numbers. The purpose is to increase performance by freeing up the CPU for other tasks. The mathematical coprocessor is also known as the floating-point unit (FPU).

SYSTEM NOTE

SINGLE-PRECISION FLOATING-POINT BINARY NUMBERS   In the standard format for a single-precision binary number, the sign bit (S) is the left-most bit, the exponent (E) includes the next eight bits, and the mantissa or fractional part (F) includes the remaining 23 bits, as shown next.

←	32 bits	→
S	Exponent (E)	Mantissa (fraction, F)
1 bit	8 bits	23 bits

In the mantissa or fractional part, the binary point is understood to be to the left of the 23 bits. Effectively, there are 24 bits in the mantissa because in any binary number the left-most (most significant) bit is always a 1. Therefore, this 1 is understood to be there although it does not occupy an actual bit position.

The eight bits in the exponent represent a *biased exponent,* which is obtained by adding 127 to the actual exponent. The purpose of the bias is to allow very large or very small numbers without requiring a separate sign bit for the exponents. The biased exponent allows a range of actual exponent values from -126 to $+128$.

To illustrate how a binary number is expressed in floating-point format, let's use 1011010010001 as an example. First, it can be expressed as 1 plus a fractional binary number by moving the binary point 12 places to the left and then multiplying by the appropriate power of two.

$$1011010010001 = 1.011010010001 \times 2^{12}$$

Assuming that this is a positive number, the sign bit (S) is 0. The exponent, 12, is expressed as a biased exponent by adding it to 127 ($12 + 127 = 139$). The biased exponent (E) is expressed as the binary number 10001011. The mantissa is the fractional part (F) of the binary number, .011010010001. Because there is always a 1 to the left of the binary point in the power-of-two expression, it is not included in the mantissa. The complete floating-point number is

S	E	F
0	10001011	01101001000100000000000

Next, let's see how to evaluate a binary number that is already in floating-point format. The general approach to determining the value of a floating-point number is expressed by the following formula:

$$\text{Number} = (-1)^S(1 + F)(2^{E-127})$$

To illustrate, let's consider the following floating-point binary number:

S	E	F
1	10010001	10001110001000000000000

The sign bit is 1. The biased exponent is $10010001 = 145$. Applying the formula, we get

$$\text{Number} = (-1)^1 (1.10001110001)(2^{145-127})$$
$$= (-1)(1.10001110001)(2^{18}) = -1100011100010000000$$

This floating-point binary number is equivalent to $-407{,}688$ in decimal. Since the exponent can be any number between -126 and $+128$, extremely large and small numbers can be expressed. A 32-bit floating-point number can replace a binary integer number having 129 bits. Because the exponent determines the position of the binary point, numbers containing both integer and fractional parts can be represented.

There are two exceptions to the format for floating-point numbers: The number 0.0 is represented by all 0s, and infinity is represented by all 1s in the exponent and all 0s in the mantissa.

EXAMPLE 2–18

Convert the decimal number 3.248×10^4 to a single-precision floating-point binary number.

SOLUTION

Convert the decimal number to binary.

$$3.248 \times 10^4 = 32480 = 111111011100000_2 = 1.11111011100000 \times 2^{14}$$

The MSB will not occupy a bit position because it is always a 1. Therefore, the mantissa is the fractional 23-bit binary number 11111011100000000000000 and the biased exponent is

$$14 + 127 = 141 = 10001101_2$$

The complete floating-point number is

0	10001101	11111011100000000000000

RELATED PROBLEM

Determine the binary value of the following floating-point binary number:

0 10011000 10000100010100110000000

SECTION 2–6 CHECKUP

1. Express the decimal number $+9$ as an 8-bit binary number in the sign-magnitude system.

2. Express the decimal number -33 as an 8-bit binary number in the 1's complement system.

3. Express the decimal number -46 as an 8-bit binary number in the 2's complement system.

4. List the three parts of a signed, floating-point number.

2-7 ARITHMETIC OPERATIONS WITH SIGNED NUMBERS

In the last section, you learned how signed numbers are represented in three different forms. In this section, you will learn how signed numbers are added, subtracted, multiplied, and divided. Because the 2's complement form for representing signed numbers is the most widely used in computers and microprocessor-based systems, the coverage in this section is limited to 2's complement arithmetic. The processes covered can be extended to the other forms if necessary.

After completing this section, you should be able to

- Add signed binary numbers
- Define *overflow*
- Explain how computers add strings of numbers
- Subtract signed binary numbers
- Multiply signed binary numbers using the direct addition method
- Multiply signed binary numbers using the partial products method
- Divide signed binary numbers

Addition

The two numbers in an addition are the **addend** and the **augend.** The result is the **sum.** There are four cases that can occur when two signed binary numbers are added.

1. Both numbers positive
2. Positive number with magnitude larger than negative number
3. Negative number with magnitude larger than positive number
4. Both numbers negative

Let's take one case at a time using 8-bit signed numbers as examples. The equivalent decimal numbers are shown for reference.

Both numbers positive:

$$\begin{array}{rr} 00000111 & 7 \\ +00000100 & +4 \\ \hline 00001011 & 11 \end{array}$$

Addition of two positive numbers yields a positive number.

The sum is positive and is therefore in true (uncomplemented) binary.

Positive number with magnitude larger than negative number:

$$\begin{array}{rr} 00001111 & 15 \\ + 11111010 & +-6 \end{array}$$

Discard carry ⟶ $\begin{array}{rr} 1\ 00001001 & 9 \end{array}$

Addition of a positive number and a smaller negative number yields a positive number.

The final carry bit is discarded. The sum is positive and therefore in true (uncomplemented) binary.

Negative number with magnitude larger than positive number:

$$\begin{array}{rr} 00010000 & 16 \\ + 11101000 & +-24 \\ \hline 11111000 & -8 \end{array}$$

Addition of a positive number and a larger negative number or two negative numbers yields a negative number in 2's complement.

The sum is negative and therefore in 2's complement form.

Both numbers negative:

$$\begin{array}{rr} 11111011 & -5 \\ + 11110111 & +-9 \end{array}$$

Discard carry ⟶ $\begin{array}{rr} 1\ 11110010 & -14 \end{array}$

The final carry bit is discarded. The sum is negative and therefore in 2's complement form.

In a computer, the negative numbers are stored in 2's complement form so, as you can see, the addition process is very simple: *Add the two numbers and discard any final carry bit.*

OVERFLOW CONDITION   When two numbers are added and the number of bits required to represent the sum exceeds the number of bits in the two numbers, an **overflow** results as indicated by an incorrect sign bit. An overflow can occur only when both numbers are positive or both numbers are negative. If the sign bit of the result is different than the sign bit of the numbers that are added, overflow is indicated. The following 8-bit example will illustrate this condition.

$$
\begin{array}{r r}
01111101 & 125 \\
+\ 00111010 & +\ 58 \\
\hline
10110111 & 183
\end{array}
$$

Sign incorrect ⟶

Magnitude incorrect ⟶

In this example the sum of 183 requires eight magnitude bits. Since there are seven magnitude bits in the numbers (one bit is the sign), there is a carry into the sign bit which produces the overflow indication.

NUMBERS ADDED TWO AT A TIME   Now let's look at the addition of a string of numbers, added two at a time. This can be accomplished by adding the first two numbers, then adding the third number to the sum of the first two, then adding the fourth number to this result, and so on. This is how computers add strings of numbers. The addition of numbers taken two at a time is illustrated in Example 2–19.

EXAMPLE 2–19

Add the signed numbers: 01000100, 00011011, 00001110, and 00010010.

SOLUTION

The equivalent decimal additions are given for reference.

68	01000100	
+ 27	+ 00011011	Add 1st two numbers
95	01011111	1st sum
+ 14	+ 00001110	Add 3rd number
109	01101101	2nd sum
+ 18	+ 00010010	Add 4th number
127	**01111111**	Final sum

RELATED PROBLEM

Add 00110011, 10111111, and 01100011. These are signed numbers.

Subtraction

Subtraction is addition with the sign of the subtrahend changed.

Subtraction is a special case of addition. For example, subtracting +6 (the **subtrahend**) from +9 (the **minuend**) is equivalent to adding −6 to +9. Basically, *the subtraction operation changes the sign of the subtrahend and adds it to the minuend.* The result of a subtraction is called the **difference**.

The sign of a positive or negative binary number is changed by taking its 2's complement.

For example, when you take the 2's complement of the positive number 00000100 (+4), you get 11111100, which is −4 as the following sum-of-weights evaluation shows:

$$-128 + 64 + 32 + 16 + 8 + 4 = -4$$

As another example, when you take the 2's complement of the negative number 11101101 (−19), you get 00010011, which is +19 as the following sum-of-weights evaluation shows:

$$16 + 2 + 1 = 19$$

Since subtraction is simply an addition with the sign of the subtrahend changed, the process is stated as follows:

> **To subtract two signed numbers, take the 2's complement of the subtrahend and add. Discard any final carry bit.**

Example 2–20 illustrates the subtraction process.

When doing binary subtraction with the 2's complement method, it is very important that both numbers have the same number of bits.

EXAMPLE 2–20

Perform each of the following subtractions of the signed numbers:

(a) 00001000 − 00000011   **(b)** 00001100 − 11110111

(c) 11100111 − 00010011   **(d)** 10001000 − 11100010

SOLUTION

Like in other examples, the equivalent decimal subtractions are given for reference.

(a) In this case, $8 - 3 = 8 + (-3) = 5$.

```
                      00001000      Minuend (+8)
                    + 11111101      2's complement of subtrahend (−3)
   Discard carry ──→ 1 00000101     Difference (+5)
```

(b) In this case, $12 - (-9) = 12 + 9 = 21$.

```
              00001100      Minuend (+12)
            + 00001001      2's complement of subtrahend (+9)
              00010101      Difference (+21)
```

(c) In this case, $-25 - (+19) = -25 + (-19) = -44$.

```
                       11100111      Minuend (−25)
                     + 11101101      2's complement of subtrahend (−19)
   Discard carry ──→  1 11010100     Difference (−44)
```

(d) In this case, $-120 - (-30) = -120 + 30 = -90$.

```
              10001000      Minuend (−120)
            + 00011110      2's complement of subtrahend (+30)
              10100110      Difference (−90)
```

RELATED PROBLEM

Subtract 01000111 from 01011000.

Multiplication

The numbers in a multiplication are the **multiplicand**, the **multiplier**, and the **product.** These are illustrated in the following decimal multiplication:

```
      8      Multiplicand
   × 3      Multiplier
     24      Product
```

Multiplication is equivalent to adding a number to itself a number of times equal to the multiplier.

The multiplication operation in most computers is accomplished using addition. As you have already seen, subtraction is done with an adder; now let's see how multiplication is done.

Direct addition and *partial products* are two basic methods for performing multiplication using addition. In the direct addition method, you add the multiplicand a number of times equal to the multiplier. In the previous decimal example (8 × 3), three multiplicands are added: 8 + 8 + 8 = 24. The disadvantage of this approach is that it becomes very lengthy if the multiplier is a large number. For example, to multiply 350 × 75, you must add 350 to itself 75 times. Incidentally, this is why the term *times* is used to mean multiply.

When two binary numbers are multiplied, both numbers must be in true (uncomplemented) form. The direct addition method is illustrated in Example 2–21 adding two binary numbers at a time.

EXAMPLE 2–21

Multiply the signed binary numbers: 01001101 (multiplicand) and 00000100 (multiplier) using the direct addition method.

SOLUTION

Since both numbers are positive, they are in true form, and the product will be positive. The decimal value of the multiplier is 4, so the multiplicand is added to itself four times as follows:

01001101	1st time
+ 01001101	2nd time
10011010	Partial sum
+ 01001101	3rd time
11100111	Partial sum
+ 01001101	4th time
100110100	Product

Since the sign bit of the multiplicand is 0, it has no effect on the outcome. All of the bits in the product are magnitude bits.

RELATED PROBLEM

Multiply 01100001 by 00000110 using the direct addition method.

The partial products method is perhaps the more common one because it reflects the way you multiply longhand. The multiplicand is multiplied by each multiplier digit beginning with the least significant digit. The result of the multiplication of the multiplicand by a multiplier digit is called a *partial product*. Each successive partial product is moved (shifted) one place to the left and when all the partial products have been produced, they are added to get the final product. Here is a decimal example.

239	Multiplicand
× 123	Multiplier
717	1st partial product (3 × 239)
478	2nd partial product (2 × 239)
+ 239	3rd partial product (1 × 239)
29,397	Final product

The sign of the product of a multiplication depends on the signs of the multiplicand and the multiplier according to the following two rules:

- **If the signs are the same, the product is positive.**
- **If the signs are different, the product is negative.**

The basic steps in the partial products method of binary multiplication are as follows:

Step 1: Determine if the signs of the multiplicand and multiplier are the same or different. This determines what the sign of the product will be.

Step 2: Change any negative number to true (uncomplemented) form. Because most computers store negative numbers in 2's complement, a 2's complement operation is required to get the negative number into true form.

Step 3: Starting with the least significant multiplier bit, generate the partial products. When the multiplier bit is 1, the partial product is the same as the multiplicand. When the multiplier bit is 0, the partial product is zero. Shift each successive partial product one bit to the left.

Step 4: Add each successive partial product to the sum of the previous partial products to get the final product.

Step 5: If the sign bit that was determined in step 1 is negative, take the 2's complement of the product. If positive, leave the product in true form. Attach the sign bit to the product.

EXAMPLE 2–22

Multiply the signed binary numbers: 01010011 (multiplicand) and 11000101 (multiplier).

SOLUTION

Step 1: The sign bit of the multiplicand is 0 and the sign bit of the multiplier is 1. The sign bit of the product will be 1 (negative).

Step 2: Take the 2's complement of the multiplier to put it in true form.

$$11000101 \longrightarrow 00111011$$

Step 3 and 4: The multiplication proceeds as follows. Notice that only the magnitude bits are used in these steps.

1010011	Multiplicand
× 0111011	Multiplier
1010011	1st partial product
+ 1010011	2nd partial product
11111001	Sum of 1st and 2nd
+ 0000000	3rd partial product
011111001	Sum
+ 1010011	4th partial product
1110010001	Sum
+ 1010011	5th partial product
100011000001	Sum
+ 1010011	6th partial product
1001100100001	Sum
+ 0000000	7th partial product
1001100100001	Final product

Step 5: Since the sign of the product is a 1 as determined in step 1, take the 2's complement of the product.

$$1001100100001 \longrightarrow 0110011011111$$

Attach the sign bit ⟶

1 0110011011111

RELATED PROBLEM

Verify the multiplication is correct by converting to decimal numbers and performing the multiplication.

Division

The numbers in a division are the **dividend,** the **divisor,** and the **quotient.** These are illustrated in the following standard division format.

$$\frac{dividend}{divisor} = quotient$$

The division operation in computers is accomplished using subtraction. Since subtraction is done with an adder, division can also be accomplished with an adder.

The result of a division is called the *quotient;* the quotient is the number of times that the divisor will go into the dividend. This means that the divisor can be subtracted from the dividend a number of times equal to the quotient, as illustrated by dividing 21 by 7.

21	Dividend
− 7	1st subtraction of divisor
14	1st partial remainder
− 7	2nd subtraction of divisor
7	2nd partial remainder
− 7	3rd subtraction of divisor
0	Zero remainder

In this simple example, the divisor was subtracted from the dividend three times before a remainder of zero was obtained. Therefore, the quotient is 3.

The sign of the quotient depends on the signs of the dividend and the divisor according to the following two rules:

- **If the signs are the same, the quotient is positive.**
- **If the signs are different, the quotient is negative.**

When two binary numbers are divided, both numbers must be in true (uncomplemented) form. The basic steps in a division process are as follows:

Step 1:   Determine if the signs of the dividend and divisor are the same or different. This determines what the sign of the quotient will be. Initialize the quotient to zero.

Step 2:   Subtract the divisor from the dividend using 2's complement addition to get the first partial remainder and add 1 to the quotient. If this partial remainder is positive, go to step 3. If the partial remainder is zero or negative, the division is complete.

Step 3:   Subtract the divisor from the partial remainder and add 1 to the quotient. If the result is positive, repeat for the next partial remainder. If the result is zero or negative, the division is complete.

Continue to subtract the divisor from the dividend and the partial remainders until there is a zero or a negative result. Count the number of times that the divisor is subtracted and you have the quotient. Example 2–23 illustrates these steps using 8-bit signed binary numbers.

EXAMPLE 2–23

Divide 01100100 by 00011001.

SOLUTION

Step 1:   The signs of both numbers are positive, so the quotient will be positive. The quotient is initially zero: 00000000.

Step 2:   Subtract the divisor from the dividend using 2's complement addition (remember that final carries are discarded).

```
  01100100      Dividend
+ 11100111      2's complement of divisor
  01001011      Positive 1st partial remainder
```

Add 1 to quotient: 00000000 + 00000001 = 00000001.

Step 3:   Subtract the divisor from the 1st partial remainder using 2's complement addition.

```
  01001011      1st partial remainder
+ 11100111      2's complement of divisor
  00110010      Positive 2nd partial remainder
```

Add 1 to quotient: 00000001 + 00000001 = 00000010.

Step 4:   Subtract the divisor from the 2nd partial remainder using 2's complement addition.

```
  00110010      2nd partial remainder
+ 11100111      2's complement of divisor
  00011001      Positive 3rd partial remainder
```

Add 1 to quotient: 00000010 + 00000001 = 00000011.

Step 5:   Subtract the divisor from the 3rd partial remainder using 2's complement addition.

```
  00011001      3rd partial remainder
+ 11100111      2's complement of divisor
  00000000      Zero remainder
```

Add 1 to quotient: 00000011 + 00000001 = **00000100** (final quotient). The process is complete.

RELATED PROBLEM

Verify that the process is correct by converting to decimal numbers and performing the division.

SECTION 2–7 CHECKUP

1. List the four cases when numbers are added.

2. Add the signed numbers 00100001 and 10111100.

3. Subtract the signed numbers 00110010 from 01110111.

4. What is the sign of the product when two negative numbers are multiplied?

5. Multiply 01111111 by 00000101.

6. What is the sign of the quotient when a positive number is divided by a negative number?

7. Divide 00110000 by 00001100.

2–8 HEXADECIMAL NUMBERS

The hexadecimal number system has sixteen characters; it is used primarily as a compact way of displaying or writing binary numbers because it is very easy to convert between binary and hexadecimal. As you are probably aware, long binary numbers are difficult to read and write because it is easy to drop or transpose a bit. Since computers and microprocessors understand only 1s and 0s, it is necessary to use these digits when you program in "machine language." Imagine writing a sixteen bit instruction for a microprocessor system in 1s and 0s. It is much more efficient to use hexadecimal or octal; octal numbers are covered in Section 2–9. Hexadecimal is widely used in computer and microprocessor applications.

After completing this section, you should be able to

- **List the hexadecimal characters**
- **Count in hexadecimal**
- **Convert from binary to hexadecimal**
- **Convert from hexadecimal to binary**
- **Convert from hexadecimal to decimal**
- **Convert from decimal to hexadecimal**
- **Add hexadecimal numbers**
- **Determine the 2's complement of a hexadecimal number**
- **Subtract hexadecimal numbers**

The hexadecimal number system consists of digits 0–9 and letters A–F.

The **hexadecimal** number system has a base of sixteen; that is, it is composed of 16 **numeric** and alphabetic **characters.** Most digital systems process binary data in groups that are multiples of four bits, making the hexadecimal number very convenient because each hexadecimal digit represents a 4-bit binary number (as listed in Table 2–3).

TABLE 2–3		
DECIMAL	**BINARY**	**HEXADECIMAL**
0	0000	0
1	0001	1
2	0010	2
3	0011	3
4	0100	4
5	0101	5
6	0110	6
7	0111	7
8	1000	8
9	1001	9
10	1010	A
11	1011	B
12	1100	C
13	1101	D
14	1110	E
15	1111	F

Ten numeric digits and six alphabetic characters make up the hexadecimal number system. The use of letters A, B, C, D, E, and F to represent numbers may seem strange at first, but keep in mind that any number system is only a set of sequential symbols. If you understand what quantities these symbols represent, then the form of the symbols themselves is less important once you get accustomed to using them. We will use the subscript 16 to designate hexadecimal numbers to avoid confusion with decimal numbers. Sometimes you may see an "h" following a hexadecimal number.

Counting in Hexadecimal

How do you count in hexadecimal once you get to F? Simply start over with another column and continue as follows:

. . ., E, F, 10, 11, 12, 13, 14, 15, 16, 17, 18, 19, 1A, 1B, 1C, 1D, 1E, 1F, 20, 21, 22, 23, 24, 25, 26, 27, 28, 29, 2A, 2B, 2C, 2D, 2E, 2F, 30, 31, . . .

With two hexadecimal digits, you can count up to FF_{16}, which is decimal 255. To count beyond this, three hexadecimal digits are needed. For instance, 100_{16} is decimal 256, 101_{16} is decimal 257, and so forth. The maximum 3-digit hexadecimal number is FFF_{16}, or decimal 4095. The maximum 4-digit hexadecimal number is $FFFF_{16}$, which is decimal 65,535.

With computer memories in the gigabyte (GB) range, specifying a memory address in binary is quite cumbersome. For example, it takes 32 bits to specify an address in a 4 GB memory. It is much easier to express a 32-bit code using 8 hexadecimal digits.

SYSTEM NOTE

Binary-to-Hexadecimal Conversion

Converting a binary number to hexadecimal is a straightforward procedure. Simply break the binary number into 4-bit groups, starting at the right-most bit and replace each 4-bit group with the equivalent hexadecimal symbol.

EXAMPLE 2–24

Convert the following binary numbers to hexadecimal:

(a) 1100101001010111   **(b)** 111111000101101001

SOLUTION

(a) 1100101001010111
         C   A   5   7 $= CA57_{16}$

(b) 0011111000101101001
         3   F   1   6   9 $= 3F169_{16}$

Two zeros have been added in part (b) to complete a 4-bit group at the left.

RELATED PROBLEM

Convert the binary number 1001111011110011100 to hexadecimal.

Hexadecimal-to-Binary Conversion

To convert from a hexadecimal number to a binary number, reverse the process and replace each hexadecimal symbol with the appropriate four bits.

Hexadecimal is a convenient way to represent binary numbers.

E X A M P L E   2 – 2 5

Determine the binary numbers for the following hexadecimal numbers:

(a) $10A4_{16}$ **(b)** $CF8E_{16}$ **(c)** 9742_{16}

SOLUTION

In part (a), the MSB is understood to have three zeros preceding it, thus forming a 4-bit group.

RELATED PROBLEM

Convert the hexadecimal number 6BD3 to binary.

Conversion between hexadecimal and binary is direct and easy.

It should be clear that it is much easier to deal with a hexadecimal number than with the equivalent binary number. Since conversion is so easy, the hexadecimal system is widely used for representing binary numbers in programming, printouts, and displays.

Hexadecimal-to-Decimal Conversion

One way to find the decimal equivalent of a hexadecimal number is to first convert the hexadecimal number to binary and then convert from binary to decimal.

E X A M P L E   2 – 2 6

Convert the following hexadecimal numbers to decimal:

(a) $1C_{16}$ **(b)** $A85_{16}$

SOLUTION

Remember, convert the hexadecimal number to binary first, then to decimal.

(a)   1   C

$$\overbrace{00011100} = 2^4 + 2^3 + 2^2 = 16 + 8 + 4 = \mathbf{28}_{10}$$

(b)   A   8   5

$$\overbrace{101010000101} = 2^{11} + 2^9 + 2^7 + 2^2 + 2^0$$
$$= 2048 + 512 + 128 + 4 + 1 = \mathbf{2693}_{10}$$

RELATED PROBLEM

Convert the hexadecimal number 6BD to decimal.

Another way to convert a hexadecimal number to its decimal equivalent is to multiply the decimal value of each hexadecimal digit by its weight and then take the sum of these products. The weights of a hexadecimal number are increasing powers of 16 (from right to left). For a 4-digit hexadecimal number, the weights are

$$16^3 \quad 16^2 \quad 16^1 \quad 16^0$$
$$4096 \quad 256 \quad 16 \quad 1$$

EXAMPLE 2–27

Convert the following hexadecimal numbers to decimal:

(a) $E5_{16}$ **(b)** $B2F8_{16}$

SOLUTION

Recall from Table 2–3 that letters A through F represent decimal numbers 10 through 15, respectively.

(a) $E5_{16} = (E \times 16) + (5 \times 1) = (14 \times 16) + (5 \times 1) = 224 + 5 = \mathbf{229}_{10}$

(b) $B2F8_{16} = (B \times 4096) + (2 \times 256) + (F \times 16) + (8 \times 1)$
$$= (11 \times 4096) + (2 \times 256) + (15 \times 16) + (8 \times 1)$$
$$= \quad 45{,}056 \quad + \quad 512 \quad + \quad 240 \quad + \quad 8 \quad = \mathbf{45{,}816}_{10}$$

RELATED PROBLEM

Convert $60A_{16}$ to decimal.

CALCULATOR TUTORIAL

Conversion of a Hexadecimal Number to a Decimal Number

EXAMPLE

Convert hexadecimal 28A to decimal.

		HEX
TI-36X	**Step 1:**	[3rd] [(]

		A
	Step 2:	[2] [8] [3rd] [1/x]

		DEC
	Step 3:	[3rd] [EE] 650

Decimal-to-Hexadecimal Conversion

Repeated division of a decimal number by 16 will produce the equivalent hexadecimal number, formed by the remainders of the divisions. The first remainder produced is the least significant digit (LSD). Each successive division by 16 yields a remainder that becomes a digit in the equivalent hexadecimal number. This procedure is similar to repeated division by 2 for decimal-to-binary conversion that was covered in Section 2–3. Example 2–28 illustrates the procedure. Note that when a quotient has a fractional part, the fractional part is multiplied by the divisor to get the remainder.

┌─ E X A M P L E   2 – 2 8 ─────────────────────────────

Convert the decimal number 650 to hexadecimal by repeated division by 16.

SOLUTION

Hexadecimal
remainder

$$\frac{650}{16} = 40 .625 \rightarrow 0.625 \times 16 = 10 = \quad A$$

$$\frac{40}{16} = 2 .5 \longrightarrow 0.5 \times 16 = 8 = \quad 8$$

$$\frac{2}{16} = 0 .125 \longrightarrow 0.125 \times 16 = 2 = \quad 2$$

└─ Stop when whole number
   quotient is zero.

2 8 A   Hexadecimal number

MSD ┘   └ LSD

RELATED PROBLEM

Convert decimal 2591 to hexadecimal.

──

Hexadecimal Addition

Addition can be done directly with hexadecimal numbers by remembering that the hexadecimal digits 0 through 9 are equivalent to decimal digits 0 through 9 and that hexadecimal digits A through F are equivalent to decimal numbers 10 through 15. When adding two hexadecimal numbers, use the following rules. (Decimal numbers are indicated by a subscript 10.)

1. In any given column of an addition problem, think of the two hexadecimal digits in terms of their decimal values. For instance, $5_{16} = 5_{10}$ and $C_{16} = 12_{10}$.

2. If the sum of these two digits is 15_{10} or less, bring down the corresponding hexadecimal digit.

3. If the sum of these two digits is greater than 15_{10}, bring down the amount of the sum that exceeds 16_{10} and carry a 1 to the next column.

┌─ C A L C U L A T O R   T U T O R I A L ─────────────────

Conversion of a Decimal Number to a Hexadecimal Number

EXAMPLE

Convert decimal 650 to hexadecimal.

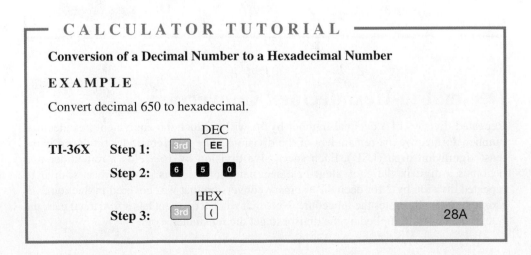

		DEC
TI-36X	**Step 1:**	[3rd] [EE]
	Step 2:	[6] [5] [0]
		HEX
	Step 3:	[3rd] [( ]

28A

──

━━ E X A M P L E   2 – 2 9 ━━

Add the following hexadecimal numbers:

(a) $23_{16} + 16_{16}$ **(b)** $58_{16} + 22_{16}$

(c) $2B_{16} + 84_{16}$ **(d)** $DF_{16} + AC_{16}$

SOLUTION

(a) 23_{16} right column: $3_{16} + 6_{16} = 3_{10} + 6_{10} = 9_{10} = 9_{16}$

    $\underline{+16_{16}}$ left column: $2_{16} + 1_{16} = 2_{10} + 1_{10} = 3_{10} = 3_{16}$

    39_{16}

(b) 58_{16} right column: $8_{16} + 2_{16} = 8_{10} + 2_{10} = 10_{10} = A_{16}$

    $\underline{+22_{16}}$ left column: $5_{16} + 2_{16} = 5_{10} + 2_{10} = 7_{10} = 7_{16}$

    $7A_{16}$

(c) $2B_{16}$ right column: $B_{16} + 4_{16} = 11_{10} + 4_{10} = 15_{10} = F_{16}$

    $\underline{+84_{16}}$ left column: $2_{16} + 8_{16} = 2_{10} + 8_{10} = 10_{10} = A_{16}$

    AF_{16}

(d) DF_{16} right column: $F_{16} + C_{16} = 15_{10} + 12_{10} = 27_{10}$

    $\underline{+AC_{16}}$ $27_{10} - 16_{10} = 11_{10} = B_{16}$ with a 1 carry

    $18B_{16}$ left column: $D_{16} + A_{16} + 1_{16} = 13_{10} + 10_{10} + 1_{10} = 24_{10}$

                  $24_{10} - 16_{10} = 8_{10} = 8_{16}$ with a 1 carry

RELATED PROBLEM

Add $4C_{16}$ and $3A_{16}$.

Hexadecimal Subtraction

As you have learned, the 2's complement allows you to subtract by adding binary numbers. Since a hexadecimal number can be used to represent a binary number, it can also be used to represent the 2's complement of a binary number.

There are three ways to get the 2's complement of a hexadecimal number. Method 1 is the most common and easiest to use. Methods 2 and 3 are alternate methods.

Method 1:   Convert the hexadecimal number to binary. Take the 2's complement of the binary number. Convert the result to hexadecimal. This is illustrated in Figure 2–4.

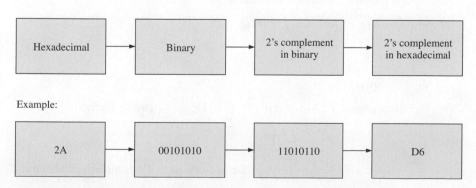

Example:

FIGURE 2–4   **Getting the 2's complement of a hexadecimal number, Method 1.**

Method 2: Subtract the hexadecimal number from the maximum hexadecimal number and add 1. This is illustrated in Figure 2–5.

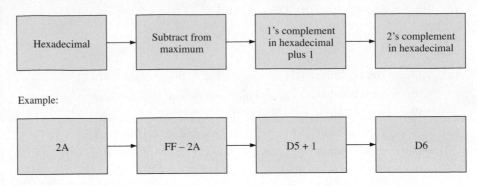

Example:

FIGURE 2–5   **Getting the 2's complement of a hexadecimal number, Method 2.**

Method 3: Write the sequence of single hexadecimal digits. Write the sequence in reverse below the forward sequence. The 1's complement of each hex digit is the digit directly below it. Add 1 to the resulting number to get the 2's complement. This is illustrated in Figure 2–6.

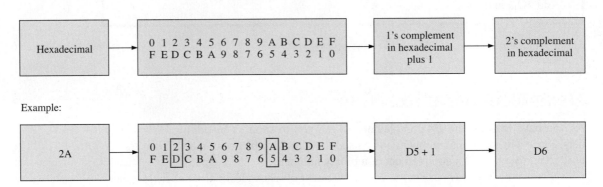

Example:

FIGURE 2–6   **Getting the 2's complement of a hexadecimal number, Method 3.**

EXAMPLE 2 – 30

Subtract the following hexadecimal numbers:

(a) $84_{16} - 2A_{16}$ **(b)** $C3_{16} - 0B_{16}$

SOLUTION

(a) $2A_{16} = 00101010$
  2's complement of $2A_{16} = 11010110 = D6_{16}$ (using Method 1)

$$
\begin{array}{r}
84_{16} \\
+D6_{16} \quad \text{Add} \\
\hline
\cancel{1}5A_{16} \quad \text{Drop carry, as in 2's complement addition}
\end{array}
$$

The difference is **$5A_{16}$**.

(b) $0B_{16} = 00001011$

2's complement of $0B_{16} = 11110101 = F5_{16}$ (using Method 1)

$$C3_{16}$$
$$\underline{+\ F5_{16}}\quad \text{Add}$$
$$\cancel{1}B8_{16}\quad \text{Drop carry}$$

The difference is $\mathbf{B8}_{16}$.

RELATED PROBLEM

Subtract 173_{16} from BCD_{16}.

SECTION 2–8 CHECKUP

1. Convert the following binary numbers to hexadecimal:

   (a) 10110011   **(b)** 110011101000

2. Convert the following hexadecimal numbers to binary:

   (a) 57_{16} **(b)** $3A5_{16}$ **(c)** $F80B_{16}$

3. Convert $9B30_{16}$ to decimal.

4. Convert the decimal number 573 to hexadecimal.

5. Add the following hexadecimal numbers directly:

   (a) $18_{16} + 34_{16}$ **(b)** $3F_{16} + 2A_{16}$

6. Subtract the following hexadecimal numbers:

   (a) $75_{16} - 21_{16}$ **(b)** $94_{16} - 5C_{16}$

2–9 OCTAL NUMBERS

Like the hexadecimal number system, the octal number system provides a convenient way to express binary numbers and codes. However, it is used less frequently than hexadecimal in conjunction with computers and microprocessors to express binary quantities for input and output purposes.

After completing this section, you should be able to

- **Write the digits of the octal number system**
- **Convert from octal to decimal**
- **Convert from decimal to octal**
- **Convert from octal to binary**
- **Convert from binary to octal**

The **octal** number system is composed of eight digits, which are

$$0, 1, 2, 3, 4, 5, 6, 7$$

To count above 7, begin another column and start over:

$$10, 11, 12, 13, 14, 15, 16, 17, 20, 21, \ldots$$

Counting in octal is similar to counting in decimal, except that the digits 8 and 9 are not used. To distinguish octal numbers from decimal numbers or hexadecimal numbers, we will use the subscript 8 to indicate an octal number. For instance, 15_8 in octal is equivalent to 13_{10} in decimal and D in hexadecimal. Sometimes you may see an "o" or a "Q" following an octal number.

The octal number system has a base of 8.

Octal-to-Decimal Conversion

Since the octal number system has a base of eight, each successive digit position is an increasing power of eight, beginning in the right-most column with 8^0. The evaluation of an octal number in terms of its decimal equivalent is accomplished by multiplying each digit by its weight and summing the products, as illustrated here for 2374_8.

$$\text{Weight:} \quad 8^3 \; 8^2 \; 8^1 \; 8^0$$

$$\text{Octal number:} \quad 2 \;\; 3 \;\; 7 \;\; 4$$

$$
\begin{aligned}
2374_8 &= (2 \times 8^3) \;\; + (3 \times 8^2) + (7 \times 8^1) + (4 \times 8^0) \\
&= (2 \times 512) + (3 \times 64) + (7 \times 8) \;\; + (4 \times 1) \\
&= \quad 1024 \quad + \quad 192 \quad + \quad 56 \quad + \quad 4 \quad = 1276_{10}
\end{aligned}
$$

Decimal-to-Octal Conversion

A method of converting a decimal number to an octal number is the repeated division-by-8 method, which is similar to the method used in the conversion of decimal numbers to binary or to hexadecimal. To show how it works, let's convert the decimal number 359 to octal. Each successive division by 8 yields a remainder that becomes a digit in the equivalent octal number. The first remainder generated is the least significant digit (LSD).

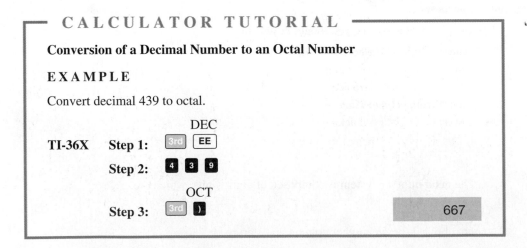

CALCULATOR TUTORIAL

Conversion of a Decimal Number to an Octal Number

EXAMPLE

Convert decimal 439 to octal.

			DEC
TI-36X	Step 1:	3rd	EE
	Step 2:	4 3 9	
			OCT
	Step 3:	3rd	)	667

Octal is a convenient way to represent binary numbers, but it is not as commonly used as hexadecimal.

Octal-to-Binary Conversion

Because each octal digit can be represented by a 3-bit binary number, it is very easy to convert from octal to binary. Each octal digit is represented by three bits as shown in Table 2–4.

TABLE 2–4 • Octal/binary conversion.								
OCTAL DIGIT	0	1	2	3	4	5	6	7
BINARY	000	001	010	011	100	101	110	111

To convert an octal number to a binary number, simply replace each octal digit with the appropriate three bits.

EXAMPLE 2–31

Convert each of the following octal numbers to binary:

(a) 13_8 **(b)** 25_8 **(c)** 140_8 **(d)** 7526_8

SOLUTION

(a) 1 3
  ↓ ↓
  001011

(b) 2 5
  ↓ ↓
  010101

(c) 1 4 0
  ↓ ↓ ↓
  001100000

(d) 7 5 2 6
  ↓ ↓ ↓ ↓
  111101010110

RELATED PROBLEM

Convert each of the binary numbers to decimal and verify that each value agrees with the decimal value of the corresponding octal number.

Binary-to-Octal Conversion

Conversion of a binary number to an octal number is the reverse of the octal-to-binary conversion. The procedure is as follows: Start with the right-most group of three bits and, moving from right to left, convert each 3-bit group to the equivalent octal digit. If there are not three bits available for the left-most group, add either one or two zeros to make a complete group. These leading zeros do not affect the value of the binary number.

EXAMPLE 2–32

Convert each of the following binary numbers to octal:

(a) 110101   **(b)** 101111001

(c) 100110011010   **(d)** 11010000100

SOLUTION

(a) 110101
  ↓ ↓
  6 5 = **65**$_8$

(b) 101111001
  ↓ ↓ ↓
  5 7 1 = **571**$_8$

(c) 100110011010
  ↓ ↓ ↓ ↓
  4 6 3 2 = **4632**$_8$

(d) 011010000100
  ↓ ↓ ↓ ↓
  3 2 0 4 = **3204**$_8$

RELATED PROBLEM

Convert the binary number 1010101000111110010 to octal.

── **SECTION 2–9 CHECKUP** ──

1. Convert the following octal numbers to decimal:

   (a) 73_8 (b) 125_8

2. Convert the following decimal numbers to octal:

   (a) 98_{10} (b) 163_{10}

3. Convert the following octal numbers to binary:

   (a) 46_8 (b) 723_8 (c) 5624_8

4. Convert the following binary numbers to octal:

   (a) 110101111   (b) 1001100010   (c) 10111111001

2–10 BINARY CODED DECIMAL (BCD)

Binary coded decimal (BCD) is a way to express each of the decimal digits with a binary code. There are only ten code groups in the BCD system, so it is very easy to convert between decimal and BCD. Because we like to read and write in decimal, the BCD code provides an excellent interface to binary systems. Examples of such interfaces are keypad inputs and digital readouts.

After completing this section, you should be able to

- **Convert each decimal digit to BCD**
- **Express decimal numbers in BCD**
- **Convert from BCD to decimal**
- **Add BCD numbers**

The 8421 BCD Code

In BCD, 4 bits represent each decimal digit.

The 8421 code is a common type of **BCD** (binary coded decimal) code. Binary coded decimal means that each decimal digit, 0 through 9, is represented by a binary code of four bits. The designation 8421 indicates the binary weights of the four bits ($2^3, 2^2, 2^1, 2^0$). The ease of conversion between 8421 code numbers and the familiar decimal numbers is the main advantage of this code. All you have to remember are the ten binary combinations that represent the ten decimal digits as shown in Table 2–5. The 8421 code is the predominant BCD code, and when we refer to BCD, we always mean the 8421 code unless otherwise stated.

TABLE 2–5 • Decimal/BCD conversion.										
DECIMAL DIGIT	0	1	2	3	4	5	6	7	8	9
BCD	0000	0001	0010	0011	0100	0101	0110	0111	1000	1001

INVALID CODES You should realize that, with four bits, sixteen numbers (0000 through 1111) can be represented but that, in the 8421 code, only ten of these are used. The six code combinations that are not used—1010, 1011, 1100, 1101, 1110, and 1111—are invalid in the 8421 BCD code.

To express any decimal number in BCD, simply replace each decimal digit with the appropriate 4-bit code, as shown by Example 2–33.

── **EXAMPLE 2–33** ──

Convert each of the following decimal numbers to BCD:

(a) 35   (b) 98   (c) 170   (d) 2469

SOLUTION

(a) 3 5   **(b)** 9 8   **(c)** 1 7 0   **(d)** 2 4 6 9
00110101   10011000   000101110000   0010010001101001

RELATED PROBLEM

Convert the decimal number 9673 to BCD.

It is equally easy to determine a decimal number from a BCD number. Start at the right-most bit and break the code into groups of four bits. Then write the decimal digit represented by each 4-bit group.

EXAMPLE 2–34

Convert each of the following BCD codes to decimal:

(a) 10000110   **(b)** 001101010001   **(c)** 1001010001110000

SOLUTION

(a) 10000110   **(b)** 001101010001   **(c)** 1001010001110000
        8   6   3   5   1   9   4   7   0

RELATED PROBLEM

Convert the BCD code 10000010001001110110 to decimal.

APPLICATIONS   Digital clocks, digital thermometers, digital meters, and other devices with seven-segment displays typically use BCD code to simplify the displaying of decimal numbers. BCD is not as efficient as straight binary for calculations, but it is particularly useful if only limited processing is required, such as in a digital thermometer.

BCD Addition

BCD is a numerical code and can be used in arithmetic operations. Addition is the most important operation because the other three operations (subtraction, multiplication, and division) can be accomplished by the use of addition. Here is how to add two BCD numbers:

Step 1:   Add the two BCD numbers, using the rules for binary addition in Section 2–4.

Step 2:   If a 4-bit sum is equal to or less than 9, it is a valid BCD number.

Step 3:   If a 4-bit sum is greater than 9, or if a carry out of the 4-bit group is generated, it is an invalid result. Add 6 (0110) to the 4-bit sum in order to skip the six invalid states and return the code to 8421. If a carry results when 6 is added, simply add the carry to the next 4-bit group.

BCD is sometimes used for arithmetic operations in computers. To represent BCD numbers in a computer, they usually are "packed," so that eight bits has two BCD digits. Normally, a computer will add numbers as if they were straight binary. Special instructions are available for computer programmers to correct the results when BCD numbers are added or subtracted. For example, in Assembly Language, the programmer will include a DAA (Decimal Adjust for Addition) instruction to automatically correct the answer to BCD following an addition.

SYSTEM NOTE

Example 2–35 illustrates BCD additions in which the sum in each 4-bit column is equal to or less than 9, and the 4-bit sums are therefore valid BCD numbers. Example 2–36 illustrates the procedure in the case of invalid sums (greater than 9 or a carry).

An alternative method to add BCD numbers is to convert them to decimal, perform the addition, and then convert the answer back to BCD.

EXAMPLE 2–35

Add the following BCD numbers:

(a) 0011 + 0100   (b) 00100011 + 00010101

(c) 10000110 + 00010011   (d) 010001010000 + 010000010111

SOLUTION

The decimal number additions are shown for comparison.

```
(a)    0011      3              (b)    0010  0011      23
     + 0100    + 4                   + 0001  0101    + 15
     ──────    ───                   ──────────────  ────
       1111      7                     0011  1000      38

(c)    1000  0110    86         (d)   0100  0101  0000    450
     + 0001  0011  + 13              + 0100  0001  0111  + 417
     ──────────────  ───             ────────────────────  ────
       1001  1001    99               1000  0110  0111    867
```

Note that in each case the sum in any 4-bit column does not exceed 9, and the results are valid BCD numbers.

RELATED PROBLEM

Add the BCD numbers: 1001000001000011 + 0000100100100101.

EXAMPLE 2–36

Add the following BCD numbers:

(a) 1001 + 0100   (b) 1001 + 1001

(c) 00010110 + 00010101   (d) 01100111 + 01010011

SOLUTION

The decimal number additions are shown for comparison.

```
(a)           1001                                     9
            + 0100                                   + 4
            ──────                                   ───
              1101      Invalid BCD number (>9)       13
            + 0110      Add 6
       ───────────────
       0001    0011     Valid BCD number
         ↓       ↓
         1       3

(b)           1001                                     9
            + 1001                                   + 9
            ──────                                   ───
         1    0010      Invalid because of carry      18
            + 0110      Add 6
       ───────────────
       0001    1000     Valid BCD number
         ↓       ↓
         1       8
```

(c)

0001	0110
+ 0001	0101

0010   1011   Right group is invalid (>9),
left group is valid.

+ 0110   Add 6 to invalid code. Add
carry, 0001, to next group.

0011   0001   Valid BCD number

⤷   ⤷
3   1

| 16 |
| + 15 |
| 31 |

(d)

	0110	0111
	+ 0101	0011

1011   1010   Both groups are invalid (>9)

+ 0110   + 0110   Add 6 to both groups.

0001   0010   0000   Valid BCD number

⤷   ⤷   ⤷
1   2   0

| 67 |
| + 53 |
| 120 |

RELATED PROBLEM

Add the BCD numbers: 01001000 + 00110100.

SECTION 2–10 CHECKUP

1. What is the binary weight of each 1 in the following BCD numbers?

   (a) 0010   **(b)** 1000

   (c) 0001   **(d)** 0100

2. Convert the following decimal numbers to BCD:

   (a) 6   **(b)** 15

   (c) 273   **(d)** 849

3. What decimal numbers are represented by each BCD code?

   (a) 10001001

   (b) 001001111000

   (c) 000101010111

4. In BCD addition, when is a 4-bit sum invalid?

2–11 DIGITAL CODES

Many specialized codes are used in digital systems. You have just learned about the BCD code; now let's look at a few others. Some codes are strictly numeric, like BCD, and others are alphanumeric; that is, they are used to represent numbers, letters, symbols, and instructions. The codes introduced in this section are the Gray code, the ASCII code, and the Unicode.

After completing this section, you should be able to

- **Explain the advantage of the Gray code**
- **Convert between Gray code and binary**
- **Use the ASCII code**
- **Discuss the Unicode**

The Gray Code

The single bit change characteristic of the Gray code minimizes the chance for error.

The **Gray code** is unweighted and is not an arithmetic code; that is, there are no specific weights assigned to the bit positions. The important feature of the Gray code is that *it exhibits only a single bit change from one code word to the next in sequence.* This property is important in many applications, such as shaft position encoders, where error susceptibility increases with the number of bit changes between adjacent numbers in a sequence.

Table 2–6 is a listing of the 4-bit Gray code for decimal numbers 0 through 15. Binary numbers are shown in the table for reference. Like binary numbers, *the Gray code can have any number of bits.* Notice the single-bit change between successive Gray code words. For instance, in going from decimal 3 to decimal 4, the Gray code changes from 0010 to 0110, while the binary code changes from 0011 to 0100, a change of three bits. The only bit change in the Gray code is in the third bit from the right: the other bits remain the same.

TABLE 2–6 • Four-bit Gray code.					
DECIMAL	**BINARY**	**GRAY CODE**	**DECIMAL**	**BINARY**	**GRAY CODE**
0	0000	0000	8	1000	1100
1	0001	0001	9	1001	1101
2	0010	0011	10	1010	1111
3	0011	0010	11	1011	1110
4	0100	0110	12	1100	1010
5	0101	0111	13	1101	1011
6	0110	0101	14	1110	1001
7	0111	0100	15	1111	1000

BINARY-TO-GRAY CODE CONVERSION   Conversion between binary code and Gray code is sometimes useful. The following rules explain how to convert from a binary number to a Gray code word:

1. The most significant bit (left-most) in the Gray code is the same as the corresponding MSB in the binary number.

2. Going from left to right, add each adjacent pair of binary code bits to get the next Gray code bit. Discard carries.

For example, the conversion of the binary number 10110 to Gray code is as follows:

The Gray code is 11101.

GRAY-TO-BINARY CODE CONVERSION   To convert from Gray code to binary, use a similar method; however, there are some differences. The following rules apply:

1. The most significant bit (left-most) in the binary code is the same as the corresponding bit in the Gray code.

2. Add each binary code bit generated to the Gray code bit in the next adjacent position. Discard carries.

For example, the conversion of the Gray code word 11011 to binary is as follows:

The binary number is 10010.

EXAMPLE 2–37

(a) Convert the binary number 11000110 to Gray code.

(b) Convert the Gray code 10101111 to binary.

SOLUTION

(a) Binary to Gray code:

(b) Gray code to binary:

$$
\begin{array}{cccccccc}
1 & 0 & 1 & 0 & 1 & 1 & 1 & 1 \\
1 & 1 & 0 & 0 & 1 & 0 & 1 & 0
\end{array}
$$

RELATED PROBLEM

(a) Convert binary 101101 to Gray code.

(b) Convert Gray code 100111 to binary.

SYSTEM EXAMPLE 2–2

SHAFT ENCODER

The concept of a 3-bit shaft position encoder is shown in Figure 2–7. Basically, there are three concentric rings that are segmented into eight sectors. The more sectors there are, the more accurately the position can be represented, but we are using only eight to illustrate. Each sector of each ring is either reflective or nonreflective. As the rings rotate with the shaft, they come under an IR emitter that produces three separate IR beams. A 1 is indicated where there is a reflected beam, and a 0 is indicated where there is no reflected beam. The IR detector senses the presence or absence of reflected beams and produces a corresponding 3-bit code. The IR emitter/detector is in a fixed position. As the shaft rotates counterclockwise through 360°, the eight sectors move under the three beams. Each beam is either reflected or absorbed by the sector surface to represent a binary or Gray code number that indicates the shaft position.

In Figure 2–7(a), the sectors are arranged in a straight binary pattern, so that the detector output goes from 000 to 001 to 010 to 011 and so on. When a beam is aligned over a reflective sector, the output is 1; when a beam is aligned over a nonreflective sector, the output is 0. If one beam is slightly ahead of the others during the transition from one sector to the next, an erroneous output can occur. Consider what happens when the beams are on the 111 sector and about to enter the 000 sector. If the MSB beam is slightly ahead, the position would be incorrectly indicated by a transitional 011 instead of a 111 or a 000. In this type of application, it is virtually impossible to maintain precise mechanical alignment

(a) Binary code

(b) Gray code

FIGURE 2–7   **A simplified illustration of how the Gray code solves the error problem in shaft position encoders. Three bits are shown to illustrate the concept, although most shaft encoders use more than 10 bits to achieve a higher resolution.**

of the IR emitter/detector beams; therefore, some error will usually occur at many of the transitions between sectors.

The Gray code is used to eliminate the error problem which is inherent in the binary code. As shown in Figure 2–7(b), the Gray code assures that only one bit will change between adjacent sectors. This means that even though the beams may not be in precise alignment, there will never be a transitional error. For example, let's again consider what happens when the beams are on the 111 sector and about to move into the next sector, 101. The only two possible outputs during the transition are 111 and 101, no matter how the beams are aligned. A similar situation occurs at the transitions between each of the other sectors.

Alphanumeric Codes

In order to communicate, you need not only numbers, but also letters and other symbols. In the strictest sense, **alphanumeric** codes are codes that represent numbers and alphabetic characters (letters). Most such codes, however, also represent other characters such as symbols and various instructions necessary for conveying information.

At a minimum, an alphanumeric code must represent 10 decimal digits and 26 letters of the alphabet, for a total of 36 items. This number requires six bits in each code combination because five bits are insufficient ($2^5 = 32$). There are 64 total combinations of six bits, so there are 28 unused code combinations. Obviously, in many applications, symbols other than just numbers and letters are necessary to communicate completely. You need spaces, periods, colons, semicolons, question marks, etc. You also need instructions to tell the receiving system what to do with the information. With codes that are six bits long, you can handle decimal numbers, the alphabet, and 28 other symbols. This should give you an idea of the requirements for a basic alphanumeric code. The ASCII is the most common alphanumeric code and is covered next.

ASCII

ASCII is the abbreviation for American Standard Code for Information Interchange. Pronounced "askee," ASCII is a universally accepted alphanumeric code used in most computers and other electronic equipment. Most computer keyboards are standardized with the ASCII. When you enter a letter, a number, or control command, the corresponding ASCII code goes into the computer.

A computer keyboard has a dedicated microprocessor that constantly scans keyboard circuits to detect when a key has been pressed and released. A unique scan code is produced by computer software representing that particular key. The scan code is then converted to an alphanumeric code (ASCII) for use by the computer.

SYSTEM NOTE

ASCII has 128 characters and symbols represented by a 7-bit binary code. Actually, ASCII can be considered an 8-bit code with the MSB always 0. This 8-bit code is 00 through 7F in hexadecimal. The first thirty-two ASCII characters are nongraphic commands that are never printed or displayed and are used only for control purposes. Examples of the control characters are "null," "line feed," "start of text," and "escape." The other characters are graphic symbols that can be printed or displayed and include the letters of the alphabet (lowercase and uppercase), the ten decimal digits, punctuation signs, and other commonly used symbols.

Table 2–7 is a listing of the ASCII code showing the decimal, hexadecimal, and binary representations for each character and symbol. The left section of the table lists the names of the 32 control characters (00 through 1F hexadecimal). The graphic symbols are listed in the rest of the table (20 through 7F hexadecimal).

EXAMPLE 2–38

Determine the binary ASCII codes that are entered from the computer's keyboard when the following C language program statement is typed in. Also express each code in hexadecimal.

$$\text{if } (x > 5)$$

SOLUTION

The ASCII code for each symbol is found in Table 2–7.

Symbol	Binary	Hexadecimal
i	1101001	69_{16}
f	1100110	66_{16}
Space	0100000	20_{16}
(	0101000	28_{16}
x	1111000	78_{16}
>	0111110	$3E_{16}$
5	0110101	35_{16}
)	0101001	29_{16}

RELATED PROBLEM

Determine the sequence of ASCII codes required for the following C program statement and express them in hexadecimal:

$$\text{if } (y < 8)$$

TABLE 2–7 • American Standard Code for Information Interchange (ASCII).

CONTROL CHARACTERS				GRAPHIC SYMBOLS											
NAME	DEC	BINARY	HEX	SYMBOL	DEC	BINARY	HEX	SYMBOL	DEC	BINARY	HEX	SYMBOL	DEC	BINARY	HEX
NUL	0	0000000	00	space	32	0100000	20	@	64	1000000	40	`	96	1100000	60
SOH	1	0000001	01	!	33	0100001	21	A	65	1000001	41	a	97	1100001	61
STX	2	0000010	02	"	34	0100010	22	B	66	1000010	42	b	98	1100010	62
ETX	3	0000011	03	#	35	0100011	23	C	67	1000011	43	c	99	1100011	63
EOT	4	0000100	04	$	36	0100100	24	D	68	1000100	44	d	100	1100100	64
ENQ	5	0000101	05	%	37	0100101	25	E	69	1000101	45	e	101	1100101	65
ACK	6	0000110	06	&	38	0100110	26	F	70	1000110	46	f	102	1100110	66
BEL	7	0000111	07	'	39	0100111	27	G	71	1000111	47	g	103	1100111	67
BS	8	0001000	08	(	40	0101000	28	H	72	1001000	48	h	104	1101000	68
HT	9	0001001	09	)	41	0101001	29	I	73	1001001	49	i	105	1101001	69
LF	10	0001010	0A	*	42	0101010	2A	J	74	1001010	4A	j	106	1101010	6A
VT	11	0001011	0B	+	43	0101011	2B	K	75	1001011	4B	k	107	1101011	6B
FF	12	0001100	0C	,	44	0101100	2C	L	76	1001100	4C	l	108	1101100	6C
CR	13	0001101	0D	–	45	0101101	2D	M	77	1001101	4D	m	109	1101101	6D
SO	14	0001110	0E	.	46	0101110	2E	N	78	1001110	4E	n	110	1101110	6E
SI	15	0001111	0F	/	47	0101111	2F	O	79	1001111	4F	o	111	1101111	6F
DLE	16	0010000	10	0	48	0110000	30	P	80	1010000	50	p	112	1110000	70
DC1	17	0010001	11	1	49	0110001	31	Q	81	1010001	51	q	113	1110001	71
DC2	18	0010010	12	2	50	0110010	32	R	82	1010010	52	r	114	1110010	72
DC3	19	0010011	13	3	51	0110011	33	S	83	1010011	53	s	115	1110011	73
DC4	20	0010100	14	4	52	0110100	34	T	84	1010100	54	t	116	1110100	74
NAK	21	0010101	15	5	53	0110101	35	U	85	1010101	55	u	117	1110101	75
SYN	22	0010110	16	6	54	0110110	36	V	86	1010110	56	v	118	1110110	76
ETB	23	0010111	17	7	55	0110111	37	W	87	1010111	57	w	119	1110111	77
CAN	24	0011000	18	8	56	0111000	38	X	88	1011000	58	x	120	1111000	78
EM	25	0011001	19	9	57	0111001	39	Y	89	1011001	59	y	121	1111001	79
SUB	26	0011010	1A	:	58	0111010	3A	Z	90	1011010	5A	z	122	1111010	7A
ESC	27	0011011	1B	;	59	0111011	3B	[	91	1011011	5B	{	123	1111011	7B
FS	28	0011100	1C	<	60	0111100	3C	\	92	1011100	5C	|	124	1111100	7C
GS	29	0011101	1D	=	61	0111101	3D	]	93	1011101	5D	}	125	1111101	7D
RS	30	0011110	1E	>	62	0111110	3E	^	94	1011110	5E	~	126	1111110	7E
US	31	0011111	1F	?	63	0111111	3F	_	95	1011111	5F	Del	127	1111111	7F

THE ASCII CONTROL CHARACTERS   The first thirty-two codes in the ASCII table (Table 2–7) represent the control characters. These are used to allow devices such as a computer and printer to communicate with each other when passing information and data. Table 2–8 lists the control characters and the control key function that allows them to be entered directly from an ASCII keyboard by pressing the control key (CTRL) and the corresponding symbol. A brief description of each control character is also given. The

TABLE 2–8 • ASCII control characters.				
NAME	**DECIMAL**	**HEX**	**KEYS**	**DESCRIPTION**
NUL	0	00	CTRL @	null character
SOH	1	01	CTRL A	start of header
STX	2	02	CTRL B	start of text
ETX	3	03	CTRL C	end of text
EOT	4	04	CTRL D	end of transmission
ENQ	5	05	CTRL E	enquire
ACK	6	06	CTRL F	acknowledge
BEL	7	07	CTRL G	bell
BS	8	08	CTRL H	backspace
HT	9	09	CTRL I	horizontal tab
LF	10	0A	CTRL J	line feed
VT	11	0B	CTRL K	vertical tab
FF	12	0C	CTRL L	form feed (new page)
CR	13	0D	CTRL M	carriage return
SO	14	0E	CTRL N	shift out
SI	15	0F	CTRL O	shift in
DLE	16	10	CTRL P	data link escape
DC1	17	11	CTRL Q	device control 1
DC2	18	12	CTRL R	device control 2
DC3	19	13	CTRL S	device control 3
DC4	20	14	CTRL T	device control 4
NAK	21	15	CTRL U	negative acknowledge
SYN	22	16	CTRL V	synchronize
ETB	23	17	CTRL W	end of transmission block
CAN	24	18	CTRL X	cancel
EM	25	19	CTRL Y	end of medium
SUB	26	1A	CTRL Z	substitute
ESC	27	1B	CTRL [	escape
FS	28	1C	CTRL /	file separator
GS	29	1D	CTRL]	group separator
RS	30	1E	CTRL ^	record separator
US	31	1F	CTRL _	unit separator

descriptions are based on obsolete teletype requirements, and the codes are generally used for different purposes than the description implies.

Extended ASCII Characters

In addition to the 128 standard ASCII characters, there are an additional 128 characters that were adopted by IBM for use in their PCs (personal computers). Because of the popularity of the PC, these particular extended ASCII characters are also used in applications other than PCs and have become essentially an unofficial standard.

The extended ASCII characters are represented by an 8-bit code series from hexadecimal 80 to hexadecimal FF and can be grouped into the following general categories: foreign (non-English) alphabetic characters, foreign currency symbols, Greek letters, mathematical symbols, drawing characters, bar graphing characters, and shading characters.

Unicode

Unicode provides the ability to encode all of the characters used for the written languages of the world by assigning each character a unique numeric value and name utilizing the universal character set (UCS). It is applicable in computer applications dealing with multilingual text, mathematical symbols, or other technical characters.

Unicode has a wide array of characters, and their various encoding forms have begun to supplant ASCII in many environments. While ASCII basically uses 7-bit codes, Unicode uses relatively abstract "code points"—non-negative integer numbers—that map sequences of one or more bytes, using different encoding forms and schemes. To permit compatibility, Unicode assigns the first 128 code points to the same characters as ASCII. One can, therefore, think of ASCII as a 7-bit encoding scheme for a very small subset of Unicode and of the UCS.

Unicode consists of about 100,000 characters, a set of code charts for visual reference, an encoding methodology and set of standard character encodings, and an enumeration of character properties such as uppercase and lowercase. It also consists of a number of related items, such as character properties, rules for text normalization, decomposition, collation, rendering, and bidirectional display order (for the correct display of text containing both right-to-left scripts, such as Arabic or Hebrew, and left-to-right scripts).

SECTION 2–11 CHECKUP

1. Convert the following binary numbers to the Gray code:

   (a) 1100   (b) 1010   (c) 11010

2. Convert the following Gray codes to binary:

   (a) 1000   (b) 1010   (c) 11101

3. What is the ASCII representation for each of the following characters? Express each as a bit pattern and in hexadecimal notation.

   (a) K   (b) r   (c) $ (d) +

2–12 ERROR DETECTION CODES

In this section, two methods for adding bits to codes to detect a single-bit error are discussed. The parity method of error detection is introduced, and the cyclic redundancy check is discussed.

After completing this section, you should be able to

- **Determine if there is an error in a code based on the parity bit**
- **Assign the proper parity bit to a code**
- **Explain the cyclic redundancy check (CRC)**

Parity Method for Error Detection

Many systems use a parity bit as a means for bit **error detection.** Any group of bits contain either an even or an odd number of 1s. A parity bit is attached to a group of bits to make the total number of 1s in a group always even or always odd. An even parity bit makes the total number of 1s even, and an odd parity bit makes the total odd.

A parity bit tells if the number of 1s in a group of bits is odd or even.

A given system operates with even or odd **parity,** but not both. For instance, if a system operates with even parity, a check is made on each group of bits received to make sure the total number of 1s in that group is even. If there is an odd number of 1s, an error has occurred.

As an illustration of how parity bits are attached to a code, Table 2–9 lists the parity bits for each BCD number for both even and odd parity. The parity bit for each BCD number is in the P column.

TABLE 2–9 • The BCD code with parity bits.

EVEN PARITY		ODD PARITY	
P	BCD	P	BCD
0	0000	1	0000
1	0001	0	0001
1	0010	0	0010
0	0011	1	0011
1	0100	0	0100
0	0101	1	0101
0	0110	1	0110
1	0111	0	0111
1	1000	0	1000
0	1001	1	1001

The parity bit can be attached to the code at either the beginning or the end, depending on system design. Notice that the total number of 1s, including the parity bit, is always even for even parity and always odd for odd parity.

DETECTING AN ERROR A parity bit provides for the detection of a single bit error (or any odd number of errors, which is very unlikely) but cannot check for two errors in one group. For instance, let's assume that we wish to transmit the BCD code 0101. (Parity can be used with any number of bits; we are using four for illustration.) The total code transmitted, including the even parity bit, is

```
          ┌──────Even parity bit
      ↓
    00101
        ↑
        └──BCD code
```

Now let's assume that an error occurs in the third bit from the left (the 1 becomes a 0).

```
          ┌──────Even parity bit
      ↓
    00001
        ↑
        └──Bit errror
```

When this code is received, the parity check circuitry determines that there is only a single 1 (odd number), when there should be an even number of 1s. Because an even number of 1s does not appear in the code when it is received, an error is indicated.

An odd parity bit also provides in a similar manner for the detection of a single error in a given group of bits.

EXAMPLE 2–39

Assign the proper even parity bit to the following code groups:

(a) 1010   (b) 111000   (c) 101101

(d) 1000111001001   (e) 101101011111

SOLUTION

Make the parity bit either 1 or 0 as necessary to make the total number of 1s even. The parity bit will be the left-most bit (color).

(a) **0**1010   (b) **1**111000   (c) **0**101101

(d) **0**100011100101   (e) **1**101101011111

RELATED PROBLEM

Add an even parity bit to the 7-bit ASCII code for the letter K.

EXAMPLE 2–40

An odd parity system receives the following code groups: 10110, 11010, 110011, 110101110100, and 1100010101010. Determine which groups, if any, are in error.

SOLUTION

Since odd parity is required, any group with an even number of 1s is incorrect. The following groups are in error: **110011** and **1100010101010.**

RELATED PROBLEM

The following ASCII character is received by an odd parity system: 00110111. Is it correct?

Cyclic Redundancy Check

The **cyclic redundancy check (CRC)** is a widely used code used for detecting one- and two-bit transmission errors when digital data is transferred on a communication link. The communication link can be between two computers that are connected to a network or between a digital storage device (such as a CD, DVD, or a hard drive) and a PC. If it is properly designed, the CRC can also detect multiple errors for a number of bits in sequence (burst errors). In CRC, a certain number of check bits, sometimes called a *checksum,* are appended to the data bits (added to end) that are being transmitted. The transmitted data is tested by the receiver for errors using the CRC. Not every possible error can be identified, but the CRC is much more efficient than just a simple parity check.

CRC is often described mathematically as the division of two polynomials to generate a remainder. A polynomial is a mathematical expression that is a sum of terms with positive exponents. When the coefficients are limited to 1s and 0s, it is called a *univariate polynomial.* An example of a univariate polynomial is $1x^3 + 0x^2 + 1x^1 + 1x^0$ or simply $x^3 + x^1 + x^0$, which can be fully described by the 4-bit binary number 1011. Most cyclic redundancy checks use a 16-bit or larger polynomial, but for simplicity the process is illustrated here with four bits.

MODULO-2 OPERATIONS   Simply put, CRC is based on the division of two binary numbers; and, as you know, division is just a series of subtractions and shifts. To do subtraction, a method called *modulo-2* addition can be used. Modulo-2 addition (or subtraction) is the same as binary addition with the carries discarded, as shown in the truth table in Table 2–10. **Truth tables** are widely used to describe the operation of

TABLE 2–10 •
Modulo-2 operation.

INPUT BITS	OUTPUT BIT
0 0	0
0 1	1
1 0	1
1 1	0

logic circuits, as you will learn in Chapter 3. With two bits, there is a total of four possible combinations, as shown in the table. This particular table describes the modulo-2 operation also known as *exclusive-OR* and can be implemented with a logic gate that will be introduced in Chapter 3. A simple rule for modulo-2 is that the output is 1 if the inputs are different; otherwise, it is 0.

CRC PROCESS   The process is as follows:

1. Select a fixed generator code; it can have fewer bits than the data bits to be checked. This code is understood in advance by both the sending and receiving devices and must be the same for both.

2. Append a number of 0s equal to the number of bits in the generator code to the data bits.

3. Divide the data bits including the appended bits by the generator code bits using modulo-2.

4. If the remainder is 0, the data and appended bits are sent as is.

5. If the remainder is not 0, the appended bits are made equal to the remainder bits in order to get a 0 remainder before data is sent.

6. At the receiving end, the receiver divides the incoming appended data bit code by the same generator code as used by the sender.

7. If the remainder is 0, there is no error detected (it is possible in rare cases for multiple errors to cancel). If the remainder is not 0, an error has been detected in the transmission and a retransmission is requested by the receiver.

Figure 2–8 illustrates the CRC process.

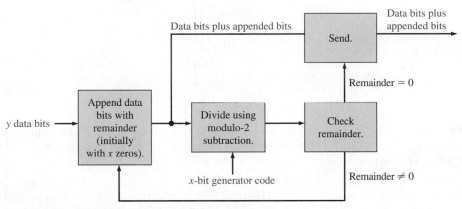

(a) Transmitting end of communication link

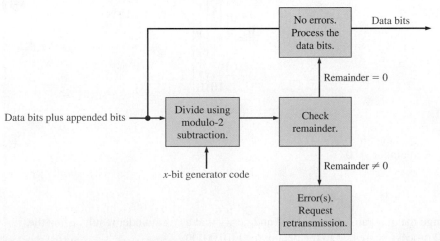

(b) Receiving end of communication link

FIGURE 2–8   **The CRC process.**

EXAMPLE 2–41

Determine the transmitted CRC for the following byte of data (D) and generator code (G). Verify that the reminder is 0.

$$D: \quad 11010011$$
$$G: \quad 1010$$

SOLUTION

Since the generator code has four data bits, add four 0s (blue) to the data byte. The appended data (D') is

$$D' = 110100110000$$

Divide the appended data by the generator code (red) using the modulo-2 operation until all bits have been used.

$$\frac{D'}{G} = \frac{110100110000}{1010}$$

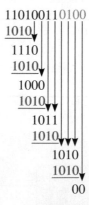

```
110100110000
1010
 1110
 1010
  1000
  1010
   1011
   1010
    1000
    1010
     100
```

Remainder = 0100. Since the remainder is not 0, append the data with the four remainder bits (blue). Then divide by the generator code (red). The transmitted CRC is **1101001 10100**.

```
110100110100
1010
 1110
 1010
  1000
  1010
   1011
   1010
    1010
    1010
      00
```

Remainder = 0

RELATED PROBLEM

Change the generator code to 1100 and verify that a 0 remainder results when the CRC process is applied to the data byte (11010011).

EXAMPLE 2–42

During transmission, an error occurs in the second bit from the left in the appended data byte generated in Example 2–41. The received data is

$$D' = 100100110100$$

Apply the CRC process to the received data to detect the error using the same generator code (1010).

SOLUTION

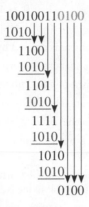

```
100100110100
1010
1100
1010
1101
1010
1111
1010
1010
1010
0100
```

Remainder = 0100. Since it is not zero, an error is indicated.

RELATED PROBLEM

Assume two errors in the data byte as follows: 10011011. Apply the CRC process to check for the errors using the same received data and the same generator code.

SECTION 2–12 CHECKUP

1. Which odd-parity code is in error?

   (a) 1011   (b) 1110   (c) 0101   (d) 1000

2. Which even-parity code is in error?

   (a) 11000110   (b) 00101000

   (c) 10101010   (d) 11111011

3. Add an even parity bit to the end of each of the following codes.

   (a) 1010100   (b) 0100000

   (c) 1110111   (d) 1000110

4. What does CRC stand for?

5. Apply modulo-2 operations to determine the following:

   (a) $1 + 1$ (b) $1 - 1$ (c) $1 - 0$ (d) $0 + 1$

SUMMARY

- A binary number is a weighted number in which the weight of each whole number digit is a positive power of two and the weight of each fractional digit is a negative power of two. The whole number weights increase from right to left—from least significant digit to most significant.

- A binary number can be converted to a decimal number by summing the decimal values of the weights of all the 1s in the binary number.

- A decimal whole number can be converted to binary by using the sum-of-weights or the repeated division-by-2 method.

- A decimal fraction can be converted to binary by using the sum-of-weights or the repeated multiplication-by-2 method.

- The basic rules for binary addition are as follows:

$$0 + 0 = 0$$
$$0 + 1 = 1$$
$$1 + 0 = 1$$
$$1 + 1 = 10$$

- The basic rules for binary subtraction are as follows:

$$0 - 0 = 0$$
$$1 - 1 = 0$$
$$1 - 0 = 1$$
$$10 - 1 = 1$$

- The 1's complement of a binary number is derived by changing 1s to 0s and 0s to 1s.
- The 2's complement of a binary number can be derived by adding 1 to the 1's complement.
- Binary subtraction can be accomplished with addition by using the 1's or 2's complement method.
- A positive binary number is represented by a 0 sign bit.
- A negative binary number is represented by a 1 sign bit.
- For arithmetic operations, negative binary numbers are represented in 1's complement or 2's complement form.
- In an addition operation, an overflow is possible when both numbers are positive or when both numbers are negative. An incorrect sign bit in the sum indicates the occurrence of an overflow.
- The hexadecimal number system consists of 16 digits and characters, 0 through 9 followed by A through F.
- One hexadecimal digit represents a 4-bit binary number, and its primary usefulness is in simplifying bit patterns and making them easier to read.
- A decimal number can be converted to hexadecimal by the repeated division-by-16 method.
- The octal number system consists of eight digits, 0 through 7.
- A decimal number can be converted to octal by using the repeated division-by-8 method.
- Octal-to-binary conversion is accomplished by simply replacing each octal digit with its 3-bit binary equivalent. The process is reversed for binary-to-octal conversion.
- A decimal number is converted to BCD by replacing each decimal digit with the appropriate 4-bit binary code.
- The ASCII is a 7-bit alphanumeric code that is widely used in computer systems for input and output of information.
- A parity bit is used to detect an error in a code.
- The CRC (cyclic redundancy check) is based on polynomial division using modulo-2 operations.

KEY TERMS

Key terms and other bold terms in the chapter are defined in the end-of-book glossary.

Alphanumeric   Consisting of numerals, letters, and other characters.

ASCII   American Standard Code for Information Interchange; the most widely used alphanumeric code.

BCD   Binary coded decimal; a digital code in which each of the decimal digits, 0 through 9, is represented by a group of four bits.

Byte   A group of eight bits.

Cyclic redundancy check (CRC)   A type of error detection code.

Floating-point number   A number representation based on scientific notation in which the number consists of an exponent and a mantissa.

Hexadecimal   Describes a number system with a base of 16.

LSB   Least significant bit; the right-most bit in a binary whole number or code.

MSB   Most significant bit; the left-most bit in a binary whole number or code.

Octal   Describes a number system with a base of eight.

Parity   In relation to binary codes, the condition of evenness or oddness of the number of 1s in a code group.

TRUE/FALSE QUIZ

Answers are at the end of the chapter.

1. The decimal number system is a weighted system with ten digits.

2. The binary number system is a weighted system with two digits.

3. LSB stands for lowest single bit.

4. In binary, $1 + 1 = 2$.

5. The 1's complement of the binary number 1010 is 0101.

6. The 2's complement of the binary number 0001 is 1110.

7. The right-most bit in a signed binary number is the sign bit.

8. The hexadecimal number system has 16 characters, six of which are alphabetic characters.

9. BCD stands for binary coded decimal.

10. ASCII stands for American standard code for information indication.

11. CRC stands for cyclic redundancy check.

12. The modulo-2 sum of 11 and 10 is 100.

SELF-TEST

Answers are at the end of the chapter.

1. $2 \times 10^1 + 8 \times 10^0$ is equal to
   (a) 10   (b) 280   (c) 2.8   (d) 28

2. The binary number 1101 is equal to the decimal number
   (a) 13   (b) 49   (c) 11   (d) 3

3. The binary number 11011101 is equal to the decimal number
   (a) 121   (b) 221   (c) 441   (d) 256

4. The decimal number 17 is equal to the binary number
   (a) 10010   (b) 11000   (c) 10001   (d) 01001

5. The decimal number 175 is equal to the binary number
   (a) 11001111   (b) 10101110   (c) 10101111   (d) 11101111

6. The sum of $11010 + 01111$ equals
   (a) 101001   (b) 101010   (c) 110101   (d) 101000

7. The difference of $110 - 010$ equals
   (a) 001   (b) 010   (c) 101   (d) 100

8. The 1's complement of 10111001 is
   (a) 01000111   (b) 01000110   (c) 11000110   (d) 10101010

9. The 2's complement of 11001000 is
   (a) 00110111   (b) 00110001   (c) 01001000   (d) 00111000

10. The decimal number $+122$ is expressed in the 2's complement form as
    (a) 01111010   (b) 11111010   (c) 01000101   (d) 10000101

11. The decimal number -34 is expressed in the 2's complement form as
    (a) 01011110   (b) 10100010   (c) 11011110   (d) 01011101

12. A single-precision floating-point binary number has a total of
    (a) 8 bits   (b) 16 bits   (c) 24 bits   (d) 32 bits

13. In the 2's complement form, the binary number 10010011 is equal to the decimal number
    (a) -19 (b) $+109$ (c) $+91$ (d) -109

14. The binary number 101100111001010100001 can be written in octal as
    (a) 5471230_8 (b) 5471241_8 (c) 2634521_8 (d) 23162501_8

15. The binary number 1000110101000110111 can be written in hexadecimal as
    (a) $AD467_{16}$ (b) $8C46F_{16}$ (c) $8D46F_{16}$ (d) $AE46F_{16}$

16. The binary number for $F7A9_{16}$ is
    (a) 1111011110101001   (b) 1110111110101001
    (c) 1111111010110001   (d) 1111011010101001

17. The BCD number for decimal 473 is
    (a) 111011010   (b) 110001110011   (c) 010001110011   (d) 010011110011

18. Refer to Table 2–7. The command STOP in ASCII is
    (a) 1010011101010010011111010000   (b) 1010010100110010011101010000
    (c) 1001010110110110011101010001   (d) 1010011101010010011101100100

19. The code that has an even-parity error is
    (a) 1010011   (b) 1101000   (c) 1001000   (d) 1110111

20. In the cyclic redundancy check, the absence of errors is indicated by
    (a) Remainder = generator code   (b) Remainder = 0
    (c) Remainder = 1   (d) Quotient = 0

PROBLEMS

Answers to odd-numbered problems are at the end of the book.

SECTION 2–1   **The Decimal Number System**

1. What is the weight of the digit 6 in each of the following decimal numbers?
   (a) 1386   (b) 54,692   (c) 671,920

2. Express each of the following decimal numbers as a power of ten:
   (a) 10   (b) 100   (c) 10,000   (d) 1,000,000

3. Give the value of each digit in the following decimal numbers:
   (a) 471   (b) 9356   (c) 125,000

4. How high can you count with four decimal digits?

SECTION 2–2   **The Binary Number System**

5. Convert the following binary numbers to decimal:
   (a) 11   (b) 100   (c) 111   (d) 1000
   (e) 1001   (f) 1100   (g) 1011   (h) 1111

6. Convert the following binary numbers to decimal:
   (a) 1110   (b) 1010   (c) 11100   (d) 10000
   (e) 10101   (f) 11101   (g) 10111   (h) 11111

7. Convert each binary number to decimal:
   (a) 110011.11   (b) 101010.01   (c) 1000001.111
   (d) 1111000.101   (e) 1011100.10101   (f) 1110001.0001
   (g) 1011010.1010   (h) 1111111.11111

8. What is the highest decimal number that can be represented by each of the following numbers of binary digits (bits)?
   (a) two   (b) three   (c) four   (d) five   (e) six
   (f) seven   (g) eight   (h) nine   (i) ten   (j) eleven

9. How many bits are required to represent the following decimal numbers?
   (a) 17   (b) 35   (c) 49   (d) 68
   (e) 81   (f) 114   (g) 132   (h) 205

10. Generate the binary sequence for each decimal sequence:
    (a) 0 through 7   (b) 8 through 15   (c) 16 through 31
    (d) 32 through 63   (e) 64 through 75

SECTION 2–3   Decimal-to-Binary Conversion

11. Convert each decimal number to binary by using the sum-of-weights method:
    (a) 10   (b) 17   (c) 24   (d) 48
    (e) 61   (f) 93   (g) 125   (h) 186

12. Convert each decimal fraction to binary using the sum-of-weights method:
    (a) 0.32   (b) 0.246   (c) 0.0981

13. Convert each decimal number to binary using repeated division by 2:
    (a) 15   (b) 21   (c) 28   (d) 34
    (e) 40   (f) 59   (g) 65   (h) 73

14. Convert each decimal fraction to binary using repeated multiplication by 2:
    (a) 0.98   (b) 0.347   (c) 0.9028

SECTION 2–4   Binary Arithmetic

15. Add the binary numbers:
    (a) 11 + 01   (b) 10 + 10   (c) 101 + 11
    (d) 111 + 110   (e) 1001 + 101   (f) 1101 + 1011

16. Use direct subtraction on the following binary numbers:
    (a) 11 − 1   (b) 101 − 100   (c) 110 − 101
    (d) 1110 − 11   (e) 1100 − 1001   (f) 11010 − 10111

17. Perform the following binary multiplications:
    (a) 11 × 11   (b) 100 × 10   (c) 111 × 101
    (d) 1001 × 110   (e) 1101 × 1101   (f) 1110 × 1101

18. Divide the binary numbers as indicated:
    (a) 100 ÷ 10   (b) 1001 : 11   (c) 1100 ÷ 100

SECTION 2–5   1's and 2's Complements of Binary Numbers

19. What are two ways of representing zero in 1's complement form?

20. How is zero represented in 2's complement form?

21. Determine the 1's complement of each binary number:
    (a) 101   (b) 110   (c) 1010
    (d) 11010111   (e) 1110101   (f) 00001

22. Determine the 2's complement of each binary number using either method:
    (a) 10   (b) 111   (c) 1001   (d) 1101
    (e) 11100   (f) 10011   (g) 10110000 (h) 00111101

SECTION 2–6   Signed Numbers

23. Express each decimal number in binary as an 8-bit sign-magnitude number:
    (a) +29   (b) −85   (c) +100   (d) −123

24. Express each decimal number as an 8-bit number in the 1's complement form:
    (a) −34   (b) +57   (c) −99   (d) +115

25. Express each decimal number as an 8-bit number in the 2's complement form:
    (a) +12   (b) −68   (c) +101   (d) −125

26. Determine the decimal value of each signed binary number in the sign-magnitude form:
    (a) 10011001   (b) 01110100   (c) 10111111

27. Determine the decimal value of each signed binary number in the 1's complement form:
    (a) 10011001   (b) 01110100   (c) 10111111

28. Determine the decimal value of each signed binary number in the 2's complement form:
    (a) 10011001   (b) 01110100   (c) 10111111

29. Express each of the following sign-magnitude binary numbers in single-precision floating-point format:
    (a) 0111110000101011   (b) 100110000011000

30. Determine the values of the following single-precision floating-point numbers:
    (a) 1 10000001 01001001110001000000000
    (b) 0 11001100 10000111110100100000000

SECTION 2–7   **Arithmetic Operations with Signed Numbers**

31. Convert each pair of decimal numbers to binary and add using the 2's complement form:
    (a) 33 and 15   (b) 56 and −27   (c) −46 and 25   (d) −110 and −84

32. Perform each addition in the 2's complement form:
    (a) 00010110 + 00110011   (b) 01110000 + 10101111

33. Perform each addition in the 2's complement form:
    (a) 10001100 + 00111001   (b) 11011001 + 11100111

34. Perform each subtraction in the 2's complement form:
    (a) 00110011 − 00010000   (b) 01100101 − 11101000

35. Multiply 01101010 by 11110001 in the 2's complement form.

36. Divide 01000100 by 00011001 in the 2's complement form.

SECTION 2–8   **Hexadecimal Numbers**

37. Convert each hexadecimal number to binary:
    (a) 38_{16} (b) 59_{16} (c) $A14_{16}$ (d) $5C8_{16}$
    (e) 4100_{16} (f) $FB17_{16}$ (g) $8A9D_{16}$

38. Convert each binary number to hexadecimal:
    (a) 1110   (b) 10   (c) 10111
    (d) 10100110   (e) 1111110000   (f) 100110000010

39. Convert each hexadecimal number to decimal:
    (a) 23_{16} (b) 92_{16} (c) $1A_{16}$ (d) $8D_{16}$
    (e) $F3_{16}$ (f) EB_{16} (g) $5C2_{16}$ (h) 700_{16}

40. Convert each decimal number to hexadecimal:
    (a) 8   (b) 14   (c) 33   (d) 52
    (e) 284   (f) 2890   (g) 4019   (h) 6500

41. Perform the following additions:
    (a) $37_{16} + 29_{16}$ (b) $A0_{16} + 6B_{16}$ (c) $FF_{16} + BB_{16}$

42. Perform the following subtractions:
    (a) $51_{16} - 40_{16}$ (b) $C8_{16} - 3A_{16}$ (c) $FD_{16} - 88_{16}$

SECTION 2–9   **Octal Numbers**

43. Convert each octal number to decimal:
    (a) 12_8 (b) 27_8 (c) 56_8 (d) 64_8 (e) 103_8
    (f) 557_8 (g) 163_8 (h) 1024_8 (i) 7765_8

44. Convert each decimal number to octal by repeated division by 8:
    (a) 15   (b) 27   (c) 46   (d) 70
    (e) 100   (f) 142   (g) 219   (h) 435

45. Convert each octal number to binary:
    (a) 13_8 (b) 57_8 (c) 101_8 (d) 321_8 (e) 540_8
    (f) 4653_8 (g) 13271_8 (h) 45600_8 (i) 100213_8

46. Convert each binary number to octal:
    (a) 111   (b) 10   (c) 110111
    (d) 101010   (e) 1100   (f) 1011110
    (g) 101100011001   (h) 10110000011   (i) 111111101111000

SECTION 2–10   **Binary Coded Decimal (BCD)**

47. Convert each of the following decimal numbers to 8421 BCD:
    (a) 10   (b) 13   (c) 18   (d) 21   (e) 25   (f) 36
    (g) 44   (h) 57   (i) 69   (j) 98   (k) 125   (l) 156

48. Convert each of the decimal numbers in Problem 47 to straight binary, and compare the number of bits required with that required for BCD.

49. Convert the following decimal numbers to BCD:
    (a) 104   (b) 128   (c) 132   (d) 150   (e) 186
    (f) 210   (g) 359   (h) 547   (i) 1051

50. Convert each of the BCD numbers to decimal:
(a) 0001   (b) 0110   (c) 1001
(d) 00011000   (e) 00011001   (f) 00110010
(g) 01000101   (h) 10011000   (i) 100001110000

51. Convert each of the BCD numbers to decimal:
(a) 10000000   (b) 001000110111
(c) 001101000110   (d) 010000100001
(e) 011101010100   (f) 100000000000
(g) 100101111000   (h) 0001011010000011
(i) 1001000000011000   (j) 0110011001100111

52. Add the following BCD numbers:
(a) 0010 + 0001   (b) 0101 + 0011
(c) 0111 + 0010   (d) 1000 + 0001
(e) 00011000 + 00010001   (f) 01100100 + 00110011
(g) 01000000 + 01000111   (h) 10000101 + 00010011

53. Add the following BCD numbers:
(a) 1000 + 0110   (b) 0111 + 0101
(c) 1001 + 1000   (d) 1001 + 0111
(e) 00100101 + 00100111   (f) 01010001 + 01011000
(g) 10011000 + 10010111   (h) 010101100001 + 011100001000

54. Convert each pair of decimal numbers to BCD, and add as indicated:
(a) 4 + 3   (b) 5 + 2   (c) 6 + 4   (d) 17 + 12
(e) 28 + 23   (f) 65 + 58   (g) 113 + 101   (h) 295 + 157

SECTION 2–11 Digital Codes

55. In a certain application a 4-bit binary sequence cycles from 1111 to 0000 periodically. There are four bit changes, and because of circuit delays, these changes may not occur at the same instant. For example, if the LSB changes first, the number will appear as 1110 during the transition from 1111 to 0000 and may be misinterpreted by the system. Illustrate how the Gray code avoids this problem.

56. Convert each binary number to Gray code:
(a) 11011   (b) 1001010   (c) 1111011101110

57. Convert each Gray code to binary:
(a) 1010   (b) 00010   (c) 11000010001

58. Convert each of the following decimal numbers to ASCII. Refer to Table 2–7.
(a) 1   (b) 3   (c) 6   (d) 10   (e) 18
(f) 29   (g) 56   (h) 75   (i) 107

59. Determine each ASCII character. Refer to Table 2–7.
(a) 0011000   (b) 1001010   (c) 0111101
(d) 0100011   (e) 0111110   (f) 1000010

60. Decode the following ASCII coded message:

     1001000 1100101 1101100 1101100 1101111 0101110
     0100000 1001000 1101111 1110111 0100000 1100001
     1110010 1100101 0100000 1111001 1101111 1110101
     0111111

61. Write the message in Problem 60 in hexadecimal.

62. Convert the following statement to ASCII:

           30 INPUT A, B

SECTION 2–12 Error Detection Codes

63. Determine which of the following even parity codes are in error:
(a) 100110010   (b) 011101010   (c) 10111111010001010

64. Determine which of the following odd parity codes are in error:
(a) 11110110   (b) 00110001   (c) 01010101010101010

65. Attach the proper even parity bit to each of the following bytes of data:
(a) 10100100   (b) 00001001   (c) 11111110

66. Apply modulo-2 to the following:
(a) 1100 + 1011   (b) 1111 + 0100   (c) 10011001 + 100011100

67. Verify that modulo-2 subtraction is the same as modulo-2 addition by adding the result of each operation in problem 66 to either of the original numbers to get the other number. This will show that the result is the same as the difference of the two numbers.

68. Apply CRC to the data bits 10110010 using the generator code 1010 to produce the transmitted CRC code.

69. Assume that the code produced in problem 68 incurs an error in the most significant bit during transmission. Apply CRC to detect the error.

ANSWERS TO SECTION CHECKUPS

SECTION 2–1   The Decimal Number System

1. (a) 1370: 10   **(b)** 6725: 100   **(c)** 7051: 1000   **(d)** 58.72: 0.1

2. (a) $51 = (5 \times 10) + (1 \times 1)$ **(b)** $137 = (1 \times 100) + (3 \times 10) + (7 \times 1)$

 (c) $1492 = (1 \times 1000) + (4 \times 100) + (9 \times 10) + (2 \times 1)$

 (d) $106.58 = (1 \times 100) + (0 \times 10) + (6 \times 1) + (5 \times 0.1) + (8 \times 0.01)$

SECTION 2–2   The Binary Number System

1. $2^8 - 1 = 255$

2. Weight is 16.

3. $10111101.011 = 189.375$

SECTION 2–3   Decimal-to-Binary Conversion

1. (a) $23 = 10111$ **(b)** $57 = 111001$ **(c)** $45.5 = 101101.1$

2. (a) $14 = 1110$ **(b)** $21 = 10101$ **(c)** $0.375 = 0.011$

SECTION 2–4   Binary Arithmetic

1. (a) $1101 + 1010 = 10111$ **(b)** $10111 + 01101 = 100100$

2. (a) $1101 - 0100 = 1001$ **(b)** $1001 - 0111 = 0010$

3. (a) $110 \times 111 = 101010$ **(b)** $1100 \div 011 = 100$

SECTION 2–5   1's and 2's Complements of Binary Numbers

1. (a) 1's comp of $00011010 = 11100101$ **(b)** 1's comp of $11110111 = 00001000$

 (c) 1's comp of $10001101 = 01110010$

2. (a) 2's comp of $00010110 = 11101010$ **(b)** 2's comp of $11111100 = 00000100$

 (c) 2's comp of $10010001 = 01101111$

SECTION 2–6   Signed Numbers

1. Sign-magnitude: $+9 = 00001001$

2. 1's comp: $-33 = 11011110$

3. 2's comp: $-46 = 11010010$

4. Sign bit, exponent, and mantissa

SECTION 2–7   Arithmetic Operations with Signed Numbers

1. Cases of addition: positive number is larger, negative number is larger, both are positive, both are negative

2. $00100001 + 10111100 = 11011101$

3. $01110111 - 00110010 = 01000101$

4. Sign of product is positive.

5. $00000101 \times 01111111 = 01001111011$

6. Sign of quotient is negative.

7. $00110000 \div 00001100 = 00000100$

SECTION 2–8   Hexadecimal Numbers

1. (a) $10110011 = B3_{16}$ (b) $110011101000 = CE8_{16}$
2. (a) $57_{16} = 01010111$ (b) $3A5_{16} = 001110100101$
   (c) $F8OB_{16} = 1111100000001011$
3. $9B30_{16} = 39{,}728_{10}$
4. $573_{10} = 23D_{16}$
5. (a) $18_{16} + 34_{16} = 4C_{16}$ (b) $3F_{16} + 2A_{16} = 69_{16}$
6. (a) $75_{16} - 21_{16} = 54_{16}$ (b) $94_{16} - 5C_{16} = 38_{16}$

SECTION 2–9   Octal Numbers

1. (a) $73_8 = 59_{10}$ (b) $125_8 = 85_{10}$
2. (a) $98_{10} = 142_8$ (b) $163_{10} = 243_8$
3. (a) $46_8 = 100110$ (b) $723_8 = 111010011$ (c) $5624_8 = 101110010100$
4. (a) $110101111 = 657_8$ (b) $1001100010 = 1142_8$ (c) $10111111001 = 2771_8$

SECTION 2–10   Binary Coded Decimal (BCD)

1. (a) 0010: 2   (b) 1000: 8   (c) 0001: 1   (d) 0100: 4
2. (a) $6_{10} = 0110$ (b) $15_{10} = 00010101$ (c) $273_{10} = 001001110011$
   (d) $849_{10} = 100001001001$
3. (a) $10001001 = 89_{10}$ (b) $001001111000 = 278_{10}$ (c) $000101010111 = 157_{10}$
4. A 4-bit sum is invalid when it is greater than 9_{10}.

SECTION 2–11   Digital Codes

1. (a) $1100_2 = 1010$ Gray   (b) $1010_2 = 1111$ Gray   (c) $11010_2 = 10111$ Gray
2. (a) 1000 Gray $= 1111_2$ (b) 1010 Gray $= 1100_2$ (c) 11101 Gray $= 10110_2$
3. (a) K: $1001011 \rightarrow 4B_{16}$ (b) r: $1110010 \rightarrow 72_{16}$
   (c) \$: $0100100 \rightarrow 24_{16}$ (d) +: $0101011 \rightarrow 2B_{16}$

SECTION 2–12   Error Detection Codes

1. (c) 0101 has an error.
2. (d) 11111011 has an error.
3. (a) 10101001   (b) 01000001   (c) 11101110   (d) 10001101
4. Cyclic redundancy check
5. (a) 0   (b) 0   (c) 1   (d) 1

ANSWERS TO RELATED PROBLEMS
FOR EXAMPLES

2–1   9 has a value of 900, 3 has a value of 30, 9 has a value of 9.

2–2   6 has a value of 60, 7 has a value of 7, 9 has a value of 9/10 (0.9), 2 has a value of 2/100 (0.02), 4 has a value of 4/1000 (0.004).

2–3 $10010001 = 128 + 16 + 1 = 145$

2–4 $10.111 = 2 + 0.5 + 0.25 + 0.125 = 2.875$

2–5 $125 = 64 + 32 + 16 + 8 + 4 + 1 = 1111101$

2–6 $39 = 100111$

2–7 $1111 + 1100 = 11011$

2–8 $111 - 100 = 011$

2–9 $110 - 101 = 001$

2–10 $1101 \times 1010 = 10000010$

2–11 $1100 \div 100 = 11$

2–12   00110101

2–13   01000000

2–14   See Table 2–11.

TABLE 2–11

	SIGN-MAGNITUDE	1'S COMP	2'S COMP
+19	00010011	00010011	00010011
−19	10010011	11101100	11101101

2–15 $01110111 = +119_{10}$

2–16 $11101011 = -20_{10}$

2–17 $11010111 = -41_{10}$

2–18   110000100010100011000000000

2–19   01010101

2–20   00010001

2–21   1001000110

2–22 $(83)(-59) = -4897$ (10110011011111 in 2's comp)

2–23 $100 \div 25 = 4$ (0100)

2–24 $4F79C_{16}$

2–25 0110101111010011_2

2–26 $6BD_{16} = 011010111101 = 2^{10} + 2^9 + 2^7 + 2^5 + 2^4 + 2^3 + 2^2 + 2^0$
$= 1024 + 512 + 128 + 32 + 16 + 8 + 4 + 1 = 1725_{10}$

2–27 $60A_{16} = (6 \times 256) + (0 \times 16) + (10 \times 1) = 1546_{10}$

2–28 $2591_{10} = A1F_{16}$

2–29 $4C_{16} + 3A_{16} = 86_{16}$

2–30 $BCD_{16} - 173_{16} = A5A_{16}$

2–31   (a) $001011_2 = 11_{10} = 13_8$ (b) $010101_2 = 21_{10} = 25_8$
(c) $001100000_2 = 96_{10} = 140_8$ (d) $111101010110_2 = 3926_{10} = 7526_8$

2–32 1250762_8 **2–33** 1001011001110011   **2–34** $82,276_{10}$

2–35   1001100101101000   **2–36** 10000010   **2–37** (a) 111011 (Gray)   (b) 111010_2

2–38   The sequence of codes for if (y < 8) is $69_{16}66_{16}20_{16}28_{16}79_{16}3C_{16}38_{16}29_{16}$

2–39   01001011

2–40   Yes

2–41   A 0 remainder results

2–42   Errors are indicated.

ANSWERS TO TRUE/FALSE QUIZ

1. T   **2.** T   **3.** F   **4.** F   **5.** T   **6.** F

7. F   **8.** T   **9.** T   **10.** F   **11.** T   **12.** F

ANSWERS TO SELF-TEST

1. (d)   **2.** (a)   **3.** (b)   **4.** (c)   **5.** (c)   **6.** (a)   **7.** (d)   **8.** (b)

9. (d)   **10.** (a)   **11.** (c)   **12.** (d)   **13.** (d)   **14.** (b)   **15.** (c)   **16.** (a)

17. (c)   **18.** (a)   **19.** (b)   **20.** (b)

CHAPTER 3

LOGIC GATES AND GATE COMBINATIONS

OUTLINE

OBJECTIVES

- Apply Boolean arithmetic
- Apply the basic laws and rules of Boolean algebra
- Describe the operation of the inverter, the AND gate, and the OR gate
- Describe the operation of the NAND gate and the NOR gate
- Express the operation of NOT, AND, OR, NAND, and NOR gates with Boolean algebra
- Describe the operation of the exclusive-OR and exclusive-NOR gates
- Recognize and use both the distinctive shape logic gate symbols and the rectangular outline logic gate symbols of ANSI/IEEE Standard 91-1984
- Construct timing diagrams showing the proper time relationships of inputs and outputs for the various logic gates
- Define *propagation delay time, power dissipation, speed-power product,* and *fan-out* in relation to logic gates
- Use each logic gate in simple applications
- Discuss the basic concepts of programmable logic and describe logic gates using VHDL and Verilog
- Describe basic troubleshooting methods

KEY TERMS

Boolean algebra	Propagation delay time
Variable	Fan-out
Complement	Unit load
Sum term	AND array
Product term	Fuse
Inverter	Antifuse
Truth table	EPROM
Timing diagram	EEPROM
AND gate	Flash
OR gate	SRAM
NAND gate	Target device
NOR gate	JTAG
Exclusive-OR gate	VHDL
Exclusive-NOR gate	Verilog

VISIT THE WEBSITE
Study aids for this chapter are available at
http://pearsonhighered.com/floyd

INTRODUCTION

The emphasis in this chapter is on the operation, application, and troubleshooting of logic gates. The relationship of input and output waveforms of a gate using timing diagrams is thoroughly covered. Boolean algebra is introduced.

Logic symbols used to represent the logic gates are in accordance with ANSI/IEEE Standard 91-1984. This standard has been adopted by private industry and the military for use in internal documentation as well as published literature.

Programmable logic is discussed in this chapter because of the widespread use of PLDs. Because integrated circuits (ICs) are used in applications, the logic function of a device is generally of much greater importance to the technician or technologist than the details of the component-level circuit operation within the IC package. Hardware description languages (HDLs) for programmable logic are introduced.

3–1 INTRODUCTION TO BOOLEAN ALGEBRA

Boolean algebra is the mathematics of digital systems. A basic knowledge of Boolean algebra is indispensable to the study and analysis of logic circuits, such as NOT, AND, and OR, which were introduced in Chapter 1.

After completing this section, you should be able to

- **Define** *variable*
- **Define** *literal*
- **Identify a sum term**
- **Evaluate a sum term**
- **Identify a product term**
- **Evaluate a product term**
- **Explain Boolean addition**
- **Explain Boolean multiplication**
- **Apply the commutative laws of addition and multiplication**
- **Apply the associative laws of addition and multiplication**
- **Apply the distributive law**
- **Apply the rules of Boolean algebra**

Boolean algebra uses variables and operators to describe a logic circuit. *Variable, complement,* and *literal* are terms used in Boolean algebra. A **variable** is a symbol (usually an italic uppercase letter or word) used to represent an action, a condition, or data. Any single variable can have only a 1 or a 0 value. The **complement** is the inverse of a variable and is indicated by a bar over the variable (overbar). For example, the complement of the variable A is $\overline{A}$. If $A = 1$, then $\overline{A} = 0$. If $A = 0$, then $\overline{A} = 1$. The complement of the variable A is read as "not A" or "A bar." Sometimes a prime symbol rather than an overbar is used to denote the complement of a variable; for example, B' indicates the complement of B. In this book, only the overbar is used. A **literal** is a constant value assigned to a variable or the complement of a variable.

In a microprocessor, the arithmetic logic unit (ALU) performs arithmetic and Boolean logic operations on digital data as directed by program instructions. Logical operations are equivalent to the basic gate operations that you are familiar with but deal with a minimum of 8 bits at a time. Examples of Boolean logic instructions are AND, OR, and NOT, which are called *mnemonics.* An assembly language program uses the mnemonics to specify an operation. Another program called an *assembler* translates the mnemonics into a binary code that can be understood by the microprocessor.

SYSTEM NOTE

Boolean Addition

FIGURE 3–1

Boolean addition is equivalent to the OR operation. The logical OR function of two variables is represented mathematically by a $+$ between the two variables, for example, $A + B$. The plus sign is read as "OR." Addition in Boolean algebra involves variables whose values are either binary 1 or binary 0. The basic rules of Boolean addition are illustrated in Figure 3–1 with their relation to the OR gate. Notice that Boolean addition differs from binary addition in the case where two 1s are added. There is no carry in Boolean addition.

The OR operation is the Boolean form of addition.

In Boolean algebra, a **sum term** is a sum of literals. In logic circuits, a sum term is produced by an OR operation with no AND operations involved. Some examples of sum terms are $A + \overline{B}, A + B + \overline{C}$, and $\overline{A} + B + C + \overline{D}$. A sum term is equal to 1 when one or more of the literals in the term are 1. A sum term is equal to 0 only if each of the literals is 0.

EXAMPLE 3–1

Determine the values of A, B, C, and D that make the sum term $A + \overline{B} + C + \overline{D}$ equal to 0.

SOLUTION

For the sum term to be 0, each of the literals in the term must be 0. Therefore, $A = \mathbf{0}, B = \mathbf{1}$ so that $\overline{B} = 0, C = \mathbf{0}$, and $D = \mathbf{1}$ so that $\overline{D} = 0$.

$$A + \overline{B} + C + \overline{D} = 0 + \overline{1} + 0 + \overline{1} = 0 + 0 + 0 + 0 = 0$$

RELATED PROBLEM*

Determine the values of A and B that make the sum term $\overline{A} + B$ equal to 0.

Answers are at the end of the chapter.

Boolean Multiplication

FIGURE 3–2

Boolean multiplication is equivalent to the AND operation. The logical AND function of two variables is represented mathematically either by placing a dot between the two variables, as $A \cdot B$, or by simply writing the adjacent letters without the dot, as AB. The variables are either binary 1 or binary 0, and the rules of binary multiplication apply to Boolean multiplication. The basic rules of Boolean multiplication are illustrated in Figure 3–2 with their relation to the AND gate.

The AND operation is the Boolean form of multiplication.

In Boolean algebra, a **product term** is the product of literals. In logic circuits, a product term is produced by an AND operation with no OR operations involved. Some examples of product terms are $A\overline{B}, ABC$, and $A\overline{B}C\overline{D}$. A product term is equal to 1 only if each of the literals in the term is 1. A product term is equal to 0 when one or more of the literals are 0.

EXAMPLE 3–2

Determine the values of A, B, C, and D that make the product term $A\overline{B}C\overline{D}$ equal to 1.

SOLUTION

For the product term to be 1, each of the literals in the term must be 1. Therefore, $A = \mathbf{1}, B = \mathbf{0}$ so that $\overline{B} = 1, C = \mathbf{1}$, and $D = \mathbf{0}$ so that $\overline{D} = 1$.

$$A\overline{B}C\overline{D} = 1 \cdot \overline{0} \cdot 1 \cdot \overline{0} = 1 \cdot 1 \cdot 1 \cdot 1 = 1$$

RELATED PROBLEM

Determine the values of A and B that make the product term $\overline{A}\,\overline{B}$ equal to 1.

Laws of Boolean Algebra

The basic laws of Boolean algebra—the **commutative laws** for addition and multiplication, the **associative laws** for addition and multiplication, and the **distributive law**—are the same as in ordinary algebra. Each of the laws is illustrated with two or three variables, but the number of variables is not limited to this.

COMMUTATIVE LAWS   The commutative law for the addition of two variables is written as

$$A + B = B + A \qquad (3\text{–}1)$$

This law states that the order in which the variables are ORed makes no difference. Figure 3–3 illustrates the commutative law as applied to the OR gate and shows that it doesn't matter to which input each variable is applied. (The symbol ≡ means "equivalent to.")

The commutative law for the multiplication of two variables is

$$AB = BA \qquad (3\text{–}2)$$

This law states that the order in which the variables are ANDed makes no difference. Figure 3–4 illustrates this law as applied to the AND gate.

FIGURE 3–3   **Application of the commutative law of addition.**

FIGURE 3–4
Application of the commutative law of multiplication.

ASSOCIATIVE LAWS   The associative law of addition is written as follows for three variables:

$$A + (B + C) = (A + B) + C \qquad (3\text{–}3)$$

This law states that when ORing more than two variables, the result is the same regardless of the grouping of the variables. Figure 3–5 illustrates this law as applied to 2-input OR gates.

MULTISIM

FIGURE 3–5   **Application of the associative law of addition. Open file F03-05 to verify.**

The associative law of multiplication is written as follows for three variables:

$$A(BC) = (AB)C \qquad (3\text{–}4)$$

This law states that it makes no difference in what order the variables are grouped when ANDing more than two variables. Figure 3–6 illustrates this law as applied to 2-input AND gates.

MULTISIM

FIGURE 3–6   **Application of the associative law of multiplication. Open file F03-06 to verify.**

DISTRIBUTIVE LAW   The distributive law is written for three variables as follows:

$$A(B + C) = AB + AC \qquad (3\text{–}5)$$

This law states that ORing two or more variables and then ANDing the result with a single variable is equivalent to ANDing the single variable with each of the two or more variables and then ORing the products. The distributive law also expresses the process of *factoring* in which the common variable A is factored out of the product terms, for example, $AB + AC = A(B + C)$. Figure 3–7 illustrates the distributive law in terms of gate implementation.

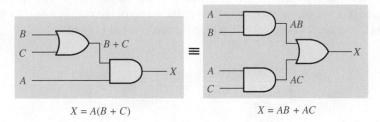

$$X = A(B + C)$$ $$X = AB + AC$$

MULTISIM

FIGURE 3–7   **Application of distributive law. Open file F03-07 to verify.**

Rules of Boolean Algebra

Table 3–1 lists 12 basic rules that are useful in manipulating and simplifying **Boolean expressions.** Rules 1 through 9 will be viewed in terms of their application to logic gates. Rules 10 through 12 will be derived in terms of the simpler rules and the laws previously discussed.

TABLE 3–1 • Basic rules of Boolean algebra.	
1. $A + 0 = A$	**7.** $A \cdot A = A$
2. $A + 1 = 1$	**8.** $A \cdot \overline{A} = 0$
3. $A \cdot 0 = 0$	**9.** $\overline{\overline{A}} = A$
4. $A \cdot 1 = A$	**10.** $A + AB = A$
5. $A + A = A$	**11.** $A + \overline{A}B = A + B$
6. $A + \overline{A} = 1$	**12.** $(A + B)(A + C) = A + BC$

A, B, or C can represent a single variable or a combination of variables.

Rule 1: $A + 0 = A$ A variable ORed with 0 is always equal to the variable. If the input variable A is 1, the output variable X is 1, which is equal to A. If A is 0, the output is 0, which is also equal to A. This rule is illustrated in Figure 3–8, where the lower input is fixed at 0.

$$X = A + 0 = A$$

FIGURE 3–8

Rule 2: $A + 1 = 1$ A variable ORed with 1 is always equal to 1. A 1 on an input to an OR gate produces a 1 on the output, regardless of the value of the variable on the other input. This rule is illustrated in Figure 3–9, where the lower input is fixed at 1.

$$X = A + 1 = 1$$

FIGURE 3–9

Rule 3: $A \cdot 0 = 0$ A variable ANDed with 0 is always equal to 0. Any time one input to an AND gate is 0, the output is 0, regardless of the value of the variable on the other input. This rule is illustrated in Figure 3–10, where the lower input is fixed at 0.

$$X = A \cdot 0 = 0$$

FIGURE 3–10

Rule 4: $A \cdot 1 = A$ A variable ANDed with 1 is always equal to the variable. If A is 0, the output of the AND gate is 0. If A is 1, the output of the AND gate is 1 because both inputs are now 1s. This rule is shown in Figure 3–11, where the lower input is fixed at 1.

$$X = A \cdot 1 = A$$

FIGURE 3–11

Rule 5: $A + A = A$ A variable ORed with itself is always equal to the variable. If A is 0, then $0 + 0 = 0$; and if A is 1, then $1 + 1 = 1$. This is shown in Figure 3–12, where both inputs are the same variable.

$$X = A + A = A$$

FIGURE 3–12

Rule 6: $A + \overline{A} = 1$ A variable ORed with its complement is always equal to 1. If A is 0, then $0 + \overline{0} = 0 + 1 = 1$. If A is 1, then $1 + \overline{1} = 1 + 0 = 1$. See Figure 3–13, where one input is the complement of the other.

$$X = A + \overline{A} = 1$$

FIGURE 3–13

Rule 7: $A \cdot A = A$ A variable ANDed with itself is always equal to the variable. If $A = 0$, then $0 \cdot 0 = 0$; and if $A = 1$, then $1 \cdot 1 = 1$. Figure 3–14 illustrates this rule.

$$X = A \cdot A = A$$

FIGURE 3–14

Rule 8: $A \cdot \overline{A} = 0$ A variable ANDed with its complement is always equal to 0. Either A or $\overline{A}$ will always be 0; and when a 0 is applied to the input of an AND gate, the output will be 0 also. Figure 3–15 illustrates this rule.

$$X = A \cdot \overline{A} = 0$$

FIGURE 3–15

Rule 9: $\overline{\overline{A}} = A$ The double complement of a variable is always equal to the variable. If you start with the variable A and complement (invert) it once, you get $\overline{A}$. If you then take $\overline{A}$ and complement (invert) it, you get A, which is the original variable. This rule is shown in Figure 3–16 using inverters.

$$X = \overline{\overline{A}} = A$$

FIGURE 3–16

Rule 10: $A + AB = A$ This rule can be proved by applying the distributive law, rule 2, and rule 4 as follows:

$$A + AB = A \cdot 1 + AB = A(1 + B) \quad \text{Factoring (distributive law)}$$
$$= A \cdot 1 \quad\quad\quad\quad\quad\quad \text{Rule 2: } (1 + B) = 1$$
$$= A \quad\quad\quad\quad\quad\quad\quad\; \text{Rule 4: } A \cdot 1 = A$$

The proof is shown in Table 3–2, which shows the truth table and the resulting logic circuit simplification.

MULTISIM

A	B	AB	A + AB	
0	0	0	0	
0	1	0	0	
1	0	0	1	
1	1	1	1	

TABLE 3–2 • Rule 10: $A + AB = A$. Open file T03-02 to verify.

equal

straight connection

Rule 11: $A + \overline{A}B = A + B$ This rule can be proved as follows:

$$
\begin{aligned}
A + \overline{A}B &= (A + AB) + \overline{A}B & \text{Rule 10: } A = A + AB \\
&= (AA + AB) + \overline{A}B & \text{Rule 7: } A = AA \\
&= AA + AB + A\overline{A} + \overline{A}B & \text{Rule 8: adding } A\overline{A} = 0 \\
&= (A + \overline{A})(A + B) & \text{Factoring} \\
&= 1 \cdot (A + B) & \text{Rule 6: } A + \overline{A} = 1 \\
&= A + B & \text{Rule 4: drop the 1}
\end{aligned}
$$

The proof is shown in Table 3–3, which shows the truth table and the resulting logic circuit simplification.

TABLE 3–3 • Rule 11: $A + \overline{A}B = A + B$. Open file T03-03 to verify.

A	B	$\overline{A}B$	$A + \overline{A}B$	$A + B$
0	0	0	0	0
0	1	1	1	1
1	0	0	1	1
1	1	0	1	1

Rule 12: $(A + B)(A + C) = A + BC$ This rule can be proved as follows:

$$
\begin{aligned}
(A + B)(A + C) &= AA + AC + AB + BC & \text{Distributive law} \\
&= A + AC + AB + BC & \text{Rule 7: } AA = A \\
&= A(1 + C) + AB + BC & \text{Factoring (distributive law)} \\
&= A \cdot 1 + AB + BC & \text{Rule 2: } 1 + C = 1 \\
&= A(1 + B) + BC & \text{Factoring (distributive law)} \\
&= A \cdot 1 + BC & \text{Rule 2: } 1 + B = 1 \\
&= A + BC & \text{Rule 4: } A \cdot 1 = A
\end{aligned}
$$

The proof is shown in Table 3–4, which shows the truth table and the resulting logic circuit simplification.

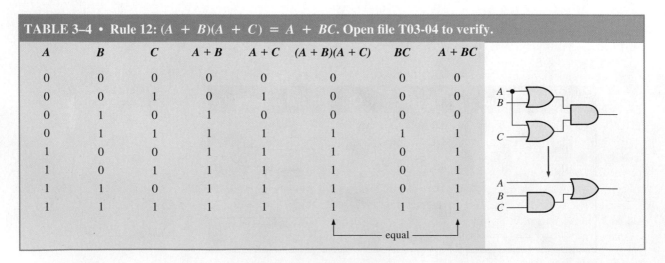

TABLE 3–4 • Rule 12: $(A + B)(A + C) = A + BC$. Open file T03-04 to verify.

A	B	C	$A + B$	$A + C$	$(A + B)(A + C)$	BC	$A + BC$
0	0	0	0	0	0	0	0
0	0	1	0	1	0	0	0
0	1	0	1	0	0	0	0
0	1	1	1	1	1	1	1
1	0	0	1	1	1	0	1
1	0	1	1	1	1	0	1
1	1	0	1	1	1	0	1
1	1	1	1	1	1	1	1

SECTION 3–1 CHECKUP*

1. If $A = 0$, what does $\overline{A}$ equal?

2. Determine the values of A, B, and C that make the sum term $\overline{A} + \overline{B} + C$ equal to 0.

3. Determine the values of A, B, and C that make the product term $A\overline{B}C$ equal to 1.

4. Apply the associative law of addition to the expression $A + (B + C + D)$.

5. Apply the distributive law to the expression $A(B + C + D)$.

*Answers are at the end of the chapter.

3–2 THE INVERTER

The inverter (NOT circuit) performs the operation called *inversion* or *complementation*. The inverter changes one logic level to the opposite level. In terms of bits, it changes a 1 to a 0 and a 0 to a 1.

After completing this section, you should be able to

- **Identify negation and polarity indicators**
- **Identify an inverter by either its distinctive shape symbol or its rectangular outline symbol**
- **Produce the truth table for an inverter**
- **Describe the logical operation of an inverter**

Standard logic symbols for the inverter are shown in Figure 3–17. Part (a) shows the *distinctive shape* symbols, and part (b) shows the *rectangular outline* symbols. In this textbook, distinctive shape symbols are generally used; however, the rectangular outline symbols are found in many industry publications, and you should become familiar with them as well. (Logic symbols are in accordance with **ANSI/IEEE** Standard 91-1984.)

(a) Distinctive shape symbols with negation indicators

(b) Rectangular outline symbols with polarity indicators

FIGURE 3–17   Standard logic symbols for the inverter (ANSI/IEEE Std. 91-1984).

The Negation and Polarity Indicators

The negation indicator is a "bubble" (o) that indicates **inversion** or *complementation* when it appears on the input or output of any logic element, as shown in Figure 3–17(a) for the inverter. Generally, inputs are on the left of a logic symbol and the output is on the right. When appearing on the input, the bubble means that a 0 is the active or *asserted* input state, and the input is called an active-LOW input. When appearing on the output, the bubble means that a 0 is the active or asserted output state, and the output is called an active-LOW output. The absence of a bubble on the input or output means that a 1 is the active or asserted state, and in this case, the input or output is called active-HIGH.

The polarity or level indicator is a "triangle" ($\searisymbol$) that indicates inversion when it appears on the input or output of a logic element, as shown in Figure 3–17(b). When appearing on the input, it means that a LOW level is the active or asserted input state. When appearing on the output, it means that a LOW level is the active or asserted output state.

Either indicator (bubble or triangle) can be used both on distinctive shape symbols and on rectangular outline symbols. Figure 3–17(a) indicates the principal inverter symbols used in this text. Note that a change in the placement of the negation or polarity indicator does not imply a change in the way an inverter operates.

TABLE 3–5 • Inverter truth table.

INPUT	OUTPUT
LOW (0)	HIGH (1)
HIGH (1)	LOW (0)

Inverter Truth Table

When a HIGH level is applied to an inverter input, a LOW level will appear on its output. When a LOW level is applied to its input, a HIGH will appear on its output. This operation is summarized in Table 3–5, which shows the output for each possible input in terms of levels and corresponding bits. A table such as this is called a **truth table.**

A timing diagram shows how two or more waveforms relate in time.

Inverter Operation

Figure 3–18 shows the output of an inverter for a pulse input, where t_1 and t_2 indicate the corresponding points on the input and output pulse waveforms.

When the input is LOW, the output is HIGH; when the input is HIGH, the output is LOW, thereby producing an inverted output pulse.

MULTISIM

FIGURE 3–18   **Inverter operation with a pulse input. Open file F03-18 to verify inverter operation.**

Timing Diagrams

FIGURE 3–19   **Timing diagram for the case in Figure 3–18.**

Recall from Chapter 1 that a **timing diagram** is basically a graph that accurately displays the relationship of two or more waveforms with respect to each other on a time basis. For example, the time relationship of the output pulse to the input pulse in Figure 3–18 can be shown with a simple timing diagram by aligning the two pulses so that the occurrences of the pulse edges appear in the proper time relationship. The rising edge of the input pulse and the falling edge of the output pulse occur at the same time (ideally). Similarly, the falling edge of the input pulse and the rising edge of the output pulse occur at the same time (ideally). This timing relationship is shown in Figure 3–19. In practice, there is a very small delay from the input transition until the corresponding output transition. Timing diagrams are especially useful for illustrating the time relationship of digital waveforms with multiple pulses.

EXAMPLE 3–3

A waveform is applied to an inverter in Figure 3–20. Determine the output waveform corresponding to the input and show the timing diagram. According to the placement of the bubble, what is the active output state?

FIGURE 3–20

SOLUTION

The output waveform is exactly opposite to the input (inverted), as shown in Figure 3–21, which is the basic timing diagram. The active or asserted output state is **0.**

FIGURE 3–21

RELATED PROBLEM

If the inverter is shown with the negative indicator (bubble) on the input instead of the output, how is the timing diagram affected?

Logic Expression for an Inverter

As you have learned, in Boolean algebra a variable is generally designated by one or two letters although there can be more. Letters near the beginning of the alphabet usually designate inputs, while letters near the end of the alphabet usually designate outputs. The complement of a variable is designated by a bar over the letter. A variable can take on a value of either 1 or 0. If a given variable is 1, its complement is 0 and vice versa.

The operation of an inverter (NOT circuit) can be expressed as follows: If the input variable is called A and the output variable is called X, then

$$X = \bar{A}$$

This expression states that the output is the complement of the input, so if $A = 0$, then $X = 1$, and if $A = 1$, then $X = 0$. Figure 3–22 illustrates this. The complemented variable $\bar{A}$ can be read as "A bar" or "not A."

FIGURE 3–22 **The inverter complements an input variable.**

An Application

Figure 3–23 shows a circuit for producing the 1's complement of an 8-bit binary number. The bits of the binary number are applied to the inverter inputs and the 1's complement of the number appears on the outputs.

FIGURE 3–23 **Example of a 1's complement circuit using inverters.**

SECTION 3–2 CHECKUP

1. When a 1 is on the input of an inverter, what is the output?

2. An active-HIGH pulse (HIGH level when asserted, LOW level when not) is required on an inverter input.

(a) Draw the appropriate logic symbol, using the distinctive shape and the negation indicator, for the inverter in this application.

(b) Describe the output when a positive-going pulse is applied to the input of an inverter.

3–3 THE AND GATE

The AND gate is one of the basic gates that can be combined to form any logic function. An AND gate can have two or more inputs and performs what is known as logical multiplication.

After completing this section, you should be able to

- **Identify an AND gate by its distinctive shape symbol or by its rectangular outline symbol**
- **Describe the operation of an AND gate**
- **Generate the truth table for an AND gate with any number of inputs**
- **Produce a timing diagram for an AND gate with any specified input waveforms**
- **Write the logic expression for an AND gate with any number of inputs**
- **Discuss examples of AND gate applications**

(a) Distinctive shape   (b) Rectangular outline with the AND (&) qualifying symbol

FIGURE 3–24   **Standard logic symbols for the AND gate showing two inputs (ANSI/IEEE Std. 91-1984).**

The term *gate* was introduced in Chapter 1 and is used to describe a circuit that performs a basic logic operation. The AND gate is composed of two or more inputs and a single output, as indicated by the standard logic symbols shown in Figure 3–24. Inputs are on the left, and the output is on the right in each symbol. Gates with two inputs are shown; however, an AND gate can have any number of inputs greater than one. Although examples of both distinctive shape symbols and rectangular outline symbols are shown, the distinctive shape symbol, shown in part (a), is used predominantly in this book.

> Logic gates are one of the fundamental building blocks of computers. Most of the functions in a computer, with the exception of certain types of memory, are implemented with logic gates used on a very large scale. For example, a microprocessor, which is the main part of a computer, is made up of hundreds of thousands or even millions of logic gates.
>
> SYSTEM NOTE

Operation of an AND Gate

An AND gate can have more than two inputs.

An **AND gate** produces a HIGH output *only* when *all* of the inputs are HIGH. When any of the inputs is LOW, the output is LOW. Therefore, the basic purpose of an AND gate is to determine when certain conditions are simultaneously true, as indicated by HIGH levels on all of its inputs, and to produce a HIGH on its output to indicate that all these conditions are true. The inputs of the 2-input AND gate in Figure 3–24 are labeled *A* and *B*, and the output is labeled *X*. The gate operation can be stated as follows:

> **For a 2-input AND gate, output *X* is HIGH only when inputs *A* and *B* are HIGH; *X* is LOW when either *A* or *B* is LOW, or when both *A* and *B* are LOW.**

AND Gate Truth Table

For an AND gate, all HIGH inputs produce a HIGH output.

The logical operation of a gate can be expressed with a truth table that lists all input combinations with the corresponding outputs, as illustrated in Table 3–6 for a 2-input AND gate. The truth table can be expanded to any number of inputs. Although the terms HIGH and LOW tend to give a "physical" sense to the input and output states, the truth table is shown with 1s and 0s; a HIGH is equivalent to a 1 and a LOW is equivalent to a 0 in positive logic. For any AND gate, regardless of the number of inputs, the output is HIGH *only* when *all* inputs are HIGH.

The total number of possible combinations of binary inputs to a gate is determined by the following formula:

$$N = 2^n \qquad (3\text{–}6)$$

where *N* is the number of possible input combinations and *n* is the number of input variables. To illustrate,

For two input variables: $N = 2^2 = 4$ combinations

For three input variables: $N = 2^3 = 8$ combinations

For four input variables: $N = 2^4 = 16$ combinations

You can determine the number of input bit combinations for gates with any number of inputs by using Equation 3–6.

TABLE 3–6 • Truth table for a 2-input AND gate.

INPUTS		OUTPUT
A	*B*	*X*
0	0	0
0	1	0
1	0	0
1	1	1

1 = HIGH, 0 = LOW

┌───┐

EXAMPLE 3–4

(a) Develop the truth table for a 3-input AND gate.

(b) Determine the total number of possible input combinations for a 4-input AND gate.

SOLUTION

(a) There are eight possible input combinations ($2^3 = 8$) for a 3-input AND gate. The input side of the truth table (Table 3–7) shows all eight combinations of three bits. The output side is all 0s except when all three input bits are 1s.

TABLE 3–7

INPUTS			OUTPUT
A	B	C	X
0	0	0	0
0	0	1	0
0	1	0	0
0	1	1	0
1	0	0	0
1	0	1	0
1	1	0	0
1	1	1	1

(b) $N = 2^4 = \mathbf{16}$. There are 16 possible combinations of input bits for a 4-input AND gate.

RELATED PROBLEM

Develop the truth table for a 4-input AND gate.

└───┘

Operation with Pulse Inputs

In most applications, the inputs to a gate are not stationary levels but are voltage waveforms that change frequently between HIGH and LOW logic levels. Now let's look at the operation of AND gates with pulse waveform inputs, keeping in mind that an AND gate obeys the truth table operation regardless of whether its inputs are constant levels or levels that change back and forth.

┌───┐

Computers can utilize all of the basic logic operations when it is necessary to selectively manipulate certain bits in one or more bytes of data. Selective bit manipulations are done with a *mask*. For example, to clear (make all 0s) the right four bits in a data byte but keep the left four bits, ANDing the data byte with 11110000 will do the job. Notice that any bit ANDed with zero will be 0 and any bit ANDed with 1 will remain the same. If 10101010 is ANDed with the mask 11110000, the result is 10100000.

SYSTEM NOTE

└───┘

Let's examine the waveform operation of an AND gate by looking at the inputs with respect to each other in order to determine the output level at any given time. In Figure 3–25, inputs A and B are both HIGH (1) during the time interval, t_1, making

FIGURE 3–25   **Example of AND gate operation with a timing diagram showing input and output relationships.**

output X HIGH (1) during this interval. During time interval t_2, input A is LOW (0) and input B is HIGH (1), so the output is LOW (0). During time interval t_3, both inputs are HIGH (1) again, and therefore the output is HIGH (1). During time interval t_4, input A is HIGH (1) and input B is LOW (0), resulting in a LOW (0) output. Finally, during time interval t_5, input A is LOW (0), input B is LOW (0), and the output is therefore LOW (0). As you know, a diagram of input and output waveforms showing time relationships is called a *timing diagram*.

EXAMPLE 3–5

If two waveforms, A and B, are applied to the AND gate inputs as in Figure 3–26, what is the resulting output waveform?

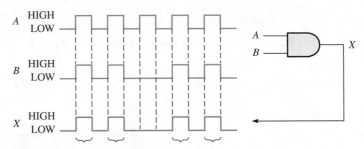

A and B are both HIGH during these four time intervals. Therefore X is HIGH.

FIGURE 3–26

SOLUTION

The output waveform X is HIGH only when both A and B waveforms are HIGH as shown in the timing diagram in Figure 3–26.

RELATED PROBLEM

Determine the output waveform and show a timing diagram if the second and fourth pulses in waveform A of Figure 3–26 are replaced by LOW levels.

Remember, when observing the waveform operation of logic gates, it is important to pay careful attention to the time relationships of all the inputs with respect to each other and to the output.

EXAMPLE 3–6

For the two input waveforms, A and B, in Figure 3–27, show the output waveform with its proper relation to the inputs.

Inputs

Output *X*

FIGURE 3–27

SOLUTION

The output waveform is HIGH only when both of the input waveforms are HIGH as shown in the timing diagram.

RELATED PROBLEM

Show the output waveform if the *B* input to the AND gate in Figure 3–27 is always HIGH.

EXAMPLE 3–7

For the 3-input AND gate in Figure 3–28, determine the output waveform in relation to the inputs.

FIGURE 3–28

SOLUTION

The output waveform *X* of the 3-input AND gate is HIGH only when all three input waveforms *A*, *B*, and *C* are HIGH.

RELATED PROBLEM

What is the output waveform of the AND gate in Figure 3–28 if the *C* input is always HIGH?

EXAMPLE 3–8

Use Multisim to simulate a 3-input AND gate with input waveforms that cycle through binary numbers 0 through 9.

SOLUTION

Use the Multisim word generator in the up counter mode to provide the combination of waveforms representing the binary sequence, as shown in Figure 3–29. The first three waveforms on the oscilloscope display are the inputs, and the bottom waveform is the output.

MULTISIM

FIGURE 3–29

RELATED PROBLEM

Use Multisim software to create the setup and simulate the 3-input AND gate as illustrated in this example. A Multisim tutorial is available at the website.

Logic Expressions for AND Gates

When variables are shown together like ABC, they are ANDed.

The operation of a 2-input AND gate can be expressed in equation form as follows: If one input variable is A, the other input variable is B, and the output variable is X, then the Boolean expression is

$$X = AB$$

Figure 3–30(a) shows the AND gate logic symbol with two input variables and the output variable indicated.

(a)   (b)   (c)

FIGURE 3–30   **Boolean expressions for AND gates with two, three, and four inputs.**

To extend the AND expression to more than two input variables, simply use a new letter for each input variable. The function of a 3-input AND gate, for example, can be expressed as $X = ABC$, where A, B, and C are the input variables. The expression for a

4-input AND gate can be $X = ABCD$, and so on. Parts (b) and (c) of Figure 3–30 show AND gates with three and four input variables, respectively.

You can evaluate an AND gate operation by using the Boolean expressions for the output. For example, each variable on the inputs can be either a 1 or a 0; so for the 2-input AND gate, make substitutions in the equation for the output, $X = AB$, as shown in Table 3–8. This evaluation shows that the output X of an AND gate is a 1 (HIGH) only when both inputs are 1s (HIGHs). A similar analysis can be made for any number of input variables.

TABLE 3–8		
A	B	AB = X
0	0	$0 \cdot 0 = 0$
0	1	$0 \cdot 1 = 0$
1	0	$1 \cdot 0 = 0$
1	1	$1 \cdot 1 = 1$

SYSTEM EXAMPLE 3–1

ENABLE/INHIBIT DEVICE

A common system application of the AND gate is to enable (that is, to allow) the passage of a signal (pulse waveform) from one point to another at certain times and to inhibit (prevent) the passage at other times. This particular use of an AND gate is shown in Figure 3–31, where the AND gate controls the passage of a signal (waveform A) to a digital counter. This system measures the frequency of waveform A. The enable pulse has a width of precisely 1 ms. When the enable pulse is HIGH, waveform A passes through the gate to the counter; and when the enable pulse is LOW, the signal is prevented from passing through the gate (inhibited).

FIGURE 3–31   The AND gate performing an enable/inhibit function for a frequency counter.

During the 1 millisecond (1 ms) interval of the enable pulse, pulses in waveform A pass through the AND gate to the counter. The number of pulses passing through during the 1 ms interval is equal to the frequency of waveform A. For example, Figure 3–31 shows six pulses in one millisecond, which is a frequency of 6 kHz. If 1000 pulses pass through the gate in the 1 ms interval of the enable pulse, there are 1000 pulses/ms, or a frequency of 1 MHz.

The counter counts the number of pulses per second and produces a binary output that goes to a decoding and display circuit to produce a readout of the frequency. The enable pulse repeats at certain intervals and a new updated count is made so that if the frequency changes, the new value will be displayed. Between enable pulses, the counter is reset so that it starts at zero each time an enable pulse occurs. The current frequency count is stored in a register so that the display is unaffected by the resetting of the counter.

SYSTEM EXAMPLE 3–2

SEAT BELT ALARM SYSTEM

In Figure 3–32, an AND gate is used in a simple automobile seat belt alarm system to detect when the ignition switch is on *and* the seat belt is unbuckled. If the ignition switch is on, a HIGH is produced on input A of the AND gate. If the seat belt is not properly buckled, a

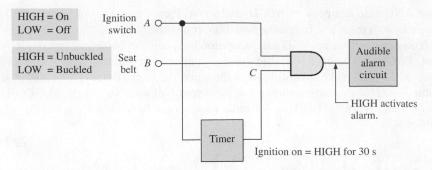

FIGURE 3–32   **Simplified seat belt alarm system using an AND gate.**

HIGH is produced on input *B* of the AND gate. Also, when the ignition switch is turned on, a timer is started that produces a HIGH on input *C* for 30 s. If all three conditions exist— that is, if the ignition is on *and* the seat belt is unbuckled *and* the timer is running—the output of the AND gate is HIGH, and an audible alarm is energized to remind the driver.

SECTION 3–3 CHECKUP

1. When is the output of an AND gate HIGH?

2. When is the output of an AND gate LOW?

3. Describe the truth table for a 5-input AND gate.

3–4 THE OR GATE

The OR gate is another of the basic gates from which all logic functions are constructed. An OR gate can have two or more inputs and performs what is known as logical addition.

After completing this section, you should be able to

- **Identify an OR gate by its distinctive shape symbol or by its rectangular outline symbol**
- **Describe the operation of an OR gate**
- **Generate the truth table for an OR gate with any number of inputs**
- **Produce a timing diagram for an OR gate with any specified input waveforms**
- **Write the logic expression for an OR gate with any number of inputs**
- **Discuss an OR gate application**

An OR gate can have more than two inputs.

An **OR gate** has two or more inputs and one output, as indicated by the standard logic symbols in Figure 3–33, where OR gates with two inputs are illustrated. An OR gate can have any number of inputs greater than one. Although both distinctive shape and rectangular outline symbols are shown, the distinctive shape OR gate symbol is used in this textbook.

(a) Distinctive shape

(b) Rectangular outline with the OR (≥ 1) qualifying symbol

FIGURE 3–33   **Standard logic symbols for the OR gate showing two inputs (ANSI/IEEE Std. 91-1984).**

Operation of an OR Gate

An OR gate produces a HIGH on the output when *any* of the inputs is HIGH. The output is LOW only when all of the inputs are LOW. Therefore, an OR gate determines when one or more of its inputs are HIGH and produces a HIGH on its output to indicate this condition. The inputs of the 2-input OR gate in Figure 3–33 are labeled *A* and *B*, and the output is labeled *X*. The operation of the gate can be stated as follows:

> **For a 2-input OR gate, output *X* is HIGH when either input *A* or input *B* is HIGH, or when both *A* and *B* are HIGH; *X* is LOW only when both *A* and *B* are LOW.**

The HIGH level is the active or asserted output level for the OR gate.

> For an OR gate, at least one HIGH input produces a HIGH output.

OR Gate Truth Table

The operation of a 2-input OR gate is described in Table 3–9. This truth table can be expanded for any number of inputs; but regardless of the number of inputs, the output is HIGH when one or more of the inputs are HIGH.

TABLE 3–9 • Truth table for a 2-input OR gate.

INPUTS		OUTPUT
A	*B*	*X*
0	0	0
0	1	1
1	0	1
1	1	1

1 = HIGH, 0 = LOW

Operation with Pulse Inputs

Now let's look at the operation of an OR gate with pulse waveform inputs, keeping in mind its logical operation. Again, the important thing in the analysis of gate operation with pulse waveforms is the time relationship of all the waveforms involved. For example, in Figure 3–34, inputs *A* and *B* are both HIGH (1) during time interval t_1, making output *X* HIGH (1). During time interval t_2, input *A* is LOW (0), but because input *B* is HIGH (1), the output is HIGH (1). Both inputs are LOW (0) during time interval t_3, so there is a LOW (0) output during this time. During time interval t_4, the output is HIGH (1) because input *A* is HIGH (1).

   In this illustration, we have applied the truth table operation of the OR gate to each of the time intervals during which the levels are nonchanging. Examples 3–9 through 3–11 further illustrate OR gate operation with waveforms on the inputs.

FIGURE 3–34 Example of OR gate operation with a timing diagram showing input and output time relationships.

EXAMPLE 3-9

If the two input waveforms, *A* and *B*, in Figure 3–35 are applied to the OR gate, what is the resulting output waveform?

When either input or both inputs are HIGH, the output is HIGH.

FIGURE 3–35

SOLUTION

The output waveform X of a 2-input OR gate is HIGH when either or both input waveforms are HIGH as shown in the timing diagram. In this case, both input waveforms are never HIGH at the same time.

RELATED PROBLEM

Determine the output waveform and show the timing diagram if input A is changed such that it is HIGH from the beginning of the existing first pulse to the end of the existing second pulse.

EXAMPLE 3–10

For the two input waveforms, A and B, in Figure 3–36, show the output waveform with its proper relation to the inputs.

FIGURE 3–36

SOLUTION

When either or both input waveforms are HIGH, the output is HIGH as shown by the output waveform X in the timing diagram.

RELATED PROBLEM

Determine the output waveform and show the timing diagram if the middle pulse of input A is replaced by a LOW level.

EXAMPLE 3–11

For the 3-input OR gate in Figure 3–37, determine the output waveform in proper time relation to the inputs.

FIGURE 3–37

SOLUTION

The output is HIGH when one or more of the input waveforms are HIGH as indicated by the output waveform X in the timing diagram.

RELATED PROBLEM

Determine the output waveform and show the timing diagram if input C is always LOW.

Logic Expressions for OR Gates

The operation of a 2-input OR gate can be expressed as follows: If one input variable is A, if the other input variable is B, and if the output variable is X, then the Boolean expression is

$$X = A + B$$

Figure 3–38(a) shows the OR gate logic symbol with two input variables and the output variable labeled.

(a)   (b)   (c)

FIGURE 3–38   **Boolean expressions for OR gates with two, three, and four inputs.**

To extend the OR expression to more than two input variables, a new letter is used for each additional variable. For instance, the function of a 3-input OR gate can be expressed as $X = A + B + C$. The expression for a 4-input OR gate can be written as $X = A + B + C + D$, and so on. Parts (b) and (c) of Figure 3–38 show OR gates with three and four input variables, respectively.

OR gate operation can be evaluated by using the Boolean expressions for the output X by substituting all possible combinations of 1 and 0 values for the input variables, as shown in Table 3–10 for a 2-input OR gate. This evaluation shows that the output X of an OR gate is a 1 (HIGH) when any one or more of the inputs are 1 (HIGH). A similar analysis can be extended to OR gates with any number of input variables.

TABLE 3–10		
A	B	$A + B = X$
0	0	$0 + 0 = 0$
0	1	$0 + 1 = 1$
1	0	$1 + 0 = 1$
1	1	$1 + 1 = 1$

Another mask operation that is used in computer programming to selectively make certain bits in a data byte equal to 1 (called setting) while not affecting any other bit is done with the OR operation. A mask is used that contains a 1 in any position where a data bit is to be set. For example, if you want to force the sign bit in an 8-bit signed number to equal 1, but leave all other bits unchanged, you can OR the data byte with the mask 10000000.

SYSTEM NOTE

SYSTEM EXAMPLE 3–3

INTRUSION DETECTION

A simplified portion of an intrusion detection and alarm system is shown in Figure 3–39. This system could be used for one room in a home—a room with two windows and a door. The sensors are magnetic switches that produce a HIGH output when open and a LOW

Open door/window sensors

HIGH = Open
LOW = Closed

HIGH activates alarm.

Alarm circuit

FIGURE 3–39   **A simplified intrusion detection system using an OR gate.**

output when closed. As long as the windows and the door are secured, the switches are closed and all three of the OR gate inputs are LOW. When one of the windows or the door is opened, a HIGH is produced on that input to the OR gate and the gate output goes HIGH. It then activates and latches an alarm circuit to warn of the intrusion.

SECTION 3–4 CHECKUP

1. When is the output of an OR gate HIGH?

2. When is the output of an OR gate LOW?

3. Describe the truth table for a 3-input OR gate.

3–5 THE NAND GATE

The NAND gate is a popular logic element because it can be used as a universal gate; that is, NAND gates can be used in combination to perform the AND, OR, and inverter operations. The universal property of the NAND gate will be examined thoroughly in Chapter 4.

After completing this section, you should be able to

- Identify a NAND gate by its distinctive shape symbol or by its rectangular outline symbol
- Describe the operation of a NAND gate
- Develop the truth table for a NAND gate with any number of inputs
- Produce a timing diagram for a NAND gate with any specified input waveforms
- Write the logic expression for a NAND gate with any number of inputs
- Describe NAND gate operation in terms of its negative-OR equivalent
- Discuss NAND gate applications

The NAND gate is the same as the AND gate except the output is inverted.

The term *NAND* is a contraction of NOT-AND and implies an AND function with a complemented (inverted) output. The standard logic symbol for a 2-input NAND gate and its equivalency to an AND gate followed by an inverter are shown in Figure 3–40(a), where the symbol ≡ means equivalent to. A rectangular outline symbol is shown in part (b).

FIGURE 3–40   Standard NAND gate logic symbols (ANSI/IEEE Std. 91-1984).

(a) Distinctive shape, 2-input NAND gate and its NOT/AND equivalent

(b) Rectangular outline, 2-input NAND gate with polarity indicator

Operation of a NAND Gate

A **NAND gate** produces a LOW output only when all the inputs are HIGH. When any of the inputs is LOW, the output will be HIGH. For the specific case of a 2-input NAND gate, as shown in Figure 3–40 with the inputs labeled A and B and the output labeled X, the operation can be stated as follows:

For a 2-input NAND gate, output X is LOW only when inputs A and B are HIGH; X is HIGH when either A or B is LOW, or when both A and B are LOW.

This operation is opposite that of the AND in terms of the output level. In a NAND gate, the LOW level (0) is the active or asserted output level, as indicated by the bubble on the output. Figure 3–41 illustrates the operation of a 2-input NAND gate for all four input

FIGURE 3–41   Operation of a 2-input NAND gate. Open file F03-41 to verify NAND gate operation.

combinations, and Table 3–11 is the truth table summarizing the logical operation of the 2-input NAND gate.

TABLE 3–11 • Truth table for a 2-input NAND gate.		
INPUTS		**OUTPUT**
A	*B*	*X*
0	0	1
0	1	1
1	0	1
1	1	0

1 = HIGH, 0 = LOW.

Operation with Pulse Inputs

Now let's look at the pulse waveform operation of a NAND gate. Remember from the truth table that the only time a LOW output occurs is when all of the inputs are HIGH.

EXAMPLE 3–12

If the two waveforms *A* and *B* shown in Figure 3–42 are applied to the NAND gate inputs, determine the resulting output waveform.

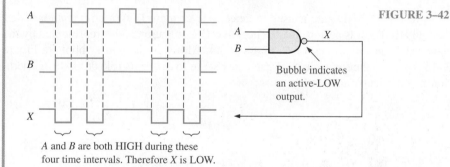

FIGURE 3–42

Bubble indicates an active-LOW output.

A and *B* are both HIGH during these four time intervals. Therefore *X* is LOW.

SOLUTION

Output waveform *X* is LOW only during the four time intervals when both input waveforms *A* and *B* are HIGH as shown in the timing diagram.

RELATED PROBLEM

Determine the output waveform and show the timing diagram if input waveform *B* is inverted.

EXAMPLE 3–13

Show the output waveform for the 3-input NAND gate in Figure 3–43 with its proper time relationship to the inputs.

FIGURE 3–43

The output waveform X is LOW only when all three input waveforms are HIGH as shown in the timing diagram.

Determine the output waveform and show the timing diagram if input waveform A is inverted.

NEGATIVE-OR EQUIVALENT OPERATION OF A NAND GATE   Inherent in a NAND gate's operation is the fact that one or more LOW inputs produce a HIGH output. Table 3–11 shows that output X is HIGH (1) when any of the inputs, A and B, is LOW (0). From this viewpoint, a NAND gate can be used for an OR operation that requires one or more LOW inputs to produce a HIGH output. This aspect of NAND operation is referred to as **negative-OR.** The term *negative* in this context means that the inputs are defined to be in the active or asserted state when LOW.

> **For a 2-input NAND gate performing a negative-OR operation, output X is HIGH when either input A or input B is LOW, or when both A and B are LOW.**

When a NAND gate is used to detect one or more LOWs on its inputs rather than all HIGHs, it is performing the negative-OR operation and is represented by the standard logic symbol shown in Figure 3–44. Although the two symbols in Figure 3–44 represent the same physical gate, they serve to define its role or mode of operation in a particular application.

NAND   Negative-OR

FIGURE 3–44   Standard symbols representing the two equivalent operations of a NAND gate.

SYSTEM EXAMPLE 3–4

STORAGE TANK HIGH-LEVEL DETECTION

Two tanks are used to store certain liquid chemicals that are required in a manufacturing process. Each tank has a sensor that detects when the chemical level drops to 25% of full. The sensors produce a HIGH level of 5 V when the tanks are more than one-quarter full. When the volume of chemical in a tank drops to one-quarter full, the sensor puts out a LOW level of 0 V.

A single green light-emitting diode (LED) on an indicator panel show when both tanks are more than one-quarter full.

Figure 3–45 shows a NAND gate with its two inputs connected to the tank level sensors and its output connected to the indicator panel. The operation can be stated as follows: If tank A *and* tank B are above one-quarter full, the LED is on.

As long as both sensor outputs are HIGH (5 V), indicating that both tanks are more than one-quarter full, the NAND gate output is LOW (0 V). The green LED circuit is arranged so that a LOW voltage turns it on. The resistor limits the LED current. The system can be modified to monitor more than two tanks by increasing the number of NAND gate inputs to correspond to the number of tanks.

FIGURE 3–45

SYSTEM EXAMPLE 3-5

STORAGE TANK LOW-LEVEL DETECTION

An alternate approach to System Example 3–4 is a system in which a red LED display comes on when at least one of the tanks falls to the quarter-full level rather than have the green LED display indicate when both are above one quarter.

  Figure 3–46 shows a NAND gate operating as a negative-OR gate to detect the occurrence of at least one LOW on its inputs. A sensor puts out a LOW voltage if the volume in its tank goes to one-quarter full or less. When this happens, the gate output goes HIGH. The red LED circuit in the panel is arranged so that a HIGH voltage turns it on. The operation can be stated as follows: If tank *A or* tank *B or* both are below one-quarter full, the LED is on.

  Notice that, in this system example and in System Example 3–4, the same 2-input NAND gate is used, but a different gate symbol is used in the schematic, illustrating the different way in which the NAND and equivalent negative-OR operations can be used.

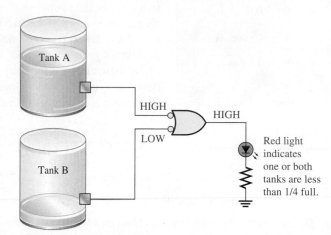

FIGURE 3–46

EXAMPLE 3-14

For the 4-input NAND gate in Figure 3–47, operating as a negative-OR gate, determine the output with respect to the inputs.

FIGURE 3–47

SOLUTION

The output waveform X is HIGH any time an input waveform is LOW as shown in the timing diagram.

RELATED PROBLEM

Determine the output waveform if input waveform A is inverted before it is applied to the gate.

A bar over a variable or variables indicates an inversion.

Logic Expressions for NAND Gates

The Boolean expression for the output of a 2-input NAND gate is

$$X = \overline{AB}$$

This expression says that the two input variables, A and B, are first ANDed and then complemented, as indicated by the bar over the AND expression. This is a description in equation form of the operation of a NAND gate with two inputs. Evaluating this expression for all possible values of the two input variables, you get the results shown in Table 3–12.

Once an expression is determined for a given logic function, that function can be evaluated for all possible values of the variables. The evaluation tells you exactly what the output of the logic circuit is for each of the input conditions, and it therefore gives you a complete description of the circuit's logic operation. The NAND expression can be extended to more than two input variables by including additional letters to represent the other variables.

TABLE 3–12

A	B	$\overline{AB} = X$
0	0	$\overline{0 \cdot 0} = \overline{0} = 1$
0	1	$\overline{0 \cdot 1} = \overline{0} = 1$
1	0	$\overline{1 \cdot 0} = \overline{0} = 1$
1	1	$\overline{1 \cdot 1} = \overline{1} = 0$

SECTION 3–5 CHECKUP

1. When is the output of a NAND gate LOW?

2. When is the output of a NAND gate HIGH?

3. Describe the functional differences between a NAND gate and a negative-OR gate. Do they both have the same truth table?

4. Write the output expression for a NAND gate with inputs A, B, and C.

3–6   THE NOR GATE

The NOR gate, like the NAND gate, is a useful logic element because it can also be used as a universal gate; that is, NOR gates can be used in combination to perform the AND, OR, and inverter operations. The universal property of the NOR gate will be examined thoroughly in Chapter 4.

After completing this section, you should be able to

- **Identify a NOR gate by its distinctive shape symbol or by its rectangular outline symbol**
- **Describe the operation of a NOR gate**
- **Develop the truth table for a NOR gate with any number of inputs**
- **Produce a timing diagram for a NOR gate with any specified input waveforms**
- **Write the logic expression for a NOR gate with any number of inputs**
- **Describe NOR gate operation in terms of its negative-AND equivalent**
- **Discuss a NOR gate application**

The NOR is the same as the OR except the output is inverted.

The term *NOR* is a contraction of NOT-OR and implies an OR function with an inverted (complemented) output. The standard logic symbol for a 2-input NOR gate and its equivalent OR gate followed by an inverter are shown in Figure 3–48(a). A rectangular outline symbol is shown in part (b).

(a) Distinctive shape, 2-input NOR gate and its NOT/OR equivalent

(b) Rectangular outline, 2-input NOR gate with polarity indicator

FIGURE 3–48   **Standard NOR gate logic symbols (ANSI/IEEE Std. 91-1984).**

Operation of a NOR Gate

A **NOR gate** produces a LOW output when *any* of its inputs is HIGH. Only when all of its inputs are LOW is the output HIGH. For the specific case of a 2-input NOR gate, as shown in Figure 3–48 with the inputs labeled *A* and *B* and the output labeled *X*, the operation can be stated as follows:

> **For a 2-input NOR gate, output *X* is LOW when either input *A* or input *B* is HIGH, or when both *A* and *B* are HIGH; *X* is HIGH only when both *A* and *B* are LOW.**

This operation results in an output level opposite that of the OR gate. In a NOR gate, the LOW output is the active or asserted output level as indicated by the bubble on the output. Figure 3–49 illustrates the operation of a 2-input NOR gate for all four possible input combinations, and Table 3–13 is the truth table for a 2-input NOR gate.

TABLE 3–13 • Truth table for a 2-input NOR gate.		
INPUTS		**OUTPUT**
A	*B*	*X*
0	0	1
0	1	0
1	0	0
1	1	0

1 = HIGH, 0 = LOW.

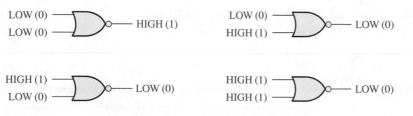

FIGURE 3–49   Operation of a 2-input NOR gate. Open file F03-49 to verify NOR gate operation.

MULTISIM

Operation with Pulse Inputs

The next two examples illustrate the operation of a NOR gate with pulse waveform inputs. Again, as with the other types of gates, we will simply follow the truth table operation to determine the output waveforms in the proper time relationship to the inputs.

EXAMPLE 3–15

If the two waveforms shown in Figure 3–50 are applied to a NOR gate, what is the resulting output waveform?

FIGURE 3–50

SOLUTION

Whenever any input of the NOR gate is HIGH, the output is LOW as shown by the output waveform *X* in the timing diagram.

RELATED PROBLEM

Invert input *B* and determine the output waveform in relation to the inputs.

EXAMPLE 3-16

Show the output waveform for the 3-input NOR gate in Figure 3–51 with the proper time relation to the inputs.

FIGURE 3–51

SOLUTION

The output X is LOW when any input is HIGH as shown by the output waveform X in the timing diagram.

RELATED PROBLEM

With the B and C inputs inverted, determine the output and show the timing diagram.

NEGATIVE-AND EQUIVALENT OF THE NOR GATE   A NOR gate, like the NAND, has another aspect of its operation that is inherent in the way it logically functions. Table 3–13 shows that a HIGH is produced on the gate output only when all of the inputs are LOW. From this viewpoint, a NOR gate can be used for an AND operation that requires all LOW inputs to produce a HIGH output. This aspect of NOR operation is called **negative-AND.** The term *negative* in this context means that the inputs are defined to be in the active or asserted state when LOW.

NOR   Negative-AND

FIGURE 3–52   Standard symbols representing the two equivalent operations of a NOR gate.

For a 2-input NOR gate performing a negative-AND operation, output X is HIGH only when both inputs A and B are LOW.

When a NOR gate is used to detect all LOWs on its inputs rather than one or more HIGHs, it is performing the negative-AND operation and is represented by the standard symbol in Figure 3–52. Remember that the two symbols in Figure 3–52 represent the same physical gate and serve only to distinguish between the two modes of its operation.

SYSTEM EXAMPLE 3-6

LANDING GEAR STATUS MONITOR

As part of an aircraft's functional monitoring system, a landing gear status monitor is required to indicate the status of the landing gears prior to landing. A green LED display turns on if all three gears are properly extended when the "gear down" switch has been activated in preparation for landing. A red LED display turns on if any of the gears fail to extend properly prior to landing. When a landing gear is extended, its sensor produces a LOW voltage. When a landing gear is retracted, its sensor produces a HIGH voltage.

Power is applied to the circuit only when the "gear down" switch is activated. A NOR gate is used for each of the two requirements as shown in Figure 3–53. One NOR gate operates as a negative-AND to detect a LOW from each of the three landing gear sensors. When all three of the gate inputs are LOW, the three landing gears are properly extended and the resulting HIGH output from the negative-AND gate turns on the green LED display. The other NOR gate operates as a NOR to detect if one or more of the

FIGURE 3–53

landing gears remain retracted when the "gear down" switch is activated. When one or more of the landing gears remain retracted, the resulting HIGH from the sensor is detected by the NOR gate, which produces a LOW output to turn on the red LED warning display.

EXAMPLE 3–17

For the 4-input NOR gate operating as a negative-AND in Figure 3–54, determine the output relative to the inputs.

FIGURE 3–54

HANDS ON TIP
When driving a load such as an LED with a logic gate, consult the manufacturer's data sheet for maximum drive capabilities (output current). A regular IC logic gate may not be capable of handling the current required by certain loads such as some LEDs. Logic gates with a buffered output, such as an open-collector (OC) or open-drain (OD) output, are available. The output current capability of typical logic gates is limited to the μA or relatively low mA range.

SOLUTION

Any time all of the input waveforms are LOW, the output is HIGH as shown by output waveform X in the timing diagram.

RELATED PROBLEM

Determine the output with input D inverted and show the timing diagram.

Logic Expressions for NOR Gates

The Boolean expression for the output of a 2-input NOR gate can be written as

$$X = \overline{A + B}$$

This equation says that the two input variables are first ORed and then complemented, as indicated by the bar over the OR expression. Evaluating this expression, you get the results shown in Table 3–14. The NOR expression can be extended to more than two input variables by including additional letters to represent the other variables.

TABLE 3–14		
A	**B**	$\overline{A + B} = X$
0	0	$\overline{0 + 0} = \overline{0} = 1$
0	1	$\overline{0 + 1} = \overline{1} = 0$
1	0	$\overline{1 + 0} = \overline{1} = 0$
1	1	$\overline{1 + 1} = \overline{1} = 0$

SECTION 3–6 CHECKUP

1. When is the output of a NOR gate HIGH?

2. When is the output of a NOR gate LOW?

3. Describe the functional difference between a NOR gate and a negative-AND gate. Do they both have the same truth table?

4. Write the output expression for a 3-input NOR gate with input variables A, B, and C.

3–7 THE EXCLUSIVE-OR AND EXCLUSIVE-NOR GATES

Exclusive-OR and exclusive-NOR gates are formed by a combination of AND gates, OR gates, and inverters. However, because of their fundamental importance in many applications, these gates are often treated as basic logic elements with their own unique symbols.

After completing this section, you should be able to

- Identify the exclusive-OR and exclusive-NOR gates by their distinctive shape symbols or by their rectangular outline symbols
- Describe the operations of exclusive-OR and exclusive-NOR gates
- Show the truth tables for exclusive-OR and exclusive-NOR gates
- Produce a timing diagram for an exclusive-OR or exclusive-NOR gate with any specified input waveforms

The Exclusive-OR Gate

Standard symbols for an exclusive-OR (XOR for short) gate are shown in Figure 3–55. The XOR gate has only two inputs. The **exclusive-OR gate** performs modulo-2 addition (introduced in Chapter 2). The output of an exclusive-OR gate is HIGH *only* when the two inputs are at opposite logic levels. This operation can be stated as follows with reference to inputs A and B and output X:

For an exclusive-OR gate, opposite inputs make the output HIGH.

For an exclusive-OR gate, output X is HIGH when input A is LOW and input B is HIGH, or when input A is HIGH and input B is LOW; X is LOW when A and B are both HIGH or both LOW.

(a) Distinctive shape   (b) Rectangular outline

FIGURE 3–55   Standard logic symbols for the exclusive-OR gate.

Exclusive-OR gates connected to form an adder circuit allow a computer to perform addition, subtraction, multiplication, and division in its Arithmetic Logic Unit (ALU). An exclusive-OR gate combines basic AND, OR, and NOT logic.

SYSTEM NOTE

The four possible input combinations and the resulting outputs for an XOR gate are illustrated in Figure 3–56. The HIGH level is the active or asserted output level and occurs only when the inputs are at opposite levels. The operation of an XOR gate is summarized in the truth table shown in Table 3–15.

TABLE 3–15 • Truth table for an exclusive-OR gate.		
INPUTS		**OUTPUT**
A	*B*	*X*
0	0	0
0	1	1
1	0	1
1	1	0

MULTISIM

FIGURE 3–56   **All possible logic levels for an exclusive-OR gate. Open file F03-56 to verify XOR gate operation.**

SYSTEM EXAMPLE 3–7

CIRCUIT FAULT DETECTION

A certain system contains two identical circuits operating in parallel. As long as both are operating properly, the outputs of both circuits are always the same. If one of the circuits fails, the outputs will be at opposite levels at some time. This method is called *redundancy*.

The outputs of the circuits are connected to the inputs of an XOR gate as shown in Figure 3–57. A failure in either one of the circuits produces differing outputs, which cause the XOR inputs to be at opposite levels. This condition produces a HIGH on the output of the XOR gate, indicating a failure in one of the circuits.

FIGURE 3–57

The Exclusive-NOR Gate

Standard symbols for an **exclusive-NOR (XNOR) gate** are shown in Figure 3–58. Like the XOR gate, an XNOR has only two inputs. The bubble on the output of the XNOR symbol indicates that its output is opposite that of the XOR gate. When the two input logic levels are opposite, the output of the exclusive-NOR gate is LOW. The operation can be stated as follows (*A* and *B* are inputs, *X* is the output):

For an exclusive-NOR gate, output *X* is LOW when input *A* is LOW and input *B* is HIGH, or when *A* is HIGH and *B* is LOW; *X* is HIGH when *A* and *B* are both HIGH or both LOW.

(a) Distinctive shape   (b) Rectangular outline

FIGURE 3–58   **Standard logic symbols for the exclusive-NOR gate.**

The four possible input combinations and the resulting outputs for an XNOR gate are shown in Figure 3–59. The operation of an XNOR gate is summarized in Table 3–16. Notice that the output is HIGH when the same level is on both inputs.

TABLE 3–16 • Truth table for an exclusive-NOR gate.

INPUTS		OUTPUT
A	B	X
0	0	1
0	1	0
1	0	0
1	1	1

MULTISIM

FIGURE 3–59 **All possible logic levels for an exclusive-NOR gate. Open file F03-59 to verify XNOR gate operation.**

Logic Expressions for the XOR and XNOR Gates

The Boolean expression for an exclusive-OR gate is

$$X = A \oplus B$$

and the expression for an exclusive-NOR gate is

$$X = \overline{A \oplus B}$$

Expressions for both the XOR and XNOR gates can also be written in terms of AND, OR, and NOT; for the XOR gate,

$$X = A\overline{B} + \overline{A}B$$

and for the XNOR gate,

$$X = A\overline{B} + \overline{A}B$$

Operation with Pulse Inputs

As we have done with the other gates, let's examine the operation of XOR and XNOR gates with pulse waveform inputs. As before, we apply the truth table operation during each distinct time interval of the pulse waveform inputs, as illustrated in Figure 3–60 for an XOR gate. You can see that the input waveforms A and B are at opposite levels during time intervals t_2 and t_4. Therefore, the output X is HIGH during these two times. Since both inputs are at the same level, either both HIGH or both LOW, during time intervals t_1 and t_3, the output is LOW during those times as shown in the timing diagram.

FIGURE 3–60 **Example of exclusive-OR gate operation with pulse waveform inputs.**

EXAMPLE 3–18

Determine the output waveforms for the XOR gate and for the XNOR gate, given the input waveforms, A and B, in Figure 3–61.

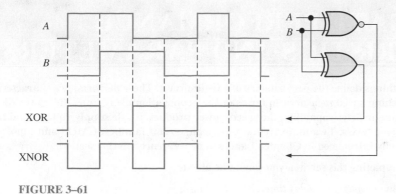

FIGURE 3–61

SOLUTION

The output waveforms are shown in Figure 3–61. Notice that the XOR output is HIGH only when both inputs are at opposite levels. Notice that the XNOR output is HIGH only when both inputs are the same.

RELATED PROBLEM

Determine the output waveforms if the two input waveforms, *A* and *B*, are inverted.

The XOR Gate as an Adder

An exclusive-OR gate can be used as a two-bit modulo-2 adder. Recall from Chapter 2 that the basic rules for binary addition are as follows: $0 + 0 = 0, 0 + 1 = 1, 1 + 0 = 1$, and $1 + 1 = 10$. An examination of the truth table for an XOR gate shows that its output is the binary sum of the two input bits. In the case where the inputs are both 1s, the output is the sum 0, but you lose the carry of 1. Later you will see how XOR gates are combined to make complete adding circuits. Table 3–17 illustrates an XOR gate used as a modulo-2 adder. It is used in CRC systems to implement the division process that was described in Chapter 2.

TABLE 3–17 • An XOR gate used to add two bits.

INPUT BITS		OUTPUT (SUM)
A	*B*	Σ
0	0	0
0	1	1
1	0	1
1	1	0 (without the 1 carry bit)

SECTION 3–7 CHECKUP

1. When is the output of an XOR gate HIGH?

2. When is the output of an XNOR gate HIGH?

3. How can you use an XOR gate to detect when two bits are different?

3–8 GATE PERFORMANCE CHARACTERISTICS AND PARAMETERS

Several things define the performance of a logic circuit. These performance characteristics are the switching speed measured in terms of the propagation delay time, the power dissipation, the fan-out or drive capability, the speed-power product, the dc supply voltage, and the input/output logic levels. Two major types of IC logic circuit families, CMOS and bipolar (TTL), were briefly introduced in Chapter 1 and will be referenced in this section.

After completing this section, you should be able to

- Define *propagation delay time*
- Define *power dissipation*
- Define *fan-out*
- Define *unit load*
- Define *speed-power product*

PROPAGATION DELAY TIME   This parameter is a result of the limitation on switching speed or frequency at which a logic circuit can operate. The terms *low speed* and *high speed,* applied to logic circuits, refer to the propagation delay time. The shorter the propagation delay, the higher the speed of the circuit and the higher the frequency at which it can operate.

High-speed logic has a short propagation delay time.

Propagation delay time, t_P, of a logic gate is the time interval between the transition of an input pulse and the occurrence of the resulting transition of the output pulse. Two different measurements of propagation delay time are associated with a logic gate and apply to all the types of basic gates:

- t_{PHL}: The time between a specified reference point on the input pulse and a corresponding reference point on the resulting output pulse, with the output changing from the HIGH level to the LOW level (HL).

- t_{PLH}: The time between a specified reference point on the input pulse and a corresponding reference point on the resulting output pulse, with the output changing from the LOW level to the HIGH level (LH).

EXAMPLE 3–19

Show the propagation delay times of the inverter in Figure 3–62(a).

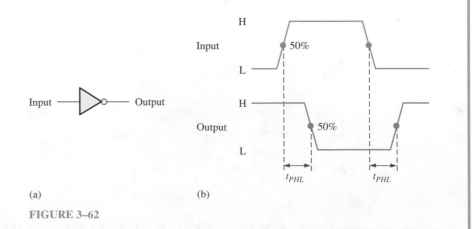

(a)　　　　　　　　　　(b)

FIGURE 3–62

SOLUTION

The propagation delay times, t_{PHL} and t_{PLH}, are indicated in part (b) of the figure. In this case, the delays are measured between the 50% points of the corresponding edges of the input and output pulses. The values of t_{PHL} and t_{PLH} are not necessarily equal but in many cases they are the same.

RELATED PROBLEM

One type of logic gate has a specified maximum t_{PLH} and t_{PHL} of 10 ns. For another type of gate the value is 4 ns. Which gate can operate at the highest frequency?

DC SUPPLY VOLTAGE (V_{CC})   The typical dc supply voltage for CMOS logic is either 5 V, 3.3 V, 2.5 V, or 1.8 V, depending on the category. An advantage of CMOS is that the supply voltages can vary over a wider range than for bipolar logic. Propagation delay time and power dissipation can be affected by variation in supply voltage. For example, the 5 V CMOS can tolerate supply variations from 2 V to 6 V and still operate properly although propagation delay time and power dissipation are significantly affected. The 3.3 V CMOS can operate with supply voltages from 2 V to 3.6 V. The typical dc supply voltage for bipolar logic is 5.0 V with a minimum of 4.5 V and a maximum of 5.5 V.

POWER DISSIPATION   The **power dissipation,** P_D, of a logic gate is the product of the dc supply voltage and the average supply current. Normally, the supply current when the gate output is LOW is greater than when the gate output is HIGH. The manufacturer's data sheet usually designates the supply current for the LOW output state as I_{CCL} and for the HIGH state as I_{CCH}. The average supply current is determined based on a 50% duty cycle (output LOW half the time and HIGH half the time), so the average power dissipation of a logic gate is

A lower power dissipation means less current from the dc supply.

$$P_D = V_{CC}\left(\frac{I_{CCH} + I_{CCL}}{2}\right) \tag{3–7}$$

CMOS gates have very low power dissipations compared to the bipolar family. However, the power dissipation of CMOS is dependent on the frequency of operation. At zero frequency the quiescent power is typically in the microwatt/gate range, and at the maximum operating frequency it can be in the low milliwatt range; therefore, power is sometimes specified at a given frequency.

INPUT AND OUTPUT LOGIC LEVELS V_{IL} is the LOW level input voltage for a logic gate, and V_{IH} is the HIGH level input voltage. The 5 V CMOS accepts a maximum voltage of 1.5 V as V_{IL} and a minimum voltage of 3.5 V as V_{IH}. Bipolar logic accepts a maximum voltage of 0.8 V as V_{IL} and a minimum voltage of 2 V as V_{IH}.

V_{OL} is the LOW level output voltage and V_{OH} is the HIGH level output voltage. For 5 V CMOS, the maximum V_{OL} is 0.33 V and the minimum V_{OH} is 4.4 V. For bipolar logic, the maximum V_{OL} is 0.4 V and the minimum V_{OH} is 2.4 V. All values depend on operating conditions as specified on the data sheet for the specific device.

SPEED-POWER PRODUCT (SPP)   This parameter (**speed-power product**) can be used as a measure of the performance of a logic circuit taking into account the propagation delay time and the power dissipation. It is especially useful for comparing the various logic gate series within the CMOS and bipolar families or for comparing a CMOS gate to a TTL gate.

The SPP of a logic circuit is the product of the propagation delay time and the power dissipation and is expressed in joules (J), which is the unit of energy. The formula is

$$SPP = t_p P_D \tag{3–8}$$

EXAMPLE 3–20

A certain gate has a propagation delay of 5 ns and $I_{CCH} = 1$ mA and $I_{CCL} = 2.5$ mA with a dc supply voltage of 5 V. Determine the speed-power product.

SOLUTION

$$P_D = V_{CC}\left(\frac{I_{CCH} + I_{CCL}}{2}\right) = 5 \text{ V}\left(\frac{1 \text{ mA} + 2.5 \text{ mA}}{2}\right) = 5 \text{ V}(1.75 \text{ mA})$$

$$= 8.75 \text{ mW}$$

$$SPP = (5 \text{ ns})(8.75 \text{ mW}) = \textbf{43.75 pJ}$$

RELATED PROBLEM

If the propagation delay of a gate is 15 ns and its *SPP* is 150 pJ, what is its average power dissipation?

FAN-OUT AND LOADING   The **fan-out** of a logic gate is the maximum number of inputs of the same series in an IC family that can be connected to a gate's output and still maintain the output voltage levels within specified limits. Fan-out is a significant parameter only for bipolar logic because of the type of circuit technology. Since very high impedances are associated with CMOS circuits, the fan-out is very high but depends on frequency because of capacitive effects.

A higher fan-out means that a gate output can be connected to more gate inputs.

Fan-out is specified in terms of **unit loads.** A unit load for a logic gate equals one input to a like circuit. For example, if the current from a LOW input (I_{IL}) of a certain gate is 0.4 mA and the current that a LOW output (I_{OL}) can accept is 8.0 mA, the number of unit loads that the gate can drive in the LOW state is

$$\text{Unit loads} = \frac{I_{OL}}{I_{IL}} = \frac{8.0 \text{ mA}}{0.4 \text{ mA}} = 20$$

Figure 3–63 shows a logic gate driving a number of other gates of the same circuit technology, where the number of gates depends on the particular circuit technology. For example, the maximum number of gate inputs (unit loads) that a certain series bipolar gate can drive is 20.

FIGURE 3–63   **The NAND gate output fans out to a maximum number of NAND gates of the same type.**

SECTION 3–8 CHECKUP

1. Which IC logic technology generally has the lowest power dissipation?

2. A positive pulse is applied to an inverter input. The time from the leading edge of the input to the leading edge of the output is 10 ns. The time from the trailing edge of the input to the trailing edge of the output is 8 ns. What are the values of t_{PLH} and t_{PHL}?

3. A certain gate has a propagation delay time of 6 ns and a power dissipation of 3 mW. Determine the speed-power product?

4. Define I_{CCL} and I_{CCH}.

5. Define V_{IL} and V_{IH}.

6. Define V_{OL} and V_{OH}.

3–9 PROGRAMMABLE LOGIC

Programmable logic was introduced in Chapter 1. In this section, the basic concept of the programmable AND array, which forms the basis for most programmable logic, is discussed, and the major process technologies are covered. A programmable logic device (PLD) is one that does not initially have a fixed-logic function but that can be programmed to implement just about any logic design. As you have learned, two types of PLD are the SPLD and CPLD. In addition to the PLD, the other major category of programmable logic is the FPGA. For simplicity, all of these devices will be referred to as PLDs. Also, some important concepts in programming are discussed.

After completing this section, you should be able to

- Describe the concept of a programmable AND array
- Discuss various process technologies
- Discuss downloading a design to a programmable logic device
- Discuss text entry and graphic entry as two methods for programmable logic design
- Explain in-system programming
- Write simple VHDL and Verilog program code for logic gates

Basic Concept of the AND Array

Most types of PLDs use some form of **AND array**. Basically, this array consists of AND gates and a matrix of interconnections with programmable links at each cross point, as shown in Figure 3–64(a). The purpose of the programmable links is to either make or break a connection between a row line and a column line in the interconnection matrix. For each input to an AND gate, only one programmable link is left intact in order to connect the desired variable to the gate input. Figure 3–64(b) illustrates an array after it has been programmed.

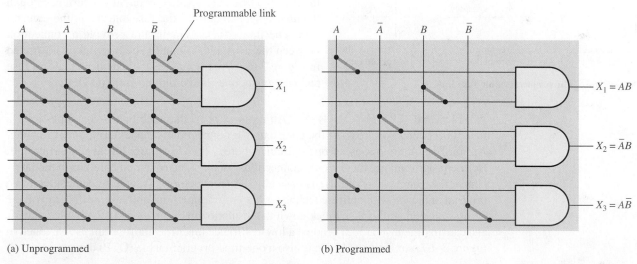

(a) Unprogrammed   (b) Programmed

FIGURE 3–64   Basic concept of a programmable AND array.

EXAMPLE 3–21

Show the AND array in Figure 3–64(a) programmed for the following outputs:
$X_1 = A\overline{B}, X_2 = \overline{A}B$, and $X_3 = \overline{A}\,\overline{B}$

SOLUTION

See Figure 3–65.

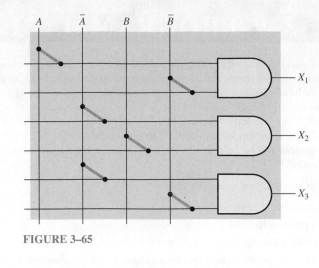

FIGURE 3–65

RELATED PROBLEM

How many rows, columns, and AND gate inputs are required for three input variables in a 3-AND gate array?

Programmable Link Process Technologies

Several different process technologies are used for programmable links in PLDs.

FUSE TECHNOLOGY   This was the original programmable link technology. It is still used in some SPLDs. The **fuse** is a metal link that connects a row and a column in the interconnection matrix. Before programming, there is a fused connection at each intersection. To program a device, the selected fuses are opened by passing a current through them sufficient to "blow" the fuse and break the connection. The intact fuses remain and provide a connection between the rows and columns. The fuse link is illustrated in Figure 3–66. Programmable logic devices that use fuse technology are one-time programmable (**OTP**).

(a) Fuse intact before programming

(b) Programming current

(c) Fuse open after programming

FIGURE 3–66   **The programmable fuse link.**

ANTIFUSE TECHNOLOGY   An **antifuse** programmable link is the opposite of a fuse link. Instead of breaking the connection, a connection is made during programming. An antifuse starts out as an open circuit whereas the fuse starts out as a short circuit. Before programming, there are no connections between the rows and columns in the interconnection matrix. An antifuse is basically two conductors separated by an insulator. To program a device with antifuse technology, a programmer tool applies a sufficient voltage across selected antifuses to break down the insulation between the two conductive materials, causing the insulator to become a low-resistance link. The antifuse link is illustrated in Figure 3–67. An antifuse device is also a one-time programmable (OTP) device.

(a) Antifuse is open before programming.

(b) Programming voltage breaks down insulation layer to create contact.

(c) Antifuse is effectively shorted after programming.

FIGURE 3–67   **The programmable antifuse link.**

EPROM TECHNOLOGY   In certain programmable logic devices, the programmable links are similar to the memory cells in **EPROMs** (electrically programmable read-only memories). This type of PLD is programmed using a special tool known as a device programmer. The device is inserted into the programmer, which is connected to a computer running the programming software. Most EPROM-based PLDs are one-time programmable (OTP). However, those with windowed packages can be erased with UV (ultraviolet) light and reprogrammed using a standard PLD programming fixture. EPROM process technology uses a special type of MOS transistor, known as a floating-gate transistor, as the programmable link. The floating-gate device utilizes a process called Fowler-Nordheim tunneling to place electrons in the floating-gate structure.

In a programmable AND array, the floating-gate transistor acts as a switch to connect the row line to either a HIGH or a LOW, depending on the input variable. For input variables that are not used, the transistor is programmed to be permanently *off* (open). Figure 3–68 shows one AND gate in a simple array. Variable A controls the state of the transistor in the first column, and variable B controls the transistor in the third column. When a transistor is *off*, like an open switch, the input line to the AND gate is at +V (HIGH). When a transistor is *on*, like a closed switch, the input line is connected to ground (LOW). When variable A or B is 0 (LOW), the transistor is *on*, keeping the input line to the AND gate LOW. When A or B is 1 (HIGH), the transistor is *off*, keeping the input line to the AND gate HIGH.

FIGURE 3–68   **A simple AND array with EPROM technology. Only one gate in the array is shown for simplicity.**

EEPROM TECHNOLOGY   Electrically erasable programmable read-only memory technology is similar to EPROM because it also uses a type of floating-gate transistor in E^2CMOS cells. The difference is that **EEPROM** can be erased and reprogrammed electrically without the need for UV light or special fixtures. An E^2CMOS device can be programmed after being installed on a printed circuit board, and many can be reprogrammed while operating in a system. This is called **in-system programming (ISP).** Figure 3–68 can also be used as an example to represent an AND array with EEPROM technology.

FLASH TECHNOLOGY   **Flash** technology is based on a single transistor link and is both nonvolatile and reprogrammable. Flash elements are a type of EEPROM but are faster and result in higher density devices than the standard EEPROM link. A detailed discussion of the flash memory element is covered later in the book.

SRAM TECHNOLOGY   Many FPGAs and some CPLDs use a process technology similar to that used in **SRAMs** (static random-access memories). The basic concept of

SRAM-based programmable logic arrays is illustrated in Figure 3–69(a). A SRAM-type memory cell is used to turn a transistor *on* or *off* to connect or disconnect rows and columns. For example, when the memory cell contains a 1 (green), the transistor is *on* and connects the associated row and column lines, as shown in part (b). When the memory cell contains a 0 (blue), the transistor is *off* so there is no connection between the lines, as shown in part (c).

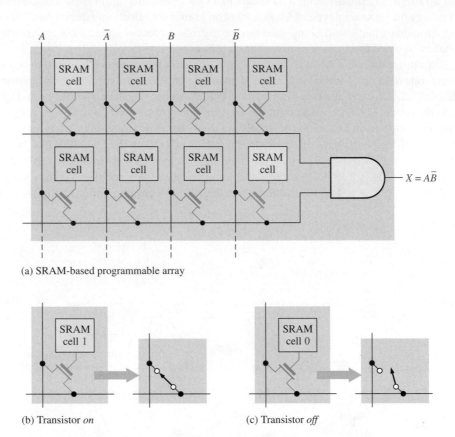

(a) SRAM-based programmable array

(b) Transistor *on*

(c) Transistor *off*

FIGURE 3–69   **Basic concept of an AND array with SRAM technology.**

SRAM technology is different from the other process technologies discussed because it is a volatile technology. This means that a SRAM cell does not retain data when power is turned *off*. The programming data must be loaded into a memory; and when power is turned *on*, the data from the memory reprograms the SRAM-based PLD.

Most system-level designs incorporate a variety of devices such as RAMs, ROMs, controllers, and processors that are interconnected by general-purpose logic devices often referred to as "glue" logic. PLDs have come to replace many of the SSI and MSI "glue" devices. The use of PLDs provides a reduction in package count.

For example, in computer memory systems, PLDs can be used for memory address decoding and to generate memory write signals as well as other functions.

SYSTEM NOTE

The fuse, antifuse, EPROM, EEPROM, and flash process technologies are nonvolatile, so they retain their programming when the power is *off*. A fuse is permanently open, an antifuse is permanently closed, and floating-gate transistors used in EPROM-based and EEPROM-based arrays can retain their *on* or *off* state indefinitely.

Device Programming

The general concept of programming was introduced in Chapter 1, and you have seen how interconnections can be made in a simple array by opening or closing the programmable links. SPLDs, CPLDs, and FPGAs are programmed in essentially the same way. The devices with OTP (one-time programmable) process technologies (fuse, antifuse, or EPROM) must be programmed with a special hardware fixture called a *programmer.* The programmer is connected to a computer by a standard interface cable. Development software is installed on the computer, and the device is inserted into the programmer socket. Most programmers have adapters that allow different types of packages to be plugged in.

EEPROM, flash, and SRAM-based programmable logic devices are reprogrammable and can be reconfigured multiple times. Although a device programmer can be used for this type of device, it is generally programmed initially on a PLD development board, as shown in Figure 3–70. A logic design can be developed using this approach because any necessary changes during the design process can be readily accomplished by simply reprogramming the PLD. A PLD to which a software logic design can be downloaded is called a **target device.** In addition to the target device, development boards typically provide other circuitry and connectors for interfacing to the computer and other peripheral circuits. Also, test points and display devices for observing the operation of the programmed device are included on the development board.

PLD development board

Programmable logic device

FIGURE 3–70 **Programming setup for reprogrammable logic devices.** (Photo courtesy of Digilent, Inc.)

DESIGN ENTRY   As you learned in Chapter 1, design entry is where the logic design is programmed using development software. The two main ways to enter a design are by text entry or graphic (schematic) entry, and manufacturers of programmable logic provide software packages to support their devices that allow for both methods.

Text entry in most development software, regardless of the manufacturer, supports two or more hardware development languages (**HDLs**). For example, all software packages support both IEEE standard HDLs, VHDL, and Verilog. Some software packages also support certain proprietary languages such as AHDL.

In **graphic (schematic) entry,** logic symbols such as AND gates and OR gates are placed on the screen and interconnected to form the desired circuit. In this method you use the familiar logic symbols, but the software actually converts each symbol and interconnections to a text file for the computer to use; you do not see this process. A simple example of two text entry screens and a graphic entry screen for an AND gate is shown in Figure 3–71. Graphic entry is generally used for less-complex logic circuits; and text entry, although it can also be used for very simple logic, is generally used for larger, more complex implementation.

Verilog1.v

```verilog
module Verilog1(A, B, X);
    output X;
    input  A,  B;
        assign X = A && B;
endmodule
```

Vhdl1.vhd

```vhdl
entity VHDL1 is
    port(A, B: in bit; X: out bit);
end entity VHDL1;
architecture ANDfunction of VHDL1 is
begin
    X <= A and B;
end ANDfunction;
```

(a) Verilog and VHDL text entry

Graphic1.bsf

(b) Equivalent graphic (schematic) entry

FIGURE 3–71   **Examples of design entry for an AND gate.**

In-System Programming (ISP)

Certain CPLDs and FPGAs can be programmed after they have been installed on a system printed circuit board (PCB). After a logic design has been developed and fully tested on a development board, it can then be programmed into a "blank" device that is already soldered onto a system board in which it will be operating. Also, if a design change is required, the device on the system board can be reconfigured to incorporate the design modifications.

In a production situation, programming a device on the system board minimizes handling and eliminates the need for keeping stocks of preprogrammed devices. It also rules out the possibility of wrong parts being placed in a product. Unprogrammed (blank) devices can be kept in the warehouse and programmed on-board as needed. This minimizes the capital a business needs for inventories and enhances the quality of its products.

JTAG   The standard established by the Joint Test Action Group is the commonly used name for IEEE Std. 1149.1. The JTAG standard was developed to provide a simple method, called boundary scan, for testing programmable devices for functionality as well as testing circuit boards for bad connections—shorted pins, open pins, bad traces, and the like. Also, JTAG has been used as a convenient way of configuring programmable devices in-system. As the demand for field-upgradable products increases, the use of JTAG as a convenient way of reprogramming CPLDs and FPGAs increases.

JTAG-compliant devices have internal dedicated hardware that interprets instructions and data provided by four dedicated signals. These signals are defined by the JTAG standard to be TDI (Test Data In), TDO (Test Data Out), TMS (Test Mode Select), and TCK (Test Clock). The dedicated JTAG hardware interprets instructions and data on the TDI and TMS signals, and drives data out on the TDO signal. The TCK signal is used to clock the process.

EMBEDDED PROCESSOR   Another approach to in-system programming is the use of an embedded microprocessor and memory. The processor is embedded within the system along with the CPLD or FPGA and other circuitry, and it is dedicated to the purpose of in-system configuration of the programmable device.

As you have learned, SRAM-based devices are volatile and lose their programmed data when the power is turned *off*. It is necessary to store the programming data in a PROM (programmable read-only memory), which is nonvolatile. When power is turned *on,* the embedded processor takes control of transferring the stored data from the PROM to the CPLD or FPGA.

Also, an embedded processor is sometimes used for reconfiguration of a programmable device while the system is running. In this case, design changes are done with software, and the new data are then loaded into a PROM without disturbing the operation of the system. The processor controls the transfer of the data to the device "on-the-fly" at an appropriate time.

Introduction to VHDL and Verilog

Hardware description languages (HDLs), such as VHDL and Verilog, differ from software programming languages because they include ways of describing the propagation of time and other characteristics of the logic. An HDL implements a new logic design in hardware (PLD), whereas a software programming language such as C or BASIC instructs existing hardware what to do. The two standard hardware description languages (HDLs) used for programming PLDs are VHDL and Verilog. Both of these HDLs have their advocates, so we will show examples of both throughout this book. *Tutorials for both VHDL and Verilog are found on the website.*

HDL DESCRIPTIONS OF LOGIC GATES   The logic gates covered in this chapter are described with VHDL and with Verilog in Figures 3–72 through 3–76. Two

FIGURE 3–72   **Inverter.**

FIGURE 3–73   AND gate.

FIGURE 3–74   OR gate.

FIGURE 3–75   3-input NAND gate.

FIGURE 3–76   Exclusive-NOR gate.

gates are left as Checkup exercises. VHDL describes a function with an **entity/architecture** structure. Verilog uses a **module** structure to describe a function. Verilog logical operators are ! for NOT, && for AND, and ‖ for OR. Keywords that are part of the language syntax are shown bold for clarity in both VHDL and Verilog.

SECTION 3–9 CHECKUP

1. List six process technologies used for programmable links in programmable logic.

2. What does the term *volatile* mean in relation to PLDs; which process technology is volatile?

3. What are two design entry methods for programming PLDs and FPGAs?

4. Define JTAG.

5. Write VHDL and Verilog descriptions of a NOR gate with three inputs.

6. Write VHDL and Verilog descriptions of an XOR gate.

3–10 TROUBLESHOOTING

Troubleshooting is the process of recognizing, isolating, and correcting a fault or failure in a system. To be an effective troubleshooter, you must understand how the system works and be able to recognize incorrect performance. Troubleshooting can be at the system level, the circuit board level, or the component level. Today, troubleshooting down to the board level is usually sufficient. Once a board is determined to be defective, it is usually replaced with a new one. However, if the circuit board is to be saved, component-level troubleshooting may be necessary.

After completing this section, you should be able to

- **Describe the steps in a troubleshooting procedure**
- **Discuss the half-splitting method**
- **Discuss the signal-tracing method**
- **Describe the troubleshooting of a particular process control system**

Basic Hardware Troubleshooting Methods

Troubleshooting at a system level requires good detective work. When a problem occurs, the list of potential causes is usually quite large. You must gather a sufficient amount of detailed information and systematically narrow the list of potential causes to determine the problem. As a general guide to troubleshooting a system, the following steps should be followed:

1. Gather information on the problem.
2. Identify the symptoms and possible failures.
3. Isolate point(s) of failure.
4. Apply proper tools to determine the cause of the problem.
5. Fix the problem.

CHECK THE OBVIOUS   After collecting information on the problem, make sure to first check for obvious faults: absence of DC power, blown fuses, tripped circuit breakers, faulty burned out indicators such as lamps, loose connectors, broken or loose wires, switches in the wrong position, physical damages, boards not properly inserted, wire fragments or solder splashes shorting components, and poor quality contacts on printed circuit boards. For any troubleshooting task, you must have a system/circuit diagram. Other useful documents are a table of signal characteristics and a prewritten troubleshooting guide for the specific system.

REPLACEMENT   Assume that a given system has multiple circuit boards. The simplest and quickest way to fix a problem is by replacing the circuit boards one by one with a known good board until the problem is corrected. This approach, of course, requires that duplicate boards be available. Another drawback to this approach is that an outside source may be causing the fault, such as a short in a connector; and by replacing the board, the fault is transferred to the new board.

REPRODUCING THE SYMPTOMS   Once the symptoms of a faulty system are identified, find a way to reproduce the problem. If the problem can be reproduced, it can be isolated and resolved. In some systems, the symptom may be self-evident, but in others it may have to be induced by application of a level or signal at a given point. Once this is done, then a systematic approach can be used to isolate the cause or causes of a problem. You should always consider the possibility that there is more than one fault.

If the symptoms are intermittent, the task of troubleshooting becomes more difficult. For example, in some cases a component may be temperature sensitive and fail only when the temperature is too high or too low. In these cases, the temperature can be varied by the simple process of blowing cool air on the component of concern to lower the temperature or using a heat gun to raise it, while monitoring the operation of the system.

HALF-SPLITTING METHOD   In this procedure, you check for the presence or absence of a signal at a point halfway between input and output. If the signal is present, you know the fault is in the second half. If the signal is absent, you know the fault is in the first half. Then you split the defective half in half and check for a signal. The process is continued until a certain area of the system has been isolated. This may be a single circuit board in a system with many circuit boards or a component on a given circuit board. In a large system, this procedure can save a lot of time over moving down the line checking each block or stage as you go. This method is usually best applied in large complex systems. Figure 3–77 is a simple illustration of this method. The system is represented with the four green blocks. Additional steps are added to left or right for additional blocks.

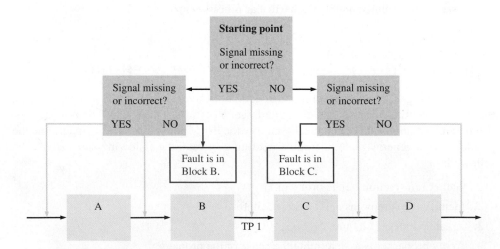

FIGURE 3–77   **Concept of the half-splitting method. The blue arrows indicate the test points.**

SIGNAL-TRACING METHOD   Signal tracing is the procedure of tracking signals as they progress through a system from input to output. Signal tracing can be used with half-splitting, where you check for a signal at each point from where the absence of a signal was detected. Signal tracing can also begin at the output where there is an incorrect or absent signal and go back toward the input from point to point until a correct signal is found. Also, you can begin at the input and check the signal and move toward the output from point to point until the correct signal is lost. In both cases, the fault would be between the point and the output. Of course, you must know what the signal is supposed to look like in order to know if anything is wrong. Figure 3–78 illustrates the concept of signal tracing.

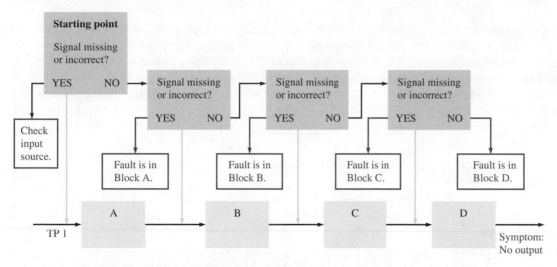

FIGURE 3–78   **Concept of the signal-tracing method. Input to output is shown. The same applies if you start at the output and go toward the input.**

SIGNAL SUBSTITUTION AND INJECTION   Signal substitution is used when the system being tested has been separated from its signal source. A generator signal is used to replace the normal signal that comes from the source when the system or portion of a system is recombined with the part that normally produces the input signal. Signal injection can be used to insert a signal at certain points in the system using the half-splitting approach.

E X A M P L E   3 – 2 2

A simple generic self-contained system that is not software driven is used to illustrate the basic approach to troubleshooting. This system, which is a type of state machine, is used to control a sequential process in a variety of specific applications. A block diagram of the system is shown in Figure 3–79. It consists of four interdependent functional blocks. The outputs consist of four signals that sequentially initiate four processes in a given sequence. A sensor detects specified conditions in the process and causes the outputs to be altered to accommodate the condition.

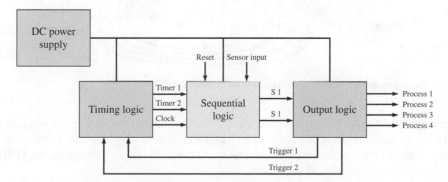

FIGURE 3–79   **Block diagram of a process control system.**

In a particular application, the system controls the packaging of golf balls into a tubular canister on an automated assembly line. A pulse on the Process 1 output moves the next canister into position, which starts the process. The pulse on the Process 2 output enables six balls to drop into the canister. When the canister is filled, a pulse on the Process 3 output causes the canister lid to be attached and sealed. A pulse on the Process 4 output initiates the attachment of the label to the canister and sends it on its way down the line.

You observe that the canisters get jammed up on the assembly line and do not move past a certain point. As the troubleshooter, how would you proceed to troubleshoot this problem?

SOLUTION

Identify the symptom as follows: When the system reaches Process 3, the lid is not attached and the canister waits in position, keeping the next canister from moving into position. The cause is most likely either the absence of a pulse on the Process 3 output or the sequence is hanging up at Process 2 and not going to Process 3. It is difficult to narrow the cause down to one specific block because certain failures in any one block could result in the problem. The specification document for the system shows the proper operation in terms of a timing diagram, which is shown in Figure 3–80.

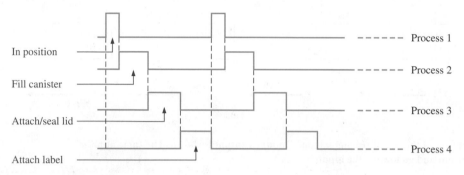

FIGURE 3–80 Correct timing diagram for the process control system.

The next step is to use an oscilloscope to look at the output waveforms. The setup is shown in Figure 3–81 with the physical system plugged into an electrical outlet.

FIGURE 3–81 Test setup for troubleshooting the process control system.

After checking that DC power is going to all circuits and inspecting the system for obvious faults, connect the outputs and observe them on the scope. Process 1, 3, and 4 outputs are at a constant LOW level, and Process 2 is a constant HIGH level. This measurement confirms that Process 2 output is stuck in the HIGH state, so the system can't advance through its normal sequence.

A fault in the Sequential Logic board is causing the state machine to lock up in the Process 2 state. To verify this, observe the system sequencing through the first two states. In order to do this, the system must be reset periodically so that the incorrect sequence repeats itself. A pulse generator can generate reset pulses with the period set to exceed the normal cycle time of the system. The scope display shown in Figure 3–82 is observed during this test procedure. This shows that the state machine is recycling through states 1 and 2 instead of states 1, 2, 3, and 4. Without a reset, the system gets to state 2 and locks up. The reset forces it back to state 1 each time.

Process 1 out

Process 2 out

Process 3 out

Process 4 out

FIGURE 3–82 Repetitive waveforms when faulty system is periodically reset.

The final step is to replace the Sequential Logic board (middle board) in Figure 3–81 and run the system without the reset input. After doing the swap out, you find that the system works properly. The next decision to be made is whether to repair or discard the faulty board. If the board is to be saved, you would then begin the troubleshooting process at the board level to determine which component(s) is defective.

SECTION 3–10 CHECKUP

1. List five steps in the troubleshooting procedure.

2. Name two troubleshooting methods.

3. List five obvious things to look for first in a failed system.

4. Is it important to know about the relationship between a cause and a symptom?

SUMMARY

- Gate symbols and Boolean expressions for the outputs of an inverter and 2-input gates are shown in Figure 3–83.

FIGURE 3–83

- Commutative laws: $A + B = B + A$

  $AB = BA$

- Associative laws: $A + (B + C) = (A + B) + C$

  $A(BC) = (AB)C$

- Distributive law: $A(B + C) = AB + AC$

- Boolean rules:

  1. $A + 0 = A$ **7.** $A \cdot A = A$

  2. $A + 1 = 1$ **8.** $A \cdot \overline{A} = 0$

  3. $A \cdot 0 = 0$ **9.** $\overline{\overline{A}} = A$

  4. $A \cdot 1 = A$ **10.** $A + AB = A$

  5. $A + A = A$ **11.** $A + \overline{A}B = A + B$

  6. $A + \overline{A} = 1$ **12.** $(A + B)(A + C) = A + BC$

- The inverter output is the complement of the input.

- The AND gate output is HIGH only when all the inputs are HIGH.

- The OR gate output is HIGH when any of the inputs is HIGH.
- The NAND gate output is LOW only when all the inputs are HIGH.
- The NAND can be viewed as a negative-OR whose output is HIGH when any input is LOW.
- The NOR gate output is LOW when any of the inputs is HIGH.
- The NOR can be viewed as a negative-AND whose output is HIGH only when all the inputs are LOW.
- The exclusive-OR gate output is HIGH when the inputs are not the same.
- The exclusive-NOR gate output is LOW when the inputs are not the same.
- Distinctive shape symbols and truth tables for various logic gates (limited to 2 inputs) are shown in Figure 3–84.

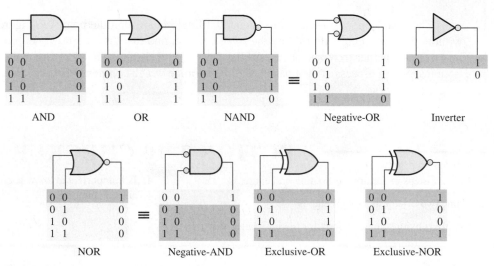

Note: Active states are shown in yellow.

FIGURE 3–84

- The average power dissipation of a logic gate is

$$P_D = V_{CC}\left(\frac{I_{CCH} + I_{CCL}}{2}\right)$$

- The speed-power product of a logic gate is

$$SPP = t_p P_D$$

- Most programmable logic devices (PLDs) are based on some form of AND array.
- Programmable link technologies are fuse, antifuse, EPROM, EEPROM, flash, and SRAM.
- A PLD can be programmed in a hardware fixture called a programmer or mounted on a development printed circuit board.
- PLDs have an associated software development package for programming.
- Two methods of design entry using programming software are text entry (HDL) and graphic (schematic) entry.
- ISP PLDs can be programmed after they are installed in a system, and they can be reprogrammed at any time.
- JTAG stands for Joint Test Action Group and is an interface standard (IEEE Std. 1149.1) used for programming and testing PLDs.
- An embedded processor is used to facilitate in-system programming of PLDs.
- CMOS is made with MOS field-effect transistors.
- Bipolar (TTL) is made with bipolar junction transistors.
- As a rule, CMOS has a lower power consumption than bipolar.

- VHDL and Verilog are hardware description languages used to describe a logic function for programming into a PLD.
- Half-splitting and signal tracing are two troubleshooting methods.

KEY TERMS

Key terms and other bold terms in the chapter are defined in the end-of-book glossary.

AND array   An array of AND gates consisting of a matrix of programmable interconnections.

AND gate   A logic gate that produces a HIGH output only when all of the inputs are HIGH.

Antifuse   A type of PLD nonvolatile programmable link that can be left open or can be shorted once as directed by the program.

Boolean algebra   The mathematics of logic circuits.

Complement   The inverse or opposite of a number. In Boolean algebra, the inverse function, expressed with a bar over a variable. The complement of a 1 is 0, and vice versa.

EEPROM   A type of nonvolatile PLD reprogrammable link based on electrically erasable programmable read-only memory cells and can be turned on or off repeatedly by programming.

EPROM   A type of PLD nonvolatile programmable link based on electrically programmable read-only memory cells and can be turned either on or off once with programming.

Exclusive-NOR gate   A logic gate that produces a LOW only when the two inputs are at opposite levels.

Exclusive-OR (XOR) gate   A logic gate that produces a HIGH output only when its two inputs are at opposite levels.

Fan-out   The number of equivalent gate inputs of the same family series that a logic gate can drive.

Flash   A type of PLD nonvolatile reprogrammable link technology based on a single transistor cell.

Fuse   A type of PLD nonvolatile programmable link that can be left shorted or can be opened once as directed by the program.

Inverter   A logic circuit that inverts or complements its input.

JTAG   Joint Test Action Group; an interface standard designated IEEE Std. 1149.1.

NAND gate   A logic gate that produces a LOW output only when all the inputs are HIGH.

NOR gate   A logic gate in which the output is LOW when one or more of the inputs are HIGH.

OR gate   A logic gate that produces a HIGH output when one or more inputs are HIGH.

Propagation delay time   The time interval between the occurrence of an input transition and the occurrence of the corresponding output transition in a logic circuit.

Product term   The Boolean product of two or more literals equivalent to an AND operation.

SRAM   A type of PLD volatile reprogrammable link based on static random-access memory cells and can be turned on or off repeatedly with programming.

Sum term   The Boolean sum of two or more literals equivalent to an OR operation.

Target device   A PLD mounted on a programming fixture or development board into which a software logic design is to be downloaded.

Timing diagram   A diagram of waveforms showing the proper timing relationship of all the waveforms.

Truth table   A table showing the inputs and corresponding output(s) of a logic circuit.

Unit load   A measure of fan-out. One gate input represents one unit load to the output of a gate within the same IC family.

Variable   A symbol used to represent an action, a condition, or data that can have a value of 1 or 0, usually designated by an italic letter or word.

Verilog   A standard hardware description language that uses a module structure to describe a function.

VHDL   A standard hardware description language that describes a function with an entity/architecture structure.

FIXED-FUNCTION LOGIC DEVICES

The diagrams shown are for CMOS and bipolar IC devices. The XX in each part number designates a particular IC family. For example, XX = HC stands for high-speed CMOS.

TRUE/FALSE QUIZ

Answers are at the end of the chapter.

1. An inverter performs the NOR operation.

2. An AND gate can have only two inputs.

3. If any input to an OR is 1, the output is 1.

4. If all inputs to an AND gate are 1, the output is 0.

5. A NAND gate has an output that is opposite the output of an AND gate.

6. A NOR gate can be considered as an OR gate followed by an inverter.

7. The output of an exclusive-OR is 0 if the inputs are opposite.

8. Two types of fixed-function logic integrated circuits are bipolar and NMOS.

9. BCD stands for binary coded decimal.

10. Fan-out is the number of similar gates that a given gate can drive.

11. Programmable logic devices are generally implemented with inverters.

12. VHDL is a type of programmable logic device.

SELF-TEST

Answers are at the end of the chapter.

1. Boolean addition represents the

    (a) inverter   **(b)** AND gate   **(c)** NOR gate   **(d)** OR gate

2. Boolean multiplication represents the

   (a) inverter   **(b)** AND gate   **(c)** NAND gate   **(d)** OR gate

3. The Boolean expression for a three-input AND gate is

   (a) $A + B + C$ **(b)** $A + BC$ **(c)** ABC **(d)** $ABC + ABC$

4. The output of an exclusive-OR gate is 1 when

   (a) inputs are both 1   **(b)** the inputs are not the same
   (c) the inputs are the same   **(d)** the inputs are both 0

5. When the input to an inverter is HIGH (1), the output is

   (a) HIGH or 1   **(b)** LOW or 1   **(c)** HIGH or 0   **(d)** LOW or 0

6. An inverter performs an operation known as

   (a) complementation   **(b)** assertion
   (c) inversion   **(d)** both answers (a) and (c)

7. The output of an AND gate with inputs A, B, and C is a 1 (HIGH) when

   (a) $A = 1, B = 1, C = 1$ **(b)** $A = 1, B = 0, C = 1$ **(c)** $A = 0, B = 0, C = 0$

8. The output of an OR gate with inputs A, B, and C is a 1 (HIGH) when

   (a) $A = 1, B = 1, C = 1$ **(b)** $A = 0, B = 0, C = 1$
   (c) $A = 0, B = 0, C = 0$ **(d)** answers (a), (b), and (c)
   (e) only answers (a) and (b)

9. A pulse is applied to each input of a 2-input NAND gate. One pulse goes HIGH at $t = 0$ and goes back LOW at $t = 1$ ms. The other pulse goes HIGH at $t = 0.8$ ms and goes back LOW at $t = 3$ ms. The output pulse can be described as follows:

   (a) It goes LOW at $t = 0$ and back HIGH at $t = 3$ ms.
   (b) It goes LOW at $t = 0.8$ ms and back HIGH at $t = 3$ ms.
   (c) It goes LOW at $t = 0.8$ ms and back HIGH at $t = 1$ ms.
   (d) It goes LOW at $t = 0.8$ ms and back LOW at $t = 1$ ms.

10. A pulse is applied to each input of a 2-input NOR gate. One pulse goes HIGH at $t = 0$ and goes back LOW at $t = 1$ ms. The other pulse goes HIGH at $t = 0.8$ ms and goes back LOW at $t = 3$ ms. The output pulse can be described as follows:

    (a) It goes LOW at $t = 0$ and back HIGH at $t = 3$ ms.
    (b) It goes LOW at $t = 0.8$ ms and back HIGH at $t = 3$ ms.
    (c) It goes LOW at $t = 0.8$ ms and back HIGH at $t = 1$ ms.
    (d) It goes HIGH at $t = 0.8$ ms and back LOW at $t = 1$ ms.

11. A pulse is applied to each input of an exclusive-OR gate. One pulse goes HIGH at $t = 0$ and goes back LOW at $t = 1$ ms. The other pulse goes HIGH at $t = 0.8$ ms and goes back LOW at $t = 3$ ms. The output pulse can be described as follows:

    (a) It goes HIGH at $t = 0$ and back LOW at $t = 3$ ms.
    (b) It goes HIGH at $t = 0$ and back LOW at $t = 0.8$ ms.
    (c) It goes HIGH at $t = 1$ ms and back LOW at $t = 3$ ms.
    (d) both answers (b) and (c)

12. A positive-going pulse is applied to an inverter. The time interval from the leading edge of the input to the leading edge of the output is 7 ns. This parameter is

    (a) speed-power product   **(b)** propagation delay, t_{PHL}
    (c) propagation delay, t_{PLH} **(d)** pulse width

13. The purpose of a programmable link in an AND array is to

    (a) connect an input variable to a gate input
    (b) connect a row to a column in the array matrix
    (c) disconnect a row from a column in the array matrix
    (d) do all of the above

14. The term OTP means

    (a) open test point   **(b)** one-time programmable
    (c) output test program   **(d)** output terminal positive

15. Types of PLD programmable link process technologies are

    (a) antifuse   **(b)** flash
    (c) ROM   **(d)** both (a) and (b)
    (e) both (a) and (c)

16. A volatile programmable link technology is
  (a) fuse   **(b)** EPROM
  (c) SRAM   **(d)** EEPROM

17. Two ways to enter a logic design using PLD development software are
  (a) text and numeric   **(b)** text and graphic
  (c) graphic and coded   **(d)** compile and sort

18. JTAG stands for
  (a) Joint Test Action Group   **(b)** Java Top Array Group
  (c) Joint Test Array Group   **(d)** Joint Time Analysis Group

19. In-system programming of a PLD typically utilizes
  (a) an embedded clock generator   **(b)** an embedded processor
  (c) an embedded PROM   **(d)** both (a) and (b)
  (e) both (b) and (c)

20. To measure the period of a pulse waveform, you must use
  (a) a DMM   **(b)** a logic probe
  (c) an oscilloscope   **(d)** a logic pulser

21. Once you measure the period of a pulse waveform, the frequency is found by
  (a) using another setting
  (b) measuring the duty cycle
  (c) finding the reciprocal of the period
  (d) using another type of instrument

PROBLEMS

Answers to odd-numbered problems are at the end of the book.

SECTION 3–1   Introduction to Boolean Algebra

1. Using Boolean notation, write an expression that is a 1 whenever one or more of its variables (A, B, C, and D) are 1s.

2. Write an expression that is a 1 only if all of its variables (A, B, C, D, and E) are 1s.

3. Write an expression that is a 1 when one or more of its variables (A, B, and C) are 0s.

4. Evaluate the following operations:
  (a) $0 + 0 + 1$ **(b)** $1 + 1 + 1$ **(c)** $1 \cdot 0 \cdot 0$
  (d) $1 \cdot 1 \cdot 1$ **(e)** $1 \cdot 0 \cdot 1$ **(f)** $1 \cdot 1 + 0 \cdot 1 \cdot 1$

5. Find the values of the variables that make each product term 1 and each sum term 0.
  (a) AB **(b)** $A\overline{B}C$ **(c)** $A + B$ **(d)** $\overline{A} + B + \overline{C}$
  (e) $\overline{A} + \overline{B} + C$ **(f)** $\overline{A} + B$ **(g)** $AB\,\overline{C}$

6. Find the value of X for all possible values of the variables.
  (a) $X = (A + B)C + B$
  (b) $X = (\overline{A + B})C$
  (c) $X = A\overline{B}C + AB$
  (d) $X = (A + B)(\overline{A} + B)$
  (e) $X = (A + BC)(\overline{B} + \overline{C})$

7. Identify the law of Boolean algebra upon which each of the following equalities is based:
  (a) $A\overline{B} + CD + A\overline{C}D + B = B + A\overline{B} + A\overline{C}D + CD$
  (b) $AB\overline{C}D + \overline{ABC} = D\overline{C}BA + \overline{CBA}$
  (c) $AB(CD + E\overline{F} + GH) = ABCD + ABE\overline{F} + ABGH$

8. Identify the Boolean rule(s) on which each of the following equalities is based:
  (a) $\overline{\overline{AB} + \overline{CD}} + \overline{EF} = AB + CD + \overline{EF}$
  (b) $A\overline{A}B + AB\overline{C} + AB\overline{B} = AB\overline{C}$
  (c) $A(BC + BC) + AC = A(BC) + AC$
  (d) $AB(C + \overline{C}) + AC = AB + AC$
  (e) $A\overline{B} + A\overline{B}C = A\overline{B}$
  (f) $ABC + \overline{AB} + \overline{AB}CD = ABC + \overline{AB} + D$

SECTION 3–2   The Inverter

9. The input waveform shown in Figure 3–85 is applied to an inverter. Draw the timing diagram of the output waveform in proper relation to the input.

FIGURE 3–85

10. A network of cascaded inverters is shown in Figure 3–86. If a HIGH is applied to point A, determine the logic levels at points B through F.

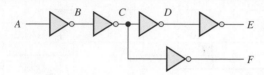

FIGURE 3–86

11. If the waveform in Figure 3–85 is applied to point A in Figure 3–86, determine the waveforms at points B through F.

SECTION 3–3   The AND Gate

12. Draw the rectangular outline symbol for a 4-input AND gate.

13. Determine the output, X, for a 2-input AND gate with the input waveforms shown in Figure 3–87. Show the proper relationship of output to inputs with a timing diagram.

FIGURE 3–87

14. Repeat Problem 13 for the waveforms in Figure 3–88.

FIGURE 3–88

15. The input waveforms applied to a 3-input AND gate are as indicated in Figure 3–89. Show the output waveform in proper relation to the inputs with a timing diagram.

FIGURE 3–89

16. The input waveforms applied to a 4-input AND gate are as indicated in Figure 3–90. Show the output waveform in proper relation to the inputs with a timing diagram.

FIGURE 3–90

SECTION 3–4   The OR Gate

17. Determine the output for a 2-input OR gate when the input waveforms are as in Figure 3–88 and draw a timing diagram.

18. Repeat Problem 15 for a 3-input OR gate.

19. Repeat Problem 16 for a 4-input OR gate.

20. For the five input waveforms in Figure 3–91, determine the output if the five signals are ANDed. Determine the output if the five signals are ORed. Draw the timing diagram for each case.

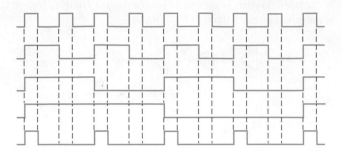

FIGURE 3–91

21. Draw the rectangular outline symbol for a 4-input OR gate.

22. Show the truth table for a 3-input OR gate.

SECTION 3–5   The NAND Gate

23. For the set of input waveforms in Figure 3–92, determine the output for the gate shown and draw the timing diagram.

FIGURE 3–92

24. Determine the gate output for the input waveforms in Figure 3–93 and draw the timing diagram.

FIGURE 3–93

25. Determine the output waveform in Figure 3–94.

FIGURE 3–94

26. As you have learned, the two logic symbols shown in Figure 3–95 represent equivalent operations. The difference between the two is strictly from a functional viewpoint. For the NAND symbol, look for two HIGHs on the inputs to give a LOW output. For the negative-OR, look for at least one LOW on the inputs to give a HIGH on the output. Using these two functional viewpoints, show that each gate will produce the same output for the given inputs.

FIGURE 3-95

SECTION 3-6   The NOR Gate

27. Repeat Problem 23 for a 2-input NOR gate.

28. Determine the output waveform in Figure 3-96 and draw the timing diagram.

FIGURE 3-96

29. Repeat Problem 25 for a 4-input NOR gate.

30. The NAND and the negative-OR symbols represent equivalent operations, but they are functionally different. For the NOR symbol, look for at least one HIGH on the inputs to give a LOW
on the output. For the negative-AND, look for two LOWs on the inputs to give a HIGH output.
Using these two functional points of view, show that both gates in Figure 3-97 will produce the
same output for the given inputs.

FIGURE 3-97

SECTION 3-7   The Exclusive-OR and Exclusive-NOR Gates

31. How does an exclusive-OR gate differ from an OR gate in its logical operation?

32. Repeat Problem 23 for an exclusive-OR gate.

33. Repeat Problem 23 for an exclusive-NOR gate.

34. Determine the output of an exclusive-OR gate for the inputs shown in Figure 3-88 and draw a
timing diagram.

SECTION 3-8   Gate Performance Characteristics and Parameters

35. Determine t_{PLH} and t_{PHL} from the oscilloscope display in Figure 3-98. The readings indicate
volts/div and sec/div for each channel.

FIGURE 3-98

36. Gate A has $t_{PLH} = t_{PHL} = 6$ ns. Gate B has $t_{PLH} = t_{PHL} = 10$ ns. Which gate can be operated at a higher frequency?

37. If a logic gate operates on a dc supply voltage of $+5$ V and draws an average current of 4 mA, what is its power dissipation?

38. The variable I_{CCH} represents the dc supply current from V_{CC} when all outputs of an IC are HIGH. The variable I_{CCL} represents the dc supply current when all outputs are LOW. For a given 5 V IC with four NAND gates, what is the average power dissipation when all the gate outputs are HIGH for half the time and LOW for half the time? $I_{CCH} = 0.5$ mA and $I_{CCL} = 1$ mA.

SECTION 3–9 Programmable Logic

39. In the simple programmed AND array with programmable links in Figure 3–99, determine the Boolean output expressions.

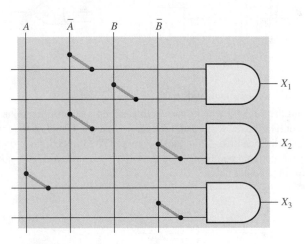

FIGURE 3–99

40. Determine by row and column number which fusible links must be blown in the programmable AND array of Figure 3–100 to implement each of the following product terms: $X_1 = \overline{A}BC, X_2 = AB\overline{C}, X_3 = \overline{A}B\overline{C}$.

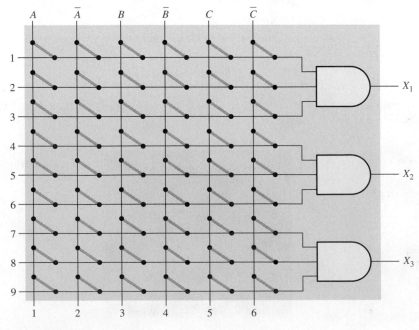

FIGURE 3–100

41. Describe a 4-input AND gate using VHDL.

42. Repeat Problem 41 using Verilog.

43. Describe a 5-input NOR gate using VHDL.

44. Repeat Problem 43 using Verilog.

SECTION 3–10   Troubleshooting

45. Define *troubleshooting*.

46. Explain the half-splitting method of troubleshooting.

47. Explain the signal-tracing method of troubleshooting.

48. Discuss signal substitution and injection.

49. Give some examples of the type of information that you look for when a system is reported to have failed.

50. If the symptom in a particular system is no output, name two possible general causes.

51. If the symptom of a particular system is an incorrect output, name two possible causes.

52. What obvious things should you look for before starting the troubleshooting process?

53. How would you isolate a fault in a system?

54. Name two common instruments used in troubleshooting.

55. Assume that you have isolated the problem down to a specific circuit board. What are your options at this point?

Special Problems

56. Sensors are used to monitor the pressure and the temperature of a chemical solution stored in a vat. The circuitry for each sensor produces a HIGH voltage when a specified maximum value is exceeded. An alarm requiring a LOW voltage input must be activated when either the pressure or the temperature is excessive. Develop a circuit for this application.

57. In a certain automated manufacturing process, electrical components are automatically inserted in a PC board. Before the insertion tool is activated, the PC board must be properly positioned, and the component to be inserted must be in the chamber. Each of these prerequisite conditions is indicated by a HIGH voltage. The insertion tool requires a LOW voltage to activate it. Develop a circuit to implement this process.

58. Modify the frequency counter in Figure 3–31 to operate with an enable pulse that is active-LOW rather than HIGH during the 1 ms interval.

59. Assume that the enable signal in Figure 3–31 has the waveform shown in Figure 3–101. Assume that waveform *B* is also available. Show how to produce an active-HIGH reset pulse to the counter only during the time that the enable signal is LOW.

FIGURE 3–101

60. Develop a device to fit in the beige block of Figure 3–102 that will cause the headlights of an automobile to be turned off automatically 15 s after the ignition switch is turned off, if the light switch is left on. Assume that a LOW is required to turn the lights off.

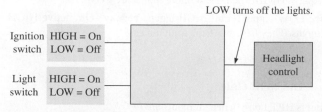

FIGURE 3–102

61. Modify the logic circuit for the intrusion alarm in Figure 3–39 so that two additional rooms, each with two windows and one door, can be protected.

62. Further modify the logic circuit from Problem 61 for a change in the input sensors where Open = LOW and Closed = HIGH.

MULTISIM TROUBLESHOOTING PRACTICE

63. Open file P03-63, and follow the instructions given there.

64. Open file P03-64, and follow the instructions given there.

65. Open file P03-65, and follow the instructions given there.

66. Open file P03-66, and follow the instructions given there.

ANSWERS TO SECTION CHECKUPS

SECTION 3–1   Introduction to Boolean Algebra

1. $\overline{A} = \overline{0} = 1$

2. $A = 1, B = 1, C = 0; \overline{A} + \overline{B} + C = \overline{1} + \overline{1} + 0 = 0 + 0 + 0 = 0$

3. $A = 1, B = 0, C = 1; A\overline{B}C = 1 \cdot \overline{0} \cdot 1 = 1 \cdot 1 \cdot 1 = 1$

4. $A + (B + C + D) = (A + B + C) + D$

5. $A(B + C + D) = AB + AC + AD$

SECTION 3–2   The Inverter

1. When the inverter input is 1, the output is 0.

   (a)

   (b)   A negative-going pulse is on the output (HIGH to LOW and back HIGH).

SECTION 3–3   The AND Gate

1. An AND gate output is HIGH only when all inputs are HIGH.

2. An AND gate output is LOW when one or more inputs are LOW.

3. Five-input AND: $X = 1$ when $ABCDE = 11111$, and $X = 0$ for all other combinations of $ABCDE$.

SECTION 3–4   The OR Gate

1. An OR gate output is HIGH when one or more inputs are HIGH.

2. An OR gate output is LOW only when all inputs are LOW.

3. Three-input OR: $X = 0$ when $ABC = 000$, and $X = 1$ for all other combinations of ABC.

SECTION 3–5   The NAND Gate

1. A NAND output is LOW only when all inputs are HIGH.

2. A NAND output is HIGH when one or more inputs are LOW.

3. NAND: active-LOW output for all HIGH inputs; negative-OR: active-HIGH output for one or more LOW inputs. They have the same truth tables.

4. $X = \overline{ABC}$

SECTION 3–6   The NOR Gate

1. A NOR output is HIGH only when all inputs are LOW.

2. A NOR output is LOW when one or more inputs are HIGH.

3. NOR: active-LOW output for one or more HIGH inputs; negative-AND: active-HIGH output for all LOW inputs. They have the same truth tables.

4. $X = \overline{A + B + C}$

SECTION 3–7   The Exclusive-OR and Exclusive-NOR Gates

1. An XOR output is HIGH when the inputs are at opposite levels.

2. An XNOR output is HIGH when the inputs are at the same levels.

3. Apply the bits to the XOR inputs; when the output is HIGH, the bits are different.

SECTION 3–8   Gate Performance Characteristics and Parameters

1. Lowest power—CMOS

2. $t_{PLH} = 10$ ns; $t_{PHL} = 8$ ns

3. 18 pJ

4. I_{CCL}—dc supply current for LOW output state; I_{CCH}—dc supply current for HIGH output state

5. V_{IL}—LOW input voltage; V_{IH}—HIGH input voltage

6. V_{OL}—LOW output voltage; V_{OH}—HIGH output voltage

SECTION 3–9   Programmable Logic

1. Fuse, antifuse, EPROM, EEPROM, flash, and SRAM

2. Volatile means that all the data are lost when power is off and the PLD must be reprogrammed; SRAM-based

3. Text entry and graphic entry

4. JTAG is Joint Test Action Group; the IEEE Std. 1149.1 for programming and test interfacing.

5.

VHDL	Verilog
entity NORgate **is**	**module** NORgate (A, B, C, X);
port (A, B, C: **in** bit; X: **out** bit);	**input** A, B, C;
end entity NORgate;	**output** X;
architecture NORfunction **of** NORgate **is**	**assign** X = !(A || B || C);
begin	**endmodule**
X <= A **nor** B **nor** C;	
end architecture NORfunction;	

6.

VHDL	Verilog
entity XORgate **is**	**module** XORgate (A, B, X);
port (A, B: **in** bit; X: **out** bit);	**input** A, B;
end entity XORgate;	**output** X;
architecture XORfunction **of** XORgate **is**	**assign** X = (A **&&** !B) || (!A **&&** B);
begin	**endmodule**
X <= A **xor** B;	
end architecture XORfunction;	

SECTION 3–10   Troubleshooting

1. Gather information, identify symptoms and possible causes, isolate point(s) of failure, apply proper tools to determine cause, and fix problem.

2. Half-splitting and signal tracing

3. Blown fuse, absence of DC power, loose connections, broken wires, loosely connected circuit board

4. Yes

ANSWERS TO RELATED PROBLEMS FOR EXAMPLES

3–1 $\overline{A} + B = 0$ when $A = 1$ and $B = 0$.

3–2 $\overline{A}\,\overline{B} = 1$ when $A = 0$ and $B = 0$.

3–3 The timing diagram is not affected.

3–4 See Table 3–18.

TABLE 3–18			
INPUTS	**OUTPUT**	**INPUTS**	**OUTPUT**
ABCD	*X*	*ABCD*	*X*
0000	0	1000	0
0001	0	1001	0
0010	0	1010	0
0011	0	1011	0
0100	0	1100	0
0101	0	1101	0
0110	0	1110	0
0111	0	1111	1

3–5 See Figure 3–103.

FIGURE 3–103

3–6 The output waveform is the same as input *A*.

3–7 See Figure 3–104.

3–8 Results are the same as example.

3–9 See Figure 3–105.

C = HIGH

FIGURE 3–104

FIGURE 3–105

3–10 See Figure 3–106.

3–11 See Figure 3–107.

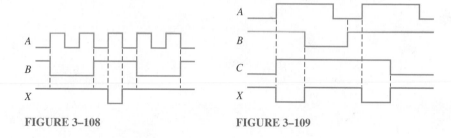

FIGURE 3–106   **FIGURE 3–107**

3–12 See Figure 3–108.

3–13 See Figure 3–109.

FIGURE 3–108   **FIGURE 3–109**

3–14 See Figure 3–110.

FIGURE 3–110

3–15 See Figure 3–111.

3–16 See Figure 3–112.

FIGURE 3–111   **FIGURE 3–112**

3–17 The output is always LOW. The output is a straight line.

3–18 The outputs are unaffected.

3–19 The gate with 4 ns t_{PLH} and t_{PHL} can operate at the highest frequency.

3–20 10 mW

3–21 6 columns, 9 rows, and 3 AND gates with three inputs each.

ANSWERS TO TRUE/FALSE QUIZ

1. F **2.** F **3.** T **4.** F **5.** T **6.** T

7. F **8.** F **9.** T **10.** T **11.** F **12.** F

ANSWERS TO SELF-TEST

1. (d) **2.** (b) **3.** (c) **4.** (b) **5.** (d) **6.** (d) **7.** (a)

8. (e) **9.** (c) **10.** (a) **11.** (d) **12.** (b) **13.** (d) **14.** (b)

15. (d) **16.** (c) **17.** (b) **18.** (a) **19.** (d) **20.** (c) **21.** (c)

CHAPTER 4

COMBINATIONAL LOGIC

OUTLINE

OBJECTIVES

- Study the operation of basic combinational logic circuits, such as AND-OR, AND-OR-Invert, exclusive-OR, and exclusive-NOR
- Write the Boolean output expression for any combinational logic circuit
- Develop a truth table from the output expression for a combinational logic circuit
- Apply DeMorgan's theorems to Boolean expressions
- Develop a combinational logic circuit for a given Boolean output expression
- Develop a combinational logic circuit for a given truth table
- Use NAND gates to implement any combinational logic function
- Use NOR gates to implement any combinational logic function
- Write VHDL and Verilog programs for simple logic circuits
- Apply combinational logic to a system application
- Troubleshoot faulty logic circuits
- Troubleshoot logic circuits by using signal tracing and waveform analysis

KEY TERMS

SOP	**Signal**
POS	**Node**
Universal gate	**Signal tracing**
Component	

INTRODUCTION

In Chapter 3, logic gates were discussed on an individual basis and in simple combinations. You will be introduced to SOP and POS implementations, which are basic forms of combinational logic. When logic gates are connected together to produce a specified output for certain specified combinations of input variables, with no storage involved, the resulting circuit is in the category of **combinational logic.** In combinational logic, the output level is at all times dependent on the combination of input levels. This chapter expands on the material introduced in earlier chapters with a coverage of the operation, implementation, and troubleshooting of various combinational logic circuits. The VHDL structural approach is introduced and applied to combinational logic. Corresponding Verilog programs are also covered.

VISIT THE WEBSITE
Study aids for this chapter are available at
http://pearsonhighered.com/floyd

4-1 BASIC COMBINATIONAL LOGIC CIRCUITS

Sum-of-product (SOP) expressions are implemented with an AND gate for each product term and one OR gate for summing all of the product terms. The SOP implementation is called AND-OR logic and is the basic form for realizing standard Boolean functions. In this section, the AND-OR and the AND-OR-Invert are examined; the exclusive-OR and exclusive-NOR gates, which are actually a form of AND-OR logic, are also covered.

After completing this section, you should be able to

- Describe the operation of AND-OR and AND-OR-Invert circuits
- Describe the operation of exclusive-OR and exclusive-NOR gates

AND-OR Logic

Figure 4–1(a) shows an AND-OR circuit using ANSI standard distinctive shape symbols consisting of two 2-input AND gates and one 2-input OR gate; Figure 4–1(b) is the ANSI standard rectangular outline symbol.

FIGURE 4–1 An example of AND-OR logic. Open file F04-01 to verify the operation.

MULTISIM

(a) Logic diagram (ANSI standard distinctive shape symbols)

(b) ANSI standard rectangular outline symbol

TABLE 4–1 • Truth table for the AND-OR logic in Figure 4–1.						
INPUTS						**OUTPUT**
A	*B*	*C*	*D*	*AB*	*CD*	*X*
0	0	0	0	0	0	0
0	0	0	1	0	0	0
0	0	1	0	0	0	0
0	0	1	1	0	1	1
0	1	0	0	0	0	0
0	1	0	1	0	0	0
0	1	1	0	0	0	0
0	1	1	1	0	1	1
1	0	0	0	0	0	0
1	0	0	1	0	0	0
1	0	1	0	0	0	0
1	0	1	1	0	1	1
1	1	0	0	1	0	1
1	1	0	1	1	0	1
1	1	1	0	1	0	1
1	1	1	1	1	1	1

The truth table for a 4-input AND-OR logic circuit is shown in Table 4–1. The intermediate AND gate outputs (the AB and CD columns) are also shown in the table. The operation of the AND-OR circuit in Figure 4–1 is stated as follows:

AND-OR logic produces an SOP expression.

> **For a 4-input AND-OR logic circuit, the output X is HIGH (1) if both input A and input B are HIGH (1) or both input C and input D are HIGH (1).**

The term **SOP** shown above the output expression in Figure 4–1 stands for sum-of-products. Recall that the AND operation is Boolean multiplication, so when two or more variables are ANDed together, the result is a product. AB and CD are Boolean product terms. Recall also that the OR operation is Boolean addition. When these two product terms are ORed together, the result is Boolean addition. Therefore, the expression $X = AB + CD$ is a sum-of-products (SOP), which is a standard form. An SOP expression can have any number of AND (product) terms ORed together.

SYSTEM EXAMPLE 4–1

STORAGE TANK MONITOR

In a certain chemical-processing plant, a liquid chemical is used in a manufacturing process. The chemical is stored in three different tanks. A level sensor in each tank produces a HIGH voltage when the level of chemical in the tank drops below a specified point. The circuit monitors the chemical level in each tank and indicates when the level in any two of the tanks drops below the specified point.

The AND-OR circuit in Figure 4–2 has inputs from the sensors on tanks A, B, and C as shown. The AND gate G_1 checks the levels in tanks A and B, gate G_2 checks tanks A and C, and gate G_3 checks tanks B and C. When the chemical level in any two of the tanks gets too low, one of the AND gates will have HIGHs on both of its inputs, causing its output to be HIGH; and so the final output X from the OR gate is HIGH. This HIGH input is then used to activate an indicator such as a lamp or audible alarm, as shown in the figure.

FIGURE 4–2

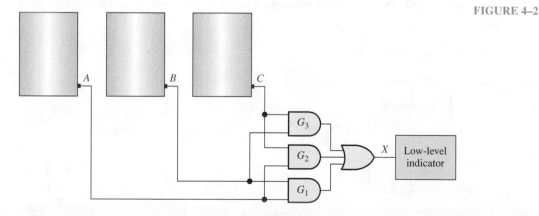

AND-OR-Invert Logic

When the output of an AND-OR circuit is complemented (inverted), it results in an AND-OR-Invert circuit. Recall that AND-OR logic directly implements SOP expressions. Product-of-sum (POS) expressions can be implemented with AND-OR-Invert logic. This is illustrated as follows, by developing the corresponding AND-OR-Invert (AOI) expression.

$$X = (\overline{A} + \overline{B})(\overline{C} + \overline{D}) = (\overline{AB})(\overline{CD}) = \overline{\overline{(AB)}\,\overline{(CD)}} = \overline{\overline{AB} + \overline{CD}} = \overline{AB + CD}$$

The logic diagram in Figure 4–3(a) shows an AND-OR-Invert circuit with four inputs. The ANSI standard rectangular outline symbol is shown in part (b).

(a)   (b)

FIGURE 4–3   An example of AND-OR-Invert logic. Open file F04-03 to verify the operation.

The operation of the AND-OR-Invert circuit in Figure 4–3 is stated as follows:

> **For a 4-input AND-OR-Invert logic circuit, the output X is LOW (0) if both input A and input B are HIGH (1) or both input C and input D are HIGH (1).**

A truth table can be developed from the AND-OR truth table in Table 4–1 by simply changing all 1s to 0s and all 0s to 1s in the output column.

The term **POS** shown above the output expression in Figure 4–3 (a) stands for product-of-sums. Recall that the OR operation is Boolean addition, so when two or more variables are ORed together, the result is a sum. $\overline{A} + \overline{B}$ and $\overline{C} + \overline{D}$ are Boolean sum terms. Recall also that the AND operation is Boolean multiplication. When these two sum terms are ANDed together, the result is Boolean multiplication. Therefore, the expression $(\overline{A} + \overline{B})(\overline{C} + \overline{D})$ is a product-of-sums (POS), which is a standard form. A POS expression can have any number of OR (sum) terms ANDed together.

Exclusive-OR Logic

The XOR gate is actually a combination of other gates.

The exclusive-OR gate was introduced in Chapter 3. Although, because of its importance, this circuit is considered a type of logic gate with its own unique symbol, it is actually a combination of two AND gates, one OR gate, and two inverters, as shown in Figure 4–4(a). The two ANSI standard exclusive-OR logic symbols are shown in parts (b) and (c).

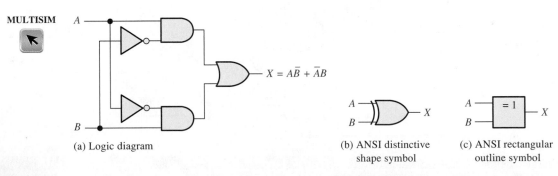

(a) Logic diagram   (b) ANSI distinctive   (c) ANSI rectangular
                                                   shape symbol   outline symbol

FIGURE 4–4   Exclusive-OR logic diagram and symbols. Open file F04-04 to verify the operation.

TABLE 4–2 • Truth table for an exclusive-OR.		
A	*B*	*X*
0	0	0
0	1	1
1	0	1
1	1	0

The output expression for the circuit in Figure 4–4 is

$$X = A\overline{B} + \overline{A}B$$

Evaluation of this expression results in the truth table in Table 4–2. Notice that the output is HIGH only when the two inputs are at opposite levels. A special exclusive-OR operator $\oplus$ is often used, so the expression $X = A\overline{B} + \overline{A}B$ can be stated as "X is equal to A exclusive-OR B" and can be written as

$$X = A \oplus B$$

Exclusive-NOR Logic

As you know, the complement of the exclusive-OR function is the exclusive-NOR, which is derived as follows:

$$X = \overline{A\overline{B} + \overline{A}B} = (\overline{A\overline{B}})(\overline{\overline{A}B}) = (\overline{A} + B)(A + \overline{B}) = \overline{A}\,\overline{B} + AB$$

Notice that the output X is HIGH only when the two inputs, A and B, are at the same level.

The exclusive-NOR can be implemented by simply inverting the output of an exclusive-OR, as shown in Figure 4–5(a), or by directly implementing the expression $\overline{A}\,\overline{B} + AB$, as shown in part (b).

(a) $X = \overline{A\overline{B} + \overline{A}B}$ (b) $X = \overline{A}\,\overline{B} + AB$

MULTISIM

FIGURE 4–5   **Two equivalent ways of implementing the exclusive-NOR. Open files F04-05 (a) and (b) to verify the operation.**

SYSTEM EXAMPLE 4–2

DATA TRANSMISSION WITH ERROR DETECTION

A certain data transmission system uses exclusive-OR gates to implement even-parity generation and checking to detect an error in a 4-bit code. Recall from Chapter 2 that a parity bit is added to a binary code in order to provide error detection. For even parity, a parity bit is added to the original code to make the total number of 1s in the code even.

The parity generator circuit in Figure 4–6 produces a 1 output when there is an odd number of 1s on the inputs in order to make the total number of 1s in the output code even.

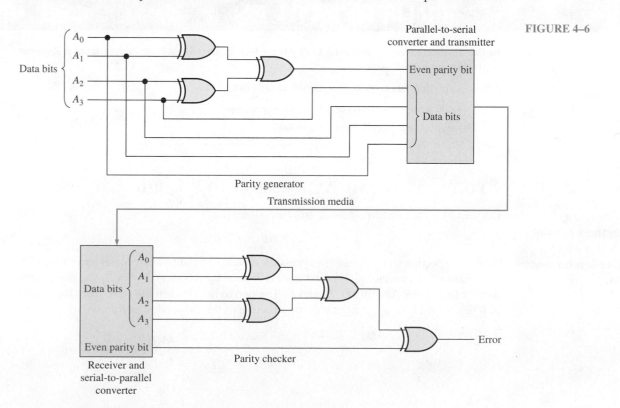

FIGURE 4–6

A 0 output is produced when there is an even number of 1s on the inputs. A total of five bits is converted to serial form and transmitted. At the receiving end, the transmitted code is converted back to parallel form. The parity checker produces a 1 output when there is an error in the five-bit code and a 0 when there is no error.

SECTION 4–1 CHECKUP*

1. Determine the output (1 or 0) of a 4-variable AND-OR-Invert circuit for each of the following input conditions:

   (a) $A = 1, B = 0, C = 1, D = 0$

   (b) $A = 1, B = 1, C = 0, D = 1$

   (c) $A = 0, B = 1, C = 1, D = 1$

2. Determine the output (1 or 0) of an exclusive-OR gate for each of the following input conditions:

   (a) $A = 1, B = 0$

   (b) $A = 1, B = 1$

   (c) $A = 0, B = 1$

   (d) $A = 0, B = 0$

3. Develop the truth table for a certain 3-input logic circuit with the output expression $X = A\overline{B}C + \overline{A}BC + \overline{A}\,\overline{B}\,\overline{C} + AB\overline{C} + ABC$.

4. Draw the logic diagram for an exclusive-NOR circuit.

*Answers are at the end of the chapter.

4–2 BOOLEAN EXPRESSIONS AND TRUTH TABLES

In this section, examples are used to illustrate how to develop a logic circuit from a Boolean expression or a truth table.

After completing this section, you should be able to

- Develop a logic circuit from a Boolean expression
- Develop a logic circuit from a truth table

From a Boolean Expression to a Logic Circuit

> For every Boolean expression there is a logic circuit, and for every logic circuit there is a Boolean expression.

Let's examine the following Boolean expression:

$$X = AB + CDE$$

A brief inspection shows that this expression is composed of two terms, AB and CDE, with a **domain** of five variables. The first term is formed by ANDing A with B, and the second term is formed by ANDing C, D, and E. The two terms are then ORed to form the output X. These operations are indicated in the structure of the expression as follows:

$$X = \underbrace{AB}_{} + \underbrace{CDE}_{} \quad \text{AND}$$

OR

Note that in this particular expression, the AND operations forming the two individual terms, *AB* and *CDE*, must be performed *before* the terms can be ORed.

To implement this Boolean expression, a 2-input AND gate is required to form the term *AB*, and a 3-input AND gate is needed to form the term *CDE*. A 2-input OR gate is then required to combine the two AND terms. The resulting logic circuit is shown in Figure 4–7.

FIGURE 4–7   **Logic circuit for** $X = AB + CDE$.

As another example, let's implement the following expression:

$$X = AB(\overline{CD} + EF)$$

A breakdown of this expression shows that the terms *AB* and $(\overline{CD} + EF)$ are ANDed. The term $\overline{CD} + EF$ is formed by first ANDing *C* and $\overline{D}$ and ANDing *E* and *F*, and then ORing these two terms. This structure is indicated in relation to the expression as follows:

Before you can implement the final expression, you must create the sum term $\overline{CD} + EF$; but before you can get this term; you must create the product terms $\overline{CD}$ and *EF;* but before you can get the term $\overline{CD}$, you must create $\overline{D}$. So, as you can see, the logic operations must be done in the proper order.

The logic gates required to implement $X = AB(\overline{CD} + EF)$ are as follows:

1. One inverter to form $\overline{D}$

2. Two 2-input AND gates to form $\overline{CD}$ and *EF*

3. One 2-input OR gate to form $\overline{CD} + EF$

4. One 3-input AND gate to form *X*

The logic circuit for this expression is shown in Figure 4–8(a). Notice that there is a maximum of four gates and an inverter between an input and output in this circuit (from input *D* to output). Often the total propagation delay time through a logic circuit is a major consideration. Propagation delays are additive, so the more gates or inverters between input and output, the greater the propagation delay time.

(a)

(b) Sum-of-products implementation of the circuit in part (a)

FIGURE 4–8   Logic circuits for $X = AB(C\overline{D} + EF) = ABC\overline{D} + ABEF$.

Unless an intermediate term, such as $C\overline{D} + EF$ in Figure 4–8(a), is required as an output for some other purpose, it is usually best to reduce a circuit to its SOP form in order to reduce the overall propagation delay time. The expression is converted to SOP as follows, and the resulting circuit is shown in Figure 4–8(b).

$$AB(C\overline{D} + EF) = ABC\overline{D} + ABEF$$

From a Truth Table to a Logic Circuit

If you begin with a truth table instead of an expression, you can write the SOP expression from the truth table and then implement the logic circuit. Table 4–3 specifies a logic function.

TABLE 4–3				
INPUTS			**OUTPUT**	
A	*B*	*C*	*X*	**PRODUCT TERM**
0	0	0	0	
0	0	1	0	
0	1	0	0	
0	1	1	1	$\overline{A}BC$
1	0	0	1	$A\overline{B}\,\overline{C}$
1	0	1	0	
1	1	0	0	
1	1	1	0	

The Boolean SOP expression obtained from the truth table by ORing the product terms for which $X = 1$ is

$$X = \overline{A}BC + A\overline{B}\,\overline{C}$$

The first term in the expression is formed by ANDing the three variables $\overline{A}$, B, and C. The second term is formed by ANDing the three variables A, $\overline{B}$, and $\overline{C}$.

The logic gates required to implement this expression are as follows: three inverters to form the $\overline{A}$, $\overline{B}$, and $\overline{C}$ variables; two 3-input AND gates to form the terms $\overline{A}BC$ and $A\overline{B}\,\overline{C}$; and one 2-input OR gate to form the final output function, $\overline{A}BC + A\overline{B}\,\overline{C}$.

The implementation of this logic function is illustrated in Figure 4–9.

FIGURE 4–9   Logic circuit for $X = \overline{A}BC + A\overline{B}\,\overline{C}$. Open file F04-09 to verify the operation.

MULTISIM

EXAMPLE 4–1

Develop the logic circuit for the operation specified in the truth table of Table 4–4.

TABLE 4–4				
INPUTS			**OUTPUT**	
A	**B**	**C**	**X**	**PRODUCT TERM**
0	0	0	0	
0	0	1	0	
0	1	0	0	
0	1	1	1	$\overline{A}BC$
1	0	0	0	
1	0	1	1	$A\overline{B}C$
1	1	0	1	$AB\overline{C}$
1	1	1	0	

SOLUTION

Notice that $X = 1$ for only three of the input conditions. Therefore, the logic expression is

$$X = \overline{A}BC + A\overline{B}C + AB\overline{C}$$

The logic gates required are three inverters, three 3-input AND gates and one 3-input OR gate. The logic circuit is shown in Figure 4–10.

FIGURE 4–10   Open file F04-10 to verify the operation.

MULTISIM

RELATED PROBLEM*

If there were no complemented variables in the Boolean expression for the logic circuit of Figure 4–10, what would be the resulting circuit?

*Answers are at the end of the chapter.

EXAMPLE 4–2

Develop a logic circuit with four input variables that will only produce a 1 output when exactly three input variables are 1s.

SOLUTION

Out of sixteen possible combinations of four variables, the combinations in which there are exactly three 1s are listed in Table 4–5, along with the corresponding product term for each.

TABLE 4–5

A	B	C	D	PRODUCT TERM
0	1	1	1	$\overline{A}BCD$
1	0	1	1	$A\overline{B}CD$
1	1	0	1	$AB\overline{C}D$
1	1	1	0	$ABC\overline{D}$

The product terms are ORed to get the following expression:

$$X = \overline{A}BCD + A\overline{B}CD + AB\overline{C}D + ABC\overline{D}$$

This expression is implemented in Figure 4–11 with AND-OR logic.

FIGURE 4–11   **Open file F04-11 to verify the operation.**

MULTISIM

RELATED PROBLEM

Show that the four cases in Table 4–5 are the only times there are three and only three 1s in the code.

Constructing a Truth Table for a Logic Circuit

A combinational logic circuit can be described by a truth table.

Once you have determined the Boolean expression for a given logic circuit, you can develop a truth table that shows the output for all possible values of the input variables. The procedure requires that you evaluate the Boolean expression for all possible combinations of values for the input variables. In the case of the circuit in Figure 4–12, there are four input variables (A, B, C, and D) and therefore sixteen ($2^4 = 16$) combinations of values are possible.

EVALUATING THE EXPRESSION   To evaluate the expression $A(B + CD)$, first find the values of the variables that make the expression equal to 1, using the rules for Boolean addition and multiplication. In this case, the expression equals 1 only if $A = 1$ and $B + CD = 1$ because

$$A(B + CD) = 1 \cdot 1 = 1$$

FIGURE 4–12   **A combinational logic circuit showing the development of the Boolean expression for the output.**

Now determine when the $B + CD$ term equals 1. The term $B + CD = 1$ if either $B = 1$ or $CD = 1$ or if both B and CD equal 1 because

$$B + CD = 1 + 0 = 1$$
$$B + CD = 0 + 1 = 1$$
$$B + CD = 1 + 1 = 1$$

The term $CD = 1$ only if $C = 1$ and $D = 1$.

To summarize, the expression $A(B + CD) = 1$ when $A = 1$ and $B = 1$ regardless of the values of C and D or when $A = 1$ and $C = 1$ and $D = 1$ regardless of the value of B. The expression $A(B + CD) = 0$ for all other value combinations of the variables.

PUTTING THE RESULTS IN TRUTH TABLE FORMAT   The first step is to list the sixteen input variable combinations of 1s and 0s in a binary sequence as shown in Table 4–6. Next, place a 1 in the output column for each combination of input variables that was determined in the evaluation. Finally, place a 0 in the output column for all other combinations of input variables. These results are shown in the truth table in Table 4–6.

TABLE 4–6 • Truth table for the logic circuit in Figure 4–12.				
INPUTS				**OUTPUT**
A	*B*	*C*	*D*	$A(B + CD)$
0	0	0	0	0
0	0	0	1	0
0	0	1	0	0
0	0	1	1	0
0	1	0	0	0
0	1	0	1	0
0	1	1	0	0
0	1	1	1	0
1	0	0	0	0
1	0	0	1	0
1	0	1	0	0
1	0	1	1	1
1	1	0	0	1
1	1	0	1	1
1	1	1	0	1
1	1	1	1	1

EXAMPLE 4–3

Use Multisim to generate the truth table for the logic circuit in Figure 4–12.

SOLUTION

Construct the circuit in Multisim and connect the Multisim Logic Converter to the inputs and output, as shown in Figure 4–13. Click on the ⬭ → 10|1 conversion bar, and the truth table appears in the display as shown.

MULTISIM

FIGURE 4–13

You can also generate the simplified Boolean expression from the truth table by clicking on ⎡10|1 SIMP A|B⎤.

RELATED PROBLEM

Open Multisim and create the setup. Do the conversions shown in this example.

SECTION 4–2 CHECKUP

1. Implement the following Boolean expressions as they are stated:

   (a) $X = ABC + AB + AC$

   (b) $X = AB(C + DE)$

2. Develop a logic circuit that will produce a 1 on its output only when all three inputs are 1s or when all three inputs are 0s.

3. Replace the AND gates with OR gates and the OR gate with an AND gate in Figure 4–12. Determine the Boolean expression for the output.

4. Construct a truth table for the circuit in Question 2.

4–3 DEMORGAN'S THEOREMS

DeMorgan's theorems provide mathematical verification of the equivalency of the NAND and negative-OR gates and the equivalency of the NOR and negative-AND gates, which were discussed in Chapter 3.

After completing this section, you should be able to

- State DeMorgan's theorems
- Relate DeMorgan's theorems to the equivalency of the NAND and negative-OR gates and to the equivalency of the NOR and negative-AND gates
- Apply DeMorgan's theorems to the simplification of Boolean expressions

DeMorgan's first theorem is stated as follows:

The complement of a product of variables is equal to the sum of the complements of the variables.

Stated another way,

The complement of two or more ANDed variables is equivalent to the OR of the complements of the individual variables.

The formula for expressing this theorem for two variables is

$$\overline{XY} = \overline{X} + \overline{Y} \qquad (4\text{–}1)$$

To apply DeMorgan's theorem, break the bar over the product of variables and change the sign from AND to OR.

DeMorgan's second theorem is stated as follows:

The complement of a sum of variables is equal to the product of the complements of the variables.

Stated another way,

The complement of two or more ORed variables is equivalent to the AND of the complements of the individual variables.

The formula for expressing this theorem for two variables is

$$\overline{X + Y} = \overline{X}\,\overline{Y} \qquad (4\text{–}2)$$

Figure 4–14 shows the gate equivalencies and truth tables for Equations 4–1 and 4–2.

INPUTS		OUTPUT	
X	Y	$\overline{XY}$	$\overline{X} + \overline{Y}$
0	0	1	1
0	1	1	1
1	0	1	1
1	1	0	0

INPUTS		OUTPUT	
X	Y	$\overline{X + Y}$	$\overline{X}\,\overline{Y}$
0	0	1	1
0	1	0	0
1	0	0	0
1	1	0	0

FIGURE 4–14 **Gate equivalencies and the corresponding truth tables that illustrate DeMorgan's theorems. Notice the equality of the two output columns in each table. This shows that the equivalent gates perform the same logic function.**

As stated, DeMorgan's theorems also apply to expressions in which there are more than two variables. The following examples illustrate the application of DeMorgan's theorems to 3-variable and 4-variable expressions.

EXAMPLE 4–4

Apply DeMorgan's theorems to the expressions $\overline{XYZ}$ and $\overline{X + Y + Z}$.

SOLUTION

$$\overline{XYZ} = \overline{X} + \overline{Y} + \overline{Z}$$
$$\overline{X + Y + Z} = \overline{X}\,\overline{Y}\,\overline{Z}$$

RELATED PROBLEM

Apply DeMorgan's theorem to the expression $\overline{\overline{X} + \overline{Y} + \overline{Z}}$.

EXAMPLE 4–5

Apply DeMorgan's theorems to the expressions $\overline{WXYZ}$ and $\overline{W + X + Y + Z}$.

SOLUTION

$$\overline{WXYZ} = \overline{W} + \overline{X} + \overline{Y} + \overline{Z}$$
$$\overline{W + X + Y + Z} = \overline{W}\,\overline{X}\,\overline{Y}\,\overline{Z}$$

RELATED PROBLEM

Apply DeMorgan's theorem to the expression $\overline{\overline{W}\,\overline{X}\,\overline{Y}\,\overline{Z}}$.

Each variable in DeMorgan's theorems as stated in Equations 4–1 and 4–2 can also represent a combination of other variables. For example, X can be equal to the term $AB + C$, and Y can be equal to the term $A + BC$. So if you can apply DeMorgan's theorem for two variables as stated by $\overline{XY} = \overline{X} + \overline{Y}$ to the expression $\overline{(AB + C)(A + BC)}$, you get the following result:

$$\overline{(AB + C)(A + BC)} = \overline{(AB + C)} + \overline{(A + BC)}$$

Notice that in the preceding result you have two terms, $\overline{AB + C}$ and $\overline{A + BC}$, to each of which you can again apply DeMorgan's theorem $\overline{X + Y} = \overline{X}\,\overline{Y}$ individually, as follows:

$$\overline{(AB + C)} + \overline{(A + BC)} = (\overline{AB})\overline{C} + \overline{A}(\overline{BC})$$

Notice that you still have two terms in the expression to which DeMorgan's theorem can again be applied. These terms are $\overline{AB}$ and $\overline{BC}$. A final application of DeMorgan's theorem gives the following result:

$$(\overline{AB})\overline{C} + \overline{A}(\overline{BC}) = (\overline{A} + \overline{B})\overline{C} + \overline{A}(\overline{B} + \overline{C})$$

Although this result can be simplified further by the use of Boolean rules and laws, DeMorgan's theorems cannot be applied further.

SECTION 4–3 CHECKUP

1. Apply DeMorgan's theorems to the following expressions:

   (a) $\overline{ABC + (\overline{D + E})}$

   (b) $\overline{(\overline{A + B})\overline{C}}$

   (c) $\overline{\overline{A} + B + C + \overline{DE}}$

4–4 THE UNIVERSAL PROPERTY OF NAND AND NOR GATES

Up to this point, you have studied combinational circuits implemented with AND gates, OR gates, and inverters. In this section, the universal property of the NAND gate and the NOR gate is discussed. The universality of the NAND gate means that it can be used as an inverter and that combinations of NAND gates can be used to implement the AND, OR, and NOR operations. Similarly, the NOR gate can be used to implement the inverter (NOT), AND, OR, and NAND operations.

After completing this section, you should be able to

- Use NAND gates to implement the inverter, the AND gate, the OR gate, and the NOR gate
- Use NOR gates to implement the inverter, the AND gate, the OR gate, and the NAND gate

The NAND Gate as a Universal Logic Element

The NAND gate is a **universal gate** because it can be used to produce the NOT, the AND, the OR, and the NOR functions. An inverter can be made from a NAND gate by connecting all of the inputs together and creating, in effect, a single input, as shown in Figure 4–15(a) for a 2-input gate. An AND function can be generated by the use of NAND gates alone, as shown in Figure 4–15(b). An OR function can be produced with only NAND gates, as illustrated in part (c). Finally, a NOR function is produced as shown in part (d).

In Figure 4–15(b), a NAND gate is used to invert (complement) a NAND output to form the AND function, as indicated in the following equation:

$$X = \overline{\overline{AB}} = AB$$

> **Combinations of NAND gates can be used to produce any logic function.**

(a) One NAND gate used as an inverter

(b) Two NAND gates used as an AND gate

(c) Three NAND gates used as an OR gate

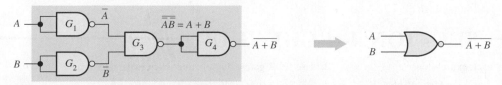

(d) Four NAND gates used as a NOR gate

FIGURE 4–15 Universal application of NAND gates. Open files F04-15(a), (b), (c), and (d) to verify each of the equivalencies.

In Figure 4–15(c), NAND gates G_1 and G_2 are used to invert the two input variables before they are applied to NAND gate G_3. The final OR output is derived as follows by application of DeMorgan's theorem:

$$X = \overline{\overline{A}\,\overline{B}} = A + B$$

In Figure 4–15(d), NAND gate G_4 is used as an inverter connected to the circuit of part (c) to produce the NOR operation $\overline{A + B}$.

The NOR Gate as a Universal Logic Element

Combinations of NOR gates can be used to produce any logic function.

Like the NAND gate, the NOR gate can be used to produce the NOT, AND, OR, and NAND functions. A NOT circuit, or inverter, can be made from a NOR gate by connecting all of the inputs together to effectively create a single input, as shown in Figure 4–16(a) with a 2-input example. Also, an OR gate can be produced from NOR gates, as illustrated in Figure 4–16(b). An AND gate can be constructed by the use of NOR gates, as shown in Figure 4–16(c). In this case the NOR gates G_1 and G_2 are used as inverters, and the final output is derived by the use of DeMorgan's theorem as follows:

$$X = \overline{\overline{A} + \overline{B}} = AB$$

Figure 4–16(d) shows how NOR gates are used to form a NAND function.

(a) One NOR gate used as an inverter

(b) Two NOR gates used as an OR gate

(c) Three NOR gates used as an AND gate

(d) Four NOR gates used as a NAND gate

MULTISIM

FIGURE 4–16   **Universal application of NOR gates. Open files F04-16(a), (b), (c), and (d) to verify each of the equivalencies.**

SECTION 4–4 CHECKUP

1. Use NAND gates to implement each expression:

   (a) $X = \overline{A} + B$

   (b) $X = A\overline{B}$

2. Use NOR gates to implement each expression:

   (a) $X = \overline{A} + B$

   (b) $X = A\overline{B}$

4–5 PULSE WAVEFORM OPERATION

General combinational logic with pulse waveform inputs is examined in this section. Keep in mind that the operation of each gate is the same for pulse waveform inputs as for constant-level inputs. The output of a logic circuit at any given time depends on the inputs at that particular time, so the relationship of the time-varying inputs is of primary importance.

After completing this section, you should be able to

- Predict behavior of combinational logic with pulse waveform inputs
- Develop a timing diagram for any given combinational logic circuit with specified inputs

---

The operation of any gate is the same regardless of whether its inputs are pulsed or constant levels. The nature of the inputs (pulsed or constant levels) does not alter the truth table of a circuit. The examples in this section illustrate the analysis of combinational logic circuits with pulse waveform inputs.

The following is a review of the operation of individual gates for use in understanding combinational circuits with pulse waveform inputs:

1. The output of an AND gate is HIGH only when all inputs are HIGH at the same time.
2. The output of an OR gate is HIGH only when at least one of its inputs is HIGH.
3. The output of a NAND gate is LOW only when all inputs are HIGH at the same time.
4. The output of a NOR gate is LOW only when at least one of its inputs is HIGH.

EXAMPLE 4–6

Determine the final output waveform X for the circuit in Figure 4–17, with input waveforms A, B, and C as shown.

$$X = \overline{A(B + C)} = \overline{AB + AC}$$

FIGURE 4–17

SOLUTION

The output expression, $\overline{AB + AC}$, indicates that the output X is LOW when both A and B are HIGH or when both A and C are HIGH or when all inputs are HIGH. The output waveform X is shown in the timing diagram of Figure 4–17. The intermediate waveform Y at the output of the OR gate is also shown.

RELATED PROBLEM

Determine the output waveform if input A is a constant HIGH level.

EXAMPLE 4–7

Draw the timing diagram for the circuit in Figure 4–18 showing the outputs of G_1, G_2, and G_3 with the input waveforms, A, and B, as indicated.

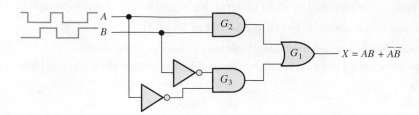

$X = AB + \overline{A}\overline{B}$

FIGURE 4–18

SOLUTION

When both inputs are HIGH or when both inputs are LOW, the output X is HIGH as shown in Figure 4–19. Notice that this is an exclusive-NOR circuit. The intermediate outputs of gates G_2 and G_3 are also shown in Figure 4–19.

FIGURE 4–19

RELATED PROBLEM

Determine the output X in Figure 4–18 if input B is inverted.

EXAMPLE 4–8

Determine the output waveform X for the logic circuit in Figure 4–20(a) by first finding the intermediate waveform at each of points Y_1, Y_2, Y_3, and Y_4. The input waveforms are shown in Figure 4–20(b).

SOLUTION

All the intermediate waveforms and the final output waveform are shown in the timing diagram of Figure 4–20(c).

RELATED PROBLEM

Determine the waveforms Y_1, Y_2, Y_3, Y_4 and X if input waveform A is inverted.

FIGURE 4–20

EXAMPLE 4–9

Determine the output waveform X for the circuit in Example 4–8, Figure 4–20(a), directly from the output expression.

SOLUTION

The output expression for the circuit is developed in Figure 4–21. The SOP form indicates that the output is HIGH when A is LOW and C is HIGH or when B is LOW and C is HIGH or when C is LOW and D is HIGH.

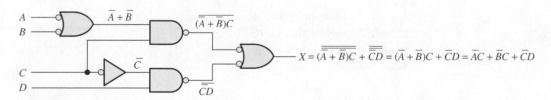

FIGURE 4–21

The result is shown in Figure 4–22 and is the same as the one obtained by the intermediate-waveform method in Example 4–8. The corresponding product terms for each waveform condition that results in a HIGH output are indicated.

FIGURE 4–22

Repeat this example if all the input waveforms are inverted.

SECTION 4–5 CHECKUP

1. One pulse with $t_W = 50 \ \mu s$ is applied to one of the inputs of an exclusive-OR circuit. A second positive pulse with $t_W = 10 \ \mu s$ is applied to the other input beginning 15 μs after the leading edge of the first pulse. Show the output in relation to the inputs.

2. The pulse waveforms A and B in Figure 4–17 are applied to the exclusive-NOR circuit in Figure 4–18. Develop a complete timing diagram.

4–6 COMBINATIONAL LOGIC WITH VHDL AND VERILOG

Three basic approaches to describing logic functions with VHDL are the data flow method, the structural method, and the behavioral method. The data flow method was introduced in Chapter 3, and we will continue it in this section for combinational logic. Recall that the purpose of using VHDL or Verilog to describe a logic function is so that it can then be programmed into a PLD. *VHDL and Verilog tutorials are available on the website*.

After completing this section, you should be able to

- **Explain the entity in VHDL**
- **Explain the architecture in VHDL**
- **Explain the module in Verilog**

A VHDL data flow description of a logic function contains an entity and an architecture. The entity identifies the inputs and outputs of the function within a port statement, and the architecture describes the logic. In writing a VHDL program, both entity and architecture must be used together in what is called an entity/architecture pair. In Verilog, both of these are done in a single module. As mentioned before, both of these HDLs have their adherents and both claim that their choice is better. Actually, both VHDL and Verilog have certain pros and cons. Some say VHDL is more versatile but that Verilog is easier to learn. Keep in mind that with this coverage, we are only touching the tip of the iceberg.

VHDL and Verilog Descriptions from a Boolean Expression

In order to describe combinational logic, you should have either the Boolean expression or the truth table. Example 4–10 shows the use of the **std_logic** data type in VHDL. The **bit** data type is shown in the remaining programs.

EXAMPLE 4–10

Write the VHDL and the Verilog descriptions of AND-OR logic with two 2-input AND gates. The expression is

$$X = AB + CD$$

SOLUTION

Name assignments can be anything all run together or separated by an underscore.

VHDL	Verilog

library ieee;
use ieee.std_logic_1164.all;
entity ANDORlogic is
    port (A, B, C, D: in std_logic; X: out std_logic);
end entity ANDORlogic;
architecture ANDORfunction of ANDORlogic is
begin
    X <= (A and B) or (C and D);
end architecture ANDORfunction;
	module ANDORlogic (A, B, C, D, X);
	    input A, B, C, D;
	    output X;
	        assign X = (A && B) || (C && D);
	endmodule

RELATED PROBLEM

If the AND-OR logic has four AND gates each with two inputs, modify the programs.

EXAMPLE 4–11

Write VHDL and Verilog programs for a logic function having the following Boolean expression:

$$X = (\overline{A}\,\overline{B} + C)(\overline{D}\,\overline{E} + F)$$

SOLUTION

VHDL	Verilog

entity Combo_Logic is
    port (A, B, C, D, E, F: in bit; X: out bit);
end entity Combo_Logic;
architecture Logic_Function of Combo_Logic is
begin
    X <= ((not A and not B) or C) and ((not D
        and not E) or F);
end architecture Logic_Function;
	module Combo_Logic (A, B, C, D, E, F, X);
	    input A, B, C, D, E, F;
	    output X;
	        assign X = ((!A && !B) || C) && ((!D
	            && !E) || F);
	endmodule

RELATED PROBLEM

How would the programs change if there were a bar over each of the parenthetical terms?

VHDL and Verilog Descriptions from a Truth Table

VHDL or Verilog descriptions can also be written directly from a truth table as Example 4–12 shows.

EXAMPLE 4–12

Write VHDL and Verilog programs for a logic function having the following truth table (Table 4–7).

TABLE 4–7				
INPUTS				OUTPUT
A	B	C	D	X
0	0	0	0	0
0	0	0	1	0
0	0	1	0	0
0	0	1	1	0
0	1	0	0	0
0	1	0	1	0
0	1	1	0	0
0	1	1	1	0
1	0	0	0	0
1	0	0	1	0
1	0	1	0	0
1	0	1	1	1
1	1	0	0	1
1	1	0	1	1
1	1	1	0	1
1	1	1	1	1

SOLUTION

As you know, the Boolean terms for the combination of inputs that produce a 1 output form the Boolean expression. In this case, there are five terms that can be written directly to the VHDL and Verilog programs.

```
                  VHDL
entity Table_4_7 is
    port (A, B, C, D: in bit; X: out bit);
end entity Table_4_7;
architecture Logic_Function of Table_4_7 is
begin
    X <= (A and not B and C and D) or (A and
         B and not C and not D) or (A and B
         and not C and D) or (A and B and C
         and not D) or (A and B and C and D);
end architecture Logic_Function;
```

```
                  Verilog
module Table_4_7 (A, B, C, D, X);
  input A, B, C, D;
  output X;
    assign X = (A && !B && C && D) ||
               (A && B && !C && !D) ||
               (A && B && !C && D) ||
               (A && B && C && !D) ||
               (A && B && C && D);
endmodule
```

RELATED PROBLEM

How would the programs change if an all 0s input also produced a 1 in the table?

Structural Approach to VHDL Programming

The structural approach to writing a VHDL description of a logic function can be compared to installing IC devices on a circuit board and interconnecting them with wires. With the structural approach, you describe logic functions and specify how they are connected together. The VHDL **component** is a way to predefine a logic function for repeated use in a program or in other programs. The component can be used to describe anything from a simple logic gate to a complex logic function. The VHDL **signal** can be thought of as a way to specify a "wire" connection between components.

A VHDL component describes predefined logic that can be stored as a package declaration in a VHDL library and called as many times as necessary in a program. You can use components to avoid repeating the same code over and over within a program. For example, you can create a VHDL component for an AND gate and then use it as many times as you wish without having to write a program for an AND gate every time you need one.

VHDL components are stored and are available for use when you write a program. This is similar to having, for example, a storage bin of ICs available when you are constructing a circuit. Every time you need to use one in your circuit, you reach into the storage bin and place it on the circuit board.

The VHDL program for any logic function can become a component and used whenever necessary in a larger program with the use of a component declaration of the following general form. **Component** is a VHDL keyword.

component name_of_component **is**

   port (port definitions);

end component name_of_component;

For simplicity, let's assume that there are predefined VHDL data flow descriptions of a 2-input AND gate with the entity name AND_gate and a 2-input OR gate with the entity name OR_gate, as shown in Figure 4–23.

```
entity AND_gate is
   port (A, B: in bit; X: out bit);
end entity AND_gate;

architecture ANDfunction of AND_gate is
begin
   X <= A and B;
end architecture ANDfunction;
```

2-input AND gate

```
entity OR_gate is
   port (A, B: in bit; X: out bit);
end entity OR_gate;

architecture ORfunction of OR_gate is
begin
   X <= A or B;
end architecture ORfunction;
```

2-input OR gate

FIGURE 4–23   **Predefined programs for a 2-input AND gate and a 2-input OR gate to be used as components in a larger program.**

Next, assume that you are writing a program for a logic circuit that has several AND gates. Instead of rewriting the program in Figure 4–23 over and over, you can use a component declaration to specify the AND gate. The port statement in the component declaration must correspond to the port statement in the entity declaration of the AND gate.

component AND_gate **is**

   port (A, B: **in** bit; X: **out** bit);

end component AND_gate;

FIGURE 4–24

USING COMPONENTS IN A PROGRAM   To use a component in a program, you must write a component instantiation statement for each instance in which the component is used. You can think of a component instantiation as a request or call for the component to be used in the main program. For example, the simple SOP logic circuit in Figure 4–24 has two AND gates and one OR gate. Therefore, the VHDL program for this circuit will have two components and three component instantiations or calls.

SIGNALS   In VHDL, signals are analogous to wires that interconnect components on a circuit board. The signals in Figure 4–24 are named OUT1 and OUT2. Signals are the *internal* connections in the logic circuit and are treated differently than the inputs and outputs. Whereas the inputs and outputs are declared in the entity declaration using the port statement, the signals are declared within the architecture using the signal statement. **Signal** is a VHDL keyword.

THE PROGRAM   The program for the logic in Figure 4–24 begins with an entity declaration as follows:

    --Program for the logic circuit in Figure 4–24
    entity AND_OR_Logic **is**
        port (IN1, IN2, IN3, IN4: **in** bit; OUT3: **out** bit);
    end entity AND_OR_Logic;

The architecture declaration contains the component declarations for the AND gate and the OR gate, the signal definitions, and the component instantiations.

    architecture LogicOperation **of** AND_OR_Logic **is**

COMPONENT INSTANTIATIONS   Let's look at the component instantiations. First, notice that the component instantiations appear between the keyword **begin** and the **end** statement. For each instantiation an identifier is defined, such as G1, G2, and G3 in this case. Then the component name is specified. The port map essentially makes all the connections for the logic function using the operator =>. For example, the first instantiation,

    G1: AND_gate **port map** (A => IN1, B => IN2, X => OUT1);

can be explained as follows: *Input A of AND gate G1 is connected to input IN1, input B of the gate is connected to input IN2, and the output X of the gate is connected to the signal OUT1.*

The three instantiation statements together completely describe the logic circuit in Figure 4–24, as illustrated in Figure 4–25.

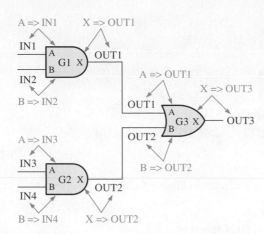

FIGURE 4–25 Illustration of the instantiation statements and port mapping applied to the AND-OR logic. Signals are shown in red.

Although the data flow approach using Boolean expressions would have been easier and probably the best way to describe this particular circuit, we have used this simple circuit to explain the concept of the structural approach. Example 4–13 compares the structural and data flow approaches to writing a VHDL program for an SOP logic circuit.

EXAMPLE 4–13

Write a VHDL program for the SOP logic circuit in Figure 4–26 using the structural approach and the data flow approach. Assume that VHDL components for a 3-input NAND gate and for a 2-input NAND are available. Notice the NAND gate G4 is shown as a negative-OR.

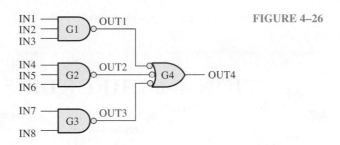

FIGURE 4–26

SOLUTION

The components and component instantiations are highlighted.

    --Program for the logic circuit in Figure 4–26

entity SOP_Logic **is**

    port (IN1, IN2, IN3, IN4, IN5, IN6, IN7, IN8: **in** bit; OUT4: **out** bit);

end entity SOP_Logic;

architecture LogicOperation **of** SOP_Logic **is**

--component declaration for 3-input NAND gate

    component NAND_gate3 **is**
      port (A, B, C: **in** bit X: **out** bit);
    end component NAND_gate3;

--component declaration for 2-input NAND gate

    component NAND_gate2 **is**
      port (A, B: **in** bit; X: **out** bit);
    end component NAND_gate;

    signal OUT1, OUT2, OUT3: bit;

begin

--component instantiations

G1: NAND_gate3 **port map** (A => IN1, B => IN2, C => IN3, X => OUT1);
G2: NAND_gate3 **port map** (A => IN4, B => IN5, C => IN6, X => OUT2);
G3: NAND_gate2 **port map** (A => IN7, B => IN8, X => OUT3);
G4: NAND_gate3 **port map** (A => OUT1, B => OUT2, C => OUT3, X => OUT4);

end architecture LogicOperation;

Using the data flow approach, the program for the logic circuit in Figure 4–26 is

entity SOP_Logic **is**

    port (IN1, IN2, IN3, IN4, IN5, IN6, IN7, IN8: **in** bit; OUT4: **out** bit);

end entity SOP_Logic;

architecture LogicOperation **of** SOP_Logic **is**

begin

    OUT4 <= (IN1 **and** IN2 **and** IN3) **or** (IN4 **and** IN5 **and** IN6) **or** (IN7 **and** IN8);

end architecture LogicOperation;

As you can see, the data flow approach results in a much simpler code for this particular logic function. However, in situations where a logic function consists of many blocks of complex logic, the structural approach might have an advantage over the data flow approach.

RELATED PROBLEM

If another NAND gate is added to the circuit in Figure 4–26 with inputs IN9 and IN10, write a component instantiation to add to the program.

SECTION 4–6 CHECKUP

1. What does an entity in VHDL do?

2. What does an architecture in VHDL do?

3. What performs the same function in Verilog as the entity/architecture pair in VHDL?

4. Name two approaches to writing a VHDL program.

5. What is a VHDL component?

6. State the purpose of a component instantiation in a program architecture.

7. How are interconnections made between components in VHDL?

8. The use of components in a VHDL program represents what approach?

4–7   A SYSTEM

A storage tank system for a pancake syrup manufacturing company is the focus of this section. The control logic allows a volume of corn syrup to be preheated to a specified temperature to achieve the proper viscosity prior to being sent to a mixing vat where ingredients such as sugar, flavoring, preservative, and coloring are added. Level and temperature sensors in the tank and the flow sensor provide the inputs for the logic.

After completing this section, you should be able to

- Explain how the system works
- Write VHDL and Verilog programs so that the system can be implemented in a PLD

System Operation

A tank that holds corn syrup for use in a pancake syrup manufacturing process is shown in Figure 4–27. In preparation for mixing, the temperature of the corn syrup when released from the tank into a mixing vat must be at a specified value for proper viscosity to produce required flow characteristics. This temperature can be selected via a keypad input. The control logic maintains the temperature at this value by turning a heater *on* and *off*. The analog output from the temperature transducer (T_{analog}) is converted to an 8-bit binary code by an analog-to-digital converter and then to an 8-bit BCD code. A temperature controller detects when the temperature falls below the specified value and turns the heater *on*. When the temperature reaches the specified value, the heater is turned *off*.

FIGURE 4–27   Tank with level and temperature sensors and controls.

The level sensors produce a HIGH when the corn syrup is at or above the minimum or at the maximum level. The valve control logic detects when the maximum level (L_{max}) or minimum level (L_{min}) has been reached and when solution is flowing into the tank (F_{inlet}). Based on these inputs, the control logic opens or closes each valve (V_{inlet} and V_{outlet}). New corn syrup can be added to the tank via the inlet valve only when the minimum level is reached. Once the inlet valve is opened, the level in the tank must reach the maximum point before the inlet valve is closed. Also, once the outlet valve is opened, the level must reach the minimum point before the outlet valve is closed. New syrup is always cooler than the syrup in the tank. Syrup cannot be released from the tank while it is being filled or its temperature is below the specified value.

INLET VALVE CONTROL   The conditions for which the inlet valve is open, allowing the tank to fill, are

- The solution level is at minimum (L_{min}).
- The tank is filling (F_{inlet}) but the maximum level has not been reached ($\overline{L}_{max}$).

Table 4–8 is the truth table for the inlet valve. A HIGH (1) is the active level for the inlet valve to be open (*on*). From the truth table, an expression for the inlet valve control output can be written.

$$V_{inlet} = \overline{L}_{max}\overline{L}_{min}\overline{F}_{inlet} + \overline{L}_{max}\overline{L}_{min}F_{inlet} + \overline{L}_{max}L_{min}F_{inlet}$$

The SOP expression can be reduced to the following simplified expression using Boolean methods:

$$V_{inlet} = \overline{L}_{min} + \overline{L}_{max}F_{inlet}$$

TABLE 4–8 • Truth table for inlet valve control.

INPUTS			OUTPUT	
L_{max}	L_{min}	F_{inlet}	V_{inlet}	DESCRIPTION
0	0	0	1	Level below minimum. No inlet flow.
0	0	1	1	Level below minimum. Inlet flow.
0	1	0	0	Level above min and below max. No inlet flow.
0	1	1	1	Level above min and below max. Inlet flow.
1	0	0	X	Invalid.
1	0	1	X	Invalid.
1	1	0	0	Level at maximum. No inlet flow.
1	1	1	0	Level at maximum. Inlet flow.

OUTLET VALVE CONTROL   The conditions for which the outlet valve is open allowing the tank to drain are

- The syrup level is above minimum and the tank is not filling.
- The temperature of the syrup is at the specified value.

Table 4–9 is the truth table for the outlet valve. A HIGH (1) is the active level for the outlet valve to be open (*on*). *Note: T* is both an input and an output (*T = Temp*).

TABLE 4–9 • Truth table for outlet valve control.

INPUTS				OUTPUT	
L_{max}	L_{min}	F_{inlet}	T	V_{outlet}	DESCRIPTION
0	0	0	0	0	Level below minimum. No inlet flow. Temp low.
0	0	0	1	0	Level below minimum. No inlet flow. Temp correct.
0	0	1	0	0	Level below minimum. Inlet flow. Temp low.
0	0	1	1	0	Level below minimum. Inlet flow. Temp correct.
0	1	0	0	0	Level above min and below max. No inlet flow. Temp low.
0	1	0	1	1	Level above min and below max. No inlet flow. Temp correct.
0	1	1	0	0	Level above min and below max. Inlet flow. Temp low.
0	1	1	1	0	Level above min and below max. Inlet flow. Temp correct.
1	0	0	0	X	Invalid.
1	0	0	1	X	Invalid.
1	0	1	0	X	Invalid.
1	0	1	1	X	Invalid.
1	1	0	0	0	Level at maximum. No inlet flow. Temp low.
1	1	0	1	1	Level at maximum. No inlet flow. Temp correct.
1	1	1	0	0	Level at maximum. Inlet flow. Temp low.
1	1	1	1	0	Level at maximum. Inlet flow. Temp correct.

From the truth table, an expression for the outlet valve control output can be written.

$$V_{outlet} = \overline{L}_{max}L_{min}\overline{F}_{inlet}\,T + L_{max}L_{min}\overline{F}_{inlet}T$$

The SOP expression can be reduced to the following simplified expression using Boolean methods:

$$V_{outlet} = L_{min}\overline{F}_{inlet}T$$

Simulation of the Valve Control Logic

The inlet and outlet valve control logic simulation screen is shown in Figure 4–28. SPDT switches are used to represent the level and flow sensor inputs and the temperature indication. Probes are used to indicate the output states.

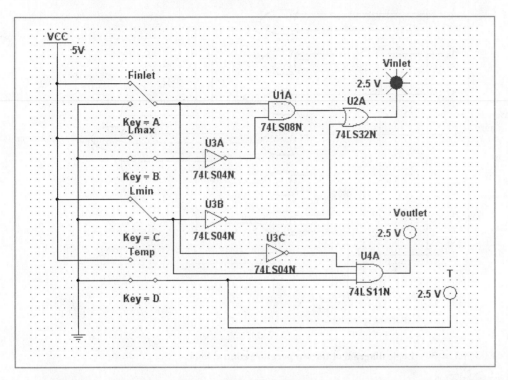

FIGURE 4–28 **Multisim circuit screen for the valve control logic. Open file F04-28 and run the simulation.**

MULTISIM

TEMPERATURE CONTROL The temperature control logic accepts an 8-bit BCD code representing the measured temperature and compares it to the BCD code for the specified temperature. A block diagram is shown in Figure 4–29. The analog-to-digital converter is covered in Chapter 12, and the binary-to-BCD converter is covered in Chapter 5.

FIGURE 4–29 **Block diagram for temperature control system.**

When the measured temperature and the specified temperature are the same, the two BCD codes are equal and the *T* output is LOW (0). When the measured temperature falls below the specified value, there is a difference in the BCD codes and the *T* output is HIGH (1), which turns on the heater. The temperature control logic can be implemented with exclusive-OR gates, as shown in Figure 4–30. Each pair of corresponding bits from the two BCD codes is applied to an exclusive-OR gate. If the bits are the same, the output of the XOR gate is 0; and if they are different, the output of the XOR gate is 1. When one or more XOR outputs equal 1, the *T* output of the OR gate equals 1, causing the heater to turn on.

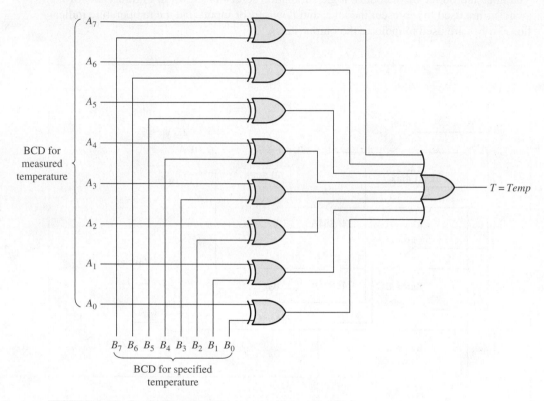

FIGURE 4–30   **Logic diagram of the temperature control logic.**

VHDL and Verilog for Inlet, Outlet, and Temperature Controls

In order to implement this control system in a PLD, a VHDL program is developed. This implementation does not include the analog-to-digital converter or the binary-to-BCD converter required for the temperature control.

First, assign input labels to the temperature control logic. The BCD for measured temperature inputs are assigned A0 through A7, as shown in Figure 4–30. The BCD for specified temperature inputs are assigned B0 through B7, as shown. The inlet and outlet valve controls in Figure 4–27 already have input label assignments. The VHDL program using the data flow approach is as follows.

entity TankControlLogic **is**

    port (Finlet, Lmax, Lmin, A0, A1, A2, A3,

        A4, A5, A6, A7, B0, B1, B2, B3, B4, B5,

        B6, B7: **in** bit; T: **buffer** bit; Vinlet, Voutlet: **out** bit);

end entity TankControlLogic;

architecture Logic_Function **of** TankControlLogic **is**

```
    begin
        T <= (A0 xor B0) or (A1 xor B1) or (A2 xor B2) or
             (A3 xor B3) or (A4 xor B4) or (A5 xor B5) or
             (A6 xor B6) or (A7 xor B7);
        Vinlet <= not Lmin or not Lmax and Finlet;
        Voutlet <= Lmin and not Finlet and Temp;
    end architecture Logic_Function;
```

Alternately, the system logic can be programmed using Verilog.

```
    module TankControlLogic (Finlet, Lmax, Lmin,
             Vinlet, Voutlet, T, A0, A1, A2, A3, A4, A5, A6, A7, B0,
          B1, B2, B3, B4, B5, B6, B7);
        input Finlet Lmax, Lmin, Temp, A0, A1, A2, A3, A4,
             A5, A6, A7, B0, B1, B2, B3, B4, B5, B6, B7;
        output Vinlet, Voutlet, T;
        assign T = ((A0 && !B0) || (!A0 && B0)) || ((A1 && !B1) || (!A1 && B1)) ||
                   ((A2 && !B2) || (!A2 && B2)) || ((A3 && !B3) || (!A3 && B3)) ||
                   ((A3 && !B3) || (!A3 && B3)) || ((A4 && !B4) || (!A4 && B4)) ||
                   ((A5 && !B5) || (!A5 && B5)) || ((A6 && !B6)) || (!A6 && B6)) ||
                   ((A7 && !B7) || (!A7 && B7))
        assign Vinlet = !Lmin || (!Lmax && Finlet);
        assign Voutlet = Lmin && !Finlet && Temp;
    endmodule
```

Alternately, the expression for T in the Verilog program can be written in a more compact form using the Verilog bit-wise operator for XOR, ^, rather than the logical operators !, &&, and || as follows:

```
    T = (A0 ^ B0) || (A1 ^ B1) || (A4 ^ B4) || (A5 ^ B5) || (A6 ^ B6) || (A7 ^ B7);
```

SECTION 4–7 CHECKUP

1. For how many input conditions is the inlet valve open?

2. For how many input conditions is the outlet valve open?

3. Why does the outlet valve control require four inputs and the inlet valve only three?

4. Once the level reaches maximum and the tank starts draining, when does the outlet valve turn off?

The preceding sections have given you some insight into the operation of combinational logic circuits and the relationships of inputs and outputs. This type of understanding is essential when you troubleshoot digital circuits and systems because you must know what logic levels or waveforms to look for throughout the circuit for a given set of input conditions. In Chapter 3, you learned two basic troubleshooting methods and studied the troubleshooting of an example system. Recall that there were two options once the defective circuit board was identified: to repair or to not repair the board. Of course, a new board replaces the defective board in either case.

In this section, an oscilloscope is used to troubleshoot at the board level when a device output is connected to several device inputs. Also, an example of signal tracing and waveform analysis methods is presented using a scope for locating a fault in combinational logic.

After completing this section, you should be able to

- **Define a circuit node**
- **Use an oscilloscope to find a faulty circuit node**
- **Use an oscilloscope to find an open gate output**
- **Use an oscilloscope to find a shorted gate input or output**
- **Use an oscilloscope for signal tracing**

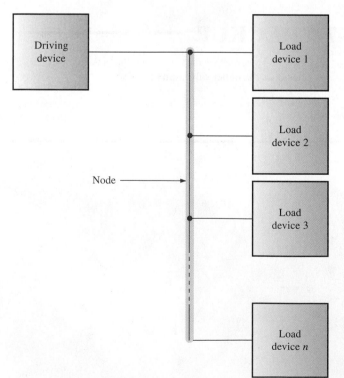

In a combinational logic circuit, the output of a driving device may be connected to two or more load devices as shown in Figure 4–31. The interconnecting paths share a common electrical point known as a **node**.

The driving device in Figure 4–31 is driving the node, and the other devices represent loads connected to the node. A driving device can drive a number of load device inputs up to its specified fan-out. Several types of failures are possible in this situation. Some of these failure modes are difficult to isolate to a single bad device because all the devices connected to the node are affected. Common types of failures are the following:

1. *Open output in driving device.* This failure will cause a loss of signal to all load devices.

2. *Open input in a load device.* This failure will not affect the operation of any of the other devices connected to the node, but it will result in loss of signal output from the faulty device.

3. *Shorted output in driving device.* This failure can cause the node to be stuck in the LOW state (short to ground) or in the HIGH state (short to V_{CC}).

4. *Shorted input in a load device.* This failure can also cause the node to be stuck in the LOW state (short to ground) or in the HIGH state (short to V_{CC}).

Troubleshooting Common Faults at the Board Level

OPEN OUTPUT IN DRIVING DEVICE   In this situation there is no pulse activity on the node. With circuit power on, an open node will normally result in a "floating" level, as illustrated in Figure 4–32.

FIGURE 4–31   Illustration of a node in a logic circuit.

There are pulses on the gate input with the other input HIGH.

No pulse activity is indicated at any point on the node. Scope may indicate "floating" level.

HIGH

IC1 IC2 IC3

If there is no pulse activity at the output pin on IC1, there is an internal open. If there is pulse activity directly on the output pin but not on the node interconnections, the connection between the pin and the board is open.

FIGURE 4–32   Open output in driving device. Assume a HIGH is on one input.

OPEN INPUT IN A LOAD DEVICE   If the check for an open driver output in IC1 is negative (there is pulse activity), then a check for an open input in a load device should be performed. Check the output of each device for pulse activity, as illustrated in Figure 4–33. If one of the inputs that is normally connected to the node is open, no pulses will be detected on that device's output.

HIGH   HIGH

IC1

HIGH

Check the output pin of each device connected to the node with other device inputs HIGH. No pulse activity on an output indicates an open input or open output.

FIGURE 4–33   Open input in a load device.

There is a LOW level at all points connected to the node.

HIGH

FIGURE 4–34   **Shorted output in the driving device or shorted input in a load.**

OUTPUT OR INPUT SHORTED TO GROUND

When the output is shorted to ground in the driving device or the input to a load device is shorted to ground, it will cause the node to be stuck LOW, as previously mentioned. A quick check with a scope probe will indicate this, as shown in Figure 4–34. A short to ground in the driving device's output or in any load input will cause this symptom, and further checks must therefore be made to isolate the short to a particular device.

Signal Tracing and Waveform Analysis

Although the methods of isolating an open or a short at a node point are very useful from time to time, the technique of **signal tracing** is of value in just about every troubleshooting situation. Waveform measurement is accomplished with an oscilloscope or a logic analyzer.

Basically, as you learned in Chapter 3, the signal tracing method requires that you observe the waveforms and their time relationships at all accessible points in the logic circuit. You can begin at the inputs and, from an analysis of the waveform timing diagram for each point, determine where an incorrect waveform first occurs. With this procedure you can usually isolate the fault to a specific device. A procedure beginning at the output and working back toward the inputs can also be used. In this approach, it is necessary to know what the signal should be based on a thorough knowledge of the circuit or on a documented troubleshooting procedure showing waveforms at various crucial points.

The general procedure for signal tracing starting at the inputs is outlined as follows:

- Within a system, define the section of logic that is suspected of being faulty.
- Start at the inputs to the section of logic under examination. We assume, for this discussion, that the input waveforms coming from other sections of the system have been found to be correct.
- For each device, beginning at the input and working toward the output of the logic circuit, observe the output waveform of the device and compare it with the input waveforms by using the oscilloscope or the logic analyzer.
- Determine if the output waveform is correct, using your knowledge of the logical operation of the device or from a troubleshooting procedure, if available.
- If the output is incorrect, the device under test may be faulty. Pull the IC device that is suspected of being faulty, and test it out-of-circuit. If the device is found to be faulty, replace the IC. If it works correctly, the fault is in the external circuitry or in another IC to which the tested one is connected.
- If the output is correct, go to the next device. Continue checking each device until an incorrect waveform is observed.

Figure 4–35 illustrates the general signal tracing procedure from inputs to output for simple combinational logic in the following steps:

Step 1:   Observe the output at test point 5 (TP5) relative to the inputs at TP1 and TP2. If it is correct, go to step 2. If the inputs are correct and the output is not correct, the gate or its connections are bad; or, if the output is LOW, the input to gate G_2 may be shorted. In this case, the fault is in IC2. Replace it and recheck the operation.

Step 2:   Observe the output of the inverter (TP6) relative to the input at TP3. If it is correct, go to Step 3. If the input is correct and the output is not correct, the

HANDS ON TIP
As you know, testing and troubleshooting logic circuits often require observing and comparing two digital waveforms simultaneously, such as an input and the output of a gate, on an oscilloscope. For digital waveforms, the scope should always be set to DC coupling on each channel input to avoid "shifting" the ground level. You should determine where the 0 V level is on the screen for both channels.

To compare the timing of the waveforms, the scope should be triggered from only one channel (don't use vertical mode or composite triggering). The channel selected for triggering should always be the one that has the lowest frequency waveform, if possible.

Step 1
☐ If correct, go to step 2.
☐ If incorrect, test IC2 and connections.

Step 2
☐ If correct, go to step 3.
☐ If incorrect, test IC1 and connections.

Step 3
☐ If correct, go to step 4.
☐ If incorrect, test IC2 and connections.

Scope is externally triggered from test point 1 (TP1).

The inverter is in IC1 and the NAND/negative-OR gates are in IC2.

Step 4
☐ If correct, go to step 5.
☐ If incorrect, test IC2 and connections.

Step 5
☐ If correct, circuit is OK.
☐ If incorrect, test IC2 and connections.

FIGURE 4–35   Example of signal tracing and waveform analysis in a portion of a printed circuit board. TP indicates test point. The correct waveforms are shown.

inverter or its connections are bad; or, if the output is LOW, the input to gate G_3 may be shorted. In this case, you can't determine which IC is bad, so both IC1 and IC2 should be replaced and the operation rechecked.

Step 3:   Observe the output of gate G_2 (TP7) relative to the inputs at TP3 and TP5. If it is correct, go to Step 4. If the output is not correct, the gate or its connections are bad; or, if the output is LOW, the input to gate G_4 may be shorted. In this case, the fault is in IC2. Replace it and recheck the operation.

Step 4:   Observe the output of gate G_3 (TP8) relative to the inputs at TP4 and TP6. If it is correct, go to Step 5. If the output is not correct, the gate or its connections are bad; or, if the output is LOW, the input to gate G_4 (TP7) may be shorted. In this case, the fault is in IC2. Replace it and recheck the operation.

Step 5:   Observe the output of gate G_4 (TP9) relative to the inputs at TP7 and TP8. If it is correct, the circuit is okay. If the output is not correct, the gate or its connections are bad or its load is shorted. Replace IC2 and see if that corrects the problem. If it doesn't, you will have to investigate further depending on where the G_4 output goes.

EXAMPLE 4–14

Determine the fault in the logic circuit of Figure 4–36(a) by using waveform analysis. You have observed the waveforms shown in green in Figure 4–36(b). The red waveforms are correct and are provided for comparison.

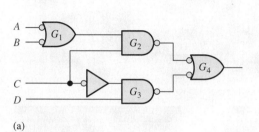

(a)

(b)

FIGURE 4–36

SOLUTION

1. Determine what the correct waveform should be for each gate. The correct waveforms are shown in red, superimposed on the actual measured waveforms, in Figure 4–36(b).

2. Compare waveforms gate by gate until you find a measured waveform that does not match the correct waveform.

   In this example, everything tested is correct until gate G_3 is checked. The output of this gate is not correct as the differences in the waveforms indicate. An analysis of the waveforms indicates that if the D input to gate G_3 is open and acting as a HIGH, you will get the output waveform measured (shown in red). Notice that the output of G_4 is also incorrect due to the incorrect input from G_3.

   Replace the IC containing G_3, and check the circuit's operation again.

RELATED PROBLEM

For the inputs in Figure 4–36(b), determine the output waveform for the logic circuit (output of G_4) if the inverter has an open output.

SECTION 4–8 CHECKUP

1. List four common internal failures in logic gates.

2. One input of a NOR gate is externally shorted to $+V_{CC}$. How does this condition affect the gate operation?

3. Determine the output of gate G_4 in Figure 4–36(a), with inputs as shown in part (b), for the following faults:

   (a) one input to G_1 shorted to ground

   (b) the inverter input shorted to ground

   (c) an open output in G_3

SUMMARY

- AND-OR logic produces an output expression in SOP form.
- AND-OR-Invert logic produces a complemented SOP form, which is actually a POS form.
- The operational symbol for exclusive-OR is $\oplus$. An exclusive-OR expression can be stated in two equivalent ways:

$$A\overline{B} + \overline{A}B = A \oplus B$$

- To describe a logic circuit, start with the logic circuit, and develop the Boolean output expression or the truth table or both.
- Implementation of a logic circuit is the process in which you start with the Boolean output expressions or the truth table and develop a logic circuit that produces the output function.
- DeMorgan's theorems:
  1. The complement of a product is equal to the sum of the complements of the terms in the product.

  $$\overline{XY} = \overline{X} + \overline{Y}$$

  2. The complement of a sum is equal to the product of the complements of the terms in the sum.

  $$\overline{X + Y} = \overline{X}\,\overline{Y}$$

- NAND and negative-OR operations are equivalent.
- NOR and negative-AND operations are equivalent.
- In the structural approach, a VHDL component is a predefined logic function stored for use throughout a program or in other programs.
- A component instantiation is used to call for a component in a program.
- A VHDL signal effectively acts as an internal interconnection in a VHDL structural description.

KEY TERMS

Key terms and other bold terms in the chapter are defined in the end-of-book glossary.

Component A VHDL feature that can be used to predefine a logic function for multiple use throughout a program or programs.

Node A common connection point in a circuit in which a gate output is connected to one or more gate inputs.

POS Product-of-sums; a form of Boolean expression that is basically the ANDing of ORed terms.

Signal A waveform; a type of VHDL object that holds data.

Signal tracing A troubleshooting technique in which waveforms are observed in a step-by-step manner beginning at the input and working toward the output or vice versa. At each point the observed waveform is compared with the correct signal for that point.

SOP Sum-of-products; a form of Boolean expression that is basically the ORing of ANDed terms.

Universal gate Either a NAND gate or a NOR gate. The term *universal* refers to the property of a gate that permits any logic function to be implemented by that gate or by a combination of that kind.

TRUE/FALSE QUIZ

Answers are at the end of the chapter.

1. AND-OR logic can have only two 2-input AND gates.
2. If the inputs of an exclusive-OR gate are the same, the output is HIGH (1).
3. If the inputs of an exclusive-NOR gate are different, the output is LOW (0).
4. A parity generator can be implemented using exclusive-OR gates.
5. NAND gates cannot be used to produce the OR function.
6. NOR gates can be used to produce the AND function.
7. Any SOP expression can be implemented using only NAND gates.

8. The dual symbol for a NAND gate is a negative-AND symbol.

9. Negative-OR is equivalent to NAND.

10. Data flow and structural are two approaches to writing VHDL programs.

SELF-TEST

Answers are at the end of the chapter.

1. The output expression for an AND-OR circuit having one AND gate with inputs A, B, C, and D and one AND gate with inputs E and F is
   (a) $ABCDEF$
   (b) $A + B + C + D + E + F$
   (c) $(A + B + C + D)(E + F)$
   (d) $ABCD + EF$

2. A logic circuit with an output $X = A\overline{B}C + A\overline{C}$ consists of
   (a) two AND gates and one OR gate
   (b) two AND gates, one OR gate, and two inverters
   (c) two OR gates, one AND gate, and two inverters
   (d) two AND gates, one OR gate, and one inverter

3. To implement the expression $\overline{A}BCD + A\overline{B}CD + ABC\overline{D}$, it takes one OR gate and
   (a) one AND gate
   (b) three AND gates
   (c) three AND gates and four inverters
   (d) three AND gates and three inverters

4. The output expression for an AND-OR-Invert circuit having one AND gate with inputs, A, B, C, and D and one AND gate with inputs E and F is
   (a) $\overline{ABCD + EF}$
   (b) $\overline{A} + \overline{B} + \overline{C} + \overline{D} + \overline{E} + \overline{F}$
   (c) $\overline{(A + B + C + D)(E + F)}$
   (d) $(\overline{A} + \overline{B} + \overline{C} + \overline{D})(\overline{E} + \overline{F})$

5. An exclusive-OR function is expressed as
   (a) $\overline{A}\,\overline{B} + AB$
   (b) $\overline{A}B + A\overline{B}$
   (c) $(\overline{A} + B)(A + \overline{B})$
   (d) $(\overline{A} + \overline{B}) + (A + B)$

6. The AND operation can be produced with
   (a) two NAND gates
   (b) three NAND gates
   (c) one NOR gate
   (d) three NOR gates

7. The OR operation can be produced with
   (a) two NOR gates
   (b) three NAND gates
   (c) four NAND gates
   (d) both answers (a) and (b)

8. All Boolean expressions can be implemented with
   (a) NAND gates only
   (b) NOR gates only
   (c) combinations of NAND and NOR gates
   (d) combinations of AND gates, OR gates, and inverters
   (e) any of these

9. A VHDL component
   (a) can be used once in each program
   (b) is a predefined description of a logic function
   (c) can be used multiple times in a program
   (d) is part of a data flow description
   (e) answers (b) and (c)

10. A VHDL component is called for use in a program by using a
   (a) signal
   (b) variable
   (c) component instantiation
   (d) architecture declaration

PROBLEMS

Answers to odd-numbered problems are at the end of the book.

SECTION 4–1   Basic Combinational Logic Circuits

1. Draw the ANSI distinctive shape logic diagram for a 3-wide, 4-input AND-OR-Invert circuit. Also draw the ANSI standard rectangular outline symbol.

2. Write the output expression for each circuit in Figure 4–37.

3. Write the output expression for each circuit as it appears in Figure 4–38.

(a)

(b)

FIGURE 4–37

(a)

(b)

(c)

(d)

(e)

(f)

FIGURE 4–38

4. Develop the truth table for each circuit in Figure 4–38.

5. Develop the truth table for each circuit in Figure 4–39 on page 212.

SECTION 4–2   Boolean Expressions and Truth Tables

6. Develop an AND-OR-Invert logic circuit for a power saw that removes power (logic 0) if the guard is not in place (logic 0) and the switch is *on* (logic 1) or the switch is *on* and the motor is too hot (logic 1).

7. An AND-OR-Invert logic chip has two 4-input AND gates connected to a 2-input NOR gate. Write the Boolean expression for the circuit (assume the inputs are labeled *A* through *H*).

8. Use AND gates, OR gates, or combinations of both to implement the following logic expressions as stated:
   (a) $X = AB$
   (b) $X = A + B$
   (c) $X = AB + C$
   (d) $X = ABC + D$
   (e) $X = A + B + C$
   (f) $X = ABCD$
   (g) $X = A(CD + B)$
   (h) $X = AB(C + DEF) + CE(A + B + F)$

9. Use AND gates, OR gates, and inverters as needed to implement the following logic expressions as stated:
   (a) $X = AB + \overline{B}C$
   (b) $X = A(B + \overline{C})$
   (c) $X = A\overline{B} + AB$
   (d) $X = \overline{ABC} + B(EF + \overline{G})$
   (e) $X = A[BC(A + B + C + D)]$
   (f) $X = B(C\overline{D}E + \overline{E}FG)(\overline{AB} + C)$

FIGURE 4–39

10. Use NAND gates, NOR gates, or combinations of both to implement the following logic expressions as stated:
(a) $X = \overline{A}B + CD + \overline{(A + B)}(ACD + \overline{B}E)$
(b) $X = AB\overline{C}\,\overline{D} + D\overline{E}F + \overline{AF}$
(c) $X = \overline{A}[B + \overline{C}(D + E)]$

11. Write the logic expression for the truth table in Table 4–10. Determine the number and the types of gates required.

TABLE 4–10			
INPUTS			**OUTPUT**
A	*B*	*C*	*X*
0	0	0	1
0	0	1	0
0	1	0	1
0	1	1	0
1	0	0	1
1	0	1	0
1	1	0	1
1	1	1	1

12. Write the logic expression for the truth table in Table 4–11. Determine the number and the types of gates required.

TABLE 4–11

INPUTS				OUTPUT
A	*B*	*C*	*D*	*X*
0	0	0	0	0
0	0	0	1	0
0	0	1	0	1
0	0	1	1	1
0	1	0	0	1
0	1	0	1	0
0	1	1	0	0
0	1	1	1	0
1	0	0	0	1
1	0	0	1	1
1	0	1	0	1
1	0	1	1	1
1	1	0	0	0
1	1	0	1	0
1	1	1	0	0
1	1	1	1	1

13. Develop a truth table for each of the following standard SOP expressions:
    (a) $A\overline{B}C + \overline{A}B\overline{C} + ABC$
    (b) $\overline{X}\,\overline{Y}\,\overline{Z} + \overline{X}\,\overline{Y}Z + XY\overline{Z} + X\,\overline{Y}Z + \overline{X}YZ$

14. Develop a truth table for each of the following standard SOP expressions:
    (a) $\overline{A}B\overline{C}D + \overline{A}BC\overline{D} + A\overline{B}\,\overline{C}D + \overline{A}\,\overline{B}\,\overline{C}\,\overline{D}$
    (b) $WXY\overline{Z} + W\overline{X}Y\overline{Z} + \overline{W}XYZ + W\overline{X}\,\overline{Y}Z + WX\overline{Y}Z$

15. For each truth table in Table 4–12 on page 214, derive a standard SOP and a standard POS expression.

SECTION 4–3 DeMorgan's Theorems

16. Apply DeMorgan's theorems to each expression:
    (a) $\overline{A + \overline{B}}$
    (b) $\overline{\overline{A}B}$
    (c) $\overline{A + B + C}$
    (d) $\overline{ABC}$
    (e) $\overline{A(B + C)}$
    (f) $\overline{AB + CD}$
    (g) $\overline{AB + CD}$
    (h) $\overline{(A + \overline{B})(\overline{C} + D)}$

17. Apply DeMorgan's theorems to each expression:
    (a) $\overline{A\overline{B}(C + \overline{D})}$
    (b) $\overline{AB(CD + EF)}$
    (c) $\overline{(A + \overline{B} + C + \overline{D}) + ABC\overline{D}}$
    (d) $\overline{(\overline{A} + B + C + D)(A\overline{B}\,\overline{C}D)}$
    (e) $\overline{AB(CD + \overline{E}F)(\overline{AB} + \overline{CD})}$

TABLE 4–12

A	B	C	X		A	B	C	X
0	0	0	0		0	0	0	0
0	0	1	1		0	0	1	0
0	1	0	0		0	1	0	0
0	1	1	0		0	1	1	0
1	0	0	1		1	0	0	0
1	0	1	1		1	0	1	1
1	1	0	0		1	1	0	1
1	1	1	1		1	1	1	1

(a) (b)

A	B	C	D	X		A	B	C	D	X
0	0	0	0	1		0	0	0	0	0
0	0	0	1	1		0	0	0	1	0
0	0	1	0	0		0	0	1	0	1
0	0	1	1	1		0	0	1	1	0
0	1	0	0	0		0	1	0	0	1
0	1	0	1	1		0	1	0	1	1
0	1	1	0	1		0	1	1	0	0
0	1	1	1	0		0	1	1	1	1
1	0	0	0	0		1	0	0	0	0
1	0	0	1	1		1	0	0	1	0
1	0	1	0	0		1	0	1	0	0
1	0	1	1	0		1	0	1	1	1
1	1	0	0	1		1	1	0	0	1
1	1	0	1	0		1	1	0	1	0
1	1	1	0	0		1	1	1	0	0
1	1	1	1	0		1	1	1	1	1

(c) (d)

18. Apply DeMorgan's theorems to the following:

(a) $\overline{\overline{(\overline{ABC})(\overline{EFG})} + \overline{(\overline{HIJ})(\overline{KLM})}}$

(b) $\overline{(A + \overline{\overline{BC}} + CD) + \overline{\overline{BC}}}$

(c) $\overline{(\overline{A} + B)(\overline{C} + D)(E + \overline{F})(\overline{G} + H)}$

SECTION 4–4 The Universal Property of NAND and NOR Gates

19. Implement the logic circuits in Figure 4–37 using only NAND gates.

20. Implement the logic circuit in Figure 4–40 using only NAND gates.

FIGURE 4–40

21. Repeat Problem 19 using only NOR gates.

22. Repeat Problem 20 using only NOR gates.

SECTION 4–5 Pulse Waveform Operation

23. Given the logic circuit and the input waveforms in Figure 4–41, draw the output waveform.

FIGURE 4–41

24. For the logic circuit in Figure 4–42, draw the output waveform in proper relationship to the inputs.

FIGURE 4–42

25. For the input waveforms in Figure 4–43, what logic circuit will generate the output waveform shown?

FIGURE 4–43

26. Repeat Problem 25 for the waveforms in Figure 4–44.

FIGURE 4–44

27. For the circuit in Figure 4–45, draw the waveforms at the numbered points in the proper relationship to each other.

FIGURE 4–45

28. Assuming a propagation delay through each gate of 10 nanoseconds (ns), determine if the *desired* output waveform X in Figure 4–46 (a pulse with a minimum $t_{PW} = 25$ ns positioned as shown) will be generated properly with the given inputs.

FIGURE 4–46

SECTION 4–6   **Combinational Logic with VHDL and Verilog**

29. Describe a 3-input AND gate with VHDL and Verilog.

30. Write a VHDL program using the data flow approach (Boolean expressions) to describe the logic circuit in Figure 4–37(b). Write a Verilog program.

31. Write VHDL programs using the data flow approach (Boolean expressions) for the logic circuits in Figure 4–38(e) and (f). Write a Verilog program for both circuits.

32. Write a VHDL program using the structural approach for the logic circuit in Figure 4–39(d). Assume component declarations for each type of gate are already available. Write a Verilog program.

33. Repeat Problem 32 for the logic circuit in Figure 4–39(f).

34. Describe the logic represented by the truth table in Table 4–10 using VHDL and Verilog.

35. Develop a VHDL program for the logic in Figure 4–47(a), using both the data flow and the structural approach. Compare the resulting programs. Write the program in Verilog.

36. Develop a VHDL program for the logic in Figure 4–47(b), using both the data flow and the structural approach. Compare the resulting programs. Write the program in Verilog.

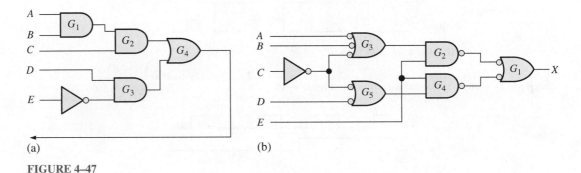

(a)   (b)

FIGURE 4–47

37. Given the following VHDL program, create the truth table that describes the logic circuit.

```
entity CombLogic is
    port (A, B, C, D: in bit; X: out bit);
end entity CombLogic;

architecture Example of CombLogic is
    begin
        X <= not((not A and not B) or (not A and not C) or (not A and not D) or
            (not B and not C) or (not B and not D) or (not D and not C));
end architecture Example;
```

38. Describe the logic circuit shown in Figure 4–48 with a VHDL program, using the data flow approach.

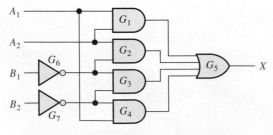

FIGURE 4–48

39. Repeat Problem 38 using the VHDL structural approach.

SECTION 4–7   **A System**

40. Implement the inlet valve logic using NOR gates and inverters.

41. Repeat Problem 40 for the outlet valve logic.

42. Implement the temperature control logic using XNOR gates.

43. Show a circuit to enable an additive to be introduced into the syrup through another inlet only when the temperature is at the specified value and the syrup is at the low-level sensor.

SECTION 4–8   Troubleshooting

44. For the logic circuit and the input waveforms in Figure 4–49, the indicated output waveform is observed. Determine if this is the correct output waveform.

FIGURE 4–49

45. The output waveform in Figure 4–50 is incorrect for the inputs that are applied to the circuit. Assuming that one gate in the circuit has failed, with its output either an apparent constant HIGH or a constant LOW, determine the faulty gate and the type of failure (output open or shorted).

FIGURE 4–50

46. Repeat Problem 45 for the circuit in Figure 4–51, with input and output waveforms as shown.

FIGURE 4–51

47. Figure 4–52(a) (p. 218) is a logic circuit under test. Figure 4–52(b) shows the waveforms as observed on a logic analyzer. The output waveform is incorrect for the inputs that are applied to the circuit. Assuming that one gate in the circuit has failed, with its output either an apparent constant HIGH or a constant LOW, determine the faulty gate and the type of failure.

(a)

(b)

FIGURE 4–52

48. The logic circuit in Figure 4–53 has the input waveforms shown.
   (a) Determine the correct output waveform in relation to the inputs.
   (b) Determine the output waveform if the output of gate G_3 is open.
   (c) Determine the output waveform if the upper input to gate G_5 is shorted to ground.

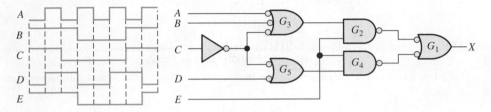

FIGURE 4–53

49. The logic circuit in Figure 4–54 has only one intermediate test point available besides the output, as indicated. For the inputs shown, you observe the indicated waveform at the test point. Is this waveform correct? If not, what are the possible faults that would cause it to appear as it does?

FIGURE 4–54

Special Problems

50. Develop a logic circuit to produce a HIGH output only if the input, represented by a 4-bit binary number, is greater than twelve or less than three. First develop the truth table and then draw the logic diagram.

51. Develop the logic circuit necessary to meet the following requirements:

A battery-powered lamp in a room is to be operated from two switches, one at the back door and one at the front door. The lamp is to be on if the front switch is on and the back switch is off, or if the front switch is off and the back switch is on. The lamp is to be off if both switches are off or if both switches are on. Let a HIGH output represent the on condition and a LOW output represent the off condition.

52. Develop the NAND logic for a hexadecimal keypad encoder that will convert each key closure to binary.

MULTISIM TROUBLESHOOTING PRACTICE

53. Open file P04-53 and follow the instructions given there.

54. Open file P04-54 and follow the instructions given there.

55. Open file P04-55 and follow the instructions given there.

56. Open file P04-56 and follow the instructions given there.

ANSWERS TO SECTION CHECKUPS

SECTION 4–1 Basic Combinational Logic Circuits

1. (a) $\overline{AB + CD} = \overline{1 \cdot 0 + 1 \cdot 0} = 1$
   (b) $\overline{AB + CD} = \overline{1 \cdot 1 + 0 \cdot 1} = 0$
   (c) $\overline{AB + CD} = \overline{0 \cdot 1 + 1 \cdot 1} = 0$

2. (a) $A\overline{B} + \overline{A}B = 1 \cdot \overline{0} + \overline{1} \cdot 0 = 1$ (b) $A\overline{B} + \overline{A}B = 1 \cdot \overline{1} + \overline{1} \cdot 1 = 0$
   (c) $A\overline{B} + \overline{A}B = 0 \cdot \overline{1} + \overline{0} \cdot 1 = 1$ (d) $A\overline{B} + \overline{A}B = 0 \cdot \overline{0} + \overline{0} \cdot 0 = 0$

3. $X = 1$ when $ABC = 000, 011, 101, 110,$ and 111; $X = 0$ when $ABC = 001, 010,$ and 100

4. $X = AB + \overline{A}\,\overline{B}$; the circuit consists of two AND gates, one OR gate, and two inverters. See Figure 4–5(b) for diagram.

SECTION 4–2 Boolean Expressions and Truth Tables

1. (a) $X = ABC + AB + AC$: three AND gates, one OR gate
   (b) $X = AB(C + DE)$: three AND gates, one OR gate

2. $X = ABC + \overline{A}\,\overline{B}\,\overline{C}$; two AND gates, one OR gate, and three inverters

3. $A + B(C + D)$

4.

A	B	C	X
0	0	0	1
0	0	1	0
0	1	0	0
0	1	1	0
1	0	0	0
1	0	1	0
1	1	0	0
1	1	1	1

SECTION 4–3 DeMorgan's Theorems

1. (a) $\overline{A} + \overline{B} + \overline{C} + D\overline{E}$
   (b) $\overline{A}\,\overline{B}\,\overline{C}$
   (c) $\overline{A}\,\overline{B}\,\overline{C} + (D + \overline{E})$

SECTION 4–4 The Universal Property of NAND and NOR Gates

1. (a) $X = \overline{A} + B$: a 2-input NAND gate with A and $\overline{B}$ on its inputs.
   (b) $X = A\overline{B}$: a 2-input NAND with A and $\overline{B}$ on its inputs, followed by one NAND used as an inverter.

2. (a) $X = \overline{A} + B$: a 2-input NOR with inputs $\overline{A}$ and B, followed by one NOR used as an inverter.
   (b) $X = A\overline{B}$: a 2-input NOR with $\overline{A}$ and B on its inputs.

SECTION 4–5   Pulse Waveform Operation

1. The exclusive-OR output is a 15 μs pulse followed by a 25 μs pulse, with a separation of 10 μs between the pulses.

2. The output of the exclusive-NOR is HIGH when both inputs are HIGH or when both inputs are LOW.

SECTION 4–6   Combinational Logic with VHDL and Verilog

1. An entity defines inputs and outputs.

2. An architecture defines type of circuit.

3. Module

4. Data flow and structural (behavioral is a third type)

5. A VHDL component is a predefined program describing a specified logic function.

6. A component instantiation is used to call for a specified component in a program architecture.

7. Interconnections between components are made using VHDL signals.

8. Components are used in the structural approach.

SECTION 4–7   A System

1. Three

2. Two

3. The outlet valve control depends on the temperature where the inlet valve does not.

4. The outlet valve turns off when there is no inlet flow and the level is above minimum.

SECTION 4–8   Troubleshooting

1. Common gate failures are input or output open; input or output shorted to ground.

2. Input shorted to V_{CC} causes output to be stuck LOW.

3. (a) G_4 output is HIGH until rising edge of seventh pulse, then it goes LOW.
   (b) G_1 output is the same as input D.
   (c) G_4 output is the inverse of the G_2 output shown in Figure 4–36(b).

ANSWERS TO RELATED PROBLEMS FOR EXAMPLES

4–1   The same as Figure 4–10 without the inverters

4–2   There is only one complemented variable for each AND gate.

4–3   Multisim results should agree with Figure 4–13.

4–4 $\overline{\overline{X} + \overline{Y} + \overline{Z}} = XYZ$

4–5 $\overline{\overline{W}\,\overline{X}\,\overline{Y}\,\overline{Z}} = W + X + Y + Z$

4–6   See Figure 4–55.

4–7   See Figure 4–56.

FIGURE 4–55   FIGURE 4–56

4–8   See Figure 4–57.

4–9   See Figure 4–58.

FIGURE 4–57

FIGURE 4–58

4–10

VHDL	Verilog
entity ANDORlogic is	module ANDORlogic (A, B, C, D, E, F, G, H, X);
port (A, B, C, D, E, F, G, H: in bit; X: out bit);	input A, B, C, D, E, F, G, H;
end entity ANDORlogic;	output X;
architecture ANDORfunction of ANDORlogic is	assign X = (A && B) || (C && D) || (E && F) ||
begin	(G && H);
X <= (A and B) or (C and D) or (E and F) or	endmodule
(G and H);	
end architecture ANDORfunction;	

4–11

VHDL	Verilog
entity Combo_Logic is	module Combo_Logic (A, B, C, D, E, F, X);
port (A, B, C, D, E, F: in bit; X: out bit);	input A, B, C, D, E, F;
end entity Combo_Logic;	output X;
architecture Logic_Function of Combo_Logic is	assign X = !(!(!A && !B) || C) && !(!(!D && !E) ||
begin	F);
X <= not(not(not A and not B) or C) and	endmodule
not(not(not D and not E) or F);	
end architecture Logic_Function;	

4–12

VHDL	Verilog
entity Table_4_7 is	module Combo_Logic (A, B, C, D, X);
port (A, B, C, D: in bit; X: out bit);	input A, B, C, D;
end entity Table_4_7;	output X;
architecture Logic_Function of Table_4_7 is	assign X = (!A && !B && !C && !D) || (A && !B
begin	&& C && D) || (A && B && !C && !D) || (A
X <= (not A and not B and not C and not D)	&& B && !C && D) || (A && B && C &&
or (A and not B and C and D) or (A and B	!D) || (A && B && C && D);
and not C and not D) or (A and B and not C	endmodule
and D) or (A and B and C and not D) or (A	
and B and C and D);	
end architecture Logic_Function;	

4–13   G5: NAND_gate2 **port map** (*A* => IN9, B => IN10, X => OUT5);

4–14   See Figure 4–59.

FIGURE 4–59

ANSWERS TO TRUE/FALSE QUIZ

1. F **2.** F **3.** T **4.** T **5.** F
6. T **7.** T **8.** F **9.** T **10.** T

ANSWERS TO SELF-TEST

1. (d) **2.** (b) **3.** (c) **4.** (d) **5.** (b)
6. (a) **7.** (d) **8.** (e) **9.** (e) **10.** (c)

CHAPTER 5

FUNCTIONS OF COMBINATIONAL LOGIC

OUTLINE

OBJECTIVES

- Explain a tablet-bottling control system
- Distinguish between half-adders and full-adders
- Use full-adders to implement multibit parallel binary adders
- Explain the differences between ripple carry and look-ahead carry parallel adders
- Use the magnitude comparator to determine the relationship between two binary numbers and use cascaded comparators to handle the comparison of larger numbers
- Implement a basic binary decoder
- Use BCD-to-7-segment decoders in display systems

- Apply a decimal-to-BCD priority encoder in a simple keyboard application
- Convert from binary to Gray code, and Gray code to binary by using logic devices
- Apply data selectors/multiplexers in multiplexed displays and as a function generator
- Use decoders as demultiplexers
- Explain the meaning of parity
- Use parity generators and checkers to detect bit errors in digital systems
- Implement a simple data communications system
- Write VHDL and Verilog programs for various combination logic functions
- Identify glitches, common bugs in digital systems

KEY TERMS

Half-adder	Priority encoder
Full-adder	Multiplexer (MUX)
Cascading	Demultiplexer
Ripple carry	(DEMUX)
Look-ahead carry	Parity bit
Decoder	Glitch
Encoder	

VISIT THE WEBSITE
Study aids for this chapter are available at
http://pearsonhighered.com/floyd

INTRODUCTION

In this chapter, several types of combinational logic circuits are introduced including adders, comparators, decoders, encoders, code converters, multiplexers (data selectors), demultiplexers, and parity generators/checkers. The tablet-bottling system that was introduced in Chapter 1 is discussed in relation to the application of the logic functions covered. Programmable logic implementation of several logic functions using VHDL and Verilog is covered. Examples of fixed-function IC devices are included at the end of the chapter.

5–1 A SYSTEM

A tablet-bottling control system was introduced in Chapter 1 and will be examined in more detail in this section. This particular system is somewhat simplified to illustrate how all the logic functions that will be covered in this chapter can be applied, although the counter and register are not covered in detail until later chapters. As with most systems, there are other ways in which a system's function can be implemented. By studying this system and how it works, you will learn how the various logic functions can be interconnected and operate as a complete system to accomplish a given task.

After completing this section, you should be able to

- **Explain the overall operation of the tablet-bottling control system**
- **Describe the purpose of each logic function**

Figure 5–1 is a general block diagram of the tablet-bottling control system. Recall from Chapter 1 that the job of this system is simply to fill each bottle with a preset number of tablets, which is accomplished by counting the number of tablets going into a bottle and limiting the quantity to the preset number. The system is set for 50 tablets per bottle but with eight bits can accommodate up to 255 tablets per bottle. The system can store a binary number representing up to 255 total tablets before sending the accumulated sum to a computer where the total will be tallied over a specified time period. This limitation is because register B can store eight bits.

KEYPAD, ENCODER, AND REGISTER A   The keypad has ten outputs, one for each of the decimal digits. When a key is pressed, the encoder changes the decimal input to the corresponding 4-bit BCD code, as shown in Figure 5–2. Register A stores 12 bits, which is three BCD digits. Because there are to be 50 tablets per bottle, a 0 is first entered on the keypad; and the BCD code for 0, which is 0000, is stored in the shift register. Next a 5 is entered, and the BCD code 0101 is stored in the register, as indicated. Finally, another 0 is entered.

DECODER A AND CODE CONVERTER A   Decoder A transforms the BCD input from register A into a 3-digit 7-segment code for the "Tablets/bottle" display. Actually, decoder A consists of three BCD-to-7-segment decoders because three digits must be displayed at the same time (remember the number can be up to 255). Code converter A converts the BCD code to a 7-bit binary number for comparison with the actual binary tablet count. Figure 5–3 illustrates these functions.

COMPARATOR, COUNTER, ADDER, AND REGISTER B   The set number of tablets per bottle (50) is applied to the comparator, which compares it to the actual number of tablets in a bottle at any point in time. The counter counts each tablet as it falls into a bottle, and the number in the counter is continuously compared to the set number. As long as the number of tablets counted is less than the set number, the valve remains open. When the count reaches the set number, the output of the comparator produces an output signal (HIGH) that closes the valve and advances the conveyer to the next bottle. The counter is reset to zero, the valve is opened when the comparator output goes back to zero, and the filling of the new bottle begins.

FIGURE 5–1 Tablet-bottling control system. Each segment is referenced to a following figure for more detail.

FIGURE 5–2

FIGURE 5–3

The number of tablets in the bottle is added to the previous total number of tablets stored in register B, and the new sum is stored back into register B. When the sum that is stored in the register reaches the limit of 250 (five bottles), as determined by decoder B, the parallel output of the register is multiplexed to the remote site where the sum is accumulated and stored in a computer. Figure 5–4 shows this operation at the point where the bottle is full with 50 tablets and the previous sum in register B is 100. The new sum of 150 replaces the previous sum of 100.

DECODER B, MUX, AND DEMUX   Decoder B decodes the contents of register B to detect the binary code for 250 and enable the multiplexer. The multiplexer acts as a parallel-to-serial converter. The multiplexing sequence (source not shown) is initiated, and the eight data bits are sent serially to the demultiplexer at a remote location. The demultiplexer converts the serial data back to parallel format and sends it to a computer that accumulates the totals over an extended period of time, as illustrated in Figure 5–5.

FIGURE 5–4

FIGURE 5–5

SECTION 5–1 CHECKUP*

1. After four bottles (50 tablets/bottle) have been filled, what are the binary numbers in Register A and Register B?

2. If the number of tablets per bottle is reset to 25, what are the binary numbers in Registers A and B after five bottles have been filled?

3. Explain the purpose of the comparator.

4. Explain the purpose of Decoder B.

*Answers are at the end of the chapter.

5–2 HALF AND FULL ADDERS

Adders are important in computers and also in other types of digital systems in which numerical data are processed, as you have seen. An understanding of the basic adder operation is fundamental to the study of digital systems. In this section, the half-adder and the full-adder are introduced.

After completing this section, you should be able to

- Describe the function of a half-adder
- Draw a half-adder logic diagram
- Describe the function of the full-adder
- Draw a full-adder logic diagram using half-adders
- Implement a full-adder using AND-OR logic

The Half-Adder

A half-adder adds two bits and produces a sum and an output carry.

Recall the basic rules for binary addition as stated in Chapter 2.

$$
\begin{aligned}
0 + 0 &= 0 \\
0 + 1 &= 1 \\
1 + 0 &= 1 \\
1 + 1 &= 10
\end{aligned}
$$

The operations are performed by a logic circuit called a **half-adder**.

The half-adder accepts two binary digits on its inputs and produces two binary digits on its outputs—a sum bit and a carry bit.

A half-adder is represented by the logic symbol in Figure 5–6.

MULTISIM

FIGURE 5–6   Logic symbol for a half-adder. Open file F05-06 to verify operation.

HALF-ADDER LOGIC   From the operation of the half-adder as stated in Table 5–1, expressions can be derived for the sum and the output carry as functions of the inputs.

A	B	C_{out}	Σ
0	0	0	0
0	1	0	1
1	0	0	1
1	1	1	0

TABLE 5–1 • Half-adder truth table.

Σ = sum
C_{out} = output carry
A and B = input variables (operands)

Notice that the output carry (C_{out}) is a 1 only when both A and B are 1s; therefore, C_{out} can be expressed as the AND of the input variables.

$$C_{out} = AB \qquad (5\text{–}1)$$

Now observe that the sum output (Σ) is a 1 only if the input variables, A and B, are not equal. The sum can therefore be expressed as the exclusive-OR of the input variables.

$$\Sigma = A \oplus B \qquad (5\text{–}2)$$

From Equations 5–1 and 5–2, the logic implementation required for the half-adder function can be developed. The output carry is produced with an AND gate with A and B on the inputs, and the sum output is generated with an exclusive-OR gate, as shown in Figure 5–7. Remember that the exclusive-OR can be implemented with AND gates, an OR gate, and inverters.

FIGURE 5–7 Half-adder logic diagram.

The Full-Adder

The second category of adder is the **full-adder**.

> **The full-adder accepts two input bits and an input carry and generates a sum output and an output carry.**

The basic difference between a full-adder and a half-adder is that the full-adder accepts an input carry. A logic symbol for a full-adder is shown in Figure 5–8, and the truth table in Table 5–2 shows the operation of a full-adder.

A full-adder has an input carry while the half-adder does not.

FIGURE 5–8 Logic symbol for a full-adder. Open file F05-08 to verify operation.

MULTISIM

FULL-ADDER LOGIC   The full-adder must add the two input bits and the input carry. From the half-adder you know that the sum of the input bits A and B is the exclusive-OR

TABLE 5–2 • Full-adder truth table.				
A	*B*	C_{in}	C_{out}	Σ
0	0	0	0	0
0	0	1	0	1
0	1	0	0	1
0	1	1	1	0
1	0	0	0	1
1	0	1	1	0
1	1	0	1	0
1	1	1	1	1

C_{in} = input carry, sometimes designated as *CI*
C_{out} = output carry, sometimes designated as *CO*
Σ = sum
A and *B* = input variables (operands)

of those two variables, $A \oplus B$. For the input carry (C_{in}) to be added to the input bits, it must be exclusive-ORed with $A \oplus B$, yielding the equation for the sum output of the full-adder.

$$\Sigma = (A \oplus B) \oplus C_{in} \qquad (5\text{--}3)$$

This means that to implement the full-adder sum function, two 2-input exclusive-OR gates can be used. The first must generate the term $A \oplus B$, and the second has as its inputs the output of the first XOR gate and the input carry, as illustrated in Figure 5–9(a).

(a) Logic required to form the sum of three bits

(b) Complete logic circuit for a full-adder (each half-adder is enclosed by a shaded area)

 MULTISIM   FIGURE 5–9   **Full-adder logic. Open file F05-09 to verify operation.**

The output carry is a 1 when both inputs to the first XOR gate are 1s or when both inputs to the second XOR gate are 1s. You can verify this fact by studying Table 5–2. The output carry of the full-adder is therefore produced by input *A* ANDed with input *B* and $A \oplus B$ ANDed with C_{in}. These two terms are ORed, as expressed in Equation 5–4. This function is implemented and combined with the sum logic to form a complete full-adder circuit, as shown in Figure 5–9(b).

$$C_{out} = AB + (A \oplus B)C_{in} \qquad (5\text{--}4)$$

Notice in Figure 5–9(b) there are two half-adders, connected as shown in the block diagram of Figure 5–10(a), with their output carries ORed. The logic symbol shown in Figure 5–10(b) will normally be used to represent the full-adder.

(a) Arrangement of two half-adders to form a full-adder

(b) Full-adder logic symbol

FIGURE 5–10   **Full-adder implemented with half-adders.**

EXAMPLE 5–1

For each of the three full-adders in Figure 5–11, determine the outputs for the inputs shown.

(a)   (b)   (c)

FIGURE 5–11

SOLUTION

(a) The input bits are $A = 1$, $B = 0$, and $C_{in} = 0$.

$$1 + 0 + 0 = 1 \text{ with no carry}$$

Therefore, $\Sigma = \mathbf{1}$ and $C_{out} = \mathbf{0}$.

(b) The input bits are $A = 1$, $B = 1$, and $C_{in} = 0$.

$$1 + 1 + 0 = 0 \text{ with a carry of } 1$$

Therefore, $\Sigma = \mathbf{0}$ and $C_{out} = \mathbf{1}$.

(c) The input bits are $A = 1$, $B = 0$, and $C_{in} = 1$.

$$1 + 0 + 1 = 0 \text{ with a carry of } 1$$

Therefore, $\Sigma = \mathbf{0}$ and $C_{out} = \mathbf{1}$.

RELATED PROBLEM*

What are the full-adder outputs for $A = 1$, $B = 1$, and $C_{in} = 1$?

*Answers are at the end of the chapter.

SECTION 5–2 CHECKUP

1. Determine the sum (Σ) and the output carry (C_{out}) of a half-adder for each set of input bits:

(a) 01   **(b)** 00   **(c)** 10   **(d)** 11

2. A full-adder has $C_{in} = 1$. What are the sum (Σ) and the output carry (C_{out}) when $A = 1$ and $B = 1$?

5–3 PARALLEL ADDERS

Two or more full-adders are connected to form parallel binary adders. In this section, you will learn the basic operation of this type of adder and its associated input and output functions.

After completing this section, you should be able to

- **Use full-adders to implement a parallel binary adder**
- **Explain the addition process in a parallel binary adder**
- **Use the truth table for a 4-bit parallel adder**
- **Expand the 4-bit adder to accommodate 8-bit or 16-bit addition**

As you saw in Section 5–2, a single full-adder is capable of adding two 1-bit numbers and an input carry. To add binary numbers with more than one bit, you must use additional full-adders. When one binary number is added to another, each column generates a sum bit and a 1 or 0 carry bit to the next column to the left, as illustrated here with 2-bit numbers.

Carry bit from right column

$$\overset{\overset{1}{\downarrow}}{11}$$
$$+\ 01$$
$$\overline{100}$$

In this case, the carry bit from second column becomes a sum bit.

Addition is performed by computers on two numbers at a time, called *operands*. The *source operand* is a number that is to be added to an existing number called the *destination operand*, which is held in an ALU register, such as the accumulator. The sum of the two numbers is then stored back in the accumulator. Addition is performed on integer numbers or floating-point numbers using ADD or FADD instructions respectively.

SYSTEM NOTE

To add two binary numbers, a full-adder is required for each bit in the numbers. So for 2-bit numbers, two adders are needed; for 4-bit numbers, four adders are used; and so on. The carry output of each adder is connected to the carry input of the next higher-order adder, as shown in Figure 5–12 for a 2-bit adder. Notice that either a half-adder can be used for the least significant position or the carry input of a full-adder can be made 0 (grounded) because there is no carry input to the least significant bit position.

General format, addition of two 2-bit numbers:

$$A_2A_1$$
$$+\ B_2B_1$$
$$\overline{\Sigma_3\Sigma_2\Sigma_1}$$

FIGURE 5–12   **Block diagram of a 2-bit parallel adder using two full-adders. Open file F05-12 to verify operation.**

In Figure 5–12 the least significant bits (LSB) of the two numbers are represented by A_1 and B_1. The next higher-order bits are represented by A_2 and B_2. The three sum bits are $\Sigma_1, \Sigma_2,$ and Σ_3. Notice that the output carry from the left-most full-adder becomes the most significant bit (MSB) in the sum, Σ_3.

EXAMPLE 5–2

Determine the sum generated by the 3-bit parallel adder in Figure 5–13 and show the intermediate carries when the binary numbers 101 and 011 are being added.

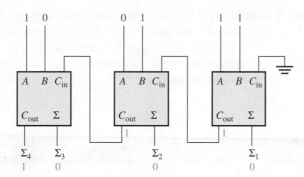

FIGURE 5–13

SOLUTION

The LSBs of the two numbers are added in the right-most full-adder. The sum bits and the intermediate carries are indicated in blue in Figure 5–13.

RELATED PROBLEM

What are the sum outputs when 111 and 101 are added by the 3-bit parallel adder?

Four-Bit Parallel Adders

A group of four bits is called a **nibble.** A basic 4-bit parallel adder is implemented with four full-adder stages as shown in Figure 5–14. Again, the LSBs (A_1 and B_1) in each number being added go into the right-most full-adder; the higher-order bits are applied as shown to the successively higher-order adders, with the MSBs (A_4 and B_4) in each number being applied to the left-most full-adder. The carry output of each adder is connected to the carry input of the next higher-order adder as indicated. These are called *internal carries.*

(a) Block diagram

(b) Logic symbol

FIGURE 5–14   A 4-bit parallel adder.

In keeping with most manufacturers' data sheets, the input labeled C_0 is the input carry to the least significant bit adder; C_4, in the case of four bits, is the output carry of the most significant bit adder; and Σ_1 (LSB) through Σ_4 (MSB) are the sum outputs. The logic symbol is shown in Figure 5–14(b).

In terms of the method used to handle carries in a parallel adder, there are two types: the *ripple carry* adder and the *carry look-ahead* adder. These are discussed in Section 5–4.

Truth Table for a 4-Bit Parallel Adder

Table 5–3 is the truth table for a 4-bit adder. On some data sheets, truth tables may be called *function tables* or *functional truth tables*. The subscript n represents the adder bits and can be 1, 2, 3, or 4 for the 4-bit adder. C_{n-1} is the carry from the previous adder. Carries C_1, C_2, and C_3 are generated internally. C_0 is an external carry input and C_4 is an output. Example 5–3 illustrates how to use Table 5–3.

TABLE 5–3 • **Truth table for each stage of a 4-bit parallel adder.**				
C_{n-1}	A_n	B_n	Σ_n	C_n
0	0	0	0	0
0	0	1	1	0
0	1	0	1	0
0	1	1	0	1
1	0	0	1	0
1	0	1	0	1
1	1	0	0	1
1	1	1	1	1

EXAMPLE 5–3

Use the 4-bit parallel adder truth table (Table 5–3) to find the sum and output carry for the addition of the following two 4-bit numbers if the input carry (C_{n-1}) is 0:

$$A_4A_3A_2A_1 = 1100 \quad \text{and} \quad B_4B_3B_2B_1 = 1100$$

SOLUTION

For $n = 1$: $A_1 = 0$, $B_1 = 0$, and $C_{n-1} = 0$. From the 1st row of the table,

$$\Sigma_1 = \mathbf{0} \quad \text{and} \quad C_1 = 0$$

For $n = 2$: $A_2 = 0$, $B_2 = 0$, and $C_{n-1} = 0$. From the 1st row of the table,

$$\Sigma_2 = \mathbf{0} \quad \text{and} \quad C_2 = 0$$

For $n = 3$: $A_3 = 1$, $B_3 = 1$, and $C_{n-1} = 0$. From the 4th row of the table,

$$\Sigma_3 = \mathbf{0} \quad \text{and} \quad C_3 = 1$$

For $n = 4$: $A_4 = 1$, $B_4 = 1$, and $C_{n-1} = 1$. From the last row of the table,

$$\Sigma_4 = 1 \quad \text{and} \quad C_4 = 1$$

C_4 becomes the output carry; the sum of 1100 and 1100 is 11000.

RELATED PROBLEM

Use the truth table (Table 5–3) to find the result of adding the binary numbers 1011 and 1010.

Adder Expansion

The 4-bit parallel adder can be expanded to handle the addition of two 8-bit numbers by using two 4-bit adders. The carry input of the low-order adder (C_0) is connected to ground because there is no carry into the least significant bit position, and the carry output of the low-order adder is connected to the carry input of the high-order adder, as shown in Figure 5–15(a). This process is known as **cascading**. Notice that, in this case, the output carry is designated C_8 because it is generated from the eighth bit position. The low-order adder is the one that adds the lower or less significant four bits in the numbers, and the high-order adder is the one that adds the higher or more significant four bits in the 8-bit numbers.

Similarly, four 4-bit adders can be cascaded to handle two 16-bit numbers as shown in Figure 5–15(b). Notice that the output carry is designated C_{16} because it is generated from the sixteenth bit position.

Adders can be expanded to handle more bits by cascading.

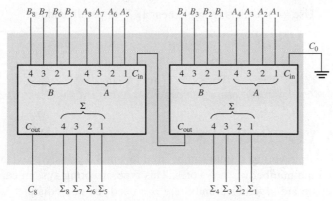

(a) Cascading of two 4-bit adders to form an 8-bit adder

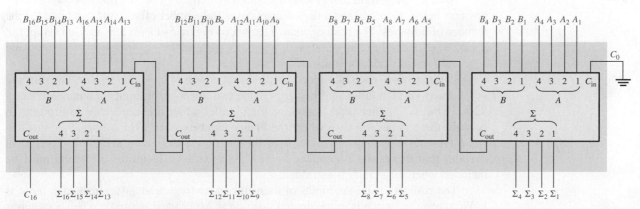

(b) Cascading of four 4-bit adders to form a 16-bit adder

FIGURE 5–15   **Examples of adder expansion.**

EXAMPLE 5–4

Show the input and output states (1s and 0s) for the 8-bit adder in Figure 5–15(a) for the following 8-bit input numbers:

$$A_8 A_7 A_6 A_5 A_4 A_3 A_2 A_1 = 10111001 \quad \text{and} \quad B_8 B_7 B_6 B_5 B_4 B_3 B_2 B_1 = 10011110$$

SOLUTION

The sum of the two 8-bit numbers is

$$\Sigma_9 \Sigma_8 \Sigma_7 \Sigma_6 \Sigma_5 \Sigma_4 \Sigma_3 \Sigma_2 \Sigma_1 = 101010111$$

The input and output states are shown in Figure 5–16.

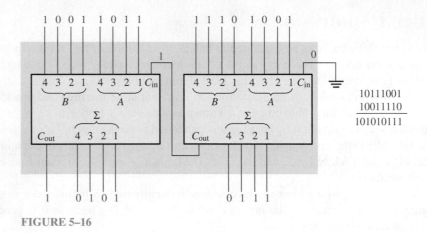

$$\begin{array}{r} 10111001 \\ 10011110 \\ \hline 101010111 \end{array}$$

FIGURE 5–16

RELATED PROBLEM

Use 4-bit adders to implement a 12-bit parallel adder.

SYSTEM EXAMPLE 5–1

A VOTING SYSTEM

A simple voting system can be used to simultaneously provide the number of "yes" votes and the number of "no" votes. This type of voting system can be used where a group of people are assembled and there is a need for immediately determining opinions (for or against), making decisions, or voting on certain issues or other matters.

In its simplest form, the system includes a switch for "yes" or "no" selection at each position in the assembly and a digital display for the number of yes votes and one for the number of no votes. The basic system is shown in Figure 5–17 for a 6-position setup, but it can be expanded to any number of positions with additional 6-position modules and additional parallel adder and display circuits.

In Figure 5–17 each full-adder can produce the sum of up to three votes. The sum and output carry of each full-adder then goes to the two lower-order inputs of a parallel binary adder. The two higher-order inputs of the parallel adder are connected to ground (0) because there is never a case where the binary input exceeds 0011 (decimal 3). For this basic 6-position system, the outputs of the parallel adder go to a BCD-to-7-segment decoder that drives the 7-segment display. As mentioned, additional circuits must be included when the system is expanded.

The resistors from the inputs of each full-adder to ground assure that each input is LOW when the switch is in the neutral position (CMOS logic is used). When a switch is moved to the "yes" or to the "no" position, a HIGH level (V_{CC}) is applied to the associated full-adder input.

FIGURE 5–17

SECTION 5–3 CHECKUP

1. Two 4-bit numbers (1101 and 1011) are applied to a 4-bit parallel adder. The input carry is 1. Determine the sum (Σ) and the output carry.

2. How many 4-bit adders would be required to add two binary numbers each representing decimal numbers up through 1000_{10}?

5–4 RIPPLE CARRY AND LOOK-AHEAD CARRY ADDERS

As mentioned in the last section, parallel adders can be placed into two categories based on the way in which internal carries from stage to stage are handled. Those categories are ripple carry and look-ahead carry. Externally, both types of adders are the same in terms of inputs and outputs. The difference is the speed at which they can add numbers. The look-ahead carry adder is much faster than the ripple carry adder.

After completing this section, you should be able to

- Discuss the difference between a ripple carry adder and a look-ahead carry adder
- State the advantage of look-ahead carry addition
- Define *carry generation* and *carry propagation* and explain the difference
- Develop look-ahead carry logic

The Ripple Carry Adder

A **ripple carry** adder is one in which the carry output of each full-adder is connected to the carry input of the next higher-order stage (a stage is one full-adder). The sum and the output carry of any stage cannot be produced until the input carry occurs; this causes a time delay in the addition process, as illustrated in Figure 5–18. The carry propagation delay for each full-adder is the time from the application of the input carry until the output carry occurs, assuming that the A and B inputs are already present.

FIGURE 5–18 **A 4-bit parallel ripple carry adder showing "worst-case" carry propagation delays.**

Full-adder 1 (FA1) cannot produce a potential output carry until an input carry is applied. Full-adder 2 (FA2) cannot produce a potential output carry until FA1 produces an output carry. Full-adder 3 (FA3) cannot produce a potential output carry until an output carry is produced by FA1 followed by an output carry from FA2, and so on. As you can see in Figure 5–18, the input carry to the least significant stage has to ripple through all the adders before a final sum is produced. The cumulative delay through all the adder stages is a "worst-case" addition time. The total delay can vary, depending on the carry bit produced by each full-adder. If two numbers are added such that no carries (0) occur between stages, the addition time is simply the propagation time through a single full-adder from the application of the data bits on the inputs to the occurrence of a sum output; however, worst-case addition time must always be assumed.

The Look-Ahead Carry Adder

The speed with which an addition can be performed is limited by the time required for the carries to propagate, or ripple, through all the stages of a parallel adder. One method of speeding up the addition process by eliminating this ripple carry delay is called **look-ahead carry** addition. The look-ahead carry adder anticipates the output carry of each stage, and based on the inputs, produces the output carry by either carry generation or carry propagation.

Carry generation occurs when an output carry is produced (generated) internally by the full-adder. A carry is generated only when both input bits are 1s. The generated carry, C_g, is expressed as the AND function of the two input bits, A and B.

$$C_g = AB \qquad (5–5)$$

Carry propagation occurs when the input carry is rippled to become the output carry. An input carry may be propagated by the full-adder when either or both of the input bits are 1s. The propagated carry, C_p, is expressed as the OR function of the input bits.

$$C_p = A + B \qquad (5–6)$$

The conditions for carry generation and carry propagation are illustrated in Figure 5–19. The three arrowheads symbolize ripple (propagation).

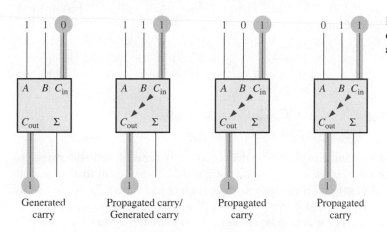

FIGURE 5–19   Illustration of conditions for carry generation and carry propagation.

Generated carry   Propagated carry/ Generated carry   Propagated carry   Propagated carry

The output carry of a full-adder can be expressed in terms of both the generated carry (C_g) and the propagated carry (C_p). The output carry (C_{out}) is a 1 if the generated carry is a 1 OR if the propagated carry is a 1 AND the input carry (C_{in}) is a 1. In other words, we get an output carry of 1 if it is generated by the full-adder ($A = 1$ AND $B = 1$) or if the adder propagates the input carry ($A = 1$ OR $B = 1$) AND $C_{in} = 1$. This relationship is expressed as

$$C_{out} = C_g + C_p C_{in} \qquad (5–7)$$

Now let's see how this concept can be applied to a parallel adder, whose individual stages are shown in Figure 5–20 for a 4-bit example. For each full-adder, the output carry is

FIGURE 5–20   Carry generation and carry propagation in terms of the input bits to a 4-bit adder.

Full-adder 4

$C_{g4} = A_4 B_4$
$C_{p4} = A_4 + B_4$

Full-adder 3

$C_{g3} = A_3 B_3$
$C_{p3} = A_3 + B_3$

Full-adder 2

$C_{g2} = A_2 B_2$
$C_{p2} = A_2 + B_2$

Full-adder 1

$C_{g1} = A_1 B_1$
$C_{p1} = A_1 + B_1$

dependent on the generated carry (C_g), the propagated carry (C_p), and its input carry (C_{in}). The C_g and C_p functions for each stage are *immediately* available as soon as the input bits A and B and the input carry to the LSB adder are applied because they are dependent only on these bits. The input carry to each stage is the output carry of the previous stage.

Based on this analysis, we can now develop expressions for the output carry, C_{out}, of each full-adder stage for the 4-bit example.

Full-adder 1:

$$C_{out1} = C_{g1} + C_{p1}C_{in1}$$

Full-adder 2:

$$C_{in2} = C_{out1}$$
$$C_{out2} = C_{g2} + C_{p2}C_{in2} = C_{g2} + C_{p2}C_{out1} = C_{g2} + C_{p2}(C_{g1} + C_{p1}C_{in1})$$
$$= C_{g2} + C_{p2}C_{g1} + C_{p2}C_{p1}C_{in1}$$

Full-adder 3:

$$C_{in3} = C_{out2}$$
$$C_{out3} = C_{g3} + C_{p3}C_{in3} = C_{g3} + C_{p3}C_{out2} = C_{g3} + C_{p3}(C_{g2} + C_{p2}C_{g1} + C_{p2}C_{p1}C_{in1})$$
$$= C_{g3} + C_{p3}C_{g2} + C_{p3}C_{p2}C_{g1} + C_{p3}C_{p2}C_{p1}C_{in1}$$

Full-adder 4:

$$C_{in4} = C_{out3}$$
$$C_{out4} = C_{g4} + C_{p4}C_{in4} = C_{g4} + C_{p4}C_{out3}$$
$$= C_{g4} + C_{p4}(C_{g3} + C_{p3}C_{g2} + C_{p3}C_{p2}C_{g1} + C_{p3}C_{p2}C_{p1}C_{in1})$$
$$= C_{g4} + C_{p4}C_{g3} + C_{p4}C_{p3}C_{g2} + C_{p4}C_{p3}C_{p2}C_{g1} + C_{p4}C_{p3}C_{p2}C_{p1}C_{in1}$$

Notice that in each of these expressions, the output carry for each full-adder stage is dependent only on the initial input carry (C_{in1}), the C_g and C_p functions of that stage, and the C_g and C_p functions of the preceding stages. Since each of the C_g and C_p functions can be expressed in terms of the A and B inputs to the full-adders, all the output carries are immediately available (except for gate delays), and you do not have to wait for a carry to ripple through all the stages before a final result is achieved. Thus, the look-ahead carry technique speeds up the addition process.

The C_{out} equations are implemented with logic gates and connected to the full-adders to create a 4-bit look-ahead carry adder, as shown in Figure 5–21.

FIGURE 5–21 **Logic diagram for a 4-stage look-ahead carry adder.**

SECTION 5–4 CHECKUP

1. The input bits to a full-adder are $A = 1$ and $B = 0$. Determine C_g and C_p.

2. Determine the output carry of a full-adder when $C_{in} = 1$, $C_g = 0$, and $C_p = 1$.

5–5 COMPARATORS

The basic function of a comparator is to compare the magnitudes of two binary quantities to determine the relationship of those quantities. In its simplest form, a comparator circuit determines whether two numbers are equal.

After completing this section, you should be able to

- Use the exclusive-NOR gate as a basic comparator
- Explain the internal logic of a magnitude comparator that has both equality and inequality outputs
- Use a comparator to compare the magnitudes of two 4-bit numbers
- Apply cascading to expand a comparator to eight or more bits

Equality

The exclusive-NOR gate can be used as a basic comparator because its output is a 0 if the two input bits are not equal and a 1 if the input bits are equal. Figure 5–22 shows the exclusive-NOR gate as a 2-bit comparator.

A comparator determines if two binary numbers are equal or unequal.

0 — ⟩⟩o— 1   The input bits are equal.
0

1 — ⟩⟩o— 0   The input bits are not equal.
0

0 — ⟩⟩o— 0   The input bits are not equal.
1

1 — ⟩⟩o— 1   The input bits are equal.
1

FIGURE 5–22   Basic comparator operation.

In order to compare binary numbers containing two bits each, an additional exclusive-NOR gate is necessary. The two least significant bits (LSBs) of the two numbers are compared by gate G_1, and the two most significant bits (MSBs) are compared by gate G_2, as shown in Figure 5–23. If the two numbers are equal, their corresponding bits are the same, and the output of each exclusive-NOR gate is a 1. If the corresponding sets of bits are not equal, a 0 occurs on that exclusive-NOR gate output.

In order to produce a single output indicating an equality or inequality of two numbers, an AND gate can be combined with XNOR gates, as shown in Figure 5–23. The output of each exclusive-NOR gate is applied to the AND gate input. When the two input bits for each exclusive-NOR are equal, the corresponding bits of the numbers are equal, producing a 1 on both inputs to the AND gate and thus a 1 on the output. When the two numbers are not equal, one or both sets of corresponding bits are unequal, and a 0 appears on at least one input to the AND gate to produce a 0 on its output. Thus, the output of the AND gate indicates equality (1) or inequality (0) of the two numbers. Example 5–5 illustrates this operation for two specific cases.

General format: Binary number $A \rightarrow A_1A_0$
Binary number $B \rightarrow B_1B_0$

MULTISIM

FIGURE 5–23   Logic diagram for equality comparison of two 2-bit numbers. Open file F05-23 to verify operation.

EXAMPLE 5–5

Apply each of the following sets of binary numbers to the comparator inputs in Figure 5–23, and determine the output by following the logic levels through the circuit.

(a) 10 and 10   **(b)** 11 and 10

SOLUTION

(a) The output is **1** for inputs 10 and 10, as shown in Figure 5–24(a).

(b) The output is **0** for inputs 11 and 10, as shown in Figure 5–24(b).

(a)

(b)

FIGURE 5–24

RELATED PROBLEM

Repeat the process for binary inputs of 01 and 10.

The basic comparator can be expanded to any number of bits. The AND gate sets the condition that all corresponding bits of the two numbers must be equal if the two numbers themselves are equal.

In a computer, the *cache* is a very fast intermediate memory between the central processing unit (CPU) and the slower main memory. The CPU requests data by sending out its *address* (unique location) in memory. Part of this address is called a *tag*. The *tag address comparator* compares the tag from the CPU with the tag from the cache directory. If the two agree, the addressed data is already in the cache and is retrieved very quickly. If the tags disagree, the data must be retrieved from the main memory at a much slower rate.

SYSTEM NOTE

Inequality

FIGURE 5–25 Logic symbol for a 4-bit comparator with inequality indication.

In addition to the equality output, many IC comparators provide additional outputs that indicate which of the two binary numbers being compared is the larger. That is, there is an output that indicates when number A is greater than number B ($A > B$) and an output that indicates when number A is less than number B ($A < B$), as shown in the logic symbol for a 4-bit comparator in Figure 5–25.

To determine an inequality of binary numbers A and B, you first examine the highest-order bit in each number. The following conditions are possible:

1. If $A_3 = 1$ and $B_3 = 0$, number A is greater than number B.

2. If $A_3 = 0$ and $B_3 = 1$, number A is less than number B.

3. If $A_3 = B_3$, then you must examine the next lower bit position for an inequality.

These three operations are valid for each bit position in the numbers. The general procedure used in a comparator is to check for an inequality in a bit position, starting with the highest-order bits (MSBs). When such an inequality is found, the relationship of the two numbers is established, and any other inequalities in lower-order bit positions must be ignored because it is possible for an opposite indication to occur; *the highest-order indication must take precedence.*

EXAMPLE 5–6

Determine the $A = B$, $A > B$, and $A < B$ outputs for the input numbers shown on the comparator in Figure 5–26.

SOLUTION

The number on the A inputs is 0110 and the number on the B inputs is 0011. The $A > B$ **output is HIGH and the other outputs are LOW.**

RELATED PROBLEM

What are the comparator outputs when $A_3A_2A_1A_0 = 1001$ and $B_3B_2B_1B_0 = 1010$?

FIGURE 5–26

SECTION 5–5 CHECKUP

1. The binary numbers $A = 1011$ and $B = 1010$ are applied to the inputs of the comparator in Figure 5–25. Determine the outputs.

2. The binary numbers $A = 11001011$ and $B = 11010100$ are applied to an 8-bit comparator. Determine the states of the outputs.

5–6 DECODERS

A **decoder** is a digital circuit that detects the presence of a specified combination of bits (code) on its inputs and indicates the presence of that code by a specified output level. In its general form, a decoder has *n* input lines to handle *n* bits and from one to 2^n output lines to indicate the presence of one or more *n*-bit combinations. In this section, several decoders are introduced. The basic principles can be extended to other types of decoders.

After completing this section, you should be able to

- **Define** *decoder*
- **Develop a logic circuit to decode any combination of bits**
- **Expand decoders to accommodate larger numbers of bits in a code**
- **Discuss zero suppression in 7-segment displays**
- **Apply decoders to specific applications**

The Basic Binary Decoder

Suppose you need to determine when a binary 1001 occurs on the inputs of a digital circuit. An AND gate can be used as the basic decoding element because it produces a HIGH output only when all of its inputs are HIGH. Therefore, you must make sure that all of the

In the representation of a binary number or other weighted code in this book, the LSB is the right-most bit in a horizontal arrangement and the topmost bit in a vertical arrangement, unless specified otherwise.

inputs to the AND gate are HIGH when the binary number 1001 occurs; this can be done by inverting the two middle bits (the 0s), as shown in Figure 5–27.

(a)   (b)

FIGURE 5–27   Decoding logic for the binary code 1001 with an active-HIGH output.

The logic equation for the decoder of Figure 5–27(a) is developed as illustrated in Figure 5–27(b). You should verify that the output is 0 except when $A_0 = 1, A_1 = 0, A_2 = 0$, and $A_3 = 1$ are applied to the inputs. A_0 is the LSB and A_3 is the MSB.

If a NAND gate is used in place of the AND gate in Figure 5–27, a LOW output will indicate the presence of the proper binary code, which is 1001 in this case.

An *instruction* tells the computer what operation to perform. Instructions are in machine code (1s and 0s) and, in order for the computer to carry out an instruction, the instruction must be decoded. Instruction decoding is one of the steps in instruction *pipelining,* which are as follows: Instruction is read from the memory (instruction fetch), instruction is decoded, operand(s) is (are) read from memory (operand fetch), instruction is executed, and result is written back to memory. Basically, pipelining allows the next instruction to begin processing before the current one is completed.

SYSTEM NOTE

EXAMPLE 5–7

Determine the logic required to decode the binary number 1011 by producing a HIGH level on the output.

SOLUTION

The decoding function can be formed by complementing only the variables that appear as 0 in the desired binary number, as follows:

$$X = A_3\overline{A}_2A_1A_0 \quad (1011)$$

This function can be implemented by connecting the true (uncomplemented) variables A_0, A_1, and A_3 directly to the inputs of an AND gate, and inverting the variable A_2 before applying it to the AND gate input. The decoding logic is shown in Figure 5–28.

FIGURE 5–28   Decoding logic for producing a HIGH output when 1011 is on the inputs.

RELATED PROBLEM

Develop the logic required to detect the binary code 10010 and produce an active-LOW output.

The 4-Bit Decoder

In order to decode all possible combinations of four bits, sixteen decoding gates are required ($2^4 = 16$). This type of decoder is commonly called either a *4-line-to-16-line decoder* because there are four inputs and sixteen outputs or a *1-of-16 decoder* because for any given code on the inputs, one of the sixteen outputs is activated. A list of the sixteen binary codes and their corresponding decoding functions is given in Table 5–4.

If an active-LOW output is required for each decoded number, the entire decoder can be implemented with NAND gates and inverters. In order to decode each of the sixteen binary codes, sixteen NAND gates are required (AND gates can be used to produce active-HIGH outputs).

A logic symbol for a 4-line-to-16-line (1-of-16) decoder with active-LOW outputs is shown in Figure 5–29. The BIN/DEC label indicates that a binary input makes the corresponding decimal output active. The input labels 8, 4, 2, and 1 represent the binary weights of the input bits ($2^3 2^2 2^1 2^0$).

MULTISIM

FIGURE 5–29 Logic symbol for a 4-line-to-16-line (1-of-16) decoder. Open file F05-29 to verify operation.

TABLE 5–4 • Decoding functions and truth table for a 4-line-to-16-line (1-of-16) decoder with active-LOW outputs.

DECIMAL DIGIT	A_3	A_2	A_1	A_0	DECODING FUNCTION	0	1	2	3	4	5	6	7	8	9	10	11	12	13	14	15
0	0	0	0	0	$\bar{A}_3\bar{A}_2\bar{A}_1\bar{A}_0$	0	1	1	1	1	1	1	1	1	1	1	1	1	1	1	1
1	0	0	0	1	$\bar{A}_3\bar{A}_2\bar{A}_1 A_0$	1	0	1	1	1	1	1	1	1	1	1	1	1	1	1	1
2	0	0	1	0	$\bar{A}_3\bar{A}_2 A_1\bar{A}_0$	1	1	0	1	1	1	1	1	1	1	1	1	1	1	1	1
3	0	0	1	1	$\bar{A}_3\bar{A}_2 A_1 A_0$	1	1	1	0	1	1	1	1	1	1	1	1	1	1	1	1
4	0	1	0	0	$\bar{A}_3 A_2\bar{A}_1\bar{A}_0$	1	1	1	1	0	1	1	1	1	1	1	1	1	1	1	1
5	0	1	0	1	$\bar{A}_3 A_2\bar{A}_1 A_0$	1	1	1	1	1	0	1	1	1	1	1	1	1	1	1	1
6	0	1	1	0	$\bar{A}_3 A_2 A_1\bar{A}_0$	1	1	1	1	1	1	0	1	1	1	1	1	1	1	1	1
7	0	1	1	1	$\bar{A}_3 A_2 A_1 A_0$	1	1	1	1	1	1	1	0	1	1	1	1	1	1	1	1
8	1	0	0	0	$A_3\bar{A}_2\bar{A}_1\bar{A}_0$	1	1	1	1	1	1	1	1	0	1	1	1	1	1	1	1
9	1	0	0	1	$A_3\bar{A}_2\bar{A}_1 A_0$	1	1	1	1	1	1	1	1	1	0	1	1	1	1	1	1
10	1	0	1	0	$A_3\bar{A}_2 A_1\bar{A}_0$	1	1	1	1	1	1	1	1	1	1	0	1	1	1	1	1
11	1	0	1	1	$A_3\bar{A}_2 A_1 A_0$	1	1	1	1	1	1	1	1	1	1	1	0	1	1	1	1
12	1	1	0	0	$A_3 A_2\bar{A}_1\bar{A}_0$	1	1	1	1	1	1	1	1	1	1	1	1	0	1	1	1
13	1	1	0	1	$A_3 A_2\bar{A}_1 A_0$	1	1	1	1	1	1	1	1	1	1	1	1	1	0	1	1
14	1	1	1	0	$A_3 A_2 A_1\bar{A}_0$	1	1	1	1	1	1	1	1	1	1	1	1	1	1	0	1
15	1	1	1	1	$A_3 A_2 A_1 A_0$	1	1	1	1	1	1	1	1	1	1	1	1	1	1	1	0

EXAMPLE 5–8

A certain application requires that a 5-bit number be decoded. Use 4-bit decoders like the one in Figure 5–30 to implement the logic. Note the EN (enable) function. The binary number is represented by the format $A_4A_3A_2A_1A_0$.

FIGURE 5–30

SOLUTION

Since each decoder can handle only four bits, two decoders must be used to decode five bits. The fifth bit, A_4, is connected to the chip select inputs, $\overline{CS_1}$ and $\overline{CS_2}$, of one decoder, and $\overline{A_4}$ is connected to the $\overline{CS_1}$ and $\overline{CS_2}$ inputs of the other decoder, as shown in Figure 5–31. When the decimal number is 15 or less, $A_4 = 0$, the low-order decoder is enabled, and the high-order decoder is disabled. When the decimal number is greater than 15, $A_4 = 1$ so $A_4 = 0$, the high-order decoder is enabled, and the low-order decoder is disabled.

FIGURE 5–31

RELATED PROBLEM

Determine the output in Figure 5–31 that is activated for the binary input 10110.

SYSTEM EXAMPLE 5-2

INPUT/OUTPUT (I/O) PORT

A decoder is used in computers and other types of systems for input/output selection as depicted in the general diagram of Figure 5–32.

FIGURE 5–32   **A simplified I/O port system with a port address decoder with only four address lines shown.**

Computers must communicate with a variety of external devices called *peripherals* by sending and/or receiving data through what is known as input/output (I/O) ports. These external devices include printers, modems, scanners, external disk drives, keyboard, video monitors, and other computers. As illustrated in Figure 5–32, a decoder can be used to select the I/O port as determined by the computer so that data can be sent or received from a specific external device.

Each I/O port has a number, called an address, which uniquely identifies it. When the computer wants to communicate with a particular device, it issues the appropriate address code for the I/O port to which that particular device is connected. This binary port address is decoded and the appropriate decoder output is activated to enable the I/O port.

As shown in Figure 5–32, binary data are transferred within the computer on a data bus, which is a set of parallel lines. For example, an 8-bit bus consists of eight parallel lines that can carry one byte of data at a time. The data bus goes to all of the I/O ports, but any data coming in or going out will only pass through the port that is enabled by the port address decoder.

The BCD-to-Decimal Decoder

The BCD-to-decimal decoder converts each BCD code (8421 code) into one of ten possible decimal digit indications. It is frequently referred as a *4-line-to-10-line decoder* or a *1-of-10 decoder*.

The method of implementation is the same as for the 1-of-16 decoder previously discussed, except that only ten decoding gates are required because the BCD code represents only the ten decimal digits 0 through 9. A list of the ten BCD codes and their corresponding decoding functions is given in Table 5–5. Each of these decoding functions is implemented with NAND gates to provide active-LOW outputs. If an active-HIGH output is required, AND gates are used for decoding. The logic is identical to that of the first ten decoding gates in the 1-of-16 decoder (see Table 5–4).

DECIMAL DIGIT	BCD CODE A_3	A_2	A_1	A_0	DECODING FUNCTION
0	0	0	0	0	$\overline{A_3}\,\overline{A_2}\,\overline{A_1}\,\overline{A_0}$
1	0	0	0	1	$\overline{A_3}\,\overline{A_2}\,\overline{A_1}A_0$
2	0	0	1	0	$\overline{A_3}\,\overline{A_2}A_1\overline{A_0}$
3	0	0	1	1	$\overline{A_3}\,\overline{A_2}A_1A_0$
4	0	1	0	0	$\overline{A_3}A_2\overline{A_1}\,\overline{A_0}$
5	0	1	0	1	$\overline{A_3}A_2\overline{A_1}A_0$
6	0	1	1	0	$\overline{A_3}A_2A_1\overline{A_0}$
7	0	1	1	1	$\overline{A_3}A_2A_1A_0$
8	1	0	0	0	$A_3\overline{A_2}\,\overline{A_1}\,\overline{A_0}$
9	1	0	0	1	$A_3\overline{A_2}\,\overline{A_1}A_0$

TABLE 5–5 • BCD decoding functions.

EXAMPLE 5–9

Figure 5–33(a) shows a BCD/DEC decoder. If the input waveforms in Figure 5–33(b) are applied to the inputs of the decoder, show the output waveforms.

SOLUTION

The output waveforms are shown in Figure 5–33(c). As you can see, the inputs are sequenced through the BCD for digits 0 through 9. The output waveforms in the timing diagram indicate that sequence on the decimal-value outputs.

RELATED PROBLEM

Construct a timing diagram showing input and output waveforms for the case where the BCD inputs sequence through the decimal numbers as follows: 0, 2, 4, 6, 8, 1, 3, 5, and 9.

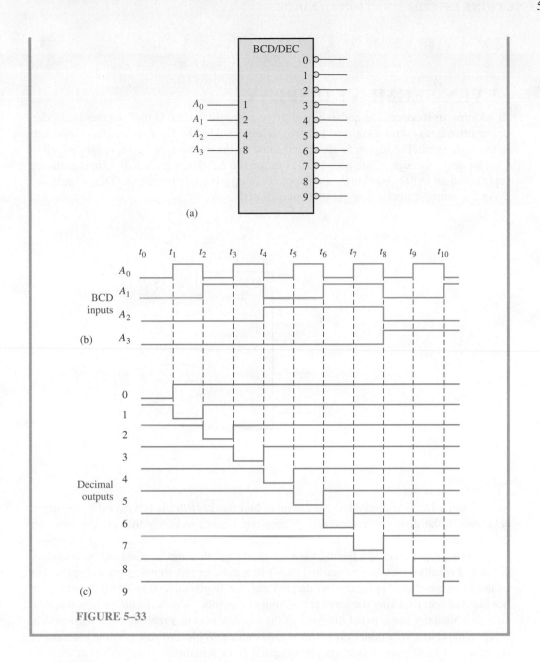

FIGURE 5–33

The BCD-to-7-Segment Decoder

The BCD-to-7-segment decoder accepts the BCD code on its inputs and provides outputs to drive 7-segment display devices to produce a decimal readout. The logic diagram for a basic 7-segment decoder is shown in Figure 5–34. The active-LOW outputs drive a common-cathode type of 7-segment display.

FIGURE 5–34   Logic symbol for a BCD-to-7-segment decoder/driver with active-LOW outputs. Open file F05-34 to verify operation.

MULTISIM

SYSTEM EXAMPLE 5–3

SEVEN-SEGMENT DISPLAY

In addition to its decoding and segment drive capability, a BCD-to-7-segment decoder/driver often has several additional features as indicated by the $\overline{LT}$, $\overline{RBI}$, $\overline{BI}/\overline{RBO}$ functions in the logic symbol of Figure 5–35. As indicated by the bubbles on the logic symbol, all of the outputs (*a* through *g*) are active-LOW as are the $\overline{LT}$ (lamp test), $\overline{RBI}$ (ripple blanking input), and $\overline{BI}/\overline{RBO}$ (blanking input/ripple blanking output) functions. The outputs can drive a common-cathode 7-segment display directly.

FIGURE 5–35

When a LOW is applied to the $\overline{LT}$ input and the $\overline{BI}/\overline{RBO}$ is HIGH, all of the seven segments in the display are turned on. Lamp test is used to verify that no segments are burned out.

Zero suppression is a feature used for multidigit displays to blank out unnecessary zeros and results in leading or trailing zeros in a number not showing on a display. For example, in a 6-digit display the number 6.4 may be displayed as 006.400 if the zeros are not blanked out. Blanking the zeros at the front of a number is called *leading zero suppression* and blanking the zeros at the back of the number is called *trailing zero suppression.* Keep in mind that only nonessential zeros are blanked. With zero suppression, the number 030.080 will be displayed as 30.08 (the essential zeros remain).

Zero suppression is accomplished using the $\overline{RBI}$ and $\overline{BI}/\overline{RBO}$ functions. $\overline{RBI}$ is the ripple blanking input and $\overline{RBO}$ is the ripple blanking output; these are used for zero suppression. $\overline{BI}$ is the blanking input that shares the same pin with $\overline{RBO}$; in other words, the $\overline{BI}/\overline{RBO}$ pin can be used as an input or an output. When used as a $\overline{BI}$ (blanking input), all segment outputs are HIGH (nonactive) when $\overline{BI}$ is LOW, which overrides all other inputs. The $\overline{BI}$ function is not part of the zero suppression capability of the device.

All of the segment outputs of the decoder are nonactive (HIGH) if a zero code (0000) is on its BCD inputs and if its $\overline{RBI}$ is LOW. This causes the display to be blank and produces a LOW $\overline{RBO}$.

The logic diagram in Figure 5–36(a) illustrates leading zero suppression for a whole number. The highest-order digit position (left-most) is always blanked if a zero code is on its BCD inputs because the $\overline{RBI}$ of the most-significant decoder is made LOW by connecting it to ground. The $\overline{RBO}$ of each decoder is connected to the $\overline{RBI}$ of the next lowest-order decoder so that all zeros to the left of the first nonzero digit are blanked. For example, in part (a) of the figure the two highest-order digits are zeros and therefore are blanked. The remaining two digits, 3 and 0 are displayed.

The logic diagram in Figure 5–36(b) illustrates trailing zero suppression for a fractional number. The lowest-order digit (right-most) is always blanked if a zero code is on its BCD inputs because the $\overline{RBI}$ is connected to ground. The $\overline{RBO}$ of each decoder is connected

(a) Illustration of leading zero suppression

(b) Illustration of trailing zero suppression

FIGURE 5–36   **Examples of zero suppression.**

to the $\overline{RBI}$ of the next highest-order decoder so that all zeros to the right of the first nonzero digit are blanked. In part (b) of the figure, the two lowest-order digits are zeros and therefore are blanked. The remaining two digits, 5 and 7 are displayed. To combine both leading and trailing zero suppression in one display and to have decimal point capability, additional logic is required.

SECTION 5–6 CHECKUP

1. A 3-line-to-8-line decoder can be used for octal-to-decimal decoding. When a binary 101 is on the inputs, which output line is activated?

2. How many 1-of-16 decoders are necessary to decode a 6-bit binary number?

3. Would you select a decoder/driver with active-HIGH or active-LOW outputs to drive a common-cathode 7-segment LED display?

An **encoder** is a combinational logic circuit that essentially performs a "reverse" decoder function. An encoder accepts an active level on one of its inputs representing a digit, such as a decimal or octal digit, and converts it to a coded output, such as BCD or binary. Encoders can also be devised to encode various symbols and alphabetic characters. The process of converting from familiar symbols or numbers to a coded format is called *encoding*.

After completing this section, you should be able to

- Determine the logic for a decimal-to-BCD encoder
- Explain the purpose of the priority feature in encoders
- Expand an encoder
- Apply the encoder to a specific system example

The Decimal-to-BCD Encoder

This type of encoder has ten inputs—one for each decimal digit—and four outputs corresponding to the BCD code, as shown in Figure 5–37. This is a basic 10-line-to-4-line encoder.

FIGURE 5–37 **Logic symbol for a decimal-to-BCD encoder.**

The BCD (8421) code is listed in Table 5–6. From this table you can determine the relationship between each BCD bit and the decimal digits in order to analyze the logic. For instance, the most significant bit of the BCD code, A_3, is always a 1 for decimal digit 8 or 9. An OR expression for bit A_3 in terms of the decimal digits can therefore be written as

$$A_3 = 8 + 9$$

Bit A_2 is always a 1 for decimal digit 4, 5, 6 or 7 and can be expressed as an OR function as follows:

$$A_2 = 4 + 5 + 6 + 7$$

Bit A_1 is always a 1 for decimal digit 2, 3, 6, or 7 and can be expressed as

$$A_1 = 2 + 3 + 6 + 7$$

TABLE 5–6				
	BCD CODE			
DECIMAL DIGIT	A_3	A_2	A_1	A_0
0	0	0	0	0
1	0	0	0	1
2	0	0	1	0
3	0	0	1	1
4	0	1	0	0
5	0	1	0	1
6	0	1	1	0
7	0	1	1	1
8	1	0	0	0
9	1	0	0	1

An *assembler* can be thought of as a software encoder because it interprets the mnemonic instructions with which a program is written and carries out the applicable encoding to convert each mnemonic to a machine code instruction (series of 1s and 0s) that the computer can understand. Examples of mnemonic instructions for a microprocessor are ADD, MOV (move data), MUL (multiply), XOR, JMP (jump), and OUT (output to a port).

SYSTEM NOTE

Finally, A_0 is always a 1 for decimal digit 1, 3, 5, 7, or 9. The expression for A_0 is

$$A_0 = 1 + 3 + 5 + 7 + 9$$

Now let's look at the logic circuitry required for encoding each decimal digit to a BCD code by using the logic expressions just developed. It is simply a matter of ORing the appropriate decimal digit input lines to form each BCD output. The basic encoder logic resulting from these expressions is shown in Figure 5–38.

The basic operation of the circuit in Figure 5–38 is as follows: When a HIGH appears on *one* of the decimal digit input lines, the appropriate levels occur on the four BCD output lines. For instance, if input line 9 is HIGH (assuming all other input lines are LOW), this condition will produce a HIGH on outputs A_0 and A_3 and LOWs on outputs A_1 and A_2, which is the BCD code (1001) for decimal 9.

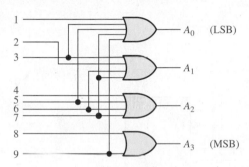

FIGURE 5–38   Logic diagram of a decimal-to-BCD encoder. A 0-digit input is not needed because the BCD outputs are all LOW when there are no HIGH inputs.

THE DECIMAL-TO-BCD PRIORITY ENCODER   This type of encoder performs the same basic encoding function as previously discussed. A **priority encoder** also offers additional flexibility in that it can be used in applications that require priority detection. The priority function means that the encoder will produce a BCD output corresponding to the *highest-order decimal digit* input that is active and will ignore any other lower-order active inputs. For instance, if the 6 and the 3 inputs are both active, the BCD output is 0110 (which represents decimal 6).

A priority encoder with active-LOW inputs (0) for decimal digits 1 through 9 and active-LOW BCD outputs is indicated in the logic symbol in Figure 5–39. A BCD zero output is represented when none of the inputs is active.

FIGURE 5–39   A decimal-to-BCD priority encoder (HPRI means highest value input has priority).

EXAMPLE 5–10

If LOW levels appear on inputs 3, 4, and 7 of the encoder shown in Figure 5–39, indicate the state of the four outputs. All other inputs are HIGH.

SOLUTION

Input 7 is the highest-order decimal digit input having a LOW level and represents decimal 7. Therefore, the output levels indicate the BCD code for decimal 7 where $\overline{A}_0$ is the LSB and $\overline{A}_3$ is the MSB. Output $\overline{A}_0$ is LOW, $\overline{A}_1$ is LOW, $\overline{A}_2$ is LOW, and $\overline{A}_3$ is HIGH.

RELATED PROBLEM

What are the outputs of the same encoder if all its inputs are LOW? If all its inputs are HIGH?

SYSTEM EXAMPLE 5–4

KEYPAD ENCODER

The ten decimal digits on a numeric keypad must be encoded for processing by the logic circuitry. In this example, when one of the keys is pressed, the decimal digit is encoded to the corresponding BCD code. Figure 5–40 shows a simple keyboard encoder arrangement using a priority encoder. The keys are represented by ten push-button switches, each with a **pull-up resistor** to $+V$. The pull-up resistor ensures that the line is HIGH when a key is not depressed. When a key is depressed, the line is connected to ground, and a LOW is applied to the corresponding encoder input. The zero key is not connected because the BCD output represents zero when none of the other keys is depressed. In systems such as the computer, the keystrokes are encoded into ASCII because both numeric and alphanumeric characters are used.

The BCD complement output of the encoder goes into a storage device, and each successive BCD code is stored until the entire number has been entered. Methods of storing BCD numbers and other binary data are covered in later chapters.

FIGURE 5–40   **A simplified keyboard encoder.**

SECTION 5–7 CHECKUP

1. Suppose the HIGH levels are applied to the 2 input and the 9 input of the circuit in Figure 5–38.

   (a) What are the states of the output lines?

   (b) Does this represent a valid BCD code?

   (c) What is the restriction on the encoder logic in Figure 5–38?

2. **(a)** What is the $\overline{A}_3\,\overline{A}_2\,\overline{A}_1\,\overline{A}_0$ output when LOWs are applied to inputs 4 and 8 of the priority encoder in Figure 5–39?

   (b) What does this output represent?

5–8 CODE CONVERTERS

In this section, we will examine some methods of using combinational logic circuits to convert from one code to another.

After completing this section, you should be able to

- **Explain the process for converting BCD to binary**
- **Use exclusive-OR gates for conversions between binary and Gray codes**

BCD-to-Binary Conversion

One method of BCD-to-binary code conversion uses adder circuits. The basic conversion process is as follows:

1. The value, or weight, of each bit in the BCD number is represented by a binary number.
2. All of the binary representations of the weights of bits that are 1s in the BCD number are added.
3. The result of this addition is the binary equivalent of the BCD number.

A more concise statement of this operation is

> **The binary numbers representing the weights of the BCD bits are summed to produce the total binary number.**

Let's examine an 8-bit BCD code (one that represents a 2-digit decimal number) to understand the relationship between BCD and binary. For instance, you already know that the decimal number 87 can be expressed in BCD as

$$\underset{8}{\underline{1000}} \quad \underset{7}{\underline{0111}}$$

The left-most 4-bit group represents 80, and the right-most 4-bit group represents 7. That is, the left-most group has a weight of 10, and the right-most group has a weight of 1. Within each group, the binary weight of each bit is as follows:

	TENS DIGIT				UNITS DIGIT			
Weight:	80	40	20	10	8	4	2	1
Bit designation:	B_3	B_2	B_1	B_0	A_3	A_2	A_1	A_0

The binary equivalent of each BCD bit is a binary number representing the weight of that bit within the total BCD number. This representation is given in Table 5–7.

BCD BIT	BCD WEIGHT	(MSB) 64	32	16	8	4	2	(LSB) 1
				BINARY REPRESENTATION				
A_0	1	0	0	0	0	0	0	1
A_1	2	0	0	0	0	0	1	0
A_2	4	0	0	0	0	1	0	0
A_3	8	0	0	0	1	0	0	0
B_0	10	0	0	0	1	0	1	0
B_1	20	0	0	1	0	1	0	0
B_2	40	0	1	0	1	0	0	0
B_3	80	1	0	1	0	0	0	0

TABLE 5–7 • Binary representations of BCD bit weights.

If the binary representations for the weights of all the 1s in the BCD number are added, the result is the binary number that corresponds to the BCD number. Example 5–11 illustrates this.

EXAMPLE 5–11

Convert the BCD numbers 00100111 (decimal 27) and 10011000 (decimal 98) to binary.

SOLUTION

Write the binary representations of the weights of all 1s appearing in the numbers, and then add them together.

RELATED PROBLEM

Show the process of converting 01000001 in BCD to binary.

Binary-to-Gray and Gray-to-Binary Conversion

The basic process for Gray-binary conversions was covered in Chapter 2. Exclusive-OR gates can be used for these conversions. Programmable logic devices (PLDs) can also be programmed for these code conversions. Figure 5–41 shows a 4-bit binary-to-Gray code converter, and Figure 5–42 illustrates a 4-bit Gray-to-binary converter.

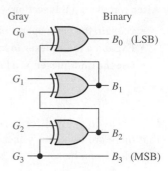

FIGURE 5–41 **Four-bit binary-to-Gray conversion logic. Open file F05-41 to verify operation.**

MULTISIM

FIGURE 5–42 **Four-bit Gray-to-binary conversion logic. Open file F05-42 to verify operation.**

MULTISIM

EXAMPLE 5–12

(a) Convert the binary number 0101 to Gray code with exclusive-OR gates.

(b) Convert the Gray code 1011 to binary with exclusive-OR gates.

SOLUTION

(a) 0101_2 is 0111 Gray. See Figure 5–43(a).

(b) 1011 Gray is 1101_2. See Figure 5–43(b).

(a)

(b)

FIGURE 5–43

RELATED PROBLEM

How many exclusive-OR gates are required to convert 8-bit binary to Gray?

SECTION 5–8 CHECKUP

1. Convert the BCD number 10000101 to binary.

2. Draw the logic diagram for converting an 8-bit binary number to Gray code.

5–9 MULTIPLEXERS (DATA SELECTORS)

A **multiplexer (MUX)** is a device that allows digital information from several sources to be routed onto a single line for transmission over that line to a common destination. The basic multiplexer has several data-input lines and a single output line. It also has data-select inputs, which permit digital data on any one of the inputs to be switched to the output line. Multiplexers are also known as data selectors.

After completing this section, you should be able to

- **Explain the basic operation of a multiplexer**
- **Expand a multiplexer to handle more data inputs**
- **Use the multiplexer as a logic function generator**

In a multiplexer, data goes from several lines to one line.

A logic symbol for a 4-input multiplexer (MUX) is shown in Figure 5–44. Notice that there are two data-select lines because with two select bits, any one of the four data-input lines can be selected.

FIGURE 5–44 Logic symbol for a 1-of-4 data selector/multiplexer.

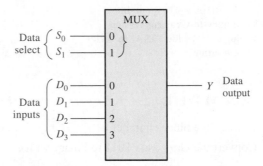

In Figure 5–44, a 2-bit code on the data-select (S) inputs will allow the data on the selected data input to pass through to the data output. If a binary 0 $(S_1 = 0$ and $S_0 = 0)$ is applied to the data-select lines, the data on input D_0 appear on the data-output line. If a binary 1 $(S_1 = 0$ and $S_0 = 1)$ is applied to the data-select lines, the data on input D_1 appear on the data output. If a binary 2 $(S_1 = 1$ and $S_0 = 0)$ is applied, the data on D_2 appear on the output. If a binary 3 $(S_1 = 1$ and $S_0 = 1)$ is applied, the data on D_3 are switched to the output line. A summary of this operation is given in Table 5–8.

TABLE 5–8 • Data selection for a 1-of-4-multiplexer.

DATA-SELECT INPUTS		INPUT SELECTED
S_1	S_0	
0	0	D_0
0	1	D_1
1	0	D_2
1	1	D_3

Now let's look at the logic circuitry required to perform this multiplexing operation. The data output is equal to the state of the *selected* data input. You can therefore, derive a logic expression for the output in terms of the data input and the select inputs.

The data output is equal to D_0 only if $S_1 = 0$ and $S_0 = 0$: $Y = D_0\overline{S}_1\overline{S}_0$.

The data output is equal to D_1 only if $S_1 = 0$ and $S_0 = 1$: $Y = D_1\overline{S}_1 S_0$.

The data output is equal to D_2 only if $S_1 = 1$ and $S_0 = 0$: $Y = D_2 S_1\overline{S}_0$.

The data output is equal to D_3 only if $S_1 = 1$ and $S_0 = 1$: $Y = D_3 S_1 S_0$.

A *bus* is a single or multiple conductor pathway along which electrical signals are sent from one part of a system to another. In computer networks, a *shared bus* is one that is connected to all the microprocessors in the system in order to exchange data. A shared bus may contain memory and input/output devices that can be accessed by all the microprocessors in the system. Access to the shared bus is controlled by a *bus arbiter* (a multiplexer of sorts) that allows only one microprocessor at a time to use the system's shared bus.

SYSTEM NOTE

When these terms are ORed, the total expression for the data output is

$$Y = D_0 \overline{S}_1 \overline{S}_0 + D_1 \overline{S}_1 S_0 + D_2 S_1 \overline{S}_0 + D_3 S_1 S_0$$

The implementation of this equation requires four 3-input AND gates, a 4-input OR gate, and two inverters to generate the complements of S_1 and S_0, as shown in Figure 5–45. Because data can be selected from any one of the input lines, this circuit is also referred to as a **data selector.**

FIGURE 5–45   **Logic diagram for a 4-input multiplexer. Open file F05-45 to verify operation.**

MULTISIM

EXAMPLE 5–13

The data-input and data-select waveforms in Figure 5–46(a) are applied to the multiplexer in Figure 5–45. Determine the output waveform in relation to the inputs.

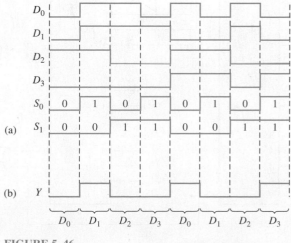

FIGURE 5–46

SOLUTION

The binary state of the data-select inputs during each interval determines which data input is selected. Notice that the data-select inputs go through a repetitive binary sequence 00, 01, 10, 11, 00, 01, 10, 11, and so on. The resulting output waveform is shown in Figure 5–46(b).

RELATED PROBLEM

Construct a timing diagram showing all inputs and the output if the S_0 and S_1 waveforms in Figure 5–46 are interchanged.

FIGURE 5–47 **Logic symbol for an 8-input data selector/multiplexer.**

An 8-input Data Selector/Multiplexer

The multiplexer shown in Figure 5–47 has eight data inputs (D_0–D_7) and, therefore, three data-select or address input lines (S_0–S_2). Three bits are required to select any one of the eight data inputs ($2^3 = 8$). A LOW on the $\overline{Enable}$ input allows the selected input data to pass through to the output. Notice that the data output and its complement are both available. The $G\frac{0}{7}$ label within the logic symbol indicates the AND relationship between the data-select inputs and each of the data inputs 0 through 7.

EXAMPLE 5–14

Use the 8-input multiplexer in Figure 5–47 and any other logic necessary to multiplex 16 data lines onto a single data-output line.

SOLUTION

An implementation of this function is shown in Figure 5–48. Four bits are required to select one of 16 data inputs ($2^4 = 16$). In this application the $\overline{Enable}$ input is used as the most significant data-select bit. When the MSB in the data-select code is LOW, the left multiplexer is enabled, and one of the data inputs (D_0 through D_7) is selected by the other three data-select bits. When the data-select MSB is HIGH, the right multiplexer is enabled, and one of the data inputs (D_8 through D_{15}) is selected. The selected input data are then passed through to the negative-OR gate and onto the single output line.

FIGURE 5–48 **A 16-input multiplexer.**

RELATED PROBLEM

Determine the codes on the select inputs required to select each of the following data inputs: D_0, D_4, D_8, and D_{13}.

SYSTEM EXAMPLE 5–5

MULTIPLEXED DISPLAY

Many types of systems require data to be displayed in readable form. Figure 5–49 shows a simplified method of multiplexing BCD numbers to a 7-segment display. Two digit numbers are displayed on the 7-segment readout by the use of a single BCD-to-7-segment decoder. This basic method of display multiplexing can be extended to displays with any number of digits.

FIGURE 5–49   Simplified 7-segment display multiplexing logic.

The basic operation is as follows. Two BCD digits ($A_3A_2A_1A_0$ and $B_3B_2B_1B_0$) are applied to the multiplexer, which consists of four separate 2-input multiplexers. Each of the four multiplexers shares a common data-select line and a common *Enable*. Because there are only two inputs to be selected in each multiplexer, a single data-select input is sufficient. Notice that the four multiplexers are indicated by the partitioned outline and that the inputs common to all four multiplexers are indicated as inputs to the notched block at the top, which is called the *common control block*. All labels within the upper MUX block apply to the other blocks below it.

Notice the 1 and $\overline{1}$ labels in the MUX blocks and the G1 label in the common control block. These labels are an example of the **dependency notation** system specified in the ANSI/IEEE Standard 91-1984. In this case G1 indicates an AND relationship between the data-select input and the data inputs with 1 or $\overline{1}$ labels. (The $\overline{1}$ means that the AND relationship applies to the complement of the G1 input.) In other words, when the data-select input is HIGH, the B inputs of the multiplexers are selected; and when the data-select input is LOW, the A inputs are selected. A "G" is always used to denote AND dependency.

A square wave is applied to the data-select line, and when it is LOW, the A bits ($A_3 A_2 A_1 A_0$) are passed through to the inputs of the BCD-to-7-segment decoder. The LOW on the data-select also puts a LOW on the A_1 input of the 2-line-to-4-line decoder, thus activating its 0 output and enabling the A-digit display by effectively connecting its common terminal to ground. The A digit is now *on* and the B digit is *off*.

When the data-select line goes HIGH, the B bits ($B_3 B_2 B_1 B_0$) are passed through to the inputs of the BCD-to-7-segment decoder. Also, the decoder's 1 output is activated, thus enabling the B-digit display. The B digit is now *on* and the A digit is *off*. The cycle repeats at the frequency of the data-select square wave. This frequency must be high enough to prevent visual flicker as the digit displays are multiplexed.

A LOGIC FUNCTION GENERATOR   A useful application of the data selector/ multiplexer is in the generation of combinational logic functions in sum-of-products form. When used in this way, the device can replace discrete gates.

To illustrate, an 8-input data selector/multiplexer can be used to implement any specified 3-variable logic function if the variables are connected to the data-select inputs and each data input is set to the logic level required in the truth table for that function. For example, if the function is a 1 when the variable combination is $\overline{A_2} A_1 \overline{A_0}$, the 2 input (selected by 010) is connected to a HIGH. This HIGH is passed through to the output when this particular combination of variables occurs on the data-select lines. Example 5–15 will help clarify this application.

EXAMPLE 5–15

Implement the logic function specified in Table 5–9 by using an 8-input data selector/multiplexer.

TABLE 5–9			
INPUTS			OUTPUT
A_2	A_1	A_0	Y
0	0	0	0
0	0	1	1
0	1	0	0
0	1	1	1
1	0	0	0
1	0	1	1
1	1	0	1
1	1	1	0

SOLUTION

Notice from the truth table that Y is a 1 for the following input variable combinations: 001, 011, 101, and 110. For all other combinations, Y is 0. For this function to be implemented with the data selector, the data input selected by each of the above-mentioned combinations must be connected to a HIGH (5 V). All the other data inputs must be connected to a LOW (ground), as shown in Figure 5–50.

$$Y = \overline{A}_2 \overline{A}_1 A_0 + \overline{A}_2 A_1 A_0 + A_2 \overline{A}_1 A_0 + A_2 A_1 \overline{A}_0$$

FIGURE 5–50   **Data selector/multiplexer connected as a 3-variable logic function generator.**

RELATED PROBLEM

Use the data selector/multiplexer to implement the following expression:

$$Y = \overline{A}_2 \overline{A}_1 \overline{A}_0 + A_2 \overline{A}_1 \overline{A}_0 + \overline{A}_2 A_1 \overline{A}_0$$

Example 5–15 illustrated how the 8-input data selector can be used as a logic function generator for three variables. Actually, this device can be also used as a 4-variable logic function generator by the utilization of one of the bits (A_0) in conjunction with the data inputs.

A 4-variable truth table has sixteen combinations of input variables. When an 8-bit data selector is used, each input is selected twice: the first time when A_0 is 0 and the second time when A_0 is 1. With this in mind, the following rules can be applied (Y is the output, and A_0 is the least significant bit):

1. If $Y = 0$ both times a given data input is selected by a certain combination of the input variables, $A_3 A_2 A_1$, connect that data input to ground (0).

2. If $Y = 1$ both times a given data input is selected by a certain combination of the input variables, $A_3 A_2 A_1$, connect the data input to $+V$ (1).

3. If Y is different the two times a given data input is selected by a certain combination of the input variables, $A_3 A_2 A_1$, and if $Y = A_0$, connect that data input to A_0.

4. If Y is different the two times a given data input is selected by a certain combination of the input variables, $A_3 A_2 A_1$, and if $Y = \overline{A}_0$, connect that data input to $\overline{A}_0$.

EXAMPLE 5–16

Implement the logic function in Table 5–10 by using an 8-input data selector/multiplexer.

SOLUTION

The data-select inputs are $A_3 A_2 A_1$. In the first row of the table, $A_3 A_2 A_1 = 000$ and $Y = A_0$. In the second row, where $A_3 A_2 A_1$ again is 000, $Y = A_0$. Thus, A_0 is connected to the 0 input. In the third row of the table, $A_3 A_2 A_1 = 001$ and $Y = \overline{A}_0$. Also, in the fourth row, when $A_3 A_2 A_1$ again is 001, $Y = \overline{A}_0$. Thus, A_0 is inverted and

TABLE 5–10

DECIMAL DIGIT	INPUTS				OUTPUT
	A_3	A_2	A_1	A_0	Y
0	0	0	0	0	0
1	0	0	0	1	1
2	0	0	1	0	1
3	0	0	1	1	0
4	0	1	0	0	0
5	0	1	0	1	1
6	0	1	1	0	1
7	0	1	1	1	1
8	1	0	0	0	1
9	1	0	0	1	0
10	1	0	1	0	1
11	1	0	1	1	0
12	1	1	0	0	1
13	1	1	0	1	1
14	1	1	1	0	0
15	1	1	1	1	1

connected to the 1 input. This analysis is continued until each input is properly connected according to the specified rules. The implementation is shown in Figure 5–51.

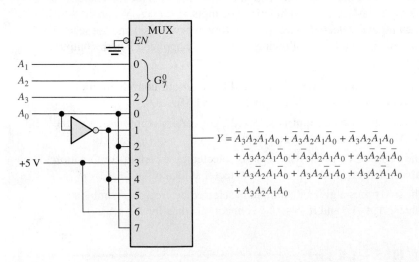

$$Y = \overline{A}_3\overline{A}_2\overline{A}_1 A_0 + \overline{A}_3\overline{A}_2 A_1 \overline{A}_0 + \overline{A}_3 A_2 \overline{A}_1 A_0$$
$$+ \overline{A}_3 A_2 A_1 \overline{A}_0 + \overline{A}_3 A_2 A_1 A_0 + A_3 \overline{A}_2 \overline{A}_1 \overline{A}_0$$
$$+ A_3 \overline{A}_2 A_1 \overline{A}_0 + A_3 A_2 \overline{A}_1 \overline{A}_0 + A_3 A_2 \overline{A}_1 A_0$$
$$+ A_3 A_2 A_1 A_0$$

FIGURE 5–51 **Data selector/multiplexer connected as a 4-variable logic function generator.**

RELATED PROBLEM

In Table 5–10, if $Y = 0$ when the inputs are all zeros and is alternately a 1 and a 0 for the remaining rows in the table, use a data selector/multiplexer to implement the resulting logic function.

SECTION 5–9 CHECKUP

SECTION 5–9 CHECKUP

1. In Figure 5–45, $D_0 = 1, D_1 = 0, D_2 = 1, D_3 = 0, S_0 = 1$, and $S_1 = 0$. What is the output?

2. How many select or address inputs are required for a 32-bit data selector/multiplexer?

3. An 8-bit multiplexer like in Figure 5–47 has alternating LOW and HIGH levels on its data inputs beginning with $D_0 = 0$. The data-select lines are sequenced through a binary count (000, 001, 010, and so on) at a frequency of 1 kHz. The enable input is LOW. Describe the data output waveform.

4. Briefly describe the purpose of each of the following devices in Figure 5–49:

   (a) Multiplexer

   (b) BCD-to-7-segment decoder/driver

   (c) 2-line-to-4-line decoder

5–10 DEMULTIPLEXERS

A **demultiplexer (DEMUX)** basically reverses the multiplexing function. It takes digital information from one line and distributes it to a given number of output lines. For this reason, the demultiplexer is also known as a **data distributor**. As you will learn, decoders can also be used as demultiplexers.

After completing this section, you should be able to

- Explain the basic operation of a demultiplexer
- Describe how the 4-line-to-16-line decoder can be used as a demultiplexer
- Develop the timing diagram for a demultiplexer with specified data and data selection inputs

Figure 5–52 shows a 1-line-to-4-line demultiplexer (DEMUX) circuit. The data-input line goes to all of the AND gates. The two data-select lines enable only one gate at a time, and the data appearing on the data-input line will pass through the selected gate to the associated data-output line.

In a demultiplexer, data goes from one line to several lines.

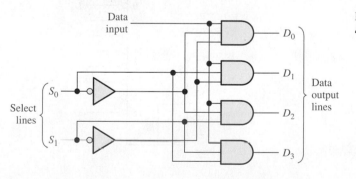

FIGURE 5–52 A 1-line-to-4-line demultiplexer.

EXAMPLE 5–17

The serial data-input waveform (Data in) and data-select inputs (S_0 and S_1) are shown in Figure 5–53. Determine the data-output waveforms on D_0 through D_3 for the demultiplexer in Figure 5–52.

SOLUTION

Notice that the select lines go through a binary sequence so that each successive input bit is routed to D_0, D_1, D_2, and D_3 in sequence, as shown by the output waveforms in Figure 5–53.

FIGURE 5–53

RELATED PROBLEM

Develop the timing diagram for the demultiplexer if the S_0 and S_1 waveforms are both inverted.

A Decoder as a Demultiplexer

The 4-line-to-16-line decoder can also be used in demultiplexing applications. The logic symbol for the decoder used as a demultiplexer is shown in Figure 5–54. In demultiplexer applications, the input lines are used as the data-select lines. One of the chip select (EN) inputs is used as the data-input line, with the other chip select input held LOW to enable the internal negative-AND gate at the bottom of the diagram.

FIGURE 5–54   **The 4-line-to-16-line decoder used as a dcmultiplexer.**

SECTION 5–10 CHECKUP

1. Generally, how can a decoder be used as a demultiplexer?

2. The demultiplexer in Figure 5–54 has a binary code of 1010 on the data-select lines, and the data-input line is LOW. What are the states of the output lines?

Errors can occur as digital codes are being transferred from one point to another within a digital system or while codes are being transmitted from one system to another. The errors take the form of undesired changes in the bits that make up the coded information; that is, a 1 can change to a 0, or a 0 to a 1, because of component malfunctions or electrical noise. In most digital systems, the probability that even a single bit error will occur is very small, and the likelihood that more than one will occur is even smaller. Nevertheless, when an error occurs undetected, it can cause serious problems in a digital system.

After completing this section, you should be able to

- Explain the concept of parity
- Implement a basic parity circuit with exclusive-OR gates
- Describe the operation of basic parity generating and checking logic
- Discuss a 9-bit parity generator/checker
- Discuss how error detection can be implemented in a data transmission

---

The parity method of error detection in which a **parity bit** is attached to a group of information bits in order to make the total number of 1s either even or odd (depending on the system) was covered in Chapter 2. In addition to parity bits, several specific codes also provide inherent error detection.

> A parity bit indicates if the number of 1s in a code is even or odd for the purpose of error detection.

Basic Parity Logic

In order to check for or to generate the proper parity in a given code, a basic principle is:

> **The sum (disregarding carries) of an even number of 1s is always 0, and the sum of an odd number of 1s is always 1.**

Therefore, to determine if a given code has **even parity** or **odd parity,** all the bits in that code are summed. As you know, the modulo-2 sum of two bits can be generated by an exclusive-OR gate, as shown in Figure 5–55(a); the modulo-2 sum of four bits can be formed by three exclusive-OR gates connected as shown in Figure 5–55(b); and so on. When the number of 1s on the inputs is even, the output X is 0 (LOW). When the number of 1s is odd, the output X is 1 (HIGH).

(a) Summing of two bits   (b) Summing of four bits

FIGURE 5–55

A 9-bit Parity Generator/Checker

The logic symbol and function table for a 9-bit parity checker/generator are shown in Figure 5–56. This particular device can be used to check for odd or even parity on a 9-bit code

FIGURE 5–56 A 9-bit parity generator/checker.

(a) Traditional logic symbol

NUMBER OF INPUTS A–I THAT ARE HIGH	OUTPUTS	
	Σ EVEN	Σ ODD
0, 2, 4, 6, 8	H	L
1, 3, 5, 7, 9	L	H

(b) Function table

(eight data bits and one parity bit), or it can be used to generate a parity bit for a binary code with up to nine bits. The inputs are *A* through *I*; when there is an even number of 1s on the inputs, the Σ Even output is HIGH and the Σ Odd output is LOW.

PARITY CHECKER   When this device is used as an even parity checker, the number of input bits should always be even; and when a parity error occurs, the Σ Even output goes LOW and the Σ Odd output goes HIGH. When it is used as an odd parity checker, the number of input bits should always be odd; and when a parity error occurs, the Σ Odd output goes LOW and the Σ Even output goes HIGH.

PARITY GENERATOR   If this device is used as an even parity generator, the parity bit is taken at the Σ Odd output because this output is a 0 if there is an even number of input bits and it is a 1 if there is an odd number. When used as an odd parity generator, the parity bit is taken at the Σ Even output because it is a 0 when the number of inputs bits is odd.

The microprocessor in a computer performs internal parity checks as well as parity checks of the external data and address buses. In a read operation, the external system can transfer the parity information together with the data bytes. The microprocessor checks whether the resulting parity is even and sends out the corresponding signal. When it sends out an address code, the microprocessor does not perform an address parity check, but it does generate an even parity bit for the address.

SYSTEM NOTE

SYSTEM EXAMPLE 5–6

A DATA TRANSMISSION SYSTEM WITH ERROR DETECTION

A simplified data transmission system is shown in Figure 5–57 to illustrate an application of parity generators/checkers, as well as multiplexers and demultiplexers, and to illustrate the need for data storage in some systems.

In this application, digital data from seven sources are multiplexed onto a single line for transmission to a distant point. The seven data bits (D_0 through D_6) are applied to the multiplexer data inputs and, at the same time, to the even parity generator inputs. The Σ Odd output of the parity generator is used as the even parity bit. This bit is 0 if the number of 1s on the inputs *A* through *I* is even and is a 1 if the number of 1s on *A* through *I* is odd. This bit is D_7 of the transmitted code.

The data-select inputs are repeatedly cycled through a binary sequence, and each data bit, beginning with D_0, is serially passed through and onto the transmission line ($\overline{Y}$). In this example, the transmission line consists of four conductors: one carries the serial data and three carry the timing signals (data selects). There are more sophisticated ways of sending the timing information, but we are using this direct method to illustrate a basic principle.

At the demultiplexer end of the system, the data-select signals and the serial data stream are applied to the demultiplexer. The data bits are distributed by the demultiplexer onto the output lines in the order in which they occurred on the multiplexer inputs. That is, D_0 comes out on the D_0 output, D_1 comes out on the D_1 output, and so on. The parity bit comes out on the D_7 output. These eight bits are temporarily stored and applied to the even parity checker. Not all of the bits are present on the parity checker inputs until the parity bit D_7 comes out and is stored. At this time, the error gate is enabled by the data-select code 111. If the parity is correct, a 0 appears on the Σ Even output, keeping the Error output at 0. If the parity is incorrect, all 1s appear on the error gate inputs, and a 1 on the Error output results.

FIGURE 5–57   **Simplified data transmission system with error detection.**

This particular system has demonstrated the need for data storage. Storage devices will be introduced in Chapter 6 and used in other later chapters.

The timing diagram in Figure 5–58 illustrates a specific case in which two 8-bit words are transmitted, one with correct parity and one with an error.

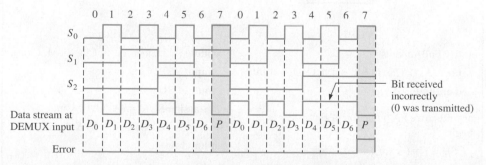

FIGURE 5–58   **Example of data transmission with and without error for the system in Figure 5–57.**

SECTION 5–11 CHECKUP

1. Add an even parity bit to each of the following codes:

   (a) 110100   (b) 01100011

2. Add an odd parity bit to each of the following codes:

   (a) 1010101   (b) 1000001

3. Check each of the even parity codes for an error.

   (a) 100010101

   (b) 1110111001

5–12 LOGIC FUNCTIONS WITH VHDL AND VERILOG

As you know, any logic function must be described with a hardware description language (HDL) in preparation for programming it into a PLD. In this section, several of the functions covered are described in both VHDL and Verilog. Keep in mind that in both languages there is more than one way to describe a given function. *Tutorials for VHDL and Verilog are on the website.*

After completing this section, you should be able to

- Use VHDL and Verilog to describe several types of combinational logic functions
- Specify variable arrays in both VHDL and Verilog
- Use comment lines

Full-Adder

The full-adder (FA) logic symbol with the Boolean expressions for the sum and output carry is shown in Figure 5–59. Using the data flow approach, you can describe the FA using VHDL. The Verilog description is also shown.

FIGURE 5–59 Logic symbol and full-adder VHDL and Verilog descriptions.

4-Bit Parallel Adder

Figure 5–60 shows a 4-bit parallel adder and its VHDL structural description. The VHDL full-adder description is used as a component. For logic functions with multiple input and/or outputs, the VHDL **bit_vector** can be used to group bits together in an array. For example, if you declare an identifier called A for a **bit_vector** data type, the first element in A would be accessed as A(0), the next element would be accessed as A(1), and so on. In the signal declaration, C1, C2, and C3 are the internal carry bits. Inputs A and B and output SUM are bit vector types with 4 bits (0–3) as specified in the entity port statement. Note that **std_logic** and **std_logic_vector** can be used in place of **bit** and **bit_vector**; however, in that case, the lines **library** ieee; and **use** ieee.**std_logic**_1164.all; must precede the **entity** line.

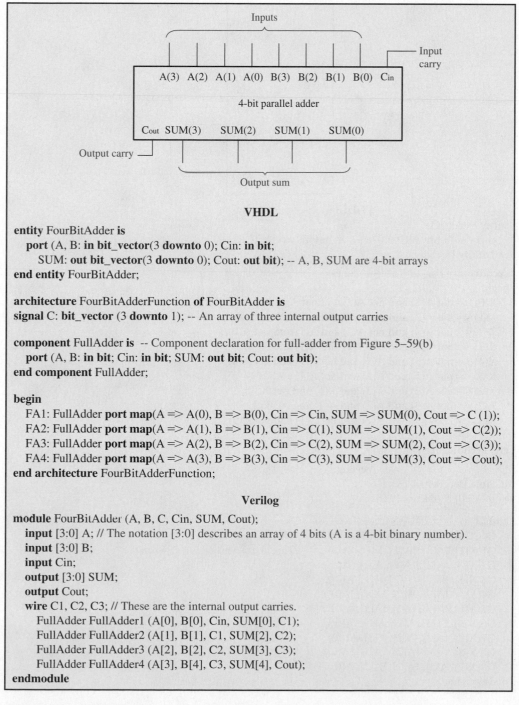

VHDL

entity FourBitAdder **is**
  port (A, B: **in bit_vector**(3 **downto** 0); Cin: **in bit**;
    SUM: **out bit_vector**(3 **downto** 0); Cout: **out bit**); -- A, B, SUM are 4-bit arrays
end entity FourBitAdder;

architecture FourBitAdderFunction **of** FourBitAdder **is**
signal C: **bit_vector** (3 **downto** 1); -- An array of three internal output carries

component FullAdder **is** -- Component declaration for full-adder from Figure 5–59(b)
  port (A, B: **in bit**; Cin: **in bit**; SUM: **out bit**; Cout: **out bit**);
end component FullAdder;

begin
  FA1: FullAdder **port map**(A => A(0), B => B(0), Cin => Cin, SUM => SUM(0), Cout => C (1));
  FA2: FullAdder **port map**(A => A(1), B => B(1), Cin => C(1), SUM => SUM(1), Cout => C(2));
  FA3: FullAdder **port map**(A => A(2), B => B(2), Cin => C(2), SUM => SUM(2), Cout => C(3));
  FA4: FullAdder **port map**(A => A(3), B => B(3), Cin => C(3), SUM => SUM(3), Cout => Cout);
end architecture FourBitAdderFunction;

Verilog

module FourBitAdder (A, B, C, Cin, SUM, Cout);
  input [3:0] A; // The notation [3:0] describes an array of 4 bits (A is a 4-bit binary number).
  input [3:0] B;
  input Cin;
  output [3:0] SUM;
  output Cout;
  wire C1, C2, C3; // These are the internal output carries.
    FullAdder FullAdder1 (A[0], B[0], Cin, SUM[0], C1);
    FullAdder FullAdder2 (A[1], B[1], C1, SUM[2], C2);
    FullAdder FullAdder3 (A[2], B[2], C2, SUM[3], C3);
    FullAdder FullAdder4 (A[3], B[4], C3, SUM[4], Cout);
endmodule

FIGURE 5–60   **Logic symbol and 4-bit parallel adder VHDL and Verilog descriptions.**

The Verilog description is shown in Figure 5–60. Notice that it also uses a special notation to specify an array. The red text in both programs are comments and do not affect the program. In VHDL, a comment line is preceded by two hyphens (--); and in Verilog, a comment line is preceded by two slashes (//).

BCD-to-Decimal Decoder

As you know, the function of the BCD-to-decimal decoder is to accept a group of four BCD bits and decode it into one of the ten outputs that represents the corresponding decimal digit. Again, the use of the **bit_vector** allows the program to reference the input or the output as indexed bits in an array with a single identifier name. Figure 5–61 shows the symbol for the decoder and the corresponding VHDL and Verilog descriptions.

VHDL
```
entity Decoder is
    port (A: in bit_vector(0 to 3); X: out bit_vector (0 to 9));
end entity Decoder;

architecture Function of Decoder is
begin
    X(0) <= not A(3) and not A(2) and not A(1) and not A(0);
    X(1) <= not A(3) and not A(2) and not A(1) and A(0);
    X(2) <= not A(3) and not A(2) and A(1) and not A(0);
    X(3) <= not A(3) and not A(2) and A(1) and A(0);
    X(4) <= not A(3) and A(2) and not A(1) and not A(0);
    X(5) <= not A(3) and A(2) and not A(1) and A(0);
    X(6) <= not A(3) and A(2) and A(1) and not A(0);
    X(7) <= not A(3) and A(2) and A(1) and A(0);
    X(8) <= A(3) and not A(2) and not A(1) and not A(0);
    X(9) <= A(3) and not A(2) and not A(1) and A(0);
end architecture Function
```

Verilog
```
module Decoder (A, X);
input [3:0] A; output [9:0] X;

and
    G0(X[0], !A[3], !A[2], !A[1], !A[0]),      //The ! means NOT.
    G1(X[1], !A[3], !A[2], !A[1], A[0]),       //This line states that X = 1 when the A inputs are 0001.
    G2(X[2], !A[3], !A[2], A[1], !A[0]),
    G3(X[3], !A[3], !A[2], A[1], A[0]),
    G4(X[4], !A[3], A[2], !A[1], !A[0]),
    G5(X[5], !A[3], A[2], !A[1], A[0]),
    G6(X[6], !A[3], A[2], A[1], !A[0]),
    G7(X[7], !A[3], A[2], A[1], A[0]),
    G8(X[8], A[3], !A[2], !A[1], !A[0]),
    G9(X[9], A[3], !A[2], !A[1], A[0]);
endmodule
```

FIGURE 5–61 Logic symbol and BCD-to-decimal decoder VHDL and Verilog descriptions.

SECTION 5–12 CHECKUP

1. Specify the input variable A as an 8-bit binary number (array) in VHDL.

2. Specify the input variable A as an 8-bit binary number (array) in Verilog.

3. Which HDL uses the module?

4. What does this line from the Verilog decoder mean:
   G6(X[6], !A[3], A[2], A[1], !A[0]);

5–13 TROUBLESHOOTING

In this section, the problem of glitches is introduced and examined from a troubleshooting standpoint, using a decoder as an example. A **glitch** is any undesired voltage or current spike (pulse) of very short duration. A glitch can erroneously be interpreted as a valid signal by a logic circuit and may cause improper operation.

After completing this section, you should be able to

- **Explain what a glitch is**
- **Determine the cause of glitches in a decoder application**
- **Use the method of output strobing to eliminate glitches**

A 3-line-to-8-line decoder (binary-to-octal) in Figure 5–62 illustrates how glitches occur and how to identify their cause. The $A_2 A_1 A_0$ inputs of the decoder are sequenced through a binary count, and the resulting waveforms of the inputs and outputs can be displayed on the screen of a logic analyzer or oscilloscope, as shown in Figure 5–62. A_2 transitions are delayed from A_1 transitions and A_1 transitions are delayed from A_0 transitions. This commonly occurs when waveforms are generated by a binary counter.

FIGURE 5–62   Decoder waveforms with output glitches.

The output waveforms are correct except for the glitches that occur on some of the output signals. A logic analyzer or an oscilloscope can be used to display glitches, which are normally very difficult to see. Generally, the logic analyzer is preferred, especially for low

HANDS ON TIP
In addition to glitches that are the result of differences in propagation delays, as you have seen in the case of a decoder, other types of unwanted noise spikes can also be a problem. Current and voltage spikes on the V_{CC} and ground lines are caused by the fast switching waveforms in digital circuits. This problem can be minimized by proper printed circuit board layout and decoupling capacitors. Switching spikes can be absorbed by decoupling the circuit board with a 1 μF capacitor from V_{CC} to ground. Also, smaller decoupling capacitors (0.022 μF to 0.1 μF) should be distributed at various points between V_{CC} and ground over the circuit board. Decoupling should be done especially near devices that are switching at higher rates or driving more loads such as oscillators, counters, buffers, and bus drivers.

repetition rates (less than 10 kHz) and/or irregular occurrence because most logic analyzers have a *glitch capture* capability. Oscilloscopes can be used to observe glitches with reasonable success, particularly if the glitches occur at a regular high repetition rate (greater than 10 kHz).

The points of interest indicated by the highlighted areas on the input waveforms in Figure 5–62 are displayed as shown in Figure 5–63. At point 1 there is a transitional state

FIGURE 5–63   **Decoder waveform displays showing how transitional input states produce glitches in the output waveforms.**

FIGURE 5–64   **Application of a strobe waveform to eliminate glitches on decoder outputs.**

of 000 due to delay differences in the waveforms. This causes the first glitch on the $\overline{0}$ output of the decoder. At point 2 there are two transitional states, 010 and 000. These cause the glitch on the $\overline{2}$ output of the decoder and the second glitch on the $\overline{0}$ output, respectively. At point 3 the transitional state is 100, which causes the first glitch on the $\overline{4}$ output of the decoder. At point 4 the two transitional states, 110 and 100, result in the glitch on the $\overline{6}$ output and the second glitch on the $\overline{4}$ output, respectively.

One way to eliminate the glitch problem is a method called **strobing,** in which the decoder is enabled by a strobe pulse only during the times when the waveforms are not in transition. This method is illustrated in Figure 5–64.

SECTION 5–13 CHECKUP

1. Define the term *glitch.*

2. Explain the basic cause of glitches in decoder logic.

3. Define the term *strobe.*

SUMMARY

- Half-adder and full-adder operations are summarized in Figure 5–65.
- Standard logic functions from the 74XX series are available for use in a programmable logic design.

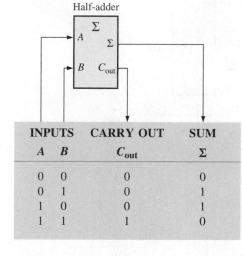

Half-adder

INPUTS		CARRY OUT	SUM
A	B	C_{out}	Σ
0	0	0	0
0	1	0	1
1	0	0	1
1	1	1	0

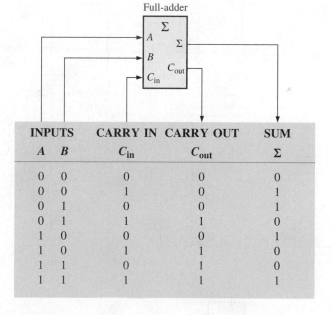

Full-adder

INPUTS		CARRY IN	CARRY OUT	SUM
A	B	C_{in}	C_{out}	Σ
0	0	0	0	0
0	0	1	0	1
0	1	0	0	1
0	1	1	1	0
1	0	0	0	1
1	0	1	1	0
1	1	0	1	0
1	1	1	1	1

FIGURE 5–65

KEY TERMS

Key terms and other bold terms in the chapter are defined in the end-of-book glossary.

Cascading Connecting two or more similar devices in a manner that expands the capability of one device.

Decoder A digital circuit that converts coded information into a familiar or noncoded form.

Demultiplexer (DEMUX) A circuit that switches digital data from one input line to several output lines in a specified time sequence.

Encoder A digital circuit that converts information to a coded form.

Full-adder A digital circuit that adds two bits and an input carry to produce a sum and an output carry.

Glitch A voltage or current spike of short duration, usually unintentionally produced and unwanted.

Half-adder A digital circuit that adds two bits and produces a sum and an output carry. It cannot handle input carries.

Look-ahead carry A method of binary addition whereby carries from preceding adder stages are anticipated, thus eliminating carry propagation delays.

Multiplexer (MUX) A circuit that switches digital data from several input lines onto a single output line in a specified time sequence.

Parity bit A bit attached to each group of information bits to make the total number of 1s odd or even for every group of bits.

Priority encoder An encoder in which only the highest value input digit is encoded and any other active input is ignored.

Ripple carry A method of binary addition in which the output carry from each adder becomes the input carry of the next higher-order adder.

FIXED-FUNCTION LOGIC DEVICES

Pin numbers are in parentheses. Go to ti.com for data sheets.

74HC42
BCD-to-decimal decoder

74LS47
BCD-to-7-segment decoder/driver

74LS85
4-bit magnitude comparator

74LS138
3-line-to-8-line decoder

74LS139
Dual 2-line-to-4-line decoder

74HC147
Decimal-to-BCD priority encoder

74LS148
Octal-to-binary encoder

74LS151
8-input data selector/multiplexer

74HC154
1-of-16-line decoder

74HC157
Quad 2-input data selector/multiplexer

74LS280
9-bit parity generator/checker

74LS283
4-bit adder

TRUE/FALSE QUIZ

Answers are at the end of the chapter.

1. A half-adder adds two binary bits.

2. A half-adder has a sum output only.

3. A full-adder adds three bits and produces two outputs.

4. Two 4-bit numbers can be added using two full-adders.

5. When the two input bits are both 1 and the carry input bit is a 1, the sum output of a full adder is 0.

6. A comparator determines when two binary numbers are equal.

7. A decoder detects the presence of a specified combination of input bits.

8. The 4-line-to-10-line decoder and the 1-of-10 decoder are two different types.

9. An encoder essentially performs a reverse decoder function.

10. A multiplexer is a logic circuit that allows digital information from a single source to be routed onto several lines.

11. Any logic function can be described using VHDL or Verilog.

SELF-TEST

Answers are at the end of the chapter.

1. A half-adder is characterized by
   (a) two inputs and two outputs   (b) three inputs and two outputs
   (c) two inputs and three outputs   (d) two inputs and one output

2. A full-adder is characterized by
   (a) two inputs and two outputs   (b) three inputs and two outputs
   (c) two inputs and three outputs   (d) two inputs and one output

3. The inputs to a full-adder are $A = 1, B = 1, C_{in} = 0$. The outputs are
   (a) $\Sigma = 1, C_{out} = 1$ (b) $\Sigma = 1, C_{out} = 0$
   (c) $\Sigma = 0, C_{out} = 1$ (d) $\Sigma = 0, C_{out} = 0$

4. A 4-bit parallel adder can add
   (a) two 4-bit binary numbers   (b) two 2-bit binary numbers
   (c) four bits at a time   (d) four bits in sequence

5. To expand a 4-bit parallel adder to an 8-bit parallel adder, you must
   (a) use four 4-bit adders with no interconnections
   (b) use two 4-bit adders and connect the sum outputs of one to the bit inputs of the other
   (c) use eight 4-bit adders with no interconnections
   (d) use two 4-bit adders with the carry output of one connected to the carry input of the other

6. If a magnitude comparator has $A = 1011$ and $B = 1001$ on its inputs, the outputs are
   (a) $A > B = 0, A < B = 1, A = B = 0$ (b) $A > B = 1, A < B = 0, A = B = 0$
   (c) $A > B = 1, A < B = 1, A = B = 0$ (d) $A > B = 0, A < B = 0, A = B = 1$

7. If a 1-of-16 decoder with active-LOW outputs exhibits a LOW on the decimal 12 output, what are the inputs?
   (a) $A_3A_2A_1A_0 = 1010$ (b) $A_3A_2A_1A_0 = 1110$
   (c) $A_3A_2A_1A_0 = 1100$ (d) $A_3A_2A_1A_0 = 0100$

8. A BCD-to-7 segment decoder has 0100 on its inputs. The active outputs are
   (a) a, c, f, g (b) b, c, f, g (c) b, c, e, f (d) b, d, e, g

9. In general, a multiplexer has
   (a) one data input, several data outputs, and selection inputs
   (b) one data input, one data output, and one selection input
   (c) several data inputs, several data outputs, and selection inputs
   (d) several data inputs, one data output, and selection inputs

10. Data selectors are basically the same as

   (a) decoders   (b) demultiplexers   (c) multiplexers   (d) encoders

11. Which of the following codes exhibit even parity?

   (a) 10011000   (b) 01111000   (c) 11111111
   (d) 11010101   (e) all   (f) both answers (b) and (c)

12. An entity and architecture are associated with

   (a) Abel   (b) Verilog   (c) VHDL   (d) PLD

PROBLEMS

Answers to odd-numbered problems are at the end of the book.

SECTION 5–1   A System

1. Assume the tablet-bottling control system is set for 36 tablets per bottle. Determine the binary numbers in register A and register B after 5 bottles have been filled.

2. After one more bottle is filled from Problem 1, determine the output of the adder.

3. For Problem 1, what is the output of the BCD-to-binary code converter A?

4. For Problem 1, determine the active outputs from the 3-digit BCD-to-7-segment decoder A.

SECTION 5–2   Half and Full Adders

5. For the full-adder of Figure 5–9, determine the logic state (1 or 0) at each gate output for the following inputs:

   (a) $A = 1, B = 1, C_{in} = 1$ (b) $A = 0, B = 1, C_{in} = 1$ (c) $A = 0, B = 1, C_{in} = 0$

6. What are the full-adder inputs that will produce each of the following outputs:

   (a) $\Sigma = 0, C_{out} = 0$ (b) $\Sigma = 1, C_{out} = 0$
   (c) $\Sigma = 1, C_{out} = 1$ (d) $\Sigma = 0, C_{out} = 1$

7. Determine the outputs of a full-adder for each of the following inputs:

   (a) $A = 1, B = 0, C_{in} = 0$ (b) $A = 0, B = 0, C_{in} = 1$
   (c) $A = 0, B = 1, C_{in} = 1$ (d) $A = 1, B = 1, C_{in} = 1$

SECTION 5–3   Parallel Adders

8. For the parallel adder in Figure 5–66, determine the complete sum by analysis of the logical operation of the circuit. Verify your result by longhand addition of the two input numbers.

9. Repeat Problem 8 for the circuit and input conditions in Figure 5–67.

FIGURE 5–66   **FIGURE 5–67**

10. The circuit shown in Figure 5–68 is a 4-bit circuit that can add or subtract numbers in a form used in computers (positive numbers in true form; negative numbers in complement form). (a) Explain what happens when the $\overline{Add}/Subt.$ input is HIGH. (b) What happens when $\overline{Add}/Subt.$ is LOW?

11. For the circuit in Figure 5–68, assume the inputs are $\overline{Add}/Subt. = 1, A = 1001$ and $B = 1100$. What is the output?

FIGURE 5–68

FIGURE 5–69

12. The input waveforms in Figure 5–69 are applied to a 2-bit adder. Determine the waveforms for the sum and the output carry in relation to the inputs by constructing a timing diagram.

13. The following sequences of bits (right-most bit first) appear on the inputs to a 4-bit parallel adder. Determine the resulting sequence of bits on each sum output.

A_1	1001
A_2	1110
A_3	0000
A_4	1011
B_1	1111
B_2	1100
B_3	1010
B_4	0010

14. In the process of checking a 4-bit parallel adder, the following logic levels are observed on the inputs: HIGH (MSB), HIGH, HIGH, HIGH, and HIGH (MSB), LOW, LOW, HIGH. The following levels are observed on the outputs: LOW (MSB), HIGH, LOW, HIGH. Determine if the adder is functioning properly.

SECTION 5–4 Ripple Carry and Look-Ahead Carry Adders

15. Each of the eight full-adders in an 8-bit parallel ripple carry adder exhibits the following propagation delays:

A to Σ and C_{out}:   40 ns
B to Σ and C_{out}:   40 ns
C_{in} to Σ: 35 ns
C_{in} to C_{out}:   25 ns

Determine the maximum total time for the addition of two 8-bit numbers.

16. Show the additional logic circuitry necessary to make the 4-bit look-ahead carry adder in Figure 5–21 into a 5-bit adder.

SECTION 5–5 Comparators

17. The waveforms in Figure 5–70 are applied to the comparator as shown. Determine the output ($A = B$) waveform.

FIGURE 5–70

18. For the 4-bit comparator in Figure 5–71, plot each output waveform for the inputs shown. The outputs are active-HIGH.

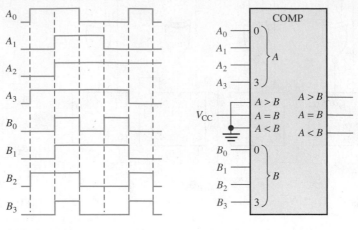

FIGURE 5–71

19. For each set of binary numbers, determine the output states for the comparator of Figure 5–25.

    (a) $A_3A_2A_1A_0 = 1100$ **(b)** $A_3A_2A_1A_0 = 1000$ **(c)** $A_3A_2A_1A_0 = 0100$

          $B_3B_2B_1B_0 = 1001$ $B_3B_2B_1B_0 = 1011$ $B_3B_2B_1B_0 = 0100$

SECTION 5–6   Decoders

20. When a HIGH is on the output of each of the decoding gates in Figure 5–72, what is the binary code appearing on the inputs? The MSB is A_3.

(a)   (b)

(c)   (d)

FIGURE 5–72

21. Show the decoding logic for each of the following codes if an active-HIGH (1) output is required:

    (a) 1101   **(b)** 1000   **(c)** 11011   **(d)** 11100

    (e) 101010   **(f)** 111110   **(g)** 000101   **(h)** 1110110

22. Solve Problem 21, given that an active-LOW (0) output is required.

23. You wish to detect only the presence of the codes 1010, 1100, 0001, and 1011. An active-HIGH output is required to indicate their presence. Develop the decoding logic with a single output that will indicate when any one of these codes is on the inputs. For any other code, the output must be LOW.

24. If the input waveforms are applied to the decoding logic as indicated in Figure 5–73, sketch the output waveform in proper relation to the inputs.

FIGURE 5–73

25. BCD numbers are applied sequentially to the BCD-to-decimal decoder in Figure 5–74. Draw a timing diagram, showing each output in the proper relationship with the others and with the inputs.

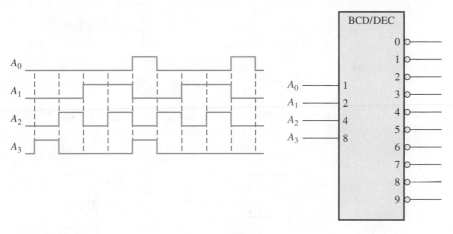

FIGURE 5–74

26. A 7-segment decoder/driver drives the display in Figure 5–75. If the waveforms are applied as indicated, determine the sequence of digits that appears on the display.

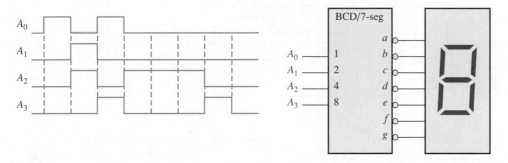

FIGURE 5–75

SECTION 5–7   Encoders

27. For the decimal-to-BCD encoder logic of Figure 5–38, assume that the 9 input and the 3 input are both HIGH. What is the output code? Is it a valid BCD (8421) code?

28. A priority encoder has active-LOW levels on inputs 1, 5, and 9. What BCD code appears on the outputs if all the other inputs are HIGH?

SECTION 5–8   Code Converters

29. Convert the following decimal numbers to BCD and then to binary.
    (a) 2   **(b)** 8   **(c)** 13   **(d)** 26   **(e)** 33

30. Show the logic required to convert a 10-bit binary number to Gray code, and use that logic to convert the following binary numbers to Gray code:
    (a) 1010101010   **(b)** 1111100000   **(c)** 0000001110   **(d)** 1111111111

31. Show the logic required to convert a 10-bit Gray code to binary, and use that logic to convert the following Gray code words to binary:

(a) 1010000000   (b) 0011001100   (c) 1111000111   (d) 0000000001

SECTION 5–9 Multiplexers (Data Selectors)

32. For the multiplexer in Figure 5–76, determine the output for the following input states: $D_0 = 0, D_1 = 1, D_2 = 1, D_3 = 0, S_0 = 1, S_1 = 0$.

33. If the data-select inputs to the multiplexer in Figure 5–76 are sequenced as shown by the waveforms in Figure 5–77, determine the output waveform with the data inputs specified in Problem 32.

FIGURE 5–76

FIGURE 5–77

34. The waveforms in Figure 5–78 are observed on the inputs of an 8-input multiplexer. Sketch the Y output waveform.

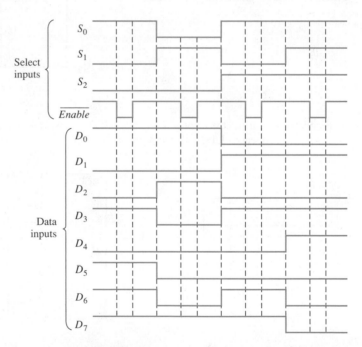

FIGURE 5–78

SECTION 5–10 Demultiplexers

35. Develop the total timing diagram (inputs and outputs) for a demultiplexer in which the inputs are as follows: The data-select inputs are repetitively sequenced through a straight binary count beginning with 0000, and the data input is a serial data stream carrying BCD data representing the decimal number 2468. The least significant digit (8) is first in the sequence, with its LSB first, and it should appear in the first 4-bit positions of the output.

SECTION 5–11 Parity Generators/Checkers

36. The waveforms in Figure 5–79 are applied to the 4-bit parity logic. Determine the output waveform in proper relation to the inputs. For how many bit times does even parity occur, and how is it indicated? The timing diagram includes eight bit times.

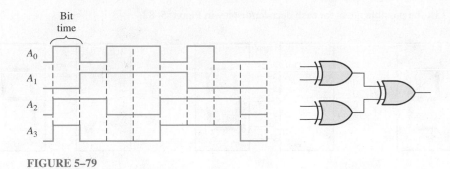

FIGURE 5–79

37. Determine the Σ Even and the Σ Odd outputs of a 9-bit parity generator/checker for the inputs in Figure 5–80. Refer to the function table in Figure 5–56.

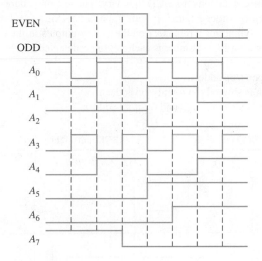

FIGURE 5–80

SECTION 5–12   Logic Functions with VHDL and Verilog

38. Write a VHDL program for an 8-bit parallel adder.

39. Write a Verilog program for an 8-bit parallel adder.

40. Write a VHDL program for a decimal-to-BCD encoder.

41. Write a Verilog program for a decimal-to-BCD encoder.

SECTION 5–13   Troubleshooting

42. The full-adder in Figure 5–81 is tested under all input conditions with the input waveforms shown. From your observation of the Σ and C_{out} waveforms, is it operating properly, and if not, what is the most likely fault?

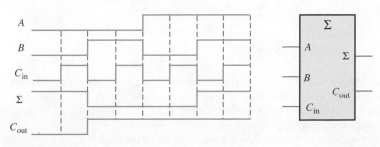

FIGURE 5–81

43. List the possible faults for each decoder/display in Figure 5–82.

(a)   (b)   (c)

FIGURE 5–82

44. Develop a systematic test procedure to check out the complete operation of the keyboard encoder in Figure 5–40.

45. You are testing a BCD-to-binary converter consisting of 4-bit adders as shown in Figure 5–83. First verify that the circuit converts BCD to binary. The test procedure calls for applying BCD numbers in sequential order beginning with 0_{10} and checking for the correct binary output. What symptom or symptoms will appear on the binary outputs in the event of each of the following faults? For what BCD number is each fault *first* detected?
   (a) The A_1 input is open (top adder).
   (b) The C_{out} is open (top adder).
   (c) The Σ_4 output is shorted to ground (top adder).
   (d) The 32 output is shorted to ground (bottom adder).

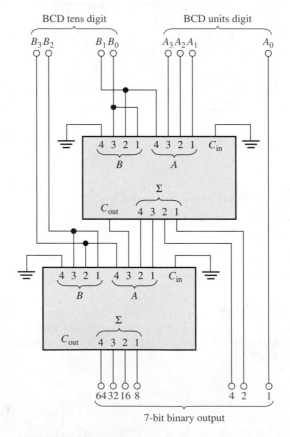

FIGURE 5–83

46. For the 7-segment display multiplexing system in Figure 5–49, determine the most likely cause or causes for each of the following symptoms:
   (a) The *B*-digit (MSD) display does not turn on at all.
   (b) Neither 7-segment display turns on.
   (c) The *f*-segment of both displays appears to be on all the time.
   (d) There is a visible flicker on the displays.

47. Develop a systematic procedure to fully test the data selector shown in Figure 5–47.

48. During the testing of the data transmission system in Figure 5–57, a code is applied to the D_0 through D_6 inputs that contains an odd number of 1s. A single bit error is deliberately introduced on the serial data transmission line between the MUX and the DEMUX, but the system does not indicate an error (error output $= 0$). After some investigation, you check the inputs to the even parity checker and find that D_0 through D_6 contain an even number of 1s, as you would expect. Also, you find that the D_7 parity bit is a 1. What are the possible reasons for the system not indicating the error?

49. In general, describe how you would fully test the data transmission system in Figure 5–57, and specify a method for the introduction of parity errors.

Special Problems

50. Modify the design of the 7-segment display multiplexing system in Figure 5–49 to accommodate two additional digits.

51. Using Table 5–2, write the SOP expressions for the Σ and C_{out} of a full-adder and show the logic diagrams. Then implement them with inverters and AND-OR logic.

52. Implement the logic function specified in Table 5–11 and write a VHDL program for the implementation.

TABLE 5–11

INPUTS				OUTPUT
A_3	A_2	A_1	A_0	Y
0	0	0	0	0
0	0	0	1	0
0	0	1	0	1
0	0	1	1	1
0	1	0	0	0
0	1	0	1	0
0	1	1	0	1
0	1	1	1	1
1	0	0	0	1
1	0	0	1	0
1	0	1	0	1
1	0	1	1	1
1	1	0	0	0
1	1	0	1	1
1	1	1	0	0
1	1	1	1	1

53. Using two of the 6-position adder modules from Figure 5–17, develop a 12-position voting system.

MULTISIM TROUBLESHOOTING PRACTICE

54. Open file P05-54 and follow the instructions given there.

55. Open file P05-55 and follow the instructions given there.

56. Open file P05-56 and follow the instructions given there.

57. Open file P05-57 and follow the instructions given there.

ANSWERS TO SECTION CHECKUPS

SECTION 5–1   A System

1. Register A: 50 = 110010; Register B: 200 = 11001000

2. Register A: 25 = 11001; Register B: 125 = 1111101

3. The comparator determines when a bottle is filled with the preset number of tablets.

4. Decoder B detects when register B contains the maximum number.

SECTION 5–2   Half and Full Adders

1. (a) $\Sigma = 1, C_{out} = 0$
   (b) $\Sigma = 0, C_{out} = 0$
   (c) $\Sigma = 1, C_{out} = 0$
   (d) $\Sigma = 0, C_{out} = 1$

2. $\Sigma = 1, C_{out} = 1$

SECTION 5–3   Parallel Adders

1. $C_{out}\Sigma_4\Sigma_3\Sigma_2\Sigma_1 = 11001$

2. Three 4-bit adders are required to add two 10-bit numbers.

SECTION 5–4   Ripple Carry and Look-Ahead Carry Adders

1. $C_g = 0, C_p = 1$

2. $C_{out} = 1$

SECTION 5–5   Comparators

1. $A > B = 1, A = B = 0, A < B = 0$, when $A = 1011$ and $B = 1010$

2. $A < B = 1; A = B = 0; A > B = 0$

SECTION 5–6   Decoders

1. Output 5 is active when 101 is on the inputs.

2. Four 1-of-16 decoders are used to decode a 6-bit binary number.

3. Active-HIGH output drives a common-cathode LED display.

SECTION 5–7   Encoders

1. (a) $A_0 = 1, A_1 = 1, A_2 = 0, A_3 = 1$
   (b) No, this is not a valid BCD code.
   (c) Only one input can be active for a valid output.

2. (a) $\overline{A}_3 = 0, \overline{A}_2 = 1, \overline{A}_1 = 1, \overline{A}_0 = 1$
   (b) The output is 0111, which is the complement of 1000 (8).

SECTION 5–8   Code Converters

1. 10000101 (BCD) = 1010101_2

2. An 8-bit binary-to-Gray converter consists of seven exclusive-OR gates in an arrangement like that in Figure 5–41 but with inputs B_0–B_7.

SECTION 5–9 Multiplexers (Data Selectors)

1. The output is 0.
2. 5 select inputs ($2^5 = 32$).
3. The data output alternates between LOW and HIGH as the data-select inputs sequence through the binary states.
4. **(a)** The multiplexer multiplexes the two BCD codes to the 7-segment decoder.
   (b) The BCD-to-7-segment decoder decodes the BCD to energize the display.
   (c) The 2-line-to-4-line decoder enables the 7-segment displays alternately.

SECTION 5–10 Demultiplexers

1. A decoder can be used as a multiplexer by using the input lines for data selection and an Enable line for data input.
2. The outputs are all HIGH except D_{10}, which is LOW.

SECTION 5–11 Parity Generators/Checkers

1. **(a)** Even parity: $\underline{1}110100$
   (b) Even parity: $\underline{0}01100011$
2. **(a)** Odd parity: $\underline{1}1010101$
   (b) Odd parity: $\underline{1}1000001$
3. **(a)** Code is correct, four 1s.
   (b) Code is in error, seven 1s

SECTION 5–12 Logic Functions with VHDL and Verilog

1. A: **in bit_vector** (7 **downto** 0);
2. **input** [7:0] A;
3. Module is used in Verilog.
4. X = 6 when the A inputs are 0110.

SECTION 5–13 Troubleshooting

1. A glitch is a very short-duration voltage spike (usually unwanted).
2. Glitches are caused by transition states.
3. Strobe is the enabling of a device for a specified period of time when the device is not in transition.

ANSWERS TO RELATED PROBLEMS FOR EXAMPLES

5–1 $\Sigma = 1, C_{\text{out}} = 1$

5–2 $\Sigma_1 = 0, \Sigma_2 = 0, \Sigma_3 = 1, \Sigma_4 = 1$

5–3 $1011 + 1010 = 10101$

5–4 See Figure 5–84.

FIGURE 5–84

5–5   See Figure 5–85.

5–6 $A > B = 0, A = B = 0, A < B = 1$

5–7   See Figure 5–86.

FIGURE 5–85   **FIGURE 5–86**

5–8   Output 22

5–9   See Figure 5–87.

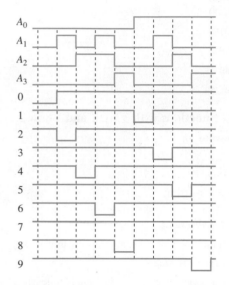

FIGURE 5–87

5–10 All inputs LOW: $\overline{A}_0 = 0, \overline{A}_1 = 1, \overline{A}_2 = 1, \overline{A}_3 = 0$

All inputs HIGH: All outputs HIGH.

5–11 BCD 01000001

$\qquad\qquad\qquad\longrightarrow\qquad$ 00000001   1

$\qquad\qquad\qquad\longrightarrow\qquad$ <u>00101000</u>   <u>40</u>

$\qquad\qquad$ Binary   00101001   41

5–12 Seven exclusive-OR gates

5–13 See Figure 5–88.

FIGURE 5–88

5–14 $D_0: S_3 = 0, S_2 = 0, S_1 = 0, S_0 = 0$

$\qquad D_4: S_3 = 0, S_2 = 1, S_1 = 0, S_0 = 0$

$\qquad D_8: S_3 = 1, S_2 = 0, S_1 = 0, S_0 = 0$

$\qquad D_{13}: S_3 = 1, S_2 = 1, S_1 = 0, S_0 = 1$

5–15   See Figure 5–89.

FIGURE 5–89

5–16   See Figure 5–90.

FIGURE 5–90

5–17   See Figure 5–91.

FIGURE 5–91

ANSWERS TO TRUE/FALSE QUIZ

1. T   **2.** F   **3.** T   **4.** F   **5.** F   **6.** T

7. T   **8.** F   **9.** T   **10.** F   **11.** T

ANSWERS TO SELF-TEST

1. (a)   **2.** (b)   **3.** (c)   **4.** (a)   **5.** (d)   **6.** (b)

7. (c)   **8.** (b)   **9.** (d)   **10.** (c)   **11.** (f)   **12.** (c)

CHAPTER 6

LATCHES, FLIP-FLOPS, AND TIMERS

OUTLINE

OBJECTIVES

- Describe a traffic signal control system
- Use logic gates to construct basic latches
- Explain the difference between an S-R latch and a D latch
- Recognize the difference between a latch and a flip-flop
- Explain how D and J-K flip-flops differ
- Understand the significance of propagation delays, set-up time, hold time, maximum operating frequency, minimum clock pulse widths, and power dissipation in the application of flip-flops
- Apply flip-flops in basic applications
- Explain how retriggerable and nonretriggerable one-shots differ
- Connect a 555 timer to operate as either an astable multivibrator or a one-shot
- Use VHDL and Verilog to implement flip-flops
- Use VHDL and Verilog to implement a traffic signal control system
- Discuss system troubleshooting

KEY TERMS

Latch	**Preset**
Bistable	**Clear**
SET	**Propagation delay time**
RESET	**Set-up time**
Clock	**Hold time**
Edge-triggered flip-flop	**Power dissipation**
D flip-flop	**Monostable**
Synchronous	**One-shot**
J-K flip-flop	**Astable**
Toggle	**Timer**

INTRODUCTION

A traffic signal control system is introduced at the beginning of this chapter. Bistable, monostable, and astable logic devices called *multivibrators* are covered. Two categories of bistable devices are the latch and the flip-flop. Bistable devices have two stable states, called SET and RESET; they can retain either of these states

VISIT THE WEBSITE
Study aids for this chapter are available at
http://pearsonhighered.com/floyd

indefinitely, making them useful as storage devices. The basic difference between latches and flip-flops is the way in which they are changed from one state to the other. The flip-flop is a basic building block for counters, registers, and other sequential control logic and is used in certain types of memories. The monostable multivibrator, commonly known as the one-shot, has only one stable state. A one-shot produces a single controlled-width pulse when activated or triggered. The astable multivibrator has no stable state and is used primarily as an oscillator, which is a self-sustained waveform generator. Pulse oscillators are used as the sources for timing waveforms in digital systems. Using VHDL and Verilog to describe bistable logic as well as the traffic signal control system is covered.

6–1 A SYSTEM

In this section, a traffic signal control system is presented to illustrate how combinational logic, flip-flops, and timers can be used to accomplish a specified function. The system controls the sequencing of traffic lights at a busy main street and an occasionally used side street.

After completing this section, you should be able to

- **Describe how the system is implemented**
- **Explain how timers are used in this system**
- **Explain how flip-flops are used in this system**

Timing Requirements

The following are the timing requirements for the traffic signal control system. These requirements are illustrated in Figure 6–1.

- The green light for the main street will stay on for a minimum of 25 s or as long as there is no vehicle on the side street.
- The green light for the side street will stay on until there is no vehicle on the side street up to a maximum of 25 s.
- The yellow caution light will stay on for 4 s between changes from green to red on both the main street and the side street.

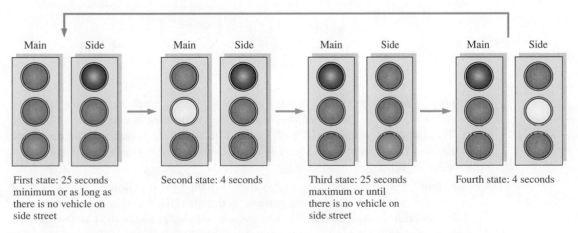

First state: 25 seconds minimum or as long as there is no vehicle on side street

Second state: 4 seconds

Third state: 25 seconds maximum or until there is no vehicle on side street

Fourth state: 4 seconds

FIGURE 6–1   Traffic light timing sequence.

The State Diagram

A state diagram graphically shows the sequence of states, the conditions for each state, and the requirements for transitions from one state to the next. Although Figure 6–1 is a basic form of state diagram, a standard type of diagram shows the logic states and the Boolean expressions for the transitions.

DEFINING THE VARIABLES   The variables that determine how the system sequences through the various states are defined as follows:

- V_s A vehicle is present on the side street.
- T_L The 25 s timer (long timer) is *on*.
- T_S The 4 s timer (short timer) is *on*.

A complemented variable indicates the opposite condition.

STATE DESCRIPTIONS   A **state diagram** is shown in Figure 6–2. Each of the four states is assigned a 2-bit Gray code as indicated. A looping arrow means that the system remains in a state, and an arrow between states means that the system transitions to the next state. The Boolean expression or variable associated with each of the arrows in the state diagram indicates the condition under which the system remains in a state or transitions to the next state.

FIGURE 6–2   State diagram for the traffic signal control system.

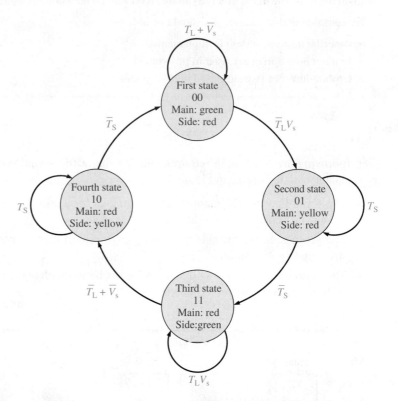

First State   The Gray code is 00. In this state (S_1), the light is green on the main street and red on the side street for 25 s when the long timer is *on* or there is no vehicle on the side street. This condition is expressed as $T_L + \overline{V}_s$. The system transitions to the next state when the long timer goes *off* and there is a vehicle on the side street. This condition is expressed as $\overline{T}_L V_s$.

Second State   The Gray code is 01. In this state (S_2), the light is yellow on the main street and red on the side street. The system remains in this state for 4 s when the short timer is *on*. This condition is expressed as T_S. The system transitions to the next state when the short timer goes *off*. This condition is expressed as $\overline{T}_S$.

Third State   The Gray code is 11. In this state (S_3), the light is red on the main street and green on the side street for 25 s when the long timer is *on* as long as there is a vehicle on the side street. This condition is expressed as $T_L V_s$. The system transitions to the next state when the long timer goes *off* or when there is no vehicle on the side street. This condition is expressed as $\overline{T}_L + \overline{V}_s$.

Fourth State The Gray code is 10. In this state (S_4), the light is red on the main street and yellow on the side street. The system remains in this state for 4 s when the short timer is *on*. This condition is expressed as T_S. The system transitions back to the first state when the short timer goes *off*. This condition is expressed as $\overline{T}_S$.

Block Diagram

The traffic signal control system consists of three parts: combinational logic, sequential logic, and timing circuits, as shown in Figure 6–3.

FIGURE 6–3 **Block diagram of the traffic signal control system.**

The combinational logic portion of the system provides outputs to turn the signal lights on and off. It also provides trigger outputs to start the long and short timers. The input sequence to this logic represents the four states described by the state diagram. The timing circuits portion of the system provides the 25 s and the 4 s timing outputs. The sequential logic produces the sequence of 2-bit Gray codes representing the four states.

The Combinational Logic

The combinational logic consists of a state decoder, light output logic, and trigger logic.

STATE DECODER This logic decodes the 2-bit Gray code from the sequential logic to determine which of the four states the system is in. The inputs to the state decoder are the two Gray code bits G_1 and G_0. There are four state outputs S_1, S_2, S_3, and S_4. For each of the four input codes, one and only one of the outputs is activated. The logic and Boolean expressions are shown in Figure 6–4.

LIGHT OUTPUT LOGIC This logic has the four state outputs of the state decoder as its inputs and produces six outputs to turn the traffic lights on and off. These outputs are designated *MR*, *MY*, *MG* (main red, main yellow, main green) and *SR*, *SY*, *SG* (side red, side yellow, side green).

Gray code state inputs { G_0, G_1

S_1
S_2 } State outputs
S_3
S_4

$S_1 = \overline{G_1}\overline{G_0}$
$S_2 = \overline{G_1}G_0$
$S_3 = G_1G_0$
$S_4 = G_1\overline{G_0}$

FIGURE 6–4 **State decoder logic and output expressions.**

Refer to the state diagram in Figure 6–2. The main red is *on* in the third state (S_3) or in the fourth state (S_4), so the Boolean expression is

$$MR = S_3 + S_4$$

The main yellow is *on* in the second state (S_2), so the expression is

$$MY = S_2$$

The main green is *on* in the first state (S_1), so the expression is

$$MG = S_1$$

Similarly, the state diagram is used to obtain the following expressions for the side street:

$$SR = S_1 + S_2$$
$$SY = S_4$$
$$SG = S_3$$

TRIGGER LOGIC   The trigger logic produces two outputs, the long trigger output and the short trigger output. The long trigger output initiates the 25 s timer on a LOW-to-HIGH transition at the beginning of the first or third states. The short trigger output initiates the 4 s timer on a LOW-to-HIGH transition at the beginning of the second or fourth states. The Boolean expressions for this logic are

$$LongTrig = S_1 + S_3$$
$$ShortTrig = S_2 + S_4$$

Timing Circuits

The timing circuits portion of the traffic signal control system in Figure 6–3 consists of a 25 s timer and a 4 s timer, as shown in Figure 6–5. The 25 s timer and the 4 s timer are triggered by the long trigger and short trigger signals, respectively, from the combinational logic. The system clock is produced by an astable multivibrator (covered later) or other type of oscillator.

FIGURE 6–5   **Block diagram of the timing circuits and the system clock generator.**

Sequential Logic

The sequential logic controls the sequencing of the traffic signals, based on inputs from the timing circuits and the side street vehicle sensor. The sequential logic produces a 2-bit Gray code sequence for each of the four states that were described in Figure 6–2.

THE COUNTER   The sequential logic consists of a 2-bit Gray code counter made up of two flip-flops and the associated logic, as shown in Figure 6–6. The counter produces the four-state sequence on outputs G_0 and G_1. Transitions from one state to the next are determined by the short timer (T_S), the long timer (T_L), and vehicle sensor for the side street (V_s) inputs. The 10 kHz clock input (CLK) comes from the system clock oscillator in the timing circuits.

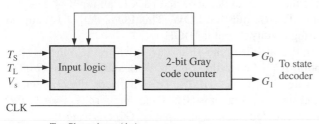

FIGURE 6–6   Block diagram of the sequential logic.

T_S: Short timer (4 s)
T_L: Long timer (25 s)
V_s: Vehicle sensor for the side street

SECTION 6–1 CHECKUP*

1. Referring to the state diagram, how long can the system remain in the first state?

2. How long can the system remain in the fourth state?

3. What determines how long the main street light stays red?

*Answers are at the end of the chapter.

6–2 LATCHES

The latch is a type of temporary storage device that has two stable states (bistable) and is normally placed in a category separate from that of flip-flops. Latches are similar to flip-flops because they are bistable devices that can reside in either of two states using a feedback arrangement, in which the outputs are connected back to the opposite inputs. The main difference between latches and flip-flops is in the method used for changing their state.

After completing this section, you should be able to

- **Explain the operation of a basic S-R latch**
- **Explain the operation of a gated S-R latch**
- **Explain the operation of a gated D latch**
- **Implement an S-R or D latch with logic gates**

The S-R (SET-RESET) Latch

A latch is a type of **bistable** logic device or **multivibrator.** An active-HIGH input S-R (SET-RESET) latch is formed with two cross-coupled NOR gates, as shown in Figure 6–7(a); an active-LOW input $\overline{S}$-$\overline{R}$ latch is formed with two cross-coupled NAND gates, as shown

A latch can reside in either of its two states, SET or RESET.

(a) Active-HIGH input S-R latch   (b) Active-LOW input $\overline{S}$-$\overline{R}$ latch

FIGURE 6–7   Two versions of SET-RESET (S-R) latches. Open files F06-07(a) and (b) and verify the operation of both latches.

MULTISIM

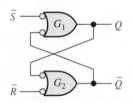

FIGURE 6–8   Negative-OR equivalent of the NAND gate $\overline{S}$-$\overline{R}$ latch in Figure 6–7(b).

SET means that the Q output is HIGH.

RESET means that the Q output is LOW.

in Figure 6–7(b). Notice that the output of each gate is connected to an input of the opposite gate. This produces the regenerative **feedback** that is characteristic of all latches and flip-flops.

To explain the operation of the latch, we will use the NAND gate $\overline{S}$-$\overline{R}$ latch in Figure 6–7(b). This latch is redrawn in Figure 6–8 with the negative-OR equivalent symbols used for the NAND gates. This is done because LOWs on the $\overline{S}$ and $\overline{R}$ lines are the activating inputs.

The latch in Figure 6–8 has two inputs, $\overline{S}$ and $\overline{R}$, and two outputs, Q and $\overline{Q}$. Let's start by assuming that both inputs and the Q output are HIGH, which is the normal latched state. Since the Q output is connected back to an input of gate G_2, and the $\overline{R}$ input is HIGH, the output of G_2 must be LOW. This LOW output is coupled back to an input of gate G_1, ensuring that its output is HIGH.

When the Q output is HIGH, the latch is in the **SET** state. It will remain in this state indefinitely until a LOW is temporarily applied to the $\overline{R}$ input. With a LOW on the $\overline{R}$ input and a HIGH on $\overline{S}$, the output of gate G_2 is forced HIGH. This HIGH on the $\overline{Q}$ output is coupled back to an input of G_1, and since the $\overline{S}$ input is HIGH, the output of G_1 goes LOW. This LOW on the Q output is then coupled back to an input of G_2, ensuring that the $\overline{Q}$ output remains HIGH even when the LOW on the $\overline{R}$ input is removed. When the Q output is LOW, the latch is in the **RESET** state. Now the latch remains indefinitely in the RESET state until a momentary LOW is applied to the $\overline{S}$ input.

In normal operation, the outputs of a latch are always complements of each other.

When Q is HIGH, $\overline{Q}$ is LOW, and when Q is LOW, $\overline{Q}$ is HIGH.

An invalid condition in the operation of an active-LOW input $\overline{S}$-$\overline{R}$ latch occurs when LOWs are applied to both $\overline{S}$ and $\overline{R}$ at the same time. As long as the LOW levels are simultaneously held on the inputs, both the Q and $\overline{Q}$ outputs are forced HIGH, thus violating the basic complementary operation of the outputs. Also, if the LOWs are released simultaneously, both outputs will attempt to go LOW. Since there is always some small difference in the propagation delay time of the gates, one of the gates will dominate in its transition to the LOW output state. This, in turn, forces the output of the slower gate to remain HIGH. In this situation, you cannot reliably predict the next state of the latch.

Figure 6–9 illustrates the active-LOW input $\overline{S}$-$\overline{R}$ latch operation for each of the four possible combinations of levels on the inputs. (The first three combinations are valid, but the last is not.) Table 6–1 summarizes the logic operation in truth table form. Operation of the active-HIGH input NOR gate latch in Figure 6–7(a) is similar but requires the use of opposite logic levels.

Table 6–1 • Truth table for an active-LOW input $\overline{S}$-$\overline{R}$ latch.				
INPUTS		**OUTPUTS**		
$\overline{S}$	$\overline{R}$	Q	$\overline{Q}$	**COMMENTS**
1	1	NC	NC	No change. Latch remains in present state.
0	1	1	0	Latch SET.
1	0	0	1	Latch RESET.
0	0	1	1	Invalid condition

Logic symbols for both the active-HIGH input and the active-LOW input latches are shown in Figure 6–10.

Example 6–1 illustrates how an active-LOW input $\overline{S}$-$\overline{R}$ latch responds to conditions on its inputs. LOW levels are pulsed on each input in a certain sequence and the resulting

Latch starts out RESET ($Q = 0$).   Latch starts out SET ($Q = 1$).

(a) Two possibilities for the SET operation

Latch starts out SET ($Q = 1$).   Latch starts out RESET ($Q = 0$).

(b) Two possibilities for the RESET operation

HIGHS on both inputs   Simultaneous LOWs on both inputs

(c) No-change condition   (d) Invalid condition

FIGURE 6–9   The three modes of basic $\overline{S}\text{-}\overline{R}$ latch operation (SET, RESET, no-change) and the invalid condition.

FIGURE 6–10   Logic symbols for the S-R and $\overline{S}\text{-}\overline{R}$ latch.

(a) Active-HIGH input
S-R latch

(b) Active-LOW input
$\overline{S}\text{-}\overline{R}$ latch

Q output waveform is observed. The $\overline{S} = 0, \overline{R} = 0$ condition is avoided because it results in an invalid mode of operation and is a major drawback of any SET-RESET type of latch.

EXAMPLE 6–1

If the $\overline{S}$ and $\overline{R}$ waveforms in Figure 6–11(a) are applied to the inputs of the latch in Figure 6–10(b), determine the waveform that will be observed on the Q output. Assume that Q is initially LOW.

(a)

FIGURE 6–11

SOLUTION

See Figure 6–11(b).

RELATED PROBLEM*

Determine the Q output of an active-HIGH input S-R latch if the waveforms in Figure 6–11(a) are inverted and applied to the inputs.

*Answers are at the end of the chapter.

The Gated S-R Latch

A gated latch requires an enable input, EN (G is also used to designate an enable input). The logic diagram and logic symbol for a gated S-R latch are shown in Figure 6–12. The S and R inputs control the state to which the latch will go when a HIGH level is applied to the EN input. The latch will not change until EN is HIGH; but as long as it remains HIGH, the output is controlled by the state of the S and R inputs. The gated latch is a level-sensitive device. In this circuit, the invalid state occurs when both S and R are simultaneously HIGH and EN is also HIGH.

FIGURE 6–12 A gated S-R latch.

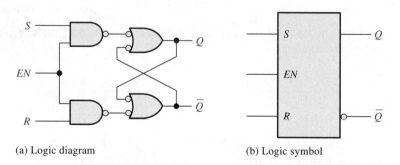

(a) Logic diagram   (b) Logic symbol

EXAMPLE 6–2

Determine the Q output waveform if the inputs shown in Figure 6–13(a) are applied to a gated S-R latch that is initially RESET.

FIGURE 6–13

SOLUTION

The Q waveform is shown in Figure 6–13(b). When S is HIGH and R is LOW, a HIGH on the EN input sets the latch. When S is LOW and R is HIGH, a HIGH on the EN input resets the latch. When both S and R are LOW, the Q output does not change from its present state.

RELATED PROBLEM

Determine the Q output of a gated S-R latch if the S and R inputs in Figure 6–13(a) are inverted.

The Gated D Latch

Another type of gated latch is called the D latch. It differs from the S-R latch because it has only one input in addition to EN. This input is called the D (data) input. Figure 6–14 contains a logic diagram and logic symbol of a D latch. When the D input is HIGH and the EN input is HIGH, the latch will set. When the D input is LOW and EN is HIGH, the latch will reset. Stated another way, the output Q follows the input D when EN is HIGH.

(a) Logic diagram   (b) Logic symbol

FIGURE 6–14   A gated D latch.

EXAMPLE 6–3

Determine the Q output waveform if the inputs shown in Figure 6–15(a) are applied to a gated D latch, which is initially RESET.

FIGURE 6–15

SOLUTION

The Q waveform is shown in Figure 6–15(b). When D is HIGH and EN is HIGH, Q goes HIGH. When D is LOW and EN is HIGH, Q goes LOW. When EN is LOW, the state of the latch is not affected by the D input.

RELATED PROBLEM

Determine the Q output of the gated D latch if the D input in Figure 6–15(a) is inverted.

Latches are sometimes used for multiplexing data onto a bus. For example, data being input to a computer from an external source have to share the data bus with data from other sources. When the data bus becomes unavailable to the external source, the existing data must be temporarily stored, and latches placed between the external source and the data bus may be used to do this. When the data bus is unavailable to the external source, the latches must be disconnected from the bus using a method known as tristating. When the data bus becomes available, the external data pass through the latches, thus the term *transparent latch*. The gated D latch performs this function because when it is enabled, the data on its input appear on the output just as though there were a direct connection. Data on the input are stored as soon as the latch is disabled.

SYSTEM NOTE

SECTION 6–2 CHECKUP

1. List three types of latches.

2. Develop the truth table for the active-HIGH input S-R latch in Figure 6–7(a).

3. What is the Q output of a D latch when $EN = 1$ and $D = 1$?

6–3 FLIP-FLOPS

Flip-flops are synchronous bistable devices, also known as *bistable multivibrators*. **In this case, the term** *synchronous* **means that the output changes state only at a specified point (leading or trailing edge) on the triggering input called the clock (CLK), which is designated as a control input, *C*; that is, changes in the output occur in synchronization with the clock. Flip-flops are edge-sensitive whereas gated latches are level-sensitive.**

After completing this section, you should be able to

- **Define** *clock*
- **Define** *edge-triggered flip-flop*
- **Explain the difference between a flip-flop and a latch**
- **Identify an edge-triggered flip-flop by its logic symbol**
- **Discuss the difference between a positive and a negative edge-triggered flip-flop**
- **Discuss and compare the operation of D and J-K edge-triggered flip-flops and explain the differences in their truth tables.**
- **Discuss the asynchronous inputs of a flip-flop**

Edge-Triggered Flip-Flops

The dynamic input indicator ▷ means the flip-flop changes state only on the edge of a clock pulse.

An **edge-triggered flip-flop** changes state either at the positive edge (rising edge) or at the negative edge (falling edge) of the clock pulse and is sensitive to its inputs only at this transition of the clock. Two types of edge-triggered flip-flops are covered in this section: D and J-K. The logic symbols for these flip-flops are shown in Figure 6–16. Notice that each type can be either positive edge-triggered (no bubble at *C* input) or negative edge-triggered (bubble at *C* input). The key to identifying an edge-triggered flip-flop by its logic symbol is the small triangle inside the block at the clock (*C*) input. This triangle is called the *dynamic input indicator*.

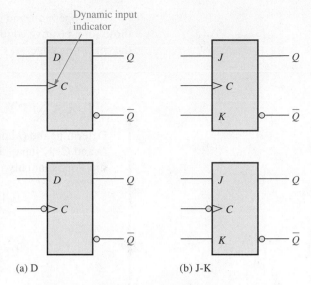

FIGURE 6–16   Edge-triggered flip-flop logic symbols (top: positive edge-triggered; bottom: negative edge-triggered).

THE EDGE-TRIGGERED D FLIP-FLOP   The *D* input of the **D flip-flop** is called a **synchronous** input because data on the input is transferred to the flip-flop's output only on the triggering edge of the clock pulse. When *D* is HIGH, the *Q* output goes HIGH on the triggering edge of the clock pulse, and the flip-flop is SET. When *D* is LOW, the *Q* output goes LOW on the triggering edge of the clock pulse, and the flip-flop is RESET.

This basic operation of a positive edge-triggered D flip-flop is illustrated in Figure 6–17, and Table 6–2 is the truth table for this type of flip-flop. Remember, *the flip-flop cannot change state except on the triggering edge of a clock pulse*. The *D* input can change at any time when the clock input is LOW or HIGH (except for a very short interval around the triggering transition of the clock) without affecting the output. Just remember that *Q* follows *D* at the triggering edge of the clock.

(a) *D* = 1; flip-flop SETS on positive clock edge. (If already SET, it remains SET.)

(b) *D* = 0; flip-flop RESETS on positive clock edge. (If already RESET, it remains RESET.)

The *Q* output of a D flip-flop assumes the state of the *D* input on the triggering edge of the clock.

FIGURE 6–17   **Operation of a positive edge-triggered D flip-flop.**

Table 6–2 • Truth table for a positive edge-triggered D flip-flop.				
INPUTS		**OUTPUTS**		
D	**CLK**	*Q*	*$\overline{Q}$*	**COMMENTS**
1	↑	1	0	SET (stores a 1)
0	↑	0	1	RESET (stores a 0)

↑ = clock transition LOW to HIGH

Semiconductor memories in computers consist of large numbers of individual cells. Each storage cell holds a 1 or a 0. One type of memory is the Static Random Access Memory or SRAM, which uses flip-flops for the storage cells because a flip-flop will retain either of its two states indefinitely as long as dc power is applied, thus the term *static*. This type of memory is classified as a *volatile* memory because all the stored data are lost when power is turned off. Another type of memory, the Dynamic Random Access Memory or DRAM, uses capacitance rather than flip-flops as the basic storage element and must be periodically refreshed in order to maintain the stored data.

SYSTEM NOTE

The operation and truth table for a negative edge-triggered D flip-flop are the same as those for a positive edge-triggered device except that the falling edge of the clock pulse is the triggering edge.

E X A M P L E   6 – 4

Determine the Q and $\overline{Q}$ output waveforms of the flip-flop in Figure 6–18 for the D and CLK inputs in Figure 6–19. (a). Assume that the positive edge-triggered flip-flop is initially RESET.

FIGURE 6–18

FIGURE 6–19

S O L U T I O N

1. At clock pulse 1, D is LOW, so Q remains LOW.
2. At clock pulse 2, D is LOW, so Q remains LOW (RESET).
3. At clock pulse 3, D is HIGH, so Q goes HIGH (SET).
4. At clock pulse 4, D is LOW, so Q goes LOW (RESET).
5. At clock pulse 5, D is HIGH, so Q goes HIGH (SET).
6. At clock pulse 6, D is HIGH, so Q stays HIGH.

Once Q is determined, $\overline{Q}$ is easily found since it is simply the complement of Q. The resulting waveforms for Q and $\overline{Q}$ are shown in Figure 6–19(b) for the input waveforms.

R E L A T E D   P R O B L E M

Determine Q and $\overline{Q}$ for the D input in Figure 6–19(a) if the flip-flop is a negative edge-triggered device.

E X A M P L E   6 – 5

Given the waveforms in Figure 6–20(a) for the D input and the clock, determine the Q output waveform if the flip-flop starts out RESET.

FIGURE 6–20

SOLUTION

The Q output goes to the state of the D input at the time of the positive-going clock edge. The resulting output is shown in Figure 6–20(b).

RELATED PROBLEM

Determine the Q output for the D flip-flop if the D input in Figure 6–20(a) is inverted.

THE EDGE-TRIGGERED J-K FLIP-FLOP   The J and K inputs of the **J-K flip-flop** are synchronous inputs because data on these inputs are transferred to the flip-flop's output only on the triggering edge of the clock pulse. When J is HIGH and K is LOW, the Q output goes HIGH on the triggering edge of the clock pulse, and the flip-flop is SET. When J is LOW and K is HIGH, the Q output goes LOW on the triggering edge of the clock pulse, and the flip-flop is RESET. When both J and K are LOW, the output does not change from its prior state. When J and K are both HIGH, the flip-flop changes state on the triggering edge of the clock. This is called the **toggle** mode.

In the toggle mode, a J-K flip-flop changes state on every clock pulse.

This basic operation of a positive edge-triggered flip-flop is illustrated in Figure 6–21, and Table 6–3 is the truth table for this type of flip-flop. Remember, *the flip-flop cannot change state except on the triggering edge of a clock pulse.* The J and K inputs can be changed at any time when the clock input is LOW or HIGH (except for a very short interval around the triggering transition of the clock) without affecting the output.

(a) $J = 1$, $K = 0$; flip-flop SETS on positive clock edge. (If already SET, it remains SET.)

(b) $J = 0$, $K = 1$; flip-flop RESETS on positive clock edge. (If already RESET, it remains RESET.)

(c) $J = 0$, $K = 0$; flip-flop does not change. (If SET, it remains SET; if RESET, it remains RESET.)

(d) $J = 1$, $K = 1$; flip-flop changes on each clock pulse (toggle).

FIGURE 6–21   **Operation of a positive edge-triggered J-K flip-flop.**

Table 6–3 • Truth table for a positive edge-triggered J-K flip-flop.					
INPUTS			**OUTPUTS**		
J	K	**CLK**	Q	$\overline{Q}$	**COMMENTS**
0	0	↑	Q_0	$\overline{Q}_0$	No change
0	1	↑	0	1	RESET
1	0	↑	1	0	SET
1	1	↑	$\overline{Q}_0$	Q_0	Toggle

↑ = clock transition LOW to HIGH
Q_0 = output level prior to clock transition

The operation and truth table for a negative edge-triggered J-K flip-flop are the same as those for a positive edge-triggered device except that the falling edge of the clock pulse is the triggering edge.

EXAMPLE 6–6

The waveforms in Figure 6–22(a) are applied to the J, K, and clock inputs as indicated. Determine the Q output, assuming that the flip-flop is initially RESET.

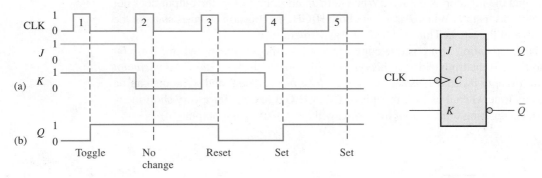

FIGURE 6–22

SOLUTION

Since this is a negative edge-triggered flip-flop, as indicated by the "bubble" at the clock input, the Q output will change only on the negative-going edge of the clock pulse.

1. At the first clock pulse, both J and K are HIGH; and because this is a toggle condition, Q goes HIGH.

2. At clock pulse 2, a no-change condition exists on the inputs, keeping Q at a HIGH level.

3. When clock pulse 3 occurs, J is LOW and K is HIGH, resulting in a RESET condition; Q goes LOW.

4. At clock pulse 4, J is HIGH and K is LOW, resulting in a SET condition; Q goes HIGH.

5. A SET condition still exists on J and K when clock pulse 5 occurs, so Q will remain HIGH.

The resulting Q waveform is indicated in Figure 6–22(b).

RELATED PROBLEM

Determine the Q output of the J-K flip-flop if the J and K inputs in Figure 6–22(a) are inverted.

EXAMPLE 6–7

Determine the Q and $\overline{Q}$ output waveforms of the flip-flop in Figure 6–23 for the J, K, and CLK inputs in Figure 6–24(a). Assume that the positive edge-triggered flip-flop is initially RESET.

FIGURE 6–23

FIGURE 6–24

SOLUTION

1. At clock pulse 1, J is LOW and K is LOW, so Q does not change.
2. At clock pulse 2, J is LOW and K is HIGH, so Q remains LOW (RESET).
3. At clock pulse 3, J is HIGH and K is LOW, so Q goes HIGH (SET).
4. At clock pulse 4, J is LOW and K is HIGH, so Q goes LOW (RESET).
5. At clock pulse 5, J is HIGH and K is HIGH, so Q toggles HIGH (SET).
6. At clock pulse 6, J is HIGH and K is HIGH, so Q toggles LOW.

Once Q is determined, $\overline{Q}$ is easily found since it is simply the complement of Q. The resulting waveforms for Q and $\overline{Q}$ are shown in Figure 6–24(b) for the input waveforms in part (a).

RELATED PROBLEM

Determine Q and $\overline{Q}$ for the J and K inputs in Figure 6–24(a) if the flip-flop is a negative edge-triggered device.

EXAMPLE 6–8

The waveforms in Figure 6–25(a) are applied to the flip-flop as shown. Determine the Q output, starting in the RESET state.

FIGURE 6–25

SOLUTION

The Q output assumes the state determined by the states of the J and K inputs at the positive-going edge (triggering edge) of the clock pulse. A change in J or K after the triggering edge of the clock has no effect on the output, as shown in Figure 6–25(b).

RELATED PROBLEM

Interchange the J and K inputs and determine the resulting Q output.

Asynchronous Preset and Clear Inputs

> **An active preset input makes the Q output HIGH (SET).**

> **An active clear input makes the Q output LOW (RESET).**

For the flip-flops just discussed, the D and J-K inputs are called *synchronous inputs* because data on these inputs are transferred to the flip-flop's output only on the triggering edge of the clock pulse; that is, the data are transferred synchronously with the clock.

Most flip-flops also have **asynchronous** inputs. These are inputs that affect the state of the flip-flop *independent of the clock.* They are normally called **preset** (*PRE*) and **clear** (*CLR*), or *direct set* (S_D) and *direct reset* (R_D). An active level on the preset input will set the flip-flop, and an active level on the clear input will reset it. A logic symbol for a J-K flip-flop with preset and clear inputs is shown in Figure 6–26. These inputs are active-LOW, as indicated by the bubbles. These preset and clear inputs must both be kept HIGH for synchronous operation. In normal operation, preset and clear would not be LOW at the same time.

FIGURE 6–26 Logic symbol for a J-K flip-flop with active-LOW preset and clear inputs.

EXAMPLE 6–9

For the positive edge-triggered D flip-flop with preset and clear inputs in Figure 6–27, determine the Q output for the inputs shown in the timing diagram in part (a) if Q is initially LOW.

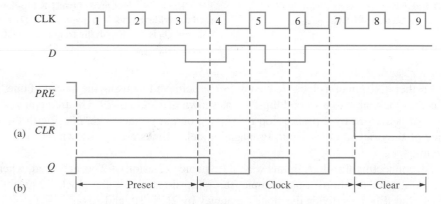

(a)

(b)

FIGURE 6–27   Open file F06-27 to verify the operation.

MULTISIM

SOLUTION

1. During clock pulses 1, 2, and 3, the preset ($\overline{PRE}$) is LOW, keeping the flip-flop SET regardless of the synchronous D input.

2. For clock pulses 4, 5, 6, and 7, the output follows the input on the triggering edge of the clock pulse because both $\overline{PRE}$ and $\overline{CLR}$ are HIGH.

3. For clock pulses 8 and 9, the clear ($\overline{CLR}$) input is LOW, keeping the flip-flop RESET regardless of the synchronous inputs.

The resulting Q output is shown in Figure 6–27(b).

RELATED PROBLEM

If you interchange the $\overline{PRE}$ and $\overline{CLR}$ waveforms in Figure 6–27(a), what will the Q output look like?

Frequency Division

One application of a flip-flop is dividing (reducing) the frequency of a periodic waveform. When a pulse waveform is applied to the clock input of a D or J-K flip-flop that is connected to toggle ($D = \overline{Q}$ or $J = K = 1$), the Q output is a square wave with one-half the frequency of the clock input. Thus, a single flip-flop can be applied as a divide-by-2 device, as is illustrated in Figure 6–28. As you can see, the flip-flop changes state on each triggering clock edge (positive edge-triggered in this case) for both a D and a J-K flip-flop. This results in an output that changes at half the frequency of the clock waveform.

FIGURE 6–28 **The D and J-K flip-flops as divide-by-2 devices.** Q **is one-half the frequency of CLK.**

FIGURE 6–29 **Example of two J-K flip-flops used to divide the clock frequency by 4.** Q_A **is one-half and** Q_B **is one-fourth the frequency of CLK.**

Further division of a clock frequency can be achieved by using the output of one flip-flop as the clock input to a second flip-flop, as shown in Figure 6–29. The frequency of the Q_A output is divided by 2 by flip-flop B. The Q_B output is, therefore, one-fourth the frequency of the original clock input. Propagation delay times are not shown on the timing diagrams.

By connecting flip-flops in this way, a frequency division of 2^n is achieved, where n is the number of flip-flops. For example, three flip-flops divide the clock frequency by $2^3 = 8$; four flip-flops divide the clock frequency by $2^4 = 16$; and so on.

EXAMPLE 6–10

Develop the f_{out} waveform for the circuit in Figure 6–30 when an 8 kHz square wave input is applied to the clock input of flip-flop A.

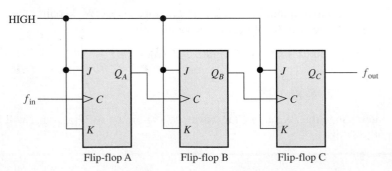

FIGURE 6–30

SOLUTION

The three flip-flops are connected to divide the input frequency by eight ($2^3 = 8$) and the f_{out} waveform is shown in Figure 6–31. Since these are positive edge-triggered flip-flops, the outputs change on the positive-going clock edge. There is one output pulse for every eight input pulses, so the output frequency is 1 kHz. Waveforms of Q_A and Q_B are also shown.

FIGURE 6–31

RELATED PROBLEM

How many flip-flops are required to divide a frequency by thirty-two?

Counting

Another important application of flip-flops is in digital counters, which are covered in detail in Chapter 8. The concept is illustrated in Figure 6–32. Negative edge-triggered J-K flip-flops are used for illustration. Both flip-flops are initially RESET. Flip-flop A toggles on the negative-going transition of each clock pulse. The Q output of flip-flop A clocks flip-flop B, so each time Q_A makes a HIGH-to-LOW transition, flip-flop B toggles. The resulting Q_A and Q_B waveforms are shown in the figure.

FIGURE 6–32   **Flip-flops used to generate a binary count sequence. Two repetitions (00, 01, 10, 11) are shown.**

Observe the sequence of Q_A and Q_B in Figure 6–32. Prior to clock pulse 1, $Q_A = 0$ and $Q_B = 0$; after clock pulse 1, $Q_A = 1$ and $Q_B = 0$; after clock pulse 2, $Q_A = 0$ and $Q_B = 1$; and after clock pulse 3, $Q_A = 1$ and $Q_B = 1$. If we take Q_A as the least significant bit, a 2-bit sequence is produced as the flip-flops are clocked. This binary sequence repeats every four clock pulses, as shown in the timing diagram of Figure 6–32. Thus, the flip-flops are counting in sequence from 0 to 3 (00, 01, 10, 11) and then recycling back to 0 to begin the sequence again.

EXAMPLE 6–11

Determine the output waveforms in relation to the clock for Q_A, Q_B, and Q_C in the circuit of Figure 6–33 and show the binary sequence represented by these waveforms.

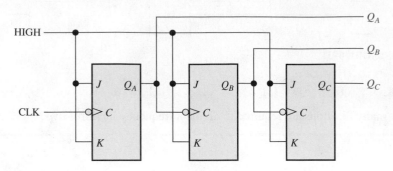

FIGURE 6–33

SOLUTION

The output timing diagram is shown in Figure 6–34. Notice that the outputs change on the negative-going edge of the clock pulses. The outputs go through the binary sequence 000, 001, 010, 011, 100, 101, 110, and 111 as indicated.

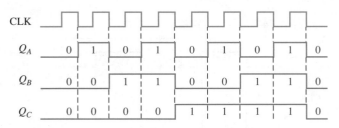

FIGURE 6–34

RELATED PROBLEM

How many flip-flops are required to produce a binary sequence representing decimal numbers 0 through 15?

SYSTEM EXAMPLE 6–1

TRAFFIC SIGNAL CONTROL SYSTEM, SEQUENTIAL LOGIC

Recall from Section 6–1 that the sequential logic is one of the component blocks of the traffic signal control system. The diagram in Figure 6–35 shows how two D flip-flops can be used to implement the Gray code counter. Outputs for the counter control logic provide the D inputs to the flip-flops, so they sequence through the proper states.

The D flip-flop transition table is shown in Table 6–4. A next-state table developed from the state diagram in Figure 6–2 is shown in Table 6–5.

THE COUNTER CONTROL LOGIC Using Tables 6–4 and 6–5, the conditions required for each flip-flop to go to the 1 state can be determined. For example, G_0 goes from 0 to 1 when the present state is 00 and the condition on input D_0 is $\overline{T}_L V_s$, as indicated

FIGURE 6–35   Sequential logic diagram with two D flip-flops used to implement a counter.

Table 6–4 • D flip-flop transition table.		
OUTPUT TRANSITIONS		**FLIP-FLOP INPUT**
Q_N	Q_{N+1}	D
0	$\rightarrow$ 0	0
0	$\rightarrow$ 1	1
1	$\rightarrow$ 0	0
1	$\rightarrow$ 1	1

Table 6–5 • Next-state table for the counter.						
PRESENT STATE		**NEXT STATE**			**FF INPUTS**	
Q_1	Q_0	Q_1	Q_0	**INPUT CONDITIONS**	D_1	D_0
0	0	0	0	$T_L + \overline{V}_s$	0	0
0	0	0	1	$\overline{T}_L V_s$	0	1
0	1	0	1	T_S	0	1
0	1	1	1	$\overline{T}_S$	1	1
1	1	1	1	$T_L V_s$	1	1
1	1	1	0	$\overline{T}_L + \overline{V}_s$	1	0
1	0	1	0	T_S	1	0
1	0	0	0	$\overline{T}_S$	0	0

on the second row of Table 6–5. D_0 must be a 1 to make G_0 go to a 1 or to remain a 1 on the next clock pulse. A Boolean expression describing the conditions that make D_0 a 1 is derived from Table 6–5 as follows, where $Q_1 = G_1$ and $Q_0 = G_0$:

$$D_0 = \overline{G}_1\overline{G}_0\overline{T}_L V_s + \overline{G}_1 G_0 T_S + \overline{G}_1 G_0 \overline{T}_S + G_1 G_0 T_L V_s$$

In the two middle terms, the T_S and the $\overline{T}_S$ variables cancel, leaving the expression

$$D_0 = \overline{G}_1\overline{G}_0\overline{T}_L V_s + \overline{G}_1 G_0 + G_1 G_0 T_L V_s$$

It can be shown that the D_0 expression can be further simplified using the **Karnaugh map** method (See the website for coverage of the Karnaugh map method).

$$D_0 = \overline{G}_1\overline{T}_L V_s + \overline{G}_1 G_0 + G_0 T_L V_s$$

Also, from Table 6–5, an expression for D_1 can be developed as follows where $Q_1 = G_1$ and $Q_0 = G_0$:

$$D_1 = \overline{G}_1 G_0 \overline{T}_S + G_1 G_0 T_L V_s + G_1 G_0 \overline{T}_L + G_1 G_0 \overline{V}_s + G_1 \overline{G}_0 T_S$$

It can be shown that the D_1 expression can be further simplified using Boolean and Karnaugh map methods.

$$D_1 = G_0 \overline{T}_S + G_1 T_S$$

Based on the minimized expressions for D_0 and D_1, the complete sequential logic diagram is shown in Figure 6–36.

FIGURE 6–36   Complete diagram of the sequential logic.

SECTION 6–3 CHECKUP

1. Describe the main difference between a gated D latch and an edge-triggered D flip-flop.

2. How does a J-K flip-flop differ from a D flip-flop in its basic operation?

3. Assume that the flip-flop in Figure 6–20 is negative edge-triggered. Describe the output waveform for the same CLK and D waveforms.

4. How must a J-K flip-flop be connected to function as a divide-by-2 device?

5. How many flip-flops are required to produce a divide-by-64 device?

6–4 FLIP-FLOP OPERATING CHARACTERISTICS

The performance, operating requirements, and limitations of flip-flops are specified by several operating characteristics or parameters found on the data sheet for the device. Generally, the specifications are applicable to all CMOS and bipolar (TTL) flip-flops.

After completing this section, you should be able to

- Define *propagation delay time*
- Explain the various propagation delay time specifications
- Define *set-up time* and discuss how it limits flip-flop operation
- Define *hold time* and discuss how it limits flip-flop operation
- Discuss the significance of maximum clock frequency
- Discuss the various pulse width specifications
- Define *power dissipation* and calculate its value for a specific device

Propagation Delay Times

A **propagation delay time** is the interval of time required after an input signal has been applied for the resulting output change to occur. Four categories of propagation delay times are important in the operation of a flip-flop:

1. Propagation delay t_{PLH} as measured from the triggering edge of the clock pulse to the LOW-to-HIGH transition of the output. This delay is illustrated in Figure 6–37(a).

2. Propagation delay t_{PHL} as measured from the triggering edge of the clock pulse to the HIGH-to-LOW transition of the output. This delay is illustrated in Figure 6–37(b).

(a)   (b)

FIGURE 6–37   **Propagation delays, clock to output.**

3. Propagation delay t_{PLH} as measured from the leading edge of the preset input to the LOW-to-HIGH transition of the output. This delay is illustrated in Figure 6–38(a) for an active-LOW preset input.

4. Propagation delay t_{PHL} as measured from the leading edge of the clear input to the HIGH-to-LOW transition of the output. This delay is illustrated in Figure 6–38(b) for an active-LOW clear input.

(a)   (b)

FIGURE 6–38   **Propagation delays, preset input to output and clear input to output.**

Set-up Time

The **set-up time** (t_s) is the minimum interval required for the logic levels to be maintained constantly on the inputs (J and K, or D) prior to the triggering edge of the clock pulse in order for the levels to be reliably clocked into the flip-flop. This interval is illustrated in Figure 6–39 for a D flip-flop.

FIGURE 6–39   Set-up time (t_s). The logic level must be present on the D input for a time equal to or greater than t_s before the triggering edge of the clock pulse for reliable data entry.

Hold Time

The **hold time** (t_h) is the minimum interval required for the logic levels to remain on the inputs after the triggering edge of the clock pulse in order for the levels to be reliably clocked into the flip-flop. This is illustrated in Figure 6–40 for a D flip-flop.

FIGURE 6–40   Hold time (t_h). The logic level must remain on the D input for a time equal to or greater than t_h after the triggering edge of the clock pulse for reliable data entry.

Maximum Clock Frequency

The maximum clock frequency (f_{max}) is the highest rate at which a flip-flop can be reliably triggered. At clock frequencies above the maximum, the flip-flop would be unable to respond quickly enough, and its operation would be impaired.

Pulse Widths

Minimum pulse widths (t_{PW}) for reliable operation are usually specified by the manufacturer for the clock, preset, and clear inputs. Typically, the clock is specified by its minimum HIGH time and its minimum LOW time.

Power Dissipation

The **power dissipation** of any digital circuit is the total power consumption of the device. For example, if the flip-flop operates on a $+5$ V dc source and draws 5 mA of current, the power dissipation is

$$P = V_{CC} \times I_{CC} = 5\,\text{V} \times 5\,\text{mA} = 25\,\text{mW}$$

HANDS ON TIP
An advantage of CMOS is that it can operate over a wider range of dc supply voltages (typically 2 V to 6 V) than bipolar and, therefore, less expensive power supplies that do not have precise regulation can be used. Also, batteries can be used as secondary or primary sources for CMOS circuits. In addition, lower voltages mean that the IC dissipates less power. The drawback is that the performance of CMOS is degraded with lower supply voltages. For example, the guaranteed maximum clock frequency of a CMOS flip-flop is much less at $V_{CC} = 2$ V than at $V_{CC} = 6$ V.

The power dissipation is very important in most applications in which the capacity of the dc supply is a concern. As an example, let's assume that you have a digital system that requires a total of ten flip-flops, and each flip-flop dissipates 25 mW of power. The total power requirement is

$$P_T = 10 \times 25 \text{ mW} = 250 \text{ mW} = 0.25 \text{ W}$$

This tells you the output capacity required of the dc supply. If the flip-flops operate on +5 V dc, then the amount of current that the supply must provide is

$$I = \frac{250 \text{ mW}}{5 \text{ V}} = 50 \text{ mA}$$

You must use a +5 V dc supply that is capable of providing at least 50 mA of current.

SECTION 6-4 CHECKUP

1. Define the following:

   (a) set-up time   (b) hold time

2. Assume one flip-flop has a $t_{PHL} = 17$ ns and another has a $t_{PHL} = 40$ ns. Which flip-flop can be operated at the highest frequency?

6-5 TIMERS

Two basic types of timing circuits commonly used in electronic systems are the one-shot or monostable multivibrator and the free-running or astable multivibrator. The one-shot is a device with only one stable state. A one-shot is normally in its stable state and will change to its unstable state only when triggered. Once it is triggered, the one-shot remains in its unstable state for a predetermined length of time and then automatically returns to its stable state. The time that the device stays in its unstable state determines the pulse width of its output. The astable multivibrator has no stable state and changes back and forth between its two unstable states at a specified rate.

After completing this section, you should be able to

- Describe the basic operation of a one-shot
- Explain how a nonretriggerable one-shot works
- Explain how a retriggerable one-shot works
- Set up specific one-shots to obtain a given output pulse width
- Describe the basic elements of a 555 timer
- Set up a 555 timer as a one-shot
- Set up a 555 timer as an astable multivibrator

Figure 6-41 shows the block diagram of a one-shot (monostable multivibrator) that is composed of a logic section and a timing section. When a pulse is applied to the **trigger** input, the output of the one-shot, Q, goes HIGH.

The timing section consists of a resistor and a capacitor. The capacitor immediately begins to charge through R. The rate at which it charges is determined by the RC time constant (the product of the values of the resistor and capacitor). When the capacitor charges to a certain voltage level, the output of the one-shot goes back LOW. The duration of the output pulse, t_{PW}, is proportional to the RC time constant by the equation $t_{PW} = kRC$, where k is a constant of proportionality and depends on the particular device.

A one-shot produces a single pulse each time it is triggered.

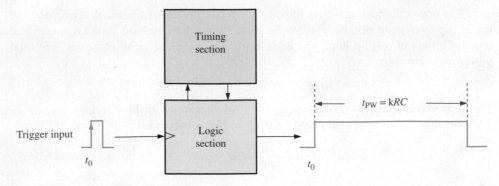

FIGURE 6–41   **Block diagram of a one-shot.**

A typical one-shot logic symbol is shown in Figure 6–42(a), and the same symbol with an external R and C is shown in Figure 6–42(b). The two basic types of one-shots are nonretriggerable and retriggerable.

FIGURE 6–42   **Basic one-shot logic symbols. *CX* and *RX* stand for external components.**

A nonretriggerable one-shot will not respond to any additional trigger pulses from the time it is triggered into its unstable state until it returns to its stable state. In other words, it will ignore any trigger pulses occurring before it times out. The time that the one-shot remains in its unstable state is the pulse width of the output.

Figure 6–43 shows the nonretriggerable one-shot being triggered at intervals greater than its pulse width and at intervals less than the pulse width. Notice that in the second case, the additional pulses are ignored.

FIGURE 6–43   **Nonretriggerable one-shot action.**

A retriggerable one-shot can be triggered before it times out. The result of retriggering is an extension of the pulse width as illustrated in Figure 6–44.

HANDS ON TIP
In normal operation, a one-shot produces only a single pulse, which can be difficult to measure on an oscilloscope because the pulse does not occur regularly. To obtain a stable display for test purposes, it is useful to trigger the one-shot from a pulse generator that is set to a longer period than the expected pulse width and trigger the oscilloscope from the same pulse. For very long pulses, either store the waveform using a digital storage oscilloscope or shorten the time constant by some known factor. For example, replace a 1000 μF capacitor with a 1 μF capacitor to shorten the time by a factor of 1000. A faster pulse is easier to see and measure in an oscilloscope.

FIGURE 6–44   Retriggerable one-shot action.

Specific One-Shots

Two specific IC one-shots are the 74121, which is nontriggerable and the 74xx122, which is retriggerable. The output pulse width formula for the 74121 is

$$t_{PW} = 0.7RC_{EXT} \qquad\qquad (6\text{--}1)$$

R can be either the internal resistor ($2\ k\Omega$) or the external resistor (R_{EXT}) with a selected value; 0.7 is the constant of proportionality, k.

The pulse width formula for the 74xx122 is

$$t_{PW} = 0.32RC_{EXT}\left(1 + \frac{0.7}{R}\right) \qquad\qquad (6\text{--}2)$$

R can be either the internal resistor ($10\ k\Omega$) or the external resistor (R_{EXT}) with a selected value; $0.32\ (1 + 0.7/R)$ is the constant of proportionality, k. Refer to manufacturer's data sheet for connection details.

SYSTEM EXAMPLE 6–2

A SEQUENTIAL TIMER

One practical one-shot application is a sequential timer that can be used to illuminate a series of lights. This type of circuit can be used, for example, in a lane change directional indicator for highway construction projects, as shown in Figure 6–45, or in sequential turn signals on automobiles.

Figure 6–46 shows three one-shots connected as a sequential timer. The labels RX/CX and CX are for the external connection of resistance and capacitance. This particular circuit produces a sequence of three 1 s pulses. The first one-shot is triggered by a switch closure or a low-frequency pulse input, producing a 1 s output pulse. When the first one-shot (OS 1) times out and the a 1 s pulse goes LOW, the second one-shot (OS 2) is triggered, also producing a 1 s output pulse. When this second pulse goes LOW, the third one-shot (OS 3) is triggered and the third 1 s pulse is produced. The output timing is illustrated in the figure. Variations of this basic arrangement can be used to produce a variety of timed outputs, such as the one shown in Figure 6–45, which would require eight one-shots. (A single one-shot controls the arrowhead when all five lights are on at the same time.) The time for each light is set with the timing components.

FIGURE 6–45   Example of a timing application.

FIGURE 6–46   Sequential timer.

The 555 Timer as a One-Shot

The 555 timer is a versatile and widely used IC device because it can be configured in two different modes as either a monostable multivibrator (one-shot) or as an astable multivibrator (pulse oscillator).

An external resistor and capacitor connected as shown in Figure 6–47 are used to set up the 555 timer as a nonretriggerable one-shot. The pulse width of the output is determined by the time constant of R_1 and C_1 according to the following formula:

$$t_{\text{PW}} = 1.1R_1C_1 \tag{6–3}$$

The control voltage input is not used and is connected to a decoupling capacitor C_2 to prevent noise from affecting the trigger and threshold levels.

FIGURE 6–47   The 555 timer connected as a one-shot.

Before a trigger pulse is applied to the one-shot, the output is LOW and an internal transistor is *on*, keeping C_1 discharged. When a negative-going trigger pulse is applied to begin the pulse, the output goes HIGH and the internal transistor turns *off*, allowing capacitor C_1 to begin charging through R_1. When C_1 charges to one-third of the supply voltage ($\frac{1}{3} V_{\text{CC}}$), the output goes back LOW and the internal transistor turns *on* immediately, discharging C_1. The charging rate of C_1 determines how long the output is HIGH.

EXAMPLE 6–12

What is the output pulse width for a 555 monostable circuit with $R_1 = 2.2\,k\Omega$ and $C_1 = 0.01\,\mu F$?

SOLUTION

From Equation 6–3 the pulse width is

$$t_{PW} = 1.1R_1C_1 = 1.1(2.2\,k\Omega)(0.01\,\mu F) = \mathbf{24.2\,\mu s}$$

RELATED PROBLEM

For $C_1 = 0.01\,\mu F$, determine the value of R_1 for a pulse width of 1 ms.

The 555 Timer as an Astable Multivibrator

An astable multivibrator is a device that has no stable states; it changes back and forth (oscillates) between two unstable states without any external triggering. The resulting output is typically a square wave that is used as a clock signal in many types of sequential logic circuits. Astable multivibrators are also known as pulse **oscillators.**

A 555 timer connected to operate as an astable multivibrator is shown in Figure 6–48. Notice that the threshold input (*THRESH*) is now connected to the trigger input (*TRIG*). The external components R_1, R_2, and C_1 form the timing network that sets the frequency of oscillation. The 0.01 μF capacitor, C_2, connected to the control (*CONT*) input is strictly for decoupling and has no effect on the operation; in some cases it can be left off.

FIGURE 6–48   The 555 timer connected as an astable multivibrator (oscillator).

Initially, when the power is turned on, the capacitor (C_1) is uncharged and thus the trigger voltage is at 0 V. This causes the internal transistor to be *off*. Now, C_1 begins charging through R_1 and R_2. When the capacitor voltage reaches $\frac{2}{3}\,V_{CC}$, the internal transistor

All computers require a timing source to provide accurate clock waveforms. The timing section controls all system timing and is responsible for the proper operation of the system hardware. The timing section usually consists of a crystal-controlled oscillator and counters for frequency division. Using a high-frequency oscillator divided down to a lower frequency provides for greater accuracy and frequency stability.

SYSTEM NOTE

is turned *on*, creating a discharge path for the capacitor through R_2 and the transistor. The capacitor now begins to discharge, and at the point where the capacitor discharges down to $1/3 V_{CC}$, the transistor turns *off*. Another charging cycle begins, and the entire process repeats. The result is a rectangular wave output whose duty cycle (ratio of pulse width to period) depends on the values of R_1 and R_2. The frequency of oscillation is given by the following formula, or it can be found using the graph in Figure 6–49.

$$f = \frac{1.44}{(R_1 + 2R_2)C_1} \qquad (6\text{-}4)$$

FIGURE 6–49   **Frequency of oscillation as a function of C_1 and $R_1 + 2R_2$. The sloped lines are values of $R_1 + 2R_2$.**

By selecting R_1 and R_2, the duty cycle of the output can be adjusted. Since C_1 charges through $R_1 + R_2$ and discharges only through R_2, duty cycles approaching a minimum of 50 percent can be achieved if $R_2 \gg R_1$ so that the charging and discharging times are approximately equal.

An expression for the duty cycle is developed as follows. The time that the output is HIGH (t_H) is how long it takes C_1 to charge from $1/3 V_{CC}$ to $2/3 V_{CC}$. It is expressed as

$$t_H = 0.7(R_1 + R_2)C_1 \qquad (6\text{-}5)$$

The time that the output is LOW (t_L) is how long it takes C_1 to discharge from $1/3 V_{CC}$ to $2/3 V_{CC}$. It is expressed as

$$t_L = 0.7R_2C_1 \qquad (6\text{-}6)$$

The period, T, of the output waveform is the sum of t_H and t_L. This is the reciprocal of f in Equation 6–4.

$$T = t_H + t_L = 0.7(R_1 + 2R_2)C_1$$

Finally, the duty cycle is

$$\text{Duty cycle} = \frac{t_H}{T} = \frac{t_H}{t_H + t_L}$$

$$\textbf{Duty cycle} = \left(\frac{R_1 + R_2}{R_1 + 2R_2}\right)100\% \qquad (6\text{-}7)$$

EXAMPLE 6–13

A 555 timer configured to run in the astable mode (pulse oscillator) is shown in Figure 6–50. Determine the frequency of the output and the duty cycle.

MULTISIM

FIGURE 6–50   **Open file F06-50 to verify operation.**

SOLUTION

Use Equations 6–4 and 6–7.

$$f = \frac{1.44}{(R_1 + 2R_2)C_1} = \frac{1.44}{(2.2\,k\Omega + 9.4\,k\Omega)0.022\,\mu F} = \mathbf{5.64\ kHz}$$

$$\text{Duty cycle} = \left(\frac{R_1 + R_2}{R_1 + 2R_2}\right)100\% = \left(\frac{2.2\,k\Omega + 4.7\,k\Omega}{2.2\,k\Omega + 9.4\,k\Omega}\right)100\% = \mathbf{59.5\%}$$

RELATED PROBLEM

What value of R_2 is required to increase the frequency to 10 kHz?

SECTION 6–5 CHECKUP

1. Describe the difference between a nonretriggerable and a retriggerable one-shot.

2. How is the output pulse width set in most IC one-shots?

3. What is the pulse width of a 555 timer one-shot when $C = 1\,\mu F$ and $R = 10\,k\Omega$?

4. Explain the difference in operation between an astable multivibrator and a monostable multivibrator.

5. For a certain astable multivibrator, $t_H = 15$ ms and $T = 20$ ms. What is the duty cycle of the output?

6–6 BISTABLE LOGIC WITH VHDL AND VERILOG

As you know, any logic function must be described with a hardware description language (HDL) in preparation for programming it into a PLD. In this section, the S-R latch and the D and J-K flip-flops are described in both VHDL and Verilog. Keep in mind that in both languages there is more than one way to describe a given function. Refer to the VHDL or Verilog tutorials on the website.

After completing this section, you should be able to

- Use VHDL and Verilog to describe several types of latches and flip-flops
- Specify variable arrays in both VHDL and Verilog
- Use comment lines

S-R Latch

The logic diagram for the programming model of an S-R latch is shown in Figure 6–51. The latch is described using VHDL and Verilog.

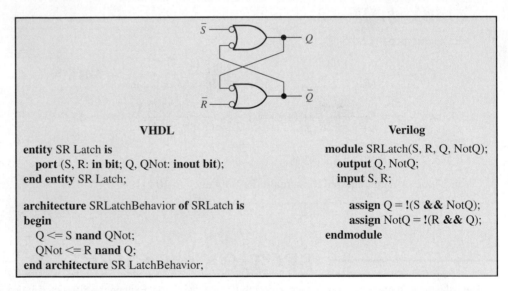

VHDL

```
entity SR Latch is
    port (S, R: in bit; Q, QNot: inout bit);
end entity SR Latch;

architecture SRLatchBehavior of SRLatch is
begin
    Q <= S nand QNot;
    QNot <= R nand Q;
end architecture SR LatchBehavior;
```

Verilog

```
module SRLatch(S, R, Q, NotQ);
    output Q, NotQ;
    input S, R;

    assign Q = !(S && NotQ);
    assign NotQ = !(R && Q);
endmodule
```

FIGURE 6–51 Logic diagram and VHDL and Verilog programs for the S-R latch. The symbol && in the Verilog program means AND and the symbol ! means NOT.

D Flip-Flop

The logic diagram for a D flip-flop is shown in Figure 6–52. The flip-flop is described using VHDL and Verilog. The pulse transition detector converts the clock pulse (CLK) to a short spike (Clock) on its triggering edge. The VHDL program includes this triggering with the **if rising_edge** term and in the Verilog program with the **@(posedge Clock)** term.

J-K Flip-Flop

The logic diagram for a J-K flip-flop is shown in Figure 6–53. The flip-flop is described using VHDL and Verilog. The pulse transition detector converts the clock pulse (CLK) to a short spike (Clock) on its triggering edge. The VHDL program includes this triggering with the **if rising_edge** term and in the Verilog program with the **@(posedge Clock)** term.

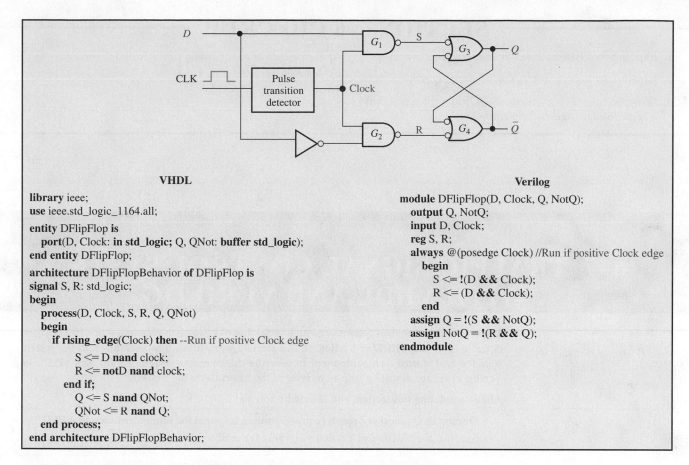

VHDL

VHDL

```
library ieee;
use ieee.std_logic_1164.all;

entity DFlipFlop is
  port(D, Clock: in std_logic; Q, QNot: buffer std_logic);
end entity DFlipFlop;

architecture DFlipFlopBehavior of DFlipFlop is
signal S, R: std_logic;
begin
  process(D, Clock, S, R, Q, QNot)
  begin
    if rising_edge(Clock) then --Run if positive Clock edge

      S <= D nand clock;
      R <= notD nand clock;
    end if;
      Q <= S nand QNot;
      QNot <= R nand Q;
  end process;
end architecture DFlipFlopBehavior;
```

Verilog

```
module DFlipFlop(D, Clock, Q, NotQ);
  output Q, NotQ;
  input D, Clock;
  reg S, R;
  always @(posedge Clock) //Run if positive Clock edge
    begin
      S <= !(D && Clock);
      R <= (D && Clock);
    end
  assign Q = !(S && NotQ);
  assign NotQ = !(R && Q);
endmodule
```

FIGURE 6–52   Logic diagram and VHDL and Verilog programs for the D flip-flop.

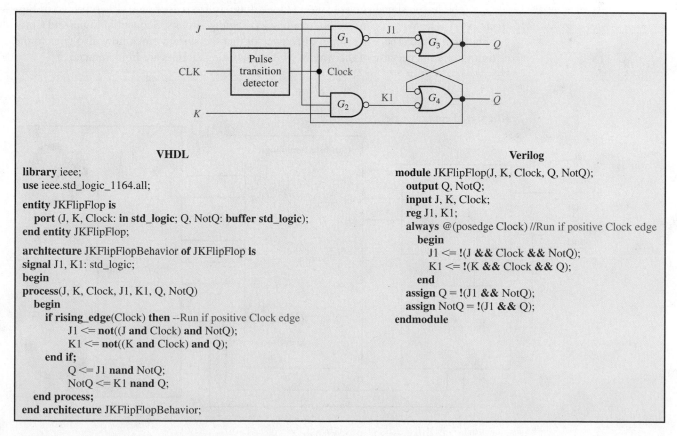

VHDL

```
library ieee;
use ieee.std_logic_1164.all;

entity JKFlipFlop is
  port (J, K, Clock: in std_logic; Q, NotQ: buffer std_logic);
end entity JKFlipFlop;

architecture JKFlipFlopBehavior of JKFlipFlop is
signal J1, K1: std_logic;
begin
process(J, K, Clock, J1, K1, Q, NotQ)
  begin
    if rising_edge(Clock) then --Run if positive Clock edge
      J1 <= not((J and Clock) and NotQ);
      K1 <= not((K and Clock) and Q);
    end if;
        Q <= J1 nand NotQ;
        NotQ <= K1 nand Q;
  end process;
end architecture JKFlipFlopBehavior;
```

Verilog

```
module JKFlipFlop(J, K, Clock, Q, NotQ);
  output Q, NotQ;
  input J, K, Clock;
  reg J1, K1;
  always @(posedge Clock) //Run if positive Clock edge
    begin
      J1 <= !(J && Clock && NotQ);
      K1 <= !(K && Clock && Q);
    end
  assign Q = !(J1 && NotQ);
  assign NotQ = !(J1 && Q);
endmodule
```

FIGURE 6–53   Logic diagram and VHDL and Verilog programs for the J-K flip-flop.

SECTION 6–6 CHECKUP

1. Explain how VHDL specifies an edge-triggered operation.

2. Explain how Verilog specifies an edge-triggered operation.

3. Write the lines of code in both VHDL and Verilog to specify a negative edge-triggered operation.

4. Define each of the following logical Verilog operators: !, ||, and &&.

6–7 TRAFFIC SIGNAL CONTROL SYSTEM WITH VHDL AND VERILOG

The traffic signal control system presented in Section 6–1 can be implemented in a PLD by first describing it with VHDL or Verilog. In this section, you will see how each block in the system is coded and is used as a component to describe the complete system. As for any VHDL or Verilog coverage, it will be helpful to refer to the tutorials on the website.

After completing this section, you should be able to

- Discuss the general approach to programming a system for implementation in a PLD
- See how the VHDL and Verilog programs are used to implement the system

The traffic signal control system block diagram from Section 6–1 is repeated in Figure 6–54. The system consists of three blocks: Timing circuits, Sequential logic, and Combinational logic. There is a small difference in that the system clock now also goes to the timing circuits because of the way in which the timing circuits are implemented.

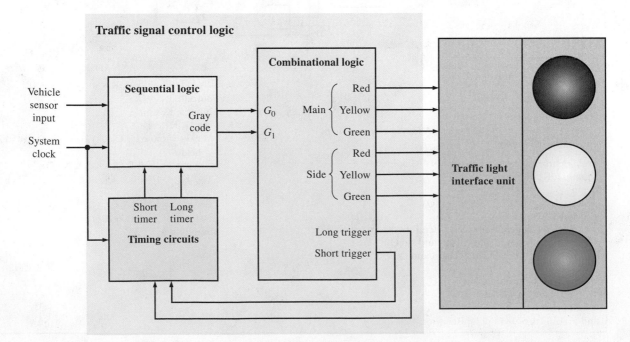

FIGURE 6–54   Block diagram of the traffic signal control system.

A programming model for the traffic signal control system is shown in Figure 6–55, where all the input and output labels are given. Notice that the Timing circuits block is split into two parts; the Frequency divider and the Timer circuits; and the Combinational logic block is divided into the State decoder and two logic sections (Light output logic and Trigger logic). This model will be used to develop VHDL and Verilog programs for the system.

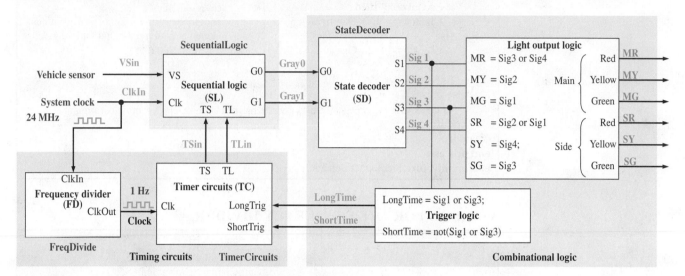

FIGURE 6–55   **Programming model for the traffic signal control system.**

The Timing Circuits Block

The two parts of the Timing circuits block are the Frequency divider (FD) and the Timer circuits (TC). A system clock frequency of 24 MHz is assumed. The Frequency divider part divides the 24 MHz system clock down to a 1 Hz clock. The Timer circuits part simulates the one-shot outputs described in Section 6–1 to produce outputs of TS = 4 s and TL = 25 s.

FREQUENCY DIVIDER   The purpose of the frequency divider is to produce a 1 Hz clock for the timer circuits. The input ClkIn in this application is a 24.00 MHz oscillator that drives the program code. SetCount is used to initialize the count for a 1 Hz interval. The program FreqDivide counts up from zero to the value assigned to SetCount (one-half the oscillator speed) and inverts the output identifier ClkOut.

The integer value Cnt is set to zero prior to operation. The clock pulses are counted and compared to the value assigned to SetCount. When the number of pulses counted reaches the value in SetCount, the output ClkOut is checked to see if it is currently set to a 1 or 0. If ClkOut is currently 0, ClkOut is assigned a 1; otherwise, ClkIn is set to 1. Cnt is assigned a value of 0 and the process repeats. Toggling the output ClkOut each time the value of SetCount is reached creates a 1 Hz clock output with a 50% duty cycle.

VHDL FOR THE FREQUENCY DIVIDER

```
library ieee;
use ieee.std_logic_1164.all;                    ClkIn: 24.00 MHz clock driver
                                                ClkOut: Output at 1 Hz
entity FreqDivide is
port(ClkIn, in std_logic;                        Cnt: Counts up to value in SetCount
    ClkOut: buffer std_logic);                  SetCount: Holds ½ timer interval value
end entity FreqDivide;

architecture FreqDivide Behavior of FreqDivide is
begin
    FreqDivide: process(ClkIn)
    variable Cnt: integer := 0;
    variable SetCount: integer;
```

```
begin
  SetCount := 12000000; -- 1/2 duty cycle
  if (ClkIn'EVENT and ClkIn = '1') then
    if (Cnt = SetCount) then
      if ClkOut = '0' then
        ClkOut <= '1'; --Output high 50%
      else
        ClkOut <= '0'; --Output Low 50%
      end if;
      Cnt := 0;
    else
      Cnt := Cnt + 1;
    end if;
  end if;
end process;
end architecture FreqDivideBehavior;
```

SetCount is assigned a value equal to half the system clock to produce a 1 Hz output. In this case, a 24 MHz system clock is used.

The if statement causes program to wait for a clock event and clock = 1 to start operation.

Check that the terminal value in SetCount has been reached at which time ClkOut is toggled and Cnt is reset to 0.

If terminal value has not been reached, Cnt is incremented.

VERILOG FOR THE FREQUENCY DIVIDER

Clkln: 24.00 MHz clock driver
ClkOut: Output at 1 Hz
Cnt: Counts up to value in SetCount
SetCount: Holds ½ timer interval value
Q: Holds output value within the always block

```
module FreqDivide (Clkln, ClkOut);
  input    Clkln;
  inout    ClkOut;
  integer  Cnt = 0;
  integer  SetCount = 12000000; //½ duty cycle
  reg [0:0] Q;

  always @(posedge Clkln)
  begin
    if (Cnt == SetCount)
    begin
      if (ClkOut == 0)
      begin
        Q = 1; //Output high 50%
      end
      else
      begin
        Q = 0; //Output Low 50%
      end
      Cnt = 0;
    end
    else
    begin
      Cnt = Cnt + 1;
    end
  end
  assign ClkOut = Q;
endmodule
```

SetCount is assigned a value equal to half the system clock to produce a 1 Hz output. In this case a 24 MHz system clock is used.

The always statement causes program to wait for a positive edge clock event.

Check that the terminal value in SetCount has been reached at which time ClkOut is toggled and Cnt is reset to 0.

If terminal value has not been reached. Cnt is incremented.

Value stored in Q is assigned to ClkOut outside the always block.

TIMER CIRCUITS The program TimerCircuits uses two one-shot instances consisting of a 25 s timer (TLong) and a 4 s timer (TShort). The 25 s and the 4 s timers are triggered by long trigger (LongTrig) and short trigger (ShortTrig). In the VHDL and Verilog programs, countdown timers driven by a 1 Hz clock input (Clk) replicate the one-shot components TLong and TShort. The values stored in SetCountLong and SetCountShort are assigned to the Duration inputs of one-shot components TLong and TShort, setting the

25-second and 4-second timeouts. When Enable is set LOW, the one-shot timer is initiated and output QOut is set HIGH. When the one-shot timers time out, QOut is set LOW. The output of one-shot component TLong is sent to TimerCircuits identifier TL. The output of one-shot component TShort is sent to TimerCircuits identifier TS.

VHDL FOR THE TIMING CIRCUITS

```
library ieee;
use ieee.std_logic_1164.all;                    LongTrig: Long timeout timer enable input
                                                ShortTrig: Short timeout timer enable input
entity TimerCircuits is                            Clk: 1 Hz Clock input
  port(LongTrig, ShortTrig, Clk: in std_logic;     TS: Short timer timeout signal
       TS, TL: buffer std_logic);                   TL: Long timer timeout signal
end entity TimerCircuits;

architecture TimerBehavior of TimerCircuits is
component OneShot is
  port(Enable, Clk: in std_logic;
       Duration :in integer range 0 to 25;    }  Component declaration for OneShot.
       QOut     :buffer std_logic);
end component OneShot;

signal SetCountLong, SetCountShort: integer range 0 to 25;   SetCountLong: Holds long timer duration
begin                                                        SetCountShort: Holds short timer duration
  SetCountLong <= 25; ]  Long and short count times are hard-coded
  SetCountShort <= 4;  } to 25 and 4 based on a 1 Hz clock.
  TLong:OneShot port map(Enable=>LongTrig, Clk=>Clk, Duration=>SctCountLong, QOut=>TL);  <— Instantiation TLong
  TShort:OneShotport map(Enable=>ShortTrig, Clk=>Clk, Duration=>SetCountShort, QOut=>TS); <— Instantiation TShort
end architecture TimerBehavior;
```

VERILOG FOR THE TIMING CIRCUITS

```
                                                LongTrig: Long timeout timer enable input
                                                ShortTrig: Short timeout timer enable input
                                                   Clk: 1 Hz Clock input
                                                    TS: Short timer timeout signal
                                                    TL: Long timer timeout signal

module TimerCircuits(LongTrig, ShortTrig, Clk, TS, TL);   SetCountLong: Holds long timer duration
  input LongTrig, ShortTrig, Clk;                         SetCountShort: Holds short timer duration
  inout TS, TL;

  wire[8:0] SetCountLong = 25; ]  Long and short count times are hard-coded
  wire[8:0] SetCountShort = 4; }  to 25 and 4 based on a 1 Hz clock.
  OneShot TLong(.Enable(LongTrig), .Clk(Clk), .Duration(SetCountLong), .QOut(TL));  <— Instantiation TLong
  OneShot TShort(.Enable(ShortTrig), .Clk(Clk), .Duration(SetCountShort), .QOut(TS)); <— Instantiation TShort
endmodule
```

The Sequential Logic Block

The Sequential logic (SL) determines the sequence of the traffic lights. The program SequentialLogic provides the Gray code logic needed to drive the traffic signal system based on input from the program TimerCircuits and the side street vehicle sensor. The sequential logic code produces a 2-bit Gray code sequence for each of the four sequence states defined in the program. The component definition dff is used to instantiate two DFlip-Flop instances DFF0 and DFF1. DFF0 and DFF1 drive the two-bit Gray code. The Gray code output sequences the traffic light system through each of four states. Internal variables D0 and D1 store the results of the D0 and D1 Boolean expressions developed in System Example 6–1. The stored results in D0 and D1 are assigned to DFlipFlops DFF0 and DFF1 along with the system clock to drive outputs G0 and G1 from the DFlipFlop Q outputs.

VHDL FOR THE SEQUENTIAL LOGIC

```
library ieee;
use ieee.std_logic_1164.all;

entity SequentialLogic is
port(VS, TL, TS, Clk: in std_logic; G0, G1: inout std_logic);
end entity SequentialLogic;

architecture SequenceBehavior of SequentialLogic is

component dff is
port(D, Clk: in std_logic;
Q: out std_logic);
end component dff;

signal D0, D1: std_logic;
begin
D1 <= (G0 and not TS) or (G1 and TS);
D0 <= (not G1 and not TL and VS) or (not G1 and G0)
or (G0 and TL and VS);

DFF0: dff port map(D=> D0, Clk => Clk, Q => G0);
DFF1: dff port map(D=> D1, Clk => Clk, Q => G1);

end architecture SequenceBehavior;
```

VS: Vehicle sensor input
TL: Long timer input
TS: Short timer input
Clk: System clock
G0: Gray code output bit 0
G1: Gray code output bit 1
D0: Logic for DFlipFlop DFF0
D1: Logic for DFlipFlop DFF1

Component declaration for D flip-flop (dff)

Logic definitions for D flip-flop inputs D0 and D1 derived from Boolean expressions developed in System Example 6–1.

Component instantiations

VERILOG FOR THE SEQUENTIAL LOGIC

```
module SequentialLogic(VS, TL, TS, Clk, G0, G1);
input VS, TL, TS, Clk;
inout G0, G1;

wire D0, D1;

assign D1 = (G0 && !TS) || (G1 && TS);
assign D0 = (!G1 && !TL && VS) || (!G1 && G0) || (G0 && TL && VS);

dff DFF0(.d(D0), .Clk(Clk), .q(G0));
dff DFF1(.d(D1), .Clk(Clk), .q(G1));

endmodule
```

VS: Vehicle sensor input
TL: Long timer input
TS: Short timer input
Clk: System clock
G0: Gray code output bit 0
G1: Gray code output bit 1
D0: Logic for DFlip-Flop DFF0
D1: Logic for DFlipFlop DFF1

Logic definitions for D flip-flop inputs D0 and D1 derived from Boolean expressions developed in System Example 6–1.

Component instantiations

The Combinational Logic Block

The combinational logic block consists of three parts: State decoder, Light output logic, and Trigger logic.

STATE DECODER   The state decoder (SD) is part of the Combinational logic block. The source program StateDecoder decodes the 2-bit Gray code from the program SequentialLogic to determine which of four states the system is in. The inputs to the state decoder are the two Gray code bits G1 and G0. The Boolean expressions for the four state outputs S1, S2, S3 and S4, were developed in Section 6–1. For each of the four input codes, one and only one of the outputs is activated.

VHDL FOR THE STATE DECODER

```
library ieee;
use ieee.std_logic_1164.all;

entity StateDecoder is
port(G0, G1: in std_logic; S1, S2, S3, S4: out std_logic);
end entity StateDecoder;

architecture StateDecoderBehavior of StateDecoder is
begin
S1 <= not G1 and not G0;
S2 <= not G1 and G0;
S3 <= G1 and G0;
S4 <= G1 and not G0;

end architecture StateDecoderBehavior;
```

G0,G1 :Gray code input
S1–S4 :State output

Logic definitions for state decoder developed in Section 6–1

2-Bit Gray Code		
	G1	G0
State 1	0	0
State 2	0	1
State 3	1	1
State 4	1	0

VERILOG FOR THE STATE DECODER

```
module StateDecoder(G0, G1, S1, S2, S3, S4);
input G0, G1;
output S1, S2, S3, S4;

assign S1 = !G1 && !G0;
assign S2 = !G1 && G0;
assign S3 = G1 && G0;
assign S4 = G1 && !G0;

endmodule
```

G0,G1 :Gray code input
S1–S4 :State output

Logic definitions for state decoder developed in Section 6–1

2-Bit Gray Code		
	G1	G0
State 1	0	0
State 2	0	1
State 3	1	1
State 4	1	0

LIGHT OUTPUT LOGIC   This block of logic accepts inputs Sig1–Sig4 from the state decoder and produces the traffic light outputs MR, MY, MG, SR, SY, and SG for the traffic light interface unit. The Boolean expressions were developed in Section 6–1 and are implemented in the program TrafficLights, which describes the complete system.

TRIGGER LOGIC   This block of logic accepts input Sig1 and Sig3 from the state decoder and produces the triggers, LongTime and ShortTime, for the Timer Circuits. The Boolean expressions were developed in Section 6–1 and are implemented in the program TrafficLights.

The Complete Traffic Signal Control System

The program TrafficLights completes the traffic signal control system. Components Freq-Divide, TimerCircuits, SequentialLogic, and StateDecoder are used to compose the completed system. Signal CLKin from the TrafficLights program source code is the clock input to the FreqDivide component. The frequency divided output ClkOut is stored as local variable Clock and is the divided clock input to the TimerCircuits and Sequential-Logic components. TimerCircuits is controlled by local variables LongTime and Short-Time, which are controlled by the outputs Sig1 and Sig3 from component StateDecoder. StateDecoder also provides outputs Sig1 through Sig4 to control the traffic lights MG, SG, MY, SY, MR, and SR. TimerCircuit timeout signals TS and TL are stored in variables TLin (timer long in) and TSin (timer short in).

Signals TSin and TLin from TimerCircuits are used along with vehicle sensor VSin as inputs to the SequentialLogic component. The outputs from SequentialLogic G0 and G1 are stored in variables Gray0 and Gray1 as inputs to component StateDecoder. Component StateDecoder returns signals S1 through S4 which are in turn passed to variables Sig1 through Sig4. The values stored in variables Sig1 through Sig4 provide the logic for outputs MG, SG, MY, SY, MR, SR; and local timer triggers LongTime and ShortTime are sent to TimerCircuits.

VHDL FOR THE TRAFFIC SIGNAL CONTROL SYSTEM

```vhdl
                                              VSin   : Vehicle sensor input
library ieee;                                 CLKin  : System Clock
use ieee.std_logic_1164.all;                  MR     : Main red light output
                                              SR     : Side red light output
entity TrafficLights is                       MY     : Main yellow light output
port(VSin, ClkIn: in std_logic; MR, SR, MY, SY, MG, SG: out std_logic);   SY : Side yellow light output
end entity TrafficLights;                     MG     : Main green light output
                                              SG     : Side green light output


architecture TrafficLightsBehavior of TrafficLights is


component StateDecoder is
port(G0, G1: in std_logic; S1, S2, S3, S4: out std_logic);  } Component declaration for StateDecoder
end component StateDecoder;


component SequentialLogic is
port(VS, TL, TS, Clk: in std_logic; G0, G1: inout std_logic);  } Component declaration for SequentialLogic
end component SequentialLogic;


component TimerCircuits is
port(LongTrig, ShortTrig, Clk: In std_logic; TS, TL: buffer std_logic);  } Component declaration for TimerCircuits
end component TimerCircuits;


component FreqDivide is
port(Clkin: in std_logic; ClkOut: buffer std_logic);  } Component declaration for FreqDivider
end component FreqDivide;

signal Sig1, Sig2, Sig3, Sig4, Gray0, Gray1: std_logic;
signal LongTime, ShortTime, TLin, TSin, Clock: std_logic;


begin                                         Sig1-4   : Return values from StateDecoder
MR <= Sig3 or Sig4;                           Gray0-1  : SequentialLogic Gray code return
SR <= Sig2 or Sig1;                           LongTime : Trigger input to TimerCircuits
MY <= Sig2;           Logic definitions for the   ShortTime : Trigger input to TimerCircuits
SY <= Sig4;           light output logic           TLin    : Store TimerCircuits long timeout
MG <= Sig1;                                   TSin    : Store TimerCircuits Short timeout
SG <= Sig3;                                   Clock   : Divided clock from FreqDivide

LongTime <= Sig1 or Sig3;       }
ShortTime <= not(Sig1 or Sig3);  }  Logic definitions for the trigger logic


SD: StateDecoder    port map(G0 => Gray0, G1 => Gray1, S1 => Sig1, S2 => Sig2, S3 => Sig3, S4 => Sig4);
SL: SequentialLogic port map(VS => VSin, TL => TLin, TS => TSin, Clk => Fout, G0 => Gray0, G1 => Gray1);   } Component
TC: TimerCircuits   port map(LongTrig=>LongTime, ShortTrig=>ShortTime, Clk=>Clock, TS=>TSin, TL=>TLin);    } instantiations
FD: FreqDivide      port map (Clkln => CLKin, ClkOut =>-Clock);


end architecture TrafficLightsBehavior;
```

VERILOG FOR THE TRAFFIC SIGNAL CONTROL SYSTEM

```verilog
module TrafficLights(VSin, CLKin, MR, SR, MY, SY, MG, SG);
input VSin, CLKin;
output MR, SR, MY, SY, MG, SG;


wire Sig1, Sig2, Sig3, Sig4, Gray0, Gray1;
wire LongTime, ShortTime, TLin, TSin, Clk, Clock;


assign MR = Sig3 || Sig4;
assign SR = Sig2 || Sig1;
assign MY = Sig2;
assign SY = Sig4;
assign MG = Sig1;
assign SG = Sig3;
assign LongTime = Sig1 || Sig3
assign ShortTime = !(Sig1 || Sig3);


StateDecoder SD(.G0(Gray0), .G1(Gray1), .S1(Sig1), .S2(Sig2), .S3(Sig3), .S4(Sig4));
SequentialLogic SL(.VS(VSin), .TL(TLin), .TS(TSin), .Clk(Clock), .G0(Gray0), .G1(Gray1));
TimerCircuits TC(.LongTrig(LongTime), .ShortTrig(ShortTime), .Clk(Clock), .TS(TSin), .TL(TLin));
FreqDivide FD(.Clkin(CLKin), .ClkOut(Clock));

endmodule
```

- *Logic definitions for the light output logic* (MR through SG assignments)
- *Logic definitions for the trigger logic* (LongTime, ShortTime assignments)
- *Component instantiations* (StateDecoder, SequentialLogic, TimerCircuits, FreqDivide)

VSin	: Vehicle sensor input
CLKin	: System Clock
MR	: Main red light output
SR	: Side red light output
MY	: Main yellow light output
SY	: Side yellow light output
MG	: Main green light output
SG	: Side green light output
Sig1-4	: Return values from StateDecoder
Gray0-1	: SequentialLogic Gray code return
LongTime	: Trigger input to TimerCircuits
ShortTime	: Trigger input to TimerCircuits
TLin	: Store TimerCircuits long timeout
TSin	: Store TimerCircuits short timeout
Clock	: Divided clock from FreqDivide

SECTION 6–7 CHECKUP

1. How are the programs broken down in terms of system blocks?

2. How is the complete system program formed?

3. Describe the observable difference between VHDL and Verilog programs.

6–8 TROUBLESHOOTING

It is standard practice to test a new circuit or system design to be sure that it is operating as specified. This process is sometimes referred to as debugging. A circuit or system can be debugged in two ways: software simulation and hardware testing. In software simulation, a program such as Multisim is used to construct and test the new design. In hardware testing, designs are "breadboarded" and tested before the design is finalized. The term *breadboard* refers to a method of temporarily hooking up a circuit so that its operation can be verified and any design flaws worked out before a prototype unit is built. Once a design has been thoroughly tested, the prototype is implemented on one or more circuit boards. A complete system test is then done on the boards. In this section, the traffic signal control system is the focus, and several aspects of testing, debugging, and troubleshooting will be discussed.

After completing this section, you should be able to

- Discuss software simulation applied to system debugging
- Describe prototype breadboarding as a method of testing a design.
- Explain several methods of hardware troubleshooting

Multisim Software Simulation

COMBINATIONAL LOGIC   Figure 6–56 shows the Multisim simulation screen for the complete combinational logic portion of the traffic signal control system including the state decoder, light output logic, and trigger logic. Switches are used to simulate the Gray code inputs, and probes are used to indicate the activity on the outputs of the combinational logic. The switches and probes are for simulation testing only and are not part of the logic circuit.

MULTISIM

FIGURE 6–56   Multisim circuit screen for the combinational logic showing the first state. Open Multisim file F06-56 on the website and run the simulation for the combinational logic portion of the traffic signal control system. Observe the operation for each of the four states in the light sequence.

TIMING CIRCUITS   Figure 6–57 shows the simulation screen for the timing circuits portion of the traffic signal control system including the 25 s timer, 4 s timer, and oscillator. This is an alternate version that differs from the HDL description shown earlier in the chapter and represents the implementation using 555 timers.

SIMULATION TIME VERSUS REAL TIME   Simulation time is based on the software program and the computer it is running on; it is much greater than real time that will occur in the actual circuit. For the particular computer used to generate the simulation shown in Figure 6–57, the ratio of simulation time to real time is 1000 but may vary depending on the speed of the computer used to run the software as well as other factors. Because of this ratio of simulation time to real time, the *RC* time constant values for the one-shots are 1000 times smaller than the values you would use in an actual hardware circuit. By connecting a Multisim probe to a one-shot output and using the switch to trigger that one-shot, you can observe the output in real time. A virtual oscilloscope connected to that output will record a much shorter time than the real time observation. For example, a scope connected to the output of the 25 s long timer (U1) as shown in Figure 6–58(a) will

FIGURE 6–57   Multisim screen for a version of the timing circuits. The switches are for test purposes only.

display an output pulse with a duration of 24.2 ms when the timer is triggered. However, a probe connected to that output will stay on for approximately 24 s (depending on the computer and the number of time steps used in the simulation). The output of the 555 timer configured as an oscillator is shown in Figure 6–58(b) using the virtual scope.

Grace Hopper, a mathematician and pioneer programmer, developed considerable troubleshooting skills as a naval officer working with the Harvard Mark I computer in the 1940s. She found and documented in the Mark I's log the first real computer bug. It was a moth that had been trapped in one of the electromechanical relays inside the machine, causing the computer to malfunction. From then on, when asked if anything was being accomplished, those working on the computer would reply that they were "debugging" the system. The term stuck, and finding problems in a computer (or other electronic device), particularly the software, would always be known as debugging.

SYSTEM NOTE

PROTOTYPE CHECK   The computer simulation of timers is more likely to have an undetected error because of the simulation versus real time issue previously discussed. It is usually advisable to verify a block like this prior to combining it into the complete system. For this application, the timing circuits are constructed on a prototyping system, and the results are measured carefully. The prototyping system that is selected is one that is closely linked to Multisim and can later be used to develop a circuit board. It is the National Instruments ELVIS system. The prototype circuit and measured data are shown in Figure 6–59.

Probe light stays on for
approximately 24 s

Scope indicates 24.2 ms

(a) Illustration of one-shot simulation.

Switch is moved to ground
momentarily to trigger one-shot.

(b) 10 kHz oscillator output

MULTISIM

FIGURE 6–58   **Testing the simulated timer circuit. Open Multisim file F06-58 on the website. Run the timing circuit simulation using your Multisim software and observe the operation.**

SEQUENTIAL LOGIC   Figure 6–60 shows the simulation screen for the sequential logic portion of the traffic signal control system.

THE COMPLETE TRAFFIC SIGNAL CONTROL SYSTEM   Now that all three parts of the system have been simulated and tested, they are combined to form the complete system. A simulation screen for the complete system is shown in Figure 6–61. Each part of the system has been converted to a Multisim subcircuit.

The main green light will stay on continuously as long as there is no vehicle on the side street. When a vehicle appears on the side street, the main green light will time out and change to red for 25 s to allow side street traffic to move. If there is continuous traffic on the side street, the main red and green lights will alternate every 25 s.

CIRCUIT	OUTPUT
U1 Long timer	29 s pulse
U2 Short timer	4.6 s pulse
U3 Oscillator	9.6 kHz

FIGURE 6–59

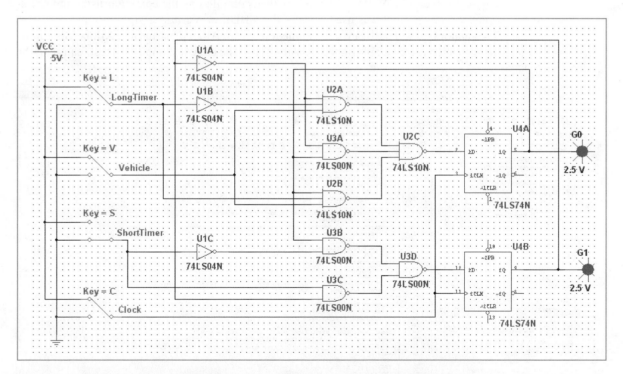

FIGURE 6–60   Multisim screen for sequential logic simulation. The switches are for test purposes only. When a probe is red, a 1 is indicated. Open file F06-60 on the website using your Multisim software. All of the inputs are shown as switch inputs for simulation purposes. Set up the input conditions for T_L, T_S, and V_s. Then momentarily make the clock input go LOW (ground) and then back HIGH (V_{CC}). Follow the sequence indicated by the state diagram in Figure 6–2.

MULTISIM

FIGURE 6–61 Simulation of the traffic signal control system with each part represented by a subcircuit. Open File F06-61 on the website. Run the traffic signal controller simulation using your Multisim software and observe the operation. Lights will appear randomly when first turned on. Simulation times may vary.

Development Software Simulation

Once a new design has been verified by Multisim simulation and prototyping, it is described with VHDL or Verilog code. The code is then entered into a computer using development software with a *text editor,* as shown in Figure 6–62(a) for the traffic signal control system with VHDL. The code is then compiled by the computer using the development software that was used to enter the VHDL or Verilog design into the computer. Two major types of software development tools are Altera Quartus and Xilinx ISE, tutorials for which are available on the website. When the code is compiled, a test feature of the

(a) VHDL code is entered on text editor.

FIGURE 6–62 Traffic signal control design entered in VHDL and simulated using PLD development software (Quartus II in this case).

(b) Simulation waveforms on the waveform editor.

FIGURE 6–62   (Continued)

development software is used to specify inputs and run a simulation to produce the output waveform(s) on a screen, generally called the *waveform editor*, as shown in Figure 6–62(b).

Hardware Testing

Let's assume that the traffic signal control system has been described using the VHDL or Verilog code from Section 6–7 and downloaded into a programmable logic device (PLD) using a development software such as Altera Quartus or Xilinx ISE. The PLD is mounted on a development board such as the one shown in Figure 6–63. There are several methods for testing system circuit boards with PLDs on them.

PLD

FIGURE 6–63   A typical development board used for programming and testing a PLD. A design specified in VHDL or Verilog is implemented in the PLD.

TRADITIONAL TESTING   Depending on the complexity of the system, standard laboratory or field instruments can be used, as shown in Figure 6–64. For systems with a large number of waveforms to observe, a logic analyzer instead of an oscilloscope would be used. You can apply input signals to input pins and check the output pins for the proper waveforms. This traditional method is practical for one-of-a kind development boards and for the preproduction testing of prototype systems.

BED-OF-NAILS TESTING   The testing of printed circuit boards at production levels must be done automatically. The **bed-of-nails** (BON) method was one of the first approaches to automated testing. The pc board is placed on a fixture with an array of small nail-like test probes that make contact with test points on the board. The signals at each test point are checked simultaneously by automated test equipment. The concept is illustrated in Figure 6–65.

Basically, the purpose of automated production testing is to find any manufacturing flaws, such as open or shorted pins and wrong, missing, or misaligned components. This automated process does not primarily test for functionality of the logic. It is assumed that each component has been tested for functionality prior to installation on the circuit board and that the only flaws should be those created during manufacturing.

FLYING PROBE TESTING   Another method for testing printed circuit boards is called the **flying probe** method. A typical flying probe and its basic operation are shown in Figure 6–66. A test probe is positioned above a circuit board that is to be tested. The probe can be automatically moved in three axes—along the *x*-axis, the *y*-axis, and the *z*-axis of the board—to make contact with any specified test points. The movement of the probe is controlled by software that uses the physical layout of the board to determine the coordinates. Many flying probe testers have multiple probes for one board.

The flying probe method of testing overcomes some of the limitations of the bed-of-nails method. First, the BON method requires a different fixture for each type of circuit board, but the flying probe method requires no fixture. Also, the flying probe can access

HANDS ON TIP
Glitches that occur in digital systems are very fast (extremely short in duration) and can be difficult to see on an oscilloscope, particularly at lower sweep rates. A logic analyzer, however, can show a glitch easily. To look for glitches using a logic analyzer, select "latch" mode or (if available) transitional sampling. In the latch mode, the analyzer looks for a voltage level change. When a change occurs, even if it is of extremely short duration (a few nanoseconds), the information is "latched" into the analyzer's memory as another sampled data point. When the data are displayed, the glitch will show as an obvious change in the sampled data, making it easy to identify.

FIGURE 6–64 Traditional testing using standard instruments. An oscilloscope or a logic analyzer can be used. (Photo used with permission from Tektronix, Inc.)

OR

FIGURE 6–65 Bed-of-nails automated testing.

To ATE
(automatic test equipment)

(a) 3-axis movement

(b) Movement from point to point

FIGURE 6–66   **Flying probe testing of a circuit board.**

more points on a board because the probe can be moved to any position, and it can access the top of the board where the components are.

BOUNDARY SCAN TESTING   Limited access to test points led to the concept of placing the test points within the integrated circuit devices themselves. Most PLDs (CPLDs and FPGAs) include boundary scan logic as part of their internal structure independent of the functionality of the logic programmed into the PLD. These devices are JTAG compliant (JTAG is a connection standard).

A circuit, known as a boundary scan cell, is placed between the programmable logic and each input and output pin of the PLD. The cells are basically memory cells that store a 1 or a 0. The cells connected to the programmable logic inputs are called input cells, and those connected to the programmable logic outputs are called output cells. **Boundary scan** testing is based on the JTAG standard (IEEE Std. 1149.1).

TROUBLESHOOTING IN THE FIELD   Once a system has been installed on site, faults that may occur generally require on-site troubleshooting. For example, let's assume that the traffic signal control system is in a housing mounted on a pole at the intersection. In this situation, there would not be time for extensive troubleshooting because the traffic light must be back up and working quickly. Unless the fault is obvious and easy to remedy, such as a loose wire or blown fuse, the approach is to replace the circuit board. The faulty board can then be tested and repaired off-site.

SECTION 6–8 CHECKUP

1. What are the two basic approaches to debugging or testing?

2. How would you test a one-of-a kind or prototype circuit board?

3. List three methods for production testing a circuit board.

SUMMARY

- Latches are bistable devices whose state normally depends on asynchronous inputs.
- Edge-triggered flip-flops are bistable devices with synchronous inputs whose state depends on the inputs only at the triggering transition of a clock pulse. Changes in the outputs occur at the triggering transition of the clock.
- Monostable multivibrators (one-shots) have one stable state. When the one-shot is triggered, the output goes to its unstable state for a time determined by an *RC* circuit.

- Astable multivibrators have no stable states and are used as oscillators to generate timing waveforms in digital systems.
- VHDL and Verilog can be used to describe a complete system.
- Two methods of debugging are software simulation and hardware testing.

KEY TERMS

Key terms and other bold terms in the chapter are defined in the end-of-book glossary.

Astable   Having no stable state. An astable multivibrator oscillates between two quasi-stable states.

Bistable   Having two stable states. Flip-flops and latches are bistable multivibrators.

Clear   An input used to reset a flip-flop (make the Q output 0).

Clock   The triggering input of a flip-flop.

D flip-flop   A type of bistable multivibrator in which the output assumes the state of the D input on the triggering edge of a clock pulse.

Edge-triggered flip-flop   A type of flip-flop in which the data are entered and appear on the output on the same clock edge.

Hold time   The time interval required for the control levels to remain on the inputs to a flip-flop after the triggering edge of the clock in order to reliably activate the device.

J-K flip-flop   A type of flip-flop that can operate in the SET, RESET, no-change, and toggle modes.

Latch   A bistable digital circuit used for storing a bit.

Monostable   Having only one stable state. A monostable multivibrator, commonly called a *one-shot,* produces a single pulse in response to a triggering input.

One-shot   A monostable multivibrator.

Power dissipation   The amount of power required by a circuit.

Preset   An asynchronous input used to set a flip-flop (make the Q output 1).

Propagation delay time   The interval of time required after an input signal has been applied for the resulting output change to occur.

RESET   The state of a flip-flop or latch when the output is 0; the action of producing a RESET state.

SET   The state of a flip-flop or latch when the output is 1; the action of producing a SET state.

Set-up time   The time interval required for the control levels to be on the inputs to a digital circuit, such as a flip-flop, prior to the triggering edge of a clock pulse.

Synchronous   Having a fixed time relationship.

Timer   A circuit that can be used as a one-shot or as an oscillator.

Toggle   The action of a flip-flop when it changes state on each clock pulse.

TRUE/FALSE QUIZ

Answers are at the end of the chapter.

1. A latch has two stable states.

2. A latch is considered to be in the SET state when the Q output is LOW.

3. A gated D latch must be enabled in order to change state.

4. Flip-flops and latches are both bistable devices.

5. An edge-triggered D flip-flop changes state whenever the D input changes.

6. A clock input is necessary for an edge-triggered flip-flop.

7. When both the J and K inputs are HIGH, an edge-triggered J-K flip-flop changes state on each clock pulse.

8. A one-shot is also known as an astable multivibrator.

9. When triggered, a one-shot produces a single pulse.

10. The 555 timer can be used as a one-shot or as a pulse oscillator.

SELF-TEST

Answers are at the end of the chapter.

1. If an S-R latch has a 1 on the S input and a 0 on the R input and then the S input goes to 0, the latch will be

   (a) set   (b) reset   (c) invalid   (d) clear

2. The invalid state of an S-R latch occurs when

   (a) $S = 1, R = 0$ (b) $S = 0, R = 1$
   (c) $S = 1, R = 1$ (d) $S = 0, R = 0$

3. For a gated D latch, the Q output always equals the D input

   (a) before the enable pulse   (b) during the enable pulse
   (c) immediately after the enable pulse   (d) answers (b) and (c)

4. Like the latch, the flip-flop belongs to a category of logic circuits known as

   (a) monostable multivibrators   (b) bistable multivibrators
   (c) astable multivibrators   (d) one-shots

5. The purpose of the clock input to a flip-flop is to

   (a) clear the device
   (b) set the device
   (c) always cause the output to change states
   (d) cause the output to assume a state dependent on the controlling (J-K or D) inputs.

6. For an edge-triggered D flip-flop,

   (a) a change in the state of the flip-flop can occur only at a clock pulse edge
   (b) the state that the flip-flop goes to depends on the D input
   (c) the output follows the input at each clock pulse
   (d) all of these answers

7. A feature that distinguishes the J-K flip-flop from the D flip-flop is the

   (a) toggle condition   (b) preset input
   (c) type of clock   (d) clear input

8. A flip-flop is in the toggle condition when

   (a) $J = 1, K = 0$ (b) $J = 1, K = 1$
   (c) $J = 0, K = 0$ (d) $J = 0, K = 1$

9. A J-K flip-flop with $J = 1$ and $K = 1$ has a 10 kHz clock input. The Q output is

   (a) constantly HIGH   (b) constantly LOW
   (c) a 10 kHz square wave   (d) a 5 kHz square wave

10. A one-shot is a type of

    (a) monostable multivibrator   (b) astable multivibrator   (c) timer
    (d) answers (a) and (c)   (e) answers (b) and (c)

11. The output pulse width of a nonretriggerable one-shot depends on

    (a) the trigger intervals   (b) the supply voltage
    (c) a resistor and capacitor   (d) the threshold voltage

12. An astable multivibrator

    (a) requires a periodic trigger input   (b) has no stable state
    (c) is an oscillator   (d) produces a periodic pulse output
    (e) answers (a), (b), (c), and (d)   (f) answers (b), (c), and (d) only

PROBLEMS

Answers to odd-numbered problems are at the end of the book.

SECTION 6–1   A System

1. How long can the traffic signal control system remain in the first state?

2. How long can the system remain in the fourth state?

3. Write the Boolean expressions for the light outputs MR, MY, MG, SR, SY, and SG in terms of G_0 and G_1.

4. Using the expressions developed in Problem 3, show the logic diagram for the light outputs.

SECTION 6–2   Latches

5. If the waveforms in Figure 6–67 are applied to an active-LOW input S-R latch, draw the resulting Q output waveform in relation to the inputs. Assume that Q starts LOW.

FIGURE 6–67

6. Solve Problem 5 for the input waveforms in Figure 6–68 applied to an active-HIGH S-R latch.

FIGURE 6–68

7. Solve Problem 5 for the input waveforms in Figure 6–69.

FIGURE 6–69

8. For a gated S-R latch, determine the Q and $\overline{Q}$ outputs for the inputs in Figure 6–70. Show them in proper relation to the enable input. Assume that Q starts LOW.

9. Solve Problem 8 for the inputs in Figure 6–71.

FIGURE 6–70

FIGURE 6–71

10. Solve Problem 8 for the inputs in Figure 6–72.

11. For a gated D latch, the waveforms shown in Figure 6–73 are observed on its inputs. Draw the timing diagram showing the output waveform you would expect to see at Q if the latch is initially RESET.

FIGURE 6–72

FIGURE 6–73

SECTION 6–3   Flip-Flops

12. Two edge-triggered D flip-flops are shown in Figure 6–74. If the inputs are as shown, draw the Q output of each flip-flop relative to the clock, and explain the difference between the two. The flip-flops are initially RESET.

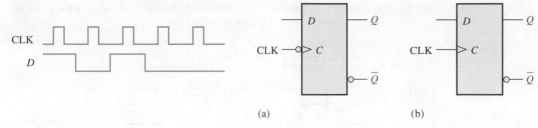

FIGURE 6–74

13. The Q output of an edge-triggered D flip-flop is shown in relation to the clock signal in Figure 6–75. Determine the input waveform on the D input that is required to produce this output if the flip-flop is a positive edge-triggered type.

FIGURE 6–75

14. Draw the Q output relative to the clock for a D flip-flop with the inputs as shown in Figure 6–76. Assume positive edge-triggering and Q initially LOW.

15. Solve Problem 14 for the inputs in Figure 6–77.

FIGURE 6–76   **FIGURE 6–77**

16. For a positive edge-triggered J-K flip-flop with inputs as shown in Figure 6–78, determine the Q output relative to the clock. Assume that Q starts LOW.

17. Solve Problem 16 for the inputs in Figure 6–79.

FIGURE 6–78   **FIGURE 6–79**

18. Determine the Q waveform relative to the clock if the signals shown in Figure 6–80 are applied to the inputs of the J-K flip-flop. Assume that Q is initially LOW.

FIGURE 6–80

19. For a negative edge-triggered J-K flip-flop with the inputs in Figure 6–81, develop the Q output waveform relative to the clock. Assume that Q is initially LOW.

FIGURE 6–81

20. The following serial data are applied to the flip-flop through the AND gates as indicated in Figure 6–82. Determine the resulting serial data that appear on the Q output. There is one clock pulse for each bit time. Assume that Q is initially 0 and that $\overline{PRE}$ and $\overline{CLR}$ are HIGH. Rightmost bits are applied first.

J_1: 1 0 1 0 0 1 1; J_2: 0 1 1 1 0 1 0; J_3: 1 1 1 1 0 0 0; K_1: 0 0 0 1 1 1 0; K_2: 1 1 0 1 1 0 0; K_3: 1 0 1 0 1 0 1

21. For the circuit in Figure 6–82, complete the timing diagram in Figure 6–83 by showing the Q output (which is initially LOW). Assume $\overline{PRE}$ and $\overline{CLR}$ remain HIGH.

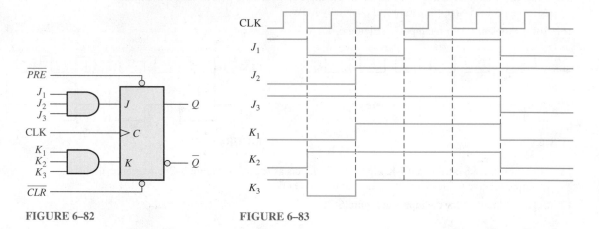

FIGURE 6–82   **FIGURE 6–83**

22. Solve Problem 21 with the same J and K inputs but with the $\overline{PRE}$ and $\overline{CLR}$ inputs as shown in Figure 6–84 in relation to the clock.

FIGURE 6–84

23. A D flip-flop is connected as shown in Figure 6–85. Determine the Q output in relation to the clock. What specific function does this device perform?

FIGURE 6–85

24. For the circuit in Figure 6–86, develop a timing diagram for eight clock pulses, showing the Q_A and Q_B outputs in relation to the clock.

HIGH

FIGURE 6–86

SECTION 6–4   Flip-Flop Operating Characteristics

25. What determines the power dissipation of a flip-flop?

26. Typically, a manufacturer's data sheet specifies four different propagation delay times associated with a flip-flop. Name and describe each one.

27. The data sheet of a certain flip-flop specifies that the minimum HIGH time for the clock pulse is 30 ns and the minimum LOW time is 37 ns. What is the maximum operating frequency?

28. The flip-flop in Figure 6–87 is initially RESET. Show the relation between the Q output and the clock pulse if propagation delay t_{PLH} (clock to Q) is 8 ns.

29. The direct current required by a particular flip-flop that operates on a +5 V dc source is found to be 10 mA. A certain digital device uses 15 of these flip-flops. Determine the current capacity required for the +5 V dc supply and the total power dissipation of the system.

30. For the circuit in Figure 6–86, determine the maximum frequency of the clock signal for reliable operation if the set-up time for each flip-flop is 2 ns and the propagation delays (t_{PLH} and t_{PHL}) from clock to output are 5 ns for each flip-flop.

FIGURE 6–87

SECTION 6–5   Timers

31. Determine the pulse width of a 555 timer configured as a one-shot if the external resistor is 3.3 kΩ and the external capacitor is 2000 pF.

32. An output pulse of 5 μs duration is to be generated by a 555 operating as a one-shot. Using a capacitor of 10,000 pF, determine the value of external resistance required.

33. Create a one-shot, using a 555 timer that will produce a 0.25 s output pulse.

34. A 555 timer is configured to run as an astable multivibrator as shown in Figure 6–88. Determine its frequency.

35. Determine the values of the external resistors for a 555 timer used as an astable multivibrator with an output frequency of 20 kHz, if the external capacitor C is 0.002 μF and the duty cycle is to be approximately 75%.

FIGURE 6–88

SECTION 6–6   Bistable Logic with VHDL and Verilog

36. Given the following VHDL statement:

Q <= J1 **nand** NotQ;

Create the functionally equivalent statement replacing the **nand** operator with the **and** operator.

37. In Verilog, the **reg** data type is used to describe the internal identifiers S and R in the DFlipFlop description in Figure 6–52. Name an equivalent VHDL data type.

38. In the VHDL JKFlipFlop code in Figure 6–53, a **process** block structure contains the functional code defining the behavior of the flip flop. What Verilog code structure performs this functionality?

39. Name the Verilog keyword required to set the result of a Boolean equation to an output identifier.

SECTION 6–7   Traffic Signal Control System with VHDL and Verilog

40. In the programming model in Figure 6–55, what does the FreqDivide block do?

41. List the statements in the VHDL program for the TimerCircuits that result in the 4 s and 25 s time intervals.

42. Repeat Problem 41 for the Verilog program.

43. Write the lines of code in both VHDL and Verilog that describe the sequential logic.

44. What numerical value would be assigned to SetCount in the FreqDivide component of the traffic signal control system if the system clock were reduced to 12 Hz?

45. We would like to add an additional state to the StateDecoder component. Explain the modification required to introduce a fifth output state S5.

46. Can the program TrafficLights be simplified if a 1 Hz clock replaced the 24 MHz system clock? If so, explain the required TrafficLights program source modifications.

47. In the traffic signal control system, a "Walk/Don't Walk" sign is added to the main and side streets. The "Walk" and "Don't" displays are to have separate inputs; one for "Walk" and a second for "Don't". What changes to the TrafficLights program are needed to operate the new signs?

SECTION 6–8   Troubleshooting

48. The flip-flop in Figure 6–89 is tested under all input conditions as shown. Is it operating properly? If not, what is the most likely fault?

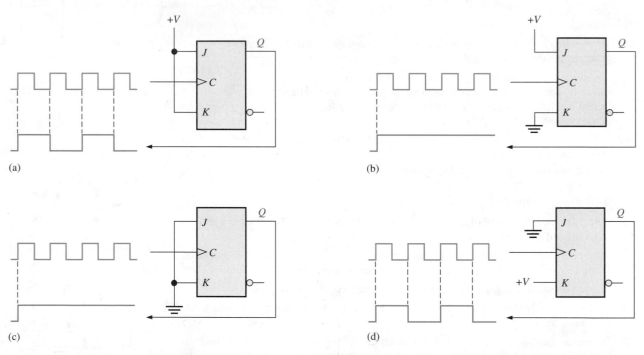

FIGURE 6–89

49. A quad NAND gate IC is used to construct a gated S-R latch on a protoboard in the lab as shown in Figure 6–90. The schematic in part (a) is used to connect the circuit in part (b). When you try to operate the latch, you find that the Q output stays HIGH no matter what the inputs are. Determine the problem.

50. Determine if the flip-flop in Figure 6–91 is operating properly, and if not, identify the most probable fault.

51. The parallel data storage circuit in Figure 6–92 does not operate properly. To check it out, you first make sure that V_{CC} and ground are connected, and then you apply LOW levels to all the D inputs and pulse the clock line. You check the Q outputs and find them all to be LOW; so far, so good. Next you apply HIGHs to all the D inputs and again pulse the clock line. When you check the Q outputs, they are still all LOW. What is the problem, and what procedure will you use to isolate the fault to a single device?

(a)

(b)

FIGURE 6–90

FIGURE 6–91

(a)

FIGURE 6–92 Example of flip-flops used in a basic register for parallel data storage.

(b)

52. The flip-flop circuit in Figure 6–93(a) is used to generate a binary count sequence. The gates form a decoder that is supposed to produce a HIGH when a binary zero or a binary three state occurs (00 or 11). When you check the Q_A and Q_B outputs, you get the display shown in part (b), which reveals glitches on the decoder output (X) in addition to the correct pulses. What is causing these glitches, and how can you eliminate them?

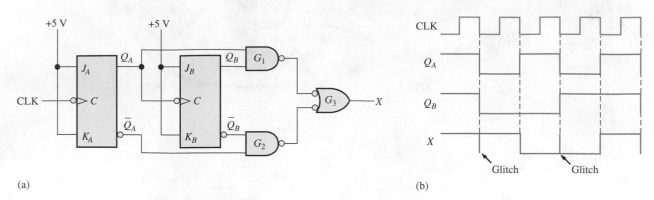

(a)   (b)

FIGURE 6–93

53. Determine the Q_A, Q_B and X outputs over six clock pulses in Figure 6–93(a) for each of the following faults in the bipolar (TTL) circuits. Start with both Q_A and Q_B LOW.
    (a) J_A input open
    (b) K_B input open
    (c) Q_B output open
    (d) clock input to flip-flop B shorted
    (e) gate G_2 output open

54. Two one-shot ICs are connected on a circuit board as shown in Figure 6–94. After observing the oscilloscope display, do you conclude that the circuit is operating properly? If not, what is the most likely problem?

FIGURE 6–94

Special Problems

55. Implement a basic counting circuit that produces a binary sequence from zero through seven by using negative edge-triggered J-K flip-flops.

56. In the shipping department of a softball factory, the balls roll down a conveyor and through a chute single file into boxes for shipment. Each ball passing through the chute activates a switch circuit that produces an electrical pulse. The capacity of each box is 32 balls. Implement a logic circuit to indicate when a box is full so that an empty box can be moved into position.

57. List the changes that would be necessary in the traffic signal control system to add a 15 s left turn arrow for the main street. The turn arrow will occur after the red light and prior to the green light. Modify the state diagram in Figure 6–2 to show these changes.

MULTISIM TROUBLESHOOTING PRACTICE

58. Open file P06-58 and follow the instructions given there.

59. Open file P06-59 and follow the instructions given there.

60. Open file P06-60 and follow the instructions given there.

61. Open file P06-61 and follow the instructions given there.

62. Open file P06-62 and follow the instructions given there.

ANSWERS TO SECTION CHECKUPS

SECTION 6–1   A System

1. For 25 s (T_L) or as long as there is no vehicle on the side street ($\overline{V}_s$)
2. For 4 s (T_S)
3. Main will stay red for up to 25 s if there is a vehicle on the side street ($T_L V_s$).

SECTION 6–2   Latches

1. Three types of latches are S-R, gated S-R, and gated D.
2. $SR = 00$, NC; $SR = 01$, $Q = 0$; $SR = 10$, $Q = 1$; $SR = 11$, invalid
3. $Q = 1$

SECTION 6–3   Flip-Flops

1. The output of a gated D latch can change any time the gate enable (EN) input is active. The output of an edge-triggered D flip-flop can change only on the triggering edge of a clock pulse.
2. The output of the D flip-flop follows the D input on each clock pulse. The output of the J-K flip-flop depends on both of the states of the J and K inputs.
3. Output Q goes HIGH on the trailing edge of the first clock pulse, LOW on the trailing edge of the second pulse, HIGH on the trailing edge of the third pulse, and LOW on the trailing edge of the fourth pulse.
4. For divide-by-2 operation, the flip-flop must toggle ($J = 1$, $K = 1$).
5. Six flip-flops are used in a divide-by-64 device.

SECTION 6–4   Flip-Flop Operating Characteristics

1. **(a)** Set-up time is the time required for input data to be present before the triggering edge of the clock pulse.
   (b) Hold time is the time required for data to remain on the inputs after the triggering edge of the clock pulse.
2. The flip-flop with $t_{PHL} = 17$ ns

SECTION 6–5   Timers

1. A nonretriggerable one-shot times out before it can respond to another trigger input. A retriggerable one-shot responds to each trigger input.
2. Pulse width is set with external R and C components.
3. 11 ms.
4. An astable multivibrator has no stable state. A monostable multivibrator has one stable state.
5. Duty cycle $= (15 \text{ ms}/20 \text{ ms})100\% = 75\%$

SECTION 6–6   Bistable Logic with VHDL and Verilog

1. VHDL specifies edge-triggered operation with the **if rising_edge** term.
2. Verilog specifies edge-triggered operation with **@(posedge Clock).**
3. VHDL; **if falling_edge;** Verilog: **@(negedge Clock)**
4. ! is NOT, ∥ is OR, and && is AND.

SECTION 6–7   **Traffic Signal Control System with VHDL and Verilog**

1. The program blocks are TimerCircuits, FreqDivide, SequentialLogic, and StateDecoder.

2. The program blocks listed in answer 1 are specified as components in the complete system program. The two timer triggers and the six light control outputs are specified in terms of the state decoder output using Boolean expressions.

3. A noticeable difference is that Verilog generally provides for shorter programs.

SECTION 6–8   **Troubleshooting**

1. Multisim software simulation; development software simulation.

2. Test one-of-a-kind circuit board using traditional test methods with lab instruments.

3. Production test using bed-of-nails, flying probe, or boundary scan techniques.

ANSWERS TO RELATED PROBLEMS FOR EXAMPLES

6–1   The Q output is the same as shown in Figure 6–11(b).

6–2   See Figure 6–95.

6–3   See Figure 6–96.

FIGURE 6–95

FIGURE 6–96

6–4   See Figure 6–97.

6–5   See Figure 6–98.

FIGURE 6–97

FIGURE 6–98

6–6   See Figure 6–99.

6–7   See Figure 6–100.

FIGURE 6–99

FIGURE 6–100

6–8   See Figure 6–101.

6–9   See Figure 6–102.

FIGURE 6–101

FIGURE 6–102

6–10 $2^5 = 32$. Five flip-flops are required.

6–11   Sixteen states require four flip-flops ($2^4 = 16$).

6–12 $R_1 = 91\,k\Omega$

6–13 $R_2 \cong 2.2\,k\Omega$

ANSWERS TO TRUE/FALSE QUIZ

1. T   **2.** F   **3.** T   **4.** T   **5.** F   **6.** T

7. T   **8.** F   **9.** T   **10.** T

ANSWERS TO SELF-TEST

1. (a)   **2.** (c)   **3.** (d)   **4.** (b)   **5.** (d)   **6.** (d)

7. (a)   **8.** (b)   **9.** (d)   **10.** (d)   **11.** (c)   **12.** (f)

CHAPTER 7

SHIFT REGISTERS

OUTLINE

OBJECTIVES

- Use shift registers in a system
- Identify the basic forms of data movement in shift registers
- Explain how serial in/serial out, serial in/parallel out, parallel in/serial out, and parallel in/parallel out shift registers operate
- Describe how a bidirectional shift register operates
- Determine the sequence of a Johnson counter
- Set up a ring counter to produce a specified sequence
- Construct a ring counter from a shift register
- Use a shift register as a time-delay device
- Use a shift register to implement a serial-to-parallel data converter
- Describe a basic shift-register-controlled keyboard encoder
- Describe a system using VHDL and Verilog

KEY TERMS

Register	**Load**
Stage	**Bidirectional**

INTRODUCTION

Shift registers are a type of sequential logic circuit closely related to digital counters. Registers are used primarily for the storage of digital data and typically do not possess a characteristic internal sequence of states as do counters. There are exceptions, however, and these are covered in Section 7–5.

In this chapter a system using shift registers is introduced. The basic types of shift registers are studied and several applications are presented. Also, troubleshooting methods are discussed, and implementation of a system using VHDL and Verilog is presented.

VISIT THE WEBSITE
Study aids for this chapter are available at
http://pearsonhighered.com/floyd

This section provides an introduction to the use of shift registers for temporarily storing and shifting data in a system. A security system that provides coded access to a secured area is developed. The system consists of a keypad, security code logic, and code-selection logic. When security code is stored in the system, access is achieved by entering the correct code on a keypad.

After completing this section, you should be able to

- **Explain the overall operation of the system**
- **Describe how shift registers are used for both parallel and serial storage and movement of data**
- **Explain how other devices such as the one-shot, the encoder, and the comparator are used in the system**

A block diagram for the security system is shown in Figure 7–1. The system consists of the security code logic, the code-selection logic, and the keypad. The keypad is a standard numeric keypad.

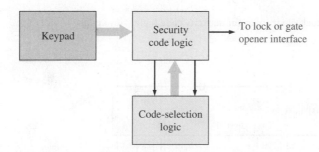

FIGURE 7–1 **Block diagram of the security system.**

Basic Operation

A 4-digit entry code is set with user-accessible DIP switches. Initially pressing the # key sets up the system for the first digit in the code. For entry, the code is entered one digit at a time on the keypad and converted to a BCD code for processing by the security code logic. If the entered code agrees with the stored code, the output activates the access mechanism and allows the door or gate, depending on the type of area that is secured, to be opened.

The Security Code Logic

The security code logic compares the code entered on the keypad with the predetermined code. A logic diagram of the security code logic with keypad is shown in Figure 7–2.

In order to gain entry, first the # key on the keypad is pressed to enable the one-shots to be triggered, thus initializing the 8-bit shift register C with a preset pattern (00010000). Next the four digits of the code are entered in proper sequence on the keypad. As each digit is entered, it is converted to BCD by the decimal-to-BCD encoder, and a clock pulse is produced by one-shot A that shifts the 4-bit code into register A. The one-shot is triggered by a transition on the output of the OR gate when a key is pressed. At the same time, the corresponding BCD digit from the code-selection logic is shifted into register B. Also, one-shot B is triggered after one-shot A to provide a delayed clock pulse for register C to serially shift the preloaded pattern (00010000). The left-most three 0s are simply "fillers" and serve no purpose in the operation of the system. The outputs of registers A and B are applied to the comparator; if the codes are the same, the output of the comparator goes HIGH, placing shift register C in the SHIFT mode.

FIGURE 7–2   Block diagram of the security code logic with keypad.

FIGURE 7–3   Logic diagram of the code-selection logic.

Each time an entered digit agrees with the preset digit, the 1 in shift register C is shifted right one position. On the fourth code agreement, the 1 appears on the output of the shift register and activates the mechanism to unlock the door or open the gate. If the code digits do not agree, the output of the comparator goes LOW, placing shift register C in the LOAD mode so the shift register is reinitialized to the preset pattern (00010000).

The Code-Selection Logic

A logic diagram of the code-selection logic is shown in Figure 7–3. This part of the system includes a set of DIP switches into which a 4-digit entry code is set. Initially pressing the # key sets up the system for the first digit in the code by causing a preset pattern to be loaded into shift register C (0001). Initially, the four bits in the first code digit are selected by a HIGH on the Q_0 output of the shift register, enabling the four AND gates labeled A1–A4. As each digit of the code is entered on the keypad, the clock from the security code logic shifts the 1 in the shift register to sequentially enable each set of four AND gates. As a result, the BCD digits in the security code appear sequentially on the outputs. In the security code logic, each of the code digits is compared to the digit entered on the keypad.

SECTION 7–1 CHECKUP*

1. Explain the purpose of the OR gate in Figure 7–2.

2. If the digit 4 is entered on the keypad, what appears on the output of register A?

3. Explain the purpose of one-shot B in the security code logic.

*Answers are at the end of the chapter.

7–2 BASIC SHIFT REGISTER OPERATIONS

Shift registers consist of arrangements of flip-flops and are important in applications involving the temporary storage and the transfer of data in a digital system. A register has no specified sequence of states, except in certain very specialized applications. A register, in general, is used solely for storing and shifting data (1s and 0s) entered into it from an external source and typically possesses no characteristic internal sequence of states.

After completing this section, you should be able to

- **Explain how a flip-flop stores a data bit**
- **Define the storage capacity of a shift register**
- **Describe the shift capability of a register**

A **register** is a digital circuit with two basic functions: data storage and data movement. The storage capability of a register makes it an important type of memory device. Figure 7–4 illustrates the concept of storing a 1 or a 0 in a D flip-flop. A 1 is applied to the data input as shown, and a clock pulse is applied that stores the 1 by *setting* the flip-flop. When the 1 on the input is removed, the flip-flop remains in the SET state, thereby storing the 1. A similar procedure applies to the storage of a 0 by *resetting* the flip-flop, as also illustrated in Figure 7–4.

The *storage capacity* of a register is the total number of bits (1s and 0s) of digital data it can retain. Each **stage** (flip-flop) in a shift register represents one bit of storage capacity; therefore, the number of stages in a register determines its storage capacity.

The *shift capability* of a register permits the movement of data from stage to stage within the register or into or out of the register upon application of clock pulses. Figure 7–5 illustrates the types of data movement in shift registers. The block represents any arbitrary 4-bit register, and the arrows indicate the direction of data movement.

A register can consist of one or more flip-flops used to store and shift data.

FIGURE 7–4 The flip-flop as a storage element.

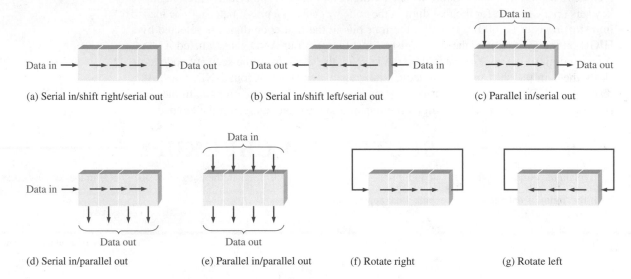

(a) Serial in/shift right/serial out

(b) Serial in/shift left/serial out

(c) Parallel in/serial out

(d) Serial in/parallel out

(e) Parallel in/parallel out

(f) Rotate right

(g) Rotate left

FIGURE 7–5 Basic data movement in shift registers. (Four bits are used for illustration. The bits move in the direction of the arrows.)

SECTION 7–2 CHECKUP

1. What two principal functions are performed by a shift register?

2. What indicates when a flip-flop in a register is storing a 1?

7–3 TYPES OF SHIFT REGISTERS

In this section, four types of shift registers are discussed: serial in/serial out, serial in/parallel out, parallel in/serial out, and parallel in/parallel out.

After completing this section, you should be able to

- Describe the operation of four types of shift register
- Explain how data bits are entered into a shift register
- Describe how data bits are shifted through the register
- Explain how data bits are taken out of a shift register
- Develop and analyze timing diagrams for shift registers

Serial In/Serial Out Shift Registers

The serial in/serial out shift register accepts data serially—that is, one bit at a time on a single line. It produces the stored information on its output also in serial form. Let's first look at the serial entry of data into a typical shift register. Figure 7–6 shows a 4-bit device implemented with D flip-flops. With four stages, this register can store up to four bits of data.

FIGURE 7–6   **Serial in/serial out shift register.**

Table 7–1 shows the entry of the four bits, 1010, into the shift register in Figure 7–6, beginning with the least significant bit. The register is initially clear. The 0 is put onto the serial data input line, making the data bit $D = 0$ for FF0. When the first clock pulse is applied, FF0 is reset, thus storing the 0.

CLK	FF0 (Q_0)	FF1 (Q_1)	FF2 (Q_2)	FF3 (Q_3)
Initial	0	0	0	0
1	0	0	0	0
2	1	0	0	0
3	0	1	0	0
4	1	0	1	0

TABLE 7–1 • Shifting a 4-bit code into the shift register in Figure 7–6. Data bits are indicated by a beige screen.

Next the second bit, which is a 1, is applied to the serial data input, making $D = 1$ for FF0 and $D = 0$ for FF1 because the D input of FF1 is connected to the Q_0 output. When the second clock pulse occurs, the 1 on the data input is shifted into FF0, causing FF0 to set; and the 0 that was in FF0 is shifted into FF1.

The third bit, a 0, is now put onto the serial data input line, and a third clock pulse is applied. The 0 is entered into FF0, the 1 stored in FF0 is shifted into FF1, and the 0 stored in FF1 is shifted into FF2.

The last bit, a 1, is now applied to the serial data input, and a fourth clock pulse is applied. This time the 1 is entered into FF0, the 0 stored in FF0 is shifted into FF1, the 1 stored in FF1 is shifted into FF2, and the 0 stored in FF2 is shifted into FF3. This completes the serial entry of the four bits into the shift register, where they can be stored for any length of time as long as the flip-flops have dc power.

If you want to get the data out of the register, the bits must be shifted out serially and taken off the Q_3 output, as Table 7–2 illustrates. After CLK4 in the data-entry operation just

For serial data, one bit at a time is transferred.

Frequently, it is necessary to *clear* an internal register in a computer. For example, a register may be cleared prior to an arithmetic or other operation. One way that registers in a computer are cleared is using software to subtract the contents of the register from itself. The result, of course, will always be zero. For example, a computer instruction that performs this operation is SUB AL,AL. With this instruction, the register named AL is cleared.

SYSTEM NOTE

CLK	FF0 (Q_0)	FF1 (Q_1)	FF2 (Q_2)	FF3 (Q_3)
Initial	1	0	1	0
5	0	1	0	1
6	0	0	1	0
7	0	0	0	1
8	0	0	0	0

TABLE 7–2 • Shifting a 4-bit code out of the shift register in Figure 7–6. Data bits are indicated by a beige screen.

described (Table 7–1), the LSB, 0, appears on the Q_3 output. When clock pulse CLK5 is applied, the second bit appears on the Q_3 output. Clock pulse CLK6 shifts the third bit to the serial data output, and CLK7 shifts the fourth bit to the output. While the original four bits are being shifted out, more bits can be shifted in. All zeros have been shifted in after CLK 8.

EXAMPLE 7–1

Show the states of the 5-bit register in Figure 7–7(a) for the specified data input and clock waveforms. Assume that the register is initially cleared (all 0s).

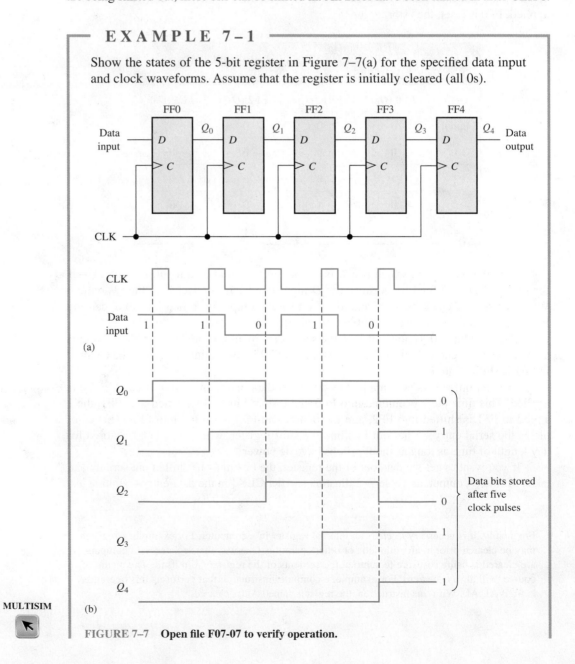

MULTISIM

FIGURE 7–7   Open file F07-07 to verify operation.

SOLUTION

The first data bit (1) is entered into the register on the first clock pulse and then shifted from left to right as the remaining bits are entered and shifted. The register contains $Q_4Q_3Q_2Q_1Q_0 = 11010$ after five clock pulses. See Figure 7–7(b).

RELATED PROBLEM*

Show the states of the register if the data input is inverted. The register is initially cleared.

*Answers are at the end of the chapter.

A traditional logic block symbol for an 8-bit serial in/serial out shift register is shown in Figure 7–8. The "SRG 8" designation indicates a shift register (SRG) with an 8-bit capacity.

FIGURE 7–8   Logic symbol for an 8-bit serial in/serial out shift register.

All microprocessors have special instructions that can emulate a serial shift register. The accumulator register can shift data to the left or right. A right shift is equivalent to a divide-by-2 operation and a left shift is equivalent to a multiply-by-2 operation. Data in the accumulator can be shifted left or right with the rotate instructions; ROR is the rotate right instruction, and ROL is the rotate left instruction. Two other instructions treat the carry flag bit as an additional bit for the rotate operation. These are the RCR for rotate carry right and RCL for rotate carry left.

SYSTEM NOTE

Serial In/Parallel Out Shift Registers

Data bits are entered serially (least-significant bit first) into a serial in/parallel out shift register in the same manner as in serial in/serial out shift registers. The difference is the way in which the data bits are taken out of the register; in the parallel out register, the output of each stage is available. Once the data bits are stored, each bit appears on its respective output line, and all bits are available simultaneously, rather than on a bit-by-bit basis as with the serial output. Figure 7–9 shows a 4-bit serial in/parallel out shift register and its logic block symbol.

(a)   (b)

FIGURE 7–9   A serial in/parallel out shift register.

EXAMPLE 7–2

Show the states of the 4-bit register (SRG 4) for the data input and clock waveforms in Figure 7–10(a). The register initially contains all 1s.

(a)

(b)

FIGURE 7–10

SOLUTION

The register contains 0110 after four clock pulses. See Figure 7–10(b).

RELATED PROBLEM

If the data input remains 0 after the fourth clock pulse, what is the state of the register after three additional clock pulses?

SYSTEM EXAMPLE 7–1

TIME DELAY

The serial in/serial out shift register can be used to provide a time delay from input to output that is a function of both the number of stages (n) in the register and the clock frequency.

When a data pulse is applied to the serial input as shown in Figure 7–11 (A and B connected together), it enters the first stage on the triggering edge of the clock pulse. It is then shifted from stage to stage on each successive clock pulse until it appears on the serial output n clock periods later. This time-delay operation is illustrated in Figure 7–11, in which an 8-bit serial in/serial out shift register is used with a clock frequency of 1 MHz to achieve a time delay (t_d) of 8 μs (8 $\times$ 1 μs).

FIGURE 7–11   **The serial in/serial out shift register as a time-delay device.**

The time delay can be adjusted up or down by changing the clock frequency. The time delay can also be increased by cascading shift registers and decreased by taking the outputs from successively lower stages in the register if the outputs are available, as illustrated in Figure 7–12, using an 8-bit serial in/parallel out shift register with a clock period of 2 μs.

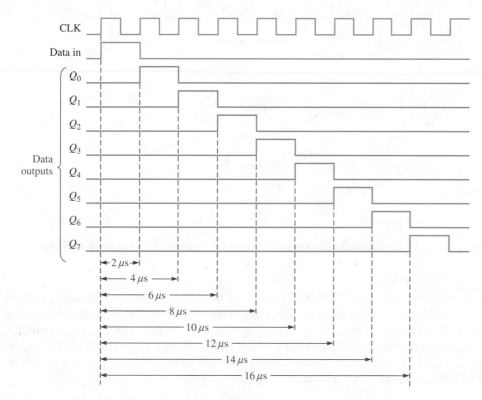

FIGURE 7–12   **Shift register and timing diagram showing time delays.**

Parallel In/Serial Out Shift Registers

For a register with parallel data inputs, the bits are entered simultaneously into their respective stages on parallel lines rather than on a bit-by-bit basis on one line as with serial data inputs. The serial output is the same as in serial in/serial out shift registers, once the data are completely stored in the register.

Figure 7–13 illustrates a 4-bit parallel in/serial out shift register and a typical logic symbol. Notice that there are four data-input lines, D_0, D_1, D_2, and D_3, and a $SHIFT/\overline{LOAD}$ input, which allows four bits of data to **load** in parallel into the register. When $SHIFT/\overline{LOAD}$ is LOW, gates G_1 through G_4 are enabled, allowing each data bit to be applied to the D input of its respective flip-flop. When a clock pulse is applied, the flip-flops with $D = 1$ will set and those with $D = 0$ will reset, thereby storing all four bits simultaneously.

For parallel data, multiple bits are transferred at one time.

MULTISIM

(b) Logic symbol

FIGURE 7–13   A 4-bit parallel in/serial out shift register. Open file F07-13 to verify operation.

When $SHIFT/\overline{LOAD}$ is HIGH, gates G_1 through G_4 are disabled and gates G_5 through G_7 are enabled, allowing the data bits to shift right from one stage to the next. The OR gates allow either the normal shifting operation or the parallel data-entry operation, depending on which AND gates are enabled by the level on the $SHIFT/\overline{LOAD}$ input. Notice that FF0 has a single AND to disable the parallel input, D_0. It does not require an AND/OR arrangement because there is no serial data in.

EXAMPLE 7–3

Show the data-output waveform for a 4-bit register with the parallel input data and the clock and $SHIFT/\overline{LOAD}$ waveforms given in Figure 7–14(a). Refer to Figure 7–13(a) for the logic diagram.

SOLUTION

On clock pulse 1, the parallel data ($D_0D_1D_2D_3 = 1010$) are loaded into the register, making Q_3 a 0. On clock pulse 2, the 1 from Q_2 is shifted onto Q_3; on clock pulse 3, the 0 is shifted onto Q_3; on clock pulse 4, the last data bit (1) is shifted onto Q_3; and on clock pulse 5, all data bits have been shifted out, and only 1s remain in the register (assuming the D_0 input remains a 1). See Figure 7–14(b).

RELATED PROBLEM

Show the data-output waveform for the clock and $SHIFT/\overline{LOAD}$ inputs shown in Figure 7–14(a) if the parallel data are $D_0D_1D_2D_3 = 0101$.

FIGURE 7–14

Parallel In/Parallel Out Shift Registers

Both parallel entry and parallel output of data have already been discussed. The parallel in/parallel out register employs both methods. Immediately following the simultaneous entry of all data bits, the bits appear on the parallel outputs. Figure 7–15 shows the logic diagram of a parallel in/parallel out register.

A universal shift register has both serial and parallel input and output capability.

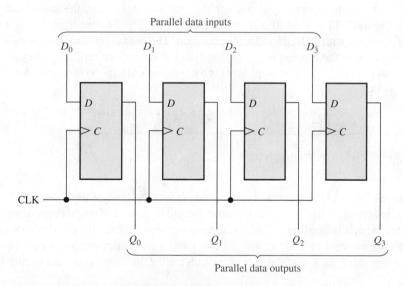

FIGURE 7–15   **A parallel in/parallel out register.**

SYSTEM EXAMPLE 7–2

SERIAL-TO-PARALLEL DATA CONVERTER

Serial data transmission from one digital system to another is commonly used to reduce the number of wires in the transmission line. For example, eight bits can be sent serially over one wire, but it takes eight wires to send the same data in parallel.

Serial data transmission is widely used by peripherals to pass data back and forth to a computer. For example, USB (*universal serial bus*) is used to connect keyboards, printers, scanners, and more to the computer. All computers process data in parallel form, thus the requirement for serial-to-parallel conversion. A simplified serial-to-parallel data converter, in which two types of shift registers are used, is shown in Figure 7–16.

FIGURE 7–16

To illustrate the operation of this serial-to-parallel converter, the serial data format shown in Figure 7–17 is used. It consists of eleven bits. The first bit (start bit) is always 0 and always begins with a HIGH-to-LOW transition. The next eight bits (D_7 through D_0) are the data bits (one of the bits can be parity), and the last one or two bits (stop bits) are always 1s. When no data bits are being sent, there is a continuous HIGH on the serial data line.

FIGURE 7–17 Serial data format.

The HIGH-to-LOW transition of the start bit sets the control flip-flop, which enables the clock generator. After a fixed delay time, the clock generator begins producing a pulse waveform, which is applied to the data-input register and to the divide-by-8 counter. (Counters are covered in Chapter 8.) The clock has a frequency precisely equal to that of the incoming serial data, and the first clock pulse after the start bit occurs during the first data bit.

The timing diagram in Figure 7–18 illustrates the following basic operation: The eight data bits (D_7 through D_0) are serially shifted into the data-input register. Shortly after the eighth clock pulse, the terminal count (TC) goes from LOW to HIGH, indicating the counter is at the last state. This rising edge is ANDed with the clock pulse, which is still HIGH, producing a rising edge at $TC \cdot$ CLK. This parallel loads the eight data bits from the data-input shift register to the data-output register. A short time later, the clock pulse goes LOW and this HIGH-to-LOW transition triggers the one-shot, which produces a short-duration pulse to clear the counter and reset the control flip-flop and thus disable the clock generator. The system is now ready for the next group of eleven bits, and it waits for the next HIGH-to-LOW transition at the beginning of the start bit.

By reversing the process just stated, parallel-to-serial data conversion can be accomplished. Since the serial data format must be produced, start and stop bits must be added to the sequence.

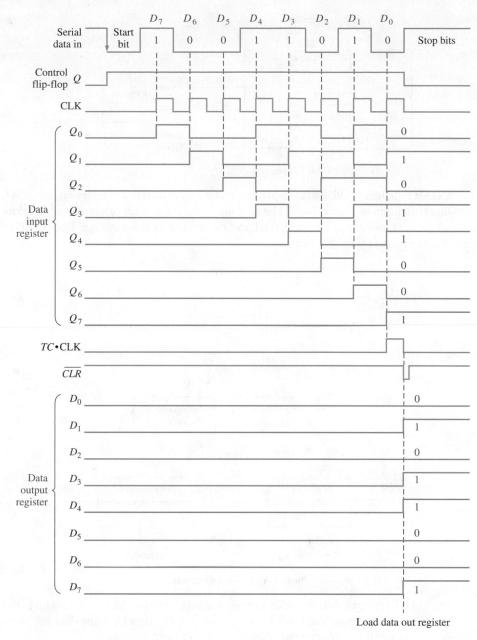

FIGURE 7–18

SYSTEM EXAMPLE 7–3

UNIVERSAL ASYNCHRONOUS RECEIVER TRANSMITTER (UART)

Computers and microprocessor-based systems also send and receive data in a parallel format. Frequently, these systems must communicate with external devices that send and/or receive serial data. An interfacing device used to accomplish these conversions is the UART (Universal Asynchronous Receiver Transmitter). Figure 7–19 illustrates the UART in a general microprocessor-based system application.

FIGURE 7–19 UART interface.

A UART includes both serial-to-parallel and parallel-to-serial conversion, as shown in the block diagram in Figure 7–20. The data bus is basically a set of parallel conductors along which data move between the UART and the microprocessor system. Buffers interface the data registers with the data bus.

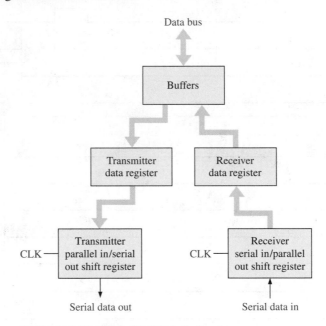

FIGURE 7–20 Basic UART block diagram.

The UART receives data in serial format, converts the data to parallel format, and places them on the data bus. The UART also accepts parallel data from the data bus, converts the data to serial format, and transmits them to an external device.

SECTION 7–3 CHECKUP

1. Develop the logic diagram for the shift register in Figure 7–6, using J-K flip-flops to replace the D flip-flops.

2. How many clock pulses are required to enter a byte of data serially into an 8-bit shift register?

3. The bit sequence 1101 is serially entered (least-significant bit first) into a 4-bit parallel out shift register that is initially clear. What are the Q outputs after two clock pulses?

4. How can a serial in/parallel out register be used as a serial in/serial out register?

5. Explain the function of the $SHIFT/\overline{LOAD}$ input.

6. In Figure 7–15, $D_0 = 1$, $D_1 = 0$, $D_2 = 0$, and $D_3 = 1$. After three clock pulses, what are the data outputs?

7–4 BIDIRECTIONAL SHIFT REGISTERS

A **bidirectional** shift register is one in which the data can be shifted either left or right. It can be implemented by using gating logic that enables the transfer of a data bit from one stage to the next stage to the right or to the left, depending on the level of a control line.

After completing this section, you should be able to

- Explain the operation of a bidirectional shift register
- Develop and analyze timing diagrams for bidirectional shift registers

A 4-bit bidirectional shift register is shown in Figure 7–21. A HIGH on the $RIGHT/\overline{LEFT}$ control input allows data bits inside the register to be shifted to the right, and a LOW enables data bits inside the register to be shifted to the left. An examination of the gating logic will make the operation apparent. When the $RIGHT/\overline{LEFT}$ control input is HIGH, gates G_1 through G_4 are enabled, and the state of the Q output of each flip-flop is passed through to the D input of the *following* flip-flop. When a clock pulse occurs, the data bits are shifted one place to the *right*. When the $RIGHT/\overline{LEFT}$ control input is LOW, gates G_5 through G_8 are enabled, and the Q output of each flip-flop is passed through to the D input of the *preceding* flip-flop. When a clock pulse occurs, the data bits are then shifted one place to the *left*.

FIGURE 7–21 Four-bit bidirectional shift register. Open file F07-21 to verify the operation.

MULTISIM

EXAMPLE 7–4

Determine the state of the shift register of Figure 7–21 after each clock pulse for the given $RIGHT/\overline{LEFT}$ control input waveform in Figure 7–22(a). Assume that $Q_0 = 1$, $Q_1 = 1$, $Q_2 = 0$, and $Q_3 = 1$ and that the serial data-input line is LOW.

SOLUTION

See Figure 7–22(b).

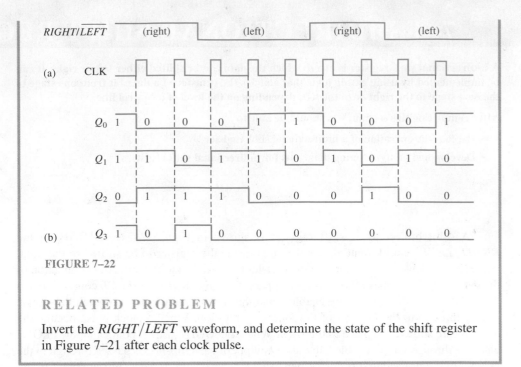

FIGURE 7–22

RELATED PROBLEM

Invert the $RIGHT/\overline{LEFT}$ waveform, and determine the state of the shift register in Figure 7–21 after each clock pulse.

SECTION 7–4 CHECKUP

1. Assume that the 4-bit bidirectional shift register in Figure 7–21 has the following contents: $Q_0 = 1$, $Q_1 = 1$, $Q_2 = 0$, and $Q_3 = 0$. There is a 1 on the serial data-input line. If $RIGHT/\overline{LEFT}$ is HIGH for three clock pulses and LOW for two more clock pulses, what are the contents after the fifth clock pulse?

7–5 SHIFT REGISTER COUNTERS

A shift register counter is basically a shift register with the serial output connected back to the serial input to produce special sequences. These devices are often classified as counters because they exhibit a specified sequence of states. Two of the most common types of shift register counters, the Johnson counter and the ring counter, are introduced in this section.

After completing this section, you should be able to

- Discuss how a shift register counter differs from a basic shift register
- Explain the operation of a Johnson counter
- Specify a Johnson sequence for any number of bits
- Explain the operation of a ring counter and determine the sequence of any specific ring counter

The Johnson Counter

In a **Johnson counter** the complement of the output of the last flip-flop is connected back to the D input of the first flip-flop (it can be implemented with other types of flip-flops as well). If the counter starts at 0, this feedback arrangement produces a characteristic sequence of states, as shown in Table 7–3 for a 4-bit device and in Table 7–4 for a 5-bit device. Notice that the 4-bit sequence has a total of eight states, or bit patterns, and that the 5-bit sequence has a total of ten states. In general, a Johnson counter will produce a modulus of $2n$, where n is the number of stages in the counter.

TABLE 7–3 • Four-bit Johnson sequence.				
CLOCK PULSE	Q_0	Q_1	Q_2	Q_3
0	0	0	0	0
1	1	0	0	0
2	1	1	0	0
3	·1	1	1	0
4	1	1	1	1
5	0	1	1	1
6	0	0	1	1
7	0	0	0	1

TABLE 7–4 • Five-bit Johnson sequence.					
CLOCK PULSE	Q_0	Q_1	Q_2	Q_3	Q_4
0	0	0	0	0	0
1	1	0	0	0	0
2	1	1	0	0	0
3	1	1	1	0	0
4	1	1	1	1	0
5	1	1	1	1	1
6	0	1	1	1	1
7	0	0	1	1	1
8	0	0	0	1	1
9	0	0	0	0	1

The implementations of the 4-stage and 5-stage Johnson counters are shown in Figure 7–23. The implementation of a Johnson counter is very straightforward and is the same regardless of the number of stages. The Q output of each stage is connected to the D input of the next stage (assuming that D flip-flops are used). The single exception is that the $\overline{Q}$ output of the last stage is connected back to the D input of the first stage. As the sequences in Table 7–3 and 7–4 show, if the counter starts at 0, it will "fill up" with 1s from left to right, and then it will "fill up" with 0s again.

Diagrams of the timing operations of the 4-bit and 5-bit Johnson counters are shown in Figures 7–24 and 7–25, respectively.

(a) Four-bit Johnson counter

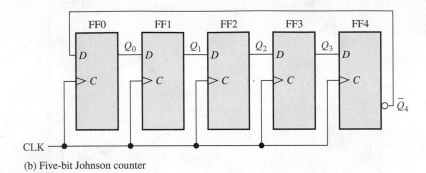

(b) Five-bit Johnson counter

FIGURE 7–23 **Four-bit and 5-bit Johnson counters.**

FIGURE 7–24 **Timing sequence for a 4-bit Johnson counter.**

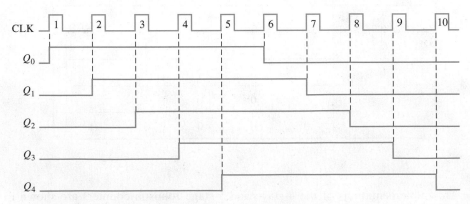

FIGURE 7–25 **Timing sequence for a 5-bit Johnson counter.**

The Ring Counter

The **ring counter** utilizes one flip-flop for each state in its sequence. It has the advantage that decoding gates are not required. In the case of a 10-bit ring counter, there is a unique output for each decimal digit.

A logic diagram for a 10-bit ring counter is shown in Figure 7–26. The sequence for this ring counter is given in Table 7–5. Initially, a 1 is preset into the first flip-flop, and the rest of the flip-flops are cleared. Notice that the interstage connections are the same as those for a Johnson counter, except that Q rather than $\overline{Q}$ is fed back from the last stage. The ten outputs of the counter indicate directly the decimal count of the clock pulse. For instance, a 1 on Q_0 represents a zero, a 1 on Q_1 represents a one, a 1 on Q_2 represents a two, a 1 on Q_3 represents a three, and so on. You should verify for yourself that the 1 is always retained in the counter and simply shifted "around the ring," advancing one stage for each clock pulse.

Modified sequences can be achieved by having more than a single 1 in the counter, as illustrated in Example 7–5.

FIGURE 7–26   A 10-bit ring counter. Open file F07-26 to verify operation.

MULTISIM

TABLE 7–5 • Ten-bit ring counter sequence.										
CLOCK PULSE	Q_0	Q_1	Q_2	Q_3	Q_4	Q_5	Q_6	Q_7	Q_8	Q_9
0	1	0	0	0	0	0	0	0	0	0 ←
1	0	1	0	0	0	0	0	0	0	0
2	0	0	1	0	0	0	0	0	0	0
3	0	0	0	1	0	0	0	0	0	0
4	0	0	0	0	1	0	0	0	0	0
5	0	0	0	0	0	1	0	0	0	0
6	0	0	0	0	0	0	1	0	0	0
7	0	0	0	0	0	0	0	1	0	0
8	0	0	0	0	0	0	0	0	1	0
9	0	0	0	0	0	0	0	0	0	1

EXAMPLE 7–5

If a 10-bit ring counter similar to Figure 7–26 has the initial state 1010000000, determine the waveform for each of the Q outputs.

SOLUTION

See Figure 7–27.

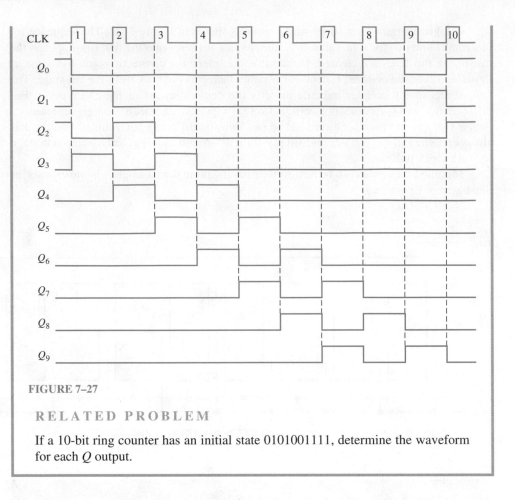

FIGURE 7–27

RELATED PROBLEM

If a 10-bit ring counter has an initial state 0101001111, determine the waveform for each Q output.

SYSTEM EXAMPLE 7–4

KEYBOARD ENCODER

The keyboard encoder is a good example of the application of a shift register used as a ring counter in conjunction with other devices.

Figure 7–28 shows a simplified keyboard encoder for encoding a key closure in a 64-key matrix organized in eight rows and eight columns. Two 4-bit shift registers are connected as an 8-bit ring counter with a fixed bit pattern of seven 1s and one 0 preset into it when the power is turned on. Two priority encoders are used as eight-line-to-three-line encoders (9 input HIGH, 8 output unused) to encode the ROW and COLUMN lines of the keyboard matrix. The parallel in/parallel out register holds the ROW/COLUMN code from the priority encoders.

The basic operation of the keyboard encoder in Figure 7–28 is as follows: The ring counter "scans" the rows for a key closure as the clock signal shifts the 0 around the counter at a 5 kHz rate. The 0 (LOW) is sequentially applied to each ROW line, while all other ROW lines are HIGH. All the ROW lines are connected to the ROW encoder inputs, so the 3-bit output of the ROW encoder at any time is the binary representation of the ROW line that is LOW. When there is a key closure, one COLUMN line is connected to one ROW line. When the ROW line is taken LOW by the ring counter, that particular COLUMN line is also pulled LOW. The COLUMN encoder produces a binary output corresponding to the COLUMN in which the key is closed. The 3-bit ROW code plus the 3-bit COLUMN code uniquely identifies the key that is closed. This 6-bit code is applied to the inputs of the key code register. When a key is closed, the two one-shots produce a delayed clock pulse to parallel-load the 6-bit code into the key code register. This delay allows the contact bounce to die out. Also, the first one-shot output inhibits the ring counter to prevent it from scanning while the data are being loaded into the key code register.

FIGURE 7–28

The 6-bit code in the key code register is now applied to a ROM (read-only memory) to be converted to an appropriate alphanumeric code that identifies the keyboard character. ROMs are studied in Chapter 10.

SECTION 7–5 CHECKUP

1. How many states are there in an 8-bit Johnson counter sequence?

2. Write the sequence of states for a 3-bit Johnson counter starting with 000.

3. In the keyboard encoder, how many times per second does the ring counter scan the keyboard?

4. What is the 6-bit ROW/COLUMN code (key code) for the top row and the left-most column in the keyboard encoder?

5. What do you believe is the purpose of the diodes in the keyboard encoder? What do you believe is the purpose of the resistors?

The security system that was introduced in Section 7–1 can be described using VHDL or Verilog for implementation in a PLD. The three blocks of the system (keypad, security code logic, and code-selection logic) are combined in the program code to describe the complete system.

After completing this section, you should be able to

- Discuss the general approach to programming a system for implementation in a PLD
- See how the VHDL and Verilog programs are used to implement the system

---

The security system block diagram is shown in Figure 7–29 as a programming model. Six program components perform the logical operations of the security system. Each component corresponds to a block or blocks in the figure. The security system program SecuritySystem contains the code that defines how the components interact.

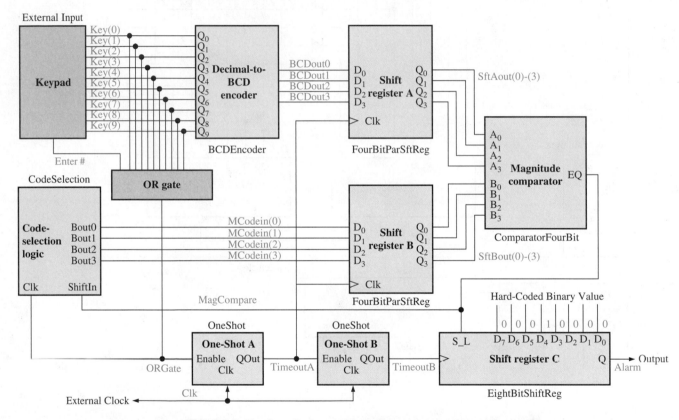

FIGURE 7–29   Security system block diagram as a programming model.

The security system includes a ten-bit input vector Key—one input bit for each decimal digit—and an input Enter, representing a typical numeric keypad. Once a key is pressed, the data stored in input array Key is sent to the decimal-to-BCD encoder (BCDEncoder). Its 4-bit output is then sent to the inputs of the 4-bit parallel in/parallel out shift register A (FourBitParSftReg). An external system clock applied to input Clk drives the overall security system. The Alarm output signal is set HIGH upon a successful arming operation.

Pressing the Enter key sends an initial HIGH clock signal to the code-selection logic block (CodeSelection), which loads an initial binary value of 1000 to shift register B. At this time, a binary 0000 is stored in shift register A, and the output of the magnitude comparator (ComparatorFourBit) is set LOW. The code-selection logic is now ready to present the first stored code value that is to be compared to the value of the first numeric keypad entry. At this time a LOW on the 8-bit parallel in/serial out shift register C (EightBitShiftReg) S_L input loads an initial value of 00010000.

When a numeric key is pressed, the output of the OR gate (ORGate) clocks the first stored value to the inputs of shift register B, and the output of the decimal-to-BCD encoder is sent to the inputs of shift register A. If the values in shift registers A and B match, the output of the magnitude comparator is set HIGH; and the code-selection logic is ready to clock in the next stored code value.

At the conclusion of four successful comparisons of stored code values against four correct keypad entries, the value 00010000 initially in shift register C will shift four places to the right, setting the Alarm output to a HIGH. An incorrect keypad entry will not match the stored code value in shift register B and the magnitude comparator will output a LOW. With the comparator output LOW, the code-selection logic will reset to the first stored code value; and the value 00010000 is reloaded into shift register C, starting the process over again.

To clock the keypad and the stored code values through the system, two one-shots (OneShot) are used. The one-shots allow data to stabilize before acting on it. One-shot A receives an Enable signal from the keypad OR gate, which initiates the first timed process. The OR gate output is also sent to the code-selection logic, and the first code value from the code-selection logic is sent to the inputs of shift register A. When one-shot A times out, the selected keypad entry and the current code from the code-selection logic are stored in shift registers A and B for comparison by the magnitude comparator, and an Enable is sent to one-shot B. If the codes in shift registers A and B match, the value stored in shift register C shifts one place to the right after one-shot B times out.

The six components used in the security system program SecuritySystem are shown in Figure 7–30. The VHDL and Verilog programs for each component are available in Appendix B.

Components

FIGURE 7–30 Security system components (VHDL/Verilog codes are available in Appendix B.)

VHDL FOR THE SECURITY SYSTEM

```
library ieee;
use ieee.std_logic_1164.all;

entity SecuritySystem is
port (key: in std_logic_vector(0 to 9); Enter: in std_logic;
      Clk: in std_logic; Alarm: out std_logic);
end entity SecuritySystem;

architecture SecuritySystemBehavior of SecuritySystem is
```

Key   : 10 - Key input
Enter   : # - Key input
Clk   : System clock
Alarm   : Alarm output

```
component BCDEncoder is
port(D: in std_logic_vector(0 to 9);
     Q: out std_logic_vector(0 to 3));
end component BCDEncoder;
```
Component declaration for BCDEncoder

```
component FourBitParSftReg is
port(D: in std_logic_vector(0 to 3);
     Clk: in std_logic;
     Q: out std_logic_vector(0 to 3));
end component FourBitParSftReg;
```
Component declaration for FourBitParSftReg

```
component ComparatorFourBit is
port(A, B: in std_logic_vector(0 to 3);
     EQ: out std_logic);
end component ComparatorFourBit;
```
Component declaration for ComparatorFourBit

```
component OneShot is
port(Enable, Clk: in std_logic;
     QOut: buffer std_logic);
end component OneShot;
```
Component declaration for OneShot

```
component EightBitShiftReg is
port(S_L, Clk: in std_logic;
     D: in std_logic_vector(0 to 7);
     Q: buffer std_logic);
end component EightBitShiftReg;
```
Component declaration for EightBitShiftReg

```
component CodeSelection is
port(ShiftIn, Clk: in std_logic;
     Bout: out std_logic_vector(1 to 4));
end component CodeSelection;
```
Component declaration for CodeSelection

```
signal BCDout: std_logic_vector(0 to 3);
signal SftAout: std_logic_vector(0 to 3);
signal SftBout: std_logic_vector(0 to 3);
signal MCodein: std_logic_vector(0 to 3);
signal ORgate: std_logic;
signal MagCompare: std_logic;
signal TimeoutA, TimeoutB: std_logic;
```
BDCout : BCD encoder return
SftAout : Shift Register A return
SftBout : Shift Register B return
MCodein : Security Code value
ORgate : OR output from 10-keypad
MagCompare : Key entry to code compare
TimeoutA/B : One-shot timer variables

```
begin
ORgate <= (Key(0) or Key(1) or Key(2) or Key(3) or Key(4)
           or key(5) or Key(6) or Key(7) or Key(8) or Key(9));
```
Logic definition for ORGate

```
BCD: BCDEncoder
port map(D(0)=>Key(0),D(1)=>Key(1),D(2)=>Key(2),D(3)=>Key(3),
         D(4)=>Key(4),D(5)=>Key(5),D(6)=>Key(6),D(7)=>Key(7),D(8)=>Key(8),D(9)=>Key(9),
         Q(0)=>BCDout(0),Q(1)=>BCDout(1),Q(2)=>BCDout(2),Q(3)=>BCDout(3));

ShiftRegisterA: FourBitParSftReg
port map(D(0)=>BCDout(0),D(1)=>BCDout(1),D(2)=>BCDout(2),D(3)=>BCDout(3),
         Clk=>not TimeoutA,Q(0)=>SftAout(0),Q(1)=>SftAout(1),Q(2)=>SftAout(2),Q(3)=>SftAout(3));

ShiftRegisterB: FourBitParSftReg
port map(D(0)=>MCodein(0),D(1)=>MCodein(1),D(2)=>MCodein(2),D(3)=>MCodein(3),
         Clk=>not TimeoutA,Q(0)=>SftBout(0),Q(1)=>SftBout(1),Q(2)=>SftBout(2),Q(3)=>SftBout(3));
Magnitude Comparator: ComparatorFourBit port map(A=>SftAout,B=>SftBout,EQ=>MagCompare);

OSA:OneShot port map(Enable=>Enter or ORgate,Clk=>Clk,QOut=>TimeoutA);
OSB:OneShot port map(Enable=>not TimeoutA,Clk=>Clk,QOut=>TimeoutB);

ShiftRegisterC:EightBitShiftReg
port map(S_L=>MagCompare,Clk=> TimeoutB,D(0)=>'0',D(1)=>'0',
         D(2)=>'0',D(3)=>'1',D(4)=>'0',D(5)=>'0',D(6)=>'0',D(7)=>'0',Q=>Alarm);
CodeSelectionA: CodeSelection
port map(ShiftIn=>MagCompare,Clk=>Enter or ORGate,Bout=>MCodein);
end architecture SecuritySystemBehavior;
```
Component instantiations

VERILOG FOR THE SECURITY SYSTEM

```
module SecuritySystem(Key, Enter, Clk, Alarm);          Key   : 10 - Key input
  input [9:0] Key;                                      Enter : # - Key input
  input Enter;                                          Clk   : System clock
  input Clk;                                            Alarm : Alarm output
  output Alarm;

  wire[3:0] BCDout;               BDCout : BCD encoder return
  wire[3:0] SftAout;              SftAout  : 4-bit shift register A
  wire[3:0] SftBout;             ˙SftBout  : 4-bit shift register B
  wire[3:0] MCodein;             MCodein  : Security Code value
  wire ORGate;                   ORgate   : OR output from 10-keypad
  wire MagCompare;               MagCompare : Key entry to code compare
  wire TimeoutA, TimeoutB;       TimeoutA/B : One-shot timer variables

  assign ORGate = ( Key[0] || Key[1] || Key[2] || Key[3] || Key[4] ||   } Logic definition for
                    Key[5] || Key[6] || Key[7] || Key[8] || Key[9] ||);  } OR Gate

  BCDEncoder BCD(.D(Key),.Q(BCDout));
  FourBitParSftReg ShiftRegisterA(.D(BCDout),.Clk(!TimeoutA),.Q(SftAout));
  FourBitParSftReg ShiftRegisterB(.D(MCodein),.Clk(!TimeoutA),.Q(SftBout));
  ComparatorFourBit MagnitudeComparator(.A(SftAout),.B(SftBout),.EQ(MagCompare));      } Component instantiations
  OneShot OSA(.Enable(Enter || ORGate),.Clk(Clk),.QOut(TimeoutA));
  OneShot OSB(.Enable(!TimeoutA),.Clk(Clk),.QOut(TimeoutB));
  EightBitShiftReg ShiftRegisterC(.S_L(MagCompare),.Clk(TimeoutB),.D(0010),.Q(Alarm));
  CodeSelection CodeSelectionA(.ShiftIn(MagCompare),.Clk(Enter || ORGate),.Bout(MCodein));
endmodule
```

SECTION 7–6 CHECKUP

1. List the types of components used in the VHDL program?

2. What is the only block in the security system not defined as a component? How is it described?

7–7 TROUBLESHOOTING

In this section, in addition to Multisim debugging and software testing, a traditional method of troubleshooting sequential logic and other more complex system hardware is also discussed. This method uses a procedure of "exercising" the circuit under test with a known input wave-form (stimulus) and then observing the output for the correct bit pattern.

After completing this section, you should be able to

- Use Multisim to debug a system design before implementation
- Describe how development software is used to implement and test a system
- Explain the procedure of "exercising" as a troubleshooting technique
- Discuss troubleshooting of a serial-to-parallel converter

Multisim Simulation

Figure 7–31 shows the simulation screen for the security code logic that was covered in Section 7–1. A DIP switch (J6) is used to simulate the 10-digit keypad and switch J1 simulates the # key. Switches J2–J5 are used for test purposes to enter the code that is produced by the code selection logic in the complete system. Probe lights are used only for test purposes to indicate the states of registers A and B, the output of the comparator, and the output of register C.

MULTISIM **FIGURE 7–31** **Multisim screen for the security code logic. The switches and probe lights are for test purposes only. When a probe light is *on*, a 1 is indicated. Open F07-31 and run the simulation to verify operation.**

Figure 7–32 shows the simulation screen for the code-selection logic. Switches are selected for this application as the most convenient and cost-effective way to enter and store the code, thus eliminating the need for a semiconductor memory.

Software Development and Testing

A VHDL or Verilog code for a device or system can be tested in software before being committed to hardware (PLD). The 4-bit bidirectional shift register is used for illustration. A development software simulation run with Altera Quartus II for the following 4-bit bidirectional shift register VHDL code is shown in Figure 7–33.

FIGURE 7–32 Multisim screen for the code-selection logic. The probe lights are for test purposes only. When a probe light is *on*, a 1 is indicated. Open F07-32 and run the code-selection logic simulation to observe the operation.

MULTISIM

```
library ieee;                              R_L : Right-Left Shift
use ieee.std_logic_1164.all;               Datain : Serial data in
                                           Clk : System clock
entity FourBitBiDirSftReg is               Q0–Q3 : Shift register output
port(R_L, Dataln, Clk: in std_logic);
    Q0, Q1, Q2, Q3: buffer std_logic);     D0: Logic for DFlipFlop DFF0
end entity FourBitBiDirSftReg;             D1: Logic for DFlipFlop DFF1
                                           D2: Logic for DFlipFlop DFF2
architecture FourBitBehavior of FourBitBiDirSftReg is  D3: Logic for DFlipFlop DFF3
component dff is
  port(D,Clk: in std_logic;        ⎫ Component declaration
       Q: out std_logic);          ⎬ for D flipflop (dff)
end component dff;                  ⎭

signal D0, D1, D2, D3: std_logic;
begin
  D0<=(Datain and R_L) or (not R_L and Q1);
  D1<=(Q0 and R_L) or (not R_L and Q2);   ⎫ Bidirectional shift register logic
  D2<=(Q1 and R_L) or (not R_L and Q3);   ⎬ defined for DFlipFlop stages
  D3<=(Q2 and R_L) or (not R_L and Datain); ⎭ DFF0 through DFF3

DFF0:dff port map(D=> D0, Clk => Clk, Q => Q0);  ⎫
DFF1:dff port map(D=> D1, Clk => Clk, Q => Q1);  ⎬ Component instantiations
DFF2:dff port map(D=> D2, Clk => Clk, Q => Q2);  ⎭
DFF3:dff port map(D=> D3, Clk => Clk, Q => Q3);
end architecture FourBitBehavior;
```

FIGURE 7–33   **Development software simulation of 4-bit bidirectional shift register operation.**

The difference in this simulation and a Multisim simulation is as follows. Multisim is used to debug and verify a design before it is described with an HDL such as VHDL or Verilog. The development software simulation is used to test the HDL code before programming it into a PLD. In this simulation Datain is set to a logic HIGH and shifted to Q0, given a logic HIGH on input shift direction R_L. A logic LOW is now assigned to the direction bit R_L and the logic HIGH on Q0 is shifted to Q3. After three clock cycles, the direction bit R_L is set to a logic HIGH and the value shifted to Q0 is shifted back to Q3, demonstrating the action of the 4-bit bidirectional shift register.

Hardware Testing

The serial-to-parallel data converter in Figure 7–16 in System Example 7–2 is used to illustrate the "exercising" procedure. The main objective in exercising the circuit is to force all elements (flip-flops and gates) into all of their states to be certain that nothing is stuck in a given state as a result of a fault. The input test pattern, in this case, must be designed to force each flip-flop in the registers into both states; to clock the counter through all of its eight states; and to take the control flip-flop, the clock generator, the one-shot, and the AND gate through their paces.

The input test pattern that accomplishes this objective for the serial-to-parallel data converter is based on the serial data format in Figure 7–17. It consists of the pattern 10101010 in one serial group of data bits followed by 01010101 in the next group, as shown in Figure 7–34. These patterns are generated on a repetitive basis by a special test-pattern generator. The basic test setup is shown in Figure 7–35.

After both patterns have been run through the circuit under test, all the flip-flops in the data-input register and in the data-output register have resided in both SET and RESET states, the counter has gone through its sequence (once for each bit pattern), and all the other devices have been exercised.

To check for proper operation, each of the parallel data outputs is observed for an alternating pattern of 1s and 0s as the input test patterns are repetitively shifted into the data-input register and then loaded into the data-output register. The proper timing diagram is shown in Figure 7–36. The outputs can be observed in pairs with a dual-trace oscilloscope, or all eight outputs can be observed simultaneously with a logic analyzer configured for timing analysis.

If one or more outputs of the data-output register are incorrect, then you must back up to the outputs of the data-input register. If these outputs are correct, then the problem is associated with the data-output register. Check the inputs to the data-output register directly on the pins of the PLD package for an open input line. Check that power and ground are correct (look for the absence of noise on the ground line). Verify that the load line is a solid LOW and that there are clock pulses on the clock input of the correct amplitude.

HANDS ON TIP
When measuring digital signals with an oscilloscope, you should always use dc coupling, rather than ac coupling. The reason that ac coupling is not best for viewing digital signals is that the 0 V level of the signal will appear at the *average* level of the signal, not at true ground or 0 V level. It is much easier to find a "floating" ground or incorrect logic level with dc coupling. If you suspect an open ground in a digital circuit, increase the sensitivity of the scope to the maximum possible. A good ground will never appear to have noise under this condition, but an open will likely show some noise, which appears as a random fluctuation in the 0 V level.

FIGURE 7–34   **Sample test pattern.**

FIGURE 7–35   **Serial-to-parallel data converter implemented in a PLD being exercised.**
(Instrument photos courtesy of Tektronix, Inc.; board photo courtesy of Digilent, Inc.)

FIGURE 7–36   **Proper outputs for the circuit under test in Figure 7–35.**
The input test pattern is shown.

Make sure that the connection to the logic analyzer did not short two output lines together. If all of these checks pass inspection, then it is likely that the output register is defective. If the data-input register outputs are also incorrect, the fault could be associated with the input register itself or with any of the other logic, and additional investigation is necessary to isolate the problem.

SECTION 7–7 CHECKUP

1. What is the purpose of providing a test input to a sequential logic circuit?

2. Generally, when an output waveform is found to be incorrect, what is the next step to be taken?

SUMMARY

- The basic types of data movement in shift registers are
    1. Serial in/shift right/serial out
    2. Serial in/shift left/serial out
    3. Parallel in/serial out
    4. Serial in/parallel out
    5. Parallel in/parallel out
    6. Rotate right
    7. Rotate left
- Shift register counters are shift registers with feedback that exhibit special sequences. Examples are the Johnson counter and the ring counter.
- The Johnson counter has $2n$ states in its sequence, where n is the number of stages.
- The ring counter has n states in its sequence.

KEY TERMS

Key terms and other bold terms in the chapter are defined in the end-of-book glossary.

Bidirectional   Having two directions. In a bidirectional shift register, the stored data can be shifted right or left.

Load   To enter data into a shift register.

Register   One or more flip-flops used to store and shift data.

Stage   One storage element in a register.

FIXED-FUNCTION LOGIC

8-bit serial in/parallel out shift register

8-bit parallel load shift register

4-bit parallel access shift register   **4-bit bidirectional universal shift register**

TRUE/FALSE QUIZ

Answers are at the end of the chapter.

1. Shift registers consist of an arrangement of flip-flops.

2. Two functions of a shift register are data storage and data movement.

3. In a serial shift register, several data bits are entered at the same time.

4. All shift registers are defined by specified sequences.

5. A shift register can have both parallel and serial outputs.

6. A shift register with four stages can store a maximum count of fifteen.

7. The Johnson counter is a special type of shift register.

8. The modulus of an 8-bit Johnson counter is eight.

9. A ring counter uses one flip-flop for each state in its sequence.

10. A shift register can be used as a time delay device.

SELF-TEST

Answers are at the end of the chapter.

1. A stage in a shift register consists of
   - **(a)** a latch
   - **(b)** a flip-flop
   - **(c)** a byte of storage
   - **(d)** four bits of storage

2. To serially shift a byte of data into a shift register, there must be
   - **(a)** one clock pulse
   - **(b)** one load pulse
   - **(c)** eight clock pulses
   - **(d)** one clock pulse for each 1 in the data

3. To parallel load a byte of data into a shift register with a synchronous load, there must be
   - **(a)** one clock pulse
   - **(b)** one clock pulse for each 1 in the data
   - **(c)** eight clock pulses
   - **(d)** one clock pulse for each 0 in the data

4. The group of bits 10110101 is serially shifted (right-most bit first) into an 8-bit parallel output shift register with an initial state of 11100100. After two clock pulses, the register contains
   - **(a)** 01011110
   - **(b)** 10110101
   - **(c)** 01111001
   - **(d)** 00101101

5. With a 100 kHz clock frequency, eight bits can be serially entered into a shift register in

   (a) 80 μs
   (b) 8 μs
   (c) 80 ms
   (d) 10 μs

6. With a 1 MHz clock frequency, eight bits can be parallel entered into a shift register

   (a) in 8 μs
   (b) in the propagation delay time of eight flip-flops
   (c) in 1 μs
   (d) in the propagation delay time of one flip-flop

7. A modulus-10 Johnson counter requires

   (a) ten flip-flops
   (b) four flip-flops
   (c) five flip-flops
   (d) twelve flip-flops

8. A modulus-10 ring counter requires a minimum of

   (a) ten flip-flops
   (b) five flip-flops
   (c) four flip-flops
   (d) twelve flip-flops

9. When an 8-bit serial in/serial out shift register is used for a 24 μs time delay, the clock frequency must be

   (a) 41.67 kHz
   (b) 333 kHz
   (c) 125 kHz
   (d) 8 MHz

10. The purpose of the ring counter in the keyboard encoding circuit of Figure 7–28 is

    (a) to sequentially apply a HIGH to each row for detection of key closure
    (b) to provide trigger pulses for the key code register
    (c) to sequentially apply a LOW to each row for detection of key closure
    (d) to sequentially reverse bias the diodes in each row

PROBLEMS

Answers to odd-numbered problems are at the end of the book.

SECTION 7–1   A System

1. Write the BCD code sequence produced by the encoder in Figure 7–2 if a 4-digit access number 4739 is entered on the keypad.

2. What is the state of shift register C in Figure 7–2 after two correct code digits are entered for the entry code 4739?

3. Assume the entry code is 1939. Determine the states of shift register A and shift register C after the second correct digit has been entered in Figure 7–2.

4. If the digit 4 is entered on the keypad, what appears on the outputs of shift register A in Figure 7–2?

5. Initially, what is the state of the Q outputs of the shift register in Figure 7–3.

6. What is the state of the shift register in Figure 7–3 after three clock pulses?

7. Assume the entry code is 7646 and the digits 7645 are entered. Determine the states of shift register A and shift register C after each of the digits is entered.

8. To increase the entry code to 5 digits, what would have to be added to the logic in Figure 7–3?

SECTION 7–2   Basic Shift Register Operations

9. Why are shift registers considered basic memory devices?

10. What is the storage capacity of a register that can retain two bytes of data?

11. Name two functions of a shift register.

SECTION 7–3   Types of Shift Registers

12. The sequence 1011 is applied to the input of a 4-bit serial shift register that is initially cleared. What is the state of the shift register after three clock pulses?

13. For the data input and clock in Figure 7–37, determine the states of each flip-flop in the shift register of Figure 7–6 and show the Q waveforms. Assume that the register contains all 1s initially.

FIGURE 7–37

14. Solve Problem 13 for the waveforms in Figure 7–38.

FIGURE 7–38

15. What is the state of the register in Figure 7–39 after each clock pulse if it starts in the 101001111000 state?

FIGURE 7–39

16. For the serial in/serial out shift register, determine the data-output waveform for the data-input and clock waveforms in Figure 7–40. Assume that the register is initially cleared.

FIGURE 7–40

17. Solve Problem 16 for the waveforms in Figure 7–41.

FIGURE 7–41

18. A leading-edge clocked serial in/serial out shift register has a data-output waveform as shown in Figure 7–42. What binary number is stored in the 8-bit register if the first data bit out (leftmost) is the LSB?

FIGURE 7–42

19. Show a complete timing diagram including the parallel outputs for the shift register in Figure 7–9. Use the waveforms in Figure 7–40 with the register initially clear.

20. Solve Problem 19 for the input waveforms in Figure 7–41.

21. Develop the Q_0 through Q_7 outputs for an 8-bit serial in/parallel out shift register with the input waveform shown in Figure 7–43.

FIGURE 7–43

22. The shift register in Figure 7–44(a) has $SHIFT/\overline{LOAD}$ and CLK inputs as shown in part (b). The serial data input (SER) is a 0. The parallel data inputs are $D_0 = 1, D_1 = 0, D_2 = 1$, and $D_3 = 0$ as shown. Develop the data-output waveform in relation to the inputs.

FIGURE 7–44

23. The waveforms in Figure 7–45 are applied to an 8-bit parallel in/serial out shift register. The parallel inputs are all 0. Determine the Q_7 waveform.

FIGURE 7–45

24. Solve Problem 23 if the parallel inputs are all 1.

25. Solve Problem 23 if the serial input is inverted.

26. Determine all the Q output waveforms for a 4-bit parallel in/parallel out shift register when the inputs are as shown in Figure 7–46.

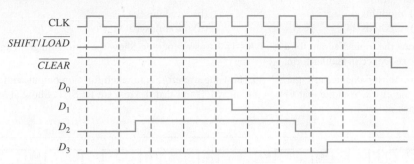

FIGURE 7–46

27. Solve Problem 26 if the $SHIFT/\overline{LOAD}$ input is inverted and the register is initially clear.

28. Use two 4-bit shift registers to form an 8-bit shift register. Show the required connections.

SECTION 7–4   Bidirectional Shift Registers

29. For the 8-bit bidirectional register in Figure 7–47, determine the state of the register after each clock pulse for the $RIGHT/\overline{LEFT}$ control waveform given. A HIGH on this input enables a shift to the right, and a LOW enables a shift to the left. Assume that the register is initially storing the decimal number seventy-six in binary, with the right-most position being the LSB. There is a LOW on the data-input line.

FIGURE 7–47

30. Solve Problem 29 for the waveforms in Figure 7–48.

FIGURE 7–48

31. Use two 4-bit bidirectional shift registers to create an 8-bit bidirectional shift register. Show the connections. Assume each shift register has a shift left (SL) data input and a shift right (SR) data input.

32. Determine the Q outputs of a 4-bit serial in/parallel out bidirectional shift register with the inputs shown in Figure 7–49.

FIGURE 7–49

SECTION 7–5   Shift Register Counters

33. How many flip-flops are required to implement each of the following in a Johnson counter configuration:

   (a) modulus-6   (b) modulus-10
   (c) modulus-14   (d) modulus-16

34. Draw the logic diagram for a modulus-18 Johnson counter. Show the timing diagram and write the sequence in tabular form.

35. For the ring counter in Figure 7–50, show the waveforms for each flip-flop output with respect to the clock. Assume that FF0 is initially SET and that the rest are RESET. Show at least ten clock pulses.

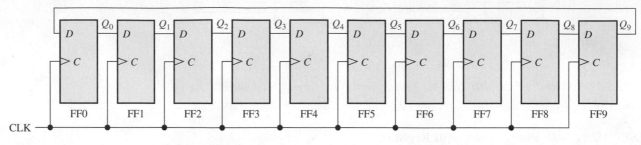

FIGURE 7–50

36. The waveform pattern in Figure 7–51 is required. Devise a ring counter, and indicate how it can be preset to produce this waveform on its Q_9 output. At CLK16 the pattern begins to repeat.

FIGURE 7–51

SECTION 7–6   Security System with VHDL and Verilog

37. If the code selection component of the security system is modified to store a five-digit code, what changes are needed to support the additional digit?

38. Explain the need for component one-shot B in the security system.

39. What would happen if two keypad keys were pressed before one-shot A could complete its timeout? How would the system respond?

SECTION 7–7   Troubleshooting

40. Based on the waveforms in Figure 7–52(a), determine the most likely problem with the register in part (b) of the figure.

FIGURE 7–52

41. Refer to the parallel in/serial out shift register in Figure 7–13. The register is in the state where $Q_0Q_1Q_2Q_3 = 1001$, and $D_0D_1D_2D_3 = 1010$ is loaded in. When the $SHIFT/\overline{LOAD}$ input is taken HIGH, the data shown in Figure 7–53 are shifted out. Is this operation correct? If not, what is the most likely problem?

FIGURE 7–53

42. You have found that the bidirectional register in Figure 7–21 will shift data right but not left. What is the most likely fault?

43. For the keyboard encoder in Figure 7–28, list the possible faults for each of the following symptoms:
   (a) The state of the key code register does not change for any key closure.
   (b) The state of the key code register does not change when any key in the third row is closed. A proper code occurs for all other key closures.
   (c) The state of the key code register does not change when any key in the first column is closed. A proper code occurs for all other key closures.
   (d) When any key in the second column is closed, the left three bits of the key code ($Q_0Q_1Q_2$) are correct, but the right three bits are all 1s.

44. Develop a test procedure for exercising the keyboard encoder in Figure 7–28. Specify the procedure on a step-by-step basis, indicating the output code from the key code register that should be observed at each step in the test.

45. What symptoms are observed for the following failures in the serial-to-parallel converter in Figure 7–16:
   (a) AND gate output stuck in HIGH state
   (b) clock generator output stuck in LOW state
   (c) third stage of data-input register stuck in SET state
   (d) terminal count output of counter stuck in HIGH state

Special Problems

46. Modify the serial-to-parallel converter in Figure 7–16 to provide 16-bit conversion.

47. Develop an 8-bit parallel-to-serial data converter that produces the data format in Figure 7–17. Show a logic diagram and specify the devices.

48. Develop a power-on $\overline{LOAD}$ circuit for the keyboard encoder in Figure 7–28. This circuit must generate a short-duration LOW pulse when the power switch is turned on.

MULTISIM TROUBLESHOOTING PRACTICE

MULTISIM

49. Open file P07-49 and follow the instructions given there.

50. Open file P07-50 and follow the instructions given there.

51. Open file P07-51 and follow the instructions given there.

52. Open file P07-52 and follow the instructions given there.

53. Open file P07-53 and follow the instructions given there.

ANSWERS TO SECTION CHECKUPS

SECTION 7–1   A System

   1. The OR gate detects a key closure and triggers one-shot A.

   2. 0100

   3. OS B, provides a delayed clock to register C.

SECTION 7–2   Basic Shift Register Operations

   1. Storage and data movement are two functions of a shift register.

   2. When the Q output of a flip-flop is HIGH, a 1 is stored.

SECTION 7–3   Types of Shift Registers

1. FF0: data input to J_0, $\overline{\text{data input}}$ to K_0; FF1: Q_0 to J_1, $\overline{Q}_0$ to K_1; FF2: Q_1 to J_2, $\overline{Q}_1$ to K_2; FF3: Q_2 to J_3, $\overline{Q}_2$ to K_3

2. Eight clock pulses

3. 0100 after 2 clock pulses

4. Take the serial output from the right-most flip-flop for serial out operation.

5. When $SHIFT/\overline{LOAD}$ is HIGH, the data are shifted right one bit per clock pulse. When $SHIFT/\overline{LOAD}$ is LOW, the data on the parallel inputs are loaded into the register.

6. The data outputs are 1001.

SECTION 7–4   Bidirectional Shift Registers

1. 1111 after the fifth clock pulse

SECTION 7–5   Shift Register Counters

1. Sixteen states are in an 8-bit Johnson counter sequence.

2. For a 3-bit Johnson counter: 000, 100, 110, 111, 011, 001, 000

3. 625 scans/second

4. $Q_5Q_4Q_3Q_2Q_1Q_0 = 011011$

5. The diodes provide unidirectional paths for pulling the ROWs LOW and preventing HIGHs on the ROW lines from being connected to the switch matrix. The resistors pull the COLUMN lines HIGH.

SECTION 7–6   Security System with VHDL and Verilog

1. Decimal-to-BCD encoder, 4-bit parallel-in/parallel-out shift register, 8-bit parallel-in/serial-out shift register, 4-bit magnitude comparator, code-selection logic, one-shot.

2. The OR gate is not treated as a component. It is described with a Boolean expression.

SECTION 7–7   Troubleshooting

1. A test input is used to sequence the circuit through all of its states.

2. Check the input to that portion of the circuit. If the signal on that input is correct, the fault is isolated to the circuitry between the good input and the bad output.

ANSWERS TO RELATED PROBLEMS FOR EXAMPLES

7–1 See Figure 7–54.

The output is $Q_4Q_3Q_2Q_1Q_0 = 00101$ after 5 clock pulses.

FIGURE 7–54

7–2 The state of the register after three additional clock pulses is 0000.

7–3 See Figure 7–55.

FIGURE 7–55

7–4 See Figure 7–56.

FIGURE 7–56

7–5 See Figure 7–57.

FIGURE 7–57

ANSWERS TO TRUE/FALSE QUIZ

1. T **2.** T **3.** F **4.** F **5.** T **6.** T

7. T **8.** F **9.** T **10.** T

ANSWERS TO SELF-TEST

1. (b) **2.** (c) **3.** (a) **4.** (c) **5.** (a)

6. (d) **7.** (c) **8.** (a) **9.** (b) **10.** (c)

CHAPTER 8

COUNTERS

OUTLINE

OBJECTIVES

- Explain how a digital clock operates
- Discuss the types of state machines
- Describe the difference between an asynchronous and a synchronous counter
- Analyze counter timing diagrams
- Analyze counter circuits
- Explain how propagation delays affect the operation of a counter
- Determine the modulus of a counter
- Modify the modulus of a counter
- Recognize the difference between a 4-bit binary counter and a decade counter
- Use an up/down counter to generate forward and reverse binary sequences
- Determine the sequence of a counter
- Use cascaded counters to achieve a higher modulus
- Use logic gates to decode any given state of a counter
- Eliminate glitches in counter decoding
- Troubleshoot counters for various types of faults

KEY TERMS

Counter	Recycle
State machine	Modulus
Moore state machine	Decade
Mealy state machine	Synchronous
Asynchronous	Cascade

INTRODUCTION

As you have learned, flip-flops can be connected together to perform counting operations. Such a group of flip-flops is a counter. The number of flip-flops used and the way in which they are connected determine the number of states (called the modulus) and also the specific sequence of states that the counter goes through during each complete cycle.

Counters are classified into two broad categories according to the way they are clocked: asynchronous and synchronous. In asynchronous counters, commonly called

VISIT THE WEBSITE
Study aids for this chapter are available at
http://pearsonhighered.com/floyd

ripple counters, the first flip-flop is clocked by the external clock pulse and then each successive flip-flop is clocked by the output of the preceding flip-flop. In synchronous counters, the clock input is connected to all of the flip-flops so that they are clocked simultaneously. Within each of these two categories, counters are classified primarily by the type of sequence, the number of states, or the number of flip-flops in the counter.

8–1 A SYSTEM

A **counter** is a sequential logic device that is used to divide signal frequency, count events, and generate specified sequences of bits. Counters are made up of flip-flops connected in various ways to produce the desired results. A common example of an application of counters is in timekeeping systems.

After completing this section, you should be able to

- **Explain the overall operation of a digital clock**
- **Describe how counters are used for counting down a frequency to achieve an output that indicates time of day**

Figure 8–1 is a simplified logic diagram of a digital clock that displays seconds, minutes, and hours. First, a 60 Hz sinusoidal ac voltage is converted to a 60 Hz pulse waveform and divided down to a 1 Hz pulse waveform by a divide-by-60 counter formed by a divide-by-10 counter followed by a divide-by-6 counter.

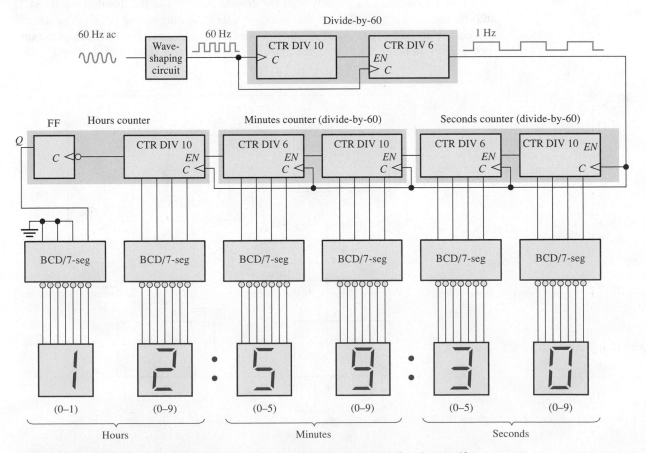

FIGURE 8–1 Simplified logic diagram for a 12-hour digital clock. Logic details using specific devices are shown in Figures 8–2 and 8–3.

Both the *seconds* and *minutes* counts are also produced by divide-by-60 counters, the details of which are shown in Figure 8–2. These counters count from 0 to 59 and then recycle to 0; synchronous decade counters are used in this particular implementation. Notice that the divide-by-6 portion is formed with a decade counter with a truncated sequence achieved by using the decoder count 6 to asynchronously clear the counter. The terminal count (TC), 59, is also decoded to enable the next counter in the chain. The **terminal count** is the final state in a counter's sequence.

FIGURE 8–2 Logic diagram of typical divide-by-60 counter using synchronous decade counters. Note that the outputs are in binary order (the right-most bit is the LSB).

The *hours* counter is implemented with a decade counter and a flip-flop as shown in Figure 8–3. Consider that initially both the decade counter and the flip-flop are RESET, and the decode-12 gate (G_2) and the decode-9 gate (G_1) outputs are HIGH. The decade counter advances through all of its states from zero to nine, and on the clock pulse that recycles it from nine back to zero, the flip-flop goes to the SET state ($J = 1, K = 0$). This

FIGURE 8–3 Logic diagram for hours counter and decoders. Note that on the counter inputs and outputs, the right-most bit is the LSB.

illuminates a 1 on the tens-of-hours display. The total count is now ten (the decade counter is in the zero state and the flip-flop is SET).

Next, the total count advances to eleven and then to twelve. In state 12 the Q_2 output of the decade counter is HIGH, the flip-flop is still SET, and thus the decode-12 gate output is LOW. This activates the $\overline{LOAD}$ input of the decade counter. On the next clock pulse, the decade counter is preset to 0001 from the data inputs, and the flip-flop is RESET ($J = 0, K = 1$). As you can see, this logic always causes the counter to recycle from twelve back to one rather than back to zero.

SECTION 8–1 CHECKUP*

1. Explain the purpose of each NAND gate in Figure 8–3.

2. Identify the two recycle conditions for the hours counter in Figure 8–1. Explain the reason for each.

*Answers are at the end of the chapter.

8–2 FINITE STATE MACHINES

A state machine is a sequential circuit having a limited (finite) number of states occuring in a prescribed order. A counter is an example of a state machine; the number of states is called the *modulus*. Two types of state machines are the Moore and the Mealy. The Moore state machine is one where the outputs depend only on the internal present state. The Mealy state machine is one where the outputs depend on both the internal present state and on the inputs. Both types have a timing input (clock) that is not considered a controlling input.

After completing this section, you should be able to

- **Describe a Moore state machine**
- **Describe a Mealy state machine**
- **Discuss examples of Moore and Mealy state machines**

General Models of Finite State Machines

A Moore state machine consists of combinational logic that determines the sequence and memory (flip-flops), as shown in Figure 8–4(a). A Mealy state machine is shown in part (b). In the Moore machine, the combinational logic is a gate array with outputs that determine the next state of the flip-flops in the memory. There may or may not be inputs to the combinational logic. There may also be output combinational logic, such as a decoder. If there is an input(s), it does not affect the outputs because they always correspond to and are dependent only on the present state of the memory. For the Mealy machine, the present state affects the outputs, just as in the Moore machine; but in addition, the inputs also affect the outputs. The outputs come directly from the combinational logic and not the memory.

(a) Moore machine

(b) Mealy machine

FIGURE 8–4   Two types of sequential logic.

EXAMPLE OF A MOORE MACHINE   Figure 8–5(a) shows a Moore machine (modulus-26 binary counter with states 0 through 25) that is used to control the number of tablets (25) that go into each bottle in an assembly line. When the binary number in the memory (flip-flops) reaches binary 25 (11001), the counter recycles to 0 and the tablet flow and clock is cut off until the next bottle is in place. The combinational logic for state transitions sets the modulus of the counter so that it sequences from binary state 0 to binary state 25, where 0 is the reset or rest state and the output combinational logic decodes binary state 25. There is no input in this case, other than the clock, so the next state is determined only by the present state, which makes this a Moore machine. One tablet is bottled for each clock pulse. Once a bottle is in place, the first tablet is inserted at binary state 1, the second at binary state 2, and the twenty-fifth tablet when the binary state is 25. Count 25 is decoded and used to stop the flow of tablets and the clock. The counter recycles to binary state 0 and waits until the next bottle is in position. Then the clock resumes, the count goes to binary state 1, and the cycle repeats, as illustrated by the state diagram in Figure 8–5(b).

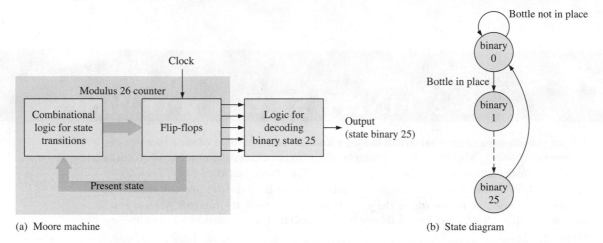

(a) Moore machine

(b) State diagram

FIGURE 8–5   A fixed-modulus binary counter as an example of a Moore state machine. The dashed line in the state diagram means the states between binary 1 and 25 are not shown for simplicity.

EXAMPLE OF A MEALY MACHINE   Let's assume that the tablet-bottling system uses three different sizes of bottles: a 25-tablet bottle, a 50-tablet bottle, and a 100-tablet bottle. This operation requires a state machine with three different terminal counts: 25, 50, and 100. One approach is illustrated in Figure 8–6(a). The combinational logic sets the

(a) Mealy machine

(b) State diagram

FIGURE 8–6   A variable-modulus binary counter as an example of a Mealy state machine. The red arrows in the state diagram represent the recycle paths that depend on the input number. The black dashed lines mean the interim states are not shown for simplicity.

modulus of the counter depending on the modulus-select inputs. The output of the counter depends on both the present state and the modulus-select inputs, making this a Mealy machine. The state diagram is shown in part (b).

SECTION 8–2 CHECKUP

1. What characterizes a finite state machine?

2. Name the types of finite state machines.

3. Explain the difference between the two types of state machines.

8–3 ASYNCHRONOUS COUNTERS

The term **asynchronous** refers to events that do not have a fixed time relationship with each other and, generally, do not occur at the same time. An *asynchronous counter* is one in which the flip-flops (FF) within the counter do not change states at exactly the same time because they do not have a common clock pulse.

After completing this section, you should be able to

- **Describe the operation of a 2-bit asynchronous binary counter**
- **Describe the operation of a 3-bit asynchronous binary counter**
- **Define** *ripple* **in relation to counters**
- **Describe the operation of an asynchronous decade counter**
- **Develop counter timing diagrams**

A 2-Bit Asynchronous Binary Counter

Figure 8–7 shows a 2-bit counter connected for asynchronous operation. Notice that the clock (CLK) is applied to the clock input (C) of *only* the first flop-flop, FF0, which is always the least significant bit (LSB). The second flip-flop, FF1, is triggered by the $\overline{Q}_0$ output of FF0. FF0 changes state at the positive-going edge of each clock pulse, but FF1 changes only when triggered by a positive-going transition of the $\overline{Q}_0$ output of FF0. Because of the inherent propagation delay time through a flip-flop, a transition of the input clock pulse (CLK) and a transition of the $\overline{Q}_0$ output of FF0 can never occur at exactly the same time. Therefore, the two flip-flops are never simultaneously triggered, so the counter operation is asynchronous.

The clock input of an asynchronous counter is always connected only to the LSB flip-flop.

FIGURE 8–7 A 2-bit asynchronous binary counter. Open file F08-07 to verify operation.

MULTISIM

THE TIMING DIAGRAM   Let's examine the basic operation of the asynchronous counter of Figure 8–7 by applying four clock pulses to FF0 and observing the Q output of each flip-flop. Figure 8–8 illustrates the changes in the state of the flip-flop outputs in response to the clock pulses. Both flip-flops are connected for toggle operation ($J = 1, K = 1$) and are assumed to be initially RESET (Q LOW).

FIGURE 8–8   Timing diagram for the counter of Figure 8–7. As in previous chapters, output waveforms are shown in green.

Asynchronous counters are also known as ripple counters.

The positive-going edge of CLK1 (clock pulse 1) causes the Q_0 output of FF0 to go HIGH, as shown in Figure 8–8. At the same time the $\overline{Q}_0$ output goes LOW, but it has no effect on FF1 because a positive-going transition must occur to trigger the flip-flop. After the leading edge of CLK1, $Q_0 = 1$ and $Q_1 = 0$. The positive-going edge of CLK2 causes Q_0 to go LOW. Output $\overline{Q}_0$ goes HIGH and triggers FF1, causing Q_1 to go HIGH. After the leading edge of CLK2, $Q_0 = 0$ and $Q_1 = 1$. The positive-going edge of CLK3 causes Q_0 to go HIGH again. Output $\overline{Q}_0$ goes LOW and has no effect on FF1. Thus, after the leading edge of CLK3, $Q_0 = 1$ and $Q_1 = 1$. The positive-going edge of CLK4 causes Q_0 to go LOW, while $\overline{Q}_0$ goes HIGH and triggers FF1, causing Q_1 to go LOW. After the leading edge of CLK4, $Q_0 = 0$ and $Q_1 = 0$. The counter has now recycled to its original state (both flip-flops are RESET).

In the timing diagram, the waveforms of the Q_0 and Q_1 outputs are shown relative to the clock pulses as illustrated in Figure 8–8. For simplicity, the transitions of Q_0, Q_1, and the clock pulses are shown as simultaneous even though this is an asynchronous counter. There is, of course, some small delay between the CLK and the Q_0 transition and between the $\overline{Q}_0$ transition and the Q_1 transition.

Q_0 is always the LSB unless otherwise specified.

Note in Figure 8–8 that the 2-bit counter exhibits four different states, as you would expect with two flip-flops ($2^2 = 4$). Also, notice that if Q_0 represents the least significant bit (LSB) and Q_1 represents the most significant bit (MSB), the sequence of counter states represents a sequence of binary numbers as listed in Table 8–1.

CLOCK PULSE	Q_1	Q_0
Initially	0	0
1	0	1
2	1	0
3	1	1
4 (recycles)	0	0

TABLE 8–1 • Binary state sequence for the counter in Figure 8–7.

Since it goes through a binary sequence, the counter in Figure 8–7 is a binary counter. It actually counts the number of clock pulses up to three, and on the fourth pulse it recycles to its original state ($Q_0 = 0$, $Q_1 = 0$). The term **recycle** is commonly applied to counter operation; it refers to the transition of the counter from its final state back to its original state.

A 3-BIT ASYNCHRONOUS BINARY COUNTER   The state sequence for a 3-bit binary counter is listed in Table 8–2, and a 3-bit asynchronous binary counter is shown in Figure 8–9(a). The basic operation is the same as that of the 2-bit counter except that the 3-bit counter has eight states, due to its three flip-flops. A timing diagram is shown in Figure 8–9(b) for eight clock pulses. Notice that the counter progresses through a binary count of zero through seven and then recycles to the zero state. This counter can be easily expanded for higher count, by connecting additional toggle flip-flops.

TABLE 8–2 • State sequence for a 3-bit binary counter.			
CLOCK PULSE	Q_2	Q_1	Q_0
Initially	0	0	0
1	0	0	1
2	0	1	0
3	0	1	1
4	1	0	0
5	1	0	1
6	1	1	0
7	1	1	1
8 (recycles)	0	0	0

(a)

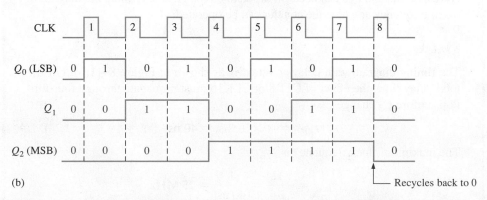

(b)

└── Recycles back to 0

FIGURE 8–9   **Three-bit asynchronous binary counter and its timing diagram for one cycle. Open file F08-09 to verify operation.**

MULTISIM

PROPAGATION DELAY   Asynchronous counters are commonly referred to as **ripple counters** for the following reason: The effect of the input clock pulse is first "felt" by FF0. This effect cannot get to FF1 immediately because of the propagation delay through FF0. Then there is the propagation delay through FF1 before FF2 can be triggered.

Thus, the effect of an input clock pulse "ripples" through the counter, taking some time, due to propagation delays, to reach the last flip-flop.

To illustrate, notice that all three flip-flops in the counter of Figure 8–9 change state on the leading edge of CLK4. This ripple clocking effect is shown in Figure 8–10 for the first four clock pulses, with the propagation delays indicated. The LOW-to-HIGH transition of Q_0 occurs one delay time (t_{PLH}) after the positive-going transition of the clock pulse. The LOW-to-HIGH transition of Q_1 occurs one delay time (t_{PLH}) after the positive-going transition of $\overline{Q}_0$. The LOW-to-HIGH transition of Q_2 occurs one delay time (t_{PLH}) after the positive-going transition of $\overline{Q}_1$. As you can see, FF2 is not triggered until two delay times after the positive-going edge of the clock pulse, CLK4. Thus, it takes three propagation delay times for the effect of the clock pulse, CLK4, to ripple through the counter and change Q_2 from LOW to HIGH.

FIGURE 8–10 Propagation delays in a 3-bit asynchronous (ripple-clocked) binary counter.

This cumulative delay of an asynchronous counter is a major disadvantage in many applications because it limits the rate at which the counter can be clocked and creates decoding problems. The maximum cumulative delay in a counter must be less than the period of the clock waveform.

EXAMPLE 8–1

A 4-bit asynchronous binary counter is shown in Figure 8–11(a). Each flip-flop is negative edge-triggered and has a propagation delay for 10 nanoseconds (ns). Develop a timing diagram showing the Q output of each flip-flop, and determine the total propagation delay time from the triggering edge of a clock pulse until a corresponding change can occur in the state of Q_3. Also determine the maximum clock frequency at which the counter can be operated.

SOLUTION

The timing diagram with delays omitted is as shown in Figure 8–11(b). For the total delay time, the effect of CLK8 or CLK16 must propagate through four flip-flops before Q_3 changes, so

$$t_{p(tot)} = 4 \times 10 \text{ ns} = \textbf{40 ns}$$

The maximum clock frequency is

$$f_{max} = \frac{1}{t_{p(tot)}} = \frac{1}{40 \text{ ns}} = \textbf{25 MHz}$$

The counter should be operated below this frequency to avoid problems due to the propagation delay.

RELATED PROBLEM*

Show the timing diagram if all of the flip-flops in Figure 8–11(a) are positive edge-triggered.

(a)

(b)

FIGURE 8–11   **Four-bit asynchronous binary counter and its timing diagram. Open file F08-11 and verify the operation.**

Answers are at the end of the chapter.

MULTISIM

Asynchronous Decade Counters

The **modulus** of a counter is the number of unique states through which the counter will sequence. The maximum possible number of states (maximum modulus) of a counter is 2^n, where n is the number of flip-flops in the counter. Counters can be designed to have a number of states in their sequence that is less than the maximum of 2^n. This type of sequence is called a *truncated sequence.*

One common modulus for counters with truncated sequences is ten (called MOD10). Counters with ten states in their sequence are called **decade** counters. A **decade counter** with a count sequence of zero (0000) through nine (1001) is a BCD decade counter because its ten-state sequence produces the BCD code. This type of counter is useful in display applications in which BCD is required for conversion to a decimal readout.

To obtain a truncated sequence, it is necessary to force the counter to recycle before going through all of its possible states. For example, the BCD decade counter must recycle back to the 0000 state after the 1001 state. A decade counter requires four flip-flops (three flip-flops are insufficient because $2^3 = 8$).

Let's use a 4-bit asynchronous counter such as the one in Example 8–1 and modify its sequence to illustrate the principle of truncated counters. One way to make the counter recycle after the count of nine (1001) is to decode count ten (1010) with a NAND gate and connect the output of the NAND gate to the clear ($\overline{CLR}$) inputs of the flip-flops, as shown in Figure 8–12(a).

PARTIAL DECODING   Notice in Figure 8–12(a) that only Q_1 and Q_3 are connected to the NAND gate inputs. This arrangement is an example of *partial decoding,* in which the two unique states ($Q_1 = 1$ and $Q_3 = 1$) are sufficient to decode the count of ten because none of the other states (zero through nine) have both Q_1 and Q_3 HIGH at the same time. When the counter goes into count ten (1010), the decoding gate output goes LOW and asynchronously resets all the flip-flops.

A counter can have 2^n states, where n is the number of flip-flops.

FIGURE 8–12   An asynchronously clocked decade counter with asynchronous recycling.

(a)

(b)

The resulting timing diagram is shown in Figure 8–12(b). Notice that there is a glitch on the Q_1 waveform. The reason for this glitch is that Q_1 must first go HIGH before the count of ten can be decoded. Not until several nanoseconds after the counter goes to the count of ten does the output of the decoding gate go LOW (both inputs are HIGH). Thus, the counter is in the 1010 state for a short time before it is reset to 0000, thus producing the glitch on Q_1 and the resulting glitch on the $\overline{CLR}$ line that resets the counter.

Other truncated sequences can be implemented in a similar way, as Example 8–2 shows.

EXAMPLE 8–2

Show how an asynchronous counter can be implemented having a modulus of twelve with a straight binary sequence from 0000 through 1011.

SOLUTION

Since three flip-flops can produce a maximum of eight states, four flip-flops are required to produce any modulus greater than eight but less than or equal to sixteen.

When the counter gets to its last state, 1011, it must recycle back to 0000 rather than going to its normal next state of 1100, as illustrated in the following sequence chart:

Q_3	Q_2	Q_1	Q_0	
0	0	0	0	←
.	.	.	.	
.	.	.	.	Recycles
.	.	.	.	
1	0	1	1	←
1	1	0	0	← Normal next state

Observe that Q_0 and Q_1 both go to 0 anyway, but Q_2 and Q_3 must be forced to 0 on the twelfth clock pulse. Figure 8–13(a) shows the modulus-12 counter. The NAND gate partially decodes count twelve (1100) and resets flip-flop 2 and flip-flop 3. Thus, on the twelfth clock pulse, the counter is forced to recycle from count eleven to count zero, as shown in the timing diagram of Figure 8–13(b). (It is in count twelve for only a few nanoseconds before it is reset by the glitch on $\overline{CLR}$.)

(a)

(b)

FIGURE 8–13   Asynchronously clocked modulus-12 counter with asynchronous recycling.

RELATED PROBLEM

How can the counter in Figure 8–13(a) be modified to make it a modulus-13 counter?

SECTION 8–3 CHECKUP

1. What does the term *asynchronous* mean in relation to counters?

2. How many states does a modulus-14 counter have? What is the minimum number of flip-flops required?

8–4 SYNCHRONOUS COUNTERS

The term synchronous refers to events that have a fixed time relationship with each other. A *synchronous counter* is one in which all the flip-flops in the counter are clocked at the same time by a common clock pulse.

After completing this section, you should be able to

- Describe the operation of a 2-bit synchronous binary counter
- Describe the operation of a 3-bit synchronous binary counter
- Describe the operation of a 4-bit synchronous binary counter
- Describe the operation of a synchronous decade counter
- Develop counter timing diagrams

A 2-Bit Synchronous Binary Counter

Figure 8–14 shows a 2-bit synchronous binary counter. Notice that an arrangement different from that for the asynchronous counter must be used for the J_1 and K_1 inputs of FF1 in order to achieve a binary sequence.

FIGURE 8–14   A 2-bit synchronous binary counter.

The clock input goes to each flip-flop in a synchronous counter.

The operation of this synchronous counter is as follows: First, assume that the counter is initially in the binary 0 state; that is, both flip-flops are RESET. When the positive edge of the first clock pulse is applied, FF0 will toggle and Q_0 will therefore go HIGH. What happens to FF1 at the positive-going edge of CLK1? To find out, let's look at the input conditions of FF1. Inputs J_1 and K_1 are both LOW because Q_0, to which they are connected, has not yet gone HIGH. Remember, there is a propagation delay from the triggering edge of the clock pulse until the Q output actually makes a transition. So, $J = 0$ and $K = 0$ when the leading edge of the first clock pulse is applied. This is a no-change condition, and therefore FF1 does not change state. A timing detail of this portion of the counter operation is shown in Figure 8–15(a).

After CLK1, $Q_0 = 1$ and $Q_1 = 0$ (which is the binary 1 state). When the leading edge of CLK2 occurs, FF0 will toggle and Q_0 will go LOW. Since FF1 has a HIGH ($Q_0 = 1$) on its J_1 and K_1 inputs at the triggering edge of this clock pulse, the flip-flop toggles and Q_1 goes HIGH. Thus, after CLK2, $Q_0 = 0$ and $Q_1 = 1$ (which is a binary 2 state). The timing detail for this condition is shown in Figure 8–15(b).

FIGURE 8–15   Timing details for the 2-bit synchronous counter operation (the propagation delays of both flip-flops are assumed to be equal).

When the leading edge of CLK3 occurs, FF0 again toggles to the SET state ($Q_0 = 1$), and FF1 remains SET ($Q_1 = 1$) because its J_1 and K_1 inputs are both LOW ($Q_0 = 0$). After this triggering edge, $Q_0 = 1$ and $Q_1 = 1$ (which is a binary 3 state). The timing detail is shown in Figure 8–15(c).

Finally, at the leading edge of CLK4, Q_0 and Q_1 go LOW because they both have a toggle condition on their J and K inputs. The timing detail is shown in Figure 8–15(d). The counter has now recycled to its original state, binary 0.

The complete timing diagram for the counter in Figure 8–14 is shown in Figure 8–16. Notice that all the waveform transitions appear coincident; that is, the propagation delays are not indicated. Although the delays are an important factor in the synchronous counter operation, in an overall timing diagram they are normally omitted for simplicity. Major waveform relationships resulting from the normal operation of a circuit can be conveyed completely without showing small delay and timing differences. However, in high-speed digital circuits, these small delays are an important consideration in design and troubleshooting.

FIGURE 8–16   **Timing diagram for the counter of Figure 8–14.**

A 3-Bit Synchronous Binary Counter

A 3-bit synchronous binary counter is shown in Figure 8–17, and its timing diagram is shown in Figure 8–18. You can understand this counter operation by examining its sequence of states as shown in Table 8–3.

MULTISIM

FIGURE 8–17   **A 3-bit synchronous binary counter. Open file F08-17 to verify the operation.**

FIGURE 8–18   **Timing diagram for the counter of Figure 8–17.**

CLOCK PULSE	Q_2	Q_1	Q_0
TABLE 8–3 • State sequence for a 3-bit binary counter.			
Initially	0	0	0
1	0	0	1
2	0	1	0
3	0	1	1
4	1	0	0
5	1	0	1
6	1	1	0
7	1	1	1
8 (recycles)	0	0	0

First, let's look at Q_0. Notice that Q_0 changes on each clock pulse as the counter progresses from its original state to its final state and then back to its original state. To produce this operation, FF0 must be held in the toggle mode by constant HIGHs on its J_0 and K_0 inputs. Notice that Q_1 goes to the opposite state following each time Q_0 is a 1. This change occurs at CLK2, CLK4, CLK6, and CLK8. The CLK8 pulse causes the counter to recycle. To produce this operation, Q_0 is connected to the J_1 and K_1 inputs of FF1. When Q_0 is a 1 and a clock pulse occurs, FF1 is in the toggle mode and therefore changes state. The other times, when Q_0 is a 0, FF1 is in the no-change mode and remains in its present state.

Next, let's see how FF2 is made to change at the proper times according to the binary sequence. Notice that both times Q_2 changes state, it is preceded by the unique condition in which both Q_0 and Q_1 are HIGH. This condition is detected by the AND gate and applied to the J_2 and K_2 inputs of FF2. Whenever both Q_0 and Q_1 are HIGH, the output of the AND gate makes the J_2 and K_2 inputs of FF2 HIGH, and FF2 toggles on the following clock pulse. At all other times, the J_2 and K_2 inputs of FF2 are held LOW by the AND gate output, and FF2 does not change state.

The analysis of the counter in Figure 8–17 is summarized in Table 8–4.

The TSC or *time stamp counter* in some microprocessors is used for performance monitoring, which enables a number of parameters important to the overall performance of a system to be determined exactly. By reading the TSC before and after the execution of a procedure, the precise time required for the procedure can be determined based on the processor cycle time. In this way, the TSC forms the basis for all time evaluations in connection with optimizing system operation. For example, it can be accurately determined which of two or more programming sequences is more efficient. This is a very useful tool for compiler developers and system programmers in producing the most effective code.

SYSTEM NOTE

A 4-Bit Synchronous Binary Counter

Figure 8–19(a) shows a 4-bit synchronous binary counter, and Figure 8–19(b) shows its timing diagram. This particular counter is implemented with negative edge-triggered flip-flops. The reasoning behind the J and K input control for the first three flip-flops is the

TABLE 8–4 • Summary of the analysis of the counter in Figure 8–17.

CLOCK PULSE	OUTPUTS			J-K INPUTS						AT THE NEXT CLOCK PULSE		
	Q_2	Q_1	Q_0	J_2	K_2	J_1	K_1	J_0	K_0	FF2	FF1	FF0
Initially	0	0	0	0	0	0	0	1	1	NC*	NC	Toggle
1	0	0	1	0	0	1	1	1	1	NC	Toggle	Toggle
2	0	1	0	0	0	0	0	1	1	NC	NC	Toggle
3	0	1	1	1	1	1	1	1	1	Toggle	Toggle	Toggle
4	1	0	0	0	0	0	0	1	1	NC	NC	Toggle
5	1	0	1	0	0	1	1	1	1	NC	Toggle	Toggle
6	1	1	0	0	0	0	0	1	1	NC	NC	Toggle
7	1	1	1	1	1	1	1	1	1	Toggle	Toggle	Toggle
										Counter recycles back to 000.		

*NC indicates *No Change*.

(a)

(b)

FIGURE 8–19 A 4-bit synchronous binary counter and timing diagram. Points where the AND gate outputs are HIGH are indicated by the shaded areas.

same as previously discussed for the 3-bit counter. The fourth stage, FF3, changes only twice in the sequence. Notice that both of these transitions occur following the times that Q_0, Q_1, and Q_2 are all HIGH. This condition is decoded by AND gate G_2 so that when a clock pulse occurs, FF3 will change state. For all other times the J_3 and K_3 inputs of FF3 are LOW, and it is in a no-change condition.

A 4-Bit Synchronous Decade Counter

A decade counter has ten states.

As you know, a BCD decade counter exhibits a truncated binary sequence and goes from 0000 through the 1001 state. Rather than going from the 1001 state to the 1010 state, it recycles to the 0000 state. A synchronous BCD decade counter is shown in Figure 8–20. The timing diagram for the decade counter is shown in Figure 8–21.

MULTISIM

FIGURE 8–20 A synchronous BCD decade counter. Open file F08-20 to verify operation.

FIGURE 8–21 Timing diagram for the BCD decade counter (Q_0 is the LSB).

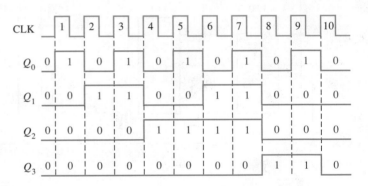

You can understand the counter operation by examining the sequence of states in Table 8–5 and by following the implementation in Figure 8–20. First, notice that FF0 (Q_0) toggles on each clock pulse, so the logic equation for its J_0 and K_0 inputs is

$$J_0 = K_0 = 1$$

This equation is implemented by connecting J_0 and K_0 to a constant HIGH level.

Next, notice in Table 8–5 that FF1 (Q_1) changes on the next clock pulse each time $Q_0 = 1$ and $Q_3 = 0$, so the logic equation for the J_1 and K_1 inputs is

$$J_1 = K_1 = Q_0\overline{Q}_3$$

This equation is implemented by ANDing Q_0 and $\overline{Q}_3$ and connecting the gate output to the J_1 and K_1 inputs of FF1.

Flip-flop 2 (Q_2) changes on the next clock pulse each time both $Q_0 = 1$ and $Q_1 = 1$. This requires an input logic equation as follows:

$$J_2 = K_2 = Q_0Q_1$$

This equation is implemented by ANDing Q_0 and Q_1 and connecting the gate output to the J_2 and K_2 inputs of FF2.

Finally, FF3 (Q_3) changes to the opposite state on the next clock pulse each time $Q_0 = 1$, $Q_1 = 1$, and $Q_2 = 1$ (state 7), or when $Q_0 = 1$ and $Q_3 = 1$ (state 9). The equation for this is as follows:

$$J_3 = K_3 = Q_0Q_1Q_2 + Q_0Q_3$$

TABLE 8–5 • States of a BCD decade counter.				
CLOCK PULSE	Q_3	Q_2	Q_1	Q_0
Initially	0	0	0	0
1	0	0	0	1
2	0	0	1	0
3	0	0	1	1
4	0	1	0	0
5	0	1	0	1
6	0	1	1	0
7	0	1	1	1
8	1	0	0	0
9	1	0	0	1
10 (recycles)	0	0	0	0

This function is implemented with the AND/OR logic connected to the J_3 and K_3 inputs of FF3 as shown in the logic diagram in Figure 8–20. Notice that the differences between this decade counter and the modulus-16 binary counter in Figure 8–19 are the $Q_0\overline{Q}_3$ AND gate, the Q_0Q_3 AND gate, and the OR gate; this arrangement detects the occurrence of the 1001 state and causes the counter to recycle properly on the next clock pulse.

Computers contain an internal counter that can be programmed for various frequencies and tone durations, thus producing "music." To select a particular tone, the programmed instruction selects a divisor that is sent to the counter. The divisor sets the counter up to divide the basic peripheral clock frequency to produce an audio tone. The duration of a tone can also be set by a programmed instruction; thus, a basic counter is used to produce melodies by controlling the frequency and duration of tones.

SYSTEM NOTE

SYSTEM EXAMPLE 8–1

PARALLEL-TO-SERIAL DATA CONVERSION (MULTIPLEXING)

A simplified example of a data transmission system using multiplexing and demultiplexing techniques was introduced in Chapter 5. Essentially, the parallel data bits on the multiplexer inputs are converted to serial data bits on the single transmission line. A group of bits appearing simultaneously on parallel lines is called *parallel data.* A group of bits appearing on a single line in a time sequence is called *serial data.*

Parallel-to-serial conversion is normally accomplished by the use of a counter to provide a binary sequence for the data-select inputs of a data selector/multiplexer, as illustrated in Figure 8–22. The Q outputs of the modulus-8 counter are connected to the data-select inputs of an 8-bit multiplexer.

Figure 8–23 is a timing diagram illustrating the operation of this circuit. The first byte (eight-bit group) of parallel data is applied to the multiplexer inputs. As the counter goes through a binary sequence from zero to seven, each bit, beginning with D_0, is sequentially

FIGURE 8–22

FIGURE 8–23

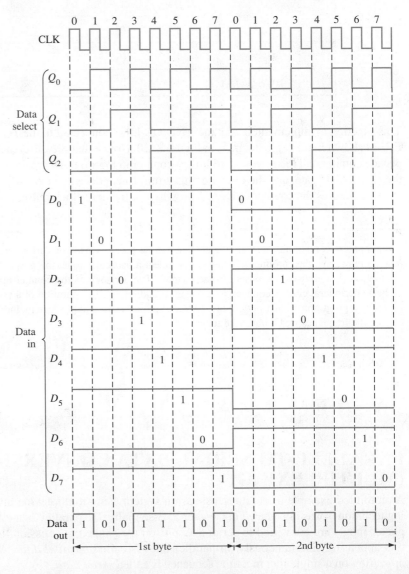

selected and passed through the multiplexer to the output line. After eight clock pulses the data byte has been converted to a serial format and sent out on the transmission line. When the counter recycles back to 0, the next byte is applied to the data inputs and is sequentially converted to serial form as the counter cycles through its eight states. This process continues repeatedly as each parallel byte is converted to a serial byte.

SECTION 8–4 CHECKUP

1. How does a synchronous counter differ from an asynchronous counter?

2. How many states does an 8-bit synchronous binary counter have?

8–5 UP/DOWN SYNCHRONOUS COUNTERS

An up/down counter is one that is capable of progressing in either direction through a certain sequence. An up/down counter, sometimes called a bidirectional counter, can have any specified sequence of states. A 3-bit binary counter that advances upward through its sequence (0, 1, 2, 3, 4, 5, 6, 7) and then can be reversed so that it goes through the sequence in the opposite direction (7, 6, 5, 4, 3, 2, 1, 0) is an illustration of up/down sequential operation.

After completing this section, you should be able to

- Explain the basic operation of an up/down counter

In general, most up/down counters can be reversed at any point in their sequence. For instance, the 3-bit binary counter can be made to go through the following sequence:

$$\overbrace{\qquad}^{\text{UP}} \qquad \overbrace{\qquad}^{\text{UP}}$$
$$0, 1, 2, 3, 4, 5, \underbrace{4, 3, 2,}_{\text{DOWN}} 3, 4, 5, 6, 7, \underbrace{6, 5,}_{\text{DOWN}} \text{etc.}$$

Table 8–6 shows the complete up/down sequence for a 3-bit binary counter. The arrows indicate the state-to-state movement of the counter for both its UP and its DOWN modes of operation. An examination of Q_0 for both the up and down sequences shows that FF0 toggles on each clock pulse. Thus, the J_0 and K_0 inputs of FF0 are

$$J_0 = K_0 = 1$$

For the up sequence, Q_1 changes state on the next clock pulse when $Q_0 = 1$. For the down sequence, Q_1 changes on the next clock pulse when $Q_0 = 0$. Thus, the J_1 and K_1 inputs of FF1 must equal 1 under the conditions expressed by the following equation:

$$J_1 = K_1 = (Q_0 \cdot \text{UP}) + (\overline{Q_0} \cdot \text{DOWN})$$

TABLE 8–6 • Up/Down sequence for a 3-bit binary counter.

CLOCK PULSE	UP	Q_2	Q_1	Q_0	DOWN
0		0	0	0	
1		0	0	1	
2		0	1	0	
3		0	1	1	
4		1	0	0	
5		1	0	1	
6		1	1	0	
7		1	1	1	

For the up sequence, Q_2 changes state on the next clock pulse when $Q_0 = Q_1 = 1$. For the down sequence, Q_2 changes on the next clock pulse when $Q_0 = Q_1 = 0$. Thus, the J_2 and K_2 inputs of FF2 must equal 1 under the conditions expressed by the following equation:

Incrementing a counter increases its count by one.

$$J_2 = K_2 = (Q_0 \cdot Q_1 \cdot \text{UP}) + (\overline{Q}_0 \cdot \overline{Q}_1 \cdot \text{DOWN})$$

Decrementing a counter decreases its count by one.

Each of the conditions for the J and K inputs of each flip-flop produces a toggle at the appropriate point in the counter sequence.

Figure 8–24 shows a basic implementation of a 3-bit up/down binary counter using the logic equations just developed for the J and K inputs of each flip-flop. Notice that the $UP/\overline{DOWN}$ control input is HIGH for UP and LOW for DOWN.

MULTISIM

FIGURE 8–24 A basic 3-bit up/down synchronous counter. Open file F08-24 to verify operation.

EXAMPLE 8–3

Show the timing diagram and determine the sequence of a 4-bit synchronous binary up/down counter if the clock and $UP/\overline{DOWN}$ control inputs have waveforms as shown in Figure 8–25(a). The counter starts in the all-0s state and is positive edge-triggered.

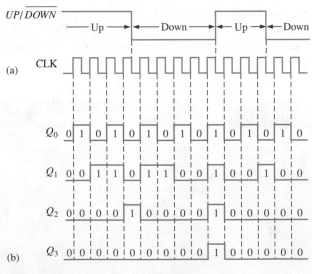

(a)

(b)

FIGURE 8–25

SOLUTION

The timing diagram showing the Q outputs is shown in Figure 8–25(b). From these waveforms, the counter sequence is as shown in Table 8–7.

TABLE 8–7

Q_3	Q_2	Q_1	Q_0	
0	0	0	0	
0	0	0	1	
0	0	1	0	UP
0	0	1	1	
0	1	0	0	
0	0	1	1	
0	0	1	0	
0	0	0	1	DOWN
0	0	0	0	
1	1	1	1	
0	0	0	0	
0	0	0	1	UP
0	0	1	0	
0	0	0	1	DOWN
0	0	0	0	

RELATED PROBLEM

Show the timing diagram if the *UP/$\overline{DOWN}$* control waveform in Figure 8–25(a) is inverted.

SYSTEM EXAMPLE 8–2

AUTOMOBILE PARKING CONTROL

This system example illustrates the use of an up/down counter to solve an everyday problem. The problem is to devise a means of monitoring available spaces in a one-hundred-space parking garage and provide for an indication of a full condition by illuminating a display sign and lowering a gate bar at the entrance.

A system that solves this problem consists of optoelectronic sensors at the entrance and exit of the garage, an up/down counter and associated circuitry, and an interface circuit that uses the counter output to turn the FULL sign on or off as required and lower or raise the gate bar at the entrance. A general block diagram of this system is shown in Figure 8–26.

FIGURE 8–26

A logic diagram of the up/down counter is shown in Figure 8–27. It consists of two cascaded up/down decade counters. The operation is described in the following paragraphs.

FIGURE 8–27

The counter is initially preset to 0 using the parallel data inputs, which are not shown. Each automobile entering the garage breaks a light beam, activating a sensor that produces an electrical pulse. This positive pulse sets the S-R latch on its leading edge. The LOW on the $\overline{Q}$ output of the latch puts the counter in the UP mode. Also, the sensor pulse goes through the NOR gate and clocks the counter on the LOW-to-HIGH transition of its trailing edge. Each time an automobile enters the garage, the counter is advanced by one **(incremented).** When the one-hundredth automobile enters, the counter goes to its last state (100_{10}). The *MAX/MIN* output goes HIGH and activates the interface circuit (no detail), which lights the FULL sign and lowers the gate bar to prevent further entry.

When an automobile exits, an optoelectronic sensor produces a positive pulse, which resets the S-R latch and puts the counter in the DOWN mode. The trailing edge of the clock decreases the count by one **(decremented).** If the garage is full and an automobile leaves, the *MAX/MIN* output of the counter goes LOW, turning off the FULL sign and raising the gate.

SECTION 8–5 CHECKUP

1. A 4-bit up/down binary counter is in the DOWN mode and in the 1010 state. On the next clock pulse, to what state does the counter go?

2. What is the terminal count of a 4-bit binary counter in the UP mode? In the DOWN mode? What is the next state after the terminal count in the DOWN mode?

8–6   CASCADED COUNTERS

Counters can be connected in cascade to achieve higher-modulus operation. In essence, cascading means that the last-stage output of one counter drives the input of the next counter.

After completing this section, you should be able to

- **Determine the overall modulus of cascaded counters**
- **Analyze the timing diagram of a cascaded counter configuration**
- **Use cascaded counters as a frequency divider**
- **Use cascaded counters to achieve specified truncated sequences**

ASYNCHRONOUS CASCADING   An example of two asynchronous counters connected in cascade is shown in Figure 8–28 for a 2-bit and a 3-bit ripple counter. The timing diagram is shown in Figure 8–29. Notice that the final output of the modulus-8 counter, Q_4, occurs once for every 32 input clock pulses. The overall modulus of the two cascaded counters is $4 \times 8 = 32$; that is, they act as a divide-by-32 counter.

The overall modulus of cascaded counters is equal to the product of the individual moduli.

Modulus-4 counter   Modulus-8counter

FIGURE 8–28   **Two cascaded asynchronous counters (all *J* and *K* inputs are HIGH).**

FIGURE 8–29   **Timing diagram for the cascaded counter configuration of Figure 8–28.**

SYNCHRONOUS CASCADING   When operating synchronous counters in a cascaded configuration, it is necessary to use the count enable and the terminal count functions to achieve higher-modulus operation. On some devices the count enable is labeled simply *CTEN* (or some other designation such as *G*), and terminal count (*TC*) is analogous to ripple clock output (*RCO*) on some IC counters.

Figure 8–30 shows two decade counters connected in cascade. The terminal count (*TC*) output of counter 1 is connected to the count enable (*CTEN*) input of counter 2. Counter 2 is inhibited by the LOW on its *CTEN* input until counter 1 reaches its last, or terminal, state and its terminal count output goes HIGH. This HIGH now enables counter 2, so that when the first clock pulse after counter 1 reaches its terminal count (CLK10), counter

FIGURE 8–30   **A modulus-100 counter using two cascaded decade counters.**

The time stamp counter (TSC), mentioned in a previous system note, is a 64-bit counter. It is interesting to observe that if this counter (or any full-modulus 64-bit counter) is clocked at a frequency of 1 GHz, it will take 585 years for it to go through all of its states and reach its terminal count. In contrast, a 32-bit full-modulus counter will exhaust all of its states in approximately 4.3 seconds when clocked at 1 GHz. The difference is astounding.

SYSTEM NOTE

2 goes from its initial state to its second state. Upon completion of the entire second cycle of counter 1 (when counter 1 reaches terminal count the second time), counter 2 is again enabled and advances to its next state. This sequence continues. Since these are decade counters, counter 1 must go through ten complete cycles before counter 2 completes its first cycle. In other words, for every ten cycles of counter 1, counter 2 goes through one cycle. Thus, counter 2 will complete one cycle after one hundred clock pulses. The overall modulus of these two cascaded counters is $10 \times 10 = 100$.

When viewed as a frequency divider, the circuit of Figure 8–30 divides the input clock frequency by 100. Cascaded counters are often used to divide a high-frequency clock signal to obtain highly accurate pulse frequencies. Cascaded counter configurations used for such purposes are sometimes called *countdown chains*. For example, suppose that you have a basic clock frequency of 1 MHz and you wish to obtain 100 kHz, 10 kHz, and 1 kHz; a series of cascaded decade counters can be used. If the 1 MHz signal is divided by 10, the output is 100 kHz. Then if the 100 kHz signal is divided by 10, the output is 10 kHz. Another division by 10 produces the 1 kHz frequency. The general implementation of this countdown chain is shown in Figure 8–31.

FIGURE 8–31 **Three cascaded decade counters forming a divide-by-1000 frequency divider with intermediate divide-by-10 and divide-by-100 outputs.**

EXAMPLE 8–4

Determine the overall modulus of the two cascaded counter configurations in Figure 8–32.

FIGURE 8–32

In Figure 8–32(a), the overall modulus for the 3-counter configuration is

$$8 \times 12 \times 16 = \mathbf{1536}$$

In Figure 8–32(b), the overall modulus for the 4-counter configuration is

$$10 \times 4 \times 7 \times 5 = \mathbf{1400}$$

RELATED PROBLEM

How many cascaded decade counters are required to divide a clock frequency by 100,000?

EXAMPLE 8–5

Use up/down decade counters connected in the UP mode to obtain a 10 kHz waveform from a 1 MHz clock. Show the logic diagram.

SOLUTION

To obtain 10 kHz from a 1 MHz clock requires a division factor of 100. Two decade counters must be cascaded as shown in Figure 8–33. The left counter produces a terminal count (*TC*) pulse for every 10 clock pulses. The right counter produces a terminal count (*TC*) pulse for every 100 clock pulses.

FIGURE 8–33   **A divide-by-100 counter using two decade up/down decade counters connected for the up sequence.**

RELATED PROBLEM

Determine the frequency of the waveform at the Q_0 output of the second counter (the one on the right) in Figure 8–33.

Cascaded Counters with Truncated Sequences

The preceding discussion has shown how to achieve an overall modulus (divide-by-factor) that is the product of the individual moduli of all the cascaded counters. This can be considered *full-modulus cascading*.

Often an application requires an overall modulus that is less than that achieved by full-modulus cascading. That is, a truncated sequence must be implemented with cascaded counters. To illustrate this method, we will use the cascaded counter configuration in Figure 8–34. This particular circuit uses four 4-bit synchronous binary counters. If these four counters (sixteen bits total) were cascaded in a full-modulus arrangement, the modulus would be

$$2^{16} = 65,536$$

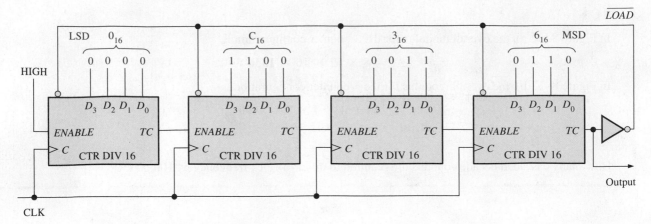

FIGURE 8–34 A divide-by-40,000 counter using 4-bit binary counters. Note that each of the parallel data inputs is shown in binary order (the right-most bit D_0 is the LSB in each counter).

Let's assume that a certain application requires a divide-by-40,000 counter (modulus 40,000). The difference between 65,536 and 40,000 is 25,536, which is the number of states that must be *deleted* from the full-modulus sequence. The technique used in the circuit of Figure 8–34 is to preset the cascaded counter to 25,536 (63C0 in hexadecimal) each time it recycles, so that it will count from 25,536 up to 65,535 on each full cycle. Therefore, each full cycle of the counter consists of 40,000 states.

Notice in Figure 8–34 that the *TC* (terminal count) output of the right-most counter is inverted and applied to the $\overline{LOAD}$ input of each 4-bit counter. Each time the count reaches its terminal value of 65,535, which is 1111111111111111_2, *TC* goes HIGH and causes the number on the parallel data inputs ($63C0_{16}$) to be synchronously loaded into the counter with the clock pulse. Thus, there is one *TC* pulse from the right-most 4-bit counter for every 40,000 clock pulses.

With this technique any modulus can be achieved by synchronous loading of the counter to the appropriate initial state on each cycle.

SECTION 8–6 CHECKUP

1. How many decade counters are necessary to implement a divide-by-1000 (modulus-1000) counter? A divide-by-10,000?

2. Show with general block diagrams how to achieve each of the following, using a flip-flop, a decade counter, and a 4-bit binary counter, or any combination of these:

   (a) Divide-by-20 counter   (b) Divide-by-32 counter

   (c) Divide-by-160 counter   (d) Divide-by-320 counter

8–7 COUNTER DECODING

In many applications, it is necessary that some or all of the counter states be decoded. The decoding of a counter involves using decoders or logic gates to determine when the counter is in a certain binary state in its sequence. For instance, the terminal count function previously discussed is a single decoded state (the last state) in the counter sequence.

After completing this section, you should be able to

- Implement the decoding logic for any given state in a counter sequence
- Explain why glitches occur in counter decoding logic
- Use the method of strobing to eliminate decoding glitches

Suppose that you wish to decode binary state 6 (110) of a 3-bit binary counter. When $Q_2 = 1$, $Q_1 = 1$, and $Q_0 = 0$, a HIGH appears on the output of the decoding gate, indicating that the counter is at state 6. This can be done as shown in Figure 8–35. This is called *active-HIGH decoding*. Replacing the AND gate with a NAND gate provides active-LOW decoding.

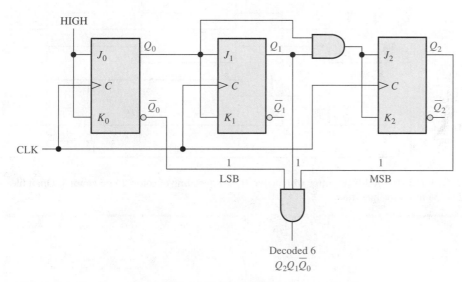

FIGURE 8–35 Decoding of state 6 (110). Open file F08-35 to verify operation.

MULTISIM

EXAMPLE 8–6

Implement the decoding of binary state 2 and binary state 7 of a 3-bit synchronous counter. Show the entire counter timing diagram and the output waveforms of the decoding gates. Binary 2 = $\overline{Q}_2 Q_1 \overline{Q}_0$ and binary 7 = $Q_2 Q_1 Q_0$.

SOLUTION

See Figure 8–36. The 3-bit counter was originally discussed in Section 8–4 (Figure 8–17).

RELATED PROBLEM

Show the logic for decoding state 5 in the 3-bit counter.

FIGURE 8–36 **A 3-bit counter with active-HIGH decoding of count 2 and count 7. Open file F08-36 to verify operation.**

Decoding Glitches

A glitch is an unwanted spike of voltage.

The problem of glitches produced by the decoding process was discussed in Chapter 5. As you have learned, the propagation delays due to the ripple effect in asynchronous counters create transitional states in which the counter outputs are changing at slightly different times. These transitional states produce undesired voltage spikes of short duration (glitches) on the outputs of a decoder connected to the counter. The glitch problem can also occur to some degree with synchronous counters because the propagation delays from the clock to the Q outputs of each flip-flop in a counter can vary slightly.

Figure 8–37 shows a basic asynchronous BCD decade counter connected to a BCD-to-decimal decoder. To see what happens in this case, let's look at a timing diagram in which the propagation delays are taken into account, as shown in Figure 8–38. Notice that these delays cause false states of short duration. The value of the false binary state at each critical transition is indicated on the diagram. The resulting glitches can be seen on the decoder outputs.

FIGURE 8–37   A basic BCD decade counter and decoder.

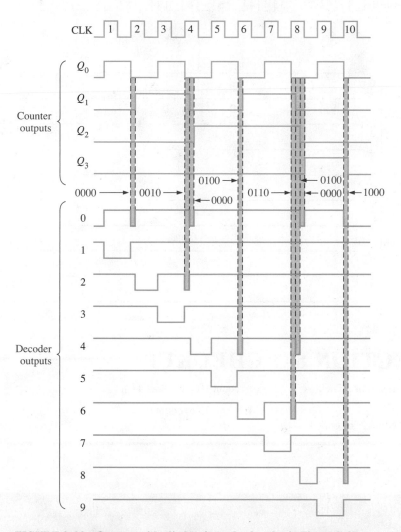

FIGURE 8–38   Outputs with glitches from the decoder in Figure 8–37.
Glitch widths are exaggerated for illustration and are usually only a few
nanoseconds wide.

One way to eliminate the glitches is to enable the decoded outputs at a time after the
glitches have had time to disappear. This method is known as *strobing* and can be accomplished in the case of an active-HIGH clock by using the LOW level of the clock to enable
the decoder, as shown in Figure 8–39. The resulting improved timing diagram is shown in
Figure 8–40.

FIGURE 8–39 The basic BCD decade counter and decoder with strobing to eliminate glitches.

FIGURE 8–40 Strobed decoder outputs for the circuit of Figure 8–39.

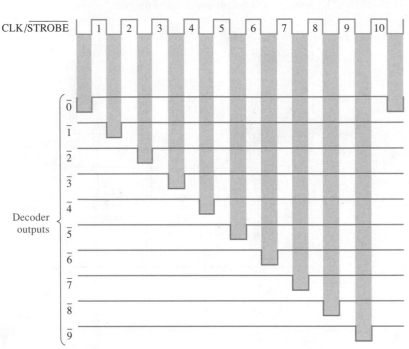

SECTION 8–7 CHECKUP

1. What transitional states are possible when a 4-bit asynchronous binary counter changes from

   (a) count 2 to count 3   **(b)** count 3 to count 4

   (c) count 10_{10} to count 11_{10}

   (d) count 15 to count 0

8–8 COUNTERS WITH VHDL AND VERILOG

In this section, the VHDL and Verilog programs for two types of counters are presented. A 4-bit synchronous binary counter is shown in Figure 8–41 and a decade counter is shown in Figure 8–42.

After completing this section, you should be able to

- Discuss the VHDL and Verilog descriptions of a 4-bit synchronous binary counter
- Discuss the VHDL and Verilog descriptions of a decade counter

Four-Bit Binary Counter

FIGURE 8–41   4-bit synchronous binary counter.

VHDL FOR FOUR-BIT BINARY COUNTER

```
library ieee;
use ieee.std_logic_1164.all;

entity FourBitBinaryCounter is
port (Clk: in std_logic;
    QOut: out std_logic);
end entity FourBitBinaryCounter;

architecture FourBitCounterBehavior of FourBitBinaryCounter is
  component jkff is
    port (J, K, Clk: in std_logic;
        Q: out std_logic);
end component jkff;

signal G1, G2: std_logic;
signal Q0, Q1, Q2: std_logic;
begin
  G1 <= Q0 and Q1;
  G2 <= G1 and Q2;

  JKFF0: jkff port map (J =>'1',K =>'1',   Clk => Clk, Q => Q0);
  JKFF1: jkff port map (J => Q0,K => Q0, Clk => Clk, Q => Q1);
  JKFF2: jkff port map (J => G1,K => G1, Clk => Clk, Q => Q2);
  JKFF3: jkff port map (J => G2,K => G2, Clk => Clk, Q => QOut);
end architecture FourBitCounterBehavior;
```

VERILOG FOR FOUR-BIT BINARY COUNTER

```
module FourBitBinaryCounter (Clk, QOut);
  input Clk;
  output QOut;

  wire G1, G2;
  wire Q0, Q1, Q2;

  assign G1 = Q0 && Q1;
  assign G2 = G1 && Q2;

  jkff JKFF0(.J(1),    .K(1),    .Clk(Clk), .Prn(1), .clrn(1), .Q(Q0));
  jkff JKFF1(.J(Q0), .K(Q0), .Clk(Clk), .Prn(1), .clrn(1), .Q(Q1));
  jkff JKFF2(.J(G1), .K(G1), .Clk(Clk), .Prn(1), .clrn(1), .Q(Q2));
  jkff JKFF3(.J(G2), .K(G2), .Clk(Clk), .Prn(1), .clrn(1), .Q(QOut));
endmodule
```

Decade Counter

FIGURE 8–42   **Decade counter.**

VHDL FOR FOUR-BIT DECADE COUNTER

```
library ieee;
use ieee.std_logic_1164.all;

entity FourBitDecadeCounter is
port(Clk: in std_logic;
     QOut: buffer std_logic);
end entity FourBitDecadeCounter;

architecture FourBitCounterBehavior of FourBitDecadeCounter is
component jkff is
  port (J, K, Clk: in std_logic;
        Q: out std_logic);
end component jkff;

signal G1, G2, G3: std_logic;
signal Q0, Q1, Q2: std_logic;
begin
  G1 <= Q0 and not QOut;
  G2 <= Q0 and Q1;
  G3 <= (Q0 and QOut) or (G2 and Q2);

  JKFF0: jkff port map (J => '1', K => '1', Clk => Clk, Q => Q0);
  JKFF1: jkff port map (J => G1, K => G1, Clk => Clk, Q => Q1);
  JKFF2: jkff port map (J => G2, K => G2, Clk => Clk, Q => Q2);
  JKFF3: jkff port map (J => G3, K => G3, Clk => Clk, Q => QOut);
end architecture FourBitCounterBehavior;
```

VERILOG FOR FOUR-BIT DECADE COUNTER

```
module FourBitDecadeCounter (Clk, QOut);
    input Clk;
    output QOut;

  wire G1, G2, G3;
  wire Q0, Q1, Q2;

  assign G1 = (Q0 && !QOut);
  assign G2 = (Q0 && Q1);
  assign G3 = (Q0 && QOut) || (G2 && Q2);

  jkff JKFF0(.J(1),   .K(1),   .Clk(Clk), .Prn(1), .clrn(1), .Q(Q0));
  jkff JKFF1(.J(G1), .K(G1), .Clk(Clk), .Prn(1), .clrn(1), .Q(Q1));
  jkff JKFF2(.J(G2), .K(G2), .Clk(Clk), .Prn(1), .clrn(1), .Q(Q2));
  jkff JKFF3(.J(G3), .K(G3), .Clk(Clk), .Prn(1), .clrn(1), .Q(QOut));
endmodule
```

SECTION 8-8 CHECKUP

1. In the VHDL program for the 4-bit synchronous binary counter, what is the purpose of the lines:

   G1 <= Q0 and Q1;
   G2 <= G1 and Q2;

2. What is the purpose of four jkff lines in the Verilog program for the decade counter?

8–9 TROUBLESHOOTING

The troubleshooting of counters can be simple or quite involved, depending on the type of counter and the type of fault. This section will give you some insight into how to approach the troubleshooting of sequential circuits.

After completing this section, you should be able to

- **Detect a faulty counter**
- **Isolate faults in maximum-modulus cascaded counters**
- **Isolate faults in cascaded counters with truncated sequences**
- **Determine faults in counters implemented with individual flip-flops**

Counters

The symptom for a faulty counter is usually that it does not advance its count. If this is the case, then check power and ground on the chip. Look at these lines with a scope to make sure there is no noise present (a noisy ground may actually be open). Check that there are clock pulses and that they have the correct amplitude and rise time and that there is not extraneous noise on the line. (Sometimes clock pulses can be loaded down by other ICs, making it appear that the counter is faulty when it is not). If power, ground, and the clock pulses are okay, check all inputs (including enable, load, and clear inputs), to see that they are connected correctly and that the logic is correct. An open input can cause a counter to work correctly some of the time—inputs should never be left open, even if they are not used. (An unused input should be connected to an inactive level). If the counter is stuck in a state and the clock is present, determine what input should be present to advance the counter. This may point to a faulty input (including clear or load inputs), which can be caused by logic elsewhere in the circuit. If inputs are all checked okay, an output may be pulled LOW or HIGH by an external short or open (or another faulty IC), keeping the output from advancing.

Cascaded Counters with Maximum Modulus

A failure in one of the counters in a chain of cascaded counters can affect all the counters that follow it. For example, if a count enable input opens, it effectively acts as a HIGH (for TTL), and the counter is always enabled. This type of failure in one of the counters will cause that counter to run at the full clock rate and will also cause all the succeeding counters to run at higher than normal rates. This is illustrated in Figure 8–43 for a divide-by-1000 cascaded counter arrangement where an open enable (*CTEN*) input acts as a TTL HIGH and continuously enables the second counter. Other faults that can affect "downstream" counter stages are open or shorted clock inputs or terminal count outputs. In some of these situations, pulse activity can be observed, but it may be at the wrong frequency. Exact frequency or frequency ratio measurements must be made.

(a) Normal operation

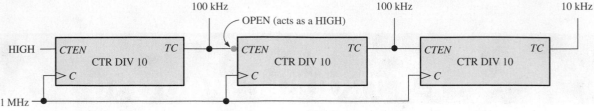

(b) Count Enable (*CTEN*) input of second counter open

FIGURE 8–43 Example of a failure that affects following counters in a cascaded arrangement.

Cascaded Counters with Truncated Sequences

The count sequence of a cascaded counter with a truncated sequence, such as that in Figure 8–44, can be affected by other types of faults in addition to those mentioned for maximum-modulus cascaded counters. For example, a failure in one of the parallel data inputs, the $\overline{LOAD}$ input, or the inverter can alter the preset count and thus change the modulus of the counter.

For example, suppose the D_3 input of the most significant counter in Figure 8–44 is open and acts as a HIGH. Instead of 6_{16} (0110) being preset into the counter, E_{16} (1110) is preset in. So, instead of beginning with $63C0_{16}$ ($25{,}536_{10}$) each time the counter recycles, the sequence will begin with $E3C0_{16}$ ($58{,}304_{10}$). This changes the modulus of the counter from 40,000 to $65{,}536 - 58{,}304 = 7232$.

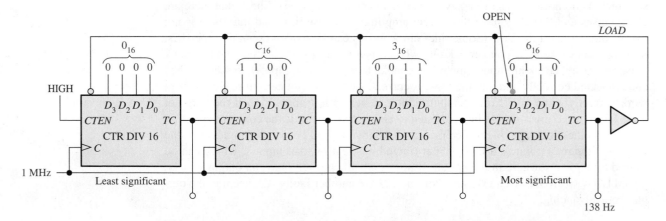

FIGURE 8–44 Example of a failure in a cascaded counter with a truncated sequence.

To check this counter, apply a known clock frequency, for example 1 MHz, and measure the output frequency at the final terminal count output. If the counter is operating properly, the output frequency is

$$f_{\text{out}} = \frac{f_{\text{in}}}{\text{modulus}} = \frac{1 \text{ MHz}}{40{,}000} = 25 \text{ Hz}$$

In this case, the specific failure described in the preceding paragraph will cause the output frequency to be

$$f_{\text{out}} = \frac{f_{\text{in}}}{\text{modulus}} = \frac{1 \text{ MHz}}{7232} \cong 138 \text{ Hz}$$

EXAMPLE 8–7

Frequency measurements are made on the truncated counter in Figure 8–45 as indicated. Determine if the counter is working properly, and if not, isolate the fault.

FIGURE 8–45

SOLUTION

Check to see if the frequency measured at TC 4 is correct. If it is, the counter is working properly.

$$\text{truncated modulus} = \text{full modulus} - \text{preset count}$$
$$= 16^4 - 82C0_{16}$$
$$= 65{,}536 - 33{,}472 = 32{,}064$$

The correct frequency at TC 4 is

$$f_4 = \frac{10\,\text{MHz}}{32{,}064} \cong 312\,\text{Hz}$$

There is a problem. The measured frequency of 637.8 Hz does not agree with the correct calculated frequency of 312 Hz.

To find the faulty counter, determine the actual truncated modulus as follows:

$$\text{modulus} = \frac{f_{\text{in}}}{f_{\text{out}}} = \frac{10\,\text{MHz}}{637.8\,\text{Hz}} = 15{,}679$$

Because the truncated modulus should be 32,064, most likely the counter is being preset to the wrong count when it recycles. The actual preset count is determined as follows:

$$\text{truncated modulus} = \text{full modulus} - \text{preset count}$$
$$\text{preset count} = \text{full modulus} - \text{truncated modulus}$$
$$= 65{,}536 - 15{,}679$$
$$= 49{,}857$$
$$= C2C0_{16}$$

This shows that the counter is being preset to $C2C0_{16}$ instead of $82C0_{16}$ each time it recycles.

Counters 1, 2, and 3 are being preset properly but counter 4 is not. Since $C_{16} = 1100_2$, the D_2 input to counter 4 is HIGH when it should be LOW. This is most likely caused by an **open input.** Check for an external open caused by a bad solder connection, a broken conductor, or a bent pin on the IC. If none can be found, replace the IC and the counter should work properly.

RELATED PROBLEM

Determine what the output frequency at TC 4 would be if the D_3 input of counter 3 were open.

Counters Implemented with Individual Flip-Flops

Counters implemented with individual flip-flop and gate ICs are sometimes more difficult to troubleshoot because there are many more inputs and outputs with external connections than there are in an IC counter. The sequence of a counter can be altered by a single open or short on an input or output, as Example 8–8 illustrates.

EXAMPLE 8–8

Suppose that you observe the output waveforms (green) that are indicated for the counter in Figure 8–46. Determine if there is a problem with the counter.

FIGURE 8–46

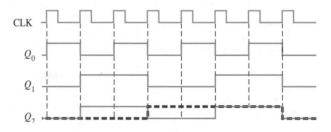

SOLUTION

The Q_2 waveform is incorrect. The correct waveform is shown as a red dashed line. You can see that the Q_2 waveform looks exactly like the Q_1 waveform, so whatever is causing FF1 to toggle appears to also be controlling FF2.

Checking the J and K inputs to FF2, you find a waveform that looks like Q_0. This result indicates that Q_0 is somehow getting through the AND gate. The only way this can happen is if the Q_1 input to the AND gate is always HIGH. However, you have seen that Q_1 has a correct waveform. This observation leads to the conclusion that the lower input to the AND gate must be internally open and acting as a HIGH. Replace the AND gate and retest the circuit.

RELATED PROBLEM

Describe the Q_2 output of the counter in Figure 8–46 if the Q_1 output of FF1 is open.

HANDS ON TIP
To observe the time relationship between two digital signals with a dual-trace analog oscilloscope, the proper way to trigger the scope is with the slower of the two signals. The reason for this is that the slower signal has fewer possible trigger points than the faster signal and there will be no ambiguity for starting the sweep. Vertical mode triggering uses a composite of both channels and should never be used for determining absolute time information. Since clock signals are usually the fastest signal in a digital system, they should not be used for triggering.

SECTION 8–9 CHECKUP

1. What failures can cause the counter in Figure 8–43 to have no pulse activity on any of the TC outputs?

2. What happens if the inverter in Figure 8–45 develops an open output?

SUMMARY

- Asynchronous and synchronous counters differ only in the way in which they are clocked. Synchronous counters can run at faster clock rates than asynchronous counters.

- The maximum modulus of a counter is the maximum number of possible states and is a function of the number of stages (flip-flops). Thus,

$$\text{Maximum modulus} = 2^n$$

where n is the number of stages in the counter. The modulus of a counter is the *actual* number of states in its sequence and can be equal to or less than the maximum modulus.

- The overall modulus of cascaded counters is equal to the product of the moduli of the individual counters.

KEY TERMS

Key terms and other bold terms in the chapter are defined in the end-of-book glossary.

Asynchronous   Not occurring at the same time.

Cascade   To connect "end-to-end" as when several counters are connected from the terminal count output of one counter to the enable input of the next counter.

Counter   A sequential logic device that is used to divide signal frequency, count events, and generate specified sequences of bits.

Decade   Characterized by ten states or values.

Mealy state machine   A state machine in which the outputs depend on both the internal present state and on the inputs.

Modulus   The number of unique states through which a counter will sequence.

Moore state machine   A state machine in which the outputs depend only on the internal present state.

Recycle   To undergo transition (as in a counter) from the final or terminal state back to the initial state.

State machine   A logic system exhibiting a sequence of states conditioned by internal logic and external inputs; any sequential circuit exhibiting a specified sequence of states.

Synchronous   Occurring at the same time.

FIXED-FUNCTION LOGIC

74XX163
4-bit synchronous binary counter

74XX190
Up-down synchronous decade counter

TRUE/FALSE QUIZ

Answers are at the end of the chapter.

1. In an asynchronous counter, all flip-flops change state at the same time.

2. In a synchronous counter, all flip-flops are clocked simultaneously.

3. An asynchronous counter is also known as a ripple counter.

4. A decade counter has sixteen states.

5. A counter with four stages has a maximum modulus of sixteen.

6. To achieve a maximum modulus of 32, sixteen stages are required.

7. If the present state is 1000, the next state of a 4-bit up/down counter in the DOWN mode is 0111.

8. Two cascaded decade counters divide the clock frequency by 20.

9. A counter with a truncated sequence has less than its maximum number of states.

10. To achieve a modulus of 100, ten decade counters are required.

SELF-TEST

Answers are at the end of the chapter.

1. Asynchronous counters are known as
   (a) ripple counters   (b) multiple clock counters
   (c) decade counters   (d) modulus counters

2. An asynchronous counter differs from a synchronous counter in
   (a) the number of states in its sequence   (b) the method of clocking
   (c) the type of flip-flops used   (d) the value of the modulus

3. The modulus of a counter is
   (a) the number of flip-flops
   (b) the actual number of states in its sequence
   (c) the number of times it recycles in a second
   (d) the maximum possible number of states

4. A 3-bit binary counter has a maximum modulus of
   (a) 3   (b) 6   (c) 8   (d) 16

5. A 4-bit binary counter has a maximum modulus of
   (a) 16   (b) 32   (c) 8   (d) 4

6. A modulus-12 counter must have
   (a) 12 flip-flops   (b) 3 flip-flops   (c) 4 flip-flops   (d) synchronous clocking

7. Which one of the following is an example of a counter with a truncated modulus?
   (a) Modulus 8   (b) Modulus 14   (c) Modulus 16   (d) Modulus 32

8. A 4-bit ripple counter consists of flip-flops that each have a propagation delay from clock to Q output of 12 ns. For the counter to recycle from 1111 to 0000, it takes a total of
   (a) 12 ns   (b) 24 ns   (c) 48 ns   (d) 36 ns

9. A BCD counter is an example of
   (a) a full-modulus counter   (b) a decade counter
   (c) a truncated-modulus counter   (d) answers (b) and (c)

10. Which of the following is an invalid state in an 8421 BCD counter?
    (a) 1100   (b) 0010   (c) 0101   (d) 1000

11. Three cascaded modulus-10 counters have an overall modulus of
    (a) 30   (b) 100   (c) 1000   (d) 10,000

12. A 10 MHz clock frequency is applied to a cascaded counter consisting of a modulus-5 counter, a modulus-8 counter, and two modulus-10 counters. The lowest output frequency possible is
    (a) 10 kHz   (b) 2.5 kHz   (c) 5 kHz   (d) 25 kHz

13. A 4-bit binary up/down counter is in the binary state of zero. The next state in the DOWN mode is
    (a) 0001   (b) 1111   (c) 1000   (d) 1110

14. The terminal count of a modulus-13 binary counter is
    (a) 0000   (b) 1111   (c) 1101   (d) 1100

PROBLEMS

Answers to odd-numbered problems are at the end of the book.

SECTION 8–1   A System

1. Assume that the digital clock of Figure 8–1 is initially reset to 12 o'clock. Determine the binary state of each counter after sixty-two 60 Hz pulses have occurred.

2. What is the output frequency of each counter in the digital clock circuit of Figure 8–1?

SECTION 8–2   Finite State Machines

3. Represent a decade counter with the terminal state decoded as a state machine. Identify the type and show the block diagram and the state diagram.

4. Identify the type of state machine for the traffic signal control system in Chapter 6. State the reason why it is the type you specified.

SECTION 8–3   Asynchronous Counters

5. For the ripple counter shown in Figure 8–47, show the complete timing diagram for eight clock pulses, showing the clock, Q_0, and Q_1 waveforms.

FIGURE 8–47

6. For the ripple counter in Figure 8–48, show the complete timing diagram for sixteen clock pulses. Show the clock, Q_0, Q_1, and Q_2 waveforms.

FIGURE 8–48

7. In the counter of Problem 6, assume that each flip-flop has a propagation delay from the triggering edge of the clock to a change in the Q output of 8 ns. Determine the worst-case (longest) delay time from a clock pulse to the arrival of the counter in a given state. Specify the state or states for which this worst-case delay occurs.

8. Show how to connect a 4-bit asynchronous counter for each of the following moduli using the Clear input:
   (a) 9   (b) 11   (c) 13   (d) 14   (e) 15

SECTION 8–4   Synchronous Counters

9. If the counter of Problem 7 were synchronous rather than asynchronous, what would be the longest delay time?

10. Show the complete timing diagram for the 5-stage synchronous binary counter in Figure 8–49. Verify that the waveforms of the Q outputs represent the proper binary number after each clock pulse.

FIGURE 8–49

11. By analyzing the *J* and *K* inputs to each flip-flop prior to each clock pulse, prove that the decade counter in Figure 8–50 progresses through a BCD sequence. Explain how these conditions in each case cause the counter to go to the next proper state.

FIGURE 8–50

12. The waveforms in Figure 8–51 are applied to the count enable, clear, and clock inputs as indicated. Show the counter output waveforms in proper relation to these inputs. The clear input is asynchronous.

FIGURE 8–51

13. A BCD decade counter is shown in Figure 8–52. The waveforms are applied to the clock and clear inputs as indicated. Determine the waveforms for each of the counter outputs (Q_0, Q_1, Q_2, and Q_3). The clear is synchronous, and the counter is initially in the binary 1000 state.

FIGURE 8–52

SECTION 8–5   Up/Down Synchronous Counters

14. Show a complete timing diagram for a 3-bit up/down counter that goes through the following sequence. Indicate when the counter is in the UP mode and when it is in the DOWN mode. Assume positive edge-triggering.

$$0, 1, 2, 3, 2, 1, 2, 3, 4, 5, 6, 5, 4, 3, 2, 1, 0$$

15. Develop the Q output waveforms for the binary up/down counter with the input waveforms shown in Figure 8–53. Start with a count of 0000.

FIGURE 8–53

16. Repeat Problem 15 if the $D/\overline{U}$ input signal is inverted with the $\overline{CLR}$ the same.

17. Repeat Problem 15 if the $\overline{CLR}$ is inverted with the $D/\overline{U}$ inputs the same.

SECTION 8–6   Cascaded Counters

18. For each of the cascaded counter configurations in Figure 8–54, determine the frequency of the waveform at each point indicated by a circled number, and determine the overall modulus.

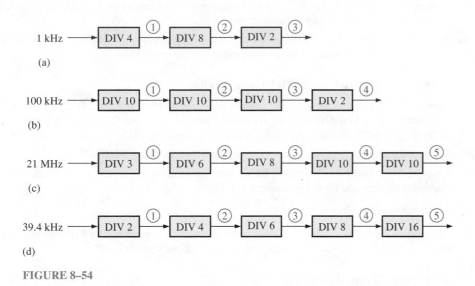

FIGURE 8–54

19. Expand the counter in Figure 8–31 to create a divide-by-10,000 counter and a divide-by-100,000 counter.

20. With general block diagrams, show how to obtain the following frequencies from a 10 MHz clock by using single flip-flops, modulus-5 counters, and decade counters:
    (a) 5 MHz   (b) 2.5 MHz   (c) 2 MHz   (d) 1 MHz   (e) 500 kHz
    (f) 250 kHz   (g) 62.5 kHz   (h) 40 kHz   (i) 10 kHz   (j) 1 kHz

SECTION 8–7   Counter Decoding

21. Given a BCD decade counter with only the Q outputs available, show what decoding logic is required to decode each of the following states and how it should be connected to the counter. A HIGH output indication is required for each decoded state. The MSB is to the left.
    (a) 0001   (b) 0011   (c) 0101   (d) 0111   (e) 1000

22. For the 4-bit binary counter connected to the decoder in Figure 8–55, determine each of the decoder output waveforms in relation to the clock pulses.

FIGURE 8–55

23. If the counter in Figure 8–55 is asynchronous, determine where the decoding glitches occur on the decoder output waveforms.

24. Modify the circuit in Figure 8–55 to eliminate decoding glitches.

25. Analyze the counter in Figure 8–35 for the occurrence of glitches on the decode gate output. If glitches occur, suggest a way to eliminate them.

26. Analyze the counter in Figure 8–36 for the occurrence of glitches on the outputs of the decoding gates. If glitches occur, make a design change that will eliminate them.

SECTION 8–8   Counters with VHDL and Verilog

27. Expand the 4-bit synchronous binary counter in Figure 8–41 to five bits. Describe the input requirements for a third AND gate G_3, required to provide the J_4 and K_4 inputs to the additional JK flip-flop stage. Write the assignment statement for G3 using VHDL or Verilog.

28. What are three advantages of implementing a counter system using VHDL?

29. Referring to the four-bit synchronous decade counter in Figure 8–42, draw the resulting timing diagram if the equation for G3 is modified as follows:

    G3 <= (Q0 and QOut) or (G1 and Q1);

30. Referring to the 4-bit synchronous binary counter timing diagram in Figure 8–19, the output of the JKFF2 flip-flop represents the binary value 2 raised to what power?

SECTION 8–9   Troubleshooting

31. For the counter in Figure 8–7, show the timing diagram for the Q_0 and Q_1 waveforms for each of the following faults (assume Q_0 and Q_1 are initially LOW):
    (a) clock input to FF0 shorted to ground
    (b) Q_0 output open
    (c) clock input to FF1 open
    (d) J input to FF0 open
    (e) K input to FF1 shorted to ground

32. Solve Problem 31 for the counter in Figure 8–14.

33. Isolate the fault in the counter in Figure 8–9(a) by analyzing the waveforms in Figure 8–56.

34. From the waveform diagram in Figure 8–57, determine the most likely fault in the counter of Figure 8–17.

35. Solve Problem 34 if the Q_2 output has the waveform observed in Figure 8–58. Outputs Q_0 and Q_1 are the same as in Figure 8–57.

FIGURE 8–56

FIGURE 8–57

FIGURE 8–58

36. You apply a 5 MHz clock to the cascaded counter in Figure 8–34 and measure a frequency of 76.2939 Hz at the last *TC* output. Is this correct, and if not, what is the most likely problem?

37. Develop a table for use in testing the counter in Figure 8–34 that will show the frequency at the final *TC* output for all possible open failures of the parallel data inputs (D_0, D_1, D_2, and D_3) taken one at a time. Use 10 MHz as the test frequency for the clock.

38. The tens-of-hours 7-segment display in the digital clock system of Figure 8–1 continuously displays a 1. All the other digits work properly. What could be the problem?

39. What would be the visual indication of an open Q_1 output in the tens portion of the minutes counter in Figure 8–1? Also see Figure 8–2.

40. One day complaints begin flooding in from patrons of a parking garage that uses the control system depicted in Figures 8–26 and 8–27. The patrons say that they enter the garage because the gate is up and the FULL sign is off but that, once in, they can find no empty space. As the technician in charge of this facility, what do you think the problem is, and how will you trouble-shoot and repair the system as quickly as possible?

Special Problems

41. For the automobile parking control system in Figure 8–26, a pattern of entrance and exit sensor pulses during a given 24-hour period are shown in Figure 8–59. If there were 53 cars already in the garage at the beginning of the period, what is the state of the counter at the end of the 24 hours?

FIGURE 8–59

42. The binary number for decimal 57 appears on the parallel data inputs of the parallel-to-serial converter in Figure 8–22 (D_0 is the LSB). The counter initially contains all zeros and a 10 kHz clock is applied. Develop the timing diagram showing the clock, the counter outputs, and the serial data output.

43. Develop a modulus-1000 counter by using decade counters.

44. Modify the counter in Figure 8–34 to achieve a modulus of 30,000.

45. Repeat Problem 44 for a modulus of 50,000.

46. Modify the digital clock in Figures 8–1, 8–2, and 8–3 so that it can be preset to any desired time.

47. Implement an alarm circuit for the digital clock that can detect a predetermined time (hours and minutes only) and produce a signal to activate an audio alarm.

48. Modify the circuit in Figure 8–27 for a 1000-space parking garage and a 3000-space parking garage.

MULTISIM TROUBLESHOOTING PRACTICE

49. Open file P08-49 and follow the instructions given there.
50. Open file P08-50 and follow the instructions given there.
51. Open file P08-51 and follow the instructions given there.
52. Open file P08-52 and follow the instructions given there.
53. Open file P08-53 and follow the instructions given there.

ANSWERS TO SECTION CHECKUPS

SECTION 8–1   A System

1. Gate G_1 resets flip-flop on first clock pulse after count 9. Gate G_2 decodes count 12 to preset counter to 0001.

2. The hours decade counter advances through each state from zero to nine, and as it recycles from nine back to zero, the flip-flop is toggled to the SET state. This produces a ten (10) on the display. When the hours decade counter is in state 12, the decode NAND gate causes the counter to recycle to state 1 on the next clock pulse. The flip-flop resets. This results in a one (01) on the display.

SECTION 8–2   Finite State Machines

1. A finite state machine has a limited number of states that occur in a prescribed order.

2. Moore and Mealy are two types of finite state machines.

3. In a Moore machine, only the present state determines the outputs. In a Mealy machine, not only the present state but the input(s) determine the outputs.

SECTION 8–3   Asynchronous Counters

1. Asynchronous means that each flip-flop after the first one is enabled by the output of the preceding flip-flop.

2. A modulus-14 counter has fourteen states requiring four flip-flops.

SECTION 8–4   Synchronous Counters

1. All flip-flops in a synchronous counter are clocked simultaneously.

2. An 8-bit synchronous binary counter has 64 states.

SECTION 8–5   Up/Down Synchronous Counters

1. The counter goes to 1001.

2. UP: 1111: DOWN: 0000; the next state is 1111.

SECTION 8–6   Cascaded Counters

1. Three decade counters produce $\div$ 1000; 4 decade counters produce $\div$ 10,000.

2. (a) $\div$ 20: flip-flop and DIV 10
   (b) $\div$ 32: flip-flop and DIV 16
   (c) $\div$ 160: DIV 16 and DIV 10
   (d) $\div$ 320: DIV 16 and DIV 10 and flip-flop

SECTION 8–7   Counter Decoding

1. (a) No transitional states because there is a single bit change
   (b) 0000, 0001, 0010, 0101, 0110, 0111
   (c) No transitional states because there is a single bit change
   (d) 0001, 0010, 0011, 0100, 0101, 0110, 0111, 1000, 1001, 1010, 1011, 1100, 1101, 1110

SECTION 8–8   Counters with VHDL and Verilog

1. G1 <= Q0 **and** Q1; Defines the logic for gate $G_1 = Q_0Q_1$

   G2 <= G1 **and** Q2; Defines the logic for gate $G_2 = Q_0Q_1Q_2$

2. These four lines define how the flip-flops are connected.

SECTION 8–9   Troubleshooting

1. No pulses on *TC* outputs: *CTEN* of first counter shorted to ground or to a LOW; clock input of first counter open; clock line shorted to ground or to a LOW; *TC* output of first counter shorted to ground or to a LOW.

2. With the inverter output open, the counter does not recycle at the preset count but acts as a full-modulus counter.

ANSWERS TO RELATED PROBLEMS FOR EXAMPLES

8–1 See Figure 8–60.

FIGURE 8–60

8–2 Connect Q_0 to the NAND gate as a third input (Q_2 and Q_3 are two of the inputs). Connect the $\overline{CLR}$ line to the $\overline{CLR}$ input of FF0 as well as FF2 and FF3.

8–3 See Figure 8–61.

FIGURE 8–61

8–4 Five decade counters are required. $10^5 = 100,000$

8–5 $f_{Q0} = 1 \text{ MHz}/[(10)(2)] = 50 \text{ kHz}$

8–6 See Figure 8–62.

FIGURE 8–62

8–7 $8AC0_{16}$ would be loaded. $16^4 - 8AC0_{16} = 65,536 - 32,520 = 30,016$

   $f_{TC4} = 10 \text{ MHz } 30,016 = 333.2 \text{ Hz}$

8–8 See Figure 8–63.

FIGURE 8–63

ANSWERS TO TRUE/FALSE QUIZ

1. F **2.** T **3.** T **4.** F **5.** T
6. F **7.** T **8.** F **9.** T **10.** F

ANSWERS TO SELF-TEST

1. (a) **2.** (b) **3.** (b) **4.** (c) **5.** (a) **6.** (c) **7.** (b)
8. (c) **9.** (d) **10.** (a) **11.** (c) **12.** (b) **13.** (b) **14.** (d)

CHAPTER 9

PROGRAMMABLE LOGIC

OUTLINE

OBJECTIVES

- Discuss the types of programmable logic, SPLDs and CPLDs, and explain their basic structure
- Describe the basic architecture of two types of SPLDs—the PAL and the GAL
- Explain the basic structure of a programmable logic array (PLA)
- Discuss the operation of macrocells
- Distinguish between CPLDs and FPGAs
- Explain the basic operation of a look-up table (LUT)
- Define *intellectual property* and *platform FPGA*
- Discuss embedded functions
- Show a basic software design flow for a programmable device
- Explain the design flow elements of design entry, functional simulation, synthesis, implementation, timing simulation, and downloading
- Discuss several methods of testing a programmable logic device, including boundary scan logic

KEY TERMS

PAL	**Design flow**
GAL	**Target device**
Macrocell	**Schematic entry**
Registered	**Text entry**
CPLD	**Functional simulation**
LAB	**Compiler**
LUT	**Timing simulation**
FPGA	**Downloading**
CLB	**Break point**
Intellectual property	**Boundary scan**

INTRODUCTION

The distinction between hardware and software is hazy. Today, new digital circuits are programmed into hardware using languages like VHDL and Verilog. The density (number of equivalent gates on a single chip) has increased dramatically over the past few years. The maximum number of gates in an FPGA (a type of PLD known as a field-programmable gate array) is over 500,000 and doubling every 18 months, according to Moore's law. At the same time, the price for a PLD is decreasing.

PLDs, such as the FPGA, can be used in conjunction with processors and software in an embedded system, or the FPGA can be the sole component with all the

VISIT THE WEBSITE
Study aids for this chapter are available at
http://pearsonhighered.com/floyd

logic functions programmed in. An embedded system is one that is dedicated to a single task or a very limited number of tasks unlike the computer, which is multipurpose and can be programmed to perform just about any task. With PLDs, logic is described with software and then implemented with the internal gates of the PLD.

In this chapter, the basic architecture (internal structure and organization) of SPLDs, CPLDs, and FPGAs is discussed. A discussion of software development tools covers the generic design flow for programming a device, including design entry, functional simulation, synthesis, implementation, timing simulation, and downloading.

9–1 SIMPLE PROGRAMMABLE LOGIC DEVICES (SPLDs)

Two major types of simple programmable logic devices (SPLDs) are the PAL and the GAL. *PAL* stands for programmable array logic, and *GAL* stands for generic array logic. Generally, a PAL is one-time programmable (OTP), and a GAL is a type of PAL that is reprogrammable. The term *GAL* is a designation originally used by Lattice Semiconductor and later licensed to other manufacturers. The basic structure of both PALs and GALs is a programmable AND array and a fixed OR array, which is a basic sum-of-products architecture.

After completing this section, you should be able to

- **Describe SPLD operation**
- **Show how a sum-of-products expression is implemented in a PAL or GAL**
- **Explain simplified PAL/GAL logic diagrams**
- **Describe a basic PAL/GAL macrocell**

SPLD: The PAL

A **PAL** (programmable array logic) consists of a programmable array of AND gates that connects to a fixed array of OR gates. Generally, PALs are implemented with fuse process technology and are, therefore, one-time programmable (OTP).

The PAL structure allows any sum-of-products (SOP) logic expression with a defined number of variables to be implemented. As you have learned, any combinational logic function can be expressed in SOP form. A simple PAL structure is shown in Figure 9–1 for two input variables and one output; most PALs have many inputs and many outputs. As you learned in Chapter 3, a programmable array is essentially a grid or matrix of conductors that form rows and columns with a programmable link at each cross point. Each programmable

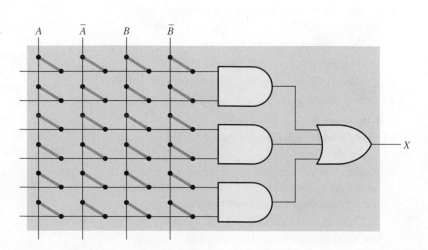

FIGURE 9–1 Basic AND/OR structure of a PAL.

link, which is a fuse in the case of a PAL, is called a *cell*. Each row is connected to the input of an AND gate, and each column is connected to an input variable or its complement. By programming the presence or absence of a fuse connection, any combination of input variables or complements can be applied to an AND gate to form any desired product term. The AND gates are connected to an OR gate, creating a sum-of-products (SOP) output.

IMPLEMENTING A SUM-OF-PRODUCTS EXPRESSION   An example of a simple PAL is programmed as shown in Figure 9–2 so that the product term AB is produced by the top AND gate, $A\overline{B}$ is produced by the middle AND gate, and $\overline{A}\,\overline{B}$ is produced by the bottom AND gate. As you can see, the fuses are left intact to connect the desired variables or their complements to the appropriate AND gate inputs. The fuses are opened where a variable or its complement is not used in a given product term. The final output from the OR gate is the SOP expression,

$$X = AB + A\overline{B} + \overline{A}\,\overline{B}$$

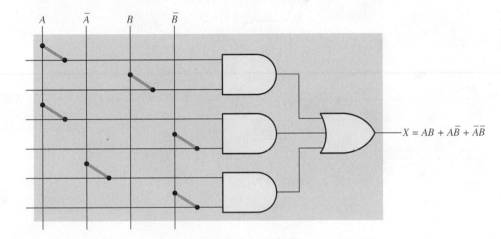

FIGURE 9–2   **PAL implementation of a sum-of-products expression.**

$$X = AB + A\overline{B} + \overline{A}\overline{B}$$

SPLD: The GAL

The **GAL** is essentially a PAL that can be reprogrammed. It has the same type of AND/OR organization that the PAL does. The basic difference is that a GAL uses a reprogrammable process technology, such as EEPROM (E^2CMOS), instead of fuses, as shown in Figure 9–3.

FIGURE 9–3   **Simplified GAL array.**

Simplified Notation for PAL/GAL Diagrams

Actual PAL and GAL devices have many AND and OR gates in addition to other elements and are capable of handling many variables and their complements. Most PAL and GAL diagrams that you may see on a data sheet use simplified notation, as illustrated in Figure 9–4, to keep the schematic from being too complicated.

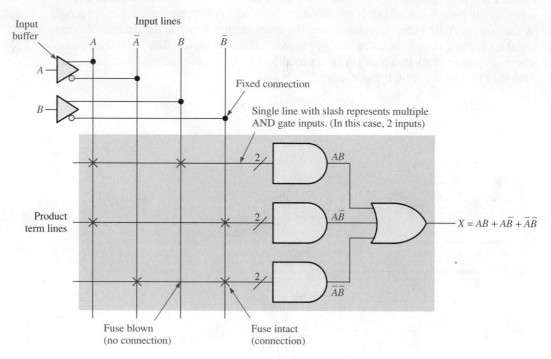

FIGURE 9–4   A portion of a programmed PAL/GAL.

The input variables to a PAL or GAL are usually buffered to prevent loading by a large number of AND gate inputs to which they are connected. On the diagram, the triangle symbol represents a buffer that produces both the variable and its complement. The fixed connections of the input variables and buffers are shown using standard dot notation.

PALs and GALs have a large number of programmable interconnection lines, and each AND gate has multiple inputs. Typical PAL and GAL logic diagrams represent a multiple-input AND gate with an AND gate symbol having a single input line with a slash and a digit representing the actual number of inputs. Figure 9–4 illustrates this for the case of 2-input AND gates.

Programmable links in an array are indicated in a diagram by a red X at the cross point for an intact fuse or other type of link and the absence of an X for an open fuse or other type of link. In Figure 9–4, the 2-variable logic function $AB + A\overline{B} + \overline{A}\,\overline{B}$ is programmed.

EXAMPLE 9–1

Show how a PAL is programmed for the following 3-variable logic function:

$$X = A\overline{B}C + \overline{A}B\overline{C} + \overline{A}\,\overline{B} + AC$$

SOLUTION

The programmed array is shown in Figure 9–5. The intact fusible links are indicated by small red Xs. The absence of an X means that the fuse is open.

FIGURE 9–5

RELATED PROBLEM*

Write the expression for the output if the fusible links connecting input A to the top row and to the bottom row in Figure 9–5 are open.

*Answers are at the end of the chapter.

PAL/GAL General Block Diagram

A block diagram of a PAL or GAL is shown in Figure 9–6. Remember, the basic difference is that a GAL has a reprogrammable array and the PAL is one-time programmable. The programmable AND array outputs go to fixed OR gates that are connected to additional

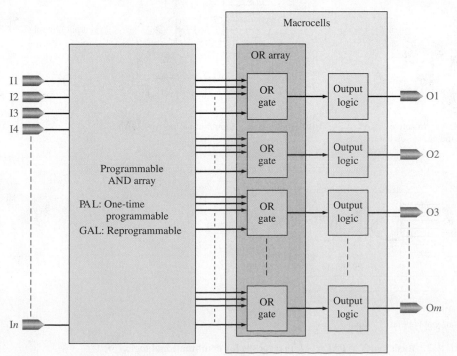

FIGURE 9–6   **General block diagram of a PAL or GAL.**

output logic. An OR gate combined with its associated output logic is typically called a *macrocell*. The complexity of the macrocell depends on the particular device, and in GALs it is often reprogrammable.

Generally, SPLD package configurations range from 20 pins to 28 pins. Two factors that you can use to help determine whether a certain PAL or GAL is adequate for a given logic design are the number of inputs and outputs and the number of equivalent gates or density. Other parameters to consider are the maximum operating frequency, delay times, and dc supply voltage. Lattice, Actel, Atmel, and Cypress are among several companies that produce SPLDs. Various SPLD manufacturers may have different ways of defining density, so you have to use the specified number of equivalent gates with this in mind.

Macrocells

A **macrocell** generally consists of one OR gate and some associated output logic. The macrocells vary in complexity, depending on the particular type of PAL or GAL. A macrocell can be configured for combinational logic, registered logic, or a combination of both. **Registered** logic means that there is a flip-flop in the macrocell to provide for sequential logic functions. The registered operation of macrocells is covered in Section 9–3.

Figure 9–7 illustrates three basic types of macrocells with combinational logic. Part (a) shows a simple macrocell with the OR gate and an inverter with a tristate control that can make the inverter like an open circuit to completely disconnect the output. The output of the tristate inverter can be either LOW, HIGH, or disconnected. Part (b) is a macrocell that can be either an input or an output. When it is used as an input, the tristate inverter is disconnected, and the input goes to the buffer that is connected to the AND array. Part (c) is a macrocell that can be programmed to have either an active-HIGH or an active-LOW output, or it can be used as an input. One input to the exclusive-OR (XOR) gate can be programmed to be either HIGH or LOW. When the programmable XOR input is HIGH, the OR gate output is inverted because $0 \oplus 1 = 1$ and $1 \oplus 1 = 0$. Similarly, when the programmable XOR input is LOW, the OR gate output is not inverted because $0 \oplus 0 = 0$ and $1 \oplus 0 = 1$.

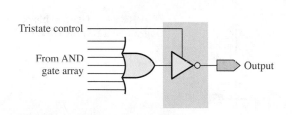

(a) Combinational output (active-LOW). An active-HIGH output would be shown without the bubble on the tristate gate symbol.

(b) Combinational input/output (active-LOW)

(c) Programmable polarity output

FIGURE 9–7 **Basic types of PAL/GAL macrocells for combinational logic.**

SECTION 9–1 CHECKUP*

1. What does PAL stand for?

2. What does GAL stand for?

3. What is the difference between a PAL and a GAL?

4. Basically, what does a macrocell contain?

*Answers are at the end of the chapter.

9–2 COMPLEX PROGRAMMABLE LOGIC DEVICES (CPLDs)

The complex programmable logic device (CPLD) is basically a single device with multiple SPLDs that provides more capacity for larger logic designs. In this section, the focus is the concepts of traditional CPLD architecture, keeping in mind that CPLDs may vary somewhat in architecture and/or in parameters such as density, process technology, power consumption, supply voltage, and speed.

After completing this section, you should be able to

- **Describe a typical CPLD**
- **Discuss the basic CPLD architecture**
- **Explain how product terms are generated in CPLDs**

The CPLD

A **CPLD** (complex programmable logic device) consists basically of multiple SPLD arrays with programmable interconnections. Although the way CPLDs are internally organized varies with the manufacturer, Figure 9–8 illustrates a generic CPLD. We will refer to each

FIGURE 9–8   Basic block diagram of a generic CPLD.

SPLD array in a CPLD as a **LAB** (logic array block). Other designations are sometimes used, such as *function block, logic block,* or *generic block.* The programmable interconnections are generally called the *PIA* (**programmable interconnect array**) although some manufacturers, such as Xilinx, use the term *AIM* (advanced interconnect matrix) or a similar designation. The LABs and the interconnections between LABs are programmed using software. A CPLD can be programmed for complex logic functions based on the SOP structure of the individual LABs (actually SPLDs). Inputs can be connected to any of the LABs, and their outputs can be interconnected to any other LABs via the PIA.

Most programmable logic manufacturers make a series of CPLDs that range in density, process technology, power consumption, supply voltage, and speed. Manufacturers usually specify CPLD density in terms of macrocells or logic array blocks. Densities can range from tens of macrocells to over 2000 macrocells in packages with up to several hundred pins. As PLDs become more complex, maximum densities will increase. Most CPLDs are reprogrammable and use EEPROM or SRAM process technology for the programmable links. Power consumption can range from a few milliwatts to a few hundred milliwatts. DC supply voltages are typically from 2.5 V to 5 V, depending on the specific device.

Several manufacturers, (for example, Altera, Xilinx, Lattice, and Cypress) produce CPLDs. As you will learn, CPLDs and other programmable logic devices are really a combination of hardware and software.

Classic CPLD Architecture

The **architecture** of a CPLD is the way in which the internal elements are organized and arranged. The architecture of specific CPLDs is similar to the block diagram of a generic CPLD (shown in Figure 9–8). It has the classic PAL/GAL structure that produces SOP functions. The density ranges from 2 LABs to 16 LABs, depending on the particular device in the series. Remember, a LAB is roughly equivalent to one SPLD, and package sizes for CPLDs vary from 44 pins to 208 pins. Typically, a series of CPLDs uses the EEPROM-based process technology. In-system programmable (ISP) versions use the JTAG standard interface.

Figure 9–9 shows a general block diagram of a typical CPLD. Four LABs are shown, but there can be up to sixteen, depending on the particular device in a series. Each of the four LABs consists of sixteen macrocells, and multiple LABs are linked together via the PIA, which is a programmable global (goes to all LABs) bus structure to which the general-purpose inputs, the I/Os, and the macrocells are connected.

THE MACROCELL   A simplified diagram of a typical macrocell is shown in Figure 9–10. The macrocell contains a small programmable AND array with five AND gates, an OR gate, a product-term selection matrix for connecting the AND gate outputs to the OR gate, and associated logic that can be programmed for input, combinational logic output, or registered output. This macrocell is covered in more detail in Section 9–3.

Although based on the same concept, this macrocell differs somewhat from the macrocell discussed in Section 9–1 in relation to SPLDs because it contains a portion of the programmable AND array and a product-term selection matrix. As shown in Figure 9–10, five AND gates feed product terms from the PIA into the product-term selection matrix. The product term from the bottom AND gate can be fed back inverted into the programmable array as a shared expander for use by other macrocells. The parallel expander inputs allow borrowing of unused product terms from other macrocells to expand an SOP expression. The product-term selection matrix is an array of programmable connections that is used to connect selected outputs from the AND array and from the expander inputs to the OR gate.

SHARED EXPANDERS   A complemented product term that can be used to increase the number of product terms in an SOP expression is available from each macrocell in a LAB. Figure 9–11 illustrates how a shared expander term from another macrocell can be used to create additional product terms. In this case, each of the five AND gates in a macrocell array is limited to four inputs and, therefore, can produce up to a 4-variable product term, as illustrated in part (a). Figure 9–11(b) shows the expansion to two product terms.

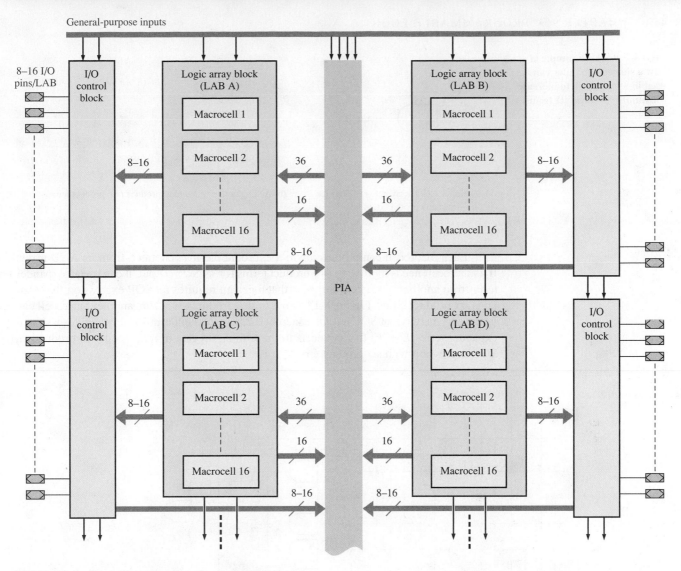

FIGURE 9–9   Basic block diagram of a typical CPLD.

FIGURE 9–10   Simplified diagram of a macrocell in a typical CPLD.

FIGURE 9–11 **Example of how a shared expander can be used in a macrocell to increase the number of product terms.**

(a) A 4-input AND array gate can produce one 4-variable product term.

(b) AND gate is expanded to produce two product terms.

Each macrocell can produce up to five product terms generated from its AND array. If a macrocell needs more than five product terms for its SOP output, it can use an expander term from another macrocell. Suppose that a design requires an SOP expression that contains six product terms. Figure 9–12 shows how a product term from another macrocell can be used to increase an SOP output. Macrocell 2, which is underutilized, generates a shared expander term $(E + F)$ that connects to the fifth AND gate in macrocell 1 to produce an SOP expression with six product terms.

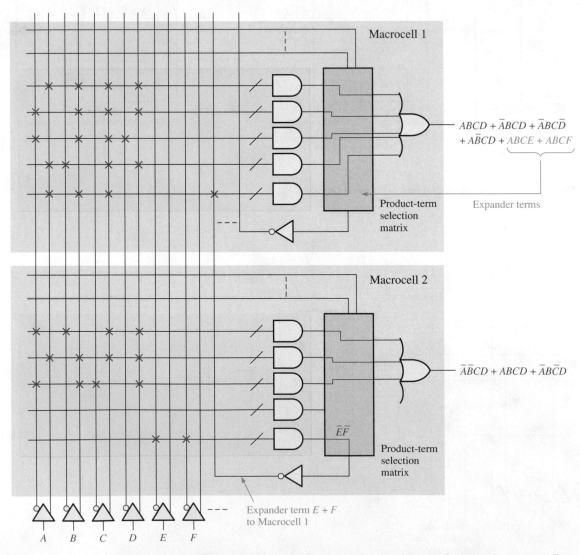

FIGURE 9–12 **Simplified illustration of using a shared expander term from another macrocell to increase an SOP expression. The red Xs and lines represent the connections produced in the hardware by the software compiler running the programmed design.**

PARALLEL EXPANDERS   Another way to increase the number of product terms for a macrocell is by using parallel expanders in which additional product terms are ORed with the terms generated by a macrocell instead of being combined in the AND array, as in the shared expander. A given macrocell can borrow unused product terms from neighboring macrocells. The basic concept is illustrated in Figure 9–13 where a simplified circuit that can produce two product terms borrows three additional product terms.

$\overline{A}BCD + ABC\overline{D} + EFG\overline{H}$ ← Parallel expander terms

$ABCD + EFGH + \overline{A}BCD + ABC\overline{D} + EFG\overline{H}$

FIGURE 9–13   **Basic concept of the parallel expander.**

Figure 9–14 shows how one macrocell can borrow parallel expander terms from another macrocell to increase the SOP output. Macrocell 2 uses three product terms from macrocell 1 to produce an eight-term SOP expression.

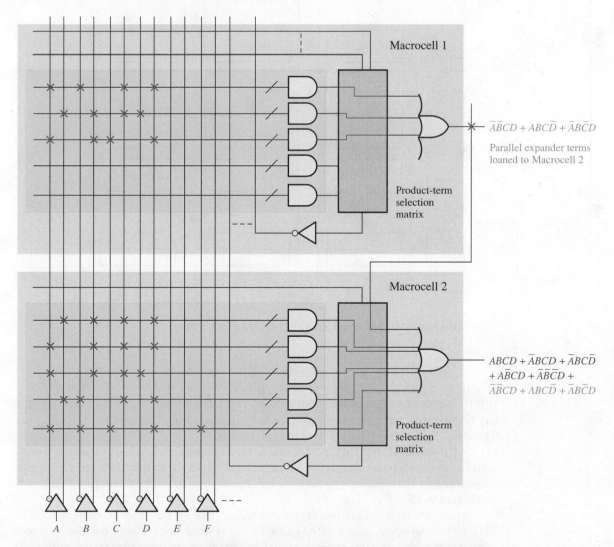

Macrocell 1

$\overline{A}\overline{B}CD + ABC\overline{D} + \overline{A}B\overline{C}D$

Parallel expander terms loaned to Macrocell 2

Product-term selection matrix

Macrocell 2

$ABCD + \overline{A}BCD + \overline{A}BC\overline{D}$
$+ A\overline{B}CD + \overline{A}\overline{B}C\overline{D} +$
$\overline{A}\overline{B}CD + ABC\overline{D} + \overline{A}B\overline{C}D$

Product-term selection matrix

A   B   C   D   E   F

FIGURE 9–14   **Simplified illustration of using parallel expander terms from another macrocell to increase an SOP expression. The red Xs and lines represent the connections produced in the hardware by the software compiler running the programmed design.**

LUT CPLD Architecture

This architecture differs from the classic CPLD previously discussed. As shown by the block diagram in Figure 9–15, this device contains logic array blocks (LABs) each with multiple **logic elements** (LEs). An LE is the basic logic design unit and is analogous to the macrocell. The programmable interconnects are arranged in a row and column arrangement running between the LABs, and input/output elements (IOEs) are oriented around the perimeter. The architecture of this type of CPLD is similar to that of FPGAs, which we discuss in Section 9–4.

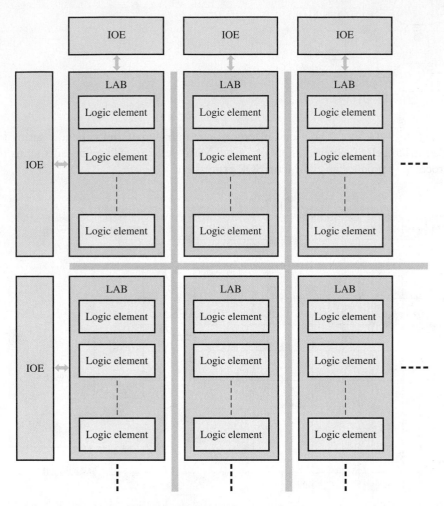

FIGURE 9–15   **Simplified block diagram of an LUT CPLD.**

A main difference between this type of CPLD and the classic AND/OR array CPLD previously discussed is the way in which a logic function is developed. Look-up tables (LUTs) are used instead of AND/OR arrays. An **LUT** is basically a type of memory that can be programmed to produce SOP functions (discussed in more detail in Section 9–4). These two approaches are contrasted in Figure 9–16.

As mentioned, the LUT CPLD has a row/column arrangement of interconnects instead of the channel-type interconnects found in most classic CPLDs. These two approaches are contrasted in Figure 9–17 and can be understood by comparing Figure 9–9 and Figure 9–15.

Most CPLDs use a nonvolatile process technology for the programmable links. The LUT CPLD, however, uses a SRAM-based process technology that is **volatile**—all programmed logic is lost when power is turned off. The memory embedded on the chip stores the program data using nonvolatile memory technology and reconfigures the CPLD on power up.

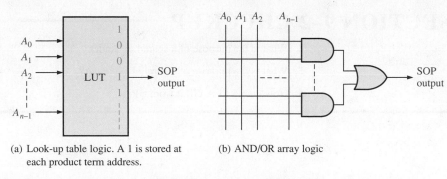

(a) Look-up table logic. A 1 is stored at each product term address.

(b) AND/OR array logic

FIGURE 9–16   **Two types of logic function generation in CPLDs.**

(a) Row/column interconnects

(b) Channel-type interconnect

FIGURE 9–17   **LUT CPLDs have row/column interconnects. Classic CPLDs have channel-type interconnects.**

PLA (Programmable Logic Array)

As you have learned, the architecture of a CPLD is the way in which the internal elements are organized and arranged. The architecture of some PLDs is based on a PLA (programmable logic array) structure rather than on a PAL (programmable array logic) structure, which we have discussed. Figure 9–18 compares a simple PAL structure with a simple PLA structure. As you know, the PAL has a programmable AND array followed by a fixed OR array and produces an SOP expression, as shown by the example in Figure 9–18(a). The **PLA** has a programmable AND array followed by a programmable OR array, as shown by the example in Figure 9–18(b).

(a) PAL-type array

$AB + A\bar{B} + \bar{A}B + \bar{A}\bar{B}$

$A \quad \bar{A} \quad B \quad \bar{B}$

$A \quad \bar{A} \quad B \quad \bar{B}$

$AB + \bar{A}\bar{B}$
$\bar{A}B + \bar{A}\bar{B}$
$A\bar{B} + \bar{A}B$
$AB + A\bar{B}$

(b) PLA-type array

FIGURE 9–18   **Comparison of a basic PLA to a basic PAL type PLD architecture.**

SECTION 9–2 CHECKUP

1. What is a CPLD?
2. What does LAB stand for?
3. Describe a LAB in a typical CPLD.

4. What is the purpose of a shared expander?
5. What is the purpose of a parallel expander?
6. How does a PLA differ from a PAL?

9–3 MACROCELL MODES

CPLD macrocells were introduced previously. A macrocell can be configured for combinational logic or registered logic outputs and inputs by programming. The term *registered* refers to the use of flip-flops. In this section, you will learn about the typical macrocell, including the combinational and the registered modes of operation. Although macrocell architecture varies among different CPLDs, a typical macrocell architecture is used for illustration.

After completing this section, you should be able to

- **Describe the operation of a typical CPLD macrocell**

Logic diagrams often use the symbol shown in Figure 9–19 to represent a multiplexer. In this case, the multiplexer has two data inputs and a select input that provides for programmable selection; the select input is usually not shown on a logic diagram.

FIGURE 9–19 Commonly used symbol for a multiplexer. It can have any number of inputs.

Data inputs $\{ D_0, D_1 \}$ Data output

Select (0 selects D_0, 1 selects D_1)

Figure 9–20 shows a complete macrocell including the flip-flop (register). The XOR gate provides for complementing the SOP function from the OR gate to produce a function in POS form. A 1 on the top input of the XOR gate complements the OR output, and a 0 lets the OR output pass uncomplemented (in SOP form). MUX 1 provides for selection of either the XOR output or an input from the I/O. MUX 2 can be programmed to select either the global clock or a clock signal based on a product term. MUX 3 can be programmed to select either a HIGH (V_{CC}) or a product-term enable for the flip-flop. MUX 4 can select the global clear or a product-term clear. MUX 5 is used to bypass the flip-flop and connect the combinational logic output to the I/O or to connect the registered output to the I/O. The flip-flop can be programmed as a D, T (toggle), or J-K flip-flop.

The Combinational Mode

When a macrocell is programmed to produce an SOP combinational logic function, the logic elements in the data path are as shown in red in Figure 9–21. As you can see, only one mux is used and the register (flip-flop) is bypassed.

FIGURE 9–20   **A CPLD macrocell.**

FIGURE 9–21   **A macrocell configured for generation of an SOP logic function. Red indicates data path.**

The Registered Mode

When a macrocell is programmed for the registered mode with the SOP combinational logic output providing the data input to the register and clocked by the global clock, the elements in the data path are as shown in red in Figure 9–22. As you can see, four multiplexers (mux) are used and the register (flip-flop) is active.

FIGURE 9–22 **A macrocell configured for generation of a registered logic function. Red indicates data path.**

SECTION 9–3 CHECKUP

1. Explain the purpose of the XOR gate in the macrocell.

2. What are the two major modes of a macrocell?

3. What does the term *registered* refer to?

4. Besides the OR gate, XOR gate, and flip-flop, what other logic element is commonly used in a macrocell?

9–4 FIELD-PROGRAMMABLE GATE ARRAYS (FPGAs)

As you have learned, the classic CPLD architecture consists of PAL/GAL or PLA-type logic blocks with programmable interconnections. Basically, the FPGA (field-programmable gate array) differs in architecture, does not use PAL/PLA type arrays, and has much greater densities than CPLDs. A typical FPGA has many times more equivalent gates than a typical CPLD. The logic-producing elements in FPGAs are generally much smaller than in CPLDs, and there are many more of them. Also, the programmable interconnections are generally organized in a row and column arrangement in FPGAs.

After completing this section, you should be able to

- Describe the basic structure of a field-programmable gate array (FPGA)
- Compare an FPGA to a CPLD
- Discuss look-up tables (LUTs)
- Discuss the SRAM-based FPGA
- Define the FPGA core

The three basic elements in an **FPGA** are the configurable logic block (CLB), the interconnections, and the input/output (I/O) blocks, as illustrated in Figure 9–23. The configurable logic blocks (CLBs) in an FPGA are not as complex as the LABs or function blocks (FBs) in a CPLD, but generally there are many more of them. When the CLBs are relatively simple, the FPGA architecture is called *fine grained*. When the CLBs are larger and more complex, the architecture is called *coarse grained*. The I/O blocks around the perimeter of the structure provide individually selectable input, output, or bidirectional access to the outside world. The distributed matrix of programmable interconnections provide for interconnection of the CLBs and connection to inputs and outputs. Large FPGAs can have tens of thousands of CLBs in addition to memory and other resources.

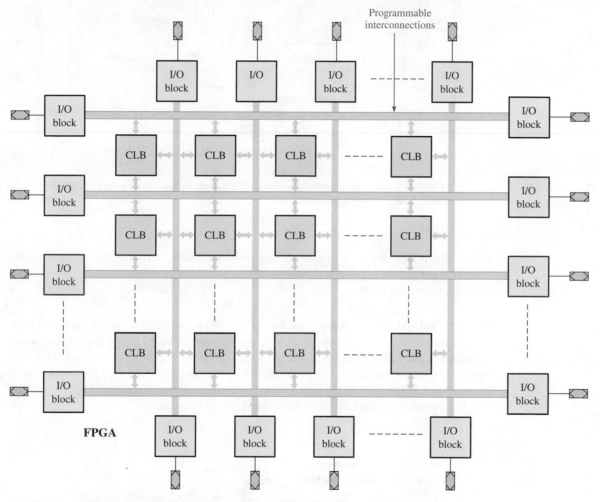

FIGURE 9–23   **Basic structure of an FPGA. CLB is configurable logic block, also known as logic array block (LAB).**

Most programmable logic manufacturers make a series of FPGAs that range in density, power consumption, supply voltage, speed, and to some degree vary in architecture. FPGAs are reprogrammable and use SRAM or antifuse process technology for the programmable links. Densities can range from hundreds of logic modules to approximately 180,000 logic modules in packages with up to over 1,000 pins. DC supply voltages are typically 1.2 V to 2.5 V, depending on the specific device.

Configurable Logic Blocks

Typically, an FPGA logic block consists of several smaller logic modules that are the basic building units, somewhat analogous to macrocells in a CPLD. Figure 9–24 shows the fundamental configurable logic blocks (CLBs) within the global row/column programmable

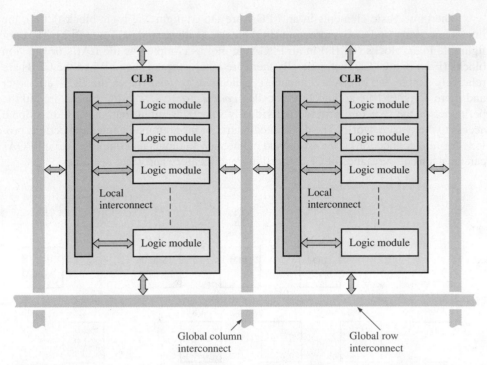

FIGURE 9–24   **Basic configurable logic blocks (CLBs) within the global row/column programmable interconnects.**

interconnects that are used to connect logic blocks. Each **CLB** is made up of multiple smaller logic modules and a local programmable interconnect that is used to connect logic modules within the CLB.

LOGIC MODULES   A logic module in an FPGA logic block can be configured for combinational logic, registered logic, or a combination of both. A flip-flop is part of the associated logic and is used for registered logic. A block diagram of a typical LUT-based logic module is shown in Figure 9–25. As you know, an LUT (look-up table) is a type of memory that is programmable and used to generate SOP combinational logic functions. The LUT essentially does the same job as the PAL or PLA does.

FIGURE 9–25   **Basic block diagram of a logic module in an FPGA.**

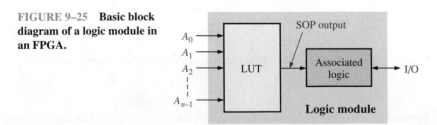

Generally, the organization of an LUT consists of a number of memory cells equal to 2^n, where n is the number of input variables. For example, three inputs can select up to eight memory cells, so an LUT with three input variables can produce an SOP expression with up to eight product terms. A pattern of 1s and 0s can be programmed into the LUT memory cells, as illustrated in Figure 9–26 for a specified SOP function. Each 1 means the associated product term appears in the SOP output, and each 0 means that the associated product term does not appear in the SOP output. The resulting SOP output expression is

$$\overline{A_2}\,\overline{A_1}\,\overline{A_0} + \overline{A_2}A_1A_0 + A_2\overline{A_1}A_0 + A_2A_1A_0$$

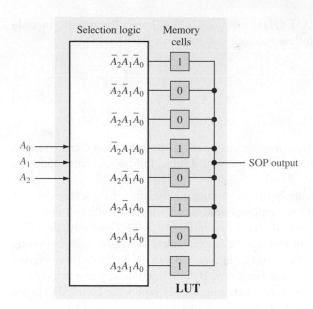

EXAMPLE 9–2

Show a basic 3-variable LUT programmed to produce the following SOP function:

$$A_2 A_1 \bar{A}_0 + A_2 \bar{A}_1 \bar{A}_0 + \bar{A}_2 A_1 A_0 + A_2 \bar{A}_1 A_0 + \bar{A}_2 \bar{A}_1 A_0$$

SOLUTION

A 1 is stored for each product term in the SOP expression, as shown in Figure 9–27.

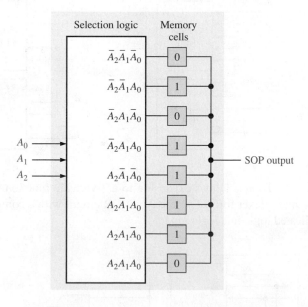

FIGURE 9–27

RELATED PROBLEM

How many memory cells would be in an LUT with four input variables? What would be the maximum possible number of product terms in the SOP output?

OPERATING MODES OF A LOGIC MODULE Typically, a logic module (LM) can be programmed for the following modes of operation:

- Normal mode
- Extended LUT mode
- Arithmetic mode
- Shared arithmetic mode

In addition to these four modes, a logic module can be utilized as a register chain to create counters and shift registers. In this section, we will discuss the normal mode and the extended LUT mode.

The *normal mode* is used primarily for generating combinational logic functions. A logic module can implement one or two combinational output functions with its two LUTs. Examples of four LUT configurations are illustrated in Figure 9–28. Generally, two SOP functions, each with four variables or less, can be implemented in an LM without sharing inputs. For example, you can have two 4-variable functions, one 4-variable function and one 3-variable function, or two 3-variable functions. By sharing inputs, you can have any combination of a total of eight inputs up to a maximum of six inputs for each LUT. In the normal mode, you are limited to 6-variable SOP functions.

FIGURE 9–28 **Examples of possible LUT configurations in a logic module (LM) in the normal mode.**

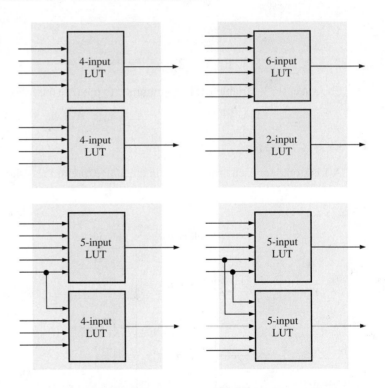

The *extended LUT mode* allows expansion to a 7-variable function, as illustrated in Figure 9–29. The multiplexer formed by the AND-OR circuit with a complemented input is part of the dedicated logic in a logic module.

FIGURE 9–29 **Expansion of a logic module (LM) to produce a 7-variable SOP function in the extended LUT mode.**

EXAMPLE 9–3

A logic module is configured in the extended LUT mode, as shown in Figure 9–30. For the specific LUT outputs shown, determine the final SOP output.

FIGURE 9–30

SOLUTION

The SOP output expression is as follows:

$$\overline{A}_5 A_4 A_3 A_2 A_1 A_0 + A_5 \overline{A}_4 A_3 A_2 A_1 A_0 + A_5 A_4 A_3 A_2 A_1 A_0 + A_6 A_5 A_4 A_3 \overline{A}_2 \overline{A}_0 + A_6 A_5 \overline{A}_4 A_3 A_2 \overline{A}_0 + A_6 A_5 A_4 A_3 A_2 \overline{A}_0$$

RELATED PROBLEM

Show an LM configured in the normal mode to produce one SOP function with five product terms from one LUT and three product terms from the other LUT.

SRAM-Based FPGAs

FPGAs are either nonvolatile because they are based on antifuse technology or they are volatile because they are based on SRAM technology. (The term *volatile* means that all the data programmed into the configurable logic blocks are lost when power is turned off.) Therefore, SRAM-based FPGAs include either a nonvolatile configuration memory embedded on the chip to store the program data and reconfigure the device each time power is turned back on or they use an external memory with data transfer controlled by a host processor. The concept of on-the-chip memory is illustrated in Figure 9–31(a). The concept of the host processor configuration is shown in part (b).

FPGA Cores

FPGAs, as we have discussed, are essentially like "blank slates" that the end user can program for any logic design. FPGAs are available that also contain hard-core logic. A **hard core** is a portion of logic in an FPGA that is put in by the manufacturer to provide a specific function and that cannot be reprogrammed. For example, if a customer needs a small microprocessor as part of a system design, it can be programmed into the FPGA by the customer or it can be provided as hard core by the manufacturer. If the embedded function has some programmable features, it is known as a **soft-core** function. An advantage of the hard-core approach is that the same design can be implemented using much less of the available capacity of the FPGA than if the user programmed it in the field, resulting in less

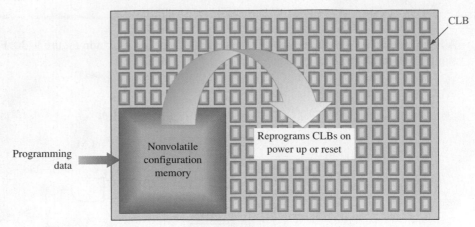

(a) Volatile FPGA with on-the-chip nonvolatile configuration memory

(b) Volatile FPGA with on-board memory and host processor

FIGURE 9–31   **Basic concepts of volatile FPGA configurations.**

space on the chip ("real estate") and less development time for the user. Also, hard-core functions have been thoroughly tested. The disadvantage of the hard core is that the specifications are fixed during manufacturing and the customer must be able to use the hard-core logic "as is." It cannot be changed later.

Hard cores are generally available for functions that are commonly used in digital systems, such as a microprocessor, standard input/output interfaces, and digital signal processors. More than one hard-core function can be programmed in an FPGA. Figure 9–32 illustrates the concept of a hard core surrounded by configurable logic programmed by the user. This is a basic embedded system because the hard-core function is embedded in the user-programmed logic.

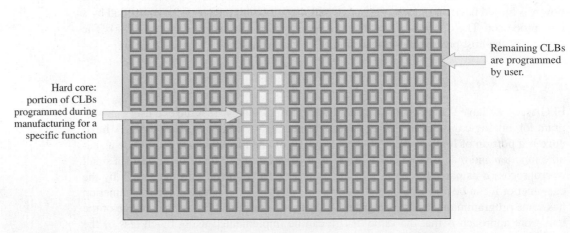

FIGURE 9–32   **Basic idea of a hard-core function embedded in an FPGA.**

Hard core designs are generally developed by and are the property of the FPGA manufacturer. Designs owned by the manufacturer are termed **intellectual property** (IP). A company usually lists the types of intellectual property that are available on its website. Some intellectual properties are a mix of hard core and soft core. A processor that has some flexibility in the selection and adjustment of certain parameters by the user is an example.

Those FPGAs containing either or both hard-core and soft-core embedded processors and other functions are known as **platform FPGAs** because they can be used to implement an entire system without the need for external support devices.

Embedded Functions

A block diagram of a typical FPGA is shown in Figure 9–33. The FPGA contains embedded memory functions as well as digital signal processing (DSP) functions. DSP functions, such as digital filters, are commonly used in many systems. As you can see in the block diagram, the embedded blocks are arranged throughout the FPGA interconnection matrix and input/output elements (IOEs) are placed around the FPGA perimeter.

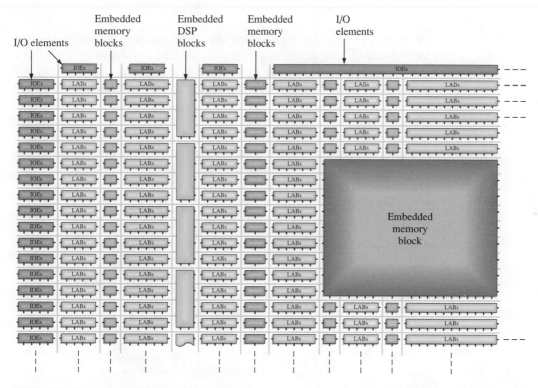

FIGURE 9–33   Example FPGA block diagram.

SECTION 9–4 CHECKUP

1. How does an FPGA differ from a CPLD?

2. What does CLB stand for?

3. Describe an LUT and discuss its purpose.

4. What is the difference between a local interconnect and a global interconnect in an FPGA?

5. What is an FPGA core?

6. Define the term *intellectual property* in relation to an FPGA manufacturer.

7. What produces combinational logic functions in an LM?

8. Name the two types of embedded functions.

9–5 PROGRAMMABLE LOGIC SOFTWARE

In order to be useful, programmable logic must have both hardware and software components combined into a functional unit. All manufacturers of SPLDs, CPLDs, and FPGAs provide software support for each hardware device. These software packages are in a category of software known as computer-aided design (CAD). PLD programming was introduced in Chapter 1 and covered further in Chapter 3. In this section, programmable logic software is presented in a generic way using the traffic signal control system for illustration. Tutorials for Altera Quartus II and Xilinx ISE are provided on the website.

After completing this section, you should be able to

- Explain the programming process in terms of design flow
- Describe the design entry phase
- Describe the functional simulation phase
- Describe the synthesis phase
- Describe the implementation phase
- Describe the timing simulation phase
- Describe the download phase

The programming process is generally referred to as **design flow**. A basic design flow diagram for implementing a logic design in a programmable device is shown in Figure 9–34. Most specific software packages incorporate these elements in one form or another and process them automatically. The device being programmed is usually referred to as the **target device**.

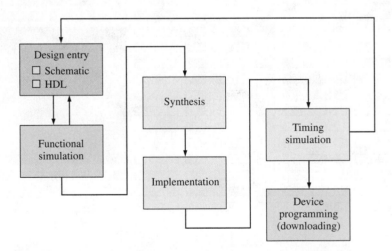

FIGURE 9–34   General design flow diagram for programming a SPLD, CPLD, or FPGA.

You must have four things to get started programming a device: a computer, development software, a programmable logic device (SPLD, CPLD, or FPGA), and a way to connect the device to the computer. These essentials are illustrated in Figure 9–35. Part (a) shows a computer that meets the system requirements for the particular software you are using. Part (b) shows the software acquired either on a CD from the device manufacturer or downloaded from the device manufacturer's website. Most manufacturers provide free software that can be downloaded and used for a limited time (Examples are Altera Quertus II and Xilinx ISE.). Part (c) shows a programmable logic device. Part (d) illustrates two means of physically connecting the device to the computer via cable by using either the programming fixture into which the device is inserted or the development board on which the device is mounted. After the software has been installed on your computer,

(a) Computer

(b) Software (CD or Website download)

(c) Device

(d) Programming hardware (programming fixture or development board with cable for
connection to computer port)

FIGURE 9–35   Essential elements for programming an SPLD, CPLD, or FPGA.   **((b) Screenshot
courtesy of Altera Corp. (d) photo courtesy of Digilent, Inc.)**

you must become familiar with the particular software tools before attempting to connect
and program a device.

Design Entry

Assume that you have a logic circuit design that you wish to implement in a programmable
device. You can enter the design on your computer in either of two basic ways: **schematic
entry** or **text entry**. In order to use text entry, you must be familiar with an HDL such as
VHDL, Verilog, or AHDL. Most programmable logic manufacturers provide software
packages that support VHDL and Verilog because they are standard HDLs. Some also
support AHDL, ABEL, or other proprietary HDLs. Schematic entry allows you to place
symbols of logic gates and other logic functions from a library on the screen and connect
them as required by your design. A knowlege of an HDL is not required for schematic
entry.

BUILDING A LOGIC DESIGN   In addition to programming languages such as
VHDL and Verilog, schematic capture can also be used in PLD development. When you
enter a complete logic circuit schematic on the screen, it is called a "flat" schematic. Complex logic circuits may be hard to fit onto the screen and difficult to read. You can enter
logic circuits in segments, save each segment as a block symbol, and then connect the
block symbols graphically to form a complex circuit, as shown in Figure 9–36 for the traffic signal control system (Chapter 6), which we will use for illustration of the process. This
is called a hierarchical approach.

The sequential logic section of the traffic signal control system is created using schematic capture and compared to the same application created using VHDL and Verilog.

FIGURE 9–36   **Block schematic diagram for the traffic signal control system.**

Figure 9–37 shows the use of VHDL and Verilog to create the sequential logic component of the system. The sequential logic portion of the traffic light application was developed in Chapter 6. The code for the expressions assigned to D0 and D1 are created straight from the Boolean expressions.

$$D_1 = G_0\overline{T}_S + G_1T_S$$
$$D_0 = \overline{G}_1\overline{T}_LV_s + \overline{G}_1G_0 + G_0T_LV_s$$

```
1    library ieee;
2    use ieee.std_logic_1164.all;
3
4    entity SequentialLogic is
5    port(VS, TL, TS, Clk: in std_logic;
6         G0, G1: inout std_logic);
7    end entity SequentialLogic;
8
9    architecture SequenceBehavior of SequentialLogic is
10   component dff is
11   port(D,Clk: in std_logic;
12        Q: out std_logic);
13   end component dff;
14
15   signal D0, D1: std_logic;
16   begin
17     D1 <= (G0 and not TS) or
18           (G1 and TS);
19
20     D0 <= (not G1 and not TL and VS)  or
21           (not G1 and G0) or
22           (G0 and TL and VS);
23
24   DFF0: dff port map(D => D0, Clk => Clk, Q => G0);
25   DFF1: dff port map(D => D1, Clk => Clk, Q => G1);
26   end architecture SequenceBehavior;
```

```
1    module SequentialLogic(VS, TL, TS, Clk, G0, G1);
2       input VS, TL, TS, Clk;
3       inout G0, G1;
4
5       wire  D0, D1;
6
7       assign D1 = (G0 && !TS) || (G1 && TS);
8       assign D0 = (!G1 && !TL && VS)|| (!G1 && G0) || (G0 && TL && VS);
9       dff DFF0(.d(D0), .Clk(Clk), .q(G0));
10      dff DFF1(.d(D1), .Clk(Clk), .q(G1));
11   endmodule
```

(a) VHDL   (b) Verilog

FIGURE 9–37   **Text entry with VHDL and Verilog descriptions of the sequential logic for the traffic signal control system.**

Figure 9–38(a) shows the sequential logic block created using schematic entry (also known as schematic capture) techniques. Breaking the schematic into separate logic circuits allows for functional compartmentalization and easier development. The Boolean expressions are implemented using separate logic gates with graphical representation of wires and I/O components needed to connect them. The completed and tested module is reduced to a simple block symbol, as shown in Figure 9–38(b), and can be inserted as a

(a)

(b)

FIGURE 9–38   The sequential logic using schematic entry.

component, as shown in Figure 9–36. A block symbol can also be created using VHDL and Verilog code.

Functional Simulation

The purpose of the **functional simulation** in the design flow is to make sure that the design you entered works as it should in terms of its logic operation, before synthesizing into a hardware design. Basically, after a logic circuit is compiled, it can then be simulated by applying input waveforms and checking the output for all possible input combinations. Functional simulation is accomplished graphically using a waveform editor or programmatically using a test bench. Graphical waveform editors allow drawing of test stimulus using waveform drawing features and drag and drop techniques.

GRAPHICAL APPROACH   Graphical generation tools allow for the easy creation of drawn stimulus waveforms for simple testing applications. Graphical waveforms are created to provide the input stimulus for the sequential logic component of the traffic signal control system as an example. Inputs Clk, TL, TS, and VS will be created using graphical tools. Output identifiers G0 and G1 require no input stimulus and are simply dragged and dropped into the Wave window. The clock definition is created using the Define Clock feature to drive the system clock Clk and limit the simulation run time. The offset, duty cycle, period, logic values, cancel, and initial edge are provided. Inputs VS, TL, and TS are created using the same graphical techniques. You can view the drawn stimulus waveforms prior to simulation. Typical windows are shown in Figure 9–39.

After you have specified the input waveforms, the simulation is ready to run. When the simulation is started, the output waveforms for G0 and G1 will be displayed as shown in Figure 9–40. This allows you to verify that the design is good or that it is

FIGURE 9–39   Functional simulation.

FIGURE 9–40   After the functional simulation is run, the output waveform should indicate that the logic is functioning properly.

working properly. In this case, the output waveform is corrected to the selected input waveforms. An incorrect output waveform would indicate a flaw in the functionality of the logic; you would have to go back, check the original design, and then re-enter a revised design.

TEST BENCH APPROACH   A programmatic approach to design simulation is to create an additional program file called a *test bench*. A test bench is similar in construction to the program code and is typically written in the same HDL as the original program. The test bench program can be as complex as the original program. In this example, a test bench program is written to provide the input stimulus for the sequential logic component of the traffic signal control system. The following test bench program is written in VHDL to simulate input waveforms for the sequential logic module.

```
library IEEE;
use IEEE.std_logic_1164.all;

entity TestSL is          } Input stimulus for the SequentialLogic unit under test (UUT) is created
end entity TestSL;        } programmatically within the program so the entity is left blank.

architecture TestSLBehavior of TestSL is

component SequentialLogic is
port(VS, TL, TS, CLK: in std_logic;
    G0, G1: inout std_logic);
end component SequentialLogic;

signal VS, TL, TS, Clk, G0, G1: std_logic;
begin
  Clk_process:process
  begin
    for iterate in 1 to 10000
        loop                      Stimulus process for
        CLK<='1';                 CLKin input. A loop
        wait for 50 ns;           structure is used to limit
        CLK<='0';                 the number clock cycles
        wait for 50 ns;           to 10000.
      end loop;
      wait;
    end process;
TLS_proccss: process
begin
      TL <='0';
      TS <= '1';
      wait for 100 us;            Stimulus process for TL
      TL <= '1';                  and TS input.
      TS <= '0';
      wait for 100 us;
    end process;
    stim_proc: process
    begin
      VS <='0';
      wait for 100 us;            Stimulus process for VSin
      VS <= '1';                  input.
      wait;
    end process;
UUT: SequentialLogic port map
(VS => VS, TL => TL, TS => TS, Clk => Clk, G0 => G0, G1 => G1);
end architecture TestSLBehavior;
```

> Creating a separate stimulus process for inputs CLK, TL, TS, and VS allows for independent control of the input identifiers.

After the test bench simulation is run, the output waveform on the waveform editor screen should indicate that the logic is functioning properly.

Synthesis

Once the design has been entered and functionally simulated to verify that its logical operation is correct, the compiler automatically goes through several phases to prepare the design to be downloaded to the target device. During this synthesis phase of the design flow, the design is optimized in terms of minimizing the number of gates, replacing logic elements with other logic elements that can perform the same function more efficiently, and eliminating any redundant logic. The final output from the synthesis phase is a netlist that describes the optimized version of the logic circuit.

To demonstrate the process of design optimization, the schematic capture version of the sequential logic section of the system is presented with redundant ORGates and NotGates, as shown in Figure 9–41(a). The AND-OR logic that was entered in the design entry phase, shown in Figure 9–41(a), could result in the optimized circuit shown in Figure 9–41(b). In this illustration, the compiler removed two 2-input OR gates and replaced them with a single 3-input OR gate. Also, one of the redundant inverters was eliminated.

(a) Original logic design

(b) Logic after synthesis

FIGURE 9–41   Example of logic optimization during synthesis. The final version is reduced by eliminating one inverter and combining two 2-input OR gates into a single 3-input OR gate.

NETLIST   A **netlist** is a connectivity list that describes components and how they are connected together. Generally, a netlist contains references to descriptions of the components or elements used. Each time a component, such as a logic gate, is used in a netlist, it is called an *instance*. Each instance has a definition that lists the connections that can be made to that kind of component and some basic properties of that component. These connection points are called *ports or pins*. Usually, each instance will have a unique name; for example, if you have two instances of AND gates, one might be "and1" and the other "and2". Aside from their names, they might otherwise be identical. Nets are the "wires" that connect things together in the circuit. Net-based netlists usually describe all the instances and their attributes, then describe each net, and specify which port they are connected to on each instance. The synthesis software generates a netlist, as shown in Figure 9–42(a). The netlist indicates the type of information that is necessary to describe a circuit. One format used for netlists is **EDIF** (Electronic Design Interchange Format). Using the netlist, the software creates a schematic representation of the net assignments, as shown in Figure 9–42(b).

```
Netlist(SequentialLogic)
Net<name>: instance<name>,<from>,<to>;
Instances: and0,and1,and2,and3,and4,or0,or1,inv0,inv1,inv2,DFF0,DFF1;
Input/outputs:l1,l2,l3,l4,O1,O2
net1:   DFF0, inport2; DFF1, inport2; l1;
net2:   and0, inport2; inv1, outport1; l2;
net3:   inv0, outport1; and4, inport2; l3;
net4:   and2, inport3, and4, inport4; l4;
net5:   and2, inport2;
net6:   DFF1, outport1; and0, inport1; inv2, output1; O0;
net7:   DFF0, outport1; and1, inport2; and3, inport2; and4, inport1; O1;
net8:   and1, inport1;
net9:   and2, inport1; and3, inport1;
net10: or0, inport1;
net11: or0, inport2;
net12: or1, inport1;
net13: or1, inport2;
net14: or1, inport3;
net15: DFF1, inport1;
net16: DFF0, inport1;
end;
```

(a) Netlist

(b) Schematic representation of netlist

FIGURE 9–42   **Synthesis produces netlist and schematic for the optimized sequential logic.**

Implementation (Software)

After the design has been synthesized, the **compiler** implements the design, which is basically a "mapping" of the design so that it will fit in the specific target device based on its architecture and pin configurations. This process is called *fitting* or *place and routing*. To accomplish the implementation phase of the design flow, the software must "know" about the specific device and have detailed pin information. Complete data on all potential target devices are generally stored in the software library.

Timing Simulation

This part of the design flow occurs after the implementation and before downloading to the target device. The **timing simulation** verifies that the circuit works at the design frequency and that there are no timing problems that will affect the overall operation.

Since a functional simulation has already been done, the circuit should work properly from a logic point of view. The development software uses information about the specific target device, such as propagation delays of the gates, to perform a timing simulation of the design. For the functional simulation, the specification of the target device was not required; but for the timing simulation, the target device must be chosen. The Waveform Editor can be used to view the result of the timing simulation just as with the functional simulation, as illustrated in Figure 9–43. If there are no problems with the timing, as shown in part (a), the design is ready to download. However, suppose that the timing simulation reveals a "glitch" due to propagation delay, as shown in Figure 9–43(b). A glitch is a very short duration spike in the waveform. In this event, you would need to carefully analyze the design for the cause, then re-enter the modified design, and repeat the design flow process. Remember, you have not committed the design to hardware at this point.

(a) Good result

(b) Timing problem

FIGURE 9–43 **Hypothetical examples of timing simulation results.**

Device Programming (Downloading)

Once the functional and timing simulations have verified that the design is working properly, you can initiate the download sequence. A **bitstream** is generated that represents the final design, and it is sent to the target device to automatically configure it. Upon completion, the design is actually in hardware and can be tested in-circuit. Figure 9–44 shows the basic concept of **downloading**.

FIGURE 9–44   Downloading a design to the target device.   (Photo courtesy of Digilent, Inc.)

SECTION 9–5 CHECKUP

1. List the phases of the design flow for programmable logic.

2. List the essential elements for programming a CPLD or FPGA.

3. What is the purpose of a netlist?

4. Which comes first in the design flow, the functional simulation or the timing simulation?

9–6 BOUNDARY SCAN LOGIC

Boundary scan is used for both the testing and the programming of the internal logic of a programmable device. The JTAG standard for boundary scan logic is specified by IEEE Std. 1149.1. Most programmable logic devices are JTAG compliant. In this section, the basic architecture of a JTAG IEEE Std. 1149.1 device is introduced and discussed in terms of the details of its boundary scan register and control logic structure.

After completing this section, you should be able to

- **Describe the required elements of a JTAG-compliant device**
- **List the mandatory JTAG inputs and outputs**
- **State the purpose of the boundary scan register**
- **State the purpose of the instruction register**
- **Explain what the bypass register is for**

IEEE Std. 1149.1 Registers

All programmable logic devices that are compliant with IEEE Std. 1149.1 require the elements shown in the simplified diagram in Figure 9–45. These are the boundary scan register, the bypass register, the instruction register, and the TAP (test access port) logic. Another register, the identification register, is optional and not shown in the figure.

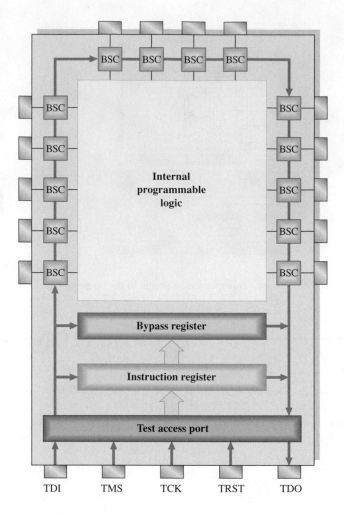

FIGURE 9–45 Greatly simplified diagram of a JTAG compliant (IEEE Std. 1149.1) programmable logic device (CPLD or FPGA). The BSCs (boundary scan cells) form the boundary scan register. Only a small number of BSCs are shown for illustration.

BOUNDARY SCAN (BS) REGISTER   The interconnected BSCs (boundary scan cells) form the boundary scan register. The serial input to the register is the TDI (test data in), and the serial output is TDO (test data out). Data from the internal logic and the input and output pins of the device can also be parallel shifted into the BS register. The BS register is used to test connections between PLDs and the internal logic that has been programmed into the device.

BYPASS (BP) REGISTER   This required data register (typically only one flip-flop) optimizes the shifting process by shortening the path between the TDI and the TDO in case the BS register or other data register is not used.

INSTRUCTION REGISTER   This required register stores instructions for the execution of various boundary scan operations.

IDENTIFICATION (ID) REGISTER   An identification register is an optional data register that is not required by IEEE Std. 1149.1. However, it is used in some boundary scan architectures to store a code that identifies the particular programmable device.

IEEE Std. 1149.1 Boundary Scan Instructions

Several standard instructions are used to control the boundary scan logic. In addition to these, other optional instructions are available.

- **BYPASS** This instruction switches the BP register into the TDI/TDO path.
- **EXTEST** This instruction switches the BS register into the TDI/TDO path and allows external pin tests and interconnection tests between the output of one programmable logic device and the input of another.

- **INTEST** This instruction switches the BS register into the TDI/TDO path and allows testing of the internal programmed logic.

- **SAMPLE/PRELOAD** This instruction is used to sample data at the device input pins and apply the data to the internal logic. Also, it is used to apply data (preload) from the internal logic to the device output pins.

- **IDCODE** This instruction switches the optional identification register into the TDI/TDO path so the ID code can be shifted out to the TDO.

IEEE Std. 1149.1 Test Access Port (TAP)

The Test Access Port (TAP) consists of control logic, four mandatory inputs and outputs, and one defined optional input, Test Reset (TRST).

- **Test Data In (TDI)** The TDI provides for serially shifting test and programming data as well as instructions into the boundary scan logic.

- **Test Data Out (TDO)** The TDO provides for serially shifting test and programming data as well as instructions out of the boundary scan logic.

- **Test Mode Select (TMS)** The TMS switches between the states of the TAP controller.

- **Test Clock (TCK)** The TCK provides timing for the TAP controller which generates control signals for the data registers and the instruction register.

A block diagram of the boundary scan logic is shown in Figure 9–46. Both instructions and data are shifted in on the TDI line. The TAP controller directs instructions into the instruction register or data into the appropriate data register. A decoded instruction from the instruction decoder selects which data register is to be accessed via MUX 1 and also if an instruction or data are to be shifted out on the TDO line via MUX 2. Also, a decoded instruction provides for setting up the boundary scan register in one of five basic modes. The boundary scan cell and its modes of operation are described next.

FIGURE 9–46 **Boundary scan logic diagram.**

The Boundary Scan Cell (BSC)

The boundary scan register is made up of boundary scan cells. A block diagram of a basic bidirectional BSC is shown in Figure 9–47. As indicated, data can be serially shifted in and out of the BSC. Also, data can be shifted into the BSC from the internal programmable logic, from a device input pin, or from the previous BSC. Additionally, data can be shifted out of the BSC to the internal programmable logic, to a device output pin, or to the next BSC.

FIGURE 9–47 **A basic bidirectional BSC.**

The architecture of a generic boundary scan cell is shown in Figure 9–48. The cell consists of two identical logic circuits, each containing two flip-flops and two multiplexers.

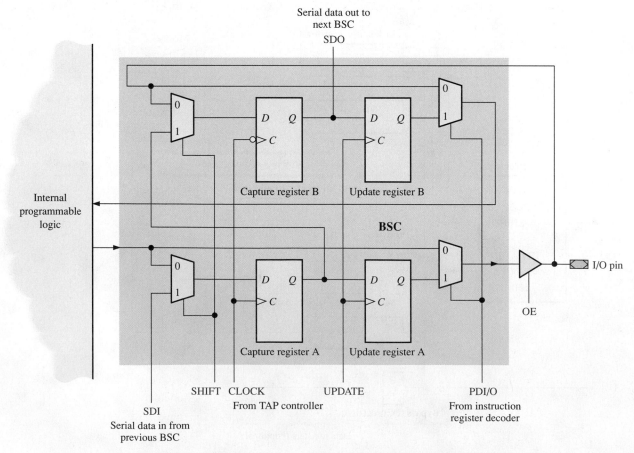

FIGURE 9–48 **Representative architecture of a typical boundary scan cell.**

Essentially, one circuit allows data to be shifted from the internal programmable logic or to a device output pin. The other circuit allows data to be shifted from a device input pin or to the internal programmable logic.

There are five modes in which the BSC can operate in terms of data flow. The first BSC mode allows data to flow serially from the previous BSC to the next BSC, as illustrated in Figure 9–49. A 1 on the SHIFT input selects the SDI. The data on the SDI line are clocked into Capture register A on the positive edge of the CLOCK. The data are then clocked into Capture register B on the negative edge of the CLOCK and appear on the SDO line. This is equivalent to serially shifting data through the boundary scan register.

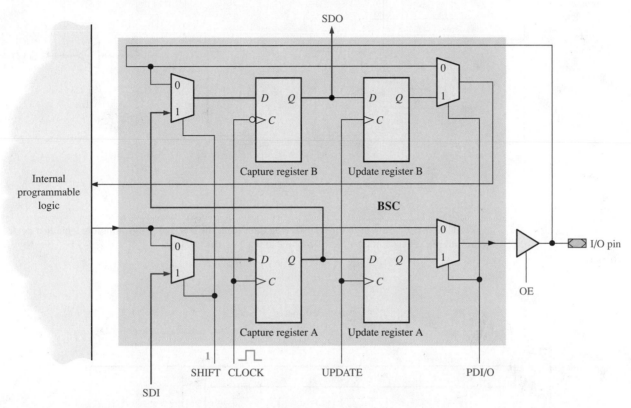

FIGURE 9–49 **Data path for serially shifting data from one BSC to the next. There is a 1 on the SHIFT input and a CLOCK pulse is applied. The red lines indicate data flow.**

The second BSC mode allows data to flow directly from the internal programmable logic to a device output pin, as illustrated in Figure 9–50. The 0 on the PDI/O (parallel data I/O) control line selects the data from the internal programmable logic. The 1 on the OE (output enable) line enables the output buffer.

The third BSC mode allows data to flow directly from a device input pin to the internal programmable logic, as illustrated in Figure 9–51. The 0 on the PDI/O (parallel data I/O) control line selects the data from the input pin. The 0 on the OE (output enable) line disables the output buffer.

The fourth BSC mode allows data to flow from the SDI to the internal programmable logic, as illustrated in Figure 9–52. A 1 on the SHIFT input selects the SDI. The data on the SDI line are clocked into Capture register A on the positive edge of the CLOCK. The data are then clocked into Capture register B on the negative edge of the CLOCK and appear on the SDO line. A pulse on the UPDATE line clocks the data into Update register B. A 1 on the PDI/O line selects the output of Update register B and applies it to the internal programmable logic. The data also appear on the SDO line.

The fifth BSC mode allows data to flow from the SDI to a device output pin and to the SDO line, as illustrated in Figure 9–53. A 1 on the SHIFT input selects the SDI. The data on the SDI line are clocked into Capture register A on the positive edge of the CLOCK.

FIGURE 9–50   Data path for transferring data from the internal programmable logic to a device output pin. There is a 0 on the PDI/O line and a 1 on the OE line.

FIGURE 9–51   Data path for transferring data from a device input pin to the internal programmable logic. There is a 0 on the PDI/O line and a 0 on the OE line.

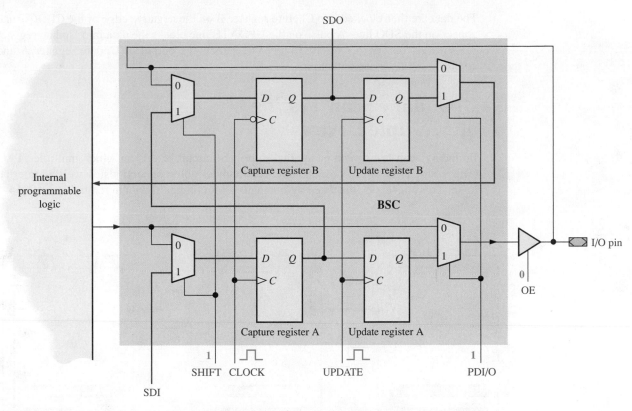

FIGURE 9–52   Data path for transferring data from the SDI to the internal programmable logic and the SDO. There is a 1 on the SHIFT line, a 1 on the PDI/O line, and a 0 on the OE line. A pulse is applied to the CLOCK line followed by a pulse on the UPDATE line.

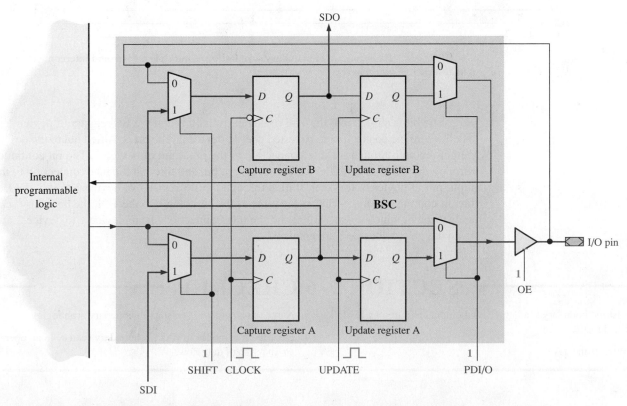

FIGURE 9–53   Data path for transferring data from the SDI to a device output pin and the SDO. There is a 1 on the SHIFT line, a 1 on the PDI/O line and a 1 on the OE line. A pulse is applied to the CLOCK line followed by a pulse on the UPDATE line.

The data are then clocked into Capture register B on the negative edge of the CLOCK and appear on the SDO line. A pulse on the UPDATE line clocks the data into Update register A. With a 1 on OE, a 1 on the PDI/O line selects the output of Update register A and applies it to the device output pin.

Boundary Scan Testing of Multiple Devices

Boundary scan testing can be applied to printed circuit boards on which multiple JTAG (IEEE Std. 1149.1) devices are mounted to check interconnections as well as internal logic. This concept is illustrated by tracing the path of data shown in red in Figure 9–54.

FIGURE 9–54 Basic concept of boundary scan testing of multiple devices and interconnections. The test path is shown in red.

The bit is shifted into the TDI of device 1 and through the BS register of device 1 to a cell where the connection to be tested goes to device 2. The bit is shifted out to the device output pin and through the interconnection to the input pin of device 2. The bit continues through the BS register of device 2 to an output pin and through the interconnection to the input pin of device 3. It is then shifted through the BS register of device 3 to the TDO. If the bit coming out of the TDO is the same as the bit going into the TDI, the boundary scan cells through which it was shifted and the interconnections from device 1 to device 2 and from device 2 to device 3 are good.

SECTION 9–6 CHECKUP

1. List the boundary scan inputs and outputs required by IEEE Std.1149.1.

2. What is the TAP?

3. Name the mandatory registers in boundary scan logic.

4. Describe five modes in which a boundary scan cell can operate in terms of data flow.

9–7 TROUBLESHOOTING

During program code development, simulation tools can be used to validate logic modules for proper operation prior to PLD programming. Two basic ways to test a device that has been programmed with a logic design are traditional and automated. In the traditional method, common laboratory test instruments can be used to check the operation. In the automated method, three fundamental approaches can be used for testing: bed-of-nails, flying probe, and boundary scan. Bed-of-nails and flying probe were described in Chapter 6, and boundary scan is introduced now. In this section, the focus is on simulation prior to device programming and boundary scan testing once the PLD has been programmed.

After completing this section, you should be able to

- Explain troubleshooting techniques using waveform simulation
- Define *break point*
- Discuss boundary scan testing

Troubleshooting with Waveform Simulation

As discussed, simulation waveform stimulus can be accomplished using a test bench program or graphically using a waveform editor. The following illustration demonstrates simulation troubleshooting techniques applied to the SequentialLogic section of the traffic signal control system created in VHDL.

FUNCTIONAL SIMULATION   Prior to download to the target device, simulation tools are useful to identify unexpected behavior. In the following illustration, the waveform output in Figure 9–55 shows that the sequential logic Gray code output from identifiers g0 and g1 does not respond to the waveform test stimulus as expected. In a timing simulation, the PLD chip libraries are loaded, and testing is conducted against a model of the target device where typically outputs start at a zero state. In the functional simulation, the basic logic is tested. Since functional simulation does not make assumptions about initial states, a circular dependency could exist where the output of one function is used to determine the outcome of a second where neither may be resolved. A break point can be inserted in the program code to determine where undetermined states may exist, so they can be addressed in the program code if needed. A **break point** is a flag placed within the program source code where the application is stopped temporarily, allowing investigation of program identifiers and the status of the I/O.

FIGURE 9–55

To investigate this behavior, you can insert a break point into the program code, so you can view the condition of identifiers G0 and G1 as the simulation progresses.

In the sequential logic component of the traffic signal control system, identifiers D0 and D1 are dependent on the output of flip-flops DFF0 and DFF1. Since D0 feeds DFF0, for example, D0 could be in an undetermined state at startup, causing G0 to also be in an undetermined state. The functional simulation would point this out as shown since G0 and G1 are left in an undetermined state. As shown in Figure 9–56, in this case, a break point is set by right-clicking line number 22 and selecting "Set Breakpoint 22". Multiple break points may be defined as needed to investigate the behavior of the program under simulation.

```
1    library ieee;
2    use ieee.std_logic_1164.all;
3
4    entity SequentialLogic is
5      port(VS, TL, IS, Clk: in std_logic;
6           G0, G1: inout std_logic);
7    end entity SequentialLogic;
8
9    architecture SequenceBehavior of SequentialLogic is
10   component dff is
11     port(D,Clk: in std_logic;
12          Q: out std_logic);
13   end component dff;
14                                          Hover over
15   signal D0 : std_logic;                 identifiers to
16   signal D1 : std_logic;                 view current
17                                          status
18   begin
19     D1 <= (G0 and not TS) or
20          (G1 and TS);
21
22     D0 <= (not G1 and not TL and VS) or
23          (not G1 and    /test1a/uut/tl    /test1a/uut/vs      /test1a/uut/g1
24                   and TL Value: 0          Value: 0            Value: U
25     /test1a/uut/d0     Time: 0 ps         Time: 0 ps          Time: 0 ps
26   DFF Value: U    map(D => D0, Clk => Clk, Q => G0);
27   DFF1 Time: 0 ps port map(D => D1, Clk => Clk, Q => G1);
28   end architecture SequenceBehavior;
```

Simulation run stopped at **break point** defined for identifier D0.

FIGURE 9–56

The simulation has stopped at the predefined break point inserted at identifier D0. By examining the condition of the supporting identifiers D0, TL, VS and G1, you determine the problem to be related to the D flip-flop components whose output value G1 is listed as "U" or undefined. D0 is dependent on identifier G1 and the flip-flop. DFF1 is in turn dependent on D1. The output of the flip-flop does not allow resolution of the Boolean expressions assigned to D0 or D1.

Examining the D flip-flop definition, you see that the flip-flop simply writes the value of the D input to output Q upon a rising clock edge. Figure 9–57(a) shows that the output

```
1    library ieee;
2    use ieee.std_logic_1164.all;
3
4    entity dff is
5      port(D,Clk: in std_logic;
6           Q: out std_logic);
7    end dff;
8
9    architecture dffbehavior of dff is
10   begin
11     process (Clk)
12     begin
13       if (Clk'event and Clk='1') then
14         Q <= D;
15       end if;
16     end process;
17   end dffbehavior;
```

(a)

```
1    library ieee;
2    use ieee.std_logic_1164.all;
3    entity dff is
4      port(D,Clk: in std_logic;
5           Q: out std_logic);
6    end dff;
7
8    architecture dffbehavior of dff is
9    signal QT :std_logic:= '0';
10   begin
11     process (Clk)
12     begin
13       if (Clk'event and Clk='1') then
14         QT <= D;
15       end if;
16     end process;
17     Q <= QT;
18   end dffbehavior;
```

(b)

Signal QT allows for the pre-initialization of a 0 to output Q

FIGURE 9–57

Q is not preinitialized, causing the output to start in an uninitialized state. To correct this problem, a new signal QT is created and initialized to 0 in Figure 9–57(b). The value of identifier D is written to signal QT upon a rising clock edge and QT is written to output Q.

Initializing the D flip-flop output to 0 allows the Boolean expressions for D0 and D1 to resolve to a value of 1 or 0. A second simulation shows that the sequential logic portion of the traffic signal control system is now able to output a valid Gray code, as shown in Figure 9–58.

FIGURE 9–58

Boundary Scan Testing

Limited access to test points led to the concept of placing the test points within the integrated circuit devices themselves. Most CPLDs and FPGAs include boundary scan logic as part of their internal structure independent of the functionality of the logic programmed into the device. These devices are JTAG compliant.

A circuit, known as a boundary scan cell, is placed between the programmable logic and each input and output pin of the device, as shown in Figure 9–59. The cells are basically memory cells that store a 1 or a 0. The cells connected to the programmable logic inputs are called input cells, and those connected to the programmable logic outputs are called output cells. **Boundary scan** testing is based on the JTAG standard

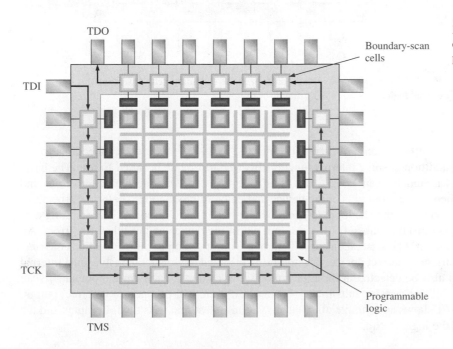

FIGURE 9–59 Basic concept of boundary scan logic in a programmable logic device.

Boundary-scan cells

Programmable logic

(IEEE Std. 1149.1). The four JTAG inputs and outputs—TDI (test data in), TDO (test data out), TCK (test clock), and TMS (test mode select)—are known as the test access port (TAP).

INTEST   When boundary scan cells are used to test the internal functionality of the device, the test mode is called Intest. The basic concept of boundary scan using Intest is as follows: A software-driven pattern of 1s and 0s is shifted in via the TDI pin and is placed on the programmable logic inputs. As a result of these applied input bits, the logic will produce output bit(s) in response. The resulting output bit(s) is (are) then shifted out on the TDO pin and checked for errors. An incorrect output, of course, indicates a fault in the programmed logic, I/O cells, or boundary scan cells.

Figure 9–60 shows a boundary scan Intest pattern 1011 for an AND-OR logic circuit that has been programmed into a device. Sixteen combinations of four TDI bits would test the circuit in all possible states according to the list in Table 9–1. The 4-bit combinations are serially shifted into the boundary scan cells, and the corresponding output is shifted out on TDO for checking. This process is controlled by boundary scan test software.

TABLE 9–1 • Boundary scan test bit pattern for the programmed device in Figure 9–60.

TDI	TDO
0000	1
0001	1
0010	0
0011	1
0100	1
0101	1
0110	1
0111	1
1000	1
1001	1
1010	0
1011	1
1100	1
1101	1
1110	1
1111	1

FIGURE 9–60   Example of a bit pattern in the boundary scan Intest for the internal logic.

EXTEST   When boundary scan cells are used to test the external connections to the device in addition to some internal functionality, the test mode is called Extest. The basic concept of boundary scan using Extest is as follows: A software-driven pattern of 1s and 0s is applied to the input pins of the device and entered into the input cells. As a result of these applied input bits, the logic will produce output bit(s) in response. The resulting output bit(s) is (are) then taken from the output pin of the device and checked for errors. An incorrect output, of course, indicates a fault in the input or output pin connections or interconnections, an incorrect device, or improperly installed device. Obviously, some internal faults can also be detected in the Extest mode. For example, faults in the boundary scan cells, I/O cells or certain faults in the programmed logic will produce an incorrect output. Figure 9–61 shows an example of a boundary scan Extest that tests the four inputs and the output of the logic circuit.

FIGURE 9–61   **Example of a bit pattern in the boundary scan Extest for external faults.**

If a fault is detected in the Extest mode, it can be either external (a bad pin connection) or internal (a faulty connection, boundary scan cell, or logic element) to the device. Therefore, in order to isolate an Extest detected fault, an Intest should be run following the Extest. If both tests show a fault, then it is internal to the device.

In the Extest mode, it is necessary to probe contacts to the input and output pins of the device. These pins have to be available at a connector to the circuit board or on test pads so they can be checked by the automatic test equipment. Pins that are not brought through a JTAG board connector can be probed at a test pad using the bed-of-nails method and/or the flying probe method. These combined approaches are illustrated in Figure 9–62.

FIGURE 9–62   A combination of boundary scan, bed-of-nails, and flying probe testing.

BOUNDARY SCAN DESCRIPTION LANGUAGE (BSDL)   This test software is part of the JTAG standard IEEE 1149.1 and uses VHDL to describe how the boundary scan logic is implemented in a specific device and how it operates. BSDL provides a standard data format for describing how IEEE 1149.1 is implemented in a JTAG-compliant device. When you use boundary scan test software tools that support BSDL, you can usually obtain BSDL from the device manufacturer.

Each device that contains dedicated boundary scan logic is supported by a BSDL file that describes that particular device. Certain things that are described in the BSDL file are the device type and port descriptions that name the I/O pins and test access port (TAP) pins and denote their nature such as input, output, or bidirectional. BSDL also provides a mapping of logical signals onto the physical pins and a description of the boundary scan logic architecture contained in the device. A bit test pattern for testing the device can be defined using BSDL.

SECTION 9–7 CHECKUP

1. Describe the purpose of a programmer-defined break point.
2. Explain the basic concept of boundary scan.
3. What are the two modes of boundary scan test?
4. Name four JTAG signals used with boundary scan.
5. What is BSDL?

SUMMARY

- A PAL is a one-time programmable (OTP) SPLD consisting of a programmable array of AND gates that connects to a fixed array of OR gates.
- The PAL structure allows any sum-of-products (SOP) logic expression with a defined number of variables to be implemented.
- The GAL is essentially a PAL that can be reprogrammed.
- In a PAL or GAL, a macrocell generally consists of one OR gate and some associated output logic.
- A CPLD is a complex programmable logic device that consists basically of multiple SPLD arrays with programmable interconnections.
- Each SPLD array in a CPLD is called a logic array block (LAB).
- A macrocell can be configured for either of two modes: the combinational mode or the registered mode.
- An FPGA (field-programmable gate array) differs in architecture, does not use PAL/PLA type arrays, and has much greater densities than typical CPLDs.
- Most FPGAs use either antifuse or SRAM-based process technology.
- Each configurable logic block (CLB) in an FPGA is made up of multiple smaller logic modules and a local programmable interconnect that is used to connect logic modules within the CLB.
- FPGAs are based on LUT architecture.
- LUT stands for *look-up table,* which is a type of memory that is programmable and used to generate SOP combinational logic functions.
- A hard core is a portion of logic embedded in an FPGA that is put in by the manufacturer to provide a specific function and which cannot be reprogrammed.
- A soft core is a portion of logic embedded in an FPGA that has some programmable features.
- Designs owned by the manufacturer are termed *intellectual property* (IP).
- The programming process is generally referred to as design flow.
- The device being programmed is usually referred to as the target device.
- In software packages for programmable logic, the operations are controlled by an application program called the compiler.
- During downloading, a bitstream is generated that represents the final design, and it is sent to the target device to automatically configure it.

- A method of internally testing a programmable device is called boundary scan, which is based on the JTAG standard (IEEE Std. 1149.1).
- The boundary scan logic in a CPLD consists of a boundary scan register, a bypass register, an instruction register, and a test access port (TAP).

KEY TERMS

Key terms and other bold terms in the chapter are defined in the end-of-book glossary.

Boundary scan   A method for internally testing a PLD based on the JTAG standard (IEEE Std. 1149.1).

Break point   A flag placed within a program source code to stop a program for investigation.

CLB   Configurable logic block; a unit of logic in an FPGA that is made up of multiple smaller logic modules and a local programmable interconnect that is used to connect logic modules within the CLB.

Compiler   An application program in development software packages that controls the operation of the software.

CPLD   A complex programmable logic device that consists basically of multiple SPLD arrays with programmable interconnections.

Design flow   The process or sequence of operations carried out to program a target device.

Downloading   The final step in a design flow in which the logic design is implemented in the target device.

FPGA   Field-programmable gate array; a programmable logic device that uses the LUT as the basic logic element and generally employs either antifuse or SRAM-based process technology.

Functional simulation   A software process that tests the logical or functional operation of a design.

GAL   A reprogrammable type of SPLD that is similar to a PAL except that it uses a reprogrammable process technology, such as EEPROM (E^2CMOS), instead of fuses.

Intellectual property (IP)   Designs owned by a manufacturer of programmable logic devices.

LAB   Logic array block; an SPLD array in a CPLD.

LUT   Look-up table; a type of memory that can be programmed to produce SOP functions.

Macrocell   Part of a PAL, GAL, or CPLD that generally consists of one OR gate and some associated output logic.

PAL   A type of one-time programmable SPLD that consists of a programmable array of AND gates that connects to a fixed array of OR gates.

Registered   A macrocell operational mode that uses a flip–flop.

Schematic entry   A method of placing a logic design into software using schematic symbols.

Target device   The programmable logic device that is being programmed.

Text entry   A method of placing a logic design into software using a hardware description language (HDL).

Timing simulation   A software process that uses information on propagation delays and netlist data to test both the logical operation and the worst-case timing of a design.

TRUE/FALSE QUIZ

Answers are at the end of the chapter.

1. A PAL consists of a programmable array of AND gates connected to a fixed array of OR gates.

2. SPLD stands for *synchronous programmable logic device*.

3. Typically, a macrocell consists of an OR gate and its associated output logic.

4. CPLD stands for *complex programmable logic device*.

5. An FPGA is a fast programmable gate array.

6. A typical FPGA has a greater gate density than a CPLD.

7. Logic array blocks are found in CPLDs.

8. The process of programming a PLD is known as design flow.

9. The device being programmed is called a target device.

10. Two types of programmable design entry are schematic and HDL.

SELF-TEST

Answers are at the end of the chapter.

1. Two types of SPLDs are
   (a) CPLD and PAL
   (b) PAL and FPGA
   (c) PAL and GAL
   (d) GAL and SRAM

2. A PAL consists of a
   (a) programmable AND array and a programmable OR array
   (b) programmable AND array and a fixed OR array
   (c) fixed AND array and a programmable OR array
   (d) fixed AND/OR array

3. A macrocell consists of a
   (a) fixed OR gate and other associated logic
   (b) programmable OR array and other associated logic
   (c) fixed AND gate and other associated logic
   (d) fixed AND/OR array with a flip-flop

4. A device that is not a type of PLD is the
   (a) CPLD   (b) GAL   (c) ROM   (d) FPGA

5. The basic AND/OR structure of SPLDs and CPLDs produces types of Boolean expressions known as
   (a) POS
   (b) SOP
   (c) product of complements
   (d) sum of complements

6. The term *LAB* stands for
   (a) logic AND block
   (b) logic array block
   (c) last asserted bit
   (d) logic assembly block

7. Two modes of macrocell operation are
   (a) input and output
   (b) registered and sequential
   (c) combinational and registered
   (d) parallel and shared

8. When a macrocell is configured to produce an SOP function, it is in the
   (a) combinational mode
   (b) parallel mode
   (c) registered mode
   (d) shared mode

9. A typical macrocell consists of
   (a) gates, multiplexers, and a flip-flop
   (b) gates and a shift register
   (c) a Gray code counter
   (d) a fixed logic array

10. Based on the complexity of its configurable logic blocks (CLBs), an FPGA can be classified as either
    (a) volatile or nonvolatile
    (b) programmable or reprogrammable
    (c) fine grained or coarse grained
    (d) platform or embedded

11. Nonvolatile FPGAs are generally based on
    (a) fuse technology
    (b) antifuse technology
    (c) EEPROM technology
    (d) SRAM technology

12. An FPGA with an embedded logic function that cannot be programmed is said to be
    (a) nonvolatile
    (b) platform
    (c) hard core
    (d) soft core

13. Hard core designs are generally developed by and are the property of the FPGA manufacturer. These designs are called
    (a) intellectual property
    (b) proprietary logic
    (c) custom designs
    (d) IEEE standards

14. For text entry of a logic design,

(a) logic symbols must be used   (b) an HDL must be used

(c) only Boolean algebra is used   (d) a special code must be used

15. In a functional simulation, the user must specify the

(a) specific target device   (b) output waveform

(c) input waveforms   (d) HDL

16. The final output of the synthesis phase of a design flow is the

(a) netlist   (b) bitstream

(c) timing simulation   (d) device pin numbers

17. EDIF stands for

(a) electronic device interchange format

(b) electrical design integrated fixture

(c) electrically destructive input function

(d) electronic design interchange format

18. The boundary scan TAP stands for

(a) test access point   (b) test array port

(c) test access port   (d) terminal access path

19. A typical boundary scan cell contains

(a) flip–flops only

(b) flip–flops and multiplexer logic

(c) latches and flip–flops

(d) latches and an encoder

20. The JTAG standard has the following inputs and outputs

(a) Intest, extest, TDI, TDO   (b) TDI, TDO, TCK, TMS

(c) ENT, CLK, SHF, CLR   (d) TCK, TMS, TMO, TLF

21. The acronym BSDL stands for

(a) board standard digital logic   (b) boundary scan down load

(c) bistable digital latch   (d) boundary scan description language

PROBLEMS

Answers to odd-numbered problems are at the end of the book.

SECTION 9–1 Simple Programmable Logic Devices (SPLDs)

1. Determine the Boolean output expression for the simple PAL array shown in Figure 9–63. The Xs represent connected links.

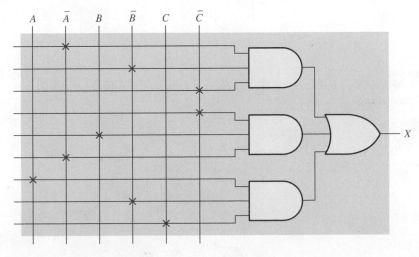

FIGURE 9–63

2. Show how the PAL-type array in Figure 9–64 should be programmed to implement each of the following SOP expressions. Use an X to indicate a connected link.
   (a) $Y = A\overline{B}C + \overline{A}B\overline{C} + ABC$
   (b) $Y = A\overline{B}C + \overline{A}\,BC + \overline{A}BC$

FIGURE 9–64

3. Modify the array in Figure 9–64 for the expression

$$Y = A\overline{B}CD + \overline{A}B\overline{C}D + A\overline{B}C\overline{D} + \overline{A}\,\overline{B}\,\overline{C}\,\overline{D}$$

4. Explain how a programmed polarity output in a PAL works.

SECTION 9–2 Complex Programmable Logic Devices (CPLDs)

5. Describe how a CPLD differs from an SPLD.

6. Refer to the block diagram in Figure 9–9 and determine the number of
   (a) inputs from the PIA to a LAB
   (b) outputs from a LAB to the PIA
   (c) inputs from an I/O control block to the PIA
   (d) outputs from a LAB to an I/O control block

7. Determine the product term for the AND gate in a CPLD array shown in Figure 9–65(a). If the AND gate is expanded, as shown in Figure 9–65(b), determine the SOP output.

FIGURE 9–65

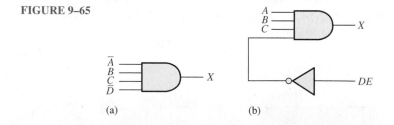

(a)   (b)

8. Determine the output of the macrocell logic in Figure 9–66 if $AB\overline{C}D + \overline{A}BCD$ is applied to the parallel expander input.

FIGURE 9–66   Parallel expander input

9. Determine the output of the array in Figure 9–67. The Xs represent connected links.

FIGURE 9–67

10. Modify the array in Figure 9–67 to produce an output $X = \overline{A}\,\overline{B}\,\overline{C} + \overline{A}\,B\,\overline{C} + ABC + A\,\overline{B}\,C$

11. Determine the output expressions for X_1 and X_2 from macrocells 1 and 2 in Figure 9–68.

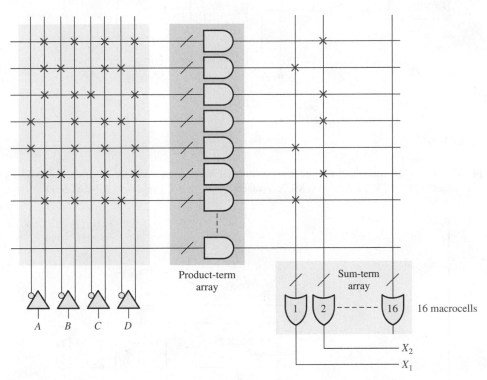

FIGURE 9–68

SECTION 9–3   Macrocell Modes

12. Determine the data output for the multiplexer in Figure 9–69 for each of the following conditions:

    (a) $D_0 = 1, D_1 = 0$, Select $= 0$ **(b)** $D_0 = 1, D_1 = 0$, Select $= 1$

FIGURE 9–69

D_0 ———

——— Data output

D_1 ———

Select

13. Determine how the macrocell in Figure 9–70 is configured (combinational or registered) and the data bit that is on the output (to I/O) for each of the following conditions. The flip-flop is a D type. Refer to Figure 9–69 for MUX data input arrangement.

    (a) XOR output = 1, flip-flop Q output = 1, from I/O input = 1, MUX 1 select = 1, MUX 2 select = 0, MUX 3 select = 0, MUX 4 select = 0, and MUX 5 select = 0.

    (b) XOR output = 0, flip-flop Q output = 0, from I/O input = 1, MUX 1 select = 1, MUX 2 select = 0, MUX 3 select = 1, MUX 4 select = 0, and MUX 5 select = 1.

FIGURE 9–70

14. For the CPLD macrocell in Figure 9–71, the following conditions are programmed: MUX 1 select = 1, MUX 2 select = 1, MUX 3 selects = 01, MUX 4 select = 0, MUX 5 select = 1,

FIGURE 9–71

MUX 6 selects $= 11$, MUX 7 selects $= 11$, MUX 8 select $= 1$, and the OR output $= 1$. The flip-flop is a D type and the MUX inputs are from D_0 at the top to D_n at the bottom.

(a) Is the macrocell configured for combinational or registered logic?
(b) Which clock is applied to the flip-flop?
(c) What is the data bit on the D input to the flip-flop?
(d) What is the output of MUX 8?

15. Repeat Problem 14 for MUX 1 select $= 0$.

SECTION 9–4 Field-Programmable Gate Arrays (FPGAs)

16. Generally, what elements make up a configurable logic block (CLB) in an FPGA? What elements make up a logic module?

17. Determine the output expression of the LUT for the internal conditions shown in Figure 9–72.

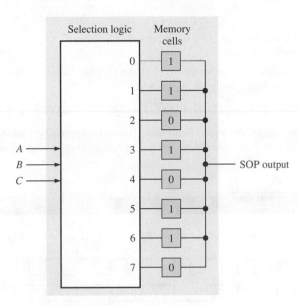

FIGURE 9–72

18. Show how to reprogram the LUT in Figure 9–72 to produce the following SOP output:

$$\overline{AB}\overline{C} + A\overline{B}\,\overline{C} + ABC$$

19. Show a logic module configured in the normal mode to produce one 4-variable SOP function and one 2-variable SOP function.

20. Determine the final SOP output function for the logic module shown in Figure 9–73.

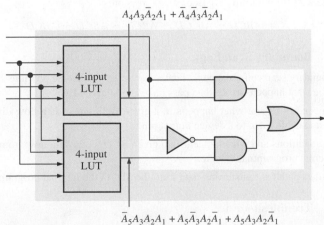

FIGURE 9–73

SECTION 9–5   **Programmable Logic Software**

21. Show the logic diagram that you would enter in the Graphic Editor for the circuit described by each of the VHDL programs.

    (a) entity AND_OR is

        port (A0, A1, A2, A3: **in bit**; X: **out bit**);

    end entity AND_OR;

    architecture LogicFunction of AND_OR is

    begin

        X <= (A0 and A1) or (A2 and not A3);

    end architecture LogicFunction;

    (b) entity LogicCircuit is

        port (A, B, C, D: **in bit**; X: **out bit**);

    end entity LogicCircuit;

    architecture Function of LogicCircuit is

    begin

        X <= (A and B) or (C and D) and

            (A and not B) and (not C and not D);

    end architecture Function;

22. Show the logic circuit that you would enter in the Graphic Editor for the following Boolean expression. Simplify before entering, if possible.

$$X = \overline{A}BCD + A\overline{B}CD + AB\overline{C}D + ABC\overline{D} + ABCD + \overline{A}\ \overline{B}\ \overline{C}\ \overline{D}$$

23. The input waveforms for the logic circuit described in Problem 22 are as shown in the Waveform Editor of Figure 9–74. Determine the output waveform that is produced after running a simulation.

FIGURE 9–74

24. Repeat Problem 23 for the following Boolean expression:

$$X = \overline{A}BC\overline{D} + A\overline{B}\ \overline{C}D + ABCD + A\overline{B}C\overline{D} + \overline{A}B\overline{C}D$$

SECTION 9–6   **Boundary Scan Logic**

25. In a given boundary scan cell, assume that data flow serially from the previous BCS to the next BSC. Describe what happens as the data pass through the given BCS.

26. Describe the conditions and what happens in a given BCS when data flow directly from the internal programmable logic to a device output pin.

27. Describe the conditions and what happens in a given BCD when data flow from a device input pin to the internal programmable logic.

28. Describe the data path for transferring data from the SDI to the internal programmable logic.

SECTION 9–7   **Troubleshooting**

29. Develop a boundary scan test bit pattern to test the logic that is programmed into the device shown in Figure 9–75 for all possible input combinations.

FIGURE 9–75

ANSWERS TO SECTION CHECKUPS

SECTION 9–1   Simple Programmable Logic Devices (SPLDs)

1. PAL: Programmable Array Logic

2. GAL: Generic Array Logic

3. A GAL is reprogrammable. A PAL is one-time programmable.

4. Basically, a macrocell consists of an OR gate and associated output logic including a flip-flop.

SECTION 9–2   Complex Programmable Logic Devices (CPLDs)

1. CPLD: Complex Programmable Logic Device

2. LAB: Logic Array Block

3. A LAB consists of 16 macrocells in a typical CPLD.

4. A shared expander is used to increase the number of product terms from a macrocell by AND-ing additional sum terms (complemented product terms) from other macrocells.

5. A parallel expander is used to increase the number of product terms from a macrocell by ORing unused product terms from other macrocells in a LAB.

6. A PLA has a programmable AND array and a programmable OR array. A PAL has a fixed OR array.

SECTION 9–3   Macrocell Modes

1. The XOR gate is used as a programmable inverter for the data. It can be programmed to invert or not invert.

2. Combinational and registered

3. Registered refers to the use of a flip-flop.

4. Multiplexer

SECTION 9–4   Field-Programmable Gate Arrays (FPGAs)

1. Generally, an FPGA is organized with a row/column interconnect structure and uses LUTs rather than AND/OR logic for generating combinational logic functions.

2. CLB: Configurable Logic Block

3. LUT: Look-Up Table. A programmable type of memory that is used to store and generate combinational logic functions.

4. A local interconnect is used to connect logic modules within a CLB. A global interconnect is used to connect a CLB with other CLBs.

5. A core is a portion of logic embedded in an FPGA to provide a specific function.

6. *Intellectual property* refers to the hard-core designs that are developed and owned by the FPGA manufacturer.

7. An LUT produces combinational logic functions in an LM.

8. Memory and DSP (digital signal processing)

SECTION 9–5   Programmable Logic Software

1. Design entry, functional simulation, synthesis, implementation, timing simulation, downloading

2. Computer running PLD development software, a programming fixture or a development board, and an interface cable

3. A netlist provides information necessary to describe a circuit.

4. The functional simulation comes before the timing simulation.

SECTION 9–6   Boundary Scan Logic

1. TDI, TMS, TCK, TDO

2. TAP: Test access port

3. Boundary scan register, bypass register, instruction register, and TAP

4. Transfer of data from SDI to SDO, transfer of data from internal programmable logic to device output pin, transfer of data from device input pin to internal programmable logic, transfer of data from SDI to internal programmable logic, and transfer of data from SDI to device output pin and to the SDO line.

SECTION 9–7   Troubleshooting

1. A break point is a user-defined location in a program where the simulation is stopped temporarily.

2. Boundary scan enables the internal testing and programming of a programmable logic device and testing of interconnections between two or more devices. It is based on the JTAG IEEE Std. 1149.1. Boundary scan uses specific logic internal to the device for testing.

3. Intest and Extest

4. TDI, TDO, TCK, TMS

5. BSDL: Boundary Scan Description Language

ANSWERS TO RELATED PROBLEMS FOR EXAMPLES

9–1 $X = \overline{B}C + \overline{A}B\overline{C} + \overline{A}\,\overline{B} + C$

9–2 Sixteen; sixteen

9–3 See Figure 9–76.

FIGURE 9–76

ANSWERS TO TRUE/FALSE QUIZ

1. T   **2.** F   **3.** T   **4.** T   **5.** F   **6.** T   **7.** F
8. T   **9.** T   **10.** T

ANSWERS TO SELF-TEST

1. (c)   **2.** (b)   **3.** (a)   **4.** (c)   **5.** (b)   **6.** (b)   **7.** (c)
8. (a)   **9.** (b)   **10.** (c)   **11.** (b)   **12.** (b)   **13.** (a)   **14.** (a)
15. (c)   **16.** (a)   **17.** (d)   **18.** (c)   **19.** (b)   **20.** (b)   **21.** (d)

CHAPTER 10

MEMORY AND STORAGE

OUTLINE

OBJECTIVES

- Explain memory hierarchy
- Define the basic memory characteristics
- Explain what a RAM is and how it works
- Explain the difference between static RAMs (SRAMs) and dynamic RAMs (DRAMs)
- Explain what a ROM is and how it works
- Describe the various types of PROMs
- Discuss the characteristics of a flash memory
- Describe the expansion of ROMs and RAMs to increase word length and word capacity
- Discuss special types of memories such as FIFO and LIFO
- Describe the basic organization of magnetic disks and magnetic tapes
- Describe the basic operation of magneto-optical disks and optical disks
- Describe basic methods for memory testing
- Develop flowcharts for memory testing

KEY TERMS

Memory hierarchy	SRAM
Memory	DRAM
Byte	Bus
Word	DDR
Cell	PROM
Address	EPROM
Capacity	Flash memory
Write	FIFO
Read	LIFO
RAM	Hard disk
ROM	

INTRODUCTION

Chapter 7 covered shift registers, which are a type of storage device; in fact, a shift register is essentially a small-scale memory. The memory devices covered in this chapter are generally used for longer-term storage of larger amounts of data than registers can provide.

Computers and other types of systems require the permanent or semipermanent storage of large amounts of binary data. Microprocessor-based systems rely on storage devices and memories for their operation because of

VISIT THE WEBSITE
Study aids for this chapter are available at
http://pearsonhighered.com/floyd

the necessity for storing programs and for retaining data during processing.

In computer terminology, *memory* usually refers to registers, cache, main (RAM and ROM), and hard disk.

Storage refers to tape, optical disk, and magnetic disk. In this chapter semiconductor memories and magnetic and optical storage media are covered.

10–1 MEMORY SYSTEM HIERARCHY

A memory system performs the data storage function in a computer. The memory system holds data temporarily during processing and also stores data and programs on a long-term basis. A computer has several types of memory, such as registers, cache, main, and hard disk. Other types of storage can also be used, such as magnetic tape, optical disk, and magnetic disk. Memory hierarchy as well as the system processor determines the processing speed of a computer.

After completing this section, you should be able to

- **Discuss several types of memory**
- **Define memory hierarchy**
- **Describe key elements in a memory hierarchy**

Three key characteristics of memory are cost, capacity, and access time. Memory cost is usually specified in cost per bit. The capacity of a memory is measured in the amount of data (bits or bytes) it can store. The access time is the time it takes to acquire a specified unit of data from the memory. The greater the capacity, the smaller the cost and the greater the access time. The smaller the access time, the greater the cost. The goal of using a memory hierarchy is to obtain the shortest possible average access time while minimizing the cost.

The speed with which data can be processed depends both on the processor speed and on the time it takes to access stored data. **Memory hierarchy** is the arrangement of various memory system elements within the computer architecture to maximize processing speed and minimize cost. Memory can be classified according to its "distance" from the processor in terms of the number of machine cycles or access time required to get data for processing. Distance is measured in time, not in physical location. Faster memory elements are considered closer to the processor compared to slower types of memory elements. Also, the cost per bit is much greater for the memory close to the processor than for the memory that is further from the processor. Figure 10–1 illustrates the arrangement of elements in a typical memory hierarchy.

A primary distinction between the memory and storage elements in Figure 10–1 is the time required for the processor to access data and programs. This access time is known as **memory latency.** The greater the latency, the further from the processor a memory or storage element is considered to be. For example, typical register latency can be up to 1 or 2 ns, cache latency can be up to about 50 ns, main memory latency can be up to about 90 ns, and hard disk latency can be up to about 20 ms. Auxiliary memory latency can range up to several seconds.

FIGURE 10–1 **Typical memory hierarchy.**

Registers

Registers are memory elements that are located within the processor. They have a very small latency as well as a low capacity (number of bits that can be stored). One goal of programming is to keep as much frequently used data in the registers as possible. The number of registers in a processor can vary from the tens to hundreds.

Caches

The next level in the hierarchy is the memory cache, which provides temporary storage. The L1 cache is located in the processor, and the L2 cache is outside of the processor. A programming goal is to keep as much of a program as possible in the cache, especially the parts of a program that are most extensively used. There can be more than two caches in a memory system.

Main Memory

Main memory generally consists of two elements: RAM (random-access memory) and ROM (read-only memory). The RAM is a working memory that temporarily stores less frequently used data and program instructions. The RAM is volatile, which means that the stored contents are lost when the power is turned off. The ROM is for permanent storage of frequently used programs and data; ROM is nonvolatile. Registers, caches, and main memory are considered primary storage.

Hard Disk

The hard disk has a very high latency and is used for mass storage of data and programs on a permanent basis. The hard disk is also used for virtual memory, space allocated for data when the primary memory fills up. In effect, virtual memory simulates primary memory with the disadvantage of high latency. Capacities range up to about 1 terabyte (TB).

$$1 \text{ TB} = 1{,}000{,}000{,}000{,}000 \text{ B} = 10^{12} \text{ B}$$

In addition to the internal hard disk, secondary storage can also include off-line storage. Off-line storage includes DVDs, CD-ROM, CD-RW, and USB flash drive. Off-line storage is removable storage.

Auxiliary Storage

Auxiliary storage, also called *tertiary storage*, includes magnetic tape libraries and optical jukeboxes. A **tape library** can store immense amounts of data (up to hundreds of petabytes). A petabyte (PB) is

$$1 \text{ PB} = 1{,}000{,}000{,}000{,}000{,}000 \text{ B} = 10^{15} \text{ B}$$

An **optical jukebox** is a robotic storage device that automatically loads and unloads optical disks. It may have as many as 2,000 slots for disks and can store hundreds of petabytes.

Relationship of Cost, Capacity, and Access Time

Figure 10–2 shows how capacity (the amount of data a memory can store) and cost per unit of storage varies as the distance from the processor, in terms of access time or latency, increases. The capacity increases and the cost decreases as access time increases.

Memory Hierarchy Performance

In a computer system, the overall processing speed is usually limited by the memory, not the processor. Programming determines how well a particular memory hierarchy is utilized. The goal is to process data at the fastest rate possible. Two key factors in establishing maximum processor performance are locality and hit rate.

If a block of data is referenced, it will tend to be either referenced again soon or a nearby data block will be referenced soon. Frequent referencing of the same data block is

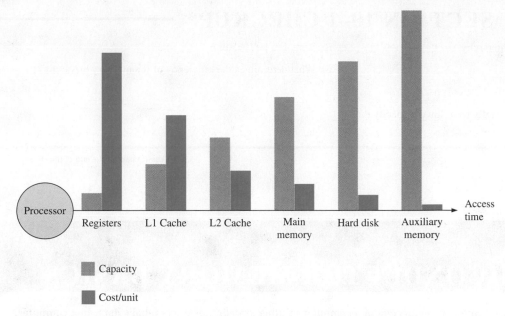

FIGURE 10–2 **Changes in memory capacity and cost per unit of data as latency (access time) increases.**

known as *temporal locality,* and the program code should be arranged so that the piece of the data in the cache is reused frequently. Referencing an adjacent data block is known as *spatial locality*, and the program code should be arranged to use consecutive pieces of data on a frequent basis.

A **miss** is a failed attempt by the processor to read or write a block of data in a given level of memory (such as the cache). A miss causes the processor to have to go to a lower level of memory (such as main memory), which has a longer latency. The three types of misses are instruction read miss, data read miss, and data write miss. A successful attempt to read or write a block of data in a given level of memory is called a *hit*. Hits and misses are illustrated in Figure 10–3, where the processor is requesting data from the cache.

FIGURE 10–3 **Illustration of a cache hit and a miss.**

The **hit rate** is the percentage of memory accesses that find the requested data in the given level of memory. The *miss rate* is the percentage of memory accesses that fail to find the requested data in the given level of memory and is equal to 1 – hit rate. The time required to access the requested information in a given level of memory is called the *hit time*. The higher the hit rate (hit to miss ratio), the more efficient the memory hierarchy is.

SECTION 10–1 CHECKUP*

1. State the purpose of memory hierarchy.

2. What is access time?

3. How does memory capacity affect the cost per bit?

4. Does higher level memory generally have lower capacity than lower level memory?

5. What is a hit? A miss?

6. What determines the efficiency of the memory hierarchy?

*Answers are at the end of the chapter.

10–2 SEMICONDUCTOR MEMORY BASICS

Memory is the portion of a computer or other system that stores binary data. In a computer, memory is accessed millions of times per second, so the requirement for speed and accuracy is paramount. Very fast semiconductor memory is available today in modules with over 1 GB (a gigabyte is one billion bytes) of capacity. These large-memory modules use exactly the same operating principles as smaller units, so we will use smaller ones for illustration in this chapter to simplify the concepts.

After completing this chapter, you should be able to

- Explain how a memory stores binary data
- Discuss the basic organization of a memory
- Describe the write operation
- Describe the read operation
- Describe the addressing operation
- Explain what RAMs and ROMs are

Units of Binary Data: Bits, Bytes, Nibbles, and Words

As a rule, memories store data in units that have from one to eight bits. The smallest unit of binary data, as you know, is the **bit.** In many applications, data are handled in an 8-bit unit called a **byte** or in multiples of 8-bit units. The byte can be split into two 4-bit units that are called **nibbles.** Bytes can also be grouped into words. The term **word** can have two meanings in computer terminology. In memories, it is defined as a group of bits or bytes that acts as a single entity that can be stored in one memory location. In assembly language, a word is specifically defined as two bytes.

The general definition of *word* is a complete unit of information consisting of a unit of binary data. When applied to computer instructions, a word is more specifically defined as two bytes (16 bits). As a very important part of assembly language used in computers, the DW (Define Word) directive means to define data in 16-bit units. This definition is independent of the particular microprocessor or the size of its data bus. Assembly language also allows definitions of bytes (8 bits) with the DB directive, double words (32 bits) with the DD directive, and quad-words (64 bits) with the QD directive.

SYSTEM NOTE

The Basic Memory Array

Each storage element in a memory can retain either a 1 or a 0 and is called a **cell.** Memories are made up of arrays of cells, as illustrated in Figure 10–4 using 64 cells as an example. Each block in the **memory array** represents one storage cell, and its location can be identified by specifying a row and a column.

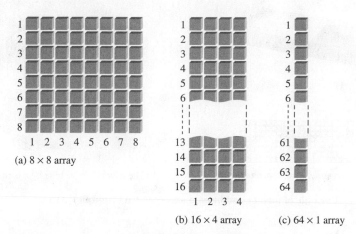

(a) 8 × 8 array

(b) 16 × 4 array

(c) 64 × 1 array

FIGURE 10–4   A 64-cell memory array organized in three different ways.

The 64-cell array can be organized in several ways based on units of data. Figure 10–4(a) shows an 8 × 8 array, which can be viewed as either a 64-bit memory or an 8-byte memory. Part (b) shows a 16 × 4 array, which is a 16-nibble memory, and part (c) shows a 64 × 1 array, which is a 64-bit memory. A memory is identified by the number of words it can store times the word size. For example, a 16k × 8 memory can store 16,384 words of eight bits each. The inconsistency here is common in memory terminology. The actual number of words is always a power of 2, which, in this case, is $2^{14} = 16,384$. However, it is common practice to state the number to the nearest thousand, in this case, 16k.

Memory Address and Capacity

A representation of a small 8 × 8 memory chip is shown in Figure 10–5(a). The location of a unit of data in a memory array is called its **address.** For example, in Figure 10–5(b), the address of a bit in the 2-dimensional array is specified by the row and column as shown. In Figure 10–5(c), the address of a byte is specified only by the row. So, as you can see, the address depends on how the memory is organized into units of data. Laptop computers have random-access memories organized in bytes. This means that the smallest group of bits that can be addressed is eight.

(a) Physical structure of 64-bit memory.

(b) The address of the blue bit is row 5, column 4.

(c) The address of the blue byte is row 3.

FIGURE 10–5   Examples of memory address in a 2-dimensional memory array.

Figure 10–6 illustrates the expansion of the 8 × 8 (64-bit) array to a 64-byte memory. The address of a byte in the array is specified by the row and column, as shown. In this case, the smallest group of bits that can be accessed is eight (1 byte). This can be viewed as a 3-dimensional array as shown in part (b).

(a) The 8 × 8 bit array expanded to a 64 × 8 bit array. This array forms a memory module.

(b) The address of the blue byte is row 5, column 8.

FIGURE 10–6   **Example of memory address in an expanded (multiple) array.**

The **capacity** of a memory is the total number of data units that can be stored. For example, in the bit-organized memory array in Figure 10–5(b), the capacity is 64 bits. In the byte-organized memory array in Figure 10–5(c), the capacity is 8 bytes, which is also 64 bits. In Figure 10–6(b), the capacity is 64 bytes. Computer memories typically have 256 MB (megabyte is one million bytes) or more of internal memory. Computers usually transfer and store data as 64-bit words, in which case all eight bits of row five in each chip in Figure 10–6(a) would be accessed.

MEMORY BANKS AND RANKS   A **bank** is a section of memory within a single memory array (chip). A memory chip can have one or more banks. Memory banks can be used for storing frequently used information. When the section of memory is identified, the data can be accessed more quickly. A **rank** is a group of chips that make up a memory module that stores data in units such as words or bytes. These terms are illustrated in Figure 10–7.

FIGURE 10–7   **Simple illustration of memory bank and memory rank.**

Basic Memory Operations

Addressing is the process of accessing a specified location in memory. Since a memory stores binary data, data must be put into the memory and data must be copied from the memory when needed. The **write** operation puts data into a specified address in the memory, and the **read** operation copies data out of a specified address in the memory. The addressing operation, which is part of both the write and the read operations, selects the specified memory address.

Data units go into the memory during a write operation and come out of the memory during a read operation on a set of lines called the *data bus*. As indicated in Figure 10–8, the data bus is bidirectional, which means that data can go in either direction (into the memory or out of the memory). In this case of byte-organized memories, the data bus has at least eight lines so that all eight bits in a selected address are transferred in parallel. For a write or a read operation, an address is selected by placing a binary code representing the desired address on a set of lines called the *address bus*. The address code is decoded internally, and the appropriate address is selected. In the case of the expanded (multiple) array memory in Figure 10–8(b) there are two decoders, one for the rows and one for the columns. The number of lines in the address bus depends on the capacity of the memory. For example, a 15-bit address code can select 32,768 locations (2^{15}) in the memory, a 16-bit address code can select 65,536 locations (2^{16}) in the memory, and so on. In personal computers a 32-bit address bus can select 4,294,967,296 locations (2^{32}), expressed as 4G.

(a) Single array memory

(b) Multiple array memory

FIGURE 10–8   **Block diagram of a single-array memory and a multiple-array memory showing address bus, address decoder(s), bidirectional data bus, and read/write inputs.**

THE WRITE OPERATION   A simplified write operation is illustrated in Figure 10–9. To store a byte of data in the memory, a code held in the address register is placed on the address bus. Once the address code is on the bus, the address decoder decodes the address and selects the specified location in the memory. The memory then gets a write command, and the data byte held in the data register is placed on the data bus and stored in the selected memory address, thus completing the write operation. When a new data byte is written into a memory address, the current data byte stored at that address is overwritten (replaced with a new data byte).

1. Address code 101 is placed on the address bus and address 5 is selected.
2. Data byte is placed on the data bus.
3. Write command causes the data byte to be stored in address 5, replacing previous data.

FIGURE 10–9 **Illustration of the write operation.**

THE READ OPERATION A simplified read operation is illustrated in Figure 10–10. Again, a code held in the address register is placed on the address bus. Once the address code is on the bus, the address decoder decodes the address and selects the specified location in the memory. The memory then gets a read command, and a "copy" of the data byte that is stored in the selected memory address is placed on the data bus and loaded into the data register, thus completing the read operation. When a data byte is read from a memory address, it also remains stored at that address. This is called *nondestructive read.*

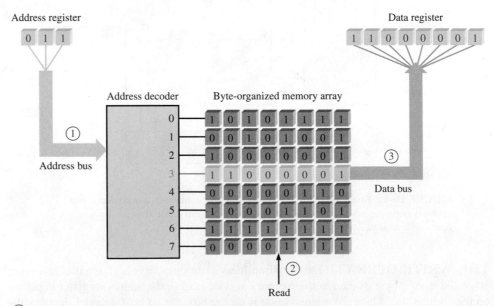

1. Address code 011 is placed on the address bus and address 3 is selected.
2. Read command is applied.
3. The contents of address 3 is placed on the data bus and shifted into data register. The contents of address 3 is not erased by the read operation.

FIGURE 10–10 **Illustration of the read operation.**

RAMs and ROMs

The two major categories of semiconductor memories are the RAM and the ROM. **RAM** (random-access memory) is a type of memory in which all addresses are accessible in an equal amount of time and can be selected in any order for a read or write operation. All RAMs have both *read* and *write* capability. Because RAMs lose stored data when the power is turned off, they are **volatile** memories.

   ROM (read-only memory) is a type of memory in which data are stored permanently or semipermanently. Data can be read from a ROM, but there is no write operation as in the RAM. The ROM, like the RAM, is a random-access memory but the term *RAM* traditionally means a random-access *read/write* memory. Several types of RAMs and ROMs will be covered in this chapter. Because ROMs retain stored data even if power is turned off, they are **nonvolatile** memories.

SECTION 10–2 CHECKUP

1. What is the smallest unit of data that can be stored in a memory?
2. What is the bit capacity of a memory that can store 256 bytes of data?
3. What is a write operation?
4. What is a read operation?
5. How is a given unit of data located in a memory?
6. Describe the difference between a RAM and a ROM.

10–3 THE RANDOM-ACCESS MEMORY (RAM)

A RAM is a read/write memory in which data can be written into or read from any selected address in any sequence. When a data unit is written into a given address in the RAM, the data unit previously stored at that address is replaced by the new data unit. When a data unit is read from a given address in the RAM, the data unit remains stored and is not erased by the read operation. This nondestructive read operation can be viewed as copying the content of an address while leaving the content intact. A RAM is typically used for short-term data storage because it cannot retain stored data when power is turned off.

After completing this section, you should be able to

- **Name the two categories of RAM**
- **Explain what a SRAM is**
- **Describe the SRAM storage cell**
- **Explain the difference between an asynchronous SRAM and a synchronous burst SRAM**
- **Explain the purpose of a cache memory**
- **Explain what a DRAM is**
- **Describe the DRAM storage cells**
- **Discuss the types of DRAM**
- **Compare the SRAM with the DRAM**

The RAM Family

The two major categories of RAM are the *static RAM* (SRAM) and the *dynamic RAM* (DRAM). **SRAMs** generally use latches as storage elements and can therefore store data indefinitely *as long as dc power is applied*. **DRAMs** use capacitors as storage elements and cannot retain data very long without the capacitors being recharged by a process called

refreshing. Both SRAMs and DRAMs will lose stored data when dc power is removed and, therefore, are classified as volatile memories.

Data can be read much faster from SRAMs than from DRAMs. However, DRAMs can store much more data than SRAMs for a given physical size and cost because the DRAM cell is much simpler, and more cells can be crammed into a given chip area than in the SRAM.

The basic types of SRAM are the *asynchronous SRAM* and the *synchronous SRAM* with a burst feature. The basic types of DRAM are the *Fast Page Mode DRAM* (FPM DRAM), the *Extended Data Out DRAM* (EDO DRAM), the *Burst EDO DRAM* (BEDO DRAM), and the *synchronous DRAM* (SDRAM). These are shown in Figure 10–11.

FIGURE 10–11   The RAM family.

Static RAMs (SRAMs)

MEMORY CELL   All SRAMs are characterized by latch memory cells. As long as dc power is applied to a **static memory** cell, it can retain a 1 or 0 state indefinitely. If power is removed, the stored data bit is lost.

Figure 10–12 shows a basic SRAM gated D latch memory cell. The cell is selected by an active level on the Select line and a data bit (1 or 0) is written into the cell by placing it on the Data in line. A data bit is read by taking it off the Data out line.

BASIC STATIC MEMORY CELL ARRAY   The memory cells in a SRAM are organized in rows and columns, as illustrated in Figure 10–13 for the case of an $n \times 4$ array. All the cells in a row share the same Row Select line. Each set of Data in and Data out lines go to each cell in a given column and are connected to a single data line that serves as both an input and output (Data I/O) through the data input and data output buffers.

To write a data unit, for example a nibble, into a given row of cells in the memory array, the Row Select line is taken to its active state and four data bits are placed on the Data I/O lines. The Write line is then taken to its active state, which causes each data bit to be stored in a selected cell in the associated column. To read a data unit, the Read line is taken to its active state, which causes the four data bits stored in the selected row to appear on the Data I/O lines.

FIGURE 10–12   A typical SRAM gated D latch memory cell.

Select

Data in   Data out

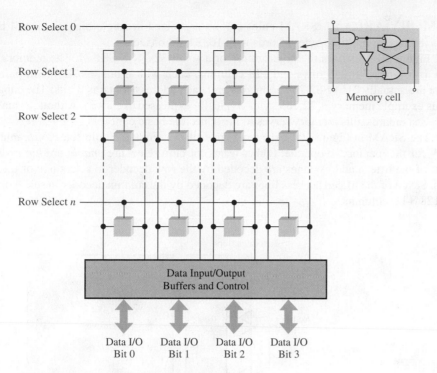

Memory cell

FIGURE 10–13   Basic SRAM array.

Basic Asynchronous SRAM Organization

An asynchronous SRAM is one in which the operation is not synchronized with a system clock. To illustrate the general organization of a SRAM, a 32k × 8 bit memory is used. A logic symbol for this memory is shown in Figure 10–14.

In the READ mode, the eight data bits that are stored in a selected address appear on the data output lines. In the WRITE mode, the eight data bits that are applied to the data input lines are stored at a selected address. The data input and data output lines (I/O_0 through I/O_7) share the same lines. During READ, they act as output lines (O_0 through O_7) and during WRITE they act as input lines (I_0 through I_7).

TRISTATE OUTPUTS AND BUSES   Tristate buffers in a memory allow the data lines to act as either input or output lines and connect the memory to the data bus in a computer. These buffers have three output states: HIGH (1), LOW (0), and HIGH-Z (open). Tristate outputs are indicated on logic symbols by a small inverted triangle (∇), as shown in Figure 10–14, and are used for compatibility with bus structures such as those found in microprocessor-based systems.

Physically, a **bus** is one or more conductive paths that serve to interconnect two or more functional components of a system or several diverse systems. Electrically, a bus is a collection of specified voltage levels and/or current levels and signals that allow various devices to communicate and work properly together.

FIGURE 10–14   Logic diagram for an asynchronous 32k × 8 SRAM.

A microprocessor is connected to memories and input/output devices by certain bus structures. An address bus allows the microprocessor to address the memories, and a data bus provides for transfer of data between the microprocessor, the memories, and the input/output devices such as monitors, printers, keyboards, and modems. A control bus allows the microprocessor to control data transfers and timing for the various components.

MEMORY ARRAY SRAM chips can be organized in single bits, nibbles (4 bits), bytes (8 bits), or multiple bytes (words with 16, 24, 32 bits, etc.).

Figure 10–15 shows the organization of a small 32k × 8 SRAM. The memory cell array is arranged in 256 rows and 128 columns, each with 8 bits, as shown in part (a). There are actually $2^{15} = 32{,}768$ addresses and each address contains 8 bits. The capacity of this example memory is 32,768 bytes (typically expressed as 32 kB). Although small by today's standards, this memory serves to introduce the basic concepts.

The SRAM in Figure 10–15(b) works as follows. First, the chip select, $\overline{CS}$, must be LOW for the memory to operate. (Other terms for chip select are *enable* or *chip enable*.) Eight of the fifteen address lines are decoded by the row decoder to select one of the 256 rows. Seven of the fifteen address lines are decoded by the column decoder to select one of the 128 8-bit columns.

(a) Memory array configuration

(b) Memory block diagram

FIGURE 10–15 **Basic organization of an asynchronous 32k × 8 SRAM.**

READ In the READ mode, the write enable input, $\overline{WE}$, is HIGH and the output enable, $\overline{OE}$, is LOW. The input tristate buffers are disabled by gate G_1, and the column output tristate buffers are enabled by gate G_2. Therefore, the eight data bits from the selected address are routed through the column I/O to the data lines (I/O_0 though I/O_7), which are acting as data output lines.

WRITE In the WRITE mode, $\overline{WE}$ is LOW and $\overline{OE}$ is HIGH. The input buffers are enabled by gate G_1, and the output buffers are disabled by gate G_2. Therefore, the eight input data bits on the data lines are routed through the input data control and the column I/O to the selected address and stored.

READ AND WRITE CYCLES Figure 10–16 shows typical timing diagrams for a memory read cycle and a write cycle. For the read cycle shown in part (a), a valid address code is applied to the address lines for a specified time interval called the *read cycle time,* t_{RC}. Next, the chip select ($\overline{CS}$) and the output enable ($\overline{OE}$) inputs go LOW. One time interval after the $\overline{OE}$ input goes LOW, a valid data byte from the selected address appears on the data lines. This time interval is called the *output enable access time,* t_{GQ}. Two other access times for the read cycle are the *address access time,* t_{AQ}, measured from the beginning of a valid address to the appearance of valid data on the data lines and the *chip enable*

(a) Read cycle ($\overline{WE}$ HIGH)

(b) Write cycle ($\overline{WE}$ LOW)

FIGURE 10–16   **Basic read and write cycle timing for the SRAM in Figure 10–15.**

access time, t_{EQ}, measured from the HIGH-to-LOW transition of $\overline{CS}$ to the appearance of valid data on the data lines.

During each read cycle, one unit of data, a byte in this case, is read from the memory.

For the write cycle shown in Figure 10–16(b), a valid address code is applied to the address lines for a specified time interval called the *write cycle time, t_{WC}.* Next, the chip select ($\overline{CS}$) and the write enable ($\overline{WE}$) inputs go LOW. The required time interval from the beginning of a valid address until the $\overline{WE}$ input goes LOW is called the *address setup time,* $t_{s(A)}$. The time that the $\overline{WE}$ input must be LOW is the write pulse width. The time that the input $\overline{WE}$ must remain LOW after valid data are applied to the data inputs is designated t_{WD}; the time that the valid input data must remain on the data lines after the $\overline{WE}$ input goes HIGH is the *data hold time, $t_{h(D)}$.*

During each write cycle, one unit of data is written into the memory.

Basic Synchronous SRAM with Burst Feature

Unlike the asynchronous SRAM, a synchronous SRAM is synchronized with the system clock. For example, in a computer system, the synchronous SRAM operates with the same clock signal that operates the microprocessor so that the microprocessor and memory are synchronized for faster operation.

The fundamental concept of the synchronous feature of a SRAM can be shown with Figure 10–17, which is a simplified block diagram of a 32k × 8 memory for purposes of illustration. The synchronous SRAM is similar to the asynchronous SRAM in terms of the memory array, address decoder, and read/write and enable inputs. The basic difference is that the synchronous SRAM uses clocked registers to synchronize all inputs with the system clock. The address, the read/write input, the chip enable, and the input data are all

FIGURE 10–17   A basic block diagram of a synchronous SRAM with burst feature.

latched into their respective registers on an active clock pulse edge. Once this information is latched, the memory operation is in sync with the clock.

For the purpose of simplification, a notation for multiple parallel lines or bus lines is introduced in Figure 10–17, as an alternative to drawing each line separately. A set of parallel lines can be indicated by a single heavy line with a slash and the number of separate lines in the set. For example, the following notation represents a set of 8 parallel lines:

The address bits A_0 through A_{14} are latched into the Address register on the positive edge of a clock pulse. On the same clock pulse, the state of the write enable ($\overline{WE}$) line and chip select ($\overline{CS}$) are latched into the Write register and the Enable register respectively. These are one-bit registers or simply flip-flops. Also, on the same clock pulse the input data are latched into the Data input register for a Write operation, and data in a selected memory address are latched into the Data output register for a Read operation, as determined by the Data I/O control based on inputs from the Write register, Enable register, and the Output enable ($\overline{OE}$).

Two basic types of synchronous SRAM are the *flow-through* and the *pipelined*. The flow-through synchronous SRAM does not have a Data output register, so the output data flow asynchronously to the data I/O lines through the output buffers. The **pipelined** synchronous SRAM has a Data output register, as shown in Figure 10–17, so the output data are synchronously placed on the data I/O lines.

THE BURST FEATURE   As shown in Figure 10–17, synchronous SRAMs normally have an address burst feature, which allows the memory to read or write up to four sequential locations using a single address. When an external address is latched in the address register, the two lowest-order address bits, A_0 and A_1, are applied to the burst logic. This produces a sequence of four internal addresses by adding 00, 01, 10, and 11 to the two lowest-order address bits on successive clock pulses. The sequence always begins with the base address, which is the external address held in the address register.

The address burst logic in a typical synchronous SRAM consists of a binary counter and exclusive-OR gates, as shown in Figure 10–18. For 2-bit burst logic, the internal burst address sequence is formed by the base address bits $A_2 - A_{14}$ plus the two burst address bits A_1' and A_0'.

To begin the burst sequence, the counter is in its 00 state and the two lowest-order address bits are applied to the inputs of the XOR gates. Assuming that A_0 and A_1 are both 0, the internal address sequence in terms of its two lowest-order bits is 00, 01, 10, and 11.

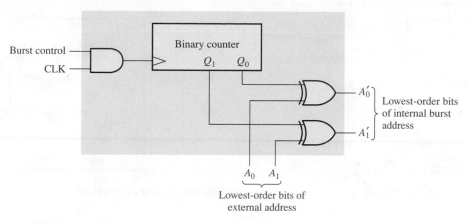

FIGURE 10–18 **Address burst logic.**

Cache Memory

One of the major applications of SRAMs is in cache memories in computers. **Cache memory** is a relatively small, high-speed memory that stores the most recently used instructions or data from the larger but slower main memory. Cache memory can also use dynamic RAM (DRAM), which is covered next. Typically, SRAM is several times faster than DRAM. Overall, a cache memory gets stored information to the microprocessor much faster than if only high-capacity DRAM is used. Cache memory is basically a cost-effective method of improving system performance without having to resort to the expense of making all of the memory faster.

The concept of cache memory is based on the idea that computer programs tend to get instructions or data from one area of main memory before moving to another area. Basically, the cache controller "guesses" which area of the slow dynamic memory the CPU (central-processing unit) will need next and moves it to the cache memory so that it is ready when needed. If the cache controller guesses right, the data are immediately available to the microprocessor. If the cache controller guesses wrong, the CPU must go to the main memory and wait much longer for the correct instructions or data. Fortunately, the cache controller is right most of the time.

CACHE ANALOGY There are many analogies that can be used to describe a cache memory, but comparing it to a home refrigerator is perhaps the most effective. A home refrigerator can be thought of as a "cache" for certain food items while the supermarket is the main memory where all foods are kept. Each time you want something to eat or drink, you can go to the refrigerator (cache) first to see if the item you want is there. If it is, you save a lot of time. If it is not there, then you have to spend extra time to get it from the supermarket (main memory).

L1 AND L2 CACHES A first-level cache (L1 cache) is usually integrated into the processor chip and has a very limited storage capacity. L1 cache is also known as *primary cache*. A second-level cache (L2 cache) is a separate memory chip or set of chips external to the processor and usually has a larger storage capacity than an L1 cache. L2 cache is also known as *secondary cache*. Some systems may have higher-level caches (L3, L4, etc.), but L1 and L2 are the most common. Also, some systems use a disk cache to enhance the performance of the hard disk because DRAM, although much slower than SRAM, is much faster than the hard disk drive. Figure 10–19 illustrates L1 and L2 cache memories in a computer system.

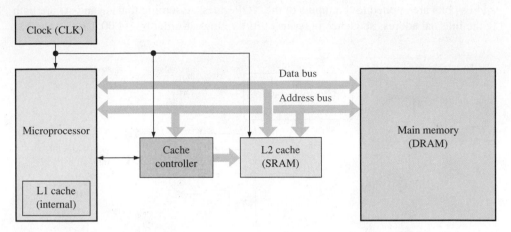

FIGURE 10–19 Block diagram showing L1 and L2 cache memories in a computer system.

Dynamic RAM (DRAM) Memory Cells

Dynamic memory cells store a data bit in a small capacitor rather than in a latch. The advantage of this type of cell is that it is very simple, thus allowing very large memory arrays to be constructed on a chip at a lower cost per bit. The disadvantage is that the storage capacitor cannot hold its charge over an extended period of time and will lose the stored data bit unless its charge is refreshed periodically. To refresh requires additional memory circuitry and complicates the operation of the DRAM. Figure 10–20 shows a typical DRAM cell consisting of a single MOS transistor (MOSFET) and a capacitor.

FIGURE 10–20 A MOS DRAM cell.

In this type of cell, the transistor acts as a switch. The basic simplified operation is illustrated in Figure 10–21 and is as follows. A LOW on the $R/\overline{W}$ line (WRITE mode) enables the tristate input buffer and disables the output buffer. For a 1 to be written into the cell, the D_{IN} line must be HIGH, and the transistor must be turned on by a HIGH on the row line. The transistor acts as a closed switch connecting the capacitor to the bit line. This connection allows the capacitor to charge to a positive voltage, as shown in Figure 10–21(a). When a 0 is to be stored, a LOW is applied to the D_{IN} line. If the capacitor is storing a 0, it remains uncharged, or if it is storing a 1, it discharges as indicated in Figure 10–21(b). When the row line is taken back LOW, the transistor turns off and disconnects the capacitor from the bit line, thus "trapping" the charge (1 or 0) on the capacitor.

To read from the cell, the $R/\overline{W}$ (Read/$\overline{Write}$) line is HIGH, enabling the output buffer and disabling the input buffer. When the row line is taken HIGH, the transistor turns on and connects the capacitor to the bit line and thus to the output buffer (sense amplifier), so the data bit appears on the data-output line (D_{OUT}). This process is illustrated in Figure 10–21(c).

For refreshing the memory cell, the $R/\overline{W}$ line is HIGH, the row line is HIGH, and the refresh line is HIGH. The transistor turns on, connecting the capacitor to the bit line. The output buffer is enabled, and the stored data bit is applied to the input of the refresh buffer, which is enabled by the HIGH on the refresh input. This produces a voltage on the bit line corresponding to the stored bit, thus replenishing the capacitor. This is illustrated in Figure 10–21(d).

(a) Writing a 1 into the memory cell

(b) Writing a 0 into the memory cell

(c) Reading a 1 from the memory cell

(d) Refreshing a stored 1

FIGURE 10–21   Basic operation of a DRAM cell.

Basic DRAM Organization

The major application of DRAMs is in the main memory of computers. The difference between DRAMs and SRAMs is the type of memory cell. As you have seen, the DRAM memory cell consists of one transistor and a capacitor and is much simpler than the SRAM cell. This allows much greater densities in DRAMs and results in greater bit capacities for a given chip area, although much slower access time.

Again, because charge stored in a capacitor will leak off, the DRAM cell requires a frequent refresh operation to preserve the stored data bit. This requirement results in more complex circuitry than in a SRAM. Several features common to most DRAMs are now discussed, using a generic 1M × 1 bit DRAM as an example.

ADDRESS MULTIPLEXING   DRAMs use a technique called *address multiplexing* to reduce the number of address lines. Figure 10–22 shows the block diagram of a 1,048,576-bit (1 Mb) DRAM with a 1M × 1 organization. We will focus on the blue blocks to illustrate address multiplexing. The green blocks represent the refresh logic.

The ten address lines are time multiplexed at the beginning of a memory cycle by the row address select ($\overline{RAS}$) and the column address select ($\overline{CAS}$) into two separate 10-bit address fields. First, the 10-bit row address is latched into the row address register. Next, the 10-bit column address is latched into the column address register. The row address and

FIGURE 10–22 **Simplified block diagram of a 1M × 1 DRAM.**

the column address are decoded to select one of the 1,048,576 addresses (2^{20} = 1,048,576) in the memory array. The basic timing for the address multiplexing operation is shown in Figure 10–23.

FIGURE 10–23 **Basic timing for address multiplexing.**

READ AND WRITE CYCLES   At the beginning of each read or write memory cycle, $\overline{RAS}$ and $\overline{CAS}$ go active (LOW) to multiplex the row and column addresses into the registers and decoders. For a read cycle, the $R/\overline{W}$ input is HIGH. For a write cycle, the $R/\overline{W}$ input is LOW. This is illustrated in Figure 10–24.

FAST PAGE MODE   In the normal read or write cycle described previously, the row address for a particular memory location is first loaded by an active-LOW $\overline{RAS}$ and then the column address for that location is loaded by an active-LOW $\overline{CAS}$. The next location is selected by another $\overline{RAS}$ followed by a $\overline{CAS}$, and so on.

FIGURE 10–24   Normal read and write cycle timing.

(a) Read cycle

(b) Write cycle

A "page" is a section of memory available at a single row address and consists of all the columns in a row. Fast page mode allows fast successive read or write operations at each column address in a selected row. A row address is first loaded by $\overline{RAS}$ going LOW and remaining LOW while $\overline{CAS}$ is toggled between HIGH and LOW. A single row address is selected and remains selected while $\overline{RAS}$ is active. Each successive $\overline{CAS}$ selects another column in the selected row. So, after a fast page mode cycle, all of the addresses in the selected row have been read from or written into, depending on $R/\overline{W}$. For example, a fast page mode cycle for the DRAM in Figure 10–22 requires $\overline{CAS}$ to go active 1024 times for each row selected by $\overline{RAS}$.

Fast page mode operation for read is illustrated by the timing diagram in Figure 10–25. When $\overline{CAS}$ goes to its nonasserted state (HIGH), it disables the data outputs. Therefore, the transition of $\overline{CAS}$ to HIGH must occur only after valid data are latched by the external system.

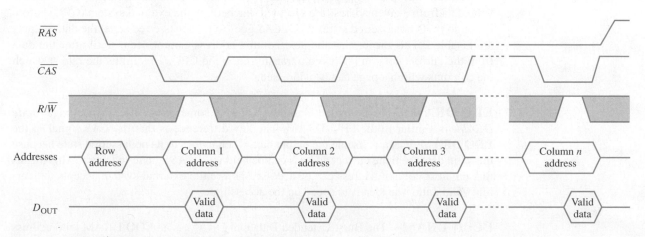

FIGURE 10–25   Fast page mode timing for a read operation.

REFRESH CYCLES   As you know, DRAMs are based on capacitor charge storage for each bit in the memory array. This charge degrades (leaks off) with time and temperature, so each bit must be periodically refreshed (recharged) to maintain the correct bit state. Typically, a DRAM must be refreshed every 8 ms to 16 ms, although for some devices the refresh period can exceed 100 ms.

A read operation automatically refreshes all the addresses in the selected row. However, in typical applications, you cannot always predict how often there will be a read cycle, and so you cannot depend on a read cycle to occur frequently enough to prevent data loss. Therefore, special refresh cycles must be implemented in DRAM systems.

Burst refresh and *distributed refresh* are the two basic refresh modes for refresh operations. In burst refresh, all rows in the memory array are refreshed consecutively each refresh period. For a memory with a refresh period of 8 ms, a burst refresh of all rows occurs once every 8 ms. The normal read and write operations are suspended during a burst refresh cycle. In distributed refresh, each row is refreshed at intervals interspersed between normal read or write cycles. For example, the memory in Figure 10–22 has 1024 rows. As an example, for an 8 ms refresh period, each row must be refreshed every 8 ms/1024 = 7.8 μs when distributed refresh is used.

The two types of refresh operations are $\overline{RAS}$ *only refresh* and $\overline{CAS}$ *before* $\overline{RAS}$ *refresh.* $\overline{RAS}$-only refresh consists of a $\overline{RAS}$ transition to the LOW (active) state, which latches the address of the row to be refreshed while $\overline{CAS}$ remains HIGH (inactive) throughout the cycle. An external counter is used to provide the row addresses for this type of operation.

The $\overline{CAS}$ before $\overline{RAS}$ refresh is initiated by $\overline{CAS}$ going LOW before $\overline{RAS}$ goes LOW. This sequence activates an internal refresh counter that generates the row address to be refreshed. This address is switched by the data selector into the row decoder.

Types of DRAMs

Now that you have learned the basic concept of a DRAM, let's briefly look at some major types. These are the *Fast Page Mode (FPM) DRAM,* the *Extended Data Out (EDO) DRAM,* the *Burst Extended Data Out (BEDO) DRAM,* the *Synchronous (S) DRAM,* and the *Double Data Note* (DDR) SDRAM.

FPM DRAM   Fast page mode operation was described earlier. This type of DRAM traditionally has been the most common and has been the type used in computers until the development of the EDO DRAM. Recall that a page in memory is all of the column addresses contained within one row address.

The idea of the **FPM DRAM** is based on the probability that the next several memory addresses to be accessed are in the same row (on the same page). Fortunately, this happens a large percentage of the time. FPM saves time over pure random accessing because in FPM the row address is specified only once for access to several successive column addresses whereas for pure random accessing, a row address is specified for each column address.

Recall that in a fast page mode read operation, the $\overline{CAS}$ signal has to wait until the valid data from a given address are accepted (latched) by the external system (CPU) before it can go to its nonasserted state. When $\overline{CAS}$ goes to its nonasserted state, the data outputs are disabled. This means that the next column address cannot occur until after the data from the current column address are transferred to the CPU. This limits the rate at which the columns within a page can be addressed.

EDO DRAM   The Extended Data Out DRAM, sometimes called *hyper page mode DRAM,* is similar to the FPM DRAM. The key difference is that the $\overline{CAS}$ signal in the **EDO DRAM** does not disable the output data when it goes to its nonasserted state because the valid data from the current address can be held until $\overline{CAS}$ is asserted again. This means that the next column address can be accessed before the external system accepts the current valid data. The idea is to speed up the access time.

BEDO DRAM   The Burst Extended Data Out DRAM is an EDO DRAM with address burst capability. Recall from the discussion of the synchronous burst SRAM that the address burst feature allows up to four addresses to be internally generated from a single external address, which saves some access time. This same concept applies to the **BEDO DRAM.**

SDRAM   Faster DRAMs are needed to keep up with the ever-increasing speed of microprocessors. The Synchronous DRAM is one way to accomplish this. Like the synchronous SRAM discussed earlier, the operation of the **SDRAM** is synchronized with the system clock, which also runs the microprocessor in a computer system. The same basic ideas described in relation to the synchronous burst SRAM, also apply to the SDRAM.

This synchronized operation makes the SDRAM totally different from the other asynchronous DRAM types. With asynchronous memories, the microprocessor must wait for the DRAM to complete its internal operations. However, with synchronous operation, the DRAM latches addresses, data, and control information from the processor under control of the system clock. This allows the processor to handle other tasks while the memory read or write operations are in progress, rather than having to wait for the memory to do its thing as is the case in asynchronous systems.

DDR SDRAM   **DDR** stands for double data rate. A DDR SDRAM is clocked on both edges of a clock pulse, whereas a SDRAM is clocked on only one edge. Because of the double clocking, a DDR SDRAM is theoretically twice as fast as an SDRAM. Sometimes the SDRAM is referred to as an SDR SDRAM (single data rate SDRAM) for contrast with the DDR SDRAM.

SECTION 10–3 CHECKUP

1. List two types of SRAM.

2. What is a cache?

3. Explain how SRAMs and DRAMs differ.

4. Describe the refresh operation in a DRAM.

5. List four types of DRAM.

10–4 THE READ-ONLY MEMORY (ROM)

A ROM contains permanently or semipermanently stored data, which can be read from the memory but either cannot be changed at all or cannot be changed without specialized equipment. A ROM stores data that are used repeatedly in system applications, such as tables, conversions, or programmed instructions for system initialization and operation. ROMs retain stored data when the power is off and are therefore nonvolatile memories.

After completing this section, you should be able to

- **List the types of ROMs**
- **Describe a basic mask ROM storage cell**
- **Explain how data are read from a ROM**
- **Discuss internal organization of a typical ROM**

The ROM Family

Figure 10–26 shows how semiconductor ROMs are categorized. The mask ROM is the type in which the data are permanently stored in the memory during the manufacturing process. The **PROM,** or programmable ROM, is the type in which the data are electrically stored by

FIGURE 10–26   **The ROM family.**

the user with the aid of specialized equipment. Both the mask ROM and the PROM can be of either MOS or bipolar technology. The **EPROM,** or erasable PROM, is strictly a MOS device. The **UV EPROM** is electrically programmable by the user, but the stored data must be erased by exposure to ultraviolet light over a period of several minutes. The electrically erasable PROM (**EEPROM** or E^{2}PROM) can be erased in a few milliseconds.

The Mask ROM

The mask ROM is usually referred to simply as a ROM. It is permanently programmed during the manufacturing process to provide widely used standard functions, such as popular conversions, or to provide user-specified functions. Once the memory is programmed, it cannot be changed. Most IC ROMs utilize the presence or absence of a transistor connection at a row/column junction to represent a 1 or a 0.

Figure 10–27 shows MOS ROM cells. The presence of a connection from a row line to the gate of a transistor represents a 1 at that location because when the row line is taken HIGH, all transistors with a gate connection to that row line turn on and connect the HIGH (1) to the associated column lines. At row/column junctions where there are no gate connections, the column lines remain LOW (0) when the row is addressed.

FIGURE 10–27   **ROM cells.**

To illustrate the ROM concept, Figure 10–28 shows a small, simplified ROM array. The blue squares represent stored 1s, and the gray squares represent stored 0s. The basic read operation is as follows. When a binary address code is applied to the address input lines, the corresponding row line goes HIGH. This HIGH is connected to the column lines through the transistors at each junction (cell) where a 1 is stored. At each cell where a 0 is stored, the column line stays LOW because of a terminating resistor. The column lines form the data output. The eight data bits stored in the selected row appear on the output lines.

As you can see, the example ROM in Figure 10–28 is organized into 16 addresses, each of which stores 8 data bits. Thus, it is a 16 × 8 (16-by-8) ROM, and its total capacity is 128 bits or 16 bytes. ROMs can be used as look-up tables (LUTs) for code conversions and logic function generation.

FIGURE 10–28 A representation of a 16 × 8-bit ROM array.

Cell storing a 1   Cell storing a 0

Address decoder

Address input lines
1
2
4
8

0 — Row 0
1 — Row 1
2 — Row 2
14 — Row 14
15 — Row 15

Data output lines
0 1 2 6 7

EXAMPLE 10–1

Show a basic ROM, similar to the one in Figure 10–28, programmed for a 4-bit binary-to-Gray conversion.

SOLUTION

Review Chapter 2 for the Gray code. Table 10–1 is developed for use in programming the ROM.

TABLE 10–1

	BINARY				GRAY		
B_3	B_2	B_1	B_0	G_3	G_2	G_1	G_0
0	0	0	0	0	0	0	0
0	0	0	1	0	0	0	1
0	0	1	0	0	0	1	1
0	0	1	1	0	0	1	0
0	1	0	0	0	1	1	0
0	1	0	1	0	1	1	1
0	1	1	0	0	1	0	1
0	1	1	1	0	1	0	0
1	0	0	0	1	1	0	0
1	0	0	1	1	1	0	1
1	0	1	0	1	1	1	1
1	0	1	1	1	1	1	0
1	1	0	0	1	0	1	0
1	1	0	1	1	0	1	1
1	1	1	0	1	0	0	1
1	1	1	1	1	0	0	0

The resulting 16 × 4 ROM array is shown in Figure 10–29. You can see that a binary code on the address input lines produces the corresponding Gray code on the output lines (columns). For example, when the binary number 0110 is applied to the address input lines, address 6, which stores the Gray code 0101, is selected.

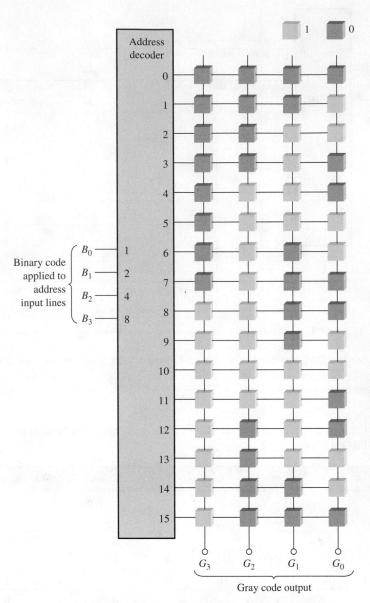

FIGURE 10–29   **Representation of a ROM programmed as a binary-to-Gray code converter.**

RELATED PROBLEM*

Using Figure 10–29, determine the Gray code output when a binary code of 1011 is applied to the address input lines.

*Answers are at the end of the chapter.

INTERNAL ROM ORGANIZATION   Most IC ROMs have a more complex internal organization than that in the basic simplified example just presented. To illustrate how an IC ROM is structured, let's use a 1024-bit device with a 256 × 4 organization. The logic symbol is shown in Figure 10–30. When any one of 256 binary codes (eight bits) is applied to the address lines, four data bits appear on the outputs if the chip select inputs are LOW. (256 addresses require eight address lines.)

Although the 256×4 organization of this device implies that there are 256 rows and 4 columns in the memory array, this is not actually the case. The memory cell array is actually a 32×32 matrix (32 rows and 32 columns), as shown in the block diagram in Figure 10–31.

The ROM in Figure 10–31 works as follows. Five of the eight address lines (A_0 through A_4) are decoded by the row decoder (often called the Y decoder) to select one of the 32 rows. Three of the eight address lines (A_5 through A_7) are decoded by the column decoder (often called the X decoder) to select four of the 32 columns. Actually, the column decoder consists of four 1-of-8 decoders (data selectors), as shown in Figure 10–31.

The result of this structure is that when an 8-bit address code (A_0 through A_7) is applied, a 4-bit data word appears on the data outputs when the chip select lines ($\overline{CS_0}$ and $\overline{CS_1}$) are LOW to enable the output buffers. This type of internal organization (architecture) is typical of IC ROMs of various capacities.

FIGURE 10–30 **A 256×4 ROM logic symbol. The $A\frac{0}{255}$ designator means that the 8-bit address code selects addresses 0 through 255.**

FIGURE 10–31 **A 1024-bit ROM with a 256×4 organization based on a 32×32 array.**

ROM is used in a computer to store the BIOS (Basic Input/Output System). These are programs that are used to perform fundamental supervisory and support functions for the computer. For example, BIOS programs stored in the ROM control certain video monitor functions, provide for disk formatting, scan the keyboard for inputs, and control certain printer functions.

SYSTEM NOTE

ROM ACCESS TIME A typical timing diagram that illustrates ROM access time is shown in Figure 10–32. The **access time,** t_a, of a ROM is the time from the application of a valid address code on the input lines until the appearance of valid output data. Access time can also be measured from the activation of the chip select ($\overline{CS}$) input to the occurrence of valid output data when a valid address is already on the input lines.

FIGURE 10–32 **ROM access time (t_a) from address change to data output with chip select already active.**

SECTION 10–4 CHECKUP

1. What is the bit storage capacity of a ROM with a 512 × 8 organization?

2. List the types of read-only memories.

3. How many address bits are required for a 2048-bit memory organized as a 256 × 8 memory?

10–5 PROGRAMMABLE ROMS

Programmable ROMs (PROMs) are basically the same as mask ROMs once they have been programmed. As you have learned, ROMs are a type of programmable logic device. The difference is that PROMs come from the manufacturer unprogrammed and are custom programmed in the field to meet the user's needs.

After completing this section, you should be able to

- **Distinguish between a mask ROM and a PROM**
- **Describe a basic PROM memory cell**
- **Discuss EPROMs including UV EPROMs and EEPROMs**
- **Analyze an EPROM programming cycle**

PROMs

A **PROM** uses some type of fusing process to store bits, in which a memory *link* is burned open or left intact to represent a 0 or a 1. The fusing process is irreversible; once a PROM is programmed, it cannot be changed.

Figure 10–33 illustrates a MOS PROM array with fusible links. The fusible links are manufactured into the PROM between the source of each cell's transistor and its column line. In the programming process, a sufficient current is injected through the fusible link to burn it open to create a stored 0. The link is left intact for a stored 1.

FIGURE 10–33   MOS PROM array with fusible links. (All drains are commonly connected to V_{DD}.)

Three basic fuse technologies used in PROMs are metal links, silicon links, and *pn* junctions. A brief description of each of these follows.

1. Metal links are made of a material such as nichrome. Each bit in the memory array is represented by a separate link. During programming, the link is either "blown" open or left intact. This is done basically by first addressing a given cell and then forcing a sufficient amount of current through the link to cause it to open.

2. Silicon links are formed by narrow, notched strips of polycrystalline silicon. Programming of these fuses requires melting of the links by passing a sufficient amount of current through them. This amount of current causes a high temperature at the fuse location that oxidizes the silicon and forms an insulation around the now-open link.

3. Shorted junction, or avalanche-induced migration, technology consists basically of two semiconductor *pn* junctions arranged back-to-back. During programming, one of the diode junctions is avalanched, and the resulting voltage and heat cause aluminum ions to migrate and short the junction. The remaining junction is then used as a forward-biased diode to represent a data bit.

EPROMs

An **EPROM** is an erasable PROM. Unlike an ordinary PROM, an EPROM can be reprogrammed if an existing program in the memory array is erased first.

An EPROM uses an NMOSFET array with an isolated-gate structure. The isolated transistor gate has no electrical connections and can store an electrical charge for indefinite periods of time. The data bits in this type of array are represented by the presence or absence of a stored gate charge. Erasure of a data bit is a process that removes the gate charge.

A typical EPROM is represented in Figure 10–34 by a logic diagram. Its operation is representative of that of other typical EPROMs of various sizes. As the logic symbol

FIGURE 10–34 **The logic symbol for a 2048 × 8 EPROM.**

shows, this device has 2048 addresses ($2^{11} = 2048$), each with eight bits. Notice that the eight outputs are tristate (∇).

Two basic types of erasable PROMs are the ultraviolet erasable PROM (UV EPROM) and the electrically erasable PROM (EEPROM).

UV EPROMS You can recognize the UV EPROM device by the UV transparent window on the package. The isolated gate in the **FET** of an ultraviolet EPROM is "floating" within an oxide insulating material. The programming process causes electrons to be removed from the floating gate. Erasure is done by exposure of the memory array chip to high-intensity ultraviolet radiation through the UV window on top of the package. The positive charge stored on the gate is neutralized after several minutes to an hour of exposure time.

To read from the memory, the output enable input ($\overline{OE}$) must be LOW and the power-down/program ($\overline{CE}/PGM$) input LOW.

To program or write to the device, a high dc voltage is applied to V_{PP} and $\overline{OE}$ is HIGH. The eight data bits to be programmed into a given address are applied to the outputs (O_0 through O_7), and the address is selected on inputs A_0 through A_{10}. Next, a HIGH level pulse is applied to the $\overline{CE}/PGM$ input. The addresses can be programmed in any order.

A timing diagram for the programming is shown in Figure 10–35. These signals are normally produced by an EPROM programmer.

FIGURE 10–35 Timing diagram for a 2048 × 8 UV EPROM programming cycle, with critical setup times (t_s) and hold times (t_h) indicated.

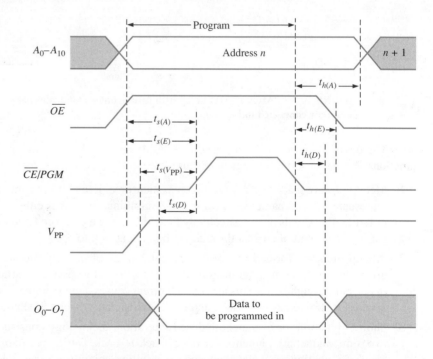

EEPROMS An electrically erasable PROM can be both erased and programmed with electrical pulses. Since it can be both electrically written into and electrically erased, the EEPROM can be rapidly programmed and erased in-circuit for reprogramming.

Two types of EEPROMs are the floating-gate MOS and the metal nitride-oxide silicon (MNOS). The application of a voltage on the control gate in the floating-gate structure permits the storage and removal of charge from the floating gate.

SECTION 10–5 CHECKUP

1. How do PROMs differ from ROMs?

2. After erasure, all bits are (1s, 0s) in a typical EPROM.

3. What is the normal mode of operation for a PROM?

The ideal memory has high storage capacity, nonvolatility, in-system read and write capability, comparatively fast operation, and cost effectiveness. The traditional memory technologies such as ROM, PROM, EPROM, EEPROM, SRAM, and DRAM individually exhibit one or more of these characteristics. Flash memory has all of the desired characteristics.

After completing this section, you should be able to

- Discuss the basic characteristics of a flash memory
- Describe the basic operation of a flash memory cell
- Compare flash memories with other types of memories
- Discuss the USB flash drive

Flash memories are high-density read/write memories (high-density translates into large bit storage capacity) that are nonvolatile, which means that data can be stored indefinitely without power. They are sometimes used in place of small-capacity hard disk drives in laptop computers.

High-density means that a large number of cells can be packed into a given surface area on a chip; that is, the higher the density, the more bits that can be stored on a given size chip. This high density is achieved in flash memories with a storage cell that consists of a single floating-gate MOS transistor. A data bit is stored as charge or the absence of charge on the floating gate depending if a 0 or a 1 is stored.

Flash Memory Cell

A single-transistor cell in a flash memory is represented in Figure 10–36. The stacked gate MOS transistor consists of a control gate and a floating gate in addition to the drain and source. The floating gate stores electrons (charge) as a result of a sufficient voltage applied to the control gate. A *0 is stored when there is more charge* and a *1 is stored when there is less or no charge.* The amount of charge present on the floating gate determines if the transistor will turn on and conduct current from the drain to the source when a control voltage is applied during a read operation.

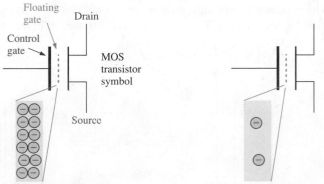

Many electrons = more charge = stored 0.   Few electrons = less charge = stored 1.

FIGURE 10–36   The storage cell in a flash memory.

Basic Flash Memory Operation

There are three major operations in a flash memory: the *programming* operation, the *read* operation, and the *erase* operation.

PROGRAMMING   Initially, all cells are at the 1 state because charge was removed from each cell in a previous erase operation. The programming operation adds electrons (charge) to the floating gate of those cells that are to store a 0. No charge is added to those cells that are to store a 1. Application of a sufficient positive voltage to the control gate with respect to the source during programming attracts electrons to the floating gate, as indicated in Figure 10–37. Once programmed, a cell can retain the charge for up to 100 years without any external power.

To store a 0, a sufficient positive voltage is applied to the control gate with respect to the source to add charge to the floating gate during programming.

To store a 1, no charge is added and the cell is left in the erased condition.

FIGURE 10–37   Simplified illustration of storing a 0 or a 1 in a flash cell during the programming operation.

READ   During a read operation, a positive voltage is applied to the control gate. The amount of charge present on the floating gate of a cell determines whether or not the voltage applied to the control gate will turn on the transistor. If a 1 is stored, the control gate voltage is sufficient to turn the transistor on. If a 0 is stored, the transistor will not turn on because the control gate voltage is not sufficient to overcome the negative charge stored in the floating gate. Think of the charge on the floating gate as a voltage source that opposes the voltage applied to the control gate during a read operation. So the floating gate charge associated with a stored 0 prevents the control gate voltage from reaching the turn-on threshold, whereas the small or zero charge associated with a stored 1 allows the control gate voltage to exceed the turn-on threshold.

When the transistor turns on, there is current from the drain to the source of the cell transistor. The presence of this current is sensed to indicate a 1, and the absence of this current is sensed to indicate a 0. This basic idea is illustrated in Figure 10–38.

When a 0 is read, the transistor remains off because the charge on the floating gate prevents the read voltage from exceeding the turn-on threshold.

When a 1 is read, the transistor turns on because the absence of charge on the floating gate allows the read voltage to exceed the turn-on threshold.

FIGURE 10–38   The read operation of a flash cell in an array.

ERASE   During an erase operation, charge is removed from all the memory cells. A sufficient positive voltage is applied to the transistor source with respect to the control gate. This is opposite in polarity to that used in programming. This voltage attracts electrons from the floating gate and depletes it of charge, as illustrated in Figure 10–39. A flash memory is always erased prior to being reprogrammed.

Basic Flash Memory Array

A simplified array of flash memory cells is shown in Figure 10–40. Only one row line is accessed at a time. When a cell in a given bit line turns on (stored 1) during a read operation, there is current through the bit line, which produces a voltage drop across the active load. This voltage drop is compared to a reference voltage with a comparator circuit and an output level indicating a 1 is produced. If a 0 is stored, then there is no current or little current in the bit line and an opposite level is produced on the comparator output.

To erase a cell, a sufficient positive voltage is applied to the source with respect to the control gate to remove charge from the floating gate during the erase operation.

FIGURE 10–39 **Simplified illustration of removing charge from a cell during erase.**

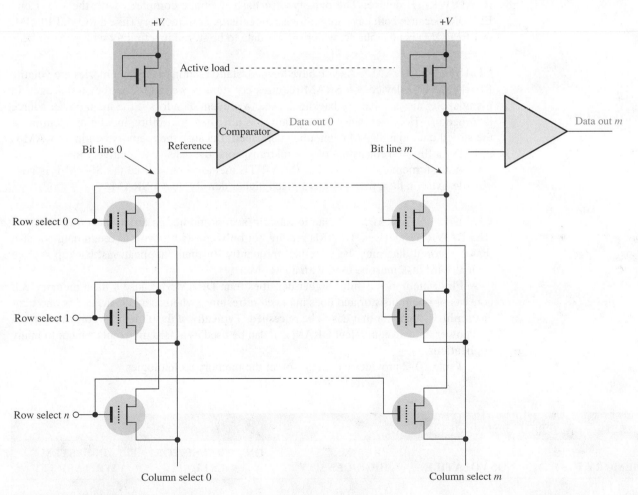

FIGURE 10–40   **Basic flash memory array.**

The memory stick is a storage medium that uses flash memory technology in a physical configuration smaller than a stick of chewing gum. Memory sticks are typically available in 128 MB to 32 GB capacities and as a kit with a PC card adaptor. Because of its compact design, it is ideal for use in small digital electronics products, such as laptop computers and digital cameras.

Comparison of Flash Memories with Other Memories

Let's compare flash memories with other types of memories with which you are already familiar.

FLASH VS. ROM, EPROM, AND EEPROM   Read-only memories are high-density, nonvolatile devices. However, once programmed the contents of a ROM can never be altered. Also, the initial programming is a time-consuming and costly process.

Although the EPROM is a high-density, nonvolatile memory, it can be erased only by removing it from the system and using ultraviolet light. It can be reprogrammed only with specialized equipment.

The EEPROM has a more complex cell structure than either the ROM or EPROM and so the density is not as high, although it can be reprogrammed without being removed from the system. Because of its lower density, the cost/bit is higher than ROMs or EPROMs.

A flash memory can be reprogrammed easily in the system because it is essentially a READ/WRITE device. The density of a flash memory compares with the ROM and EPROM because both have single-transistor cells. A flash memory (like a ROM, EPROM, or EEPROM) is nonvolatile, which allows data to be stored indefinitely with power off.

FLASH VS. SRAM   As you have learned, static random-access memories are volatile READ/WRITE devices. A SRAM requires constant power to retain the stored data. In many applications, a battery backup is used to prevent data loss if the main power source is turned off. However, since battery failure is always a possibility, indefinite retention of the stored data in a SRAM cannot be guaranteed. Because the memory cell in a SRAM is basically a flip-flop consisting of several transistors, the density is relatively low.

A flash memory is also a READ/WRITE memory, but unlike the SRAM it is nonvolatile. Also, a flash memory has a much higher density than a SRAM.

FLASH VS. DRAM   Dynamic random-access memories are volatile high-density READ/ WRITE devices. DRAMs require not only constant power to retain data but also that the stored data must be refreshed frequently. In many applications, backup storage such as hard disk must be used with a DRAM.

Flash memories exhibit higher densities than DRAMs because a flash memory cell consists of one transistor and does not need refreshing, whereas a DRAM cell is one transistor plus a capacitor that has to be refreshed. Typically, a flash memory consumes much less power than an equivalent DRAM and can be used as a hard disk replacement in many applications.

Table 10–2 provides a comparison of the memory technologies.

TABLE 10–2 • Comparison of types of memories.				
MEMORY TYPE	**NONVOLATILE**	**HIGH-DENSITY**	**ONE-TRANSISTOR CELL**	**IN-SYSTEM WRITABILITY**
Flash	Yes	Yes	Yes	Yes
SRAM	No	No	No	Yes
DRAM	No	Yes	Yes	Yes
ROM	Yes	Yes	Yes	No
EPROM	Yes	Yes	Yes	No
EEPROM	Yes	No	No	Yes

USB Flash Drive

A USB flash drive consists of a flash memory connected to a standard USB connector housed in a small case about the size of a cigarette lighter. The USB connector can be plugged into a port on a personal computer and obtains power from the computer. These memories are usually rewritable and can have a storage capacity up to 256 GB (a number which is constantly increasing), with most ranging from 2 GB to 64 GB. A typical USB flash drive is shown in Figure 10–41(a), and a basic block diagram is shown in part (b).

(a) Typical USB flash drive   (b) Basic block diagram

FIGURE 10–41   The USB flash drive.

The USB flash drive uses a standard USB A-type connector for connection to the computer, as shown in Figure 10–42(a). Peripherals, such as printers, use the USB B-type connector, which has a different shape and physical pin configuration. The USB icon is shown in part (b).

(b) Type A USB connector   (b) USB icon

FIGURE 10–42

SECTION 10–6 CHECKUP

1. What types of memories are nonvolatile?

2. What is a major advantage of a flash memory over a SRAM or DRAM?

3. List the three modes of operation of a flash memory.

10–7 MEMORY EXPANSION

Available memory can be expanded to increase the word length (number of bits in each address) or the word capacity (number of different addresses) or both. Memory expansion is accomplished by adding an appropriate number of memory chips to the address, data, and control buses. SIMMs and DIMMs, which are types of memory expansion modules, are introduced.

After completing this section, you should be able to

- Define *word-length expansion*
- Show how to expand the word length of a memory
- Define *word-capacity expansion*
- Show how to expand the word capacity of a memory

Word-Length Expansion

To increase the **word length** of a memory, the number of bits in the data bus must be increased. For example, an 8-bit word length can be achieved by using two memories, each with 4-bit words as illustrated in Figure 10–43(a). As you can see in part (b), the 16-bit address bus is commonly connected to both memories so that the combination memory still has the same number of addresses ($2^{16} = 65,536$) as each individual memory. The 4-bit data buses from the two memories are combined to form an 8-bit data bus. Now when an address is selected, eight bits are produced on the data bus—four from each memory. Example 10–2 shows the details of $65,536 \times 4$ to $65,536 \times 8$ expansion.

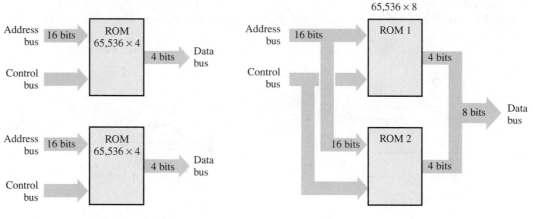

(a) Two separate 65,536 × 4 ROMs   (b) One 65,536 × 8 ROM from two 65,536 × 4 ROMs

FIGURE 10–43   Expansion of two **65,536 × 4 ROMs into a 65,536 × 8 ROM to illustrate word-length expansion.**

EXAMPLE 10–2

Expand the $65,536 \times 4$ ROM ($64k \times 4$) in Figure 10–44 to form a $64k \times 8$ ROM. Note that "64k" is the accepted shorthand for 65,536. Why not "65k"? Maybe it's because 64 is also a power-of-two.

SOLUTION

Two $64k \times 4$ ROMs are connected as shown in Figure 10–45.

FIGURE 10–44   **A 64k × 4 ROM.**

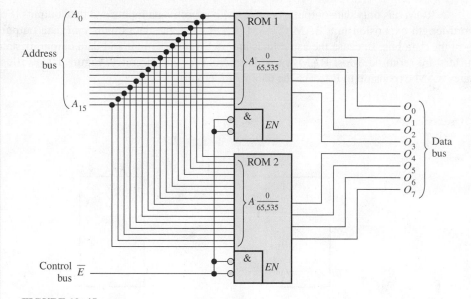

FIGURE 10–45

Notice that a specific address is accessed in ROM 1 and ROM 2 at the same time. The four bits from a selected address in ROM 1 and the four bits from the corresponding address in ROM 2 go out in parallel to form an 8-bit word on the data bus. Also notice that a LOW on the enable line, $\overline{E}$, which forms a simple control bus, enables *both* memories.

RELATED PROBLEM

Describe how you would expand a 64k $\times$ 1 ROM to a 64k $\times$ 8 ROM.

EXAMPLE 10–3

Use the memories in Example 10–2 to form a 64k $\times$ 16 ROM.

SOLUTION

In this case you need a memory that stores 65,536 16-bit words. Four 64k $\times$ 4 ROMs are required to do the job, as shown in Figure 10–46.

FIGURE 10–46

RELATED PROBLEM

How many 64k $\times$ 1 ROMs would be required to implement the memory shown in Figure 10–46?

A ROM has only data outputs, but a RAM has both data inputs and data outputs. For word-length expansion in a RAM (SRAM or DRAM), the data inputs *and* data outputs form the data bus. Because the same lines are used for data input and data output, tristate buffers are required. Most RAMs provide internal tristate circuitry. Figure 10–47 illustrates RAM expansion to increase the data word length.

FIGURE 10–47 Illustration of word-length expansion with two $2^m \times n$ RAMs forming a $2^m \times 2n$ RAM.

EXAMPLE 10-4

Use $1M \times 4$ SRAMs to create a $1M \times 8$ SRAM.

SOLUTION

Two $1M \times 4$ SRAMs are connected as shown in the simplified block diagram of Figure 10–48.

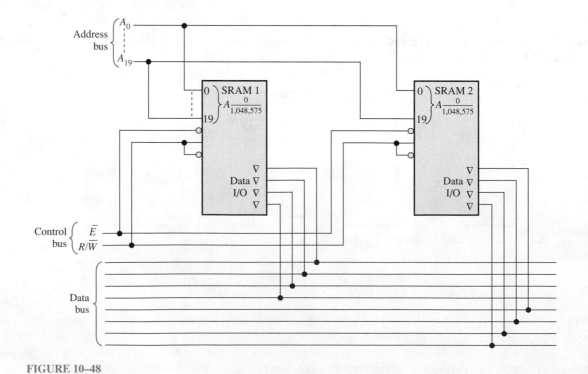

FIGURE 10–48

Use 1M × 8 SRAMs to create a 1M × 16 SRAM.

Word-Capacity Expansion

When memories are expanded to increase the **word capacity,** the *number of addresses is increased.* To achieve this increase, the number of address bits must be increased, as illustrated in Figure 10–49, (where two 1M × 8 RAMs are expanded to form a 2M × 8 memory).

(a) Individual memories each store 1,048,576 8-bit words

(b) Memories expanded to form a 2M × 8 RAM requiring a 21-bit address bus

FIGURE 10–49   Illustration of word-capacity expansion.

Each individual memory has 20 address bits to select its 1,048,576 addresses, as shown in part (a). The expanded memory has 2,097,152 addresses and therefore requires 21 address bits, as shown in part (b). The twenty-first address bit is used to enable the appropriate memory chip. The data bus for the expanded memory remains eight bits wide. Details of this expansion are illustrated in Example 10–5.

EXAMPLE 10–5

Use 512k × 4 RAMs to implement a 1M × 4 memory.

SOLUTION

The expanded addressing is achieved by connecting the enable ($\overline{E}_0$) input to the twentieth address bit (A_{19}), as shown in Figure 10–50. Input $\overline{E}_1$ is used as an enable input common to both memories. When the twentieth address bit (A_{19}) is LOW, RAM 1 is selected (RAM 2 is disabled), and the nineteen lower-order address bits ($A_0 - A_{18}$) access each of the addresses in RAM 1. When the twentieth address bit (A_{19}) is HIGH, RAM 2 is enabled by a LOW on the inverter output (RAM 1 is disabled), and the nineteen lower-order address bits ($A_0 - A_{18}$) access each of the RAM 2 addresses.

FIGURE 10–50

RELATED PROBLEM

What are the ranges of addresses in RAM 1 and in RAM 2 in Figure 10–50?

Memory Modules

SDRAMs are available in modules consisting of multiple memory ICs arranged on a printed circuit board. The most common type of SDRAM memory module is called a **DIMM** (dual in-line memory module). Another version of the DIMM is the SODIMM (small-outline DIMM). Another type of memory module, generally found in older equipment and essentially obsolete, is the SIMM (single in-line memory module). Whereas a SIMM has connection pins on one side of the pc board, a DIMM has connection pins on both sides of the board. DIMMs plug into a socket on the system motherboard for memory expansion. A generic representation of a memory module is shown in Figure 10–51 with the system board connectors into which the modules are inserted.

FIGURE 10–51 A memory module with connectors.

DIMMs generally contain DDR SDRAM memory chips. A DDR SDRAM transfers two blocks of data for each clock cycle rather than one like a standard SDRAM. There are currently three basic types of modules: DDR, DDR2 and DDR3 as follows:

- DDR modules have 184 pins and require a 2.5 voltage source.
- DDR2 modules have 240 pins and require a 1.8 voltage source.
- DDR3 modules have 240 pins and require a 1.5 voltage source.

The DDR, DDR2, and DDR3 have transfer data rates of 1600 MBps, 3200 MBps, and 6400 MBps, respectively.

When installing DIMMs, follow these steps:

1. Line up the notches on the DIMM board with the notches in the memory socket.
2. Push firmly on the module until it is securely seated in the socket.
3. Generally, the latches on both sides of the socket will snap into place when the module is completely inserted. These latches also release the module, so it can be removed from the socket.

SECTION 10–7 CHECKUP

1. How many 16k × 1 RAMs are required to achieve a memory with a word capacity of 16k and a word length of eight bits?

2. To expand the 16k × 8 memory in question 1 to a 32k × 8 organization, how many more 16k × 1 RAMs are required?

3. What does DIMM stand for?

4. What does DDR mean?

10–8 SPECIAL TYPES OF MEMORIES

In this section, the first in–first out (FIFO) memory, the last in–first out (LIFO) memory, the memory stack, and the charge-coupled device memory are covered.

After completing this section, you should be able to

- **Describe a FIFO memory**
- **Describe a LIFO memory**
- **Discuss memory stacks**
- **Explain how to use a portion of RAM as a memory stack**
- **Describe a basic CCD memory**

First In–First Out (FIFO) Memories

This type of memory is formed by an arrangement of shift registers. The term **FIFO** refers to the basic operation of this type of memory, in which the first data bit written into the memory is the first to be read out.

One important difference between a conventional shift register and a FIFO register is illustrated in Figure 10–52. In a conventional register, a data bit moves through the register only as new data bits are entered; in a FIFO register, a data bit immediately goes through the register to the right-most bit location that is empty.

Figure 10–53 is a block diagram of a FIFO serial memory. This particular memory has four serial 64-bit data registers and a 64-bit control register (marker register). When data are entered by a shift-in pulse, they move automatically under control of the marker register to the empty location closest to the output. Data cannot advance into occupied positions. However, when a data bit is shifted out by a shift-out pulse, the data bits remaining in the registers automatically move to the next position toward the output. In an asynchronous FIFO, data are shifted out independent of data entry, with the use of two separate clocks.

CONVENTIONAL SHIFT REGISTER					
INPUT	X	X	X	X	OUTPUT
0	0	X	X	X	→
1	1	0	X	X	→
1	1	1	0	X	→
0	0	1	1	1	→

FIFO SHIFT REGISTER					
INPUT	—	—	—	—	OUTPUT
0	—	—	—	0	→
1	—	—	1	0	→
1	—	1	1	0	→
0	0	1	1	0	→

X = unknown data bits.
In a conventional shift register, data stay to the left until "forced" through by additional data.

— = empty positions.
In a FIFO shift register, data "fall" through (go right).

FIGURE 10–52 Comparison of conventional and FIFO register operation.

FIGURE 10–53 Block diagram of a typical FIFO serial memory.

FIFO Applications

One important application area for the FIFO register is the case in which two systems of differing data rates must communicate. Data can be entered into a FIFO register at one rate and taken out at another rate. Figure 10–54 illustrates how a FIFO register might be used in these situations.

FIGURE 10–54 Examples of the FIFO register in data-rate buffering applications.

(a) Irregular telemetry data can be stored and retransmitted at a constant rate.

(b) Data input at a slow keyboard rate can be stored and then transferred at a higher rate for processing.

(c) Data input at a constant rate can be stored and then output in bursts.

(d) Data in bursts can be stored and reformatted into a constant-rate output.

Last In–First Out (LIFO) Memories

The **LIFO** (last in–first out) memory is found in applications involving microprocessors and other computing systems. It allows data to be stored and then recalled in reverse order; that is, the last data byte to be stored is the first data byte to be retrieved.

FIGURE 10–55   Register stack.

REGISTER STACKS   A LIFO memory is commonly referred to as a push-down stack. In some systems, it is implemented with a group of registers as shown in Figure 10–55. A stack can consist of any number of registers, but the register at the top is called the *top-of-stack*.

To illustrate the principle, a byte of data is loaded in parallel onto the top of the stack. Each successive byte pushes the previous one down into the next register. This process is illustrated in Figure 10–56. Notice that the new data byte is always loaded into the top register and the previously stored bytes are pushed deeper into the stack. The name *push-down stack* comes from this characteristic.

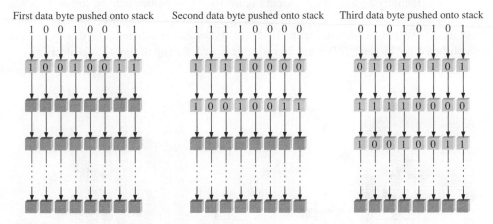

FIGURE 10–56   Simplified illustration of pushing data onto the stack.

Data bytes are retrieved in the reverse order. The last byte entered is always at the top of the stack, so when it is pulled from the stack, the other bytes pop up into the next higher locations. This process is illustrated in Figure 10–57.

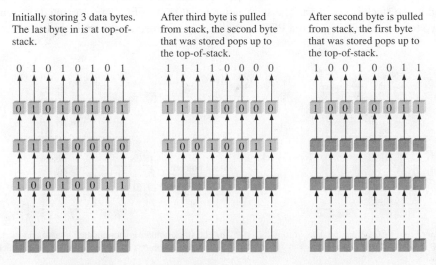

FIGURE 10–57   Simplified illustration of pulling data from the stack.

16-bit address
(hexadecimal)

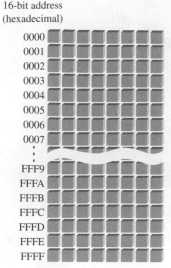

0000
0001
0002
0003
0004
0005
0006
0007

FFF9
FFFA
FFFB
FFFC
FFFD
FFFE
FFFF

FIGURE 10–58 **Representation of a 64 kB memory with the 16-bit addresses expressed in hexadecimal.**

RAM STACK Another approach to LIFO memory used in microprocessor-based systems is the allocation of a section of RAM as the stack rather than the use of a dedicated set of registers. As you have seen, for a register stack the data moves up or down from one location to the next. In a RAM stack, the data itself does not move but the top-of-stack moves under control of a register called the stack pointer.

Consider a random-access memory that is byte organized—that is, one in which each address contains eight bits—as illustrated in Figure 10–58. The binary address 0000000000001111, for example, can be written as 000F in hexadecimal. A 16-bit address can have a *minimum* hexadecimal value of 0000_{16} and a *maximum* value of $FFFF_{16}$. With this notation, a 64 kB memory array can be represented as shown in Figure 10–58. The lowest memory address is 0000_{16} and the highest memory address is $FFFF_{16}$.

Now, consider a section of RAM set aside for use as a stack. A special separate register, the stack pointer, contains the address of the top of the stack, as illustrated in Figure 10–59. A 4-digit hexadecimal representation is used for the binary addresses. In the figure, the addresses are chosen for purposes of illustration.

Now let's see how data are pushed onto the stack. The stack pointer is initially at address $FFEE_{16}$, which is the top of the stack as shown in Figure 10–59(a). The stack pointer is then decremented (decreased) by two to $FFEC_{16}$. This moves the top of the stack to a lower memory address, as shown in Figure 10–59(b). Notice that the top of the stack is not stationary as in the fixed register stack but moves downward (to lower addresses) in the RAM as data words are stored. Figure 10–59(b) shows that two bytes (one data word) are then pushed onto the stack. After the data word is stored, the top of the stack is at $FFEC_{16}$.

Small section of RAM

Stack pointer

FFEE → 0 0 0 0 0 0 0 0 Top-of-stack

(a) The stack pointer is initially at FFEE before the data word 0001001000110100 (1234) is pushed onto the stack.

Stack pointer

FFEC

0 0 1 1 0 1 0 0 Top-of-stack
0 0 0 1 0 0 1 0
0 0 0 0 0 0 0 0

(b) The stack pointer is decremented by two and the data word 0001001000110100 is placed in the two locations prior to the original stack pointer location.

FIGURE 10–59 **Illustration of the PUSH operation for a RAM stack.**

Figure 10–60 illustrates the POP operation for the RAM stack. The last data word stored in the stack is read first. The stack pointer that is at FFEC is incremented (increased) by two to address $FFEE_{16}$ and a POP operation is performed as shown in part (b). Keep in

Stack pointer

FFEC

0 0 1 1 0 1 0 0 Top-of-stack
0 0 0 1 0 0 1 0
0 0 0 0 0 0 0 0

(a) The stack pointer is at FFEC before the data word is copied (popped) from the stack.

Stack pointer

FFEE →

0 0 1 1 0 1 0 0
0 0 0 1 0 0 1 0
0 0 0 0 0 0 0 0 Top-of-stack

(b) The stack pointer is incremented by two and the last data word stored is copied (popped) from the stack.

FIGURE 10–60 **Illustration of the POP operation for the RAM stack.**

mind that RAMs are nondestructive when read, so the data word still remains in the memory after a POP operation. A data word is destroyed only when a new word is written over it.

A RAM stack can be of any depth, depending on the number of continuous memory addresses assigned for that purpose.

CCD Memories

The **CCD** (charge-coupled device) memory stores data as charges on capacitors. Unlike the DRAM, however, the storage cell does not include a transistor. High density is the main advantage of CCDs, and these devices are widely used in digital imaging.

The CCD memory consists of long rows of semiconductor capacitors, called *channels*. Data are entered into a channel serially by depositing a small charge for a 0 and a large charge for a 1 on the capacitors. These charge packets are then shifted along the channel by clock signals as more data are entered.

As with the DRAM, the charges must be refreshed periodically. This process is done by shifting the charge packets serially through a refresh circuit. Figure 10–61 shows the basic concept of a CCD channel. Because data are shifted serially through the channels, the CCD memory has a relatively long access time. CCD arrays are used in some modern cameras to capture video images in the form of light-induced charge.

FIGURE 10–61   **A CCD (charge-coupled device) channel.**

SECTION 10–8 CHECKUP

1. What is a FIFO memory?

2. What is a LIFO memory?

3. Explain the PUSH operation in a memory stack.

4. Explain the POP operation in a memory stack.

5. What does the term *CCD* stand for?

10–9 MAGNETIC AND OPTICAL STORAGE

In this section, the basics of magnetic disks, magnetic tape, magneto-optical disks, and optical disks are introduced. These storage media are important, particularly in computer applications, where they are used for mass nonvolatile storage of data and programs.

After completing this section, you should be able to

- **Describe a magnetic hard disk**
- **Discuss removable hard disks**
- **Explain the principle of magneto-optical disks**
- **Discuss the CD-ROM, CD-R, and CD-RW disks**
- **Describe the WORM**
- **Discuss the DVD-ROM**

Magnetic Storage

MAGNETIC HARD DISKS   Computers use hard disks as the internal mass storage media. **Hard disks** are rigid "platters" made of aluminum alloy or a mixture of glass and ceramic covered with a magnetic coating. Hard disk drives mainly come in three diameter sizes, 3.5 in., 2.5 in., and 1.8 in. Older formats of 8 in. and 5.25 in. are considered obsolete. A hard disk drive is hermetically sealed to keep the disks dust-free.

     Typically, two or more platters are stacked on top of each other on a common shaft or spindle that turns the assembly at several thousand rpm. A separation between each disk allows for a magnetic read/write head that is mounted on the end of an actuator arm, as shown in Figure 10–62. There is a read/write head for both sides of each disk since data are recorded on both sides of the disk surface. The drive actuator arm synchronizes all the read/write heads to keep them in perfect alignment as they "fly" across the disk surface with a separation of only a fraction of a millimeter from the disk. A small dust particle could cause a head to "crash," causing damage to the disk surface.

FIGURE 10–62   **A hard disk drive.**

FIGURE 10–63   **Simplified read/write head operation.**

BASIC READ/WRITE HEAD PRINCIPLES   The hard drive is a random-access device because it can retrieve stored data anywhere on the disk in any order. A simplified diagram of the magnetic surface read/write operation is shown in Figure 10–63. The direction or polarization of the magnetic domains on the disk surface is controlled by the direction of the magnetic flux lines (magnetic field) produced by the write head according to the direction of a current pulse in the winding. This magnetic flux magnetizes a small spot on the disk surface in the direction of the magnetic field. A magnetized spot of one polarity represents a binary 1, and one of the opposite polarity represents a binary 0. Once a spot on the disk surface is magnetized, it remains until written over with an opposite magnetic field.

Data are stored on a hard drive in the form of files. Keeping track of the location of files is the job of the device driver that manages the hard drive (sometimes referred to as hard drive BIOS). The device driver and the computer's operating system can access two tables to keep track of files and file names. The first table is called the FAT (File Allocation Table). The FAT shows what is assigned to specific files and keeps a record of open sectors and bad sectors. The second table is the Root Directory which has file names, type of file, time and date of creation, starting cluster number, and other information about the file. Other types of tables include NTFS (New Technology File System) and HFS (Hierarchical File System).

SYSTEM NOTE

When the magnetic surface passes a read head, the magnetized spots produce magnetic fields in the read head, which induce voltage pulses in the winding. The polarity of these pulses depends on the direction of the magnetized spot and indicates whether the stored bit is a 1 or a 0. The read and write heads are usually combined in a single unit.

HARD DISK FORMAT   A hard disk is organized or formatted into tracks and sectors, as shown in Figure 10–64(a). Each track is divided into a number of sectors, and each track and sector has a physical address that is used by the operating system to locate a particular data record. Hard disks typically have from a few hundred to thousands of tracks and are available with storage capacities of up to 1 TB or more. As you can see in the figure, there is a constant number of tracks/sector, with outer sectors using more surface area than the inner sectors. The arrangement of tracks and sectors on a disk is known as the *format*.

FIGURE 10–64   **Hard disk organization and formatting.**

A hard disk stack is illustrated in Figure 10–64(b). Hard disk drives differ in the number of platters in a stack, but there is always a minimum of two. All of the same corresponding tracks on each platter are collectively known as a cylinder, as indicated.

HARD DISK PERFORMANCE   Several basic parameters determine the performance of a given hard disk drive. A *seek* operation is the movement of the read/write head to the desired track. The **seek time** is the average time for this operation to be performed. Typically, hard disk drives have an average seek time of several milliseconds, depending on the particular drive.

The **latency period** is the time it takes for the desired sector to spin under the head once the head is positioned over the desired track. A worst case is when the desired sector is just past the head position and spinning away from it. The sector must rotate almost a full revolution back to the head position. *Average latency period* assumes that the disk must make half of a revolution. Obviously, the latency period depends on the constant rotational speed of the disk. Disk rotation speeds are different for different disk drives but typically are from 4200 rpm to 15,000 rpm. Some disk drives rotate at 10,033 rpm and have an average latency period of less than 3 ms.

The sum of the average seek time and the average latency period is the *access time* for the disk drive.

REMOVABLE HARD DISK   Removable hard disk drives with capacities of 1 TB are available. Keep in mind that the technology is changing so rapidly that there most likely will be further advancements at the time you are reading this.

MAGNETIC TAPE   Tape is used for backup data from mass storage devices and typically is slower than disks because data on tape is accessed serially rather than randomly. There are several types that are available, including QIC, DAT, 8 mm, and DLT.

> Tape is a viable alternative to disk due to its lower cost per bit. Though the density is lower than for a disk drive, the available surface on a tape is far greater. The highest-capacity tape media are generally on the same order as the largest available disk drive (about 1 TB—a terabyte is one trillion bytes.) Tape has historically offered enough advantage in cost over disk storage to make it a viable product, particularly for backup, where media removability is also important.
>
> SYSTEM NOTE

QIC is an abbreviation for quarter-inch cartridge and looks much like audio tape cassettes with two reels inside. Various QIC standards have from 28 to 72 tracks that can store from 80 MB to 1.2 GB. More recent innovations under the Travan standard have lengthened the tape and increased its width allowing storage capacities up to 4 GB. QIC tape drives use read/write heads that have a single write head with a read head on each side. This allows the tape drive to verify data just written when the tape is running in either direction. In the record mode, the tape moves past the read/write heads at approximately 100 inches/second, as indicated in Figure 10–65.

FIGURE 10–65   QIC tape.

DAT, which is an abbreviation for digital audio tape, uses a technique called helical scan recording. DATs offer storage capacities ranging up to 160 GB but is more expensive than QIC.

8 mm, a third type of tape format, was originally designed for the video industry but has been adopted by the computer industry as a reliable way to store large amounts of computer data. 8 mm is similar to DAT but offers storage capacities up to about 80 GB.

DLT is an abbreviation for digital linear tape. DLT is a half-inch wide tape, which is 60% wider than 8 mm and, of course, twice as wide as standard QIC. Basically, DLT differs in the way the tape-drive mechanism works to minimize tape wear compared to other systems. DLT offers storage capacities up to 110 GB.

Magneto-Optical Storage

As the name implies, magneto-optical (MO) storage devices use a combination of magnetic and optical (laser) technologies. A **magneto-optical disk** is formatted into tracks and sectors similar to magnetic disks.

The basic difference between a purely magnetic disk and an MO disk is that the magnetic coating used on the MO disk requires heat to alter the magnetic polarization. Therefore, the MO is extremely stable at ambient temperature, making data unchangeable. To write a data bit, a high-power laser beam is focused on a tiny spot on the disk, and the temperature of that tiny spot is raised above a temperature level called the Curie point (about 200 °C). Once heated, the magnetic particles at that spot can easily have their direction (polarization) changed by a magnetic field generated by the write head. Information is read from the disk with a less-powerful laser than used for writing, making use of the Kerr effect where the polarity of the reflected laser light is altered depending on the orientation of the magnetic particles. Magnetic spots of one polarity represent 0s and magnetic spots of the opposite polarity represent 1s. Basic MO operation is shown in Figure 10–66, which represents a small cross-sectional area of a disk.

(a) Unrecorded disk

(b) Writing: A high-power laser beam heats the spot, causing the magnetic particles to align with the electromagnetic field.

(c) Reading: A low-power laser beam reflects off of the reversed-polarity magnetic particles and its polarization shifts. If the particles are not reversed, the polarization of the reflected beam is unchanged.

(d) Erasing: The electromagnetic field is reversed as the high-power laser beam heats the spot, causing the magnetic particles to be restored to the original polarity.

FIGURE 10–66   **Basic principle of a magneto-optical disk.**

Optical Storage

CD-ROM   The basic Compact Disk–Read-Only Memory is a 120 mm diameter disk with a sandwich of three coatings: a polycarbonate plastic on the bottom, a thin aluminum sheet for reflectivity, and a top coating of lacquer for protection. The **CD-ROM** disk is formatted in a single spiral track with sequential 2 kB sectors and has a capacity of 680 MB. Data are prerecorded at the factory in the form of minute indentations called *pits* and the flat area surrounding the pits called *lands*. The pits are stamped into the plastic layer and cannot be erased.

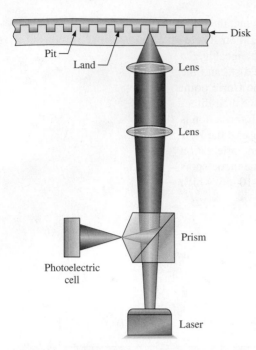

FIGURE 10–67   Basic operation of reading data from a CD-ROM.

A CD player reads data from the spiral track with a low-power infrared laser, as illustrated in Figure 10–67. The data are in the form of pits and lands as shown. Laser light reflected from a pit is 180° out-of-phase with the light reflected from the lands. As the disk rotates, the narrow laser beam strikes the series of pits and lands of varying lengths, and a photodiode detects the difference in the reflected light. The result is a series of 1s and 0s corresponding to the configuration of pits and lands along the track.

WORM   Write Once/Read Many (**WORM**) is a type of optical storage that can be written onto one time after which the data cannot be erased but can be read many times. To write data, a low-power laser is used to burn microscopic pits on the disk surface. 1s and 0s are represented by the burned and nonburned areas.

CD-R   This is essentially a type of WORM. The difference is that the CD-Recordable allows multiple write sessions to different areas of the disk. The **CD-R** disk has a spiral track like the CD-ROM, but instead of mechanically pressing indentations on the disk to represent data, the CD-R uses a laser to burn microscopic spots into an organic dye surface. When heated beyond a critical temperature with a laser during read, the burned spots change color and reflect less light than the nonburned areas. Therefore, 1s and 0s are represented on a CD-R by burned and nonburned areas, whereas on a CD-ROM they are represented by pits and lands. Like the CD-ROM, the data cannot be erased once it is written.

CD-RW   The CD-Rewritable disk can be used to read and write data. Instead of the dye-based recording layer in the CD-R, the **CD-RW** commonly uses a crystalline compound with a special property. When it is heated to a certain temperature, it becomes crystalline when it cools; but if it is heated to a certain higher temperature, it melts and becomes amorphous when it cools. To write data, the focused laser beam heats the material to the melting temperature resulting in an amorphous state. The resulting amorphous areas reflect less light than the crystalline areas, allowing the read operation to detect 1s and 0s. The data can be erased or overwritten by heating the amorphous areas to a temperature above the crystallization temperature but lower than the melting temperature that causes the amorphous material to revert to a crystalline state.

DVD-ROM   Originally DVD was an abbreviation for Digital Video Disk but eventually came to represent *Digital Versatile Disk*. Like the CD-ROM, **DVD-ROM** data are prestored on the disk. However, the pit size is smaller than for the CD-ROM, allowing more data to be stored on a track. The major difference between CD-ROM and DVD-ROM is that the CD is single-sided, while the DVD has data on both sides. Also, in addition to double-sided DVD disks, there are also multiple-layer disks that use semitransparent data layers placed over the main data layers, providing storage capacities of tens of gigabytes. To access all the layers, the laser beam requires refocusing going from one layer to the other.

SECTION 10–9 CHECKUP

1. List the major types of magnetic storage.

2. Generally, how is a magnetic disk organized?

3. How are data written on and read from a magneto-optical disk?

4. List the types of optical storage.

Because memories can contain large numbers of storage cells, testing each cell can be a lengthy and frustrating process. Fortunately, memory testing is usually an automated process performed with a programmable test instrument or with the aid of software for in-system testing. Most microprocessor-based systems provide automatic memory testing as part of their system software.

After completing this section, you should be able to

- **Discuss the checksum method of testing ROMs**
- **Discuss the checkerboard pattern method of testing RAMs**

ROM Testing

Since ROMs contain known data, they can be checked for the correctness of the stored data by reading each data word from the memory and comparing it with a data word that is known to be correct. One way of doing this is illustrated in Figure 10–68. This process requires a reference ROM that contains the same data as the ROM to be tested. A special test instrument is programmed to read each address in both ROMs simultaneously and to compare the contents. A flowchart in Figure 10–69 illustrates the basic sequence.

FIGURE 10–68   Block diagram for a complete contents check of a ROM.

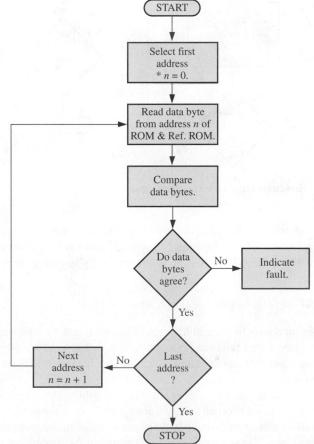

* n is the address number.

FIGURE 10–69   Flowchart for a complete contents check of a ROM.

CHECKSUM METHOD   Although the previous method checks each ROM address for correct data, it has the disadvantage of requiring a reference ROM for each different ROM to be tested. Also, a failure in the reference ROM can produce a fault indication.

In the checksum method a number, the sum of the contents of all the ROM addresses, is stored in a designated ROM address when the ROM is programmed. To test the ROM, the contents of all the addresses except the checksum are added, and the result is compared with the checksum stored in the ROM. If there is a difference, there is definitely a fault. If the checksums agree, the ROM is most likely good. However, there is a remote possibility that a combination of bad memory cells could cause the checksums to agree.

This process is illustrated in Figure 10–70 with a simple example. The checksum in this case is produced by taking the sum of each column of data bits and discarding the carries. This is actually an XOR sum of each column. The flowchart in Figure 10–71 illustrates the basic checksum test.

FIGURE 10–70   **Simplified illustration of a programmed ROM with the checksum stored at a designated address.**

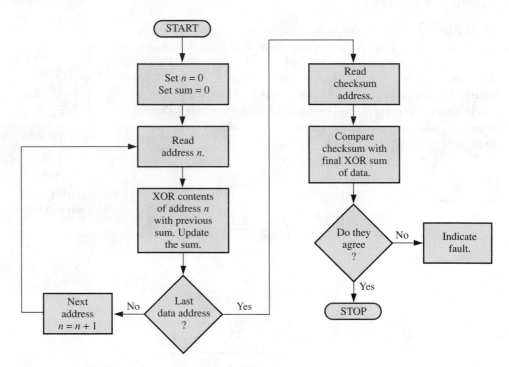

FIGURE 10–71   **Flowchart for a basic checksum test.**

The checksum test can be implemented with a special test instrument, or it can be incorporated as a test routine in the built-in (system) software or microprocessor-based systems. In that case, the ROM test routine is automatically run on system start-up.

RAM Testing

To test a RAM for its ability to store both 0s and 1s in each cell, first 0s are written into all the cells in each address and then read out and checked. Next, 1s are written into all the cells in each address and then read out and checked. This basic test will detect a cell that is stuck in either a 1 state or a 0 state.

Some memory faults cannot be detected with the all-0s–all-1s test. For example, if two adjacent memory cells are shorted, they will always be in the same state, both 0s or both 1s. Also, the all-0s–all-1s test is ineffective if there are internal noise problems such that the contents of one or more addresses are altered by a change in the contents of another address.

THE CHECKERBOARD PATTERN TEST   One way to more fully test a RAM is by using a checkerboard pattern of 1s and 0s, as illustrated in Figure 10–72. Notice that all adjacent cells have opposite bits. This pattern checks for a short between two adjacent cells; if there is a short, both cells will be in the same state.

After the RAM is checked with the pattern in Figure 10–72(a), the pattern is reversed, as shown in part (b). This reversal checks the ability of all cells to store both 1s and 0s.

A further test is to alternate the pattern one address at a time and check all the other addresses for the proper pattern. This test will catch a problem in which the contents of an address are dynamically altered when the contents of another address change.

A basic procedure for the checkerboard test is illustrated by the flowchart in Figure 10–73. The procedure can be implemented with the system software in microprocessor-based systems so that either the tests are automatic when the system is powered up or they can be initiated from the keyboard.

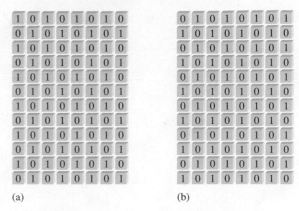

(a)   (b)

FIGURE 10–72   **The RAM checkerboard test pattern.**

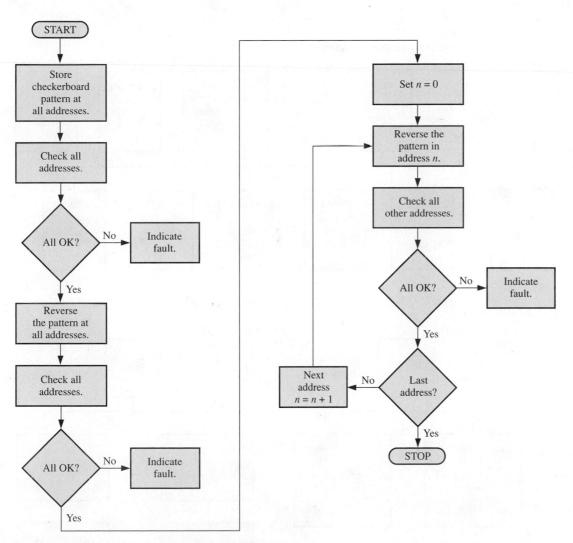

FIGURE 10–73   **Flowchart for basic RAM checkerboard test.**

SECTION 10–10 CHECKUP

1. Describe the checksum method of ROM testing.

2. Why can the checksum method not be applied to RAM testing?

3. List the three basic faults that the checkerboard pattern test can detect in a RAM.

SUMMARY

- Memory hierarchy:

- Types of semiconductor memories:

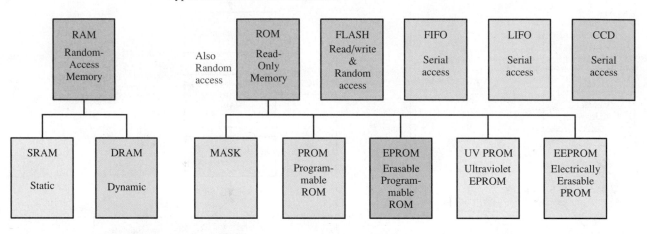

- Types of SRAMs (Static RAMs) and DRAMs (Dynamic RAMs):

- Types of magnetic storage:

Random access

Serial access

- Types of optical (laser) storage:

Magneto-Optical Disk	CD-ROM	CD-R	CD-RW	WORM	DVD-ROM
Cross between magnetic and optical	Prerecorded at factory	Recordable	Rewritable	Write once read many	Digital versatile disk

KEY TERMS

Key terms and other bold terms in the chapter are defined in the end-of-book glossary.

Address   The location of a given storage cell or group of cells in a memory.

Bus   One or more interconnections that interface one or more devices based on a standardized specification.

Byte   A group of eight bits.

Capacity   The total number of data units (bits, nibbles, bytes, words) that a memory can store.

Cell   A single storage element in a memory.

DDR   Double data rate.

DRAM   Dynamic random-access memory; a type of semiconductor memory that uses capacitors as the storage elements and is a volatile, read/write memory.

EPROM   Erasable programmable read-only memory; a type of semiconductor memory device that typically uses ultraviolet light to erase data.

FIFO   First in–first out memory.

Flash memory   A nonvolatile read/write random-access semiconductor memory in which data are stored as charge on the floating gate of a certain type of FET.

Hard disk   A magnetic storage device; typically, a stack of two or more rigid disks enclosed in a sealed housing.

LIFO   Last in–first out memory; a memory stack.

Memory   The portion of a computer or other system that stores binary data.

Memory hierarchy   The arrangement of various memory elements to maximize speed and minimize cost.

PROM Programmable read-only memory; a type of semiconductor memory.

RAM Random-access memory; a volatile read/write semiconductor memory.

Read The process of retrieving data from a memory.

ROM Read-only memory; a nonvolatile random-access semiconductor memory.

SRAM Static random-access memory; a type of volatile read/write semiconductor memory.

Word A group of bits or bytes that acts as a single entity that can be stored in one memory location; two bytes.

Write The process of storing data in a memory.

TRUE/FALSE QUIZ

Answers are at the end of the chapter.

1. A data byte consists of eight bits.

2. A memory cell can store a byte of data.

3. The write operation stores data in memory.

4. The read operation always erases the data byte.

5. RAM is a *random address memory*.

6. Stored data is lost if power is removed from a static RAM.

7. Cache is a type of memory used for intermediate or temporary storage of data.

8. Dynamic RAMs must be periodically refreshed to retain data.

9. ROM is a *random output memory*.

10. A flash memory uses a flashing beam of light to store data.

SELF-TEST

Answers are at the end of the chapter.

1. Memory hierarchy determines
   - **(a)** storage capacity
   - **(b)** processing speed
   - **(c)** access time
   - **(d)** both (b) and (c)

2. The bit capacity of a memory that has 1024 addresses and can store 8 bits at each address is
   - **(a)** 1024 **(b)** 8192 **(c)** 8 **(d)** 4096

3. A 32-bit data word consists of
   - **(a)** 2 bytes **(b)** 4 nibbles
   - **(c)** 4 bytes **(d)** 3 bytes and 1 nibble

4. Data are stored in a random-access memory (RAM) during the
   - **(a)** read operation **(b)** enable operation
   - **(c)** write operation **(d)** addressing operation

5. Data that are stored at a given address in a random-access memory (RAM) is lost when
   - **(a)** power goes off
   - **(b)** the data are read from the address
   - **(c)** new data are written at the address
   - **(d)** answers (a) and (c)

6. A ROM is a
   - **(a)** nonvolatile memory **(b)** volatile memory
   - **(c)** read/write memory **(d)** byte-organized memory

7. A memory with 256 addresses has

(a) 256 address lines   (b) 6 address lines

(c) 1 address line   (d) 8 address lines

8. A byte-organized memory has

(a) 1 data output line   (b) 4 data output lines

(c) 8 data output lines   (d) 16 data output lines

9. The storage cell in a SRAM is

(a) a flip-flop   (b) a capacitor

(c) a fuse   (d) a magnetic domain

10. A DRAM must be

(a) replaced periodically   (b) refreshed periodically

(c) always enabled   (d) programmed before each use

11. A flash memory is

(a) volatile   (b) a read-only memory

(c) a read/write memory   (d) nonvolatile

(e) answers (a) and (c)   (f) answers (c) and (d)

12. Optical storage devices employ

(a) ultraviolet light   (b) electromagnetic fields

(c) optical couplers   (d) lasers

PROBLEMS

Answers to odd-numbered problems are at the end of the book.

SECTION 10–1   Memory System Hierarchy

1. What is a cache in memory terminology and where is it generally located in a system?

2. What does main memory consist of?

3. What part of a memory system has the highest latency?

4. A certain optical jukebox can store 150 PB. Express this in terms of bytes.

5. Assume the cache in Figure 10–3 has a latency of 30 ns and the main memory a latency of 75 ns. Determine the time required for the processor to acquire a block of data for a cache hit and for a cache miss as illustrated in the figure.

6. Define the terms *temporal locality* and *spatial locality*.

SECTION 10–2   Semiconductor Memory Basics

7. Identify the ROM and the RAM in Figure 10–74.

(a)   (b)

FIGURE 10–74

8. Explain why RAMs and ROMs are both random-access memories.

9. Explain the purposes of the address bus and the data bus.

10. What memory address (0 through 256) is represented by each of the following hexadecimal numbers:
    (a) $0A_{16}$ (b) $3F_{16}$ (c) CD_{16}

SECTION 10–3 The Random-Access Memory (RAM)

11. A static memory array with four rows similar to the one in Figure 10–13 is initially storing all 0s. What is its content after the following conditions? Assume a 1 selects a row.

    $$\text{Row } 0 = 1, \text{Data in (Bit 0)} = 1$$
    $$\text{Row } 1 = 0, \text{Data in (Bit 1)} = 1$$
    $$\text{Row } 2 = 1, \text{Data in (Bit 2)} = 1$$
    $$\text{Row } 3 = 0, \text{Data in (Bit 3)} = 0$$

12. Draw a basic logic diagram for a 512×8-bit static RAM, showing all the inputs and outputs.

13. Assuming that a $64k \times 8$ SRAM has a structure similar to that of the SRAM in Figure 10–15, determine the number of rows and 8-bit columns in its memory cell array.

14. Redraw the block diagram in Figure 10–15 for a $64k \times 8$ memory.

15. Explain the difference between a SRAM and a DRAM.

16. What is the capacity of a DRAM that has twelve address lines?

SECTION 10–4 The Read-Only Memory (ROM)

17. For the ROM array in Figure 10–75, determine the outputs for all possible input combinations, and summarize them in tabular form (Blue cell is a 1, gray cell is a 0).

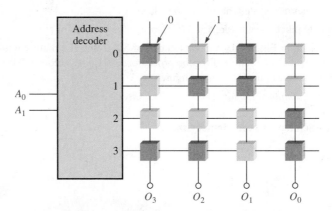

FIGURE 10–75

18. Determine the truth table for the ROM in Figure 10–76.

19. Using a procedure similar to that in Example 10–1, show a ROM for conversion of single-digit BCD to excess-3 code.

20. What is the total *bit* capacity of a ROM that has 14 address lines and 8 data outputs?

SECTION 10–5 Programmable ROMs

21. Assuming that the PROM matrix in Figure 10–77 is programmed by blowing a fuse link to create a 0, indicate the links to be blown to program an X^3 look-up table, where X is a number from 0 through 7.

FIGURE 10–76

FIGURE 10–77

22. Determine the addresses that are programmed and the contents of each address after the programming sequence in Figure 10–78 has been applied to an EPROM like the one shown in Figure 10–34.

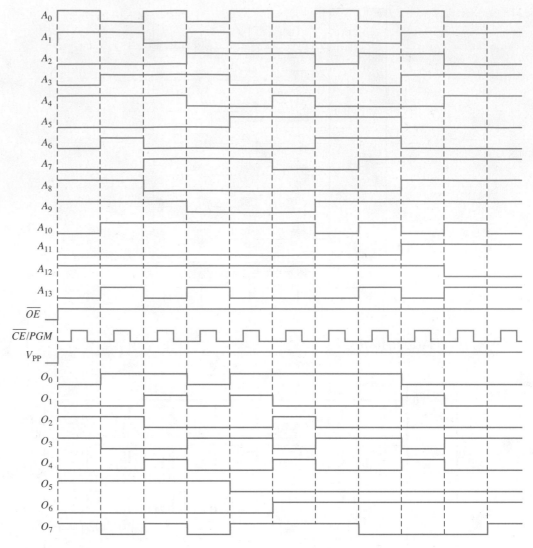

FIGURE 10–78

SECTION 10–7 Memory Expansion

23. Use 16k × 4 DRAMs to build a 64k × 8 DRAM. Show the logic diagram.

24. Using a block diagram, show how 64k × 1 dynamic RAMs can be expanded to build a 256k × 4 RAM.

25. What is the word length and the word capacity of the memory of Problem 23? Problem 24?

SECTION 10–8 Special Types of Memories

26. Complete the timing diagram in Figure 10–79 by showing the output waveforms that are initially all LOW for a FIFO serial memory like that shown in Figure 10–53.

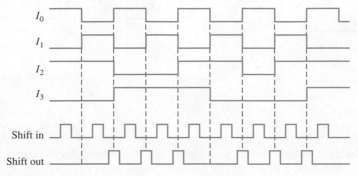

FIGURE 10–79

27. Consider a 4096 × 8 RAM in which the last 64 addresses are used as a LIFO stack. If the first address in the RAM is 000_{16}, designate the 64 addresses used for the stack.

28. In the memory of Problem 27, sixteen bytes are pushed into the stack. At what address is the first byte in located? At what address is the last byte in located?

SECTION 10–9   Magnetic and Optical Storage

29. Describe the general format of a hard disk.

30. Explain seek time and latency period in a hard disk drive.

31. Why does magnetic tape require a much longer access time than does a disk?

32. Explain the differences in a magneto-optical disk, a CD-ROM, and a WORM.

SECTION 10–10   Troubleshooting

33. Determine if the contents of the ROM in Figure 10–80 are correct.

ROM

Checksum

FIGURE 10–80

34. A 128 × 8 ROM is implemented as shown in Figure 10–81. The decoder decodes the two most significant address bits to enable the ROMs one at a time, depending on the address selected.
    (a) Express the lowest address and the highest address of each ROM as hexadecimal numbers.
    (b) Assume that a single checksum is used for the entire memory and it is stored at the highest address. Develop a flowchart for testing the complete memory system.
    (c) Assume that each ROM has a checksum stored at its highest address. Modify the flowchart developed in part (b) to accommodate this change.
    (d) What is the disadvantage of using a single checksum for the entire memory rather than a checksum for each individual ROM?

FIGURE 10–81

35. Suppose that a checksum test is run on the memory in Figure 10–81 and each individual ROM has a checksum at its highest address. What IC or ICs will you replace for each of the following error messages that appear on the system's video monitor?
    (a) ADDRESSES 40–5F FAULTY
    (b) ADDRESSES 20–3F FAULTY
    (c) ADDRESSES 00–7F FAULTY

ANSWERS TO SECTION CHECKUPS

SECTION 10–1   Memory System Hierarchy

1. The purpose of memory hierarchy is to obtain the fastest access time at the lowest cost.
2. Access time is the time it takes a processor to retrieve (read) or write a block of data stored in the memory?
3. Generally, the higher the capacity the lower the cost per bit.
4. Yes
5. A hit is when the processor finds the requested data at the first place it looks. A miss is when the processor fails to find the requested data and has to go to another level of memory to find it.
6. The hit rate

SECTION 10–2   Semiconductor Memory Basics

1. Bit is the smallest unit of data.
2. 256 bytes is 2048 bits.
3. A write operation stores data in memory.
4. A read operation takes a copy of data from memory.
5. A unit of data is located by its address.
6. A RAM is volatile and has read/write capability. A ROM is nonvolatile and has only read capability.

SECTION 10–3   The Random-Access Memory (RAM)

1. Asynchronous and synchronous with burst feature
2. A small fast temporary memory between the CPU and main memory
3. SRAMs have latch storage cells that can retain data indefinitely while power is applied. DRAMs have capacitive storage cells that must be periodically refreshed.
4. The refresh operation prevents data from being lost because of capacitive discharge. A stored bit is restored periodically by recharging the capacitor to its nominal level.
5. FPM, EDO, BEDO, Synchronous

SECTION 10–4   The Read-Only Memory (ROM)

1. 512×8 equals 4096 bits.
2. Mask ROM, PROM, EPROM, UV EPROM, EEPROM
3. Eight bits of address are required for 256 byte locations ($2^8 = 256$).

SECTION 10–5   Programmable ROMs

1. PROMs are field-programmable; ROMs are not.
2. 1s are left after EPROM erasure.
3. Read is the normal mode of operation for a PROM.

SECTION 10–6   The Flash Memory

1. Flash, ROM, EPROM, and EEPROM are nonvolatile.
2. Flash is nonvolatile; SRAM and DRAM are volatile.
3. Programming, read, erase

SECTION 10–7   Memory Expansion

1. Eight RAMs
2. Eight RAMs
3. DIMM: Dual in-line memory module
4. DDR: double data rate

SECTION 10–8   Special Types of Memories

1. In a FIFO memory the *first* bit (or word) *in* is the *first* bit (or word) *out*.
2. In a LIFO memory the *last* bit (or word) *in* is the *first* bit (or word) *out*. A stack is a LIFO.

3. The PUSH operation or instruction adds data to the memory stack.

4. The POP operation or instruction removes data from the memory stack.

5. CCD is a charge-coupled device.

SECTION 10–9   Magnetic and Optical Storage

1. Magnetic storage: hard disk, tape, and magneto-optical disk.

2. A magnetic disk is organized in tracks and sectors.

3. A magneto-optical disk uses a laser beam and an electromagnet.

4. Optical storage: CD-ROM, CD-R, CD-RW, DVD-ROM, WORM

SECTION 10–10   Troubleshooting

1. The contents of the ROM are added and compared with a prestored checksum.

2. Checksum cannot be used because the contents of a RAM are not fixed.

3. (1) a short between adjacent cells; (2) an inability of some cells to store both 1s and 0s; (3) dynamic altering of the contents of one address when the contents of another address change.

ANSWERS TO RELATED PROBLEMS FOR EXAMPLES

10–1 $G_3G_2G_1G_0 = 1110$

10–2 Connect eight 64k $\times$ 1 ROMs in parallel to form a 64k $\times$ 8 ROM.

10–3 Sixteen 64k $\times$ 1 ROMs

10–4 See Figure 10–82.

FIGURE 10–82

10–5 ROM 1: 0 to 524,287; ROM 2: 524,288 to 1,048,575

ANSWERS TO TRUE/FALSE QUIZ

1. T	**2.** F	**3.** T	**4.** F	**5.** F
6. T	**7.** T	**8.** T	**9.** F	**10.** F

ANSWERS TO SELF-TEST

1. (d)	**2.** (b)	**3.** (c)	**4.** (c)	**5.** (d)	**6.** (a)
7. (d)	**8.** (c)	**9.** (a)	**10.** (b)	**11.** (f)	**12.** (d)

CHAPTER 11

DATA TRANSMISSION

OUTLINE

OBJECTIVES

- Discuss various types of data transmission media
- Describe the methods for data transmission
- Explain data transmission modes
- Define *modulation*
- Describe the types of modulation for transmission of digital data
- Explain how digital signals are used to transmit analog information
- Define *multiplexing* and *demultiplexing*
- Discuss the types of multiplexing and demultiplexing
- Describe how the transmission medium affects the quality of data

KEY TERMS

Coaxial cable	Asynchronous
EMI	Bit rate
UTP	Baud
Optical fiber	Synchronous
Electromagnetic waves	RZ

NRZ	PAM
Manchester encoding	PWM
Simplex	PPM
Half-duplex	PCM
Full-duplex	TDM
ASK	FDM
FSK	Signal-to-noise ratio
PSK	Noise margin
QAM	

INTRODUCTION

Digital data transmission is the transfer of data over a medium or channel from one point to one or more points. Channels can be wire, cable, optical fibers, or wireless. Digital information (data) transmission falls into two basic categories, analog and digital. As you know, analog information is transmitted as a continuously varying signal. Digital data is transmitted either as a sequence of pulses, called *baseband transmission*, or by modulating an analog signal, called *broadband transmission*. Both types are generally considered digital transmission. A modem is used in broadband modulation and demodulation. Early fundamental work in data transmission and information theory was done by Harry Nyguist, Ralph Hartley, Claude Shannon, and others.

VISIT THE WEBSITE

Study aids for this chapter are available at

http://pearsonhighered.com/floyd

11-1 DATA TRANSMISSION MEDIA

All data transmission systems in their most basic form have a data source (sending device) at one end and a receiving device at the other. The two devices are connected by a transmission medium, which can be wire, coaxial cable, unshielded twisted pair cable, shielded twisted pair cable, optical fiber, or space (wireless). A digital signal is a changing electrical or electromagnetic quantity that carries information through the medium. When data is sent without modulation, usually over wires or cables, it is called *baseband transmission*. When data is modulated and sent through a wireless medium, it is called *broadband transmission*.

After completing this section, you should be able to

- Define a data transmission system
- Describe wire and cable transmission media
- Describe optical fiber transmission media
- Describe wireless transmission

The basic block diagram in Figure 11–1 illustrates the essential elements in a data transmission system.

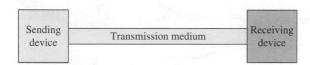

FIGURE 11–1   Basic data transmission system.

Wire Cable Connections

The simplest connection between sending and receiving devices is a wire or a conductive trace on a printed circuit (PC) board. This type of connection is typically limited to internal data transmission over very short distances within the same system or between nearby systems, such as a computer and/or peripherals. Cables, data buses, or conductive traces that connect one element to another on a PC board and between PC boards in close proximity or between parts of a system are examples of wire/cable connections, as illustrated in Figure 11–2.

FIGURE 11–2   Conductive traces on PC boards and wire interconnection between boards.

COAXIAL CABLE (COAX)   Coaxial cable consists of a center conductor within an insulating dielectric material. A copper braided or foil shield surrounds the dielectric to protect the conductor against electromagnetic interference (**EMI**). The shield is encased in a protective insulating jacket, as shown in Figure 11–3. BNC (Bayonet Neill-Concelman) connectors are typically used for coaxial connections. Coax is used in data transmission applications with data rates up to about 1 GHz. Two common applications for coax are cable TV and Internet connections.

FIGURE 11–3   Construction view of a coaxial cable.

Unshielded Twisted Pair Cable

Unshielded twisted pair (**UTP**) cable is used extensively for indoor telephone application as well as some outdoor uses. It is also found in many computer networks and video applications, such as security cameras, and also in the USB (universal serial bus) cable. UTP is color-coded according to a standard 25-pair color code. Most cables use a subset of these standard colors.

An unshielded twisted pair minimizes cross talk (discussed in Section 11–6) when bundled with other twisted pair into cables. The two wires in each pair are twisted so that they cross each other at nearly 90°, ideally cancelling any electromagnetic fields generated by the signals in the wires. UTP cables are limited to use in low-noise environments and to lower signal frequencies than coax, such as audio and other signals up to about 1 MHz. UTP cables use standard RJ-45 connectors. A common four-pair UTP cable is shown in Figure 11–4(a), and an RJ-45 connector is shown in part (b). Shielded twisted pair (**STP**) cable encased in a metal sleeve or conduit is also available and provides more protection from EMI.

(a)   (b)

FIGURE 11–4   **Example of an unshielded twisted pair (UTP) cable and connector.**

Optical Fiber Cables

The structure of a single **optical fiber** is shown in Figure 11–5(a). An optical fiber can be as small as a human hair, so many single fibers can be bundled into a cable, as shown in Figure 11–5(b).

Jacket
Cladding
Core

(a) Single optical fiber   (b) Fiber optic cable

FIGURE 11–5   **Optical fiber cables.**

Instead of using electrical pulses to transmit information through copper lines, fiber optics uses light pulses transmitted through optical fibers. Fiber optic systems have several advantages over electrical transmission media. Advantages include faster data rates, higher signal capacity (more signals at a time), and better transmission over longer distances; optical fibers are not susceptible to EMI. The main disadvantage of fiber optics is the cost, which is higher than that of coax, UTP, and STP.

Optical fiber is commonly used as a medium for telecommunication and networking. Because light propagates through the fiber with little attenuation compared to electrical cables, optical fiber is useful for long-distance transmission. Data rates from 10 GHz to 40 GHz are common, although data rates over 100 GHz are used.

When light is introduced at one end of an optical fiber called the *core*, it "bounces" along until it emerges from the other end, as shown in Figure 11–6. The fiber is typically made of pure glass, plastic, or other material that is surrounded by a highly reflective cladding, which effectively acts as a mirrored surface, using a phenomenon called *total internal reflection* to produce an almost lossless reflection. This allows the light to move around bends in the fiber.

FIGURE 11–6   **Light propagating through an optical fiber while reflecting off the internal surface.**

MODES OF LIGHT PROPAGATION   Two basic modes of light propagation in optical fibers are multimode and single mode, as illustrated in Figure 11–7. In **multimode**, light entering the fiber will tend to propagate through the core in multiple rays (modes),

(a) Multimode   (b) Single mode

FIGURE 11–7   **Modes of light propagation in an optical fiber.**

basically due to varying angles as each light ray moves along. Some of the rays will go straight down the core, while others will bounce back and forth. Still others will scatter due to the sharp angle at which they strike the cladding, resulting in attenuation in light energy. Multimode also exhibits time dispersion, which means that all the light rays do not arrive at the end of the fiber at exactly the same time. In **single mode**, the core is much smaller in diameter than in multimode. Light entering the fiber tends to propagate in a straight line as a single ray.

The diameter of the optical fiber determines the mode. Currently, there are three sizes most widely used in data transmission: 50/125, 62.5/125, and 8.3/125. The numbers are in microns (one micron is one millionth of a meter) and represent the diameters of the fiber core and cladding respectively. The 50/125 and the 62.5/125 are multimode fibers. The 8.3/125 is a single-mode fiber. Single mode results in increased bandwidth and distance for transmission, but the costs are higher than for multimode.

A Fiber-Optic Data Communications Link

A simplified block diagram of a fiber-optic communications link is shown in Figure 11–8. The source provides the electrical signal that is to be transmitted. The electrical signal is converted to a light signal and coupled to the fiber-optic cable by the transmitter. At the receiving end, the light signal is coupled out of the cable and converted back to an electrical signal. The signal is then processed and sent to the end user.

FIGURE 11–8   **Basic block diagram of a fiber-optic communications link.**

The electrical signal modulates the light intensity and produces a light signal that carries the same information as the electrical signal. Special connectors are used to connect the fiber-optic cable to various equipment.

Various types of connectors are used in fiber-optic systems. Some of these are described as follows and shown in Figure 11–9:

- ST   An AT&T trademark and one of the most widely used for multimode networks
- SC   A snap-in type multimode connector
- FC   A popular single-mode connector type
- LC   A single-mode connector
- LX-5   Similar to an LC connector except it has a shutter over the end of the fiber
- MT   A 12-fiber connector used for ribbon cables
- FDDI   All duplex, meaning the connector can accommodate two optical fibers for two-way communication.

ST   SC   FC

LC   MT   FDDI

FIGURE 11–9   Typical types of optical fiber connectors.

Wireless Transmission

The transmission of data through air and space without the use of physical connections between sending and receiving systems is known as *wireless transmission*. Wireless transmission generally can be categorized by the type of signal in terms of application, frequency, or how the data is configured.

THE ELECTROMAGNETIC SPECTRUM   The spectrum of frequencies for the range of **electromagnetic waves** is shown in Figure 11–10. Most data communications occur within the radio wave, microwave, and infrared frequencies. Another medium of wireless communications is water where sonar is used. Sonar produces low-frequency sound waves that do not fall into the electromagnetic spectrum.

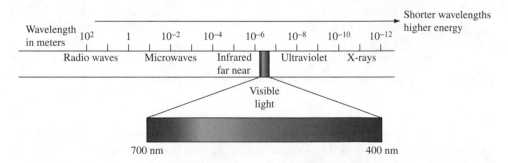

FIGURE 11–10   The electromagnetic spectrum.

RADIO FREQUENCY (RF) PROPAGATION   Three ways in which rf and microwave signals propagate through Earth's atmosphere (air) are ground wave, ionospheric, and line-of-sight. In ground wave propagation, the radio waves follow the curvature of Earth and can be up to about 2 MHz in frequency; the standard AM broadcast band is an example. Radio frequencies in the 30 MHz to 85 MHz range bounce off the ionosphere. These signals can change with time of day and weather conditions. Most ham radio bands are examples of ionospheric propagation. In line-of-sight (LOS) propagation, the receiver must be in view of the transmitter. The distance is limited to about 100 km (horizon to horizon) from a ground-based transmitter to a ground-based receiver. Long distances are achieved by placing a series of repeater towers so that each tower is within the line-of-sight of the previous tower. In the case of satellites, which use line-of-sight propagation, the distances can be extended around the world. Figure 11–11 illustrates these types of rf propagation.

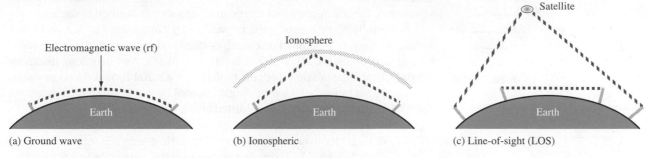

(a) Ground wave   (b) Ionospheric   (c) Line-of-sight (LOS)

FIGURE 11–11   **Ways in which rf and microwave signals can propogate.**

INFRARED (IR)   Communication in the infrared region can be line-of-sight or diffused. With LOS, the transmitter and receiver must be visible to each other with no obstacles in between. With diffusion, the IR waves reflect off nearby surfaces such as buildings, ceilings, and walls. Uses include remote control devices, weather satellites, and night vision.

SECTION 11–1 CHECKUP*

1. List the types of data transmission media.

2. What is the purpose of a coax shield?

3. Name three ways in which radio waves are propagated in wireless transmission.

4. Which type of electromagnetic radiation has the highest frequencies?

5. Generally, what is the difference between baseband and broadband?

*Answers are at the end of the chapter.

11–2 METHODS AND MODES OF DATA TRANSMISSION

Data transmission over a communication channel can be configured in several ways. A communication channel is the pathway over which data is sent and can be in the form of any of the media discussed in Section 11–1. The methods by which data is transmitted can be one bit at a time (serial) or several bits at a time (parallel), and the data can be either synchronized or unsynchronized. The modes describe the direction of the data.

After completing this section, you should be able to

- **Distinguish between serial and parallel data**
- **Distinguish between synchronous and asynchronous data**
- **Describe the three modes of data transmission**

Serial and Parallel Data

Serial data transmission is when data is transmitted one bit at a time in a bit stream, as illustrated in Figure 11–12(a). **Parallel** data transmission is when data is transmitted several bits at a time, as shown in part (b). In general, a given number of bits can be transmitted

1101 0010 ⟶
(a) Serial data

1 ⟶
1 ⟶
0 ⟶
1 ⟶
0 ⟶
0 ⟶
1 ⟶
0 ⟶
(b) Parallel data

FIGURE 11–12

faster in parallel than in series, resulting in higher data rates. However, when several bits are sent simultaneously on separate lines in parallel, slight differences in the properties of the lines can cause skewing in the data, making the data more susceptible to error, so the data rate may need to be reduced to prevent errors. Error detection and correction methods can be used in these cases.

Generally, data is processed in parallel by computers but is transmitted serially to outside systems. For example, data from a computer to a printer is typically sent over a USB, which is serial. Data that is sent over long distances via one of the transmission media is typically in a serial format. In some cases, data can be sent in parallel over a single channel by using different frequencies for each bit so the bits can be transmitted at the same time.

In most systems, serial-to-parallel and parallel-to-serial conversions are used. The basic concept is shown in Figure 11–13. A simple parallel-to-serial converter is a parallel in/serial out shift register. The parallel bits are loaded into the register and then shifted out in serial format at a specific clock frequency that determines the data rate (bits/s). Likewise, a serial in/parallel out shift register can be used as a serial-to-parallel converter.

FIGURE 11–13   Digital data conversions.

Data Synchronization

In a data transmission system, the receiving device must be able to recognize each valid group of data bits that it is receiving serially. Two types of data transmission in terms of how a receiving device knows what a sending device is transmitting are synchronous and asynchronous.

Asynchronous Data

Data is sent in short "bursts" known as **packets.** A data packet is one complete piece of information of a longer message. Typically, many packets make up the entire message. A data packet consists of data bits representing alphabetic or numeric characters, a parity bit, and start/stop bits. There is a pause between data packets so that the receiver recognizes the start bit that precedes each packet. At the end of the data packet there is one or more stop bits that tell the receiver the packet is complete.

In **asynchronous** systems, the sending and receiving devices operate with separate oscillators having the same clock frequency. Because the separate clock frequencies may drift over time, they are typically resynchronized on each data packet with the start bit. Most commonly, data is sent in small packets of perhaps 10 or 11 bits. Eight of these bits carry the information and the others are parity bits and start/stop bits. Between packets, when the channel is idle, there is a continuous logic level. A data packet always begins with a start bit with the opposite logic level as the idle period to alert the receiver that a data packet is starting. A parity bit follows the eight data bits, and a stop bit signals the end of the packet. This is illustrated in Figure 11–14.

FIGURE 11–14   Serial transmission of a data packet for a given data code.

Data Rate

Data rate is the speed of data transfer. In a serial data transmission the data rate can be stated as bit rate or baud, although bit rate is the preferred measure. The **bit rate** is the number of bits (1s and 0s) per second (bps), and the **baud** is the symbol rate or the number of data symbols (sometimes known as transitions or events) per second.

$$\text{Bit rate} = \text{bps (bit/s)}$$

$$\text{Baud} = \text{symbols per second (symbols/s)}$$

A symbol (transition) can consist of one or more bits. Therefore, bit rate is always greater than or equal to the baud. The relationship between bit rate and baud is

$$\text{Bit rate} = (\text{Number of bits per symbol})(\text{Baud})$$

or

$$\text{Baud} = \frac{\text{Bit rate}}{\text{Number of bits per symbol}}$$

Typically, data is information in the form of a modulated signal. For example, data could consist of a string of ASCII characters, voltage level values, frequency values, or phase values. If data consists of ASCII characters, each character is called a symbol and is represented by eight bits. To illustrate, assume that one symbol (8 bits in this case) is transmitted every millisecond (ms). The data rate expressed as baud is

$$\text{Baud} = (1 \text{ symbol/ms})(1000 \text{ ms/s}) = 1000 \text{ baud} = 1 \text{ kbaud}$$

The data rate in terms of bit rate is

$$\text{Bit rate} = (8 \text{ bits/symbol})(1000 \text{ symbols/s}) = 8000 \text{ bps} = 8 \text{ kbps}$$

EXAMPLE 11–1

A certain analog waveform amplitude is represented by sixteen different voltage levels. Each level is a symbol consisting of four bits ($2^4 = 16$). If eight symbols are transmitted in 1 ms, determine the baud and the bit rate.

SOLUTION

$$\text{Baud} = (8 \text{ symbols/ms})(1000 \text{ ms/s}) = \textbf{8 kbaud}$$

$$\text{Bit rate} = (4 \text{ bits/symbol})(8000 \text{ symbols/s}) = \textbf{32 kbps}$$

RELATED PROBLEM*

Determine the bit rate if a symbol is represented by eight bits and the baud rate is 5000 symbols/s.

Answers are at the end of the chapter.

Transmission Efficiency

The efficiency of a data transmission channel is the ratio of data bits to total bits. For example in Figure 11–14, there are eight data bits, a start bit, a parity bit, and a stop bit. The nondata bits are considered overhead bits. There are eleven total bits transmitted in a packet, so the efficiency of the transmission is

$$\text{Efficiency} = \frac{\text{Data bits}}{\text{Total bits}} = \frac{8 \text{ bits}}{11 \text{ bits}} = 0.727 \text{ or } 72.7\%$$

EXAMPLE 11–2

A certain system transmits a block of information containing ten packets each with eight data bits, a start bit, and a stop bit. Additional overhead bits include a 4-bit synchronization code at the beginning of the block and a parity bit at the end of the block. Determine the transmission efficiency.

SOLUTION

Data bits $= 8 \times 10 = 80$

Overhead bits $= 1 \times 10$ start bits $+ 1 \times 10$ stop bits

$+ 4$ synchronization bits $+ 1$ parity bit $= 25$

Total bits $=$ Data bits $+$ Overhead bits $= 105$

Efficiency $= \dfrac{\text{Data bits}}{\text{Total bits}} = \dfrac{80}{105} = $ **0.762 or 76.2%**

RELATED PROBLEM

Determine the efficiency if 12 data bits are included in each packet with the same number of overhead bits stated in the example.

Synchronous Data

In **synchronous** data transmission both the sender and the receiver derive timing from the same clock signal, which originates at the sender end of the system. The bits are transmitted in a continuous stream with no pauses, so the receiver must have some way to recognize where a data block starts and ends. In order for the receiver to determine when to read information bits from the channel, it must determine exactly when the data begins and the time between bits. When this timing information is determined, the receiver is synchronized with the transmitter. Unlike asynchronous transmission, the data blocks usually contain more than one character of information. Synchronous transmission is generally faster than asynchronous transmission.

One method of synchronization is by using separate channels to transmit the data and the timing information (synchronization and clock pulses). Because the transmitter originates both the data and the timing pulses, the receiver will read the data channel only when instructed to do so by the transmitter (via the timing channel) and synchronization is achieved. The disadvantage of this method is that it requires two physical lines. Two data formats that require separate data and timing are **RZ** (return to zero) and **NRZ** (nonreturn to zero). In the RZ format, a single pulse during a bit time represents a 1 and the absence of a pulse is a 0, as shown in Figure 11–15(a). In the NRZ format, a high level during a bit time represents a 1 and a low level represents a 0. A series of 1s is represented by a continuous high level, and a series of 0s is represented by a continuous low level. The waveform does not return to the low level until a zero occurs after a string of 1s and does not go back to the high level until a 1 occurs after a string of 0s. This is illustrated in Figure 11–15(b).

(a) RZ format   (b) NRZ format

FIGURE 11–15   **Data formats requiring separate timing for synchronization.**

Another more commonly used method is to embed the timing signal in the data so that only one channel is required. One technique is to use biphase or **Manchester encoding** at the transmitter. The receiver extracts the embedded timing signal and uses it to synchronize to the transmitter. Figure 11–16 illustrates Manchester encoding. A rising edge in the code is a 1 and a falling edge is a 0, as indicated by the up and down arrowheads. The edges occur at the middle of the bit time. The biphase code is sent to the receiver, and the clock is extracted from the data with a phase-locked loop. Sometimes a series of all 1s or all 0s is included in the transmission to allow the receiver to synchronize.

FIGURE 11–16   Example of Manchester-encoded data and timing.

EXAMPLE 11–3

Determine the biphase (Manchester) code for the data and clock shown in Figure 11–17(a).

FIGURE 11–17

SOLUTION

The encoded data and embedded timing are shown in Figure 11–17(b). The rising edges are 1s and the falling edges are 0s that occur in the middle of each bit time (period of the clock).

RELATED PROBLEM

If the data were all 1s, what would the Manchester code look like?

SYNCHRONOUS FRAMES   Synchronous data is sent in frames which include other bits. Frame formats vary because there are numerous standards in use. Figure 11–18 illustrates a generic synchronous protocol.

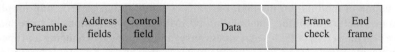

FIGURE 11–18   Basic synchronous frame structure.

Preamble   A group of bits at the beginning of a frame that is used to alert the receiver that a new frame has arrived and to synchronize the receiver's clock with the transmitted clock.

Address fields   A group of bits containing the address(s) of the sender and the receiver. One or both addresses may be present in a given protocol.

Control field   This group of bits identifies the type of data being sent, such as handshaking (establishes a connection), file transfers, or the size of the data.

Data field   This sequence is the actual information being sent and can be of a fixed or a variable length. If it is a fixed-length field, a group of bits called a *pad* is used to fill in if the actual data field is less than the fixed field.

Frame check   This field contains an error check such as parity, CRC (cyclic redundancy check), or checksum, which is a value computed by a simple algorithm of the data bits in the frame.

End frame   A group of bits that tells the receiver when the end of the frame occurs.

Transmission Modes

Three modes that characterize data channel (media) connections are simplex, half-duplex, and full-duplex. In the **simplex** connection, data flows in only one direction from the sender (transmitter) to the receiver. In a computer, for example, data flows one way from the computer to the printer. In the **half-duplex** mode, data flows both ways but not at the same time in the same channel. For example, a sender transmits information to the receiver and the receiver responds after it has received the information. In the **full-duplex** mode, the data flows both ways simultaneously in the same channel. The bandwidth of the channel is divided between the two directions. Figure 11–19 illustrates these three modes.

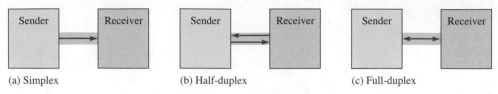

(a) Simplex   (b) Half-duplex   (c) Full-duplex

FIGURE 11–19   Data transmission modes.

SECTION 11–2 CHECKUP

1. Explain the difference between serial and parallel data.

2. What is the purpose of synchronization in a data transmission system?

3. Name three types of data formats.

4. List the modes of data transmission?

11–3 MODULATION OF ANALOG SIGNALS WITH DIGITAL DATA

Three major classes of modulation techniques for wireless transmission of digital data are amplitude-shift keying (ASK), frequency-shift keying (FSK), and phase-shift keying (PSK). Another common modulation method, basically a combination of PSK and amplitude, is known as quadrature amplitude modulation (QAM). Several variations of these main techniques are also in use. All of these techniques convey information by changing a property of an rf carrier signal, usually a sine wave, for the purpose of conveying digital information over a wireless medium.

After completing this section, you should be able to

- **Describe ASK**
- **Describe FSK**
- **Describe PSK**
- **Describe QAM**

Amplitude-Shift Keying

Amplitude-shift keying, **ASK**, is a form of modulation in which a digital signal modulates the amplitude of a higher frequency sine wave (carrier). In its simplest form, a sinusoidal carrier signal is turned on and off by the data signal and, therefore, is also known as on-off keying (OOK). When the carrier is *on*, a binary 1 is represented, and when the carrier is *off*, a binary 0 is represented. ASK is susceptible to noise interference and is not typically used for wireless data transmission. ASK is most commonly used in fiber optics where the presence of light represents a binary 1 and the absence of light represents a binary 0. Figure 11–20 illustrates the concept of ASK. The presence of a carrier signal is a 1, and the absence of a carrier signal is a 0. When modulated by digital data (1s and 0s), this method is sometimes known as BASK (binary amplitude-shift keying).

FIGURE 11–20   Illustration of amplitude-shift keying (ASK).

Frequency-Shift Keying

Frequency-shift keying, **FSK**, is a form of modulation in which a digital signal modulates the frequency of a higher frequency sine wave (carrier). A carrier signal with a lower frequency generally represents a binary 0, and a carrier signal with a higher frequency represents a binary 1. When modulated by digital data (1s and 0s), this method is sometimes known as BFSK (binary frequency-shift keying). Figure 11–21 illustrates FSK.

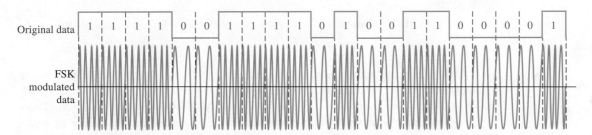

FIGURE 11–21   Illustration of frequency-shift keying (FSK).

Phase-Shift Keying

Phase-shift keying, **PSK**, is a form of modulation in which a digital signal modulates the phase of a higher frequency sine wave (carrier). A carrier signal of one phase represents a binary 1, and a carrier signal that is 180° out-of-phase represents a binary 0. When modulated by digital data (1s and 0s), this method is sometimes known as BPSK (binary phase-shift keying). In one of its many variations, PSK applications include wireless LAN and bluetooth. Figure 11–22 illustrates PSK.

FIGURE 11–22   Illustration of phase-shift keying (PSK).

FIGURE 11–23 **Eight ampli-tude/phase combinations (mod-ulation states) represent one of the eight 3-bit groups. Only one cycle of each modulation state is shown.**

Quadrature Amplitude Modulation

Quadrature amplitude modulation, **QAM**, is widely used in telecommunications and in digital cable TV. Digital QAM uses a combination of PSK and amplitude modulation to send information. Quadrature refers to a 90° phase difference. Each combination of phase and amplitude is called a *modulation state* or *symbol* and represents a combination of two or more bits. To illustrate the basic concept of QAM, we will use what is known as 8-QAM, where each of the eight modulation states (2^3) represents a unique three-bit combination. As shown in Figure 11–23, in 8-QAM, there are two different amplitudes with a quadrature phase difference between each pair. Since there are four quadratures (90°) in 360°, there are a total of eight different amplitude/phase combinations or modulation states.

A digital 8-QAM data transmission can represent any combination of three bits in any sequence. Figure 11–24 illustrates an 8-QAM transmission of the binary sequence of numbers 0 through 7. Depending on the carrier frequency and the time specified for each bit group, multiple cycles of each modulation state can represent each bit group. For simplicity, only two cycles per bit group are shown.

FIGURE 11–24 Illustration of an 8-QAM transmission of the binary sequence shown.

M-QAM There are numerous variations in QAM in terms of the number of modulation states (*M*) that can be represented. For the 8-QAM just illustrated, *M* = 8. Higher *M* values of QAM are also commonly used such as 16-QAM, 64-QAM, and 256-QAM. These higher *M* values are achieved by using more amplitude levels and/or phases. For example, a 64-QAM can have 16 amplitude levels and four phases and can represent 6-bit binary groups. For a 256-QAM, there can be 32 amplitude levels and eight phases and can represent 8-bit binary groups.

Constellation Maps

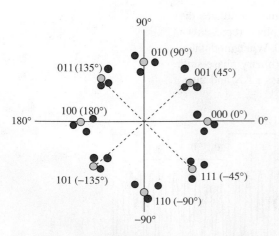

FIGURE 11–25 **Constellation map for a 3-bit PSK transmission. The phases are 0°, 45°, 90°, 135°, 180°, −45°, −90°, and −135° as indicated.**

Modulated transmission of digital data can be represented by a constellation map, which is a vector representation that graphically shows the symbol values and corresponding phases being transmitted by a system. As you have seen, when data is transmitted, a pattern is modulated into the signal, such as in PSK, where the bit pattern is represented by various phase shifts. A constellation map is useful in the design and analysis of a data transmission system and in visually understanding how the system works. Figure 11–25 shows a 4-quadrant constellation map for a 3-bit PSK transmission. Each green dot represents the ideal amplitude and phase of the modulated signal. The amplitude constant is represented by the distance of the dots from the origin. Eight different phases or symbols represent the 3-bit binary combinations.

In an actual transmission, the medium can affect both amplitude and phase shift. In the figure, the cluster of red dots around each green dot represents nonideal signal values. When the signal is received, these nonideal values can be adjusted to the ideal value (nearest green dot) as long as there is adequate separation of the clusters; there should be no confusion as to which ideal value it belongs. If there is any overlap, errors can occur.

EXAMPLE 11–4

Develop an ideal constellation map for the 8-QAM transmission represented in Figures 11–23 and 11–24.

SOLUTION

There are four phases and four amplitudes that represent a 3-bit code in this system. The ideal constellation map is shown in Figure 11–26. There are two amplitudes for each phase represented by the distance from the origin.

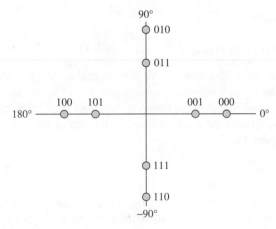

FIGURE 11–26

RELATED PROBLEM

How many phases and amplitudes could be used to represent a 4-bit code?

SECTION 11–3 CHECKUP

1. List four types of modulation techniques.

2. What parameter is changed in ASK?

3. What parameter is changed in FSK?

4. What is QAM?

5. What parameter is changed in PSK?

As you learned in the last section, analog signals are commonly used to carry digital data. In this section, you will see that digital signals can be used to carry analog information. These techniques are usually referred to as pulse modulation. A pulse parameter such as amplitude or pulse width is varied to represent an analog quantity. The details of related topics such as sampling and analog-to-digital conversion are covered in Chapter 12.

After completing this section, you should be able to

- Describe pulse amplitude modulation
- Describe pulse width modulation
- Discuss the basic concept of sampling
- Explain delta modulation

FIGURE 11–27   A simple PAM signal.

Pulse Amplitude Modulation

In pulse amplitude modulation, **PAM**, the heights or amplitudes of the pulses are varied according to the modulating analog signal, and each pulse represents a value of the analog signal. PAM is the simplest, but least used, type of pulse modulation although it is used in the Ethernet communications standard. A simple PAM sequence is shown in Figure 11–27.

Ethernet is a family of computer networking protocols described by the IEEE 802.3 standard. Systems that communicate using Ethernet divide the data into individual packets called frames. Each frame contains source and destination addresses and error-checking bits. The Ethernet standard includes several variations that specify both media and signaling standards, including type of wire or cable, data format, and data rates.

SYSTEM NOTE

GENERATION OF A PAM SIGNAL   A basic method of producing a PAM representation of an analog signal is to use a constant-amplitude pulse source to sample the analog wave that has a frequency lower than the pulses, as shown in Figure 11–28 for a sine wave input, although any form of analog signal can be converted to a PAM output. The pulses turn the switch on (closed) and off (open) to sample the waveform. When there

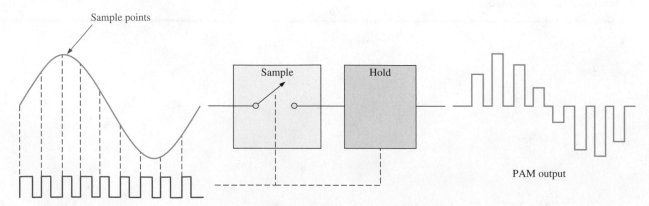

FIGURE 11–28   Basic method of pulse amplitude modulation.

is a pulse, the sample switch is closed; and the amplitude of the sine wave at that point goes to the hold element that maintains the initial analog value occurring at the beginning of each pulse for the duration of the pulse. The output goes to zero between pulses.

Pulse Width Modulation

In pulse width modulation, **PWM**, (also known as pulse duration modulation, PDM), the width or duration of the pulses and duty cycle are varied according to the modulating analog signal; each pulse represents a value of the analog signal. PWM is commonly used in various control applications. Braking systems, motor speed control, and renewable energy systems are just three examples.

Figure 11–29 illustrates one method of PWM generation, called the *intersective* method, which uses a sawtooth waveform. A triangular waveform can also be used. Again, a sine wave input is used, but the input can be any type of analog waveform. The sawtooth intersects the sinusoidal modulating signal twice during each cycle. The sawtooth is either above the sine wave or below the sine wave. When the sawtooth is above the sine wave, a low level is generated; when it is below the sine wave, a high level is generated. The resulting output is a series of pulses with widths proportional to the amplitude of the sine wave.

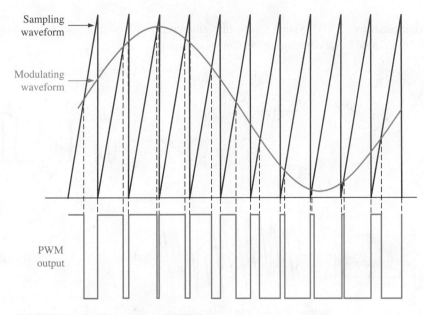

FIGURE 11–29   Illustration of PWM.

EXAMPLE 11–5

Determine the PAM signal and the PWM signal for the modulating signal in Figure 11–30. Assume ten cycles of the sampling pulse waveform or sawtooth waveform during the portion of the modulating signal shown.

FIGURE 11–30   Modulating waveform.

SOLUTION

Figure 11–31 shows the PAM and PWM results.

FIGURE 11–31

RELATED PROBLEM

How would the outputs in Figure 11–31 be affected by an increase in frequency of the pulse and sawtooth waveforms?

An intersective PWM system can be implemented simply with a sawtooth or triangular wave generator and a comparator, as shown in Figure 11–32.

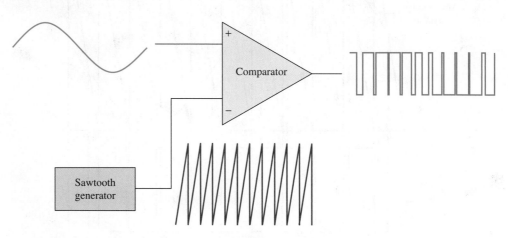

FIGURE 11–32 **A basic method of pulse width modulation.**

SYSTEM EXAMPLE 11–1

SOLAR ENERGY SYSTEM WITH BATTERY BACKUP

In this application, PWM is used in a charge controller to control the energy delivered to the storage batteries from solar modules. Figure 11–33 shows the basic concept of a PWM charge controller. In part (a), the PWM and control circuit produce pulses based on the input from the sampling circuit. The sampling circuit determines the actual battery voltage by sampling the voltage between pulses. The diode acts as a rectifier to smooth out the pulses and also blocks discharge of the battery back through the charger when the source is not generating power. Part (b) demonstrates how the battery charges during each pulse and how the width and the time between pulse change as the battery charges. At the beginning, when the battery charge is low, the wider pulses deliver more energy to the battery; as the battery charges, less energy is required so the pulses narrow.

(b) Waveforms

FIGURE 11–33   **PWM charge controller operation.**

Pulse Position Modulation

In pulse position modulation, **PPM**, (also known as pulse phase modulation) the position of each pulse relative to a reference or timing signal is varied proportional to the modulating signal waveform. The amplitude and width of the pulses in a PPM system are kept constant. An example of a PPM signal is shown in Figure 11–34, where the PPM pulses are shifted relative to the leading edges of the timing waveform.

FIGURE 11–34   **Example of a PPM signal with timing.**

As with other types of pulse modulation, there is generally more than one way to produce a modulated waveform. One method is to derive the PPM from PWM, as illustrated in Figure 11–35. Notice that the leading (positive-going) edges of the PWM signal

FIGURE 11–35 **A method of generating PPM.**

occur at fixed intervals, while the trailing (negative-going) edges vary relative to the leading edges. When the PWM signal is passed through a differentiator, short positive pulses (spikes) are generated on the leading edges and short negative pulses occur on the trailing edges. The differentiated signal is rectified to remove the positive pulses and generate the PPM waveform shown, which can be inverted to produce positive pulses.

A simplified block diagram of a PPM system is shown in Figure 11–36. PPM is widely used in optical communications, such as fiber optics, and in certain types of rf systems, such as radio control for model planes, boats, and cars. It is less sensitive to channel interference than PAM or PWM because noise can alter pulse amplitude and width but not so much the position.

FIGURE 11–36 **PPM system block diagram.**

PPM ENCODING A certain number of data bits (D) are encoded by a single pulse in one of 2^D possible positions during a specified fixed time period (T). The data rate is D/T bits per second (bps). Figure 11–37 illustrates this case of four time periods and two data bits per time period. There are $2^D = 2^2 = 4$ possible positions in each time period. As you can see, each position represents a 2-bit binary number. In the first time period, the pulse position represents 00, in the second time period the pulse position represents 10, etc. Any pulse could be in any of the four positions within each period, depending on the data being encoded. The code for this particular set of pulse positions is 00100111, and it is shown in NRZ format.

FIGURE 11–37 **Encoding of a PPM signal.**

┌───┐

EXAMPLE 11–6

For a PPM system with four data bits and a time period of 1 μs, determine the data rate. How many possible pulse positions are there in each time period?

SOLUTION

The data rate is

$$\frac{D}{T} = \frac{4 \text{ bits}}{1 \, \mu s} = \textbf{4 Mbps}$$

The number of possible pulse positions in each period is

$$2^D = 2^4 = \textbf{16}$$

RELATED PROBLEM

For eight data bits, what is the data rate ($T = 1 \, \mu$s), and what is the number of possible pulse positions in each period?

└───┘

Pulse Code Modulation

Pulse code modulation, **PCM**, involves sampling of an analog signal amplitude at regular intervals and converting the sampled values to a digital code. The sample-and-hold process was mentioned in Chapter 1 and is discussed in detail in Chapter 12. The concept of PCM is demonstrated in Figure 11–38.

FIGURE 11–38   Concept of PCM with eight levels.

The modulating signal is a sine wave in this illustration, and its amplitude is divided into eight levels as shown. Each level is represented by a 3-bit binary number ($2^3 = 8$). The sine wave is sampled at fixed intervals; and the sampled value is held until the next sample, resulting in the green stair-step waveform that approximates the sine wave. The value at each sample is converted to a 3-bit binary number, and a pulse sequence is generated where each pulse represents a 1 and the absence of a pulse represents a 0. In this case, the PCM waveform is shown in NRZ format. The higher the sample rate and the more levels used, the more accurate is the PCM representation.

PCM is used for digital audio in computers, blu-ray, CD, and DVD formats, and in digital telephone systems. A simplified block diagram of the PCM process is shown in Figure 11–39.

FIGURE 11–39   Block diagram of a PCM system.

EXAMPLE 11–7

In a PCM system with 32 levels, determine the number of code bits for each sample of an analog signal.

SOLUTION

$$\text{Code bits} = 32 = 2^5$$

Five bits represent each sample.

RELATED PROBLEM

What is the number of PCM code bits for each sample if the system has 64 levels?

DIGITAL DATA SYSTEMS   All digital data systems have certain common components and variations, depending on the type of data format. The three main data transmission combinations are digital-to-analog, analog-to-digital, and digital-to-digital. Figure 11–40 shows a general functional block diagram of a data transmission system. Each block is not used in all cases, depending on the type of data format and the type of communications channel.

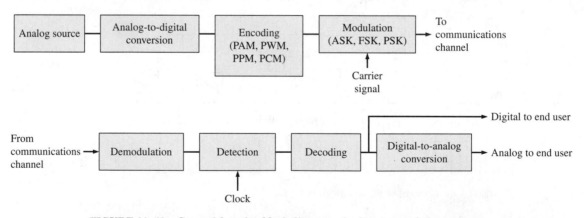

FIGURE 11–40   General function block diagram of a data transmission system.

SECTION 11–4 CHECKUP

1. List four types of pulse modulation methods.

2. What parameter is used in PAM to represent the value of the modulating signal?

3. What parameter is used in PWM to represent the value of the modulating signal?

4. What parameter is used in PPM to represent the value of the modulating signal?

5. What is used in PCM to represent the value of the modulating signal?

This topic was introduced briefly in Chapter 1. Multiplexing (also known as muxing) is a method used to transmit digital data from multiple sources over a single communication channel. Multiplexing is widely used in telecommunications and computer networks. Demultiplexing (demuxing) is the process of separating data from a single channel to multiple channels. Muxing is used on the sending end of a data communication system, and demuxing is used on the receiving end. Two major types of multiplexing are time-division and frequency-division. Time-division multiplexing is a type of baseband communications. Baseband is where digital or analog signals are sent using the entire channel bandwidth. Frequency-division multiplexing is a type of broadband communications where analog signals of different frequencies are transmitted simultaneously.

After completing this section, you should be able to

- Describe time-division multiplexing
- Discuss frequency-division multiplexing

Time-Division Multiplexing

Time-division multiplexing, **TDM**, is a technique in which data from several sources are interleaved on a time basis and sent on a single communication channel or data link. Let's say that there are three sources of digital data to be transmitted. Certain time slots are allotted for each channel so that an element of data (bits or bytes) from source 1 is sent during time slot 1, an element of data from source 2 is sent during time slot 2, and an element of data from source 3 is sent during time slot 3. This is repeated for the time slots that follow until all the data have been sent. Figure 11–41 illustrates the TDM concept.

Source 1 data	Source 2 data	Source 3 data	Source 1 data	Source 2 data	– – – – –
Time slot 1	Time slot 2	Time slot 3	Time slot 4	Time slot 5	– – – – –

→ Time

FIGURE 11–41   Basic concept of TDM.

A simplified illustration of TDM is shown in Figure 11–42 where multiple data sources are switched (multiplexed) in a time sequence (t_1 through t_4) onto a single line (communication channel), and the single stream of data is switched back onto multiple lines in a synchronized time sequence. That is, data from source 1 goes to the data 1 output during time slot t_1, data from source 2 goes to the data 2 output during time slot t_2, etc.

FIGURE 11–42   Simple illustration of TDM.

BIT-INTERLEAVED TDM   In this method, a single data bit from a source is transmitted on the channel, followed by a data bit from another source, and so on. A time slot is reserved on the channel for each input source. These time slots are synchronized with the sender and receiver so that the receiver knows to which output the data bit in each time slot should go. Bit-interleaving is demonstrated in Figure 11–43. In this case, the TDM channel data is transmitted at a rate four times greater than the data rate of the individual sources. Samples are sequentially taken of each data source (four in this case) during a bit time slot to determine if the bit is a 1 or a 0. The resulting values are sequentially placed onto the channel in 1, 2, 3, 4 order, as shown. This process is repeated for each of the bit times that follow.

FIGURE 11–43   Illustration of TDM bit interleaving.

BYTE-INTERLEAVED TDM   In this method, bytes for each input source are sequentially placed onto the data channel. As in bit-interleaving, synchronization between the mux and demux at each end of the communications channel is required. The basic concept is shown in Figure 11–44.

FIGURE 11–44   Basic idea of byte-interleaved TDM.

The byte-interleaved data is from two sources (in this case) and is sent at twice the rate as either source, as illustrated in Figure 11–45. As you can see in the figure, it is necessary

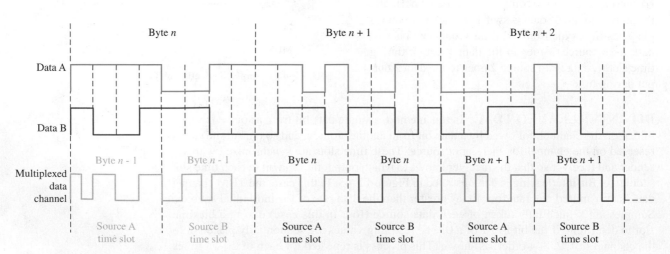

FIGURE 11–45   Byte-interleaved TDM with two data sources.

to delay the data before multiplexing until an entire byte is complete, using a process called buffering. Notice that a byte of Data A and a byte of Data B occur during the Byte n interval. These two bytes are interleaved during the Byte $n + 1$ interval, so the multiplexed data is one byte delayed from the input data (A and B). This continues for each successive data byte.

SYNCHRONOUS TDM When the time slots allotted to each source are fixed, each time slot is transmitted whether or not the source has data to send. This results in an inefficient use of the communications channel because sometimes some of the time slots are empty, as illustrated in Figure 11–46. Here, data source C is not transmitting data, so its assigned time slots are blank.

FIGURE 11–46 Example of a 3-source synchronous TDM with one data source inactive.

STATISTICAL TDM The statistical TDM approach improves channel efficiency by making use of all the time slots. Only data from active sources are transmitted, so there are no blank time slots for inactive sources. The time slot assignment is variable rather than fixed, as in synchronous TDM. This method is shown in Figure 11–47 for the case where data source C is not transmitting. If C becomes active, the time slots are reassigned to accommodate the data.

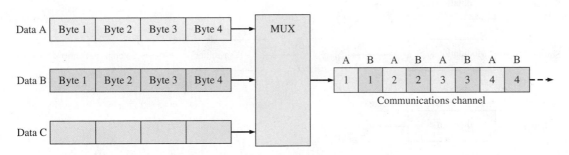

FIGURE 11–47 Example of a 3-source statistical TDM with one source inactive.

SYSTEM EXAMPLE 11–2

THE T1 SYSTEM

The telephone company in North America uses TDM for nearly all voice traffic with what is known as the T1 system. A T1 line can carry 24 digitized telephone conversations and is capable of transmitting data at a rate of 1.544 Mbps. A voice signal is sampled 8,000 times per second, and each sample is converted to a byte of digital data. A voice signal requires a transmission rate of

$$\text{Voice transmission rate} = (8000 \text{ samples/s})(8 \text{ bits/sample}) = 64 \text{ kbps}$$

The number of digitized voice signals that can be multiplexed on a T1 line is

$$\text{Voice signals} = \frac{(1.544 \text{ Mbps/s})}{(64 \text{ kbps/s})} = 24$$

A T1 transmission over the channel consists of sequential 193-bit frames, as shown in Figure 11–48. Each frame is made up of twenty-four 8-bit slots plus one signaling bit.

FIGURE 11–48   T1 channel transmission.

Frequency-Division Multiplexing

Frequency-division multiplexing, **FDM**, is a broadband technique in which the total bandwidth available to a system is divided into frequency sub-bands and information is sent in analog form. Each sub-band is assigned to a given source. The sources can transmit at the same time but at different frequencies. At the receiving end, the signals are separated using band-pass (BP) filtering. Figure 11–49 illustrates the concept of FDM.

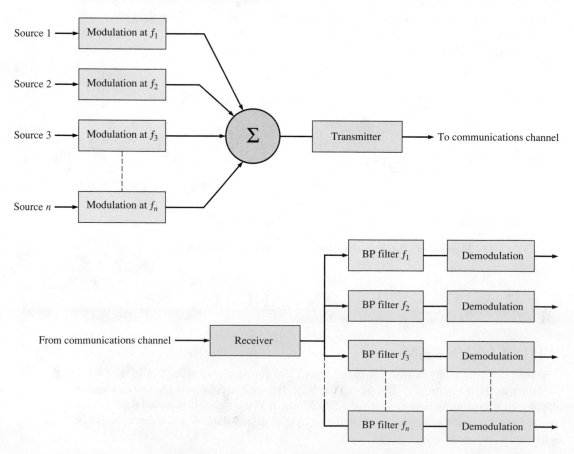

FIGURE 11–49   Basic FDM system. Σ stands for summation.

As mentioned, all sources transmit at the same time but at different frequencies. The general spectrum of a composite FDM transmission is shown in Figure 11–50. The bandwidth (*BW*) of each source is centered around the carrier frequency for that source.

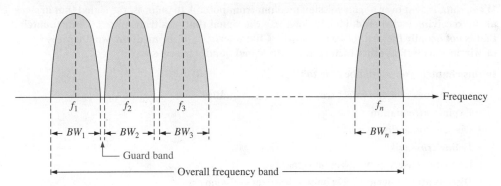

FIGURE 11–50 **Frequency spectrum of an FDM transmission.**

SYSTEM EXAMPLE 11–3

FM AND AM RADIO

FM radio uses FDM. The FM band in the frequency spectrum is 88 MHz to 108 MHz. FM stations are assigned center frequencies at 200 kHz separations beginning at 88.1 MHz with a maximum deviation of 75 kHz on either side of the center frequency. There are 25 kHz guard bands between each station to minimize interference with adjacent bands. This assignment allows up to 100 stations in a given area.

AM radio is also based on FDM and uses a band from 540 kHz to 1600 kHz with a 10 kHz bandwidth for each station. A total of 106 stations can be accommodated. Figure 11–51 shows how the FM and AM bands fit into a wider communications spectrum.

FIGURE 11–51 **Frequency spectrum for AM and FM radio.**

SECTION 11–5 CHECKUP

1. Discuss the reason that multiplexing is used in data communications.

2. What is TDM?

3. What is FDM?

4. Which has the higher efficiency, synchronous TDM or statistical TDM?

5. What is a guard band?

11-6 EFFECTS OF TRANSMISSION MEDIA ON DATA QUALITY

When data is sent over a transmission medium from point A to point B, the signal that arrives at the receiving end should, ideally, look like the signal that was transmitted. Unfortunately, that is not usually the case in the real world. Characteristics of the medium, whether it is wired or wireless, as well as other factors affect the signal quality in some way.

In this chapter, you should be able to

- List factors that can affect transmitted signal quality
- Explain attenuation
- Discuss signal distortion
- Define *cross talk*
- Describe noise and its effect on signals
- Discuss atmospheric effects on wireless transmission

Attenuation

Attenuation (also known as path loss) is the loss of energy as a signal propagates through a medium. Attenuation results in reducing the signal voltage level. Attenuation can be expressed as a linear ratio or in dB (decibels). The distance that a signal must travel is one of the factors affecting attenuation. The longer the distance, the more attenuation the signal will encounter.

WIRE MEDIA   In coax cable or twisted pairs, most of the attenuation is because of resistive losses. As you know, all wire has some resistance that causes a voltage drop when current passes through it. The longer a cable is, the more attenuation there will be. Temperature inversely affects the resistance of the material and therefore the attenuation. The signal frequency also is a factor in attenuation due to the skin effect. Alternating current tends to be much greater near the conductor surface (skin depth) than it is further inside. The skin effect increases with frequency, which causes a decrease in skin depth and thus an increase in resistance. These two causes of attenuation in a conductor are illustrated in Figure 11–52.

The formula for attenuation (A) in a metallic conductor is

$$A = 1 - \left(\frac{V_s - IR_{cond}}{V_s} \right) \qquad (11\text{--}1)$$

(a) Wire resistance produces signal reduction (attenuation)

(b) Skin depth decreases (resistance increases) with frequency

FIGURE 11–52   **Attenuation in a wire. The larger the wire diameter, the less the resistance.**

EXAMPLE 11-8

Assume that a data signal with an amplitude of 5 V is transmitted over a 2000 ft cable. The 24 gauge conductor in the cable has a resistance of 25.67 Ω/1000 ft. The load on the receiving end of the cable is 1 kΩ. Determine the attenuation of the signal and the amplitude of the signal at the receiving end of the link.

SOLUTION

First, determine the resistance of the conductor (R_{cond}) and calculate the current.

$$R_{cond} = (25.67\,\Omega/1000\ \text{ft})2000\ \text{ft} = 51.34\ \Omega \cong 51\ \Omega$$

$$I = \frac{V_s}{R_{load} + R_{cond}} = \frac{5\ \text{V}}{1\ \text{k}\Omega + 0.051\ \text{k}\Omega} = \frac{5\ \text{V}}{1.051\ \text{k}\Omega} = 4.76\ \text{mA}$$

The attenuation, A, is

$$A = 1 - \left(\frac{V_s - IR_{cond}}{V_s}\right) = 1 - \left(\frac{5\ \text{V} - (4.76\ \text{mA})(51.34\ \Omega)}{5\ \text{V}}\right) = \mathbf{0.0489}$$

The attenuation is 4.89%.

The amplitude of the signal at the receiving end (V_{rec}) is found by subtracting the voltage drop across the conductor from the source voltage.

$$V_{rec} = V_{source} - AV_{source} = 5\ \text{V} - (0.0489)(5\ \text{V}) = \mathbf{4.76\ V}$$

RELATED PROBLEM

Determine the attenuation in this example if a 20 gauge copper wire is used with a resistance of 10.36 Ω/1000 ft.

OPTICAL MEDIA Attenuation in optical fiber is the reduction in light intensity of the signal and is directly dependent on the distance that the light travels through the medium. Attenuation in an optical fiber is mainly caused by scattering, absorption, and bending of light energy. The unit of attenuation normally used is dB/km. Attenuation in fiber optics is expressed as a ratio as

$$A = 1 - \frac{\textbf{Output intensity}}{\textbf{Input intensity}} \tag{11–2}$$

Expressed in dB, the attenuation is

$$A\ \textbf{(dB)} = \textbf{10 log}\left(1 - \frac{\textbf{Output intensity}}{\textbf{Input intensity}}\right) \tag{11–3}$$

Ways to express light intensity include Watts/m^2, lumens, lux, and candela.

When light is reflected within an optical fiber, the light rays tend to be reflected in random directions (scattering), primarily due to the molecular "roughness" or irregularities of the fiber material and the inner surface of the fiber. Scattering causes an energy loss that results in attenuation. Scattering is illustrated in Figure 11–53.

FIGURE 11–53 **Illustration of light scattering and absorption in an optical fiber.**

Attenuation can be partially the result of absorption due to imperfections, intrinsic material properties, or extrinsically induced impurities. Absorption results in the conversion of optical power to another form of energy, such as heat. This is also illustrated in Figure 11–53.

Another cause of attenuation in optical fibers is bending. Macrobends occur when a fiber is curved too sharply during installation. Very small bending, microbending, can be caused by uneven coating or improper cabling procedures. Also, microbends in the fiber can be caused by deformation in the cable jacket resulting from an external force. Figure 11–54 shows examples of bending.

(a) Macrobending   (b) Microbending

FIGURE 11–54   Illustration of bending in an optical fiber.

EXAMPLE 11–9

Determine the attenuation as a ratio and in dB for a fiber optics system with an input signal intensity of 10 W/m^2 and an output signal intensity of 8.9 W/m^2.

SOLUTION

$$A = 1 - \frac{\text{Output intensity}}{\text{Input intensity}} = 1 - \frac{8.9 \text{ W/m}^2}{10 \text{ W/m}^2} = \mathbf{0.11}$$

Expressed in dB, the attenuation is

$$A = 10 \log\left(1 - \frac{\text{Output intensity}}{\text{Input intensity}} \right) = 10 \log(0.11) = \mathbf{-9.59 \text{ dB}}$$

The negative sign indicates a power loss.

RELATED PROBLEM

When a longer fiber optic cable is used, the output intensity drops to 8.1 W/m^2. What is the attenuation in dB?

WIRELESS MEDIA   Attenuation limits the range of rf (radio frequency) signals as they are transmitted through various materials. Signal strength is reduced by propagation losses and penetration loses, which cause diffraction and absorption.

As an rf wave propagates into free space, it expands in a spherical shape, resulting in some loss in energy. When travelling long distance, the rf wave may bounce off the atmosphere and the earth causing it to take multiple paths to the receiver. This is called the *multipath effect*. RF wave propagation is illustrated in Figure 11–55 for a directional type wave front.

When an rf wave encounters objects, such as trees, buildings, and rain, that are not totally transparent to rf waves, attenuation results. Some energy is lost by diffraction of the waves and some by absorption.

A formula that is used to calculate the dB attenuation of an rf signal in free space includes distance (*d*) and wavelength (λ).

Transmitter

FIGURE 11–55   Radio wave propagation.

$$A = 10 \log\left(\frac{4\pi d}{\lambda} \right)^2 \tag{11–4}$$

EXAMPLE 11–10

Data is transmitted with a carrier (rf) frequency of 50 MHz to a receiver that is 30 m away. Calculate the attenuation assuming free space conditions.

SOLUTION

The wavelength is

$$\lambda = \frac{c}{f} = \frac{299,792,458 \text{ m/s}}{50 \text{ MHz}} = 6.0 \text{ m}$$

where c is the speed of light.
   The attenuation is

$$A = 10 \log\left(\frac{4\pi d}{\lambda}\right)^2 = 20 \log\left(\frac{4\pi d}{\lambda}\right) = 20 \log\left(\frac{4\pi 30 \text{ m}}{6.0 \text{ m}}\right) = \textbf{36 dB}$$

RELATED PROBLEM

Determine the attenuation for an rf signal of 100 MHz over the same distance (30 meters).

Distortion

Distortion is the change in shape of a waveform and is usually an unwanted effect. For data transmitted over a wire cable, the capacitive and inductive characteristics of the cable are a major factor in causing waveform distortion as well as phase (delay) distortion. In combination with wire resistance, stray capacitance and inductance can result in slower rise and fall times of a data pulse as well as oscillations (called ringing), overshoot, and undershoot on the pulses. Figure 11–56 shows an example of distortion of a pulse transmitted through a coaxial cable. There is distributed capacitance between the conductor and the ground shield and between the inherent inductance of the wire and shield. In twisted-pair cables, the inductance between conductors is minimized by angles at which the wires cross each other when twisted.

Transmitted pulse   Coax cable   Received pulse

FIGURE 11–56   **Distortion of a pulse in a coaxial cable.**

Another type of distortion, cross talk, can occur due to stray mutual inductance or capacitive coupling when two or more conductors are in parallel with each other. Also, external sources of noise can induce unwanted voltages in a transmission line and cause signal distortion.

Noise

The primary cause of data error in wireless communications is due to noise. **Noise** is an unwanted electrical disturbance caused by both natural and man-made sources. Electrical storms and general background noise are examples of natural sources of noise. Electrical motors or automotive ignitions are examples of man-made sources of noise. Noise can affect electromagnetic waves by altering the modulated waves or the data that they carry.

SIGNAL-TO-NOISE RATIO (S/N)   The **signal-to-noise ratio** (also abbreviated SNR) is a measure of the signal strength relative to background noise and is usually expressed in decibels (dB). A positive S/N means that the signal voltage is greater than the noise voltage, a 0 dB S/N indicates the signal and noise are equal, and a negative S/N means the signal is less than the noise. The formula to calculate the S/N ratio is

$$S/N = 20 \log\left(\frac{V_s}{V_n}\right) \tag{11-5}$$

where V_s is the signal voltage and V_n is the noise voltage.

When the S/N is positive, the signal can usually be reliably extracted from the noise at the receiving end of a data transmission system. At an S/N of 0, the extracted signal is marginal and generally has many errors, requiring a resend request and thus a longer transmission time. For a negative S/N, the signal generally cannot be extracted from the transmission.

EXAMPLE 11–11

Determine the S/N for each case.

(a) A signal with a voltage amplitude of 500 μV is transmitted with an average background noise of 100 μV.

(b) Signal voltage and noise voltage equal at 250 μV

(c) A signal with a voltage of 100 μV and a noise voltage of 150 μV.

SOLUTION

(a) $S/N = 20 \log\left(\frac{V_s}{V_n}\right) = 20 \log\left(\frac{500\ \mu V}{100\ \mu V}\right) = \mathbf{14\ dB}$

(b) $S/N = 20 \log\left(\frac{250\ \mu V}{250\ \mu V}\right) = \mathbf{0\ dB}$

(c) $S/N = 20 \log\left(\frac{100\ \mu V}{150\ \mu V}\right) = \mathbf{-3.52\ dB}$

RELATED PROBLEM

Calculate the S/N when the signal voltage is twice the noise.

NOISE MARGIN   In communications, **noise margin** is the amount by which the actual signal level exceeds the minimum acceptable level for an error-free transmission. Noise margin is closely related to S/N and is also usually expressed in dB. Figure 11–57 illustrates the basic concept of noise margin using a digital signal.

FIGURE 11–57   **Noise margin.**

The noise margin is

$$\textbf{Noise margin (dB)} = S_T\ \textbf{(dB)} - S_m\ \textbf{(dB)} \tag{11-6}$$

where S_T is the transmitted signal level and S_m is the minimum acceptable signal level.

EXAMPLE 11–12

A certain system transmits with a signal level of 15 dB. For the given noise environment, the minimum acceptable signal level is 1.2 dB. What is the noise margin?

SOLUTION

$$\text{Noise margin} = 15 \text{ dB} - 1.2 \text{ dB} = \textbf{13.8 dB}$$

RELATED PROBLEM

Does a higher noise margin mean more or less noise?

SECTION 11–6 CHECKUP

1. Name three things that can affect the quality of a transmitted signal.

2. Define *attenuation*.

3. Define *distortion*.

4. Define *signal-to-noise ratio*.

5. Define *noise margin*.

SUMMARY

- Three essential elements in a data transmission system are sending device, transmission medium, and receiving device.

- The simplest connection between sending and receiving devices is a wire or a conductive trace on a printed circuit (PC) board.

- A coaxial cable consists of a center conductor within an insulating dielectric material surrounded by a copper braided or foil shield encased in a protective jacket.

- BNC (Bayonet Neill-Concelman) connectors are typically used for coaxial connections.

- An unshielded twisted pair minimizes cross talk when bundled with other twisted pair into cables.

- Instead of using electrical pulses to transmit information through copper lines, fiber optics uses light pulses transmitted through optical fibers.

- The transmission of data through air and space without the use of physical connections between sending and receiving systems is known as wireless transmission.

- Most data communications occur within the radio wave, microwave, and infrared frequencies.

- Three ways in which rf and microwave signals propagate through Earth's atmosphere (air) are ground wave, ionospheric, and line-of-sight.

- Serial data transmission is when data is transmitted one bit at a time in a bit stream.

- Parallel data transmission is when data is transmitted several bits at a time.

- Two types of data transmission in terms of how a receiving device knows what a sending device is transmitting are synchronous and asynchronous.

- Two data formats that require separate data and timing are RZ (return to zero) and NRZ (nonreturn to zero).

- In biphase coding, the timing and data are combined in one signal.

- Three modes that characterize data channel (media) connections are simplex, half-duplex, and full-duplex.

- Three major classes of modulation techniques for wireless transmission of digital data are amplitude-shift keying (ASK), frequency-shift keying (FSK), and phase-shift keying (PSK).

- Digital quadrature amplitude modulation (QAM) uses a combination of PSK and amplitude modulation to send information.

- Pulse amplitude modulation (PAM) is a type of modulation where the amplitudes of pulses are varied proportional to the amplitude of a modulating waveform.

- Pulse width modulation (PWM) is a type of modulation where the duration of the pulses and duty cycle are varied according to the amplitude of a modulating analog signal.

- Pulse position modulation (PPM) is a type of modulation where the position of each pulse relative to a reference is varied proportional to the amplitude of a modulating waveform.

- Pulse code modulation (PCM) is a type of modulation in which the amplitude of a modulating signal is represented by a digital code.

- Multiplexing is a method used to transmit digital data from multiple sources over a single communications channel.

- Demultiplexing is the process of separating data from a single channel to multiple channels.

- Two major types of multiplexing are time-division and frequency-division.

- Characteristics of the transmission medium, whether it is wired or wireless, as well as other factors affect the signal quality in some way.

KEY TERMS

Key terms and other bold terms in the chapter are defined in the end-of-book glossary.

ASK   Ampltude shift keying. A form of modulation in which a digital signal modulates the amplitude of a higher frequency sine wave.

Asynchronous   A condition that describes signals or systems that are not aligned or synchronized with each other in terms of timed events; two or more systems that do not have the same timing signal.

Baud   The number of symbols per second in a data transmission.

Bit rate   The number of bits per second in a data transmission.

Coaxial cable   A type of data transmission media in which a shielded conductor is used to minimize EMI.

Electromagnetic waves   Related to the electromagnetic spectrum, which includes radio waves, microwaves, infrared, visible, ultraviolet, X-rays, and gamma rays.

EMI   Electromagnetic interference.

FDM   Frequency division multiplexing. A broadband technique in which the total bandwidth available to a system is divided into frequency sub-bands and information is sent in analog form.

FSK   Frequency shift keying. A form of modulation in which a digital signal modulates the frequency of a higher frequency sine wave.

Full-duplex   A connection in which the data flows both ways simultaneously in the same channel.

Half-duplex   A connection in which the data flows both ways but not at the same time in the same channel.

Manchester encoding   A method of encoding called biphase in which a 1 is represented by a positive-going transition and a 0 is represented by a negative-going transition.

Noise margin   The amount by which the actual signal level exceeds the minimum acceptable level for an error-free transmission.

NRZ   Nonreturn to zero. A type of data format in which the signal level does not return to zero during a bit time after a high-level (1) data bit occurs.

Optical fiber   A type of data transmission media used for transmitting light signals.

PAM   Pulse amplitude modulation. A method of modulation in which the height or amplitude of the pulses are varied according to the modulating analog signal, and each pulse represents a value of amplitude of the analog signal.

PCM   Pulse code modulation. A method of modulation that involves sampling of an analog signal amplitude at regular intervals and converting the sampled values to a digital code.

PPM   Pulse position modulation. A method of modulation in which the position of each pulse relative to a reference or timing signal is varied proportional to the amplitude of the modulating signal waveform.

PSK   Phase shift keying. A form of modulation in which a digital signal modulates the phase of a higher frequency sine wave.

PWM   Pulse width modulation. A method of modulation in which the width or duration of the pulses and duty cycle are varied according to the modulating analog signal, and each pulse width represents an amplitude value of the analog signal.

QAM   Quadrature amplitude modulation. A form of modulation that uses a combination of PSK and amplitude modulation to send information.

RZ   Return to zero. A type of data format in which the signal level goes to or remains at zero after each data bit.

Signal-to-noise ratio (SNR)   A measure of the signal strength relative to background noise, usually expressed in decibels (dB).

Simplex   A connection in which data flows in only one direction from the sender (transmitter) to the receiver.

Synchronous   A condition that describes signals or systems that are aligned or synchronized with each other in terms of timed events; two or more systems that have the same timing signal.

TDM   Time division multiplexing. A technique in which data from several sources are interleaved on a time basis and sent on a single communication channel or data link.

UTP   Unshielded twisted pair. A type of transmission media.

TRUE/FALSE QUIZ

Answers are at the end of the chapter.

1. The simplest connection between sending and receiving devices in a data transmission system is a coaxial cable.

2. BNC is a type of connector used for coax.

3. The purpose of a twisted pair is to minimize cross talk.

4. Shielded twisted pair cable in conduit provides more protection from EMI than UTP does.

5. Fiber optic cables are used to transmit electrical pulses through thin optical fibers.

6. ST is a type of fiber optic connector.

7. Most data communications occur within the visible spectrum.

8. Types of rf and microwave signals that propagate through Earth's atmosphere are ground wave, ionospheric, and line-of-sight.

9. In general, a given number of bits can be transmitted faster in parallel than in series.

10. In asynchronous systems, the sending and receiving devices operate with separate oscillators having different clock frequencies.

11. Data rate is the speed of data transfer.

12. Bit rate is always less than or equal to the baud.

13. One method of synchronization is by using separate channels to transmit the data and the timing information.

14. Manchester encoding is a commonly used method to embed the timing signal in the data so that only one channel is required.

15. Biphase and Manchester are two different types of code.

16. In a simplex connection, data flows in both directions between the sender (transmitter) and the receiver.

17. BASK stands for binary amplitude-shift keying.

18. PSK applications include wireless LAN and bluetooth.

19. In quadrature amplitude modulation there are eight phase quadrants.

20. QAM is widely used in telecommunications and in digital cable TV.

21. Ethernet is a family of computer-networking protocols described by the IEEE 802.3 standard.

22. PWM and PDM are the same.

23. PPM is less sensitive to channel interference than PAM or PWM.

24. Pulse code modulation involves sampling of an analog signal amplitude at regular intervals.

25. The three main data transmission combinations are digital-to-analog, analog-to-digital, and digital-to-digital.

26. TDM stands for transmitted data multiplexing.

27. FDM is a baseband technique in which the total bandwidth available to a system is divided into frequency sub-bands, and information is sent in analog form.

28. Alternating current tends to be much greater near the conductor surface (skin depth) than it is further inside.

29. Attenuation in optical fiber is the increase in light intensity of the signal.

30. A positive S/N means that the signal voltage is greater than the noise voltage.

SELF-TEST

Answers are at the end of the chapter.

1. The main purpose of the shield in a coaxial cable is to
   (a) make the cable stronger   (b) dissipate heat
   (c) prevent EMI   (d) prevent distortion

2. UTP is color-coded according to
   (a) a 10-pair color code   (b) a 25-pair color code
   (c) the resistor color code (d) the primary colors

3. Advantages that fiber optic systems have over electrical transmission media are
   (a) higher data rate, less susceptible to noise, and longer transmission distance
   (b) lower cost, higher data rate, and simplicity
   (c) higher data rate, higher EMI, less distortion
   (d) higher baud, availability, and reliability

4. The modes of light propagation in optical fibers are
   (a) simplex and duplex   (b) multimode and single-mode
   (c) synchronous and asynchronous   (d) scatter and direct

5. The electromagnetic spectrum includes
   (a) radio waves, microwaves, and audio waves
   (b) radio waves, microwaves, and sonic waves
   (c) visible light, infrared, and alpha waves
   (d) radio waves, microwave, and infrared

6. Satellites use
   (a) ground propagation   (b) line-of-sight propagation
   (c) ionospheric propagation (d) triangulation

7. In asynchronous transmission, data is sent in short bursts known as
   (a) packets   (b) frames   (c) bundles   (d) quanta

8. For a given bit rate, the baud compared to the bit rate is
   (a) always greater   (b) always less   (c) equal or greater   (d) equal or less

9. The efficiency of a data transmission system is
   (a) the ratio of baud to bit rate   (b) the ratio of actual data rate to ideal data rate
   (c) the ratio of data bits to total bits (d) the ratio of parity bits to data bits

10. The Manchester code format is
    (a) NRZ   (b) biphase   (c) RZ   (d) FDM

11. A synchronous data frame does not contain a(n)
    (a) preamble   (b) data field   (c) address   (d) vector field

12. Three types of data channel connections in terms of data flow are
    (a) input, output, neutral   (b) simplex, half-duplex, full-duplex
    (c) simplex, duplex, triplex   (d) uniplex, diplex, biplex

13. In FSK modulation,
    (a) the frequency of a carrier signal is varied by a digital signal.
    (b) the frequency of a digital signal is varied by a carrier signal.
    (c) the phase of a carrier signal is varied by a digital signal.
    (d) the amplitude of a carrier signal is varied by a digital signal.

14. Types of modulation in which a parameter of a sine-wave carrier signal is varied by a digital signal are

    (a) PAM, PWM, PPM   (b) QAM, PAM, ASK
    (c) FSK, PSK, PPM   (d) FSK, ASK, PSK

15. QAM stands for

    (a) quadrature analysis method   (b) quadrature amplitude modulation
    (c) quasi-amplitude modulation   (d) quadratic amplitude modulation

16. In QAM, the parameters that are varied are

    (a) amplitude and frequency   (b) phase and frequency
    (c) amplitude and phase   (d) pulse width and position

17. Three methods of modulating a digital signal with analog data are

    (a) PAM, PWM, PPM   (b) PAM, ASK, PPM
    (c) FSK, QAM, PAM   (d) QAM, PAM, PWM

18. The most likely type of modulation to be used in motor speed control is

    (a) PAM   (b) PPM   (c) PWM   (d) QAM

19. A method in which an analog signal is sampled and converted to a digital code is

    (a) PCM   (b) PDM   (c) PDC   (d) PPM

20. TDM stands for

    (a) time duration modulation   (b) time division modulation
    (c) time division multiplexing   (d) time division method

21. Two methods of combining data in TDM are

    (a) fast and slow   (b) bit-interleaved and byte-interleaved
    (c) time-domain and frequency-domain   (d) simplex and duplex

22. A method by which data from multiple sources are sent simultaneously is

    (a) TDM   (b) FCM   (c) FSK   (d) FDM

23. The phenomenon where alternating current tends to be greater near the surface of a conductor is called the

    (a) skin effect   (b) Hall effect   (c) tunneling   (d) eddy current effect

24. In an optical medium, three factors that cause attenuation are

    (a) permeability, resistance, and capacitance
    (b) scattering, absorption, and bending
    (c) reflection, refraction, and absorption
    (d) intensity, luminescence, and refraction

25. A measure of the signal strength relative to background noise is called the

    (a) noise margin   (b) transmission ratio
    (c) signal-to-noise ratio   (d) distortion factor

PROBLEMS

Odd-numbered answers are at the end of the book.

SECTION 11–1   Data Transmission Media

1. List four parts of a coaxial cable.

2. What do the acronyms UTP and STP mean?

3. Name four advantages of fiber optic media over electrical transmission media.

4. List three parts of an optical fiber.

5. Describe the multimode in an optical fiber.

6. Which of the optical fiber sizes (50/125, 62.5/125 and 8.3/125) operates in multimode and which operate(s) in single mode?

7. Draw a basic block diagram of a fiber optics communications link.

8. A frequency of 100 MHz falls into what part of the electromagnetic spectrum?

9. In what frequency range does visible light fall? Use the Internet or other source to determine the conversion from wavelength to frequency.

SECTION 11–2   Methods and Modes of Data Transmission

10. If data bits are transmitted serially at a 1 MHz rate, how many bits can be transmitted in 1 ms?

11. If a byte of data is transmitted in parallel in $1 \mu s$, what is the data rate in bits per second?

12. Eight voltage levels are being transmitted by a system where each level (symbol) represents a 3-bit code. Assuming that 12 symbols are transmitted in 0.5 μs, determine the bit rate and the baud.

13. Assume a 5-bit code is used for each symbol transmitted. If the bit rate is 25 Mbs, what is the baud?

14. A certain data packet contains a total of 20 bits of which 16 are data bits. Determine the efficiency.

15. Show the data waveform for the bit sequence 101011100011 in NRZ and in RZ formats.

16. For the bit sequence in Problem 15, show the Manchester code.

17. Determine the bit sequence represented by the Manchester code in Figure 11–58.

18. List and describe each part of a typical synchronous data frame structure.

FIGURE 11–58

SECTION 11–3   Modulation of Analog Signals with Digital Data

19. Determine the binary code represented by the ASK signal in Figure 11–59. Presence of a signal is a 1 and absence of a signal is 0.

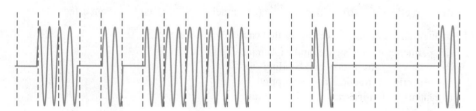

FIGURE 11–59

20. Show how you would respresent four successive bits (1010) using FSK.

21. Repeat Problem 20 for PSK.

22. Refer to Figure 11–23 and determine the sequence of bits represented by the QAM signal in Figure 11–60.

FIGURE 11–60

23. Sketch a constellation map for a 4-bit PSK system.

24. Repeat Problem 23 for a 4-bit QAM system.

SECTION 11–4   Modulation of Digital Signals with Analog Data

25. Describe the PWM intersective method.

26. Develop the PAM and PWM signals for the waveform in Figure 11–61.

(a) PAM

(b) PWM

FIGURE 11–61

27. In a certain PPM system there are 2^3 positions in each period, T. If $T - 10$ ms, determine the data rate.

28. Show the NRZ code for the PPM signal in Figure 11–62.

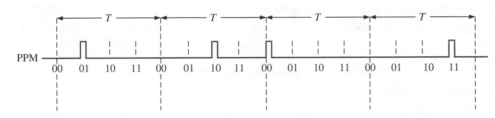

FIGURE 11–62

29. In a PCM code, how many bits are required to represent 16 voltage levels of a modulating signal?

30. Show the 4-bit PCM code in NRZ format for four successive samples of an analog waveform. The sampled values are 1, 3, 5, and 7.

SECTION 11–5   Multiplexing and Demultiplexing

31. Explain the difference between bit-interleaved and byte-interleaved TDM.

32. Describe the difference between synchronous and statistical TDM.

33. What type of filters are used in an FDM system and what is their purpose?

34. What is the frequency separation called between each source in an FDM system?

SECTION 11–6   Effects of Transmission Media on Data Quality

35. Assume that a data signal with an amplitude of 3 V is transmitted over a 500 ft cable. The 24 gauge conductor in the cable has a résistance of 25.67 Ω/1000 ft. The load on the receiving end of the cable is 600 Ω. Determine the attenuation of the signal and the amplitude of the signal at the receiving end of the link.

36. A fiber optic system has an input intensity of 100 lumens and an output intensity of 92 lumens. What is the attenuation?

37. Data is transmitted with a carrier (rf) frequency of 250 MHz to a receiver that is 50 m away. Calculate the attenuation assuming free space conditions.

38. Determine the S/N for a signal with a voltage amplitude of 900 μV that is transmitted with an average background noise of 100 μV.

39. Determine the S/N for a signal with a voltage of 100 μV and a noise voltage of 110 μV.

40. A certain system transmits with a signal level of 20 dB. For the given noise environment, the minimum acceptable signal level is 12 dB. What is the noise margin?

ANSWERS TO SECTION CHECKUPS

SECTION 11–1   Data Transmission Media

1. Wire, coaxial cable, unshielded twisted pair, shielded twisted pair, optical fiber, conductive trace, and wireless.
2. The shield protects against EMI.
3. Ground wave, ionospheric, line-of-sight
4. Gamma radiation has the highest frequencies
5. Baseband uses digital modulation (a series of pulses). Broadband uses a digitally modulated analog signal.

SECTION 11–2   Methods and Modes of Data Transmission

1. Serial data is one bit at a time in sequence. Parallel data is simultaneous multiple bits.
2. Synchronization allows the receiver to recognize the beginning and end of a data transmission.
3. RZ, NRZ, and biphase are three types of data formats.
4. Simplex, half-duplex, and full-duplex are three modes of data transmission.

SECTION 11–3   Modulation of Analog Signals with Digital Data

1. Modulation techniques are ASK, FSK, PSK, and QAM.
2. The amplitude of the analog signal is changed in ASK.
3. The frequency of the analog signal is changed in FSK.
4. QAM is quadrature amplitude modulation.
5. The phase of the analog signal is changed in PSK.

SECTION 11–4   Modulation of Digital Signals with Analog Data

1. Pulse modulation methods are PAM, PWM, PPM, and PCM.
2. Pulse amplitude
3. Pulse width
4. Pulse position
5. A binary code

SECTION 11–5   Multiplexing and Demultiplexing

1. Multiplexing sends data from several sources on a single communication channel.
2. TDM is time division multiplexing where data from multiple sources are interleaved on a time basis.
3. FDM is frequency division multiplexing where data from multiple sources are sent simultaneously at different frequencies.
4. Statistical TDM has the higher efficiency.
5. The guard band is the frequency separation between the frequency bands of the multiple sources in FDM.

SECTION 11–6   Effects of Transmission Media on Data Quality

1. Attenuation, distortion, and noise affect the quality of a transmitted signal.
2. Attenuation is the loss of energy as a signal propagates through a medium.
3. The change in shape or other parameters of a signal waveform due to characteristics and imperfections in the transmission medium.
4. Signal-to-noise ratio is a measure of the signal strength relative to background noise.
5. Noise margin is the amount by which the actual signal level exceeds the minimum acceptable level for an error-free transmission.

ANSWERS TO RELATED PROBLEMS FOR EXAMPLES

11–1   Bit rate = 40 kbps

11–2   Efficiency = 0.828 (82.8%)

11–3 See Figure 11–63.

FIGURE 11–63

11–4 Eight amplitudes and eight phases can be used to represent a 4-bit code.

11–5 There would be more pulses closer together in both cases providing a more accurate representation of the analog signal.

11–6 Data rate = 8 Mbps; 256 pulse positions

11–7 Six PCM code bits to represent 64 levels.

11–8 0.0203

11–9 −7.21 dB

11–10 −42 dB

11–11 6.02 dB

11–12 Higher noise margin means less noise for a given signal strength.

ANSWERS TO TRUE/FALSE QUIZ

1. F	**2.** T	**3.** T	**4.** T	**5.** F	**6.** T	**7.** F	**8.** T
9. T	**10.** F	**11.** T	**12.** F	**13.** T	**14.** T	**15.** F	**16.** F
17. T	**18.** T	**19.** F	**20.** T	**21.** T	**22.** T	**23.** T	**24.** T
25. T	**26.** F	**27.** F	**28.** T	**29.** F	**30.** T		

ANSWERS TO SELF-TEST

1. (c)	**2.** (b)	**3.** (a)	**4.** (b)	**5.** (d)	**6.** (b)	**7.** (a)
8. (d)	**9.** (c)	**10.** (b)	**11.** (d)	**12.** (b)	**13.** (a)	**14.** (d)
15. (b)	**16.** (c)	**17.** (a)	**18.** (c)	**19.** (a)	**20.** (c)	**21.** (b)
22. (d)	**23.** (a)	**24.** (b)	**25.** (c)			

CHAPTER 12

SIGNAL CONVERSION AND PROCESSING

OBJECTIVES

- Explain how analog signals are converted to digital form
- Discuss the purpose of filtering
- Describe the sampling process
- State the purpose of analog-to-digital conversion
- Explain how several types of ADCs operate
- State the purpose of digital-to-analog conversion
- Explain how DACs operate
- List the essential elements in a digital signal processing system
- Explain the basic concepts of a digital signal processor (DSP)
- Describe the basic architecture of a DSP
- Name some of the functions that a DSP performs

KEY TERMS

Sampling	DSP core
Nyquist frequency	MIPS
Aliasing	MFLOPS
Analog-to-digital converter (ADC)	MMACS
	Pipeline
Quantization	Fetch
Digital-to-analog converter (DAC)	Decode
	Execute
DSP	

INTRODUCTION

This chapter provides an introduction to methods of data conversion and signal processing. Data conversions and signal processing are technologies that are widely used in many applications, such as automotive, consumer, graphics/imaging, industrial, instrumentation, medical, military, telecommunications, and voice/speech applications. Digital signal processing incorporates mathematics, software programming, and processing hardware to manipulate analog signals.

VISIT THE WEBSITE
Study aids for this chapter are available at
http://pearsonhighered.com/floyd

In this section, a system for tracking the movement of the sun in order to control the position of a solar panel, Fresnel mirror array, or other type of solar collection device is discussed. The system senses the orientation of the solar panel relative to the sun's position in the sky and produces a signal that causes the solar panel to rotate until it is directly aligned with the sun. Analog-to-digital converters (ADCs), which are covered later in the chapter, are an important part of this system.

After completing this section, you should be able to

- Describe the overall system
- Discuss how ADCs are used in the system
- Use VHDL to implement a portion of the system

The Sunlight Sensors

The sun tracker system tracks the movement of the sun through the day using photosensitive devices. As illustrated in Figure 12–1, the sunlight sensor consists of two photoresistors separated by a plate that optically isolates the photoresistors from each other. The resistance of a photoresistor decreases as the light intensity increases. Other light-sensitive devices, such as the photodiode and phototransistor, could also be used. The sensor assembly is attached to the solar panel so that it is always pointing in the same direction as the panel. When the sensor arrangement and solar panel are not pointing directly at the sun, the photoresistor R_2 gets more sunlight than R_1, as illustrated in part (a) of the figure. When the sensors and solar panel are pointing directly at the sun, R_1 and R_2 get the same amount of sunlight, as shown in part (b). The voltages across the sensors depend on the orientation of the panel. After being processed through the system, the sensor voltages cause the panel to rotate to follow the sun's movement, as shown in Figure 12–2. When the solar panel and

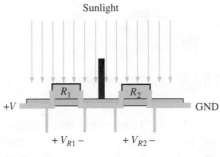

(a) $V_{R1} > V_{R2}$ (b) $V_{R1} = V_{R2}$

FIGURE 12–1 Sunlight sensor arrangement. The orientation produces voltages proportional to the incident sunlight.

Sensor assembly

Solar panel

(a) When panel and sensor are not aligned with sun, they rotate clockwise.

(b) When panel and sensor are aligned with sun, they stop rotating.

FIGURE 12–2 The sensors are mounted to the solar panel. When the panel is not pointing directly at the sun, it rotates until it is aligned.

sensor arrangement are pointed directly at the sun, the light levels for both photoresistors are balanced. The solar panel does not rotate until the sun's position changes by a certain amount, and the photoresistors begin to receive different amounts of sunlight.

The Complete System

The complete sun tracker system is shown in block diagram form in Figure 12–3. The system consists of the sunlight sensor, analog-to-digital converters, frequency divider, analog-to-digital processor, motor control, motor interface, and stepper motor. If the resistance of R_1 in Figure 12–1 is less than that of R_2, V_{R2} is less than V_{R1}, and the resulting direction signal (acting through the A/D processing block) causes the motor to turn the solar panel clockwise. If R_1 and R_2 are equal, the voltages are equal and the motor stops because the panel is directly aligned with the sun. The frequency divider divides the system clock down to a frequency compatible with the rate at which the stepper motor can be advanced.

FIGURE 12–3   **The complete sun tracker system block diagram.**

The A/D processing block receives the digitized light level inputs from the ADCs and compares the two values to determine the required direction of rotation. If the ADC1 output is greater than the ADC2 output ($V_{R1} > V_{R2}$), a start bit is issued, and the motor control initiates a bit sequence to produce the motor stepping action, which causes the solar panel to rotate clockwise. If the ADC outputs are equal ($V_{R1} = V_{R2}$), the solar panel will stop because it is pointing directly at the sun. As the sun moves further west, the system follows until the light level of both sensors decreases below a certain threshold value, and a reset signal is generated. At that point, the panel has reached its western-most position and the sunlight is greatly diminished. The motor control reverses the bit sequence to the stepper motor, and the solar panel is rotated counterclockwise back to its eastern-most position. When the sun rises the next day, the cycle repeats.

Stepper Motor Operation

The interface between the stepper motor control and the motor converts low-level currents produced by the motor control to currents that are sufficient to drive the motor windings. The stepper motor divides a single rotation of its output shaft into a fixed number of steps, as shown in Figure 12–4. Pulses (1s) are applied sequentially to the four motor windings to drive the stepping operation. For each pulse, the motor advances a set number of degrees, $D°$, clockwise. After a full cycle of four pulses, the motor has

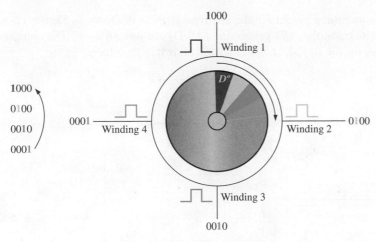

FIGURE 12–4   Basic operation of a stepper motor.

advanced $4D^\circ$. The sequential pulses continue until the stop signal occurs at a point where the solar panel is pointing directly at the sun. A construction view of a stepper motor is shown in Figure 12–5.

FIGURE 12–5   Simplified construction view of a typical stepper motor.

System Implementation

The three components shown in the blue box in Figure 12–3 are the frequency divider, the A/D processing, and the stepper motor control. This portion of the total system can be implemented in a PLD. The light sensors, ADCs, and interface are external to the PLD, as shown in Figure 12–6.

FIGURE 12–6   System block diagram with a PLD.

A programming model for the sun tracker PLD is shown in Figure 12–7. The frequency divider and the A/D processing (AtoD) are treated as VHDL components. The VHDL code for the stepper motor control is shown in the block.

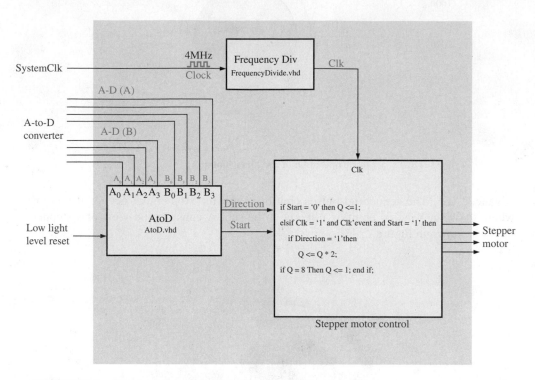

FIGURE 12–7 **Programming model of sun tracker PLD.**

FREQUENCY DIVIDER The VHDL code for the frequency divider is as follows.

```vhdl
library ieee;
use ieee.std_logic_1164.all;

entity FrequencyDivider is
port (ClockIn : in std_logic;              ClockIn : System Clock in
      ClockOut : out std_logic);           ClockOut : Divided Clock out
end FrequencyDivider;

architecture FrequencyDividerBehavior of FrequencyDivider is
begin
  process (ClockIn)
  variable DelayCount: integer range 0 to 1048575; -- Sets delay count range to 20 bits
    begin
      if (ClockIn'event and ClockIn = '1') then        Upon receiving a positive
        ClockOut <= '0';                                clock edge from ClockIn,
        if DelayCount = 300000 then                     counts to 300000 then sets
          ClockOut <= '1';                              ClockOut to 1, otherwise
          DelayCount := 0;                              ClockOut is set to zero.
        else
          DelayCount:= DelayCount +1;
        end if;
      end if;
    end process;
end FrequencyDividerBehavior;
```

AtoD   The VHDL code for the A/D processing is as follows.

```vhdl
library ieee;
use ieee.std_logic_1164.all;
entity AtoD is
port(A: in integer range 0 to 15;   -- A-D input (A)
     B: in integer range 0 to 15;   -- A-D input (B)
     Reset: in std_logic;           -- Low light level reset
     Direction: out std_logic;      -- Step Motor Direction
     Start: out std_logic);         -- Step Motor start/stop
end entity AtoD;


architecture AtoDBehavior of AtoD is
begin
   process(A, B, Reset)
   begin
     if (A < B) then
         Direction <= '1';
     else
         Direction <= '0';
     end if;
         if ((A-B) > 1) or (((B + A) < 10) and Reset = '0') then
         Start <= '1';
     else
         Start <='0';
     end if;
   end process;
end architecture AtoDBehavior;
```

Checks to see if the binary output of ADC 'A' is greater than 'B' to determine direction of rotation for sun tracking and system reset operations.

Subtracts the 8-bit binary values from ADCs and checks the value of the result against a threshold of 1 to start or stop sun tracker rotation as the sun is tracked. When the combined illumination of ADC A and B falls below a low light threshold of 10 and the low light reset switch is not closed (0), rotation is enabled in order to reset the solar panel for the next day. When the reset switch is closed (1), the system is ready for the next day's operation.

COMPLETE VHDL DESCRIPTION   The frequency divider, AtoD components, and the motor control code are combined to form a complete VHDL description of the PLD logic.

```vhdl
library IEEE;
use IEEE.std_logic_1164.all;                    Q          : Motor sequence
entity SunTracker is                            A, B       : ADC input
port (Clock: in std_logic;                      Clk        : Divided clock pulse
      Q: buffer integer range 0 to 255;         Direction  : Step Motor Direction
      A, B: in integer range 0 to 255;          Start      : Step Motor start/stop
end entity SunTracker;
architecture SunTrackerBehavior of SunTracker is
signal Clk: std_logic;         --Divided clock pulse.
signal Direction: std_logic; --Direction bit
signal Start: std_logic;       --Stepper motor start/stop

component FrequencyDivider is
port (ClockIn: in std_logic;
      ClockOut: out std_logic);
end component FrequencyDivider;

component AtoD is
port(A: in integer range 0 to 15;
     B: in integer range 0 to 15;
     Reset: in std_logic;
     Direction: out std_logic;
     Start: out std_logic);
end component AtoD;
```

Definition instantiation for component FrequencyDivider

Definition instantiation for component AtoD

```
begin
  process (Clock, Clk, Direction, Start)
begin
  if Start = '0' then Q <=1;
  elsif Clk = '1' and Clk'event and Start = '1' then
    if Direction = '1' then
      Q <= Q * 2;
    if Q = 8 then
      Q <= 1;
    end if;
  elsif Direction = '0' then
    Q <= Q/2;
    if Q = 1 then
      Q <=8;
    end if;
    end if;
  end if;
end process;
FreqDiv:FrequencyDivider port map(ClockIn => clock, ClockOut => clk);
AtoDConv:AtoD port map(A => A, B => B, Reset => Reset, Direction => Direction, Start => Start);
end architecture SunTrackerBehavior;
```

If Start is 0, output Q is set to 1 to initiate the sequence. If the direction is set to 1, output Q is multiplied by 2 to shift a 1 to the right (forwards). If the direction bit is set to 0, the value of Q is divided by 2, shifting the 1 to the left (reverse).

SECTION 12–1 CHECKUP*

1. How does the system "know" when to rotate?

2. What is the purpose of a counterclockwise rotation?

3. Describe the function of the ADCs.

4. Why is a stepper motor used?

*Answers are at the end of the chapter.

12–2 CONVERTING ANALOG SIGNALS TO DIGITAL

In order to process signals using digital techniques, the incoming analog signal must be converted into digital form.

After completing this section, you should be able to

- Explain the basic process of converting an analog signal to digital
- Describe the purpose of the sample-and-hold function
- Define the Nyquist frequency
- Define the reason for *aliasing* and discuss how it is eliminated
- Describe the purpose of an ADC

Sampling and Filtering

An anti-aliasing filter and a sample-and-hold circuit are two functions typically found in a digital signal processing system. The sample-and-hold function does two operations, the first of which is sampling. **Sampling** is the process of taking a sufficient number of discrete values at points on a waveform that will define the shape of waveform. The more samples you take, the more accurately you can define a waveform. Sampling converts an analog signal into a series of impulses, each representing the amplitude of the signal at a given instant in time. Figure 12–8 illustrates the process of sampling.

FIGURE 12–8   Illustration of the sampling process.

When an analog signal is to be sampled, there are certain criteria that must be met in order to accurately represent the original signal. All analog signals (except a pure sine wave) contain a spectrum of component frequencies. For a pure sine wave, these frequencies appear in multiples called *harmonics*. The harmonics of an analog signal are sine waves of different frequencies and amplitudes. When the harmonics of a given periodic waveform are added, the result is the original signal. Before a signal can be sampled, it must be passed through a low-pass filter (anti-aliasing filter) to eliminate harmonic frequencies above a certain value as determined by the Nyquist frequency.

THE SAMPLING THEOREM   Notice in Figure 12–8 that there are two input waveforms. One is the analog signal and the other is the sampling pulse waveform. The sampling theorem states that, in order to represent an analog signal, the sampling frequency, f_{sample}, must be at least twice the highest frequency component $f_{a(\text{max})}$ of the analog signal. Another way to say this is that the highest analog frequency can be no greater than one-half the sampling frequency. The frequency $f_{a(\text{max})}$ is known as the **Nyquist frequency** and is expressed in Equation 12–1. In practice, the sampling frequency should be more than twice the highest analog frequency.

$$f_{\text{sample}} > 2f_{a(\text{max})} \qquad\qquad (12\text{–}1)$$

To intuitively understand the sampling theorem, a simple "bouncing-ball" analogy may be helpful. Although it is not a perfect representation of the sampling of electrical signals, it does serve to illustrate the basic idea. If a ball is photographed (sampled) at one instant during a single bounce, as illustrated in Figure 12–9(a), you cannot tell anything about the path of the ball except that it is off the floor. You can't tell whether it is going up or down or the distance of its bounce. If you take photos at two equally-spaced instants during one bounce, as shown in part (b), you can obtain only a minimum amount of information about its movement and nothing about the distance of the bounce. In this particular case, you know only that the ball has been in the air at the times the two photos were taken and that the maximum height of the bounce is at least equal to the height shown in each

(a) One sample of a ball during a single bounce provides no information.

(b) Two samples of a ball during a single bounce. This is the absolute minimum required to tell anything about its movement, but generally insufficient to describe its path.

(c) Four samples of a ball during a single bounce form a rough picture of the path of the ball.

FIGURE 12–9   **Bouncing ball analogy of sampling theory.**

photo. If you take four photos, as shown in part (c), then the path that the ball follows during a bounce begins to emerge. The more photos (samples) that you take, the more accurately you can determine the path of the ball as it bounces.

THE NEED FOR FILTERING   Low-pass filtering is necessary to remove all frequency components (harmonics) of the analog signal that exceed the Nyquist frequency. If there are any frequency components in the analog signal that exceed the Nyquist frequency, an unwanted condition known as **aliasing** will occur. An alias is a signal produced when the sampling frequency is not at least twice the signal frequency. An alias signal has a frequency that is less than the highest frequency in the analog signal being sampled and therefore falls within the spectrum or frequency band of the input analog signal causing distortion. Such a signal is actually "posing" as part of the analog signal when it really isn't, thus the term *alias*.

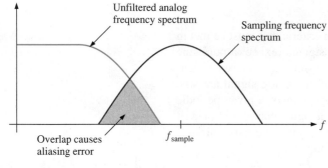

Unfiltered analog frequency spectrum

Sampling frequency spectrum

Overlap causes aliasing error

f_{sample}

FIGURE 12–10   **A basic illustration of the condition**
$f_{sample} < 2f_{a(max)}.$

Another way to view aliasing is by considering that the sampling pulses produce a spectrum of harmonic frequencies above and below the sample frequency, as shown in Figure 12–10. If the analog signal contains frequencies above the Nyquist frequency, these frequencies overlap into the spectrum of the sample waveform as shown and interference occurs. The lower frequency components of the sampling waveform become mixed in with the frequency spectra of the analog waveform, resulting in an aliasing error.

A low-pass anti-aliasing filter must be used to limit the frequency spectrum of the analog signal for a given sample frequency. To avoid an aliasing error, the filter must at least eliminate all analog frequencies above the minimum frequency in the sampling spectrum, as illustrated in Figure 12–11. Aliasing can also be avoided by sufficiently increasing the sampling frequency. However, the maximum sampling frequency is usually limited by the performance of the analog-to-digital converter (ADC) that follows it.

Filtered analog frequency spectrum

Sampling frequency spectrum

f_{sample}

FIGURE 12–11   **After low-pass filtering, the frequency spectra of the analog and the sampling signals do not overlap, thus eliminating aliasing error.**

AN APPLICATION   An example of the application of sampling is in digital audio systems. The sampling rates used are 32 kHz, 44.1 kHz, or 48 kHz (the number of samples per second). The 48 kHz rate is the most common, but the 44.1 kHz rate is used for audio CDs and prerecorded tapes. According to the Nyquist rate, the sampling frequency must be at least twice the audio signal. Therefore, the CD sampling rate of 44.1 kHz captures frequencies up to about 22 kHz, which exceeds the 20 kHz specification that is common for most audio equipment.

Many applications do not require a wide frequency range to obtain reproduced sound that is acceptable. For example, human speech contains some frequencies near 10 kHz and, therefore, requires a sampling rate of at least 20 kHz. However, if only frequencies up to 4 kHz (ideally requiring an 8 kHz minimum sampling rate) are reproduced, voice is very understandable. On the other hand, if a sound signal is not sampled at a high enough rate, the effect of aliasing will become noticeable with background noise and distortion.

Holding the Sampled Value

The holding operation is the second part of the sample-and-hold function. After filtering and sampling, the sampled level must be held constant until the next sample occurs. This is necessary for the ADC to have time to process the sampled value. This sample-and-hold operation results in a "stairstep" waveform that approximates the analog input waveform, as shown in Figure 12–12.

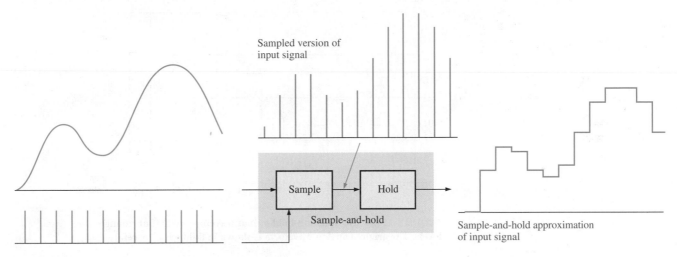

FIGURE 12–12   Illustration of a sample-and-hold operation.

Analog-to-Digital Conversion

Analog-to-digital conversion is the process of converting the output of the sample-and-hold circuit to a series of binary codes that represent the amplitude of the analog input at each of the sample times. The sample-and-hold process keeps the amplitude of the analog input signal constant between sample pulses; therefore, the analog-to-digital conversion can be done using a constant value rather than having the analog signal change during a conversion interval, which is the time between sample pulses. Figure 12–13 illustrates the basic function of an **analog-to-digital converter (ADC)**, which is a circuit that performs analog-to-digital conversion. The sample intervals are indicated by dashed lines.

FIGURE 12–13   Basic function of an analog-to-digital (ADC) converter (The binary codes and number of bits are arbitrarily chosen for illustration only). The ADC output waveform that represents the binary codes is also shown.

QUANTIZATION The process of converting an analog value to a code is called **quantization.** During the quantization process, the ADC converts each sampled value of the analog signal to a binary code. The more bits that are used to represent a sampled value, the more accurate is the representation.

To illustrate, let's quantize a reproduction of the analog waveform into four levels (0–3). Two bits are required for four levels. As shown in Figure 12–14, each quantization level is represented by a 2-bit code on the vertical axis, and each sample interval is numbered along the horizontal axis. The sampled data is held for the entire sample period. This data is quantized to the next lower level, as shown in Table 12–1 (for example, compare samples 3 and 4, which are assigned different levels).

FIGURE 12–14 Sample-and-hold output waveform with four quantization levels. The original analog waveform is shown in light gray for reference.

Table 12–1 • Two-bit quantization for the waveform in Figure 12–14.		
SAMPLE INTERVAL	**QUANTIZATION LEVEL**	**CODE**
1	0	00
2	1	01
3	2	10
4	1	01
5	1	01
6	1	01
7	1	01
8	2	10
9	3	11
10	3	11
11	3	11
12	3	11
13	3	11

If the resulting 2-bit digital codes are used to reconstruct the original waveform, you would get the waveform shown in Figure 12–15. This operation is done by **digital-to-analog converters (DACs)**, which are circuits that perform digital-to-analog conversion. As you can see, quite a bit of accuracy is lost using only two bits to represent the sampled values.

FIGURE 12–15   **The reconstructed waveform in Figure 12–14 using four quantization levels (2 bits). The original analog waveform is shown in light gray for reference.**

Now, let's see how more bits will improve the accuracy. Figure 12–16 shows the same waveform with sixteen quantization levels (4 bits). The 4-bit quantization process is summarized in Table 12–2.

FIGURE 12–16   **Sample-and-hold output waveform with sixteen quantization levels. The original analog waveform is shown in light gray for reference.**

If the resulting 4-bit digital codes are used to reconstruct the original waveform, you would get the waveform shown in Figure 12–17. As you can see, the result is much more like the original waveform than for the case of four quantization levels in Figure 12–15. This shows that greater accuracy is achieved with more quantization bits. Most integrated circuit ADCs use from 8 to 24 bits, and the sample-and-hold function is sometimes contained on the ADC chip. Several types of ADCs are introduced in the next section.

Table 12–2 • Four-bit quantization for the waveform in Figure 12–16.

SAMPLE INTERVAL	QUANTIZATION LEVEL	CODE
1	0	0000
2	5	0101
3	8	1000
4	7	0111
5	5	0101
6	4	0100
7	6	0110
8	10	1010
9	14	1110
10	15	1111
11	15	1111
12	15	1111
13	14	1110

FIGURE 12–17   **The reconstructed waveform in Figure 12–16 using sixteen quantization levels (4 bits). The original analog waveform is shown in light gray for reference.**

SECTION 12–2 CHECKUP

1. What does sampling mean?

2. Why must you hold a sampled value?

3. If the highest frequency component in an analog signal is 20 kHz, what is the minimum sample frequency?

4. What does quantization mean?

5. What determines the accuracy of the quantization process?

As you have seen, analog-to-digital conversion is the process by which an analog quantity is converted to digital form. It is necessary when measured quantities must be in digital form for processing or for display or storage. Some common types of analog-to-digital converters (ADCs) are now examined. Two important ADC parameters are *resolution*, which is the number of bits, and *throughput*, which is the sampling rate an ADC can handle in units of samples per second (sps).

After completing this section, you should be able to

- Explain what an operational amplifier is
- Show how the op-amp can be used as an inverting amplifier or a comparator
- Explain how a flash ADC works
- Discuss dual-slope ADCs
- Describe the operation of a successive-approximation ADC
- Describe a delta-sigma ADC
- Discuss testing ADCs for a missing code, incorrect code and offset

A Quick Look at an Operational Amplifier

Before getting into analog-to-digital converters (ADCs), let's look briefly at an element that is common to most types of ADCs and digital-to-analog converters (DACs). This element is the operational amplifier, or op-amp for short. This is an abbreviated coverage of the op-amp.

An **op-amp** is a linear amplifier that has two inputs (inverting and noninverting) and one output. It has a very high voltage gain and a very high input impedance, as well as a very low output impedance. The op-amp symbol is shown in Figure 12–18(a). When used as an inverting amplifier, the op-amp is configured as shown in part (b). The feedback resistor, R_f, and the input resistor, R_i, control the voltage gain according to the formula in Equation 12–2, where V_{out}/V_{in} is the closed-loop voltage gain (closed loop refers to the feedback from output to input provided by R_f). The negative sign indicates inversion.

$$\frac{V_{out}}{V_{in}} = -\frac{R_f}{R_i}$$

(12–2)

In the inverting amplifier configuration, the inverting input of the op-amp is approximately at ground potential (0 V) because feedback and the extremely high open-loop gain make the differential voltage between the two inputs extremely small. Since the noninverting input is grounded, the inverting input is at approximately 0 V, which is called *virtual ground*.

(a) Op-amp symbol

(b) Op-amp as an inverting amplifier with gain of R_f/R_i

(c) Op-amp as a comparator

FIGURE 12–18 The operational amplifier (op-amp).

When the op-amp is used as a comparator, as shown in Figure 12–18(c), two voltages are applied to the inputs. When these input voltages differ by a very small amount, the op-amp is driven into one of its two saturated output states, either HIGH or LOW, depending on which input voltage is greater.

Flash (Simultaneous) Analog-to-Digital Converter

The flash method utilizes special high-speed comparators that compare reference voltages with the analog input voltage. When the input voltage exceeds the reference voltage for a given comparator, a HIGH is generated. Figure 12–19 shows a 3-bit converter that uses seven comparator circuits; a comparator is not needed for the all-0s condition. A 4-bit converter of this type requires fifteen comparators. In general, $2^n - 1$ comparators are required for conversion to an n-bit binary code. The number of bits used in an ADC is its **resolution.** The large number of comparators necessary for a reasonable-sized binary number is one of the disadvantages of the **flash ADC.** Its chief advantage is that it provides a fast conversion time because of a high *throughput,* measured in samples per second (sps).

FIGURE 12–19   A 3-bit flash ADC.

The reference voltage for each comparator is set by the resistive voltage-divider circuit. The output of each comparator is connected to an input of the priority encoder. The encoder is enabled by a pulse on the *EN* input, and a 3-bit code representing the value of the input appears on the encoder's outputs. The binary code is determined by the highest-order input having a HIGH level.

The frequency of the enable pulses and the number of bits in the binary code determine the accuracy with which the sequence of binary codes represents the input of the ADC. The signal is sampled each time the enable pulse is active.

EXAMPLE 12–1

Determine the binary code output of the 3-bit flash ADC in Figure 12–19 for the input signal in Figure 12–20 and the encoder enable pulses shown. For this example, $V_{REF} = +8$ V.

FIGURE 12–20   Sampling of values on a waveform for conversion to binary code.

SOLUTION

The resulting digital output sequence is listed as follows and shown in the waveform diagram of Figure 12–21 in relation to the enable pulses:

$$100, 110, 111, 110, 100, 010, 000, 001, 011, 101, 110, 111$$

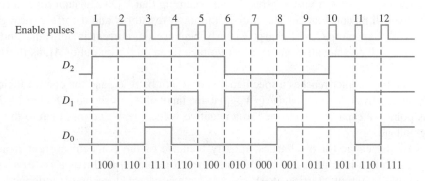

FIGURE 12–21   Resulting digital outputs for sample-and-hold values. Output D_0 is the LSB of the 3-bit binary code.

RELATED PROBLEM*

If the enable pulse frequency in Figure 12–20 were halved, determine the binary numbers represented by the resulting digital output sequence for 6 pulses. Is any information lost?

*Answers are at the end of the chapter.

Dual-Slope Analog-to-Digital Converter

A dual-slope ADC is common in digital voltmeters and other types of measurement instruments. A ramp generator (integrator) is used to produce the dual-slope characteristic. A block diagram of a dual-slope ADC is shown in Figure 12–22.

FIGURE 12–22 **Basic dual-slope ADC.**

Figure 12–23 illustrates dual-slope conversion. Start by assuming that the counter is reset and the output of the integrator is zero. Now assume that a positive input voltage is applied to the input through the switch (SW) as selected by the control logic. Since the inverting input of A_1 is at virtual ground, and assuming that V_{in} is constant for a period of time, there will be constant current through the input resistor R and therefore through the capacitor C. Capacitor C will charge linearly because the current is constant, and as a result, there will be a negative-going linear voltage ramp on the output of A_1, as illustrated in Figure 12–23(a).

When the counter reaches a specified count, it will be reset, and the control logic will switch the negative reference voltage ($-V_{REF}$) to the input of A_1, as shown in Figure 12–23(b). At this point the capacitor is charged to a negative voltage ($-V$) proportional to the input analog voltage.

Now the capacitor discharges linearly because of the constant current from the $-V_{REF}$, as shown in Figure 12–23(c). This linear discharge produces a positive-going ramp on the A_1 output, starting at $-V$ and having a constant slope that is independent of the charge voltage. As the capacitor discharges, the counter advances from its RESET state. The time it takes the capacitor to discharge to zero depends on the initial voltage $-V$ (proportional to V_{in}) because the discharge rate (slope) is constant. When the integrator (A_1) output voltage reaches zero, the comparator (A_2) switches to the LOW state and disables the clock to the counter. The binary count is latched, thus completing one conversion cycle. The binary count is proportional to V_{in} because the time it takes the capacitor to discharge depends only on $-V$, and the counter records this interval of time.

Successive-Approximation Analog-to-Digital Converter

One of the most widely used methods of analog-to-digital conversion is successive-approximation. It has a much faster conversion time than the dual-slope conversion, but it is slower than the flash method. It also has a fixed conversion time that is the same for any value of the analog input.

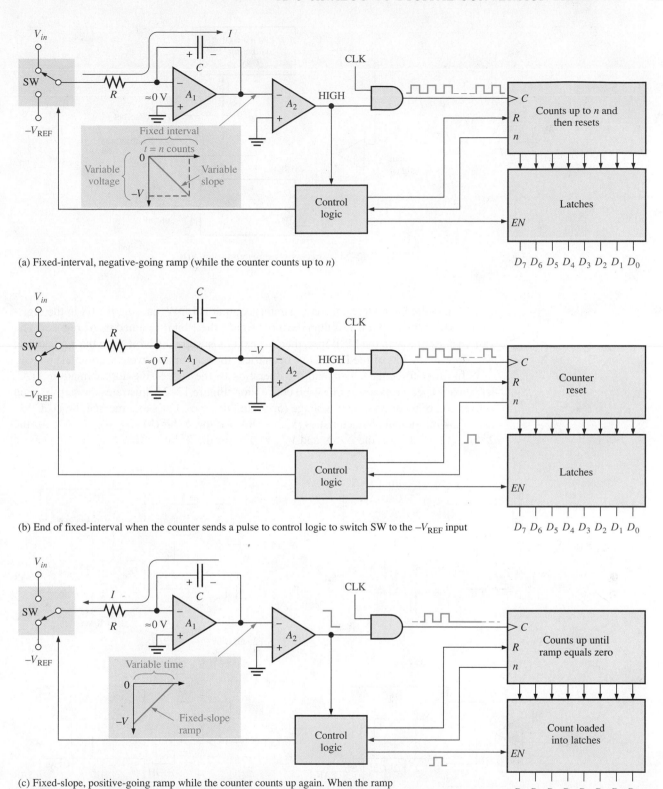

(a) Fixed-interval, negative-going ramp (while the counter counts up to n)

(b) End of fixed-interval when the counter sends a pulse to control logic to switch SW to the $-V_{REF}$ input

(c) Fixed-slope, positive-going ramp while the counter counts up again. When the ramp
    reaches 0 V, the counter stops, and the counter output is loaded into latches.

FIGURE 12–23   **Illustration of dual-slope conversion.**

Figure 12–24 shows a basic block diagram of a 4-bit successive approximation ADC.
It consists of a DAC (DACs are covered in Section 12–4), a successive-approximation
register (SAR), and a comparator. The basic operation is as follows: The input bits of the
DAC are enabled (made equal to a 1) one at a time, starting with the most significant bit
(MSB). As each bit is enabled, the comparator produces an output that indicates whether
the input signal voltage is greater or less than the output of the DAC. If the DAC output is

FIGURE 12–24   **Successive-approximation ADC.**

greater than the input signal, the comparator's output is LOW, causing the bit in the register to reset. If the output is less than the input signal, the 1 bit is retained in the register. The system does this with the MSB first, then the next most significant bit, then the next, and so on. After all the bits of the DAC have been tried, the conversion cycle is complete.

In order to better understand the operation of the successive-approximation ADC, let's take a specific example of a 4-bit conversion. Figure 12–25 illustrates the step-by-step conversion of a constant input voltage (5.1 V in this case). Let's assume that the DAC has the following output characteristics: $V_{out} = 8$ V for the 2^3 bit (MSB), $V_{out} = 4$ V for the 2^2 bit, $V_{out} = 2$ V for the 2^1 bit, and $V_{out} = 1$ V for the 2^0 bit (LSB).

FIGURE 12–25   **Illustration of the successive-approximation conversion process.**

Figure 12–25(a) shows the first step in the conversion cycle with the MSB = 1. The output of the DAC is 8 V. Since this is greater than the input of 5.1 V, the output of the comparator is LOW, causing the MSB in the SAR to be reset to a 0.

Figure 12–25(b) shows the second step in the conversion cycle with the 2^2 bit equal to a 1. The output of the DAC is 4 V. Since this is less than the input of 5.1 V, the output of the comparator switches to a HIGH, causing this bit to be retained in the SAR.

Figure 12–25(c) shows the third step in the conversion cycle with the 2^1 bit equal to a 1. The output of the DAC is 6 V because there is a 1 on the 2^2 bit input and on the 2^1 bit input; $4\text{ V} + 2\text{ V} = 6\text{ V}$. Since this is greater than the input of 5.1 V, the output of the comparator switches to a LOW, causing this bit to be reset to a 0.

Figure 12–25(d) shows the fourth and final step in the conversion cycle with the 2^0 bit equal to a 1. The output of the DAC is 5 V because there is a 1 on the 2^2 bit input and on the 2^0 bit input; $4\text{ V} + 1\text{ V} = 5\text{ V}$.

The four bits have all been tried, thus completing the conversion cycle. At this point the binary code in the register is 0101, which is approximately the binary value of the input of 5.1 V. Additional bits will produce an even more accurate result. Another conversion cycle now begins, and the basic process is repeated. The SAR is cleared at the beginning of each cycle.

Sigma-Delta Analog-to-Digital Converter

Sigma-delta is a widely used method of analog-to-digital conversion, particularly in telecommunications using audio signals. The method is based on **delta modulation** where the difference between two successive samples (increase or decrease) is quantized; other ADC methods were based on the absolute value of a sample. Delta modulation is a 1-bit quantization method.

The output of a delta modulator is a single-bit data stream where the relative number of 1s and 0s indicates the level or amplitude of the input signal. The number of 1s over a given number of clock cycles establishes the signal amplitude during that interval. A maximum number of 1s corresponds to the maximum positive input voltage. A number of 1s equal to one-half the maximum corresponds to an input voltage of zero. No 1s (all 0s) corresponds to the maximum negative input voltage. This is illustrated in a simplified way in Figure 12–26. For example, assume that 4096 1s occur during the interval when the input signal is a positive maximum. Since zero is the midpoint of the dynamic range of the input signal, 2048 1s occur during the interval when the input signal is zero. There are no 1s during the interval when the input signal is a negative maximum. For signal levels in between, the number of 1s is proportional to the level.

FIGURE 12–26   **A simplified illustration of sigma-delta analog-to-digital conversion.**

THE SIGMA-DELTA ADC FUNCTIONAL BLOCK DIAGRAM   The basic block diagram in Figure 12–27 accomplishes the conversion illustrated in Figure 12–26. The analog input signal and the analog signal from the converted quantized bit stream from the DAC in the feedback loop are applied to the summation (Σ) point. The difference (Δ) signal out of the Σ is integrated, and the 1-bit ADC increases or decreases the number of 1s depending on the difference signal. This action attempts to keep the quantized signal

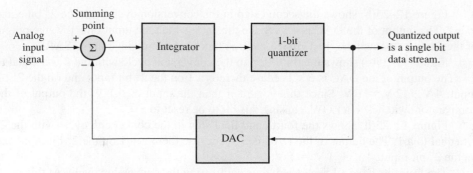

FIGURE 12–27 **Partial functional block diagram of a sigma-delta ADC.**

that is fed back equal to the incoming analog signal. The 1-bit quantizer is essentially a comparator followed by a latch.

To complete the sigma-delta conversion process using one particular approach, the single bit data stream is converted to a series of binary codes, as shown in Figure 12–28. The counter counts the 1s in the quantized data stream for successive intervals. The code in the counter then represents the amplitude of the analog input signal for each interval. These codes are shifted out into the latch for temporary storage. What comes out of the latch is a series of *n*-bit codes, which completely represent the analog signal.

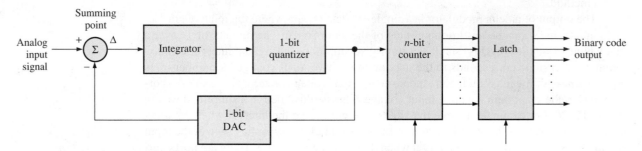

FIGURE 12–28 **One type of sigma-delta ADC.**

Testing Analog-to-Digital Converters

One method for testing ADCs is shown in Figure 12–29. A DAC is used as part of the test setup to convert the ADC output back to analog form for comparison with the test input.

FIGURE 12–29 **A method for testing ADCs.**

A test input in the form of a linear ramp is applied to the input of the ADC. The resulting binary output sequence is then applied to the DAC test unit and converted to a stairstep ramp. The input and output ramps are compared for any deviation.

Analog-to-Digital Conversion Errors

Again, a 4-bit conversion is used to illustrate the principles. Let's assume that the test input is an ideal linear ramp.

MISSING CODE The stairstep output in Figure 12–30(a) indicates that the binary code 1001 does not appear on the output of the ADC. Notice that the 1000 value stays for two intervals and then the output jumps to the 1010 value.

(a) Missing code (green) (b) Incorrect codes (green) (c) Offset

FIGURE 12–30 Illustrations of analog-to-digital conversion errors.

In a flash ADC, for example, a failure of one of the op-amp comparators can cause a missing-code error.

INCORRECT CODES The stairstep output in Figure 12–30(b) indicates that several of the binary code words coming out of the ADC are incorrect. Analysis indicates that the 2^1-bit line is stuck in the LOW (0) state in this particular case.

OFFSET Offset conditions are shown in 12–30(c). In this situation the ADC interprets the analog input voltage as greater than its actual value.

EXAMPLE 12–2

A 4-bit flash ADC is shown in Figure 12–31(a). It is tested with a setup like the one in Figure 12–29. The resulting reconstructed analog output is shown in Figure 12–31(b). Identify the problem and the most probable fault.

SOLUTION

The binary code 0011 is missing from the ADC output, as indicated by the missing step. Most likely, the output of comparator 3 is stuck in its inactive state (LOW).

RELATED PROBLEM

Reconstruct the analog output in a test setup like in Figure 12–29 if the ADC in Figure 12–31(a) has comparator 8 stuck in the HIGH output state.

(a)

(b)

FIGURE 12–31

SECTION 12–3 CHECKUP

1. What is the fastest method of analog-to-digital conversion?

2. Which analog-to-digital conversion method produces a single-bit data stream?

3. Does the successive-approximation converter have a fixed conversion time?

4. Name two types of output errors in an ADC.

12–4 DIGITAL-TO-ANALOG CONVERSION METHODS

Digital-to-analog conversion is an important part of a digital processing system. Once the digital data has been processed, it is converted back to analog form. In this section, we will examine the theory of operation of two basic types of digital-to-analog converters (DACs) and learn about their performance characteristics.

After completing this section, you should be able to

• **Explain the operation of a binary-weighted-input DAC**

• **Explain the operation of an _R/2R_ ladder DAC**

• **Discuss resolution, accuracy, linearity, monotonicity, and settling time in a DAC**

• **Discuss the testing of DACs for nonmonotonicity, differential nonlinearity, low or high gain, and offset error**

Binary-Weighted-Input Digital-to-Analog Converter

One method of digital-to-analog conversion uses a resistor network with resistance values that represent the binary weights of the input bits of the digital code. Figure 12–32 shows a 4-bit DAC of this type. Each of the input resistors will either have current or have no current, depending on the input voltage level. If the input voltage is zero (binary 0), the current is also zero. If the input voltage is HIGH (binary 1), the amount of current depends on the input resistor value and is different for each input resistor, as indicated in the figure.

FIGURE 12–32   **A 4-bit DAC with binary-weighted inputs.**

Since there is practically no current into the op-amp inverting (−) input, all of the input currents sum together and go through R_f. Since the inverting input is at 0 V (virtual ground), the drop across R_f is equal to the output voltage, so $V_{out} = I_f R_f$.

The values of the input resistors are chosen to be inversely proportional to the binary weights of the corresponding input bits. The lowest-value resistor (R) corresponds to the highest binary-weighted input (2^3). The other resistors are multiples of R (that is, $2R$, $4R$, and $8R$) and correspond to the binary weights 2^2, 2^1, and 2^0, respectively. The input currents are also proportional to the binary weights. Thus, the output voltage is proportional to the sum of the binary weights because the sum of the input currents is through R_f.

Disadvantages of this type of DAC are the number of different resistor values and the fact that the voltage levels must be exactly the same for all inputs. For example, an 8-bit converter requires eight resistors, ranging from some value of R to $128R$ in binary-weighted steps. This range of resistors requires tolerances of one part in 255 (less than 0.5%) to accurately convert the input, making this type of DAC very difficult to mass-produce.

EXAMPLE 12–3

Determine the output of the DAC in Figure 12–33(a) if the waveforms representing a sequence of 4-bit numbers in Figure 12–33(b) are applied to the inputs. Input D_0 is the least significant bit (LSB).

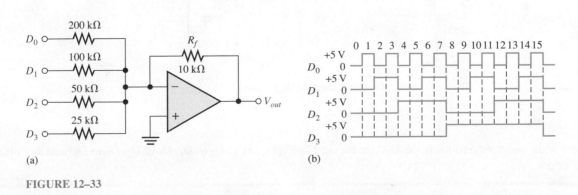

(a)   (b)

FIGURE 12–33

SOLUTION

First, determine the current for each of the weighted inputs. Since the inverting (−) input of the op-amp is at 0 V (virtual ground) and a binary 1 corresponds to +5 V, the current through any of the input resistors is 5 V divided by the resistance value.

$$I_0 = \frac{5 \text{ V}}{200 \text{ k}\Omega} = 0.025 \text{ mA}$$

$$I_1 = \frac{5 \text{ V}}{100 \text{ k}\Omega} = 0.05 \text{ mA}$$

$$I_2 = \frac{5 \text{ V}}{50 \text{ k}\Omega} = 0.1 \text{ mA}$$

$$I_3 = \frac{5 \text{ V}}{25 \text{ k}\Omega} = 0.2 \text{ mA}$$

Almost no current goes into the inverting op-amp input because of its extremely high impedance. Therefore, assume that all of the current goes through the feedback resistor R_f. Since one end of R_f is at 0 V (virtual ground), the drop across R_f equals the output voltage, which is negative with respect to virtual ground.

$$V_{out(D0)} = (10 \text{ k}\Omega)(-0.025 \text{ mA}) = -0.25 \text{ V}$$

$$V_{out(D1)} = (10 \text{ k}\Omega)(-0.05 \text{ mA}) = -0.5 \text{ V}$$

$$V_{out(D2)} = (10 \text{ k}\Omega)(-0.1 \text{ mA}) = -1 \text{ V}$$

$$V_{out(D3)} = (10 \text{ k}\Omega)(-0.2 \text{ mA}) = -2 \text{ V}$$

From Figure 12–33(b), the first binary input code is 0000, which produces an output voltage of 0 V. The next input code is 0001, which produces an output voltage of −0.25 V. The next code is 0010, which produces an output voltage of −0.5 V. The next code is 0011, which produces an output voltage of −0.25 V + −0.5 V = −0.75 V. Each successive binary code increases the output voltage by −0.25 V, so for this particular straight binary sequence on the inputs, the output is a stairstep waveform going from 0 V to −3.75 V in −0.25 V steps. This is shown in Figure 12–34.

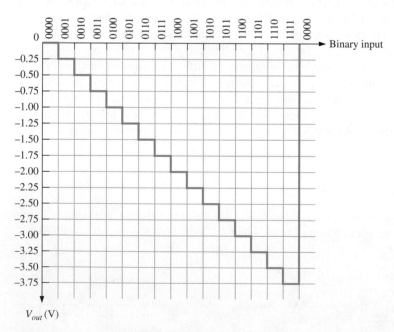

FIGURE 12–34   Output of the DAC in Figure 12–33.

RELATED PROBLEM

Reverse the input waveforms to the DAC in Figure 12–33 (D_3 to D_0, D_2 to D_1, D_1 to D_2, D_0 to D_3) and determine the output.

The *R/2R* Ladder Digital-to-Analog Converter

Another method of digital-to-analog conversion is the *R/2R* ladder, as shown in Figure 12–35 for four bits. It overcomes one of the problems in the binary-weighted-input DAC in that it requires only two resistor values.

FIGURE 12–35   An *R/2R* ladder DAC.

Start by assuming that the D_3 input is HIGH (+5 V) and the others are LOW (ground, 0 V). This condition represents the binary number 1000. A circuit analysis will show that this reduces to the equivalent form shown in Figure 12–36(a). Essentially no current goes through the 2R equivalent resistance because the inverting input is at virtual ground. Thus, all of the current ($I = 5\text{ V}/2R$) through R_7 also goes through R_f, and the output voltage is −5 V. The operational amplifier keeps the inverting (−) input near zero volts (≈ 0 V) because of negative feedback. Therefore, all current goes through R_f rather than into the inverting input.

Figure 12–36(b) shows the equivalent circuit when the D_2 input is at +5 V and the others are at ground. This condition represents 0100. If we thevenize* looking from R_8, we get 2.5 V in series with R, as shown. This results in a current through R_f of $I = 2.5\text{ V}/2R$, which gives an output voltage of −2.5 V. Keep in mind that there is no current into the op-amp inverting input and that there is no current through the equivalent resistance to ground because it has 0 V across it, due to the virtual ground.

Figure 12–36(c) shows the equivalent circuit when the D_1 input is at +5 V and the others are at ground. This condition represents 0010. Again thevenizing looking from R_8, you get 1.25 V in series with R as shown. This results in a current through R_f of $I = 1.25\text{ V}/2R$, which gives an output voltage of −1.25 V.

In part (d) of Figure 12–36, the equivalent circuit representing the case where D_0 is at +5 V and the other inputs are at ground is shown. This condition represents 0001. Thevenizing from R_8 gives an equivalent of 0.625 V in series with R as shown. The resulting current through R_f is $I = 0.625\text{ V}/2R$, which gives an output voltage of −0.625 V.

Notice that each successively lower-weighted input produces an output voltage that is halved, so that the output voltage is proportional to the binary weight of the input bits.

Performance Characteristics of Digital-to-Analog Converters

The performance characteristics of a DAC include resolution, accuracy, linearity, monotonicity, and settling time, each of which is discussed in the following list:

- **Resolution** The resolution of a DAC is the reciprocal of the number of discrete steps in the output. This, of course, is dependent on the number of input bits. For example, a 4-bit DAC has a resolution of one part in $2^4 - 1$ (one part in fifteen). Expressed as

---

*Thevenin's theorem states that any circuit can be reduced to an equivalent voltage source in series with an equivalent resistance.

(a) Equivalent circuit for $D_3 = 1, D_2 = 0, D_1 = 0, D_0 = 0$

(b) Equivalent circuit for $D_3 = 0, D_2 = 1, D_1 = 0, D_0 = 0$

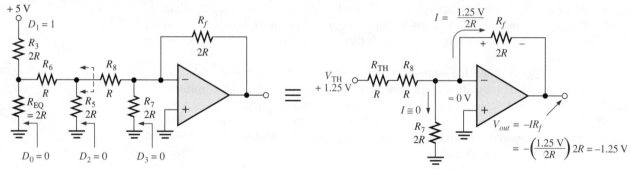

(c) Equivalent circuit for $D_3 = 0, D_2 = 0, D_1 = 1, D_0 = 0$

(d) Equivalent circuit for $D_3 = 0, D_2 = 0, D_1 = 0, D_0 = 1$

FIGURE 12–36 **Analysis of the $R/2R$ ladder DAC.**

a percentage, this is $(1/15)100 = 6.67$. The total number of discrete steps equals $2^n - 1$, where n is the number of bits. Resolution can also be expressed as the number of bits that are converted.

- **Accuracy** Accuracy is derived from a comparison of the actual output of a DAC with the expected output. It is expressed as a percentage of a full-scale, or maximum, output voltage. For example, if a converter has a full-scale output of 10 V and the accuracy is $\pm 0.1\%$, then the maximum error for any output voltage is

$(10 \text{ V})(0.001) = 10 \text{ mV}$. Ideally, the accuracy should be no worse than $\pm 1/2$ of a least significant bit. For an 8-bit converter, the least significant bit is 0.39% of full scale. The accuracy should be approximately ±0.2%.

- **Linearity** A linear error is a deviation from the ideal straight-line output of a DAC. A special case is an offset error, which is the amount of output voltage when the input bits are all zeros.

- **Monotonicity** A DAC is **monotonic** if it does not take any reverse steps when it is sequenced over its entire range of input bits.

- **Settling time** Settling time is normally defined as the time it takes a DAC to settle within ±1/2 LSB of its final value when a change occurs in the input code.

EXAMPLE 12–4

Determine the resolution, expressed as a percentage, of the following:

(a) an 8-bit DAC   **(b)** a 12-bit DAC

SOLUTION

(a) For the 8-bit converter,

$$\frac{1}{2^8 - 1} \times 100 = \frac{1}{255} \times 100 = \mathbf{0.392\%}$$

(b) For the 12-bit converter,

$$\frac{1}{2^{12} - 1} \times 100 = \frac{1}{4095} \times 100 = \mathbf{0.0244\%}$$

RELATED PROBLEM

Calculate the resolution for a 16-bit DAC.

Testing Digital-to-Analog Converters

The concept of DAC testing is illustrated in Figure 12–37. In this basic method, a sequence of binary codes is applied to the inputs, and the resulting output is observed. The binary code sequence extends over the full range of values from 0 to $2^n - 1$ in ascending order, where n is the number of bits.

FIGURE 12–37 **Basic test setup for a DAC.**

The ideal output is a straight-line stairstep as indicated. As the number of bits in the binary code is increased, the resolution is improved. That is, the number of discrete steps increases, and the output approaches a straight-line linear ramp.

Digital-to-Analog Conversion Errors

Several digital-to-analog conversion errors to be checked for are shown in Figure 12–38, which uses a 4-bit conversion for illustration purposes. A 4-bit conversion produces fifteen discrete steps. Each graph in the figure includes an ideal stairstep ramp for comparison with the faulty outputs.

NONMONOTONICITY   The step reversals in Figure 12–38(a) indicate nonmonotonic performance, which is a form of nonlinearity. In this particular case, the error occurs because the 2^1 bit in the binary code is interpreted as a constant 0. That is, a short is causing the bit input line to be stuck LOW.

(a) Nonmonotonic output (green)

(b) Differential nonlinearity (green)

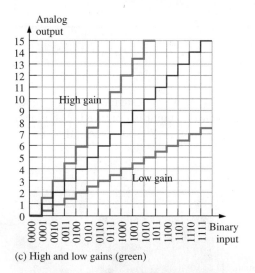

(c) High and low gains (green)

(d) Offset error (green)

FIGURE 12–38   **Illustrations of several digital-to-analog conversion errors.**

DIFFERENTIAL NONLINEARITY   Figure 12–38(b) illustrates differential nonlinearity in which the step amplitude is less than it should be for certain input codes. This particular output could be caused by the 2^2 bit having an insufficient weight, perhaps because of a faulty input resistor. We could also see steps with amplitudes greater than normal if a particular binary weight were greater than it should be.

LOW OR HIGH GAIN   Output errors caused by low or high gain are illustrated in Figure 12–38(c). In the case of low gain, all of the step amplitudes are less than ideal. In the case of high gain, all of the step amplitudes are greater than ideal. This situation may be caused by a faulty feedback resistor in the op-amp circuit.

OFFSET ERROR   An offset error is illustrated in Figure 12–38(d). Notice that when the binary input is 0000, the output voltage is nonzero and that this amount of offset is the same for all steps in the conversion. A faulty op-amp may be the culprit in this situation.

EXAMPLE 12–5

The DAC output in Figure 12–39 is observed when a straight 4-bit binary sequence is applied to the inputs. Identify the type of error, and suggest an approach to isolate the fault.

FIGURE 12–39

SOLUTION

The DAC in this case is nonmonotonic. Analysis of the output reveals that the device is converting the following sequence, rather than the actual binary sequence applied to the inputs.

0010, 0011, 0010, 0011, 0110, 0111, 0110, 0111, 1010, 1011, 1010, 1011, 1110, 1111, 1110, 1111

Apparently, the 2^1 bit is stuck in the HIGH (1) state. To find the problem, first monitor the bit input pin to the device. If it is changing states, the fault is internal to the DAC and it should be replaced. If the external pin is not changing states and is always HIGH, check for an external short to $+V$ that may be caused by a solder bridge somewhere on the circuit board.

RELATED PROBLEM

Determine the output of a DAC when a straight 4-bit binary sequence is applied to the inputs and the 2^0 bit is stuck HIGH.

The Reconstruction Filter

The output of the DAC is a "stairstep" approximation of the original analog signal after it has been processed by the digital signal processor (DSP). The purpose of the low-pass reconstruction filter (sometimes called a postfilter) is to smooth out the DAC output by eliminating the higher frequency content that results from the fast transitions of the "stairsteps," as roughly illustrated in Figure 12–40.

Output of the DAC   Reconstruction filter   Final analog output

FIGURE 12–40   **The reconstruction filter smooths the output of the DAC.**

SECTION 12–4 CHECKUP

1. What is the disadvantage of the DAC with binary-weighted inputs?

2. What is the resolution of a 4-bit DAC?

3. How do you detect nonmonotonic behavior in a DAC?

4. What effect does low gain have on a DAC output?

12–5 DIGITAL SIGNAL PROCESSING

Digital signal processing converts signals that naturally occur in analog form, such as sound, video, and information from sensors, to digital form and uses digital techniques to enhance and modify analog signal data for various applications.

After completing this section, you should be able to

- **Define *ADC***
- **Define *DSP***
- **Define *DAC***
- **Draw a basic block diagram of a digital signal processing system**

A digital signal processing system first translates a continuously varying analog signal into a series of discrete levels. This series of levels follows the variations of the analog signal and resembles a staircase, as illustrated for the case of a sine wave in Figure 12–41. The process of changing the original analog signal to a "stairstep" approximation is accomplished by a sample-and-hold circuit.

FIGURE 12–41   **An original analog signal (sine wave) and its "stairstep" approximation.**

Sample — Hold

Each held level is converted to a binary code by an ADC.

Next, the "stairstep" approximation is quantized into binary codes that represent each discrete step on the "stairsteps" by a process called analog-to-digital (A/D) conversion. The circuit that performs A/D conversion is an analog-to-digital converter (ADC).

Once the analog signal has been converted to a binary coded form, it is applied to a **DSP** (digital signal processor). The DSP can perform various operations on the incoming data, such as removing unwanted interference, increasing the amplitude of some signal frequencies and reducing others, encoding the data for secure transmissions, and detecting and correcting errors in transmitted codes. DSPs make possible, among many other things, the cleanup of sound recordings, the removal of echos from communications lines, the enhancement of images from CT scans for better medical diagnosis, and the scrambling of cellular phone conversations for privacy.

After a DSP processes a signal, the signal can be converted back to an enhanced version of the original analog signal. This is accomplished by a digital-to-analog converter (DAC). Figure 12–42 shows a basic block diagram of a typical digital signal processing system.

FIGURE 12–42   **Basic block diagram of a typical digital signal processing system.**

DSPs are actually a specialized type of microprocessor but are different from general-purpose microprocessors in a couple of significant ways. Typically, microprocessors are designed for general-purpose functions and operate with large software packages. DSPs are used for special-purpose applications; they are very fast number crunchers that must work in real time by processing information as it happens using specialized algorithms (programs). The analog-to-digital converter (ADC) in a system must take samples of the incoming analog data often enough to catch all the relevant fluctuations in the signal amplitude, and the DSP must keep pace with the sampling rate of the ADC by doing its calculations as fast as the sampled data are received. Once the digital data are processed by the DSP, they go to the digital-to-analog converter (DAC) for conversion back to analog form.

SECTION 12–5 CHECKUP

1. What does DSP stand for?

2. What does ADC stand for?

3. What does DAC stand for?

4. An analog signal is changed to a binary coded form by what circuit?

5. A binary coded signal is changed to analog form by what circuit?

12–6 THE DIGITAL SIGNAL PROCESSOR (DSP)

Essentially, a digital signal processor (DSP) is a special type of microprocessor that processes data in real time. Its applications focus on the processing of digital data that represents analog signals. A DSP, like a microprocessor, has a central processing unit (CPU) and memory units in addition to many interfacing functions.

After completing this chapter, you should be able to

- **Explain the basic concepts of a DSP**
- **List some of the applications of DSPs**
- **Describe the basic functions of a DSP in a cell phone**
- **Discuss the TMS320C6000 series DSP**

The digital signal processor (DSP) is the heart of a digital signal processing system. It takes its input from an ADC and produces an output that goes to a DAC, as shown in Figure 12–43. As you have learned, the ADC changes an analog waveform into data in the form of a series of binary codes that are then applied to the DSP for processing. After being processed by the DSP, the data go to a DAC for conversion back to analog form.

FIGURE 12–43   **The DSP has a digital input and produces a digital output.**

DSP Programming

DSPs are typically programmed in either assembly language or in C. Because programs written in assembly language can usually execute faster and because speed is critical in most DSP applications, assembly language is used much more in DSPs than in general-purpose microprocessors. Also, DSP programs are usually much shorter than traditional microprocessor programs because of their very specialized applications where much redundancy is used. In general, the instruction sets for DSPs tend to be smaller than for microprocessors.

DSP Applications

The DSP, unlike the general-purpose microprocessor, must typically process data in *real time;* that is, as it happens. Many applications in which DSPs are used cannot tolerate any noticeable delays, requiring the DSP to be extremely fast. In addition to cell phones, digital signal processors (DSPs) are used in multimedia computers, video recorders, CD players, hard disk drives, digital radio modems, and other applications to improve the signal quality. Also, DSPs are becoming more common in television applications. For example, television converters use DSP to provide compatibility with various television standards.

An important application of DSPs is in signal compression and decompression. In CD systems, for example, the music on the CD is in a compressed form so that it doesn't use as much storage space. It must be decompressed in order to be reproduced. Also signal compression is used in cell phones to allow a greater number of calls to be handled simultaneously in a local cell. Some other areas where it has had a major impact are as follows.

TELECOMMUNICATIONS   The field of telecommunications involves transferring all types of information from one location to another, including telephone conversations, television signals, and digital data. Among other functions, the DSP facilitates multiplexing many signals onto one transmission channel because information in digital form is relatively easy to multiplex and demultiplex.

At the transmitting end of a telecommunications system, DSPs are used to compress digitized voice signals for conservation of bandwidth. Compression is the process of reducing the data rate. Generally, a voice signal is converted to digital form at 8000 samples per second (sps), based on a Nyquist frequency of 4 kHz. If 8 bits are used to encode each sample, the data rate is 64 kbps. In general, reducing (compressing) the data rate from 64 kbps to 32 kbps results in no loss of sound quality. When the data are compressed to 8 kbps, the sound quality is reduced noticeably. When compressed to the minimum of 2 kbps, the sound is greatly distorted but still usable for some applications where only word recognition and not quality is important. At the receiving end of a telecommunications system, the DSP decompresses the data to restore the signal to its original form.

Sound cards used in computers use an ADC to convert sound from a microphone, audio CD player, or other source into a digital signal. The ADC sends the digital signal to a digital signal processor (DSP). Based on instructions from a ROM, one function of the DSP is to compress the digital signal so it uses less storage space. The DSP then sends the compressed data to the computer's processor which, in turn, sends the data to a hard drive or CD ROM for storage. To play a recorded sound, the stored data is retrieved by the processor and sent to the DSP where it is decompressed and sent to a DAC. The output of the DAC, which is a reproduction of the original sound signal, is applied to the speakers.

SYSTEM NOTE

Echoes, a problem in many long distance telephone connections, occur when a portion of a voice signal is returned with a delay. For shorter distances, this delay is barely noticeable; but as the distance between the transmitter and the receiver increases, so does the delay time of the echo. DSPs are used to effectively cancel the annoying echo, which results in a clear, undisturbed voice signal.

MUSIC PROCESSING   The DSP is used in the music industry to provide filtering, signal addition and subtraction, and signal editing in music preparation and recording. Also, another application of the DSP is to add artificial echo and reverberation, which are usually minimized by the acoustics of a sound studio, in order to simulate ideal listening environments from concert halls to small rooms.

SPEECH GENERATION AND RECOGNITION   DSPs are used in speech generation and recognition to enhance the quality of man/machine communication. The most common method used to produce computer-generated speech is digital recording. In digital recording, the human voice is digitized and stored, usually in a compressed form. During playback the stored voice data are uncompressed and converted back into the original analog form. Approximately an hour of speech can be stored using about 3 MB of memory.

Speech recognition is much more difficult to accomplish than speech generation. The DSP is used to isolate and analyze each word in the incoming voice signal. Certain parameters are identified in each word and compared with previous examples of the spoken word to create the closest match. Most systems are limited to a few hundred words at best. Also, significant pauses between words are usually required and the system must be "trained" for a given individual's voice. Speech recognition is an area of tremendous research effort and will eventually be applied in many commercial applications.

RADAR   In *ra*dio *d*etection *a*nd *r*anging (radar) applications, DSPs provide more accurate determination of distance using data compression techniques, decrease noise using filtering techniques, thereby increasing the range, and optimize the ability of the radar system to identify specific types of targets. DSPs are also used in similar ways in sonar systems.

IMAGE PROCESSING   The DSP is used in image-processing applications such as the computed tomography (CT) and magnetic resonance imaging (MRI), which are widely used in the medical field for looking inside the human body. In CT, X-rays are passed through a section of the body from many directions. The resulting signals are converted to digital form and stored. This stored information is used to produce calculated images that appear to be slices through the human body that show great detail and permit better diagnosis.

Instead of X-rays, MRI uses magnetic fields in conjunction with radio waves to probe inside the human body. MRI produces images, just as CT, and provides excellent discrimination between different types of tissue as well as information such as blood flow through arteries. MRI depends entirely on digital signal processing methods.

In applications such as video telephones, digital television, and other media that provide moving pictures, the DSP uses image compression to reduce the number of bits needed, making these systems commercially feasible.

FILTERING DSPs are commonly used to implement digital filters for the purposes of separating signals that have been combined with other signals or with interference and noise and for restoring signals that are distorted. Although analog filters are quite adequate for some applications, the digital filter is generally much superior in terms of the performance that can be achieved. One drawback to digital filters is that the execute time required produces a delay from the time the analog signal is applied until the time the output appears. Analog filters present no delay problems because as soon as the input occurs, the response appears on the output. Analog filters are also less expensive than digital filters. Regardless of this, the overall performance of the digital filter is far superior in many applications.

SYSTEM EXAMPLE 12–1

A CELLULAR TELEPHONE

Figure 12–44 shows a simplified block diagram of a digital cell phone. The voice **codec** (codec is the abbreviation for coder/decoder) contains, among other functions, the ADC and DAC necessary to convert between the analog voice signal and a digital voice format. Sigma-delta conversion is typically used in most cell phone applications. For transmission, the voice signal from the microphone is converted to digital form by the ADC in the codec and then it goes to the DSP for processing. From the DSP, the digital signal goes to the rf (radio frequency) section where it is modulated and changed to the radio frequency for transmission. An incoming rf signal containing voice data is picked up by the antenna, demodulated, and changed to a digital signal. It is then applied to the DSP for processing, after which the digital signal goes to the codec for conversion back to the original voice signal by the DAC. It is then amplified and applied to the speaker.

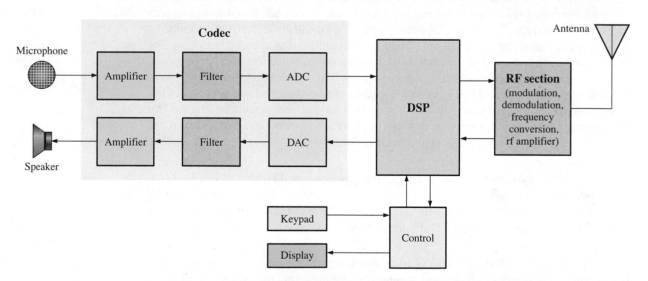

FIGURE 12–44 **Simplified block diagram of a digital cellular phone.**

The DSP performs many functions to improve and facilitate the reception and transmission of a voice signal. Some of these DSP functions are as follows:

- **Speech compression** The rate of the digital voice signal is reduced significantly for transmission in order to meet the bandwidth requirements.

- **Speech decompression** The rate of the received digital voice signal is returned to its original rate in order to properly reproduce the analog voice signal.

- **Protocol handling** The cell phone communicates with the nearest base in order to establish the location of the cell phone, allocates time and frequency slots, and arranges handover to another base station as the phone moves into another cell.

- **Error detection and correction** During transmission, error detection and correction codes are generated and, during reception, detect and correct errors induced in the rf channel by noise or interference.

- **Encryption** Converts the digital voice signal to a form for secure transmission and converts it back to original form during reception.

Basic DSP Architecture

As mentioned before, a DSP is basically a specialized microprocessor optimized for speed in order to process data in real time. Many DSPs are based on what is known as the *Harvard architecture,* which consists of a central processing unit (CPU) and two memories, one for data and the other for the program, as shown by the block diagram in Figure 12–45.

FIGURE 12–45   Many DSPs use the Harvard architecture (two memories).

Specific DSPs—The TMS320C6000 Series

DSPs are available for both fixed-point and floating-point processing. Recall from Chapter 2 that these two methods differ in the way numbers are stored and manipulated. All floating-point DSPs can also handle numbers in fixed-point format. Fixed-point DSPs are less expensive than the floating-point versions and, generally, can operate faster. The details of DSP architecture can vary significantly, even within the same family. Let's look briefly at one particular DSP series as an example of how a DSP is generally organized.

Examples of DSPs available in the TMS320C6000 series include the TMS320C62xx, the TMS320C64xx, and the TMS320C67xx, which are part of Texas Instrument's TMS320 family of devices. A general block diagram for these devices is shown in Figure 12–46.

The DSPs have a central processing unit (CPU), also known as the **DSP core**, that contains 64 general-purpose 32-bit registers in the C64xx and 32 general-purpose 32-bit registers in the C62xx and the C67xx. The C67xx can handle floating-point operations, whereas the C62xx and C64xx are fixed-point devices.

Each DSP has eight functional units that contain two 16-bit multipliers and six arithmetic logic units (ALUs). The performance of the three DSPs in the C6000 series in terms of **MIPS** (Million Instructions Per Second), **MFLOPS** (Million Floating-point Operations Per Second), and **MMACS** (Million Multiply/Accumulates per Second) is shown in Table 12–3.

DATA PATHS IN THE CPU   In the CPU, the program fetch, instruction dispatch, and instruction decode sections can provide eight 32-bit instructions to the functional units during every clock cycle. The CPU is split into two data paths, and instruction processing occurs in both data paths A and B. Each data path contains half of the general-purpose registers (16 in the C62xx and C67xx or 32 in the C64xx) and four functional units. The control register and logic are used to configure and control the various processor operations.

FUNCTIONAL UNITS   Each data path has four functional units. The M units (labeled .M1 and .M2 in Figure 12–46) are dedicated multipliers. The L units (labeled .L1 and .L2) perform arithmetic, logic, and miscellaneous operations. The S units (labeled .S1 and .S2) perform compare, shift, and miscellaneous arithmetic operations. The D units (labeled .D1 and .D2) perform load, store, and miscellaneous operations.

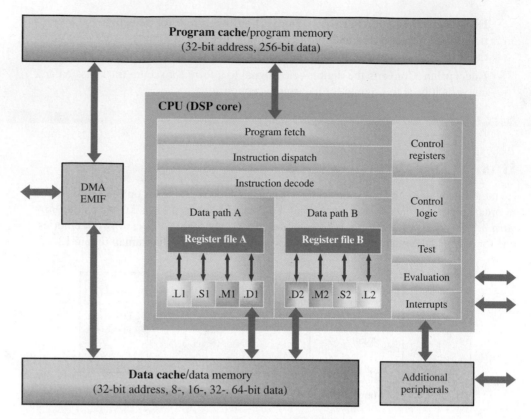

FIGURE 12–46   **General block diagram of the TMS320C6000 series DSP.**

Table 12–3 • TMS320C6000 series DSP data processing performance.				
DSP	**TYPE**	**APPLICATION**	**PROCESSING SPEED**	**MULTIPLY/ ACCUMULATE SPEED**
C62xx	Fixed-point	General-purpose	1200–2400 MIPS	300–600 MMACS
C64xx	Fixed-point	Special-purpose	3200–4800 MIPS	1600–2400 MMACS
C67xx	Floating-point	General-purpose	600–1000 MFLOPS	200–333 MMACS

PIPELINE   A **pipeline** allows multiple instructions to be processed simultaneously. A pipeline operation consists of three stages through which all instructions flow: *fetch, decode, execute.* Eight instructions at a time are first fetched from the program memory; they are then decoded, and finally they are executed.

During **fetch**, the eight instructions (called a packet) are taken from memory in four *phases,* as shown in Figure 12–47.

- **Program address generate (PG)** The program address is generated by the CPU.
- **Program address send (PS)** The program address is sent to the memory.
- **Program access ready wait (PW)** A memory read operation occurs.
- **Program fetch packet receive (PR)** The CPU receives the packet of instructions.

FIGURE 12–47   **The four fetch phases of the pipeline operation.**

Two phases make up the instruction **decode** stage of pipeline operation, as shown in Figure 12–48. The instruction dispatch (DP) phase is where the instruction packets are split into execute packets and assigned to the appropriate functional units. The instruction decode (DC) phase is where the instructions are decoded.

The **execute** stage of the pipeline operation is where the instructions from the decode stage are carried out. The execute stage has a maximum of five phases (E1 through E5), as shown in Figure 12–49. All instructions do not use all five phases. The number of phases used during execution depends on the type of instruction. Part of the execution of an instruction requires getting data from the data memory.

FIGURE 12–48   The two decode phases of the pipeline operation.

FIGURE 12–49   The five execute phases of pipeline operation.

INTERNAL DSP MEMORY AND INTERFACES   As you can see in Figure 12–46, there are two internal memories, one for data and one for program. The program memory is organized in 256 bit packets (eight 32-bit instructions) and there are 64 kB of capacity. The data memory also has a capacity of 64 kB and can be accessed in 8-, 16-, 32-, or 64-bit word lengths, depending on the specific device in the series. Both internal memories are accessed with a 32-bit address. The DMA (Direct Memory Access) is used to transfer data without going through the CPU. The EMIF (External Memory Interface) is used to support external memories when required in an application. Additional interface is provided for serial I/O ports and other external devices.

TIMERS   There are two general-purpose timers in the DSP that can be used for timed events, counting, pulse generation, CPU interrupts, and more.

PACKAGING   These particular processors are available in 352-pin ball grid array (BGA) packages, as shown in Figure 12–50, and are implemented with CMOS technology.

(a) Top view

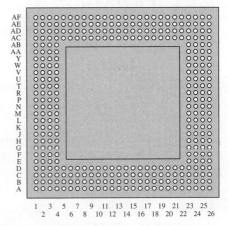

(b) Bottom view

(c) Side view

FIGURE 12–50   A 352-pin BGA package.

SECTION 12–6 CHECKUP

1. What is meant by the Harvard architecture?
2. What is a DSP core?
3. Name two categories of DSPs according to the type of numbers handled.
4. What are the two types of internal memory?

5. Define (a) MIPS   (b) MFLOPS   (c) MMACS.
6. Basically, what does pipelining accomplish?
7. Name the three stages of pipeline operation.
8. What happens during the fetch phase?

SUMMARY

- Sampling converts an analog signal into a series of impulses, each representing the signal amplitude at a given instant in time.
- The sampling theorem states that the sampling frequency must be at least twice the highest sampled frequency (Nyquist frequency).
- Analog-to-digital conversion changes an analog signal into a series of digital codes.
- Four types of analog-to-digital converters (ADCs) are flash (simultaneous), dual-slope, successive-approximation, and sigma-delta.
- Digital-to-analog conversion changes a series of digital codes that represent an analog signal back into the analog signal.
- Two types of digital-to-analog converters (DACs) are binary-weighted input and $R/2R$ ladder.
- Digital signal processing is the digital processing of analog signals, usually in real-time, for the purpose of modifying or enhancing the signal in some way.
- In general, a digital signal processing system consists of an anti-aliasing filter, a sample-and-hold circuit, an analog-to-digital converter, a DSP (digital signal processor), a digital-to-analog converter, and a reconstruction filter.
- A DSP is a specialized microprocessor optimized for speed in order to process data as it occurs (real-time).
- Most DSPs are based on the Harvard architecture, which means that there is a data memory and a program memory.
- A pipeline operation consists of fetch, decode, and execute stages.

KEY TERMS

Key terms and other bold terms in the chapter are defined in the end-of-book glossary.

Aliasing   The effect created when a signal is sampled at less than twice the signal frequency. Aliasing creates unwanted frequencies that interfere with the signal frequency when signal is recovered.

Analog-to-digital converter (ADC)   A circuit used to convert an analog signal to digital form.

Decode   A stage of the DSP pipeline operation in which instructions are assigned to functional units and are decoded.

Digital-to-analog converter (DAC)   A circuit used to convert the digital representation of an analog signal back to the analog signal.

DSP   Digital signal processor; a special type of microprocessor that processes data in real time.

DSP core   The central processing unit of a DSP.

Execute   A stage of the DSP pipeline operation in which the decoded instructions are carried out.

Fetch   A stage of the DSP pipeline operation in which an instruction is obtained from the program memory.

MFLOPS   Million floating-point operations per second.

MIPS   Million instructions per second.

MMACS   Million multiply/accumulates per second.

Nyquist frequency   The highest signal frequency that can be sampled at a specified sampling frequency; a frequency equal to or less than half the sampling frequency.

Pipeline   Part of the DSP architecture that allows multiple instructions to be processed simultaneously.

Quantization   The process whereby a binary code is assigned to each sampled value during analog-to-digital conversion.

Sampling   The process of taking a sufficient number of discrete values at points on a waveform that will define the shape of the waveform.

TRUE/FALSE QUIZ

Answers are at the end of the chapter.

1. An analog signal is converted to a digital signal by an ADC.

2. A DAC is a digital approximation computer.

3. The Nyquist frequency is twice the sampling frequency.

4. A higher sampling rate is more accurate than a lower sampling rate for a given analog signal.

5. Resolution is the number of bits used by an analog-to-digital converter.

6. Successful approximation is an analog-to-digital conversion method.

7. Delta modulation is based on the difference of two successive samples.

8. Two types of DAC are the binary-weighted input and the $R/2R$ ladder.

9. The process of converting an analog value to a code is called *quantization*.

10. A flash ADC differs from a simultaneous ADC.

SELF-TEST

Answers are at the end of the chapter.

1. An ADC is an
   - **(a)** alphanumeric data code
   - **(b)** analog-to-digital converter
   - **(c)** analog device carrier
   - **(d)** analog-to-digital comparator

2. A DAC is a
   - **(a)** digital-to-analog computer
   - **(b)** digital analysis calculator
   - **(c)** data accumulation converter
   - **(d)** digital-to-analog converter

3. Sampling of an analog signal produces
   - **(a)** a series of impulses that are proportional to the amplitude of the signal
   - **(b)** a series of impulses that are proportional to the frequency of the signal
   - **(c)** digital codes that represent the analog signal amplitude
   - **(d)** digital codes that represent the time of each sample

4. According to the sampling theorem, the sampling frequency should be
   - **(a)** less than half the highest signal frequency
   - **(b)** greater than twice the highest signal frequency
   - **(c)** less than half the lowest signal frequency
   - **(d)** greater than the lowest signal frequency

5. A hold action occurs
   - **(a)** before each sample
   - **(b)** during each sample
   - **(c)** after the analog-to-digital conversion
   - **(d)** immediately after a sample

6. The quantization process
   - **(a)** converts the sample-and-hold output to binary code
   - **(b)** converts a sample impulse to a level
   - **(c)** converts a sequence of binary codes to a reconstructed analog signal
   - **(d)** filters out unwanted frequencies before sampling takes place

7. Generally, an analog signal can be reconstructed more accurately with
   - **(a)** more quantization levels
   - **(b)** fewer quantization levels
   - **(c)** a higher sampling frequency
   - **(d)** a lower sampling frequency
   - **(e)** either answer (a) or (c)

8. A flash ADC uses

   (a) counters   (b) op-amps
   (c) an integrator   (d) flip-flops
   (e) answers (a) and (c)

9. A dual-slope ADC uses

   (a) a counter   (b) op-amps
   (c) an integrator   (d) a differentiator
   (e) answers (a) and (c)

10. The output of a sigma-delta ADC is

    (a) parallel binary codes   (b) multiple-bit data
    (c) single-bit data   (d) a difference voltage

11. In a binary-weighted DAC, the resistors on the inputs

    (a) determine the amplitude of the analog signal
    (b) determine the weights of the digital inputs
    (c) limit the power consumption
    (d) prevent loading on the source

12. In an $R/2R$ DAC, there are

    (a) four values of resistors
    (b) one resistor value
    (c) two resistor values
    (d) a number of resistor values equal to the number of inputs

13. A digital signal processing system usually operates in

    (a) real time   (b) imaginary time
    (c) compressed time   (d) computer time

14. The term *Harvard architecture* means

    (a) a CPU and a main memory
    (b) a CPU and two data memories
    (c) a CPU, a program memory, and a data memory
    (d) a CPU and two register files

15. The minimum number of general-purpose registers in the TMS320C6000 series DSPs is

    (a) 32   (b) 64   (c) 16   (d) 8

16. The two internal memories in the TMS320C6000 series each have a capacity of

    (a) 1 MB   (b) 512 kB   (c) 64 kB   (d) 32 kB

17. In the TMS320C6000 series pipeline operation, the number of instructions processed simultaneously is

    (a) eight   (b) four   (c) two   (d) one

18. The stage of the pipeline operation in which instructions are retrieved from the memory is called

    (a) execute   (b) accumulate   (c) decode   (d) fetch

PROBLEMS

Answers to odd-numbered problems are at the end of the book.

SECTION 12–1   A System

1. Assume that the voltage across a photoresistor varies from 5 V at the minimum light level to 1 V at the maximum light level. Determine the voltage increments represented by each 4-bit code produced by the ADC.

2. Referring to Problem 1, which is the voltage represented by 1001?

3. The stepper motor control generates 4-bit cycles 1000, 0100, 0010, 0001 to advance the stepper motor. If each step is 1.5°, how many cycles of the motor control is required to advance the motor 18°?

4. Explain the purpose of each functional block in the sun tracker system block diagram in Figure 12–3.

SECTION 12–2   Converting Analog Signals to Digital

5. The waveform shown in Figure 12–51 is applied to a sampling circuit and is sampled every 3 ms. Show the output of the sampling circuit. Assume a one-to-one voltage correspondence between the input and output.

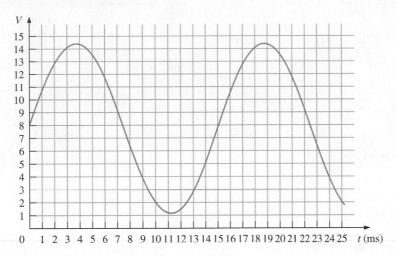

FIGURE 12–51

6. The output of the sampling circuit in Problem 5 is applied to a hold circuit. Show the output of the hold circuit.

7. If the output of the hold circuit in Problem 6 is quantized using two bits, what is the resulting sequence of binary codes?

8. Repeat Problem 7 using 4-bit quantization.

9. (a) Reconstruct the analog signal from the 2-bit quantization in Problem 7.
   (b) Reconstruct the analog signal from the 4-bit quantization in Problem 8.

10. Graph the analog function represented by the following sequence of binary numbers:

    1111, 1110, 1101, 1100, 1010, 1001, 1000, 0111, 0110, 0101, 0100, 0101, 0110, 0111, 1000, 1001, 1010, 1011, 1100, 1100, 1100, 1011, 1010, 1001.

SECTION 12–3   Analog-to-Digital Conversion Methods

11. The input voltage to a certain op-amp inverting amplifier is 10 mV, and the output is 2 V. What is the closed-loop voltage gain?

12. To achieve a closed-loop voltage gain of 330 with an inverting amplifier, what value of feedback resistor do you use if $R_i = 1.0$ kV?

13. What is the gain of an inverting amplifier that uses a 47 kV feedback resistor if the input resistor is 2.2 kV?

14. How many comparators are required to form an 8-bit flash converter?

15. Determine the binary output code of a 3-bit flash ADC for the analog input signal in Figure 12–52.

FIGURE 12–52

16. Repeat Problem 15 for the analog waveform in Figure 12–53.

FIGURE 12–53

17. For a certain 2-bit successive-approximation ADC, the maximum ladder output is +8 V. If a constant +6 V is applied to the analog input, determine the sequence of binary states for the SAR.

18. Repeat Problem 17 for a 4-bit successive-approximation ADC.

19. An ADC produces the following sequence of binary numbers when an analog signal is applied to its input: 0000, 0001, 0010, 0011, 0100, 0101, 0110, 0111, 0110, 0101, 0100, 0011, 0010, 0001, 0000.
(a) Reconstruct the input digitally.
(b) If the ADC failed so that the code 0111 were missing, what would the reconstructed output look like?

SECTION 12–4   Digital-to-Analog Conversion Methods

20. In the 4-bit DAC in Figure 12–32, the lowest-weighted resistor has a value of 10 kΩ. What should the values of the other input resistors be?

21. Determine the output of the DAC in Figure 12–54(a) if the sequence of 4-bit numbers in part (b) is applied to the inputs. The data inputs have a low value of 0 V and a high value of +5 V.

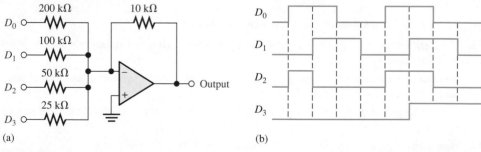

(a)   (b)

FIGURE 12–54

22. Repeat Problem 21 for the inputs in Figure 12–55.

FIGURE 12–55

23. Determine the resolution expressed as a percentage, for each of the following DACs:
    (a) 3-bit   **(b)** 10-bit   **(c)** 18-bit

24. Develop a circuit for generating an 8-bit binary test sequence for the test setup in Figure 12–37.

25. A 4-bit DAC has failed in such a way that the MSB is stuck in the 0 state. Draw the analog output when a straight binary sequence is applied to the inputs.

26. A straight binary sequence is applied to a 4-bit DAC, and the output in Figure 12–56 is observed. What is the problem?

FIGURE 12–56

SECTION 12–5   Digital Signal Processing

27. Explain the purpose of analog-to-digital conversion.

28. Fill in the appropriate functional names for the digital signal processing system block diagram in Figure 12–57.

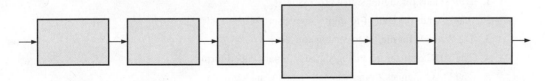

FIGURE 12–57

29. Explain the purpose of digital-to-analog conversion.

SECTION 12–6   The Digital Signal Processor (DSP)

30. A TMS320C62xx DSP has 32-bit instructions and is operating at 2000 MIPS. How many bytes per second is the DSP processing?

31. If the clock rate of a TMS320C64xx DSP is 400 MHz, how many instructions can it provide to the CPU functional units in one second?

32. How many floating-point operations can a DSP do in one second if it is specified at 1000 MFLOPS?

33. List and describe the four phases of the fetch operation in a TMS320C6000 series DSP.

34. List and describe the two phases of the decode operation in a TMS320C6000 series DSP.

ANSWERS TO SECTION CHECKUPS

SECTION 12–1 A System

1. The system rotates when one photosensor has more light than the other.

2. The system rotates counterclockwise when the level of sunlight falls below a certain level in the evening.

3. The ADCs convert the photoresister voltages to binary code.

4. To rotate the system in small increments

SECTION 12–2 Converting Analog Signals to Digital

1. Sampling is the process of converting an analog signal into a series of impulses, each representing the amplitude of the analog signal.

2. A sampled value is held to allow time to convert the value to a binary code.

3. The minimum sampling frequency is 40 kHz.

4. Quantization is the process of converting a sampled level to a binary code.

5. The number of bits determine quantization accuracy.

SECTION 12–3 Analog-to-Digital Conversion Methods

1. The simultaneous (flash) method is fastest.

2. The sigma-delta method produces a single-bit data stream.

3. Yes, successive approximation has a fixed conversion time.

4. Missing code, incorrect code, and offset are types of ADC output errors.

SECTION 12–4 Digital-to-Analog Conversion Methods

1. In a binary-weighted DAC, each resistor has a different value.

2. $(1/(2^4 - 1))100\% = 6.67\%$

3. A step reversal indicates nonmonotonic behavior in a DAC.

4. Step amplitudes in a DAC are less than ideal with low gain.

SECTION 12–5 Digital Signal Processing

1. DSP stands for digital signal processor.

2. ADC stands for analog-to-digital converter.

3. DAC stands for digital-to-analog converter.

4. The ADC changes an analog signal to binary coded form.

5. The DAC changes a binary coded signal to analog form.

SECTION 12–6 The Digital Signal Processor (DSP)

1. Harvard architecture means that there is a CPU and two memories, one for data and one for programs.

2. The DSP core is the CPU.

3. DSPs can be fixed-point or floating-point.

4. Internal memory types are data and program.

5. **(a)** MIPS—million instructions per second
   (b) MFLOPS—million floating-point operations per second
   (c) MMACS—million multiply/accumulates per second

6. Pipelining provides for the processing of multiple instructions simultaneously.

7. The stages of pipeline operation are fetch, decode, and execute.

8. During fetch, instructions are retrieved from the program memory.

ANSWERS TO RELATED PROBLEMS FOR EXAMPLES

12–1 100, 111, 100, 000, 011, 110. Yes, information is lost.

12–2 See Figure 12–58.

12–3 See Figure 12–59.

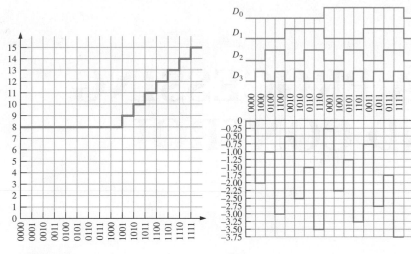

FIGURE 12–58   x   FIGURE 12–59

12–4 $(1/(2^{16} - 1))100\% = 0.00153\%$

12–5 See Figure 12–60.

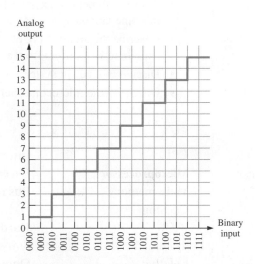

FIGURE 12–60

ANSWERS TO TRUE/FALSE QUIZ

1. T   **2.** F   **3.** F   **4.** T   **5.** T   **6.** F   **7.** T   **8.** T   **9.** T   **10.** F

ANSWERS TO SELF-TEST

1. (b)   **2.** (d)   **3.** (a)   **4.** (b)   **5.** (d)   **6.** (a)

7. (e)   **8.** (b)   **9.** (e)   **10.** (c)   **11.** (b)   **12.** (c)

13. (a)   **14.** (c)   **15.** (a)   **16.** (c)   **17.** (a)   **18.** (d)

CHAPTER 13

DATA PROCESSING AND CONTROL

OUTLINE

OBJECTIVES

- Name the basic units of a computer
- Name the computer buses and how they are used
- Discuss the considerations for a practical computer system
- Describe the purpose of buffers, decoders, and wait-state generators in a computer system
- Name the basic elements of a microprocessor
- Describe the basic architecture of a microprocessor
- Explain basic microprocessor (CPU) operation
- List and describe some microprocessor addressing modes
- Define and describe microprocessor polling, interrupts, exceptions, and bus requests
- Describe the functions of PIC and PPI devices
- Define and explain the advantage of DMA
- Discuss the operating system of a computer
- Explain pipelining, multitasking, and multiprocessing
- Describe a simple assembly language program
- List some typical microprocessor instructions
- Distinguish between assembly language and machine language
- Describe the architecture of a microcontroller and explain how it differs from a microprocessor
- Discuss embedded systems
- Discuss some microcontroller applications

KEY TERMS

CPU	Interrupt
Microprocessor	Exception
Main memory	Bus master
Caching	DMA
BIOS	Hardware
System bus	Software
Loading	Operating system
Buffer	Multitasking
Wait state	Multiprocessing
Pipelining	Machine language
ALU	Assembly language
Program	High-level language
Op-code	Microcontroller
Operand	

VISIT THE WEBSITE

Study aids for this chapter are available at
http://pearsonhighered.com/floyd

INTRODUCTION

This chapter provides a basic introduction to computers, microprocessors, and microcontrollers. It gives you a fundamental coverage of basic concepts related to data processing and control. For the most part, a generic approach is used to present basic concepts of the topics. The total computer system with practical considerations is covered. Various aspects of a microprocessor and its role as the CPU in computer systems are presented, and programming is briefly discussed. Microcontrollers are also introduced and some applications are described.

13–1   THE BASIC COMPUTER SYSTEM

General-purpose computers, with which most are familiar, and special-purpose computers are used to control various functions or perform specific tasks in areas such as medical, automotive, consumer appliances, manufacturing processes, and navigation. The general-purpose computer system, which can be programmed to do many different things, is the focus in this section.

After completing this section, you should be able to

- **Describe the basic elements of a computer**
- **Discuss what each part of a computer does**
- **Explain what a peripheral device is**

All computer systems work with information, or data, to produce a desired result. To accomplish this, computer systems must perform the following tasks:

- Acquire information from data sources, including human operators, sensors, memory and storage devices, communication networks, and other computer systems
- Process information by interpreting, evaluating, manipulating, converting, formatting, or otherwise working with acquired data in some intended fashion
- Provide information in a meaningful form to data recipients, including human operators, actuators, memory and storage devices, communication networks, and other computer systems

Specific sections and components in a computer system accomplish each of these tasks. Information processing is performed by the central processing unit, or CPU. The CPU is the brain of the computer system. The CPU acquires information through the input section of the computer system, provides information through the output section, and uses the system memory and storage to store and retrieve information as needed. The CPU transfers information to and from other sections of the computer system over special groups of signal lines called *buses*. Figure 13–1 shows a block diagram of a basic computer system. Each block will be discussed in terms of its purpose and function.

FIGURE 13–1   Basic computer block diagram.

The Central Processing Unit (CPU)

The central processing unit performs much of what is associated with the word *computer*. The **CPU** executes the instruction sequences (called *programs*) in the computer system, directly processes much of the data that passes through the computer system, and controls and coordinates the various sections that make up the computer system. To play such a large role in the computer system, the CPU consists of four separate units: the arithmetic logic unit (ALU), the instruction decoder, the timing and control unit, and the register set. The CPU is basically a **microprocessor** (or simply processor). A single IC package can contain two or more processors, forming a **multicore processor.**

Memory and Storage

Computer systems must have some means of storing and retrieving the information with which they work and use two types of devices—memory devices and storage devices—to do so. Although the usage and meanings of the terms can overlap somewhat, they primarily differ in the construction of the devices and the information they contain. Memory devices typically are semiconductor devices that store information electronically, interface with the computer system through the system buses, and contain dynamic information, such as programs and program variables, that is frequently accessed or modified. Storage devices typically store information on some physical medium, interface with the system through a peripheral interface, and contain primarily static information, such as program and data files, that is accessed or modified relatively infrequently. Memory devices are faster than storage devices; however, they have lower storage capacities and higher cost per bit than storage devices.

Memory in computer systems can be classified both by the type of memory and the function it performs. Chapter 10 discussed the different types and characteristics of memory. This section examines the functional requirements of memory in computer systems.

MAIN MEMORY   **Main memory** is the computer system memory that contains programs and data associated with them, such as program variables, the program stack, and information the operating system requires to execute the program. The earliest 8-bit processors (such as the Intel 8080, Motorola 6800, and InMOS 6502) had 16-bit address buses that could access $2^{16} = 65,536$ bytes (64 kilobytes or 64 kB) of memory. However, the main memory in 8-bit PCs was actually less than this because other devices in the system used part of the address space. The 16-bit computers that followed had 20-bit address buses that could access $2^{20} = 1,048,756$ bytes (1 megabyte or 1 MB) of memory. Modern computers require gigabytes of main memory to support the requirements of their graphical user interface (GUI), operating systems, and application programs. Main memory must meet the requirements of large storage capacity at an economical price and also allow the computer system to modify data within it. Because of these requirements, computer systems typically use some form of dynamic RAM (DRAM) for main memory, which features large capacity, low cost per bit, and read/write capability.

CACHE MEMORY   **Cache memory** is memory that computer systems use to overcome the relatively slow speed of main memory DRAM. **Caching** is a process that copies frequently accessed instructions or data from slow main memory into faster cache memory to reduce access time and improve system performance. Because of these requirements, computer systems use some form of static RAM (SRAM) for cache memory.

BASIC INPUT/OUTPUT SYSTEM (BIOS) MEMORY   The design of every computer system differs to some extent from other systems. The basic input/output system (**BIOS**) memory contains system-specific low-level code that runs the power on self-test (POST), installs specialized software called drivers to configure and provide access to the computer system hardware, and loads the operating system. The BIOS memory must retain its contents when power is removed so that the BIOS code is ready to run when the computer first powers up. This requires computer systems to use some form of nonvolatile memory for BIOS.

The earliest personal computers used read-only memory (ROM) for BIOS, so any change to the BIOS required the user to replace the ROM chip (which was often socketed) itself. Later computers used a low-power CMOS device with a back-up battery to preserve the contents when the system power was shut off. This allowed users to change and save BIOS settings when they made changes to system hardware configuration. Most recently, computers have used EEPROM and Flash devices so that users can easily upgrade the BIOS firmware to the latest revision. **Firmware** is software programs or data that has been written onto ROM.

Modern computers often use specialized types of memory in addition to those types mentioned previously. Rather than simply storing and returning program instructions and data, these memories store and return data that support and facilitate processor and system operations. Content-addressable memory and FIFO are examples.

CONTENT-ADDRESSABLE MEMORY   One specialized type of memory is the content-addressable (or associative) memory, whose operation differs from that of conventional memory. Conventional memory returns the data stored at a specified address. Content-addressable memory returns the address that contains a specified data value. Computers use content-addressable memory for special data tables that support caching and paging operations.

FIFO   Another specialized type of memory is the FIFO (first-in, first-out) memory. Conventional memory, such as SRAM and DRAM, allows computers to store data and to retrieve data from any memory location in any order. FIFO memory returns data only in the order in which the data was stored. As the acronym *FIFO* indicates, the first data stored in memory must be the first data taken out of memory. Computers use FIFO memory for special data structures called **queues**. Queues temporarily store data for which the sequence of data must be preserved, such as program instructions.

Input/Output Ports

Input/output (or I/O) ports are interfaces that allow computers to transfer data to and from external peripherals such as mice, keyboards, video monitors, scanners, printers, modems, and network adapters, as well as other computers. I/O ports vary greatly in complexity and capability. An I/O port can be serial or parallel; operate as an input, output, or both; and transfer several thousand to several billion bits per second. Many I/O ports, such as RS-232, USB 2.0, SCSI-2, FireWire, and Ethernet ports, conform to official or de facto standards to simplify computer system connections. These standards are usually developed by international organizations and typically specify not only the type of connectors, but the pin assignments, electrical signal levels, signal timing, data transmission rates, and communication protocols (i.e., the format, organization, and meaning of data patterns). For example, EIA 802, is the international standard for Ethernet communications, and IEEE 1394 is the standard for FireWire. These standards ensure that all devices that comply will be able to communicate with each other.

Processors support I/O ports and operations in one of two ways. One way is memory-mapped I/O, in which the processor treats I/O ports as memory locations and external circuitry converts standard read and write operations into I/O port accesses. The second way is direct I/O, in which specific processor pins and instructions are exclusively dedicated to data input and output operations. In either case, general-purpose processors require additional circuitry and program code to implement specific communications standards and protocols. Specialized microcontrollers like the Motorola MC68360 and NXP LPC2292 improve on this by incorporating additional circuitry and embedded firmware to support UART, I2C, Ethernet, CAN, SPI, and other popular communication standards on their I/O ports with a minimum of driver coding and external interface circuitry.

System Bus

As you have learned, the purpose of computers is to acquire, process, and provide information. To do so, computers must be able to specify where to acquire and return information, to transfer the information from its source to its destination, and to coordinate the movement of data within the computer system. The mechanism by which the computer accomplishes this is the **system bus**, which consists of three component buses: the address bus, the data bus, and the control bus.

THE ADDRESS BUS The address bus is the means by which the processor specifies the system location from which data is to be read or to which data is to be written. In early computer systems, only one processor controlled the address bus, so the address bus was unidirectional (the processor address lines were always outputs and the address lines on other system devices were always inputs); and each address line was always either LOW or HIGH. In modern computer systems, more than one device can control the bus. In addition, the address bus on some processors is bidirectional so that other devices can access registers and other internal components.

THE DATA BUS The data bus consists of signal lines over which the computer system transfers information from one device to another. Because the processor can both read data from and write data to system devices, each data line is bidirectional. The number of data lines determines the width of the data bus, which is a factor in how quickly the processor can process data. The earliest microprocessors had 4-bit and 8-bit data buses, but many modern processors have 64-bit data buses.

THE CONTROL BUS The control bus is the collection of signals that controls the transfer of data within the system and coordinates the operation of system hardware. Unlike the address and data buses, which consist of functionally identical signals that function as a group, the individual signal lines that make up the control bus vary in characteristics, nature, and function. Control signals can be unidirectional or bidirectional, can function individually or with other control signals, can be active-HIGH or active-LOW, can operate synchronously or asynchronously, and can be edge-oriented or level-oriented. Despite this individual diversity, computer systems and processor operations are similar enough that the signals that make up the control bus—read, write, interrupt, and others—are also very similar.

A Typical Computer System

The block diagram in Figure 13–2 shows the main elements in a typical computer system and how they are interconnected. Notice that the computer itself is connected to several

FIGURE 13–2 **Basic block diagram of a typical computer system including common peripherals. The computer itself is shown in the gray block.**

peripheral units. For the computer to accomplish a given task, it must communicate with the "outside world" by interfacing with people, sensing devices, or devices to be controlled through input and output ports.

Computer Software

In addition to the hardware, a major part of a computer system is the software. The software makes the hardware perform. The two major categories of software used in computers are the application software and the system software. Application software, with which most people are more familiar, includes word processors, spreadsheets, computer games, and other programs written to accomplish some specific task. The second group consists of system software, whose purpose is to manage the computer system. A major portion of the system software is the operating system. The operating system manages the system hardware, supervises the running of application software, provides a standard operating environment for programs in which they can run, and interacts with the computer hardware.

SECTION 13–1 CHECKUP*

1. What are the major functional blocks in a computer?

2. What are peripherals?

3. What is the difference between computer hardware and computer software?

4. How does content-addressable memory differ from conventional memory?

5. Compare and contrast the characteristics of the address, data, and control buses in a computer system.

*Answers are at the end of the chapter.

13–2 PRACTICAL COMPUTER SYSTEM CONSIDERATIONS

Although the general features of a computer system were discussed in the previous section, several important details that a practical computer system must include to function properly were omitted. In particular, practical computer designs incorporate special circuitry that resolves four issues that exist in real-world systems: shared signal lines, signal loading, device selection, and system timing.

After completing this section, you should be able to

- **Identify design considerations for practical computer systems**
- **Explain the role and operation of buffers, decoders, and wait-state generators in practical computer systems**

Figure 13–3 shows the block diagram of a practical computer, based on the consideration for shared signal lines, signal loading, device selection, and system timing.

FIGURE 13–3 Block diagram of a practical computer system.

Shared Signal Lines

When the outputs of two or more devices connect to the same signal line, the potential for bus contention exists. Bus contention occurs when device outputs attempt to drive a signal line to different voltage levels. This causes high current to flow from one output into the other, which can damage the devices. Typically, bus contention occurs when device outputs are at different logic levels. However, even when devices are at the same logic level, the variation for different devices will cause some device output voltages to be higher than others so that bus contention will occur. Two special types of output, the tristate output and open collector output, allow devices to share signal lines while avoiding bus contention.

The term *tristate* is a registered trademark of National Semiconductor but is often used interchangeably with the generic terms *three-state* or *3-state*. As the name suggests, the **tristate** output adds a third output state, called the high-impedance or high-Z state, to the usual logic LOW and HIGH states. The tristate switch is effectively a switch that disconnects the output of the tristate device from the signal line so that it does not interfere with other devices from driving the line. When a tristate device is enabled, it outputs a logic LOW or HIGH as other digital devices. When a tristate device is disabled, the output assumes the high-Z state and the output is said to be tristated. When tristate outputs share a signal line, only one output at a time must be enabled to ensure that bus contention will not occur. Figure 13–4 shows the operation of tristate outputs.

Devices that are designed to connect to processor buses, such as memory and interface devices, typically have tristate outputs built into them. Devices that do not have tristate outputs or open-collector outputs must use tristate buffers to connect to buses.

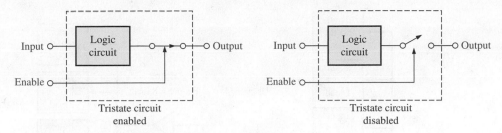

FIGURE 13–4 **Logic devices with tristate outputs.**

Signal Loading and Buffering

In most computer systems, the microprocessor buses do not connect directly to memory and other system devices. All logic outputs are affected to some extent by the inputs of the devices to which they are connected. This effect is called **loading**. In TTL logic, this loading is due to the input current that each load device requires to operate. As more loads are added to the logic output, the driving device at some point cannot provide enough current to guarantee that the load devices will operate correctly.

In CMOS logic the loading is due to the capacitance on the input of each load device. When the driving device switches states, it takes a finite amount of time for the output current to charge or discharge the input capacitance. As more loads are added to the logic output, the driving device at some point cannot charge up the total capacitance fast enough to meet the timing specifications of the logic devices so that the devices cannot function at their specified operating frequencies.

Another issue is that noise can cause erratic operation. If the input voltage changes too slowly and remains near the switching threshold for an extended amount of time, noise in the system can cause the device to switch states as the additional noise changes or moves the input voltage above and below the switching threshold.

Loading is a common issue on computer buses because the address bus and other bus lines often connect to many devices. To avoid loading problems, computer systems use special logic devices called *buffers* on system bus lines. A **buffer** is a logic gate that provides more output drive than it requires on its input to operate and isolates the output of a device from the loading effects of other devices. Inverting buffers invert the input signal and noninverting buffers do not. A special type of buffer, called a Schmitt trigger, has a special feature called *hysteresis* and is used with signals that are sensitive to noise or change so slowly that they may create switching or timing problems in the system.

One major issue with all digital outputs is that there is a limit to the number of digital inputs that they can reliably drive, called the device fan-out. When the number of inputs exceeds the fan-out of an output device, the operation of that output device may not meet the specified voltages or timing for that device. To avoid problems with signal loading, buffers are used to ensure that device fan-outs are not exceeded. Figure 13–5 shows a specific example that illustrates the use of buffers to prevent the nine input devices from exceeding the eight-load fanout of the output device.

An alternate approach to that illustrated in Figure 13–5 would be to connect seven input devices directly to the output device and use a single buffer to connect the remaining input devices. This would have reduced the parts count, but one issue with buffers is that each buffer will increase the propagation delay. If a single buffer were used, the response of the input devices connected to the buffer would be slower than that of the input devices connected directly to the output device. Using two buffers would help match the propagation delay to all the input devices. The buffers shown in Figure 13–5(b) are simple noninverting buffers, which means that the buffer output signal is identical to the buffer input signal.

There are other types of buffers to ensure that devices will not degrade the performance of a device to which they are connected. These buffers include tristate buffers like those previously mentioned, inverting buffers that invert the input signal, bidirectional buffers that can pass information through the buffer in both directions as on the data bus, and Schmitt triggers.

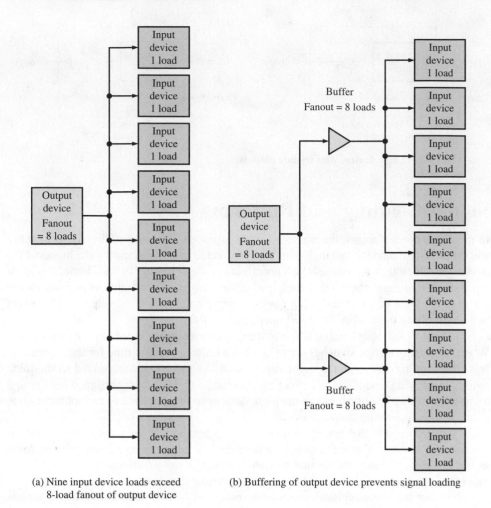

(a) Nine input device loads exceed
8-load fanout of output device

(b) Buffering of output device prevents signal loading

FIGURE 13–5   **Buffers are used to prevent overloading of driving gate.**

Device Selection

The processor uses the address bus to access ROM, RAM, hardware I/O ports, and other system devices. A question that naturally arises is how a device knows when the processor is attempting to access it rather than some other system device. The answer is that these devices have a special input, usually called a chip select (CS) or chip enable (CE), that enables the device. When the processor must access a specific device, it must assert the select line of the intended device.

  While, in theory, processors could provide separate control lines to select system devices, this is not practical for general-purpose computers because there is no way for the system designers to know what devices a system will contain. Instead, system designers use PLDs or dedicated hardware decoders, similar to that in Figure 13–6, to decode processor addresses and generate the device select lines. In this case, the processor uses a 16-bit address bus where the upper (most significant) four bits are used to generate device select outputs.

System Timing

A final issue with practical computer systems is system timing. In a computer system, the processor signals must meet the setup and hold times for each peripheral so that data is properly stored and accessed. As you have seen, decoding logic or buffers in the system can slow the processor signals. In some cases the processor runs much faster than the peripherals that are available; in other cases fast peripherals are available, but their cost prohibits designers from using them. In addition, some peripherals, such as SRAM, are

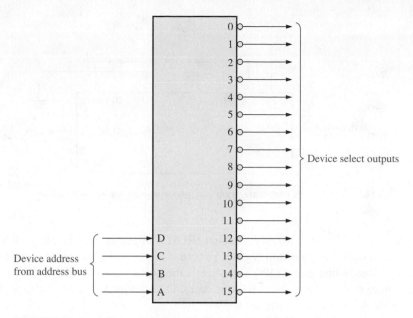

FIGURE 13–6   Address decoding for the purpose of device selection.

inherently faster than others, such as DRAM, so signal timing that meets the setup and hold times for some devices will not meet the setup and hold times for others. To resolve this issue, three different types of system buses can be used: synchronous, asynchronous, and semisynchronous.

Synchronous buses include a synchronizing clock to ensure that signals from the processor meet the setup and hold times of the peripheral. Synchronous buses are faster than asynchronous or semisynchronous buses.

Asynchronous buses will automatically insert wait states in a bus cycle until a signal indicates that the bus cycle can finish. A **wait state** holds the state of the bus signals for one processor clock so that the read or write operation is "frozen" for one clock period when the processor is accessing memory or other devices that are slow to respond. Several wait states may be necessary. Computer CPUs run at very high speeds, while memory technology does not seem to be able to catch up. Typical processors like the Intel Core 2 and the AMD Athlon 64 X2 run with a clock of several gigahertz, while the main memory clock generally is in the several hundred to over 1000 MHz range. Even some second-level CPU caches run slower than the processor core. In order to minimize the use of wait states, which slow the computer down, techniques such as CPU caches, instruction pipelines, instruction prefetch, and simultaneous multithreading are used.

Semisynchronous buses are similar to asynchronous buses except that a semisynchronous bus will complete the bus cycle unless a signal indicates that the processor should not complete the bus cycle. Until the processor can complete the cycle, it will insert wait states.

Memory and other peripheral devices do not, as a rule, have signals indicating when data is ready. The signals that instruct the processor to insert wait states must be generated by an additional logic circuit, called a *wait-state generator*, which can be basically a programmable timer or shift register. The wait-state generator is clocked by the same clock as the processor and enabled by the device select line for a specific memory or other device. After the wait-state generator is enabled by the device select line, it will generate a ready signal after a specific number of clocks. Figure 13–7 shows an 8-bit parallel-in/serial-out shift register circuit that can insert up to six wait states for an asynchronous processor by delaying the ready signal to the processor by up to six clocks.

The circuit in Figure 13–7 inserts one wait state for a single device. While the active-LOW device select line is HIGH (indicating that the device associated with the wait-state generator is not selected), the shift register loads the bit pattern 11111011 on each clock pulse and SEROUT, which corresponds to the least significant bit (D_0), is HIGH. When the device select signal goes LOW, the shift register shifts the bit pattern to the right one place with each clock. On the first clock (corresponding to zero wait states) the shifted

FIGURE 13–7   **A wait-state generator programmed for one wait state.**

pattern is 11111101 and SEROUT remains HIGH. On the second clock, which corresponds to one wait state, the shift register pattern changes to 11111110 so that SEROUT and thus the Ready line goes LOW. This signals the processor to terminate the wait state. If two or more devices have the same wait states, their device select lines can be ANDed (assuming the select lines are active-LOW).

EXAMPLE 13–1

For the wait-state generator in Figure 13–8, how many wait states will be generated when the device is selected?

FIGURE 13–8

SOLUTION

The initial pattern loaded into the shift register is 11101111_2. This shifted pattern for each clock and the corresponding number of wait states are

Clock 1 (0 wait states): 11110111_2

Clock 2 (1 wait state): 11111011_2

Clock 3 (2 wait states): 11111101_2

Clock 4 (3 wait states): 11111110_2

On the fourth clock after Device select goes LOW, the most significant bit of the SEROUT line for the shift register goes LOW. This causes the Ready output to go HIGH, terminating the bus cycle. Therefore, the wait-state generator inserts **three** wait states.

RELATED PROBLEM*

Which data input line of the shift register must be tied LOW for the wait-state generator in Figure 13–8 to insert five wait states?

*Answers are at the end of the chapter.

SECTION 13–2 CHECKUP

1. Define *bus contention* and discuss types of devices used to prevent it.

2. How does a processor enable various devices?

3. Define *wait state* and give its purpose.

4. What is the purpose of a buffer?

13–3 THE CPU: BASIC OPERATION

As you have learned, a microprocessor forms the CPU of a computer system. A microprocessor consists of several units, each designed for a specific job. The specific units, their design and organization, are called the *architecture* (do not confuse the term with the VHDL element). The architecture determines the instruction set and the process for executing those instructions. In this section, the basic concept of the microprocessor is examined at its simplest level.

After completing this section, you should be able to

- **Name the four basic elements of a microprocessor**
- **Describe the fetch/execute cycle**
- **Explain the read and write operations**

The four basic elements that are common to all microprocessors are the arithmetic logic unit (ALU), the instruction decoder, the register set, and the timing/control unit, as shown in Figure 13–9.

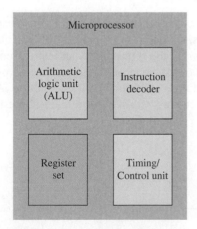

FIGURE 13–9   Elements of a microprocessor (CPU).

Figure 13–10 shows a block diagram of a microprocessor. The elements shown are common to most processors, although the internal arrangement or architecture and complexity varies. This generic block diagram of an 8-bit processor with a small register set is used to illustrate fundamental operation. Today, processors have data buses that are 64 bits, but the simplified model will show the basic operation.

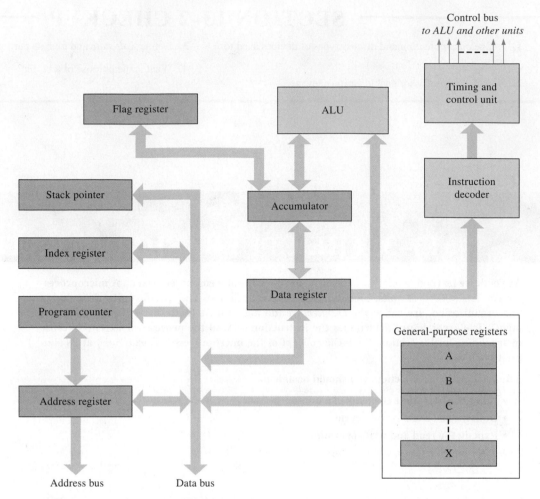

FIGURE 13–10   **Basic model of a simplified processor.**

The Fetch/Execute Cycle

When a program is being run, the processor goes through a repetitive cycle that consists of two fundamental phases, as shown in Figure 13–11. One phase is called *fetch* and the other is called *execute*. During the **fetch** phase, an instruction is read from the memory and decoded by the instruction decoder. During the **execute** phase, the processor carries out the sequence of operations called for by the instruction. As soon as one instruction has been executed, the processor returns to the fetch phase to get the next instruction from the memory.

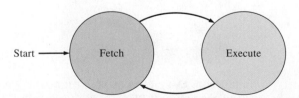

FIGURE 13–11   **The fetch/execute cycle of a processor.**

PIPELINING   A technique where the microprocessor begins to execute the next instruction in a program before the previous instruction has been completed is called **pipelining**. That is, several instructions are in the pipeline simultaneously, each at a different processing stage.

Typically, a pipeline is divided into stages or segments, and each stage can execute its operation concurrently with the other stages. When a segment completes an operation, it passes the result to the next segment in the pipeline and fetches the next operation from the preceding segment. The final results of each instruction emerge at the end of the pipeline in rapid succession. Figure 13–12 is a greatly simplified illustration of non-pipelined processing compared to pipelined processing using three stages of execution for illustration.

(a) Nonpipelined execution of a program showing three stages of execution

(b) Pipelined execution of a program showing three stages

FIGURE 13–12   Illustration of pipelining.

As shown in the figure, in the nonpipelined processing of a program, one instruction at a time is executed through all of its stages before the next instruction begins execution. As you can see, all the stages of execution are idle (gray) except the one that is active (red). In pipelined processing, as soon as one instruction has finished an execution stage, the next instruction begins that stage. Pipelining results in much shorter overall execution times. Once the pipeline is "full," there are no idle processing stages.

Processor Elements

ALU   This part of the processor contains the logic to perform arithmetic and logic operations. Data is transferred into the **ALU** from the accumulator and from the data register. For our model, the accumulator and data register are 8-bit registers that hold one byte of data. Each byte transferred into the ALU is called an *operand* because it is operated on by the ALU. As an example, Figure 13–13 shows an 8-bit number from the accumulator being added to an 8-bit number from the data register. The result of this addition operation (sum) is put back into the accumulator replacing the original operand that was stored there. When the ALU performs an operation on two operands, the result always goes into the accumulator to replace the previous operand.

FIGURE 13–13   **Example of the ALU adding two operands.**

As demonstrated in Figure 13–13, one function of the accumulator is to store an operand prior to an operation by the ALU. The other function is to store the result of the operation after it has been performed. The data register temporarily stores data that is to be put onto the data bus or that has been taken off of the data bus.

INSTRUCTION DECODER AND TIMING/CONTROL UNIT   An **instruction** is a binary code that tells the processor what it is to do. An orderly arrangement of many different instructions makes up a program. A **program** is a step-by-step procedure used by the processor to carry out a specified task.

The instruction decoder within the processor decodes an instruction code that has been transferred on the data bus from the memory. The instruction code in mnemonic form is commonly known as an **op-code**. When the op-code is decoded, the instruction decoder provides the timing and control unit with this information. The timing and control unit can produce the proper signals and timing sequence to execute the instruction.

REGISTER SET   Processors typically have two categories of registers for temporary storage of data: general-purpose registers and special-purpose registers. General-purpose registers are used to store any type data that may be required by a program. Special-purpose registers are dedicated to a specific function. Some typical special-purpose registers are described as follows.

Flag Register   This register is sometimes called a condition code register or status register. It indicates the status of the contents of the accumulator or certain other conditions within the processor. For example, it can indicate a zero result, a negative result, the occurrence of a carry, or the occurrence of an overflow from the accumulator.

Program Counter   This counter produces the sequence of memory addresses from which the program instructions are taken. The content of the program counter is always the memory address from which the next byte is to be taken. In some processors, the program counter is known as the *instruction pointer*.

Address Register   This register temporarily stores an address from the program counter in order to place it on the address bus. As soon as the program counter loads an address into the address register, it is incremented (increased by 1) to the address of the next instruction.

Stack Pointer and Index Register   The stack pointer is a register that is mainly used during program subroutines and interrupts. It is used in conjunction with the memory stack. The index register is used as one means of addressing the memory in a mode of addressing called *indexed addressing*.

The Processor and the Memory

The processor is connected to a memory with the address bus and data bus. Also, there are certain control signals that must be sent between the processor and the memory, such as the read and write controls. The address bus is unidirectional so the address bits go only one way, from the processor to the memory. The data bus is bidirectional so data bits are transferred between the processor and memory in either direction. This is illustrated in Figure 13–14.

FIGURE 13–14   **A processor and memory.**

THE READ OPERATION   To transfer data from the memory to the processor, a read operation (introduced in Chapter 10) must be performed, as shown in Figure 13–15, using an 8-bit data bus and a 16-bit address bus for illustration. To start, the program counter contains the address of the data to be read from the memory. This address is loaded into the address register and placed onto the address bus. The program counter is then incremented (advanced by one) to the next address and waits. Once the address code is on the bus, the processor timing and control unit sends a read signal to the memory. At the memory, the address bits are decoded and the desired memory location is selected. The read signal causes the contents of the selected address to be placed on the data bus. The data is then loaded into the data register to be used by the processor, completing the read operation. In this illustration, each memory location contains one byte of data. When a byte is read from memory, it is not destroyed but remains in the memory. This process of "copying" the contents of a memory location without destroying the contents is called *nondestructive read*.

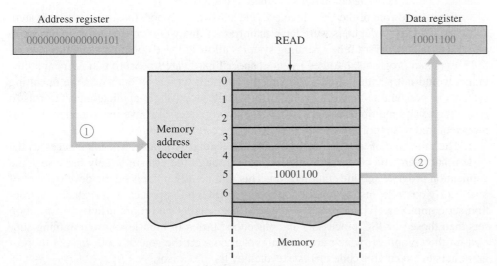

① Address 5_{10} is placed on address bus and followed by the read signal.

② Contents of address 5_{10} in memory is placed on data bus and stored in data register.

FIGURE 13–15   **Illustration of the read operation.**

THE WRITE OPERATION   To transfer data from the processor to the memory, a write operation (introduced in Chapter 10) is required, as illustrated in Figure 13–16. A data byte held in the data register is placed on the data bus, and the processor sends the

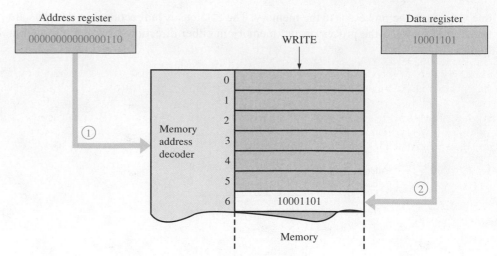

① Address code for address 6_{10} is placed on address bus.

② Data are placed on data bus and followed by the write signal. Data are stored at address 6_{10} in memory.

FIGURE 13–16   **Illustration of the write operation.**

memory a write signal. This causes the byte on the data bus to be stored at the memory location selected by the address code. The existing contents of that particular memory location are replaced by the new data. This completes the write operation.

Roles of the CPU

The CPU performs three major roles. The first role of the CPU is to control the system hardware. Specifically, the CPU determines how data moves through the computer system, which devices are active, and when specific operations and data transactions occur. In computers, some of this control is decentralized by assigning some tasks (such as peripheral access and communications and graphics processing) to devices that can perform those tasks more quickly and efficiently than the CPU itself. Even so, the CPU still coordinates the operation of the computer system as a whole.

The second role of the CPU is to provide hardware support for the operating system software. The first computers were large mainframes that were too expensive to devote to a single user or program. The operating systems allowed these computers to support multiple users and programs, but they required special hardware to ensure that users and programs would not accidentally or deliberately interfere with each other. As the operating systems in personal computers evolved from single-user single-application platforms to multitasking and multiprocessing systems, the microprocessors have incorporated the features required to support them.

The third role of the CPU is to execute application programs. The CPU accesses the system hardware and controls the flow of data through the system largely because some application program requires that it do so. This role greatly influenced the development of many early complex instruction set computing (CISC) microprocessors. Reduced instruction set computing (RISC) processors emphasize smaller and more efficient instruction sets than those in CISC processors and place the burden of high-level programming support on the compilers, which are programs that convert the source code written by programmers to executable code that is executed by the processor.

SECTION 13–3 CHECKUP

1. Describe the fetch/execute cycle.

2. Name the four elements in a microprocessor.

3. What is the ALU and what is its purpose?

4. What happens during a read operation?

5. What happens during a write operation?

As you have learned, the processor must address the memory to obtain data or to store data. There are several ways in which a processor can generate an address when an instruction is being executed. These are called addressing modes, and they provide for wide programming flexibility. Each instruction in a processor's instruction set generally has a certain addressing mode associated with it. The type and number of addressing modes vary from one processor to another. In this section, several common modes are discussed to demonstrate basic concepts and generic instructions are used for illustration.

After completing this section, you should be able to

- Explain inherent addressing
- Explain immediate addressing
- Explain direct addressing
- Explain indexed addressing
- Explain relative addressing

Inherent Addressing

Inherent addressing is sometimes known as *implied addressing*. The one-byte instructions using this mode generally require no operand, or the operand is implied by the op-code, which is a mnemonic form of an instruction. An **operand** is the object to be manipulated by the instruction. For example, an instruction used to clear the accumulator (CLRA) has an implied operand of all zeros. The implied all-zeros operand ends up in the accumulator after the instruction is executed. Another example is a halt or wait instruction (WAI), which requires no operand because it simply tells the processor to stop all operations. The sequence that the processor goes through in handling an instruction with inherent addressing is illustrated in Figure 13–17. Op-codes, similar to op-codes of a typical processor, are used for illustration.

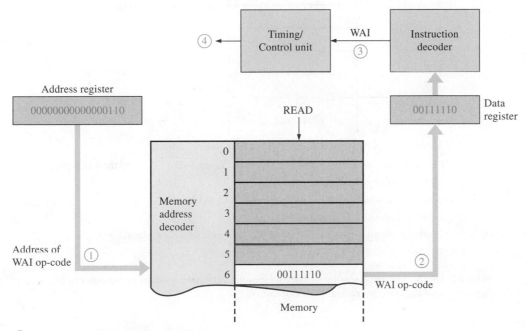

① Address code (6_{10}) is placed on address bus.

② Data are placed on data bus and stored in data register by the read signal.

③ Instruction is decoded.

④ Timing/Control unit stops processor operation.

FIGURE 13–17 Fetch/execute cycle for the wait (WAI) instruction. This illustrates inherent addressing.

Immediate Addressing

Immediate addressing is used in conjunction with two-byte instructions where the first byte is the op-code and the second byte is the operand. For example, the load accumulator (LDA) and the add to accumulator (ADDA) instructions are two examples that use immediate addressing.

The LDA immediate op-code is stored in one memory address, and the operand in the address immediately following the op-code. That is, the op-code and operand are at consecutive memory addresses. When the LDA immediate instruction is fetched and executed, it tells the processor to get the contents of the next memory location (operand) and load it into the accumulator, as illustrated in Figure 13–18.

① Address of LDA immediate op-code (1_{10}) is placed on address bus.

② LDA immediate op-code is placed on data bus and stored in data register by the read signal.

First fetch/Execute cycle

③ LDA instruction is decoded.

④ Timing/Control unit initiates a read operation to fetch the operand.

⑤ Address of operand (2_{10}) is placed on address bus.

Second fetch/Execute cycle

⑥ Operand is placed on data bus and stored in data register by the read signal.

⑦ Operand is loaded into accumulator.

FIGURE 13–18 **Illustration of immediate addressing. The process steps are numbered in sequence, and the cycle operations are color-coded.**

Direct Addressing

For an instruction using direct addressing, the first part is the op-code, and the second part is the address of the operand, not the operand itself. For example, the LDA instruction uses direct addressing as well as immediate addressing. LDA direct has a different op-code than LDA immediate. Let's assume each part is one byte for simplicity.

The LDA direct instruction is used to illustrate direct addressing. Figure 13–19 shows the LDA direct instruction in memory addresses 1 and 2. The first byte is the op-code, and the second byte is the operand address. When the LDA direct instruction is fetched and executed, it tells the processor to load the accumulator with the operand located at the memory address specified by the second byte of the instruction. The process is illustrated in Figure 13–19.

First fetch/Execute cycle
(1) Address of LDA direct op-code (1_{10}) is placed on address bus.
(2) LDA direct op-code is placed on data bus and stored in data register by the read signal.
(3) LDA instruction is decoded.
(4) Timing/Control unit initiates a read operation to fetch the address of the operand.

Second fetch/Execute cycle
(5) Address of operand address (2_{10}) is placed on address bus.
(6) Operand address is placed on data bus and stored in data register by the read signal.
(7) Operand address (6_{10}) is loaded into address register.

Third fetch/Execute cycle
(8) Operand address (6_{10}) is placed on address register.
(9) Operand is placed on data bus and loaded into data register.
(10) Operand is loaded into accumulator.

FIGURE 13–19 Illustration of direct addressing.

Indexed Addressing

Indexed addressing is used in conjunction with the index register. An instruction using indexed addressing consists of the op-code and the offset address. When an indexed instruction is executed, the offset address is added to the contents of the index register to produce an operand address. In Figure 13–20 the LDA (load accumulator) instruction is used again to illustrate indexed addressing.

(1) Address of LDA indexed op-code (0_{10}) is placed on address bus.

(2) LDA indexed op-code is placed on data bus and stored in data register by the read signal.

(3) LDA indexed instruction is decoded.

(4) Timing/Control unit initiates a read operation to fetch the address of the operand.

(5) Offset address (5_{10}) is selected at address 1_{10}.

(6) Offset address is placed on data bus and stored in data register by the read signal.

(7) Offset address is added to contents of index register (7_{10}) by the ALU to produce address of operand (12_{10}).

(8) Address of operand (12_{10}) is loaded into address register.

(9) Address of operand is placed on address bus.

(10) Operand is placed on data bus and stored in data register by the read signal.

(11) Operand is transferred into accumulator.

First fetch/Execute cycle: ① ② ③ ④

Second fetch/Execute cycle: ⑤ ⑥ ⑦

Third fetch/Execute cycle: ⑧ ⑨ ⑩ ⑪

FIGURE 13–20   Illustration of indexed addressing.

Relative Addressing

Relative addressing is used by a class of instructions known as *branch instructions*. Basically, a branch instruction allows the CPU to go back or skip ahead for a specified number of addresses in a program instead of going to the next address in sequence. Branching instructions are used to form program loops. For a relative addressing instruction (branch instruction), the first byte is the op-code and the second byte is the relative address. When a branch instruction is executed, the relative address is added to the contents of the program counter to form the address to which the program is to branch. Figure 13–21 illustrates relative addressing using a branch relative always (BRA) instruction that can branch both forward or backward. Forward branching is shown.

FIGURE 13–21   Illustration of relative addressing (branching).

The figure contains the following annotations:

First fetch/Execute cycle
1. Address of BRA relative op-code (1_{10}) is placed on address bus.
2. BRA op-code is placed on data bus and stored in data register by the read signal.
3. BRA instruction is decoded.

Second fetch/Execute cycle
4. Program count is transferred to ALU and address register.
5. Address of relative address (2_{10}) is placed on address bus.
6. Relative address (5_{10}) is placed on data bus and stored in data register by the read signal.
7. Relative address is transferred to ALU.
8. Program count and relative address are added by ALU and resulting branch address (7_{10}) is placed in address register.
9. Program branches to specified address (7_{10}).

SECTION 13–4 CHECKUP

1. List several types of addressing.

2. What is an op-code?

3. What is an operand?

4. Explain branching.

The previous sections identified and discussed the basic components, organization, and operation of microprocessors (CPUs), and computer systems. During normal operation, the CPU fetches instructions from system memory, which are decoded by the instruction decoder. Each instruction affects the operation of the timing and control unit, which in turn synchronizes the operation of the CPU, system buses, and system components to execute the instruction. In this section, specific CPU operations (interrupts, exceptions, and bus requests) that occur when special circumstances arise that preempt normal processor operation are discussed.

After completing this section, you should be able to

- Define *polling*
- Define the terms *interrupt* and *exception*
- Describe the process by which a processor responds to and services an interrupt
- Explain how an interrupt service routine differs from a subroutine
- Explain why computer systems use bus requests

A computer runs programs that limit what the computer is permitted to do and how it will respond to situations that arise. Some situations are predictable and others are not. Even when a situation is predictable, just when it will occur may not be predictable. As an example, every word processor program must respond to input from a keyboard, but the program cannot predict just when someone will press a key.

Polling

One technique to deal with unpredictable events is to have the CPU poll, or repeatedly check, the keyboard and other peripheral devices that may require attention from the CPU. Each time the CPU polls a device, it must stop the program that it is currently processing, go through the polling sequence, provide service if needed, and then return to the point where it left off in its current program. This process is very inefficient and is suitable only for devices that can be serviced at regular and predictable intervals. Figure 13–22 illustrates the concept of **polling**, where the CPU sequentially selects each peripheral device via the multiplexer to see if it needs service.

Interrupts and Exceptions

A more efficient approach than polling is to have the processor perform its normal operations and deviate from them only when some special event requires that the processor take special action to handle it. The term *exception* is sometimes used for any event that requires special handling by the processor, but most often it is used for events due to software only. The term *software interrupt* (SWI) or *trap* are also used for software events. The terms *interrupt* or *hardware interrupt* (HWI) are used for an event due to hardware. We will use **interrupt** to refer to a hardware event and **exception** for a software event that requires the CPU to deviate from its normal operation.

When the processor receives an interrupt or exception, it finishes executing the current instruction and then runs a special sequence of instructions called an *interrupt service routine* (ISR) or *exception handler*. An ISR is similar to calling a standard program subroutine but with three important exceptions. First, because the processor cannot know when an interrupt or exception will occur, it automatically saves on the register stack status information about the program that is executing at the time the interrupt or exception occurs. The information includes the contents of the condition code register as well as the address of the next instruction to be executed when the ISR is finished. Sometimes the accumulator and condition code register, which make up the program status word, are both saved. The ISR must save on the stack any other registers it may use to ensure that the interrupted process will not be affected when it resumes executing.

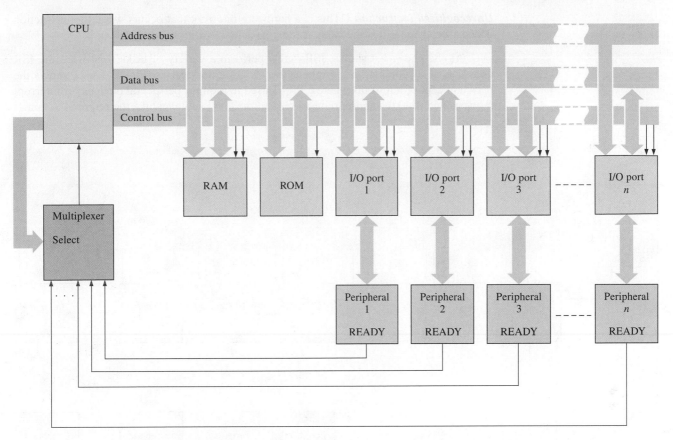

FIGURE 13–22   Basic concept of CPU polling peripheral devices.

Second, the processor obtains the address of the ISR based on the specific interrupt or exception that occurs. In some systems a programmable interrupt controller (PIC) provides the address of the ISR over the data bus when the processor acknowledges an interrupt request. Other systems use autovectored interrupts that obtain the address for each interrupt from entries in an interrupt vector table stored in memory. Each vector, or ISR address, in the table specifies the starting address of an ISR. The programmer must write the ISRs and place the starting address for each in the correct location of the interrupt vector table. If no ISR exists for an entry in the vector interrupt table, or the interrupt vector table is not properly initialized, interrupts and exceptions can cause the processor to behave erratically, "hang" (stop responding), or "crash" (abort and restart).

Third, the ISR uses a special return from interrupt (RTI) instruction, which restores the additional status information as well as the address of the next instruction. RTI is used rather than a standard return from subroutine (RET) instruction, which restores only the address of the next instruction, to exit and return processor control to the interrupted process. Before executing the RTI instruction, the ISR must restore any registers it saved on the stack. Specific interrupt and exceptions vary with each processor, but the following list describes some typical ones.

Reset   This is sometimes called a cold boot. A cold boot completely restarts the system so that the processor runs the power on self-test (POST), initializes the hardware, loads the hardware drivers and operating system, and performs all other tasks necessary to prepare the system for operation.

Software reset   This is a software exception and is sometimes called a warm boot. This also restarts the system but bypasses many of the hardware initialization tasks performed by a cold boot.

Divide by zero   This is a software exception and occurs when the processor attempts to divide a number by zero.

System timer   This is a hardware interrupt and occurs when a special timer asserts a signal indicating that a specified time interval or "time tick" (such as 1/60th of a second) has elapsed since the last occurrence.

Unrecognized instruction This is a hardware interrupt that occurs when the instruction decoder determines that the value it contains is not a valid instruction.

As the above list shows, ISRs must perform a variety of tasks. Just what the ISR does can be as complex or simple as the programmer desires. Figure 13–23 shows the basic concept of interrupts where a PIC is used to monitor peripheral devices for interrupt requests and send the appropriate address to the CPU so it can take the required action.

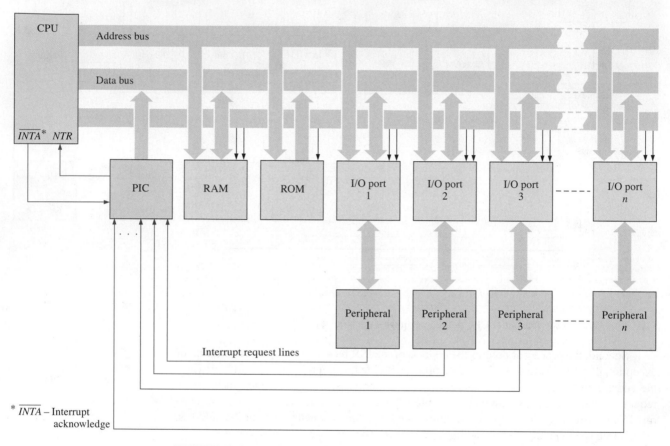

FIGURE 13–23 **Basic concept of interrupt control.**

Bus Request Operations

The device in a computer that drives the address bus and the bus control signals is called the **bus master**. In a simple computer architecture only the CPU can be bus master, which means that all communications between I/O devices must involve the CPU. More complex architectures allow other devices (or multiple CPUs) to take turns at controlling the bus. For example, a network controller card can be used to access a disk controller directly while the CPU performs other tasks that do not require the bus. Any device can place data on the data bus when the CPU reads from that device, but only the bus master drives the address bus and the bus control signals.

Although processors operate at high speeds, they are not always efficient at transferring data. When a processor transfers data from one device to another, it must use a bus cycle to read in the data from the source device and another bus cycle to write the data back out to the destination device. The overhead in reading data into the processor and writing it out again greatly slows data transfers. The bus request operation allows other bus masters to take control of the system buses and rapidly transfer data between system devices.

Bus request operations are similar to interrupts and exceptions but differ in three important ways. First, bus request operations do not complete the current instruction cycle before proceeding. Instructions can take hundreds or even thousands of clock cycles, and the circumstances that generated the bus request may be too urgent to be delayed. For example, a CD drive may be on the verge of a buffer under run and require data immediately to refill the buffer, or a memory controller may need to immediately refresh the system DRAM to prevent data from being lost. Interrupts and exceptions allow the processor to complete the current instruction cycle before processing the interrupt or exception.

Second, in a bus request operation the processor passes control of the system buses to the requesting device, which then handles all bus operations. The processor continues to execute instructions in the ISR or exception handler during interrupts. Third, once the processor grants the bus request and relinquishes the system buses, the processor cannot regain control of the system until the requesting device relinquishes control or the processor is reset. The sequence of events during a bus request operation is as follows:

1. The bus master requesting control of the system buses submits a request by asserting the processor's bus request (BR) line.

2. The processor tristates the system buses and signals that it has released control of the buses by asserting the bus grant (BG) line.

3. The requesting bus master uses the system address, data, and control lines to transfer data between system devices.

4. After completing the data transfers, the requesting bus master tristates the system buses and signals the end of the bus request operation by asserting the bus grant acknowledge (BGACK) line.

Direct Memory Access (DMA)

One important class of bus master is the **DMA** (direct memory access) controller. These devices are designed specifically to transfer large amounts of data between system devices in a fraction of the time that the system processor would require. To utilize a DMA controller, the processor first writes the starting source address, starting destination address, and number of bytes to transfer to registers within the DMA controller. The processor next enables the transfer by writing to a control register within the controller, which then initiates the bus request operation. Computer systems typically use DMA controllers to transfer data between memory and hardware peripherals, such as when loading a program or data file from a hard drive to memory or transferring a message from system memory to the transmit buffer of an Ethernet controller. DMA controllers can also move data between memory devices, such as when moving data from main memory to cache memory.

DMA speeds up data transfers between RAM and certain peripheral devices. Basically, DMA bypasses the CPU for certain types of data transfers, thus eliminating the time consumed by normal fetch and execute cycles required for each CPU read or write operation. Transfers between the disk drive and RAM are particularly suited for DMA because of the large amount of data and the serial nature of the transfers. Generally, the DMA controller can handle data transfers several times faster than the CPU. Figure 13–24 shows a comparison of a data transfer handled by the CPU (part a) and one handled by the DMA (part b).

(a) Data transfer handled by the CPU

FIGURE 13–24 Illustration of DMA vs CPU data transfer.

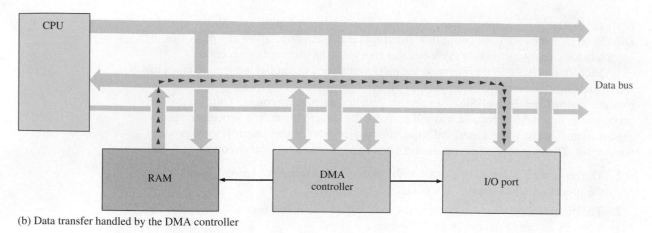

(b) Data transfer handled by the DMA controller

FIGURE 13–24   **Continued**

Bus masters other than DMA controllers also use bus request operations. Processors in multiprocessor systems use bus request operations to access shared memory and other system resources. Memory controllers also use bus request operations to perform background memory operations, such as refreshing DRAM and ensuring that the data in main memory and cache memory are consistent.

A Computer System

Figure 13–25 shows a computer system block diagram with a DMA controller and a PIC.

FIGURE 13–25   **Block diagram of a typical computer.**

SECTION 13–5 CHECKUP

1. Compare and contrast exceptions and interrupts.

2. Compare and contrast bus requests and interrupts.

3. Define and explain the purpose of direct memory accesses.

13–6 OPERATING SYSTEMS AND HARDWARE

Each computer system consists of two main components. The microprocessor, memories, interface circuits, peripherals, power supplies, and other electronic components make up what is collectively referred to as computer hardware. The programs that the microprocessor executes and that control the computer system are collectively referred to as computer software. A general rule is anything in a computer system that you can physically touch is hardware, and anything that you can't physically touch is software.

After completing this section, you should be able to

- **Explain the three basic duties of an operating system**
- **Discuss how an operating system functions in a computer system**
- **Compare and contrast the difference between multitasking and nonmultitasking operating systems**
- **Differentiate between multitasking and multiprocessing**
- **Identify and discuss the issues presented by multitasking**

Operating System Basics

The **operating system** (OS) of a computer is a special program that establishes the environment in which application programs operate. A major function of the operating system is to provide the functional interface between application programs in the system, called **processes**, and the computer hardware. Because the operating system must work closely with the computer hardware, it is often written in assembly language or programming language with low-level hardware support, such as C++.

An operating system increases the overall complexity of a computer system, but using an operating system offers a number of advantages over running stand-alone application programs. The operating system tests and initalizes hardware in the computer system, eliminating the need for each application to duplicate these functions. Operating systems also provide a standard computing environment so that applications can execute consistently. Finally, operating systems provide system services that allow applications access to commonly used system resources, such as the real-time clock, I/O ports, and data files, which simplify the code for applications programs. A drawback of operating systems is that processes may execute more slowly; accessing system resources through an operating system can take longer than a program accessing them directly. An operating system has three basic duties:

1. To schedule and allocate system resources (CPU time, memory, access to system peripherals)

2. To protect system processes and resources (preventing accidental or deliberate corruption of process code and data, preventing unauthorized access to hardware and memory)

3. To provide system services (messaging between processes, low-level hardware drivers)

Multiple Processes

Computers can run multiple processes in two basic ways. The first way, called **multitasking**, shares a single-core processor among multiple processes. The processor runs more than one process but switches between them so that each process uses only part of the processor's available time. Multitasking systems use different techniques to decide when to switch between processes. One technique allows a process to run until it must wait for some event, such as a key press, before it can continue and switches to another process that is ready to run. Another technique, called *preemptive multitasking* allows each process to run for a specific amount of time before the operating system switches to another process. A third technique, called *nonpreemptive multitasking*, allows a process to run until it voluntarily relinquishes the processor to another process. Figure 13–26 illustrates how a single-core processor multitasks.

FIGURE 13–26   Simplified model of processor multitasking.

The second way for computer system to run multiple processes is called **multiprocessing**. Multiprocessing uses multiple processors, each of which can either multitask or run a single process. Figure 13–27 illustrates the concept of multitasked multiprocessing.

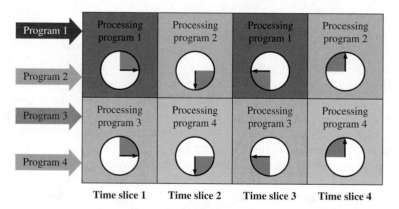

FIGURE 13–27   Multitasked multiprocessing in a multicore processor.

Supervisor and User States

It is difficult for multiple users or processes to coexist in a computer system if processes have unrestricted access to system resources. Once a process takes control of the processor and is running, it can modify or disable any software or hardware in the system that exists to control it. The only solution to this is to restrict what a process can access. Some processors use the user/supervisor state bit so that only trusted code, like the operating system, can run under certain circumstances. For multiprocess or multiuser systems, the processor executes in supervisor state when it first powers up, while the operating system is running, and when the processor responds to an interrupt. When the operating system loads and transfers control to an application program, it first clears the user/supervisor state bit. This places the process in user state and prevents it from accessing restricted parts of computer system's hardware or software.

Memory Management Unit (MMU)

One device in the computer system that has not yet been discussed is the memory management unit, or **MMU**. Memory management units are very sophisticated logic devices that handle many details associated with accessing memory in computer systems, including memory protection, wait-state generation, address translation for handling virtual memory, and cache control. As an example, consider a simplified MMU that simply provides memory protection. The processor can program the MMU with the start and end addresses of a memory range. The MMU then acts as a comparator. If the MMU detects a value on the address bus that is less than the programmed start address or greater than the programmed end address, it will generate a hardware interrupt to the processor.

System Services

Operating systems provide system services that allow applications access to commonly used system resources. This is essential for allowing processes to interact and communicate with each other to share information, coordinate operations, and otherwise function in unison. Interprocess communication uses software interrupts (also called *traps*). When one process wishes to utilize a system service, it loads specific registers with values and then invokes a specific trap to pass control to the operating system's exception handler for that trap.

When the process executes the trap, the processor enters supervisor mode and the exception handler would use the register contents to fulfill the requested service. If, for example, the requested service was to send several bytes from one process to another, the exception handler would use the starting address of the data and the number of data bytes contained in the processor registers to copy the data from the user memory of the source process to the user memory of the destination process. It would then load a condition code indicating that the service had been completed successfully (or failed) in one of the processor registers and return processor control to the requesting process. When processes are meant to interact with other processes, they each must be carefully designed to ensure that messages are passed at the right time and in the right order and that the processes can recover from communication errors. Otherwise, one process may believe that it has sent out a valid message and await a response, while the intended destination process is waiting for the first process to send a message to which it can respond. The result is that neither process can proceed.

SECTION 13–6 CHECKUP

1. What are the three basic duties of an operating system?

2. Compare and contrast multitasking and multiprocessing.

3. Describe how a memory management unit prevents one process from accessing the memory space of another process.

4. Explain how an operating system permits two processes to exchange information.

Assembly language is a way to express machine language (1s and 0s) in English-like terms, so there is a one-to-one correspondence. Assembly language has limited applications and is not portable from one processor to another, so most computer programs are written in high-level languages such as C++, JAVA and BASIC. High-level languages are portable and therefore can be used in different computers. High-level languages must be converted to the machine language for a specific microprocessor by a process called *compiling*.

After completing this section, you should be able to

- Describe some programming concepts
- Discuss the levels of programming languages

Levels of Programming Languages

A hierarchy diagram of computer programming languages relative to the computer hardware is shown in Figure 13–28. At the lowest level is the computer hardware (CPU, memory, disk drive, input/output). Next is the **machine language** that the hardware understands because it is written with 1s and 0s (remember, a logic gate can recognize only a LOW (0) or a HIGH (1). At the level above machine language is **assembly language** where the 1s and 0s are represented by English-like words. Assembly languages are considered low-level because they are closely related to machine language and are machine dependent, which means a given assembly language can only be used on a specific microprocessor.

At the level above assembly language is **high-level language,** which is closer to human language and further from machine language. An advantage of high-level language over assembly language is that it is portable, which means that a program can run on a variety of computers. Also, high-level language is easier to read, write, and maintain than assembly language. Most system software (e.g., Windows), and applications software (e.g., word processors and spreadsheets) are written with high-level languages.

High-level language
- Closer to human language
- Portable

Assembly language
- English-like terms representing binary code
- Machine dependent

Machine language
- Binary code (1s and 0s)
- Machine dependent

Computer hardware (the "machine")
- CPU
- Memory (RAM, ROM)
- Disk drives
- Input/Output

FIGURE 13–28   **Hierarchy of programming languages relative to computer hardware.**

Assembly Language

To avoid having to write out long strings of 1s and 0s to represent microprocessor instructions, English-like terms called mnemonics or op-codes are used. Each type of microprocessor has its own set of mnemonic instructions that represent binary codes for the instructions. All of the mnemonic instructions for a given microprocessor are called the instruction set. Assembly language uses the instruction set to create programs for the microprocessor; and because an assembly language is directly related to the machine language (binary code instructions), it is classified as a low-level language. Assembly language is one step removed from machine language.

Assembly language and the corresponding machine language that it represents is specific to the type of microprocessor or microprocessor family. Assembly language is not portable; that is, you cannot directly run an assembly language program written for one type of microprocessor on another type of microprocessor. For example, an assembly program for the Motorola processors will not work on the Intel processors. Even within a given family different microprocessors may have different instruction sets.

An **assembler** is a program that converts an assembly language program to machine language that is recognized by the microprocessor. Also, programs called **cross-assemblers** translate an assembly language program for one type of microprocessor to an assembly language for another type of microprocessor.

Assembly language is rarely used to create large application programs. However, assembly language is often used in a subroutine (a small program within a larger program) that can be called from a high-level language program. Assembly language is useful in subroutine applications because it usually runs faster and has none of the restrictions of a high-level language. Assembly language is also used in machine control, such as for industrial processes. Another area for assembly language is in video game programming.

Conversion of a Program to Machine Language

All programs written in either an assembly language or a high-level language must be converted into machine language in order for a particular computer to recognize the program instructions.

ASSEMBLERS   An assembler translates and converts a program written in assembly language into machine code, as indicated in Figure 13–29. The term **source program** is often used to refer to a program written in either assembly or high-level language. The term **object program** refers to a machine language translation of a source program.

FIGURE 13–29 **Assembly to machine conversion using an assembler.**

COMPILERS   A *compiler* is a program that compiles or translates a program written in a high-level language and converts it into machine code, as shown in Figure 13–30. The compiler examines the entire source program and collects and reorganizes the instructions. Every high-level language comes with a specific compiler for a specific computer, making the high-level language independent of the computer on which it is used. Some high-level languages are translated using what is called an *interpreter* that translates each line of program code to machine language.

FIGURE 13–30 **High-level to machine conversion with a compiler.**

All high-level languages, such as C++, will run on any computer. A given high-level language is valid for any computer, but the compiler that goes with it is specific to a particular type of CPU. This is illustrated in Figure 13–31, where the same high-level language program (written in C++ in this case) is converted by different machine-specific compilers.

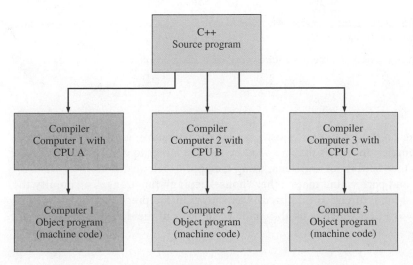

FIGURE 13–31 **Machine independence of a program written in a high-level language.**

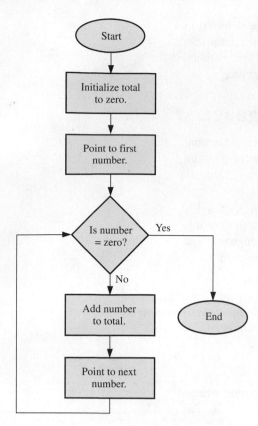

FIGURE 13–32 Flowchart for adding a list of numbers.

EXAMPLE OF AN ASSEMBLY LANGUAGE PROGRAM

For a simple assembly language program, let's say that we want the computer to add a list of numbers from the memory and place the sum of the numbers back into the memory. A zero is used as the last number in the list to indicate the end of the list of numbers. The steps required to accomplish this task are as follows:

1. Clear a register (in the microprocessor) for the total or sum of the numbers.
2. Point to the first number in the memory (RAM).
3. Check to see if the number is zero. If it is zero, all the numbers have been added.
4. If the number is not zero, add the number in the memory to the total in the register.
5. Point to the next number in the memory.
6. Repeat steps 3, 4, and 5.

A flowchart is often used to diagram the sequence of steps in a computer program. Figure 13–32 shows the flowchart for the program represented by the six steps listed above.

The following assembly language program implements the addition problem shown in the flowchart in Figure 13–32. Two of the registers in the microprocessor are named ax and bx. The comments preceded by a semicolon are not recognized by the microprocessor; they are for explanation only.

```
        mov ax,0           ;Replaces the contents of the ax register with zero.
                           ;Register ax will store the total of the addition.
        mov bx,50H         ;Places memory address hexadecimal 50 into the bx register.
next:   cmp word ptr [bx],0 ;Compares the number stored in the memory location pointed to by
                           ;the bx register to zero.
        jz done            ;If the number in the memory location is zero, jump to "done".
        add ax,[bx]        ;Add the number in the memory location pointed to by the bx register to
                           ;the number in the ax register and place the sum into the ax register.
        add bx,02          ;2 is added to the address in the bx register. Two addresses are
                           ;required to store each number which is two bytes long.
        jmp next           ;Loop back to "next" and repeat the process.
done:   mov [bx],ax        ;Replace the zero last number in the memory location pointed to by the
                           ;bx register with the total in the ax register.
        nop                ;No operation, this indicates the end of the program.
```

Depending on the assembler, most programs in assembly language will have a number of assembler directives that are used by the assembler to do a variety of tasks. These tasks include setting up segments, using the appropriate instruction set, describing data sizes, and performing many other "housekeeping" functions. To simplify the explanation, only one directive (required) was shown in the preceding program. The directive was **word ptr,** which is used to indicate the size of the data pointed to by the BX register.

EXAMPLE 13–2

Write the instructions for an assembly language program that will find the largest unsigned number in the data and place it in the last position. Assume the last data point is signaled with a zero.

SOLUTION

The flowchart is shown in Figure 13–33.

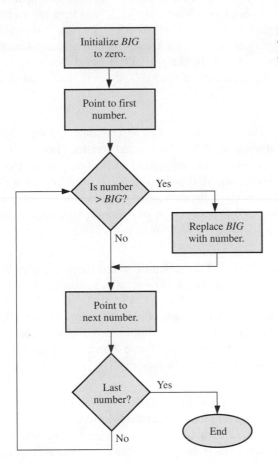

FIGURE 13–33 Flowchart.
The variable *BIG* represents
the largest value.

The data is assumed to be the same as before. The program listing (with comments) is as follows:

```
        mov ax,0000          ; initial value of BIG is in the ax register
        mov bx,0050          ; point to a location in memory (50H) where the data starts
repeat: cmp [bx],ax          ; is the data point larger than BIG?
        jbe check            ; if the data point is smaller, go to "check"
        mov ax,[bx]          ; otherwise, put new largest data point in ax
check:  add bx,02            ; point to the next number in memory (two bytes per word)
        cmp word ptr [bx],0  ; test for last data point
        jnz repeat           ; continue if the data point is not a zero
        mov [bx],ax          ; save BIG in memory
        nop                  ; no operation
```

RELATED PROBLEM

Explain how you could change the flowchart to find the smallest number in the list instead of the largest.

Types of Instructions

The programs in this section only show a few of the hundreds of variations of instructions available to programmers. To simplify learning the Intel instruction set, instructions are divided into seven categories. These categories are described here.

DATA TRANSFER   The most basic data transfer instruction MOV was introduced in the example programs. The MOV instruction, for example, can be used in several ways to copy a byte, a word (16 bits), or a double word (32 bits) between various sources and destinations such as registers, memory, and I/O ports. (A better mnemonic for MOV might have been "COPY" because this is what the instruction actually does.) Other data transfer instructions include IN (get data from a port), OUT (send data to a port), PUSH (copy data onto the stack, a separate area of memory), POP (copy data from the stack), and XCHG (exchange).

ARITHMETIC   There are a number of instructions and variations of these instructions for addition, subtraction, multiplication, and division. The ADD instruction was used in both example programs. Other arithmetic instructions include INC (increment), DEC (decrement), CMP (compare), SUB (subtract), MUL (multiply), and DIV (divide). Variations of these instructions allow for carry operations and for signed or unsigned arithmetic. These instructions allow for specification of operands located in memory, registers, and I/O ports.

BIT MANIPULATION   This group of instructions includes those used for three classes of operations: logical (Boolean) operations, shifts, and rotations. The logical instructions are NOT, AND, OR, XOR, and TEST. An example of a shift instruction is SAR (shift arithmetic right). An example of a rotate instruction is ROL (rotate left). When bits are shifted out of an operand, they are lost; but when bits are rotated out of an operand, they are looped back into the other end. These logical, shift, and rotate instructions can operate on bytes or words in registers or memory.

LOOPS AND JUMPS   These instructions are designed to alter the normal (one after the other) sequence of instructions. Most of these instructions test the processor's flags to determine which instruction should be processed next. In Example 13–2, the instructions JBE and JNZ were used to alter the path. Other instructions in this group include JMP (unconditional jump), JA (jump above), JO (jump overflow), LOOP (decrement the CX register and repeat if not zero) and many others.

STRINGS   A **string** is a **contiguous** (one after the other) sequence of bytes or words. Strings are common in computer programs. A simple example is a sentence that the programmer wishes to display on the screen. There are five basic string instructions that are designed to copy, load, store, compare, or scan a string—either as a byte at a time or a word at a time. Examples of string instructions are MOVSB (copy a string, one byte at a time) and MOVSW (copy a string, one word at a time).

SUBROUTINE AND INTERRUPTS   A **subroutine** is a miniprogram that can be used repeatedly but programmed only once. For example, if a programmer needs to convert ASCII numbers from a keyboard to a BCD format, a simple programming structure is to make the required instructions a separate process and "call" the process whenever necessary. Instructions in this group include CALL (begin the subroutine) and RET (return to the main program).

PROCESSOR CONTROL   This is a small group of instructions that allow direct control of some of the processor's flags and other miscellaneous tasks. An example is the STC (set carry flag) instruction.

High-Level Programming

The basic steps to take when you write a high-level computer program, regardless of the particular programming language that you use, are as follows:

1. Determine and specify the problem that is to be solved or task that is to be done.
2. Create an algorithm; that is, develop a series of steps to accomplish the task.
3. Express the steps using a particular programming language and enter them on the software text editor.
4. Compile (or assemble) and run the program.

A simple program will show an example of high-level programming. The following C++ program implements the same addition problem defined by the flowchart in Figure 13–32 and implemented using assembly language.

```
int total = 0;              //Initialize the total to 0.
while (*number ! = 0×00)    //Loop while the value is not found. The
                            //asterisk preceding the pointer identifier
                            //number says that the contents of the
                            //memory location pointed to by the
                            //identifier number are being evaluated.
{
total = total + *number;    //Accumulative summation of total.
number++;                   //Increment pointer to next number in memory.
}
```

This C++ program is equivalent to the assembly program that adds a series of numbers and produces a total value.

```
in total = 0;
while (*number ! = 0×00)
{
     total = total + *number;
     number++;
}
```

C++

Equivalent →

```
       mov ax, 0
       mov bx, 50H
next:  cmp word ptr [bx], 0
       jz done
       add ax, [bx]
       add bx, 02
       jmp next
done:  mov [bx], ax
       nop
```

Assembly

SECTION 13–7 CHECKUP

1. Define *program*.
2. What is an op-code?
3. What is a string?

Although a general-purpose microprocessor can interface with a variety of devices over its system buses, its ability to interface with the real world is limited. Most general-purpose microprocessors must use analog-to-digital converters (ADCs), digital-to-analog converters (DACs), universal asynchronous receivers and transmitters (UARTs) and other communication controllers, peripheral interface adapters (PIAs), external timers, and other specialized peripherals to process real-world information. A special type of processor, called a microcontroller, sometimes abbreviated as μC or MCU, combines a microprocessor core and common peripherals in a single package. Microcontrollers are used in microprocessor-controlled applications called embedded systems that perform a specific set of tasks and incorporate both the hardware and firmware required to perform them. Embedded systems include personal electronic devices such as cell phones, MP3 players, and calculators as well as consumer and industrial products such as microwave ovens, automated assembly systems, and robots.

After completing this section, you should be able to

- Describe the general architecture of microcontrollers
- Discuss the types of peripherals found in common microcontrollers
- Describe how microcontroller peripherals are configured
- Describe how microcontrollers are used in various embedded systems

Microcontroller Basics

Microcontrollers can range in complexity from simple devices with a few dozen pins to very complex devices with hundreds of pins. A common aspect of all these processors is that the design of each seeks to incorporate all the elements of a microcomputer system into a single package. A microcontroller will typically include the following functional units:

- A microprocessor (called the processor core)
- Nonvolatile memory for program code, device configuration data, and similar data that must be preserved when power is removed
- RAM for program data, internal registers, peripheral device buffers, and other data storage
- Peripheral devices such as timers, ADCs, DACs, communication controllers, and I/O ports
- Internal buses to connect the processor core to internal memory
- Internal buses to connect the processor core and internal memory to peripheral devices
- Interface circuitry to connect the microcontroller with external devices

In addition to the above list of microcontroller features, more sophisticated microcontrollers can also include the following:

- A phase-locked loop (PLL) to multiply a low-frequency external clock to a higher internal frequency, increasing the speed of microcontroller operation
- DMA controllers to improve data transfer between internal memory and peripheral devices
- Programmable logic resources, or "fabric," to implement custom functions
- A JTAG interface to support device testing and programming
- Special power modes for low-power and standby operation

Figure 13–34 shows a simplified block diagram of a typical microcontroller.

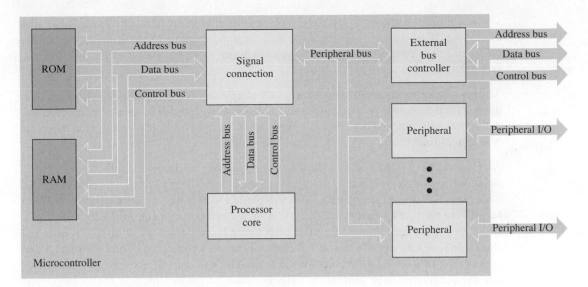

FIGURE 13–34 Simplified microcontroller block diagram.

Microcontroller Peripherals

Microcontrollers feature a wide variety of peripherals. Manufacturers select the type and number of peripherals, as well as the types and amounts of internal memory, to meet the requirements of specific applications, such as communication, automotive, and motion-control products. For example, microcontrollers that target communication applications will include a wide variety and number of communication controllers (such as Ethernet, I2C, USB, and UART) to support transmission and reception of data using multiple protocols. In contrast, microcontrollers meant for motion-control applications, such as robotic assemblies, will include ADCs, DACs, encoders, and pulse-width modulators (PWMs) for position sensing and motor feedback and control.

Many pins on microcontrollers are multifunctional. This not only helps to reduce the total pin count and cost of the device but also limits the functions that can be used at the same time. The data sheet may state that a communications microcontroller has four USB controllers, two UART controllers, an Ethernet controller, an external memory interface, and 80 general-purpose I/O (GPIO) lines, but it is unlikely that a given system can use all of these. A pin on a communications microcontroller, for example, might serve as a transmit line for USB communications, a clear-to-signal line for UART communications, a transmit line for Ethernet communications, an address line for the external bus controller, or a general-purpose I/O (GPIO) pin; but it can be configured for only one function at a time. Since applications rarely can change pin functions "on the fly," the circuit design permanently assigns the function of each microcontroller pin. If the circuit designer must use a set of pins for an external memory interface but also needs other functions that those pins provide, she either must find those functions on other pins (which is why microcontrollers offer more than one instance of a type of peripheral) or use external circuits to provide those functions.

The following describes some of the more common types of peripherals on microcontrollers.

General-Purpose I/O (GPIO) General-purpose I/O pins are typically the default function for many microcontroller pins. As the name suggests, these pins can be configured as input or outputs to read or write data, either as individual bits (for serial transfer of data) or groups of bits (for paraller transfer of data). Typical applications for GPIO lines are to read individual switches, drive LED indicators, or to select or enable latches or buffers.

Communication Controllers Communication controllers allow microcontrollers to communicate with other devices using specific communication protocols. Some standard communication protocols are universal asynchronous receive and transmit (UART), Ethernet, universal serial bus (USB), inter-integrated circuit (IIC or I2C), serial peripheral interface

(SPI), controller area network (CAN), and high-level data link control (HDLC). Because the timing, flow control, and data format of these protocols vary so widely, configuring communication controller functions is typically much more involved than for other peripheral functions.

Timers   Microcontroller timers can have multiple uses. These include setting the frequency for a communication controller, indicating when a preset time interval has elapsed, determining the elapsed time between two events, and providing a periodic time tick for a system real-time clock.

ADCs and DACs   ADCs and DACs are the means by which digital circuits interact with an analog world. As you know, digital circuits must use a limited set of values to represent a continuous range of analog data. For example, a car does not accelerate from 0 MPH to 60 MPH in sixty jerky increments of 1 MPH each, but the reading on a digital speedometer does. Similarly, the temperature of a room being heated from 65° F to 72° F does not change in seven abrupt 1° F steps, but the temperature setting on a digital thermostat that controls the furnace does. Microcontrollers use ADCs to convert analog voltage and current measurements from sensors into digital values for processing and DACs to convert digital values into analog voltages and currents to control electric and electronic circuits.

Quadrature Encoders   Quadrature encoders are used to determine the speed, direction, and position of a moving object, such as a computer mouse or a stepper motor. A quadrature encoder represents the present position of a tracked object with a Gray code sequence. When the object moves, the Gray code value changes. Each change will increment or decrement a counter to represent a positive or negative change in position. For example, a system could use the sequence $00 \rightarrow 01 \rightarrow 11 \rightarrow 10 \rightarrow 00$ to represent a positive change so that the sequence $00 \rightarrow 10 \rightarrow 11 \rightarrow 01 \rightarrow 00$ would represent a negative change. The counter value represents the position of the tracked object in the physical system relative to the starting point or origin of the tracked object; how fast the counter value changes reflects the speed of the tracked object. Quadrature encoders typically use 32-bit or larger counters to prevent an underflow or overflow condition that would make it seem that the tracked device suddenly changed from a maximum or minimum position or vice versa. Tracked objects require one encoder for each dimension of movement. An object that moves in one dimension, such a sliding door, needs only one encoder to track movement along the line of travel. An object that moves in two dimensions, such as a computer mouse, requires two encoders to track movement in the plane of travel. Objects that move in three dimensions, such as some robotic assemblies, require three encoders to track movement in the space of travel.

Pulse Width Modulators   As you know, a pulse width modulator modulates, or varies, the pulse width of a repetitive digital signal to change the signal's duty cycle (i.e., the ratio of the time that a signal is HIGH to the period of the signal). Pulse width modulators are often used in motor control. Although motor controller circuits can use the amplitude of winding current to set the speed of some motors, a more typical approach is to keep the amplitude of the applied winding current constant and vary the duty cycle to control the speed. Microcontrollers can precisely control the duty cycle to accurately set the motor's running speed. Also, microcontrollers also can change the duty cycle very quickly in response to the effects of motor speed due to line or load variations to maintain a constant running speed.

External Memory Controllers   Although most microcontrollers contain internal ROM, RAM, EEPROM, flash, and other memory for code, data, and other program information, some applications require more memory than a microcontroller contains. Consequently, many microcontrollers feature external memory controllers that permit interfacing the microcontroller to external memory devices. Some microcontrollers do not contain any internal memory so that external memory must be used. External memory controllers often feature decoded chip select lines that allow programmers to configure the size of the memory range, port size (8-bit, 16-bit, or 32-bits), and the number of wait states for each select line; they can also contain memory management units that provide memory protection for multitasking applications. External memory devices are typically limited to SRAM,

SDRAM, flash, and other memory types that do not require special bus operations, as do DRAM and EEPROM.

Configuring Peripherals

Microcontroller peripherals must be configured so that they operate the way an application requires them to do. Configuring means loading specific registers associated with the peripheral with values that control the function and operation of the peripheral. The registers and values vary with each peripheral, but the registers fall into the general categories described next. Depending upon the number of bits required to configure some aspect of a peripheral, some categories may share one register, while others may require multiple registers to contain the necessary information.

Control Registers   Control registers determine how the peripheral will function. For some peripherals a control register may select the specific peripheral as well as the characteristics for that peripheral. For example, the control registers for a communication controller could specify the specific communication protocol and the data rate, data packet size, error detection method, and operating mode (interrupt-driven or polled).

Status Registers   Status registers contain information about how the peripheral is operating and conditions associated with peripheral operation. Applications use status registers to detect errors, to determine when the peripheral has completed some task, and to decide when conditions require some special handling. The microcontroller may automatically clear some status bits when firmware corrects a detected condition, while in other cases, firmware may need to manually clear some status bits. For example, if an ADC sets the end-of-conversion status bit to indicate it has completed converting an analog value, reading the converted value from the ADC data register may automatically clear the bit; or firmware may need to specifically clear the status bit to allow the ADC to perform another conversion.

Data Registers   Data registers contain information that the peripheral processes in some way. The value in a data register can be data for the peripheral to process, data processed by the peripheral, or data currently being processed. The contents of a data register might not change unless firmware changes them, or operation of the peripheral may automatically update them. For example, the initialization register for a timer contains the initial count value that is loaded into the timer and may not change unless firmware writes a new value into the register. In contrast, the timer's count register holds the actual value of the timer and may update each time the counter is clocked. Some peripherals have only a few configuration registers.

The GPIO pins typically have only two registers: a control register that determines whether a pin is an input or output and a data register that contains the signal level of the pin. Other peripherals can have many more. A communication controller, for example, can have a control register to specify the communication parameters; a status register to monitor the operation of the controller; a transmit buffer descriptor register to specify the memory locations of data to be transmitted; a transmit length register to specify the number of bytes to transmit; a receive buffer descriptor register to specify the memory locations at which received data is to be stored; a receive length register to indicate the number of bytes received; an interrupt status register to signal communication events during reception and transmission; and an interrupt mask register to prevent or allow recognition of communication events. When configuring microcontroller peripherals, the programmer must carefully read the user manual and understand both the operation of each peripheral he intends to use as well as which configuration registers must be programmed and the configuration values to use.

As the number of products using microcontrollers has grown, manufacturers and third-party vendors have visual development and evaluation tools to simplify the process of programming microcontrollers. Many tools now allow programmers to use drop-down lists, check boxes, and other visual controls to generate C or C++ initialization code by specifying the peripherals they wish to use and how the peripherals should operate. While this is convenient and shortens development, time errors are still possible. Programmers should always review the code to verify it matches what they expected.

Microcontrollers in Embedded Systems

PERSONAL HANDHELD SYSTEMS   Smart phones, digital media devices, calculators, and portable GPS units are only a few examples of portable handheld electronic devices that are microcontroller-based embedded systems. Microcontrollers are widely used in these products because they can easily interface with the input and output hardware, rapidly process data, and consume relatively little power. Some of the most popular microcontrollers for portable handheld devices are those based on the ARM (Advanced RISC Machine) processor.

SYSTEM EXAMPLE 13–1

PROGRAMMABLE CALCULATOR

A block diagram for a microcontroller-based programmable calculator is shown in Figure 13–35.

FIGURE 13–35   **Microcontroller block diagram for programmable calculator.**

The calculator incorporates a USB communication port. The ROM contains the embedded code that implements the calculator functions and processes, and the RAM provides storage for the system stack, system data, user data, and programs. The USB controller transmits and receives data per the USB communications protocol and interfaces to the hardware that makes up the physical USB port. The calculator keypad connects to a parallel port formed by multiple GPIO lines, and the calculator LCD display interfaces with an LCD driver peripheral in the microcontroller to create the human machine interface, or HMI. A timer inside the microcontroller powers down the calculator after it has been active for a preset amount of time to save energy. Other timers in the microcontroller, which are not shown, set the communications rate for the USB controller and can provide a real-time clock and allow the user to set time-of-day alarms.

CONSUMER APPLIANCES   Virtually every electronic product today is a "smart" product that can make decisions, perform a preprogrammed sequence of events, or be manually programmed to do so. A short list of these products includes microwave ovens, coffee makers, washers and dryers, refrigerators, ovens, home entertainment components and systems, video game systems, and robotic vacuum cleaners.

AUTOMOBILE SYSTEMS   Automobiles use microcontrollers in a number of embedded systems. Embedded systems in modern automobiles monitor vehicle operation and control the engine, fuel system, brakes, air bags, door locks, environmental system, instrument display, vehicle navigation, and virtually every aspect of vehicle operation. One specific factor that can affect microcontrollers in automotive applications is the operating environment. Microcontrollers must be able to operate properly when they are exposed to the vehicle's temperature, humidity, vibration, and electrical noise they will encounter when the vehicle is operating.

AUTOMATED SYSTEMS   Two large areas of embedded applications are robotics and automation. Robotic and automated assemblies by nature must operate independently, perform repetitive tasks, process real-world data, and respond to circumstances that arise during operation. Embedded microcontroller systems can perform these tasks very well. One particular aspect of automated systems with which the microcontroller must deal is motion control. Microcontrollers must use feedback from the mechanical system to properly control the speed and acceleration of the system to ensure that it operates properly.

SYSTEM EXAMPLE 13–2

ROBOTICS SYSTEM

Figure 13–36 shows the block diagram for a basic robotics system. Although the block diagram is for a system that operates along a single axis, it can be extended to three axes for three-dimensional movement by using three microcontroller systems.

FIGURE 13–36   Basic block diagram for a robotics system.

The ROM contains the embedded code that implements the robotic functions and processes, and the RAM provides storage for the system stack and system data. The quadrature encoder receives encoded information from a motor position indicator and increments or decrements a counter depending upon how the encoded pattern changes. The pulse width modulator supplies a pulse train to a motor driver that in turn applies the voltages to the motor windings to turn the motor. GPIO lines detect when optical, magnetic, or other sensors indicate that the mechanical assembly has reached its maximum or minimum position.

When the system first powers up, the microcontroller uses the quadrature encoder and pulse width modulator to move the mechanical assembly to its minimum, or home, position and clears the counter so that zero corresponds to this home position and initializes the system. Once the system is initialized, the microcontroller then moves the mechanical assembly as programmed by driving the pulse width modulator to move the motor forward or backward and monitor the counter to determine how far and fast the mechanical assembly has moved. In most robotic systems, the microcontroller performs a complex series of calculations while monitoring the motor position and driving the motor to ensure that the mechanical assembly starts, stops, and operates smoothly.

SECTION 13–8 CHECKUP

1. How does a microcontroller differ from a microprocessor?

2. What are some common functional units found in a typical microcontroller?

3. Discuss an advantage and disadvantage of multifunctional pins on a microcontroller.

4. Which peripherals allow a microcontroller to interact with the real world?

5. How does an embedded system differ from a personal computer system?

6. Identify some types of embedded systems in which microcontrollers are found.

SUMMARY

- The basic functional components of a CPU are the ALU, register set, and timing and control unit.
- A microprocessor implements the functional components of a CPU on a single IC.
- The three basic computer buses are the address bus, data bus, and the control bus.
- Examples of peripheral computer devices include the keyboard, external disk drives, mouse, printer, modem, and scanner.
- Computer systems use I/O ports to access peripherals such as keyboards, mice, video monitors, modems, scanners, and disk drives.
- The three tasks of a computer system are to acquire, process, and return data.
- Computer systems use different types of memory for specialized functions, such as caching and queuing data.
- The functions of a CPU are to control system hardware, to provide hardware support to the system operating system, and to execute programs.
- A multicore processor has two or more microprocessors (cores), each with its own memory cache on a single chip.
- Pipelining, multitasking, and multiprocessing are techniques for decreasing the processing time.
- Buffers are used to prevent output loading but introduce propagation delays.
- Wait states are used to compensate for delays in bus signals due to bus buffers, address decoders, and mismatched timing specifications between hardware.
- Two methods for a processor to execute concurrent multiple processes are multitasking and multiprocessing.
- The three basic duties of a multitasking operating system are to allocate resources, to protect process and system resources, and to provide system services.
- The basic "language" of a computer is called machine code in which instructions are given as a series of binary codes.
- In assembly language, machine instructions are replaced with a short alphabetic English mnemonic that has a one-to-one correspondence to machine code. Assembly language also uses directives to allow the programmer to specify other parameters that are not translated directly into machine code.
- Microcontrollers integrate a microprocessor core with hardware peripherals and are well-suited for embedded applications.
- Embedded systems primarily process real-world signals and operate in real time rather than manipulate application program data.

KEY TERMS

Key terms and other bold terms in the chapter are defined in the end-of-book glossary.

ALU   Arithmetic logic unit.

Assembly language   A programming language that uses English-like words and has a one-to-one correspondence to machine language.

BIOS   The set of low-level routines that initialize the computer system hardware and allow high-level programs to interact with the system hardware.

Buffer   A circuit that prevents loading of an output.

Bus master   Any device that can control and manage the system buses in a computer system.

Caching   The process of copying frequently accessed program instructions from main memory into faster memory to increase processing speed.

CPU   Central processing unit; the "brain" of a computer that processes the program instructions.

DMA   Direct memory access.

Exception   Any software event that requires special handling by the processor.

Hardware   The circuitry and physical components of a computer system, as opposed to the instructions (called software).

High-level language   A type of computer language closest to human language that is a level above assembly language.

Interrupt   Any hardware event that requires special handling by the processor. An event that causes the current process to be temporarily stopped while a service routine is run.

Loading   The effect of the multiple inputs degrading the voltage or timing specifications of an output.

Machine language   Computer instructions written in binary code that are understood by a computer; the lowest level of programming language.

Main memory   Memory used by computer systems to store the bulk of programs and associated data.

Microcontroller   An semiconductor device that combines a microprocessor, memory, and various hardware peripherals on a single IC.

Microprocessor   A large-scale digital integrated circuit that can be programmed to perform arithmetic, logic, or other operations; the CPU of a computer.

Multiprocessing   A data processing technique that uses more than one processor.

Multitasking   A technique by which a single processor runs multiple programs concurrently.

Op-code   The mnemonic representation of a computer instruction.

Operand   The object to be manipulated by the instruction.

Operating system   The software that controls the computer system and oversees the execution of application software.

Pipelining   A technique where the microprocessor begins executing the next instruction in a program before the previous instruction has been completed.

Program   A sequential set of computer instructions designed to accomplish a given task(s).

Software   Computer programs; programs that instruct a computer what to do in order to carry out a given set of tasks.

System bus   The interconnecting paths in a computer system including the address bus, data bus, and control bus.

Wait state   A system bus delay equal to one processor clock cycle. Wait states are used to ensure that the system bus timing satisfies the address, data, and control timing specifications of a system.

TRUE/FALSE QUIZ

Answers are at the end of the chapter.

1. The three basic components of a CPU are input, memory, and output.

2. CPU stands for computer processing unit.

3. Printers, scanners, and keyboards are examples of computer peripherals.

4. The operating system of a computer is made of hardware.

5. The ALU is a key element in a microprocessor.

6. Microprocessors generally have three types of buses: address, data, and control.

7. Wait states allow other bus masters to share the system buses.

8. A multicore processor has one processor and more than one memory.

9. Pipelining allows a computer to execute a program faster.

10. Three levels of computer programming languages are machine, assembly, and high level.

11. DMA stands for digital memory access.

12. The signals associated with a DMA operation are bus request, bus grant, and bus grant acknowledge.

13. Microprocessor and microcontroller are different names for the same thing.

14. Some examples of microcontroller peripherals are GPIOs, ADCs, and quadrature encoders.

SELF-TEST

Answers are at the end of the chapter.

1. A basic computer does not include
   (a) an arithmetic logic unit   **(b)** a control unit
   (c) peripheral units   **(d)** a memory unit

2. The operation of the timing and control unit in a microprocessor is affected by the
   (a) instruction decoder   **(b)** accumulator
   (c) arithmetic logic unit   **(d)** register array

3. A 20-bit address bus supports
   (a) 100,000 memory addresses   **(b)** 1,048,576 memory addresses
   (c) 2,097,152 memory addresses   **(d)** 20,000 memory addresses

4. A bus that is used to transfer information both to and from the microprocessor is the
   (a) address bus   **(b)** data bus
   (c) both of the above   **(d)** none of the above

5. An example of a peripheral unit is the
   (a) address register   **(b)** CPU
   (c) video monitor   **(d)** interface adapter

6. In indexed addressing, the contents of the
   (a) accumulator are added to the contents of the index register
   (b) data register are added to the contents of the index register
   (c) index register are added to the contents of the program counter
   (d) index register are multiplied by the contents of the program counter

7. Polling is a method used for
   (a) determining the state of the microprocessor
   (b) establishing communications between the CPU and a peripheral
   (c) establishing a priority for communication with several peripherals
   (d) determining the next instruction

8. Polling peripherals
   (a) is less efficient than using interrupts
   (b) requires the CPU to periodically check each polled peripheral
   (c) often has the CPU checking devices that do not require service
   (d) all of the above

9. DMA stands for
   (a) digital microprocessor address
   (b) direct memory access
   (c) data multiplexed access
   (d) direct memory addressing

10. A computer program is a list of
    (a) memory addresses that contain data to be used in an operation
    (b) addresses that contain instructions to be used in an operation
    (c) instructions arranged to achieve a specific result

11. A type of assembly language instruction that alters the course of the program is called a
    (a) loop   (b) jump
    (c) both of the above   (d) none of the above

12. A wait state
    (a) terminates a bus cycle
    (b) halts the processor clock for one period
    (c) places the microprocessor in a low-power mode
    (d) delays completion of a bus cycle by one processor clock

13. A bus request differs from an interrupt in that
    (a) an interrupt requires an external bus master to complete the operation
    (b) a bus request will interrupt a current instruction cycle
    (c) an interrupt cannot be masked
    (d) a bus request does not involve a response from the processor

14. Multitasking
    (a) allows a single processor to execute more than one active process concurrently
    (b) allows a processor to share its buses with another bus master
    (c) is the same as running a terminate and stay resident (TSR) application
    (d) requires no hardware support from the CPU

15. A typical microcontroller contains
    (a) nonvolatile memory   (b) RAM
    (c) a processor core   (d) all of the above

PROBLEMS

Answers to odd numbered problems are at the end of the book.

SECTION 13–1   The Basic Computer System

1. Name the basic elements of a computer.
2. Name the functional units of a CPU.
3. What is a bus?
4. Explain how the control bus signals differ from those of the address and data buses.

SECTION 13–2   Practical Computer System Considerations

5. Name two special types of output that allow devices to share signal lines while avoiding bus contention.
6. What causes signal loading?
7. In Figure 13–5, determine the number of loads that the output is driving in parts (a) and (b).
8. In a computer system, how does the CPU select a device such as memory or Input/Output?
9. Explain the purpose of a wait state.

SECTION 13–3   The CPU: Basic Operation

10. Name the basic elements of a microprocessor.
11. List three operations that a microprocessor performs.
12. List the three microprocessor buses.
13. Explain what happens during a fetch/execute cycle.
14. Explain how pipelining works.

SECTION 13–4   The CPU: Addressing Modes

15. List the sequence of events for the inherent addressing mode.
16. List the sequence of events for the direct addressing mode.

17. List the sequence of events for the indexed addressing mode.

18. A processor is using the relative addressing mode to execute a branch instruction. If the binary number for 125_{10} is in the program counter and the number 55_{10} is in the data register, to what memory address will the processor branch?

SECTION 13–5   The CPU: Special Operations

19. Describe the purpose of an interrupt vector table.

20. How does an interrupt service routine differ from a normal subroutine?

21. Describe the sequence of events in a bus request operation.

22. Define DMA and describe its purpose in a computer system.

SECTION 13–6   Operating Systems and Hardware

23. What are the two groups of software that execute in a computer system and what do they include?

24. List the three basic duties of a multitasking operating system.

25. Identify the two ways that computers execute more than one process concurrently and describe how they differ.

26. Describe four difficulties that running multiple concurrent processes can create.

27. What are the functions of a memory management unit?

SECTION 13–7   Programming

28. What is an assembler?

29. Draw a flowchart for a program that adds the numbers from one to 10 and saves the result in a memory location named TOTAL.

30. Draw a flowchart showing how you could count the number of bytes in a string and place the count in a location in memory called COUNT. Assume the string starts at a location named START and has a 20H (hexadecimal ASCII code for a space) to signal the end. You should not count the space character.

31. Explain what happens when the instruction **mov ax,[bx]** is executed.

32. What is a compiler?

SECTION 13–8   Microcontrollers and Embedded Systems

33. What is a microcontroller?

34. Identify the functional components of the microcontroller block diagram shown in Figure 13–37.

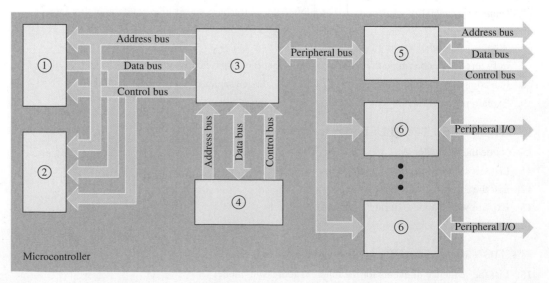

FIGURE 13–37

35. Discuss the functions of the following microcontroller peripherals.
    (a) GPIO   **(b)** ADC   **(c)** DAC   **(d)** PWM
    (e) quadrature encoder   **(f)** timer   **(g)** communication controller

36. Why are microcontrollers so widely used in embedded applications?

37. How does a microcontroller differ from a microprocessor?

ANSWERS TO SECTION CHECKUPS

SECTION 13–1   The Basic Computer System

1. The major functional blocks in a computer are CPU, memory/storage, and input/output ports.

2. Peripherals are devices external to the computer.

3. Hardware is the microprocessor, memory, hard disk, etc. Software is the program that runs the computer.

4. Conventional memory returns the data stored at a specified address. Content-addressable memory returns the address that contains a specified data value.

5. The address bus is unidirectional. The data bus is bidirectional. The control bus consists of signals that are bidirectional and operate independently rather than as a single functional group.

SECTION 13–2   Practical Computer System Considerations

1. Bus contention is the condition that results when two outputs connected to the same signal line attempt to drive it to different voltage levels. Two types of devices intended to prevent bus contention on shared signal lines are tristate devices and open collector (or open drain) devices.

2. A processor uses an address decoder to select and enable various devics.

3. Wait states in a computer system are additional clock cycles inserted in a processor bus cycle to satisfy the timing specifications of system devices.

4. A buffer prevents excess loading of a device and allows multiple devices to share the same bus.

SECTION 13–3   The CPU: Basic Operation

1. During a fetch/execute cycle, the CPU retrieves (fetches) an instruction from memory and carries out (executes) the instruction.

2. The four elements in a microprocessor are ALU, timing and control unit, register set, and instruction decoder.

3. The ALU is the arithmetic logic unit, and it carries out decoded instructions.

4. During a read operation, data is acquired from memory.

5. During a write operation, data is stored in memory.

SECTION 13–4   The CPU: Addressing Modes

1. Some addressing modes are inherent, direct, immediate, indexed, and relative.

2. An op-code is the representation of an computer program instruction.

3. An operand is an object that is manipulated by an instruction.

4. Branching occurs when the processor leaves the normal sequence in a program and branches to another place in the program.

SECTION 13–5   The CPU: Special Operations

1. Although the terms are sometimes used interchangeably, interrupts are generated by hardware sources and exceptions are generated by software sources.

2. A bus request can interrupt an instruction cycle and allows the external device to take control of the system buses. An interrupt must permit an instruction cycle to complete and is serviced by the processor which retains control of the system buses.

3. A direct memory access is a data transfer operation for which a special bus master called a DMA controller rather than the microprocessor controls the system buses. Direct memory accesses allow data transfers to occur much more rapidly than with the microprocessor because the controller does not attempt to process the data for each transfer as would the microprocessor.

SECTION 13–6 **Operating Systems and Hardware**

1. The three basic duties of an operating system are to allocate system resources, to protect processes and system resources, and to provide system services.

2. Multitasking allows a single processor to execute multiple processes concurrently by allocating processor time to each process. Multiprocessing allows a system to execute multiple processes by operating more than one processor.

3. A memory management unit prevents a process from accessing the memory space of another by comparing the contents of the address bus against a permitted range and generating an interrupt when a violation occurs so that the operating system can take appropriate action.

4. An operating system allows processes that are restricted to their own memory spaces to exchange information by providing a system service, activated by a software interrupt, that has the operating system pass the information between the processes.

SECTION 13–7 **Programming**

1. A program is a sequence of computer instructions designed to perform a specified task.

2. An op-code is an instruction expressed in mnemonic form.

3. A string is a contiguous sequence of bytes or words.

SECTION 13–8 **Microcontrollers and Embedded Systems**

1. A microcontroller basically includes many elements of a computer system where a microprocessor is only the CPU.

2. Common functional units found in a typical microcontroller are the processor core, ROM, RAM, one or more hardware peripherals, a signal connection block between the microcontroller functional blocks, and an external memory controller.

3. An advantage of multifunctional pins on a microcontroller is that it reduces the number of pins and physical size of the device. A disadvantage is that it is not always possible to have access to all the microcontroller features that an application may require so that the design requires additional external hardware.

4. Peripherals that allow microcontrollers to interact with the real world include timers, ADCs, DACs, communication controllers, GPIOs, quadrature encoders, and PWMs.

5. An embedded system is designed to interact directly with the real world and perform a specific function. A personal computer system processes data and can be configured with application software to perform a number of tasks.

6. Some embedded systems in which microcontrollers are found include personal electronics, consumer electronics, automotive systems, and communication devices.

ANSWERS TO RELATED PROBLEMS FOR EXAMPLES

13–1 D_6 must be LOW.

13–2 Change first block (initialization block) to "BIG = FFFF"; this is the largest possible unsigned number. Change first question to "Is number < BIG?"

ANSWERS TO TRUE/FALSE QUIZ

1. F	2. F	3. T	4. F	5. T	6. T	7. F
8. F	9. T	10. T	11. F	12. F	13. F	14. T

ANSWERS TO SELF-TEST

1. (c)	2. (a)	3. (b)	4. (b)	5. (c)	6. (b)	7. (b)	8. (d)
9. (b)	10. (c)	11. (c)	12. (d)	13. (c)	14. (a)	15. (d)	

CHAPTER 14

BUSES, NETWORKS, AND INTERFACING

OBJECTIVES

- Discuss the types of buses
- List bus characteristics
- Explain bus protocols
- Explain the multiplexed bus and tristate outputs
- Discuss the PCI series, AGP, ISA, IEEE-488, and SCSI types of parallel buses
- Describe the Universal Serial Bus (USB)
- Discuss the RS-232/422/423/485, SPI, I^{2}C, CAN, and Firewire types of serial buses
- Discuss the general characteristics of a network
- Describe bus, ring, star, tree, and mesh network topologies
- Define OSI
- Discuss Ethernet
- Discuss TCP/IP

KEY TERMS

System bus	SCSI
PCI	USB
ISA bus	RS-232
Local bus	SPI
Parallel bus	I^{2}C
Serial bus	Controller Area
Bus width	Network (CAN)
Bus frequency	Firewire
Bus transfer speed	Network
Bus bandwidth	Network topology
Bus protocol	OSI
Handshake	LAN
Tristate buffer	Ethernet
PCI-X	TCP/IP
PCI-Express	Client/server
GPIB	IP address

INTRODUCTION

You are already familiar with the concept of buses from Chapters 11 and 13. Selected buses, networks, and interfacing are covered in this chapter. The intent is to provide only a basic introduction. Indeed, each of the bus standards and networks that are introduced would require at least a full chapter or, in some cases, a book to

VISIT THE WEBSITE
Study aids for this chapter are available at
http://pearsonhighered.com/floyd

thoroughly develop the topic and provide all of the details. Much information on all of the bus standards, network topologies, and bus protocols is available on the Internet. With the foundation you will obtain from this chapter, you will be able to expand your knowledge in those areas for which more expertise is required.

As you know, a *bus* is the interconnection mechanism by which two or more devices, such as computers and peripheral devices, can communicate with each other. Generally, a bus is both physical and electrical, combining hardware specifications and electrical specifications. A *network* is an array of devices interconnected in such a way as to allow communication among them. Networks, as do buses, operate based on specified protocols, which are sets of rules to which they must conform. *Interfacing* is the method by which two different devices or systems are compatible in terms of transmitting and receiving data.

14–1 BUS BASICS

Essentially, a bus is a set of connections that carries digital information between two or more systems or between two or more parts of a system. For example, most computers have both internal and external buses. The internal buses connect to various internal elements to allow data transfer. External buses connect peripheral devices to the computer.

After completing this section, you should be able to

- **Name the internal computer buses**
- **Discuss the difference between internal and external buses**
- **Describe how parallel and serial buses operate**
- **List several bus characteristics**
- **Discuss bus protocol**
- **Explain how synchronous and asynchronous buses differ**

A bus allows two or more devices to communicate with each other. A bus consists of a set of physical wires and connectors and a set of electrical specifications. A bus can be either internal or external to a system.

The physical properties of a typical bus include the number of wires or PC board conductors (width), the configuration and length of the wires or conductors, and the types and configurations of the connectors. The electrical properties of a typical bus include but are not limited to some or all of the following: signal format, signal voltage levels, clock frequency, data transfer speed, bandwidth, data frame format, data rate handshaking protocol, error detection, and impedances and line termination. Each device connected to a bus must be compatible with the bus specifications in order to communicate. A sending device can also be a receiving device, and a receiving device can also be a sending device. Figure 14–1 illustrates the concept of a typical bus.

FIGURE 14–1   **Physical and electrical definition of a typical bus.**

Internal Buses

Internal computer buses were introduced in Chapter 13. As you know, internal buses in a computer carry information in the form of addresses, data, and control signals.

A **system bus** (also known as **frontside bus**) is a computer bus that connects the major components of a computer system with three component buses. It connects the CPU, memories, and I/O, as shown in Figure 14–2.

The **PCI** (peripheral component interconnect) bus, also part of the local bus system, is used to communicate with devices such as a hard disk, and with devices inserted in expansion slots on the computer motherboard, such as sound cards, network cards, and modem. A device known as a **bus bridge** interfaces two buses, usually a faster bus with a slower system bus.

Some computers also have an **ISA bus** (industry standard architecture) for communicating with motherboard slot devices that do not require the speed of the PCI bus. The system bus, PCI bus, and ISA bus form the **local bus**. The local bus is a parallel bus where multiple conductive lines form the bus.

In addition to the three local buses, some computers often use two other buses: the backside bus and the AGP bus. The **backside bus** connects the CPU directly to the cache (L2) to provide faster access to the RAM than can be achieved with the frontside bus (system bus). The **AGP** (accelerated graphics port) bus is an interface for a graphic card. The graphics controller can directly access the main memory through the AGP for faster data transfer using the PCI bus. The block diagram in Figure 14–3 illustrates the basic internal bus structure of a typical computer that uses all of the buses just discussed.

FIGURE 14–2   **The system bus.**

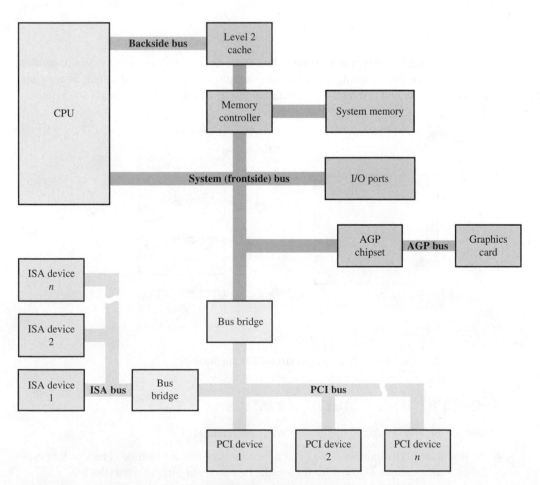

FIGURE 14–3   **Illustration of one possible configuration of internal computer buses.**

Parallel and Serial Buses

Buses can be either parallel or serial. As mentioned, the local bus in a computer is a parallel bus. A **parallel bus** carries data bits simultaneously, and a **serial bus** carries data bits sequentially one at a time. Figure 14–4 is a simple comparison of parallel and serial buses showing eight bits being transmitted.

(a) Parellel bus   (a) Serial bus

FIGURE 14–4   **Comparison of parallel and serial buses.**

It would seem that a parallel bus would transmit data faster than a serial bus because multiple data bits can be sent simultaneously. However, this is not always the case. As data rates increase, things like cross talk across parallel bus lines, timing skew between bus lines, and EMI (electromagnetic interference) become a problem that limit the speed. Serial buses are not limited by those factors and can actually transmit data at higher rates than parallel buses in many situations.

External Buses

External buses (also known as *expansion buses*) are used to connect one system to another separate system. For example, a computer connects to peripheral units such as monitor, keyboard, mouse, and printer through external buses, as illustrated in Figure 14–5.

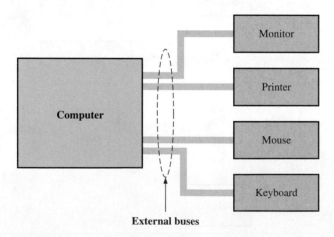

FIGURE 14–5   **Example of external bus application.**

General Bus Characteristics

A bus is typically described by the following parameters:

- **Bus width**   The number of bits that a bus can transmit at one time. The width of typical buses can vary from 1 bit for a serial bus up to 64 bits for a parallel bus.

- **Bus frequency**   The clock frequency at which the bus can operate.

- **Bus transfer speed** The number of bytes per clock cycle.

- **Bus bandwidth** The number of bytes per clock cycle times the number of clock cycles per second; that is, transfer speed times frequency. Bus bandwidth is sometimes called *throughput*.

Bus bandwidth can be specified in two ways, which depends on how the prefix M in MBps is defined. It can be defined in decimal form as a power of ten ($10^6 = 1,000,000$) or in binary form as a power of two ($2^{20} = 1,048,576$). In the decimal form, the M stands for *mega*; in the binary form, the M stands for *mebi* (*megabi*nary). This distinction can be a point of confusion in specifications, so you should be aware of the difference. The following two formulas provide for determination of the bus bandwidth. Equation 14–1 is for the decimal approach, and Equation 14–2 is for the binary approach.

$$\text{Bus bandwidth} = \frac{\text{Width (bits)} \times \text{Frequency (MHz)}}{8 \text{ bits/byte}} \tag{14–1}$$

$$\text{Bus bandwidth} = \frac{((\text{Width (bits)} \times \text{Frequency (MHz)})/8 \text{ bits/byte})10^6}{2^{20}} \tag{14–2}$$

EXAMPLE 14–1

A certain bus is specified with a width of 32 bits and a frequency of 66 MHz. Determine the bus bandwidth expressed as two different values, according to the decimal and binary definitions of M. Bps is bytes per second.

SOLUTION

Using the decimal definition of M (10^6) in the unit of MBps,

$$\text{Bandwidth} = \frac{32 \text{ bits} \times 66 \text{ MHz}}{8 \text{ bits/byte}} = \mathbf{264\ MBps}$$

Using the binary definition of M (2^{20}),

$$\text{Bandwidth} = \frac{((32 \text{ bits} \times 66 \text{ MHz})/8 \text{ bits/byte})\ 10^6}{2^{20}} = \mathbf{252\ MBps}$$

RELATED PROBLEM*

What is the frequency for a bus that has a width of 64 bits and bandwidth of 125 MBps as specified in decimal form.

*Answer is at the end of the chapter.

Table 14–1 lists some typical buses and their characteristics.

Table 14–1					
BUS	**WIDTH (BITS)**	**FREQUENCY (MHz)**	**TRANSFER SPEED (BYTES/CYCLE)**	**BANDWIDTH (MBps)**	
				Decimal	**Binary**
ISA (16 bit)	16	8.3	2	16.6	15.9
PCI	32	33	4	132	125.9
PCI-X	32	66	4	264	251.8
AGP	32	66	4	264	251.8

Bus Protocol

Bus protocol is a set of rules that allow two or more devices to communicate through a bus. Buses provide for data transfer, address selection, and control. Each device connected to a bus has an address assigned to it for identification and command. Control signals implement the protocol. These signals allow the devices to work properly together by identifying each other and communicating back and forth. One device can send a request to another device and get an acknowledgement or reply.

HANDSHAKING The **handshake** is a routine by which two devices initiate and complete a bus transfer. Figure 14–6 shows a simple handshake process, including a timing diagram, in which a requesting device (sometimes called the master) and a responding device (sometimes called the servant or slave) initiate and complete a data transfer.

① Master sends request for data (REQD) to servant.

② Servant sends an acknowledgement (ACK) and places data on bus.

③ Master receives data and removes request.

④ Servant removes acknowledge and is ready for next request.

FIGURE 14–6 **Simple example of handshake and data transfer.**

Synchronous and Asynchronous Buses

A synchronized bus includes a clock in the control lines and has a fixed protocol that is relative to the clock. A synchronous bus is fast but has the disadvantage that every device connected to it must operate at the same clock frequency. Also, the physical length of the bus may be limited because of having to carry a high-frequency clock signal.

An asynchronous bus is not clocked, so it can serve various devices with different clock rates. The asynchronous bus uses a handshake protocol to establish communication, as previously described.

Single-ended Versus Differential Buses

Data communications between devices can be classified as either single-ended or differential in terms of the physical bus configuration. In general, single-ended operation is limited in both data rate and distance (cable length). Differential operation provides much higher data rates and longer transmission distances. **Single-ended operation** uses one wire for data and one wire for ground, where the signal voltage on the wire is with respect to ground. **Differential operation** uses two wires for data and one wire for ground. The data signal is sent on one wire, and its complement (inversion) is sent on the other wire. The difference between the two data wires is the differential signal. Figure 14–7 shows both single-ended and differential operation.

(a) Single-ended

(b) Differential

FIGURE 14–7   A comparison of single-ended and differential bus operation.

A single-ended transmission is simpler and lower in cost compared to a differential transmission. Differential operation is much less sensitive to noise because of common-mode rejection. This means that when a common noise signal appears on each line, the two noise signals cancel due to the differential operation where the difference between them is zero. Since the data signals are the same but opposite in phase, they are effectively added and the data signal is reinforced while the noise signal is cancelled. Due to the twisted pair, cross talk is reduced at higher frequencies, thus allowing longer cables.

SECTION 14–1 CHECKUP*

1. Define *bus bridge*.

2. Name three internal buses often found in computer systems.

3. List two factors that may limit the speed of a parallel bus.

4. What is the basic difference between an internal and an external bus?

5. Name four bus characteristics.

6. What is bus protocol?

7. Discuss the difference between single-ended and differential bus operation.

As you know, all the components in a computer or other systems are interconnected by buses, which serve as communication paths. Physically, a bus is a set of conductive paths that serves to interconnect two or more functional components of a system or several diverse systems. Electrically, a bus is a collection of specified voltage levels and/or current levels and signals that allow the various devices connected to the bus to work properly together.

After completing this section, you should be able to

- Discuss the concept of a multiplexed bus
- Explain the reason for tristate outputs

Basic Multiplexed Buses

As you have learned, in computers the microprocessor controls and communicates with the memories and the input/output (I/O) devices via the *internal bus structure,* as indicated in Figure 14–8. A bus is multiplexed so that any of the devices connected to it can either send or receive data to or from one of the other devices. A sending device is often called a master or **source,** and a receiving device is often called a servant or **acceptor.** At any given time, there is only one source active. For example, the RAM may be sending data to the input/output (I/O) interface under control of the microprocessor.

FIGURE 14–8   The interconnection of microprocessor-based system components by a bidirectional, multiplexed bus.

Bus Signals

With synchronous bus control, the microprocessor (CPU) usually originates all control and timing signals. The other devices then synchronize their operations to those control

FIGURE 14–9   **An example of a handshaking sequence.**

and timing signals. With asynchronous bus control, the control and timing signals are generated jointly by a source and a receiver using a handshaking routine. A typical handshaking sequence is given in Figure 14–9. Handshaking routines may differ from one system to another, as you can see by comparing this sequence with the one shown in the previous section.

An important control function is called **bus arbitration.** Arbitration prevents two sources from trying to use the bus at the same time.

Connecting Devices to a Bus

Tristate buffers are normally used to interface the outputs of a source device to a bus. (Tristate buffers were introduced in Chapter 13.) Usually more than one source is connected to a bus, but only one can have access at any given time. All the other sources must be disconnected from the bus to prevent **bus contention.**

Tristate circuits are used to connect a source to a bus or disconnect it from a bus, as illustrated in Figure 14–10(a) for the case of two sources. The select input is used to connect either source A or source B but not both at the same time to the bus. When the select input is LOW, source A is connected and source B is disconnected. When the select input is HIGH, source B is connected and source A is disconnected. A switch equivalent of this action is shown in part (b) of the figure.

FIGURE 14–10   Tristate buffer interface to a bus.

(a)

(b)

When the enable input of a tristate circuit is not active, the device is in a high-impedance (**high-Z**) state and acts like an open switch. Many digital ICs provide internal tristate buffers for the output lines. A tristate output is indicated by a ∇ symbol as shown in Figure 14–11.

FIGURE 14–11   Method of indicating tristate outputs on an IC device.

(a) Active-HIGH enable   (b) Active-LOW enable

FIGURE 14–12   **Tristate buffer symbols.**

TRISTATE BUFFER OPERATION   Figure 14–12(a) shows the logic symbol for a noninverting tristate buffer with an active-HIGH enable. Part (b) of the figure shows one with an active-LOW enable.

The basic operation of a tristate buffer can be understood in terms of switching action as illustrated in Figure 14–13. When the enable input is active, the gate operates as a normal noninverting circuit. That is, the output is HIGH when the input is HIGH and LOW when the input is LOW, as shown in parts (a) and (b) respectively. The HIGH and LOW levels represent two of the states. The buffer operates in its third state when the enable input is not active. In this state, the circuit acts as an open switch, and the output is completely disconnected from the input, as shown in part (c). This is sometimes called the *high-impedance* or *high-Z* state.

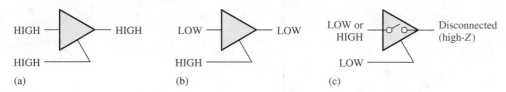

(a)   (b)   (c)

FIGURE 14–13   **Tristate buffer operation.**

Many microprocessors, memories, and other integrated circuit functions have tristate buffers that serve to interface with the buses. Such buffers are necessary when two or more devices are connected to a common bus. To prevent the devices from interfering with each other, the tristate buffers are used to disconnect all devices except the ones that are communicating at any given time.

Bus Contention

Bus contention occurs when two or more devices try to output opposite logic levels on the same common bus line. The most common form of bus contention is when one device has not completely turned off before another device connected to the bus line is turned on. This generally occurs in memory systems when switching from the READ mode to the WRITE mode or vice versa and is the result of a timing problem.

Multiplexed I/Os

Some devices that send and receive data have combined input and output lines, called I/O ports, that must be multiplexed onto the data bus. Bidirectional tristate buffers interface this type of device with the bus, as illustrated in Figure 14–14(a).

Each I/O port has a pair of tristate buffers. When the $\overline{SND/RCV}$ (Send/Receive) line is LOW, the upper tristate buffer in each pair is enabled and the lower one disabled. In this state, the device is acting as a source and sending data to the bus. When the $\overline{SND/RCV}$ line is HIGH, the lower tristate buffer in each pair is enabled so that the device is acting as an acceptor and receiving data from the bus. This operation is illustrated in Figure 14–14(b). Some devices provide for multiplexed I/O operation with internal circuitry.

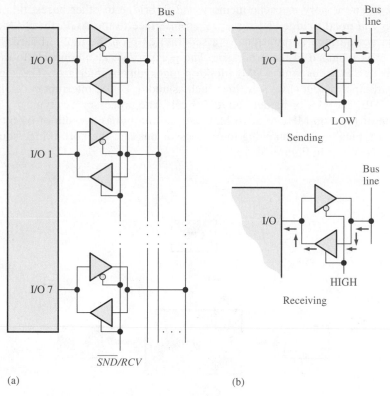

FIGURE 14–14   **Multiplexed I/O operation.**

SECTION 14–2 CHECKUP

1. Why are tristate buffers required to interface digital devices to a bus?

2. What is the purpose of a bus system?

14–3 PARALLEL BUSES

A bus is not only a set of physical connections (PC board traces or cables) but it is also a set of signals and operating parameters that are defined in the bus specification. Any devices connected to a given bus must be compatible with the bus specifications. In this section, we will briefly look at several important parallel bus standards. Further details and information on each bus can be found on the Internet.

After completing this section, you should be able to

- Discuss the PCI, PCI-X, and PCI-Express buses
- Describe the AGP and ISA
- Explain the basics of the IEEE-488 bus
- Discuss the SCSI bus and several of its variations

The Processor Bus and Bus Bridges

The processor bus, also known as the frontside bus (FSB), is internal to the system and is located on the motherboard. The CPU uses this bus to transfer information to and from the main memory or cache memory and to bridge to other buses; this bridge is known as the **northbridge** and serves as an interface to the AGP, the RAM, and the PCI. The northbridge is traditionally a separate chip on the motherboard, but more recently it is included in the processor. The northbridge also connects to the **southbridge** with a PCI bus with a DMI (desktop management interface). The southbridge handles all input/output (I/O) functions such as audio, BIOS, interrupt controller, USB devices, IDE, and ISA or its derivative devices. The processor bus typically runs at a clock rate of from 66 MHz to 1066 MHz and normally has a width of 64 bits. A diagram of a typical system showing the processor bus and the northbridge/southbridge interfaces is shown in Figure 14–15.

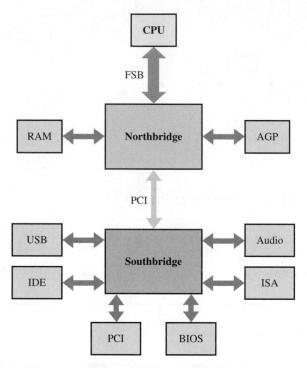

FIGURE 14–15   **Typical system layout with processor bus (FSB) and bus bridges.**

The PCI Bus

The PCI (peripheral component interconnect) bus is an internal synchronous bus for interconnecting chips, expansion boards, and processor/memory subsystems. The original PCI bus has a width of 32 bits and a frequency of 33 MHz. Another version has a width of 64 bits and a frequency of 66 MHz. Still later versions enable 64-bit data transfers using up to a 133 MHz clock to enable bandwidths of up to 1066 MBps (Mbytes/s).

The original PCI standard required 5 V power and signal levels. As the standard evolved, the option for 3.3 V was added. The latest standard provides for 3.3 V only. The 32-bit PCI connector has 62 pins and 124 contacts (62 per side). Thirty-two of the contacts are used for both a 32-bit address and 32 bits of data, which are multiplexed. The remaining contacts are used for command and control signals, power, ground, etc. A 64-bit PCI connector has an additional 32 pins for a total of 94. PIC connectors are usually white in color and are wired to the motherboard in a computer system, as shown in Figure 14–16.

The 32-bit PCI connector is shown in Figure 14–17. It has 64 pins of which 62 are used.

FIGURE 14–16   A simplified drawing of a typical computer motherboard. The white connectors are the PCI extension slots.

FIGURE 14–17   Pin layout and functional designation for a 32-bit PCI connector.

PIN	SIDE A	SIDE B	
1	−12 V	TRST#	
2	TCK	+12 V	JTAG port pins
3	Gnd	TMD	
4	TDO	TDI	
5	+5 V	+5 V	
6	+5 V	INTA#	
7	INTB#	INTC#	Interrupt lines
8	INTD#	+5 V	
9	PRSNT1#	Reserved	Indicates 7.5 or 24 W power
10	Reserved	IOPWR	+5 V or +3.3 V
11	PRSNT1#	Reserved	Indicates 7.5 or 24 V power
12	Gnd	Gnd	Key notch for 3.3 V cards
13	Gnd	Gnd	
14	Reserved	3.3 Vaux	Standby power
15	Gnd	RST#	Bus reset
16	CLK	IOPWR	33 MHz/66 MHz clock
17	Gnd	GNT#	Bus grant motherboard to card
18	REQ#	Gnd	Bus request card to motherboard
19	IOPWR	PME#	Power management event/3.3 V
20	AD[31]	AD[30]	
21	AD[29]	+3.3 V	
22	Gnd	AD[28]	
23	AD[27]	AD[26]	Address/data bus
24	AD[25]	Gnd	
25	+3.3 V	AD[24]	
26	C/BE[3]#	IDSEL	
27	AD[23]	+3.3 V	
28	Gnd	AD[22]	
29	AD[21]	AD[20]	
30	AD[19]	Gnd	
31	+3.3 V	AD[18]	
32	AD[17]	AD[16]	
33	C/BE[2]#	+3.3 V	
34	Gnd	FRAME#	Bus transfer in progress
35	IRDY#	Gnd	Initiator ready
36	+3.3 V	TRDY#	Target ready
37	DEVSEL#	Gnd	Target selected
38	Gnd	STOP#	Target halt request
39	LOCK#	+3.3 V	Locked transaction
40	PERR#	SMBCLK	Parity error/SMBus clock
41	+3.3 V	SMBDAT	SMBus data
42	SERR#	Gnd	System error
43	+3.3 V	PAR	Even parity over AD bus
44	C/BE[1]#	AD[15]	
45	AD[14]	+3.3 V	
46	Gnd	AD[13]	Address/data bus
47	AD[12]	AD[11]	
48	AD[10]	Gnd	
49	MGGEN/Gen	AD[09]	
50	Gnd	Gnd	Key notch for +5 V cards
51	Gnd	Gnd	
52	AD[08]	C/BE[0]#	
53	AD[07]	+3.3 V	
54	+3.3 V	AD[06]	
55	AD[05]	AD[04]	
56	AD[03]	Gnd	Address/data bus
57	Gnd	AD[02]	
58	AD[01]	AD[00]	
59	IOPWR	IOPWR	
60	ACK64#	REQ64#	For 64-bit expansion
61	+5 V	+5 V	
62	+5 V	+5 V	
63			
64			

The PCI-X Bus

The **PCI-X** bus is a high-performance enhancement of the PCI and is backward compatible with PCI, although it is a faster bus and has some additional features. A 64-bit bus, the PIC-X runs at a frequency of 133 MHz. The PCI-X 2.0 revision supports frequencies of 266 MHz and 533 MHz. Some of the additional features increase system reliability by minimizing errors at high transfer rates. Servers are the major application for the PCI-X.

The PCI-EXPRESS Bus

The **PCI-Express** is also designated as PCIe or PCI E. This bus differs from the PCI and PCI-X buses in that it does not use a **shared bus**. Both PCI and PCI-X use a shared bus configuration, as shown in Figure 14–18(a). Each PCIe device has a dedicated path, called a *lane*, to a single chip known as a *switch*, which provides for much faster transfer speeds, as shown in part (b).

FIGURE 14–18   Comparison of PCI and PCI-Express.

(a) PCI shared bus configuration

(b) PCI-Express lane configuration

The lanes are individual paths between the switch and the PCI devices. Each card slot can have 1, 2, 4, 8, 16, or 32 lanes. The designation PCI-Express ×1 means here is one lane, PCI-Express ×2 means there are two lanes, and so on. More lanes result in a faster data transfer. High speed makes PCI-Express ideal for video and graphics applications, but it is not compatible with PCI or PCI-X.

A single PCI-Express lane contains two pairs of conductors. One pair of conductors from a given device receives data and the other sends data in a serial format. A single lane configuration is known as x1 and is illustrated in Figure 14–19(a). A x2 configuration is shown in part (b), and a x16 configuration is shown in part (c). Other possible configurations are x4, x8, and x32.

FIGURE 14–19   PCI-Express lane configurations.

(a) x1

(b) x2

(c) x16

Bidirectional data is transferred serially on each lane. For multiple-lane configurations, the serial data on each lane is in parallel with the serial data on all the other lanes. The data is transferred at one bit per cycle on an x1 connection, two bits per cycle on an x2, and sixteen bits per cycle on the x16. The PCI-Express has a bandwidth of 4 Gbps per lane. PCI-Expess supports hot swapping in which expansion cards can be added or removed without turning off the system. PCI and PCIe are software compatible but not hardware compatible.

The AGP and ISA Buses

The AGP (accelerated graphics port) bus is intended specifically for connecting a video card to the system motherboard for 3-D computer graphics. It has a width of 32 bits and a bandwidth of up to over 2000 MBps. This bus has been superseded in many applications by PCI-Express. The ISA (industry standard architecture) bus, also known at the AT bus, has also been superseded by PCI-Express in many applications. ISA has a width of 16 bits and a maximum bandwidth of 8 MBps.

The IEEE-488 Bus

This bus standard has been around a long time and is also known as the General-Purpose Interface Bus (**GPIB**). Widely used in test and measurement applications, it was developed by Hewlett-Packard in the 1960s. The IEEE 488 specifies 24 lines that are used to transfer eight parallel data bits at a time and provide eight control signals that include three handshake lines and five bus-management lines. Also included are eight ground lines used for shielding and ground returns. The maximum data transfer rate for the IEEE 488 standard is 1 Mbps. A superset of this standard, called the HS488, has a maximum data rate of 8 Mbps.

To connect test equipment to a computer using the IEEE 488 bus, an interface card is installed in the computer, which turns the computer into a system controller. In a typical GPIB setup, up to 14 controlled devices (test and measurement instruments) can be connected to the system controller. When the system controller issues a command for a controlled device to perform a specified operation, such as a frequency measurement, it is said that the controller "talks" and the controlled device "listens."

A **listener** is an instrument capable of receiving data over the GPIB when it is addressed by the system controller (computer). Examples of listeners are printers, monitors, programmable power supplies, and programmable signal generators. A **talker** is an instrument capable of sending data over the GPIB. Examples are DMMs and frequency counters that can output bus-compatible data. Some instruments can send and receive data and are called talker/listeners; examples are computers, modems, and certain measurement instruments. The system controller can specify each of the other instruments on the bus as either a talker or a listener for the purpose of data transfer. The controller is usually a talker/listener.

A typical GPIB arrangement is shown in Figure 14–20 as an example. The three basic bus signal groupings are shown as the *data bus, data transfer control bus,* and *interface management bus.*

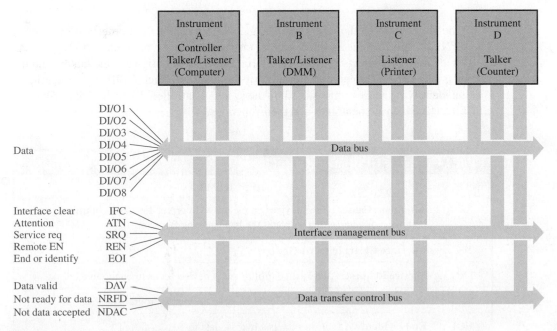

FIGURE 14–20 A Typical IEEE 488 (GPIB) connection.

The parallel data lines are designated DI/O1 through DI/O8 (data input/output). One byte of data is transferred on this bidirectional part of the bus. Every byte that is transferred undergoes a handshaking operation via the data transfer control. The three active-LOW handshaking lines indicate if data are valid ($\overline{DAV}$), if the addressed instrument is not ready for data ($\overline{NRFD}$), or if the data are not accepted ($\overline{NDAC}$). More than one instrument can accept data at the same time, and the slowest instrument sets the rate of transfer. Figure 14–21 shows the timing diagram for the GPIB handshaking sequence, and Table 14–2 describes the handshaking signals.

FIGURE 14–21 **Timing diagram for the GPIB handshaking sequence.**

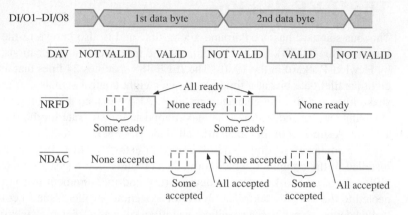

Table 14–2 • The GPIB handshaking signals.	
NAME	**DESCRIPTION**
$\overline{DAV}$	**Data Valid:** After the talker detects a HIGH on the $\overline{NRFD}$ line, a LOW is placed on this line by the talker when the data on its I/O are settled and valid.
$\overline{NRFD}$	**Not Ready for Data:** The listener places a LOW on this line to indicate that it is not ready for data. A HIGH indicates that it is ready. The $\overline{NRFD}$ line will not go HIGH until all addressed listeners are ready to accept data.
$\overline{NDAC}$	**Not Data Accepted:** The listener places a LOW on this line to indicate that it has not accepted data. When it accepts data from its I/O, it releases its $\overline{NDAC}$ line. The $\overline{NDAC}$ line to the talker does not go HIGH until the last listener has accepted data.

The five signals of the interface management bus control the orderly flow of data. The ATN (attention) line is monitored by all instruments on the bus. When ATN is active, the system controller selects the specific interface operation, designates the talkers and the listeners, and provides specific addressing for the listeners. Each GPIB instrument has a specific identifying address that is used by the system controller. Table 14–3 describes the GPIB interface management lines and their functions.

Table 14–3 • The GPIB management lines.	
NAME	**DESCRIPTION**
ATN	**Attention:** Causes all the devices on the bus to interpret data, as a controller command or address and activates the handshaking function.
IFC	**Interface Clear:** Initializes the bus.
SRQ	**Service Request:** Alerts the controller that a device needs to communicate.
REN	**Remote Enable:** Enables devices to respond to remote program control.
EOI	**End or Identify:** Indicates the last byte of data to be transferred.

The GPIB is limited to a maximum cable length of 15 meters, and there can be no more than one instrument per meter with a maximum capacitive loading of 50 pF each. The cable length limitation can be overcome by the use of bus extenders and modems. A bus extender provides for cable-interfacing of instruments that are separated by a distance greater than allowed by the GPIB specifications or for communicating over greater distances via modem-interfaced telephone lines.

The IEEE-488 connector and pin configuration are shown in Figure 14–22.

PIN	DESIGNATION	DESCRIPTION	PIN	DESIGNATION	DESCRIPTION
1	DI01	Data input/output bit.	13	DIO5	Data input/output bit.
2	DI02	Data input/output bit.	14	DIO6	Data input/output bit.
3	DI03	Data input/output bit.	15	DIO7	Data input/output bit.
4	DI04	Data input/output bit.	16	DIO8	Data input/output bit.
5	EOI	End-or-identify.	17	REN	Remote enable.
6	DAV	Data valid.	18	GND	(wire twisted with DAV)
7	NRFD	Not ready for data.	19	GND	(wire twisted with NRFD)
8	NDAC	Not data accepted.	20	GND	(wire twisted with NDAC)
9	IFC	Interface clear.	21	GND	(wire twisted with IFC)
10	SRQ	Service request.	22	GND	(wire twisted with SRQ)
11	ATN	Attention.	23	GND	(wire twisted with ATN)
12	SHIELD		24	Logic ground	

FIGURE 14–22   The IEEE-488 (GPIB) bus connector and pin assignments.

The SCSI Bus

FIGURE 14–23   SCSI symbol.

The SCSI (small computer system interface) bus, generally pronounced "skuzy", is a parallel I/O bus. SCSI is primarily used to transfer data between a computer and peripheral devices, such as hard disks, tape drives, scanners, and CD drives. Figure 14–23 shows the SCSI symbol.

The current SCSI standard is known as SCSI-3, which was preceded by SCSI-1 and SCSI-2. In addition, there are many variations of the standard with designations such as *fast, ultra,* and *wide,* which have different speeds, widths, and connectors, as shown in Table 14–4.

SCSI SIGNALS   A parallel SCSI bus contains nine control signals in addition to data, dc voltages, and ground. These signals are listed in Table 14–5.

Up to eight devices, including the host, can be connected to a SCSI bus, but only two devices can communicate at any given time. Communication is initiated when an initiating device sends a request and the target device acknowledges and performs the request. Single-ended (SE) and differential (LVD or HVD) are the three electrical specifications. Single-ended operation is limited to a cable length of six meters, and differential operation allows up to 25 meters. LVD is low-voltage differential and HVD is high-voltage differential. SCSI devices can operate either asynchronously or synchronously.

Table 14–4 • Several variations of the three basic SCSI standards.

NAME	SPECIFICATION	# OF DEVICES	BUS WIDTH	BAND WIDTH
Asynchronous SCSI	SCSI-1	8	8 bits	4 MBps
Synchronous SCSI	SCSI-1	8	8 bits	5 MBps
Wide	SCSI-2	16	16 bits	10 MBps
Fast	SCSI-2	8	8 bits	10 MBps
Fast/Wide	SCSI-2	16	16 bits	20 MBps
Ultra	SCSI-3	8	8 bits	20 MBps
Ultra/Wide	SCSI-3	8	16 bits	40 MBps
Ultra2	SCSI-3	8	8 bits	40 MBps
Ultra2/Wide	SCSI-3	16	16 bits	80 MBps
Ultra160	SCSI-3	16	16 bits	160 MBps
Ultra320	SCSI-3	16	16 bits	320 MBps

Table 14–5 • SCSI signals.

SIGNAL	DESCRIPTION
BSY	Busy. Bus in use
SEL	Select
RST	Reset
C/D	Control/Data
MSG	Message
REQ	Request
ACK	Acknowledge a request
ATN	Attention

SECTION 14–3 CHECKUP

1. What does PCI stand for?

2. List two alternate designations for the PCI-Express bus.

3. What is a *lane*?

4. What is the IEEE designation for the GPIB?

5. What does SCSI stand for?

14–4 THE UNIVERSAL SERIAL BUS (USB)

Several common serial bus standards are available. The USB is one of the most widely used and is covered in this section. Recall that a serial bus transfers data one bit at a time. As with the parallel bus, a serial bus is not only a set of physical connections but it is also a set of signals and operating parameters that are defined in the bus specification. As with other buses, only the basics are introduced here. More details and information can be found on the Internet.

After completing this section, you should be able to

- **Discuss the USB**
- **Identify and describe USB cables and connectors**
- **Discuss USB signals**

USB (universal serial bus) is a widely used standard serial bus for connecting peripherals to a computer. There are typically two or more USB ports on computers and, with USB hubs, up to 127 devices can be connected. USB allows the devices to be connected or disconnected while the computer is running (hot swapping). Figure 14–24 shows the symbol for USB.

FIGURE 14–24   USB symbol.

The original USB standard was 1.0, which was followed by 1.1. USB 2.0 replaced the two original versions and more recently USB 3.0 was introduced. The earlier versions are still in use, especially 2.0. In terms of data rate, USB has four classifications: low-speed, full-speed, high-speed, and super-speed. Table 14–6(a) shows how the data rate classifications apply to each of the USB versions, and part (b) shows the data rate values.

Table 14–6

	LOW-SPEED	FULL-SPEED	HIGH-SPEED	SUPER-SPEED	DATA RATE	MAXIMUM VALUE
USB 1.0	●	●			Low-speed	0.1875 MBps
USB 1.1	●	●			Full-speed	1.5 MBps
USB 2.0	●	●	●		High-speed	60 MBps
USB 3.0	●	●	●	●	Super-speed	625 MBps

(a)   (b)

Cable length is an important specification for buses. Table 14–7 lists maximum cable lengths for three USB versions and maximum total lengths when multiple cables are strung together using USB hubs. A **hub** is a common connection device with multiple parts.

Table 14–7

	USB 1.1	USB 2.0	USB 3.0
Max cable length	9.8 ft. (3.0 m)	16.4 ft. (5.0 m)	9.8 ft. (3.0 m)
Maximum total length	49.2 ft. (15 m)	82.0 ft. (25 m)	49.2 ft. (15 m)

USB Cable and Connectors

USB versions up to and including 2.0 have a four-wire cable that includes a twisted pair for data transmission, a +5 V wire, and a ground wire color-coded as shown in Figure 14–25(a). The standard type A and type B connectors are shown in parts (b) and (c) with pin designations.

FIGURE 14–25 USB cable and connectors for USB standards through 2.0.

(a) USB cable

(b) Type A connector

(c) Type B connector

USB hosts (computer) and devices (peripherals) have sockets, and all USB cables have a type A plug at one end and a type B plug at the other. The sockets on a host are type A, and the sockets on peripheral devices are type B. Hubs, which are common connection points with multiple parts, have both type A and type B. The USB standard also specifies smaller connectors designated mini and micro.

USB Signals

Serial data is transmitted on the twisted pair (D+ and D−) using half-duplex differential mode to minimize EMI and improve the signal-to-noise ratio. Data is sent in packets using NRZI (non-return-to-zero invert) encoding format with a 3.3 V level (differentially, there is 6.6 V between the two data lines). A packet format can contain the following fields:

Sync field   All packets start with a sync (synchronization) field. The sync field consists of eight bits for low and full speed or 32 bits for high speed and is used to synchronize the receiver clock with that of the transmitter.

PID field   The packet identification field is used to identify the type of packet that is being transmitted. There are four bits in the PID; however to ensure it is received correctly, the four bits are complemented and repeated, making an eight-bit PID code.

ADDR field   The address field specifies to which device the packet is sent. The seven bits in this field allow for 127 devices to be supported. Address 0 is invalid.

Data field   The data field contains up to 1024 bytes of data.

ENDP field   The endpoint field is made up of four bits, allowing 16 possible endpoints. An endpoint is a data source or load. Low-speed devices, however, can only have two additional endpoints on top of the default pipe. (four endpoints max). Endpoints can be described as sources or sinks of data.

CRC field   Cyclic redundancy checks are performed on the data within the packet using from 5 bits to 16 bits, depending on the type of packet.

EOP field   The end of packet field signals the end of a packet.

Four types of USB packets are token, data, handshake, and start-of-frame, as shown in Figure 14–26 with the packet format for each type. Each field is labeled and the number of bits is shown. The token packet indicates the type of transaction, the data packet contains the actual data, the handshake packet acknowledges a transaction, and the start-of-frame packet begins a new frame. The token packet, data packet, and handshake packet each have different types as specified by the PID field.

Sync 8/32	PID 8	ADDR 7	ENDP 4	CRC 5	EOP 3

(a) Token packet

Sync 8/32	PID 8	Data 0-8192	CRC 16	EOP 3

(b) Data packet

Sync 8/32	PID 8	EOP 3

(c) Handshake packet

Sync 8/32	PID 8	Frame number 11	CRC 5	EOP 3

(d) Start-of-frame packet

FIGURE 14–26 Types of USB packets.

USB 3.0

USB 3.0 is the most recent version of the USB standard following USB 1.0, 1.1, and 2.0. The 3.0 version is known as SuperSpeed USB. It is ten times faster than USB 2.0 at 4.8 Gbps. The 3.0 connectors (types A and B) are different because they now have nine contacts instead of four as in version 2.0. The 3.0 type A connector looks about the same as that for 2.0 except that the extra five pins are further inside and make contact only with a 3.0 mating connector. The connector is compatible with a 2.0 device where the front four pins are accessible. Type A and type B connectors for USB 3.0 along with the standard symbol are shown in Figure 14–27. A micro-B connector is also available.

FIGURE 14–27 USB 3.0 connectors and symbol.

(a) Type A   (b) Type B   (c) Symbol

USB 3.0 is, for the most part, backward compatible with USB 2.0, but the speed is limited to the 2.0 specification. The USB 3.0 cable consists of two additional twisted pairs of signal lines for data and an additional ground, for a total of nine wires.

The previous USB versions were half-duplex, meaning that data could be transmitted in one direction at a time. Version 3.0 is full-duplex that can send and receive data simultaneously. The two additional twisted pairs are for sending high-speed data, and the third twisted pair is for receiving data.

SYSTEM EXAMPLE 14–1

A TYPICAL COMPUTER SYSTEM

The USB in a typical computer system is shown in Figure 14–28. The computer acts as a host and uses type A connectors. The hub functions as both a host and a device.

FIGURE 14–28

---

1. What does USB stand for?

2. What are the functions of the four pins in a USB 2.0 connector?

3. Why is a twisted pair used in a USB cable?

4. Describe the basic differences between USB 2.0 and USB 3.0.

14–5 OTHER SERIAL BUSES

Although the USB is perhaps the most widely used serial bus, there are numerous others. A few of the more important bus standards are introduced in this section. For more detailed information on any bus standard, search the Internet.

After completing this section, you should be able to

- Discuss the RS-232, 422, 423, and 485 bus standards
- Describe the SPI bus
- Discuss the I²C bus
- Explain the CAN bus
- Describe the Firewire (IEEE 1394) bus

The RS-232/422/423/485 Buses

RS-232 Also known as **EIA-232**, the **RS-232** bus was once standard on computers for connection to peripheral devices. The standard provides for single-ended data transmission in either synchronous or asynchronous formats. It has been largely replaced by the USB because of its limited speed, relatively large voltage requirements, and large connector size. However, RS-232 devices are still used in industrial and telecommunication applications as well as in scientific instrumentation. The devices connected by the RS-232 are classified as DTE (data terminal equipment) or DCE (data communication equipment). Since newer computers have no RS-232 ports, USB-to-RS 232 converters can be used to connect to older RS-232 compatible peripherals, if necessary. The standard is designed for one transmitting device and one receiving device with a maximum cable length of 50 feet (15.24 meters) between them.

The maximum RS-232 data rate is 20 kbps. The data format typically consists of seven or eight bits of data, a start bit, a parity bit in some cases depending on the protocol, and a stop bit. A transmitted signal level between +5 V and +15 V represents a binary 0 and between −5 V and −15 V represents a binary 1. A received signal level between +3 V and +13 V represents a binary 0, and a signal level between −3 V and −13 V represents a binary 1. The data are transmitted in NRZ format, as the example in Figure 14–29 shows.

The standard 25-pin connector for RS-232 is shown in Figure 14–30(a). A smaller 9-pin connector is shown in part (b).

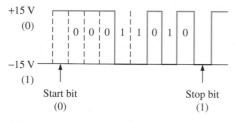

FIGURE 14–29 **Example of RS-232 transmitted data format. A parity bit is not included.**

RS-422/423/485 The **RS-422** provides for differential transmission and provides for greater distances (longer cable length) and higher data rates than the RS-232. Also, the standard defines the number of receiving devices as ten for a line with one driver (transmitting device) compared to one receiving device for the RS-232. The **RS-423** is similar to the RS-232 in that it is single-ended, but it has a higher data rate and longer cable length. The **RS-485** is a differential standard and can accommodate multilple drivers and receivers. Table 14–8 summarizes some of the features of the RS (EIA) buses.

(a) 25-pin   (b) 9-pin

FIGURE 14–30   Standard RS-232 connectors.

Table 14–8

SPECIFICATIONS	RS-232	RS-423	RS-422	RS-485
Operation	Single-ended	Single-ended	Differential	Differential
Drivers/Receivers	1/1	1/10	1/10	32/32
Cable length	50 ft	4000 ft	4000 ft	4000 ft
Max data rate	20 kbps	100 kbps	10 Mbps	10 Mbps
Driver output signal level (+/− min/max)	5 V/15 V	3.6 V/6 V	2 V/6 V	1.5 V/6 V

The SPI Bus

The **SPI** (serial-to-peripheral interface) bus is a synchronous serial communications bus that uses four wires for communication between a "master" device and a "slave" device. This standard was developed by Motorola and operates in full-duplex mode up to a data rate of 10 Mbps and can accommodate multiple slaves. The four signal wires are

1. MOSI (master out slave in) is initiated by the master and received by the slave.
2. MISO (master in slave out) is initiated by the slave and received by the master.
3. SCLK (serial clock) is generated by the master for synchronizing data transfers.
4. SS (slave select) is generated by the master to select an individual slave.

Other names are sometimes assigned to these signals such as SDI (serial data in) for MOSI and SDO (serial data out) for MISO. Figure 14–31 shows a master with a single slave.

The SPI bus is typically used in embedded systems and on PC boards for communication between microprocessors or microcontrollers and peripheral IC chips or between two processors. Much PC board "real estate" can be saved compared to using an internal parallel bus with many more connections.

Two SPI configurations that allow the master to send data to multiple slaves are the daisy-chain configuration and the parallel configuration. In the daisy-chain

FIGURE 14–31   SPI master/slave configuration.

configuration, the clock (SCLK) and the slave select (SS) are connected to all the slaves. The data is cascaded from one slave device to the next, as shown in Figure 14–32(a). The parallel configuration requires the master to generate a separate slave select signal to each slave, so the data is sent to each slave independently, as shown in part (b).

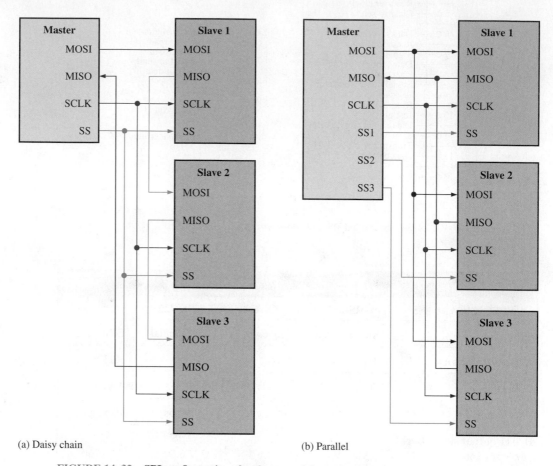

(a) Daisy chain   (b) Parallel

FIGURE 14–32   **SPI configurations for the case of three slave devices.**

SPI applications include digital audio, signal processing, and telecommunications. SPI is used to communicate with various types of peripherals, such as sensors, camera lenses, flash memory, LCD displays, video games, and many others. Many microprocessors and microcontrollers include SPI controllers that can be used as either a master or a slave.

The I²C Bus

I²C (pronounced I squared C) is also stated at I2C (I two C) and stands for *inter-integrated circuit*. It is an internal serial bus primarily for connecting ICs on a PC board. A main advantage is that it requires only two lines (plus ground) and therefore saves considerable board space compared to a parallel bus. Two signals (SDA and SCL) are used to communicate between compatible devices. Data is sent serially on the SDA line, and a clock is sent on the SCL line. Like SPI, I²C is based on the master/slave concept where the master device drives the clock line and the slaves respond to the master. Only the master can initiate a transfer over the bus, but slaves can transfer data under control of the master using clock rates up to 100 kHz in the standard mode. Two other modes, enhanced and high-speed, allow 400 kbps and 3.4 Mbps, respectively. A general bus arrangement is shown in Figure 14–33. The pull-up resistors provide for open-collector or open-drain devices.

FIGURE 14–33   An I²C bus arrangement.

DATA TRANSFER   When transferring data from master to slave, the master device sends a start bit, followed by a slave address, and a write bit. The master waits for an acknowledge (ACK) signal from the slave and then sends the data and waits for an acknowledge before sending a stop bit, as illustrated in Figure 14–34 (a). Similarly, when the master requires data from the slave, it sends a start bit followed by the address and a read bit. The slave returns an acknowledge followed by the data. When the master receives the data, it issues an acknowledge and a stop bit, as shown in Figure 14–34(b).

Start	Address	Write	ACK	Data	ACK	Stop

(a) Data transfer from master to slave

Start	Address	Read	ACK	Data	ACK	Stop

(b) Data transfer from slave to master

FIGURE 14–34   I²C data transfers. Yellow is from master. Gray is from slave.

SYSTEM EXAMPLE 14–2

DIGITAL AUDIO SYSTEM

An application of the I²C bus is shown in Figure 14–35. The digital audio system uses an I²C bus to minimize the size of the PC board in order to fit it into a limited space. The master device is a microcontroller that controls the volume, balance, tone, tuning, display, memory, and analog-to-digital converter (ADC) functions, which are the slave devices.

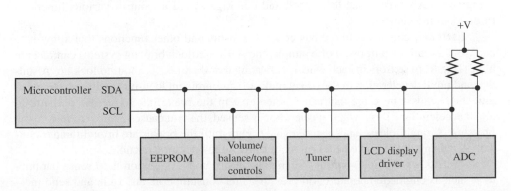

FIGURE 14–35

The CAN Bus

The **controller area network (CAN)** bus, a differential serial bus, was developed for automotive applications and is also commonly used in aerospace systems as well as other applications. The bus consists of a terminated twisted pair of signal lines, called CAN H and CAN L, plus ground. Vehicles sold in the United States are required by the SAE (Society of Automotive Engineers) to use the CAN bus protocol. The European Union has similar requirements.

Devices, called *nodes*, can be connected to the bus but are not assigned specific addresses as in the I^2C bus. There are two CAN specifications in use. The standard or basic CAN 2.0A has 11-bit message identifiers and can operate up to 250 kbps, and the full CAN has 29-bit message identifiers and can be used up to 1 Mbps. The message identifier is a label for the contents of a message and goes to each node on the bus. Each receiving node performs a test on the identifier to determine if it is relevant to that node and is used to arbitrate the bus to determine if the message is of highest priority. All of the nodes on the bus can transmit and receive messages. The bus is available to a node with a message with the highest priority (dominant) and can override a message with lower priority (recessive). When the dominant message has been processed, the recessive message is retransmitted.

DATA FRAME   The standard data frame is shown in Figure 14–36. Data is transmitted in NRZ format. The frame begins with a start-of-frame (SOF) bit followed by an arbitration field and a control field. The arbitration field contains the message identifier and a remote transmission request (RTR) bit. The control field has two reserve bits and a data length code (DLC) that specifies the length of the data field that follows and can contain up to eight bytes. The cyclic redundancy check (CRC) field provides for error detection. The acknowledge (ACK) verifies the receipt of correct data, and the frame ends with the end-of-frame field (EOF).

SOF	Arbitration field	Control field	Data field	CRC field	ACK	EOF
(1 bit)	Identifier (11 bits) RTR (1 bit)	Reserve (2 bits) DLC (4 bits)	(0–8 bytes)	(16 bits)	(2 bits)	(7 bits)

FIGURE 14–36   Standard CAN data frame format.

SYSTEM EXAMPLE 14–3

AN AUTOMOBILE CONTROL SYSTEM

A modern automobile typically has many control units (usually several dozen) for various subsystems, which include the engine control unit and other control units for transmission; ABS; cruise control; power steering; audio system; window, door, and mirror controls; airbags; and others. Figure 14–37 is a block diagram of a partial automotive control system using two CAN buses, one low-speed and one high-speed to control various functions throughout the vehicle.

Each unit connected to the bus contains sensors and other functions that allow it to carry out its unique purpose. For example, the ABS (antilock braking system) can receive a message from sensors in each wheel, indicating that the brake is about to lock up. A sudden and rapid deceleration in the wheel indicates an imminent lock-up condition. The ABS unit then sends a message that causes the valve in the brake line to release pressure to allow acceleration. Then, when acceleration is sensed, the unit causes a pump to restore the pressure. A rapid release-and-restore cycle occurs until the brakes are brought under control. A pulsing of the brake pedal can be felt when the operation occurs.

As another example, part of the engine control unit's operation is to sense parameters, such as engine temperature, oil pressure, fuel consumption, and rpm, and send messages to the driver unit. All of the units on the bus operate as a system to keep the vehicle running as smoothly and as safely as possible, while providing a comfortable environment for the driver and passengers.

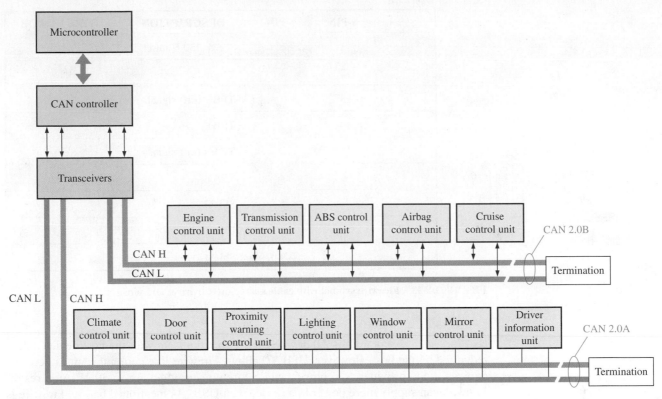

FIGURE 14–37

The Firewire Bus

Firewire, also known as IEEE-1394 and iLink, is a high-speed external serial bus developed by Apple Inc. Firewire is used in high-speed communications and real-time data transfer. It is used in professional audio and video equipment, camcorders, DVD players, external hard drives, computers, and in some auto and aircraft applications. It works very similar to the USB except that it has a higher data rate and can handle more data.

Three types of connectors are used in the Firewire standard: a 4-pin connector, a 6-pin connector, and a 9-pin connector. The cable for the 4-pin connector consists of two twisted pairs that carry data. The cable for the 6-pin connector has the two twisted pairs for data plus a power line and a ground line. The cable for the 9-pin connector has the same wires as the 6-pin configuration plus two wires that provide for a grounded shield and one wire that is currently unused. The Firewire symbol is shown in Figure 14–38(a). End views of the three connector types are shown in part (b), and the pin designations are shown in part (c).

The Firewire bus address has a total of 64 bits. Ten are for bus ID, six are for node ID, and 48 are for individual addresses. This allows up to 1023 buses each having up to 63 nodes. There are six transfer modes in the IEEE-1394 standard and its revisions: S100, S200, S400, S800, S1600, and S3200. The S100 is the base rate of 98.304 Mbps. The S200 is twice the base rate at 196.608 Mbps, and the S400 is four times the base rate at 393.216 Mbps. The S800 is 786.432 Mbps, and the S1600 and S3200 are 16 and 32 times the base rate respectively (1.6 Gbps and 3.2 Gbps). Firewire cable length cannot exceed 15 ft (4.572 m). To increase this length, up to 16 cables can be connected together.

FIREWIRE VERSUS USB   In general, any capable node can control the bus in a Firewire system, but a single host is used to control the bus in USB. USB networks use a tiered-star topology, and Firewire uses a tree topology. (Network topologies are covered in section 14–6.) A Firewire device can communicate with any node at any time if the

4-PIN	6-PIN	9-PIN	DESCRIPTION	WIRE COLOR
	1	8	+V dc (30 V max)	White
	2	6	Ground	Black
1	3	1	TPB− (diff signal)	Orange
2	4	2	TPB+	Blue
3	5	3	TPA− (diff signal)	Red
4	6	4	TPA+	Green
		5	Shield	
		7	Unused	
		9	Shield	

FIGURE 14–38   Firewire symbol with cable and connector wires and pins.

conditions allow, but a USB 2.0 device cannot communicate with the host device unless requested by the host. However, USB 3.0 allows Firewire-like communications between devices. USB provides 5 V power, while Firewire provides up to 30 V. As a result, Firewire can supply more power to a device than USB. As mentioned before, Firewire is faster than USB.

SECTION 14–5 CHECKUP

1. List all the buses introduced in this section.

2. What does SPI stand for?

3. What does I2C stand for?

4. What does CAN stand for?

5. What is another designation for Firewire?

14–6 NETWORK TOPOLOGIES

A **network** is a set of computers and associated devices that are interconnected in a specified way in order to communicate to share information and resources. When two or more computers are connected together and can communicate with each other, it is a basic network. A simple example of a network is where a printer has communication links to several computers so that all of the computer operators can use the common printer. Networks can be classified according to the communication medium (wired or wireless), the communication protocol, or the topology (pattern of interconnection).

After completing this section, you should be able to

- Distinguish between physical network topology and logical network topology
- Describe the following physical network topologies: bus, ring, star, tree, and mesh
- Describe the shared media network topology
- Describe the token-based network topology

Network topology is the layout pattern of interconnections and the protocol used to allow various devices to communicate with each other. Two classifications of network topology are physical and logical. The physical structure or the way in which devices are connected to the network with actual cables is called *physical topology*. The way in which data is moved through the network is called *logical topology* and is related to network protocol (set of rules for communication). There are several configurations of network topology in term of the physical layout. Examples of logical topologies are Ethernet (has both physical and logical characteristics), which is widely used in local area networks (LANs), and the Internet Protocol suite among others.

Physical Network Topologies

BUS NETWORK TOPOLOGY   All devices in the **bus network topology** are connected to a single common bus. This is the simplest way to connect multiple devices (usually referred to as clients, nodes, or workstations). When a device transmits a data packet, all of the other devices have access to the transmission. If two devices try to send data at the same time, a "collision" will occur and the devices must arbitrate to see which one will resend its data. A failure in the bus results in the entire network going down. Each end of a bus network must be **terminated** with a resistor to keep the signal from reflecting back when it reaches the end of the cable. Figure 14–39 illustrates the bus network topology.

FIGURE 14–39   Example of bus network topology.

STAR NETWORK TOPOLOGY   This topology is used in many home and small business networks. **Star network topology** is also known as "hub and spoke." The main feature is a central connection point called a *hub*. Each device connects to the hub through an individual cable so that if there is a problem in one of the connections, the other devices can still communicate. However, if the hub fails, the entire network goes down. The hub connects a signal from any device and passes on to all the other devices in the network. There can be any number of devices within the limitations of the hub. Figure 14–40 illustrates the star network topology.

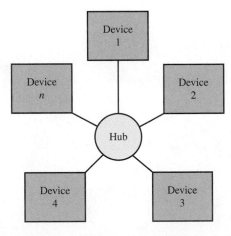

FIGURE 14–40   Example of star network topology.

RING NETWORK TOPOLOGY   The devices in the **ring network topology** are daisy-chained, and each device communicates only with its two neighboring devices. Data is sent from one device to the next device around the ring. Each device can send a data frame through the use of a special series of bits called a *token*. To send data, a device must have the token. The first device sends data that includes the token to the second device, which extracts any data addressed to it, inserts any data that it needs to send, and transmits it to the third device. This daisy-chain operation continues until the data frame returns to the original sending device. Data always moves in the same direction, either clockwise or counterclockwise. Ring networks are sometimes used in office buildings or schools. Figure 14–41 illustrates ring network topology.

TREE NETWORK TOPOLOGY   This type of network configuration is a hybrid network that combines the bus topology and the star topology and is shown in Figure 14–42. The **tree topology** allows for the expansion of a network to meet increasing needs. Another possible hybrid network is the star ring topology.

FIGURE 14-41   **Example of ring network topology.**

FIGURE 14-42   **Example of tree network topology.**

MESH NETWORK TOPOLOGY   There are multiple *routes* for data in the **mesh network topology.** Every device connects to all of the other devices so that data can take more than one possible path from one device to another. Mesh networks are used in some WANs (wide area networks), in particular the Internet. Figure 14–43 illustrates a simple mesh network topology.

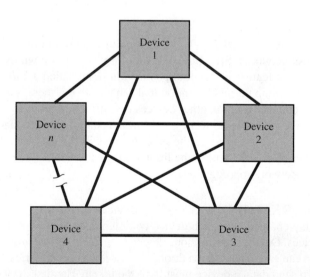

FIGURE 14-43   **Example of a mesh network topology.**

Logical Network Topology

The logical topology describes how the devices interact in order to send and receive data in a network and is based on protocols. Two categories of **logical network topology** are shared media (also known as *broadcast*) and token-based.

In the **shared media topology,** which is used in Ethernet, all devices can access the physical network at any time as long as no other device is attempting access at the same time. Each device checks to see if the data being transmitted is for it or not. If two devices transmit data at the same time, a "collision" of the data packets occurs, and the devices have to arbitrate so that one device must back off and send later.

In the **token-based topology**, there is a code called a token that travels around the network. If a device needs to send data, it grabs the token and attaches the data packet to it. Only the device that has the token can send data. If a device has no data to transmit, the token is passed along to the next device in the network.

SECTION 14–6 CHECKUP

1. Define physical network topology.

2. Define logical network topology.

3. List the major types of physical network topology.

4. Name the categories of logical network topology.

14–7 NETWORK PROTOCOL TECHNOLOGIES

A network protocol technology in data communication is a set of structures, rules, and conventions that control how computers and other network devices communicate with each other to exchange information. All of the devices in a network must conform to the standards, both physical and logical, required by the specific network technology used. There are a variety of network protocol technologies, and two of the most widely used are introduced in this section.

After completing this section, you should be able to

- **Define OSI**
- **Discuss the Ethernet**
- **Describe the TCP/IP**

Most protocols for data communication are modeled to some extent after the **OSI** (open systems interconnection) model. This is a 7-layer model in which all protocol operations and specifications for communicating on a network are broken down into seven parts called layers. Because it is loosely specified, OSI serves only as a guide for the development of network protocols and other specifications. The OSI layers are as follows:

1. **Physical**   This layer defines the actual network hardware and physical characteristics, such as type of cables or connections, voltage levels, and timing.

2. **Data**   The protocol is assigned to the data and the data packet sequencing. Also, the type of network is specified.

3. **Network**   Logical protocol, addressing, and routing are determined at this layer.

4. **Transport**   This layer provides data flow control, error checking, and data recovery.

5. **Session**   Communication with a receiving device is established, maintained, and ended by this layer.

6. **Presentation**   This layer converts data to a standard format that can be understood by all the other layers.

7. **Application**   This layer interacts with the application or operating system when files are transferred, messages are read, or any other network-related activity is required.

As mentioned, network protocol technologies use OSI only as a guide or model. Many do not exactly match the OSI structure although, generally, some or all of the specified OSI functions are implemented in some way. Protocols can be grouped in suites or stacks in accordance with their purpose. For example, a given protocol may belong to the **LAN** (local area network) or to the **MAN** (metropolitan area network). Various groups develop protocols, such as the IETF (Internet Engineering Task Force) that deals with Internet protocol and TCP/IP, the IEEE (Institute of Electrical and Electronics Engineers) that deals with LAN protocols, and the ITU-T (International Telecommunications Union-Telecommunications that deals with **WAN** (wide area network). The ISO (International Organization of Standardization) also contributes to many protocol standards.

The Ethernet

The **Ethernet** (IEEE 802.3 standard) is a widely used local area network (LAN) technology or protocol that uses a process called CSMA/CD (carrier sense multiple access/collision detection) to access the network. Ethernet is typically employed in networks with bus, star, or tree topologies and can use various media such as twisted pair, coaxial, or fiber optic cables to transfer data at rates up to 10 Mbps. Fast Ethernet has speeds up to 100 Mbps (IEEE 802.3u), Giga Ethernet (IEEE 802.3z) goes up to 1000 Mbps, and 10-Giga Ethernet (IEEE 802ae) up to 10,000 Mbps. Three basic elements in an Ethernet system are the physical medium, a set of medium-access control rules, and a data frame.

DATA FRAME   An Ethernet frame consists of 64 to 1518 bytes excluding the preamble and SOF delimiter. The data itself is from 46 to 1500 bytes, and 24 bytes make up the CRC and other fields. Figure 14–44 shows the format of an Ethernet data frame. The preamble (alternating 1s and 0s) field signals to the receiving devices that a data frame is being sent and allows for synchronization. The start-of-frame delimiter (alternating 1s and 0s with two 1s at end) field alerts the receiver that the next bit begins the destination address field, which identifies which device for which the frame is intended. The source address field identifies the source of the data. The length/type field specifies the number of data bytes or the type of frame. The data field contains the actual data, and the frame check sequence field contains a cyclic redundancy check code.

Preamble PRE	SOF delimiter SFD	Destination address DA	Source address SA	Length/type	Data	Frame check sequence FCS
7 bytes	1 byte	6 bytes	6 bytes	2 bytes	46–1500 bytes	4 bytes

FIGURE 14–44   Ethernet data frame.

THE CSMA/CD   As you know, this is the process used by Ethernet to access the network. The carrier sense (CS) part of the process means that a device that wants to send data must check to see if it has a medium (cable) connected over which to transmit. The multiple access (MA) part of the process provides for any device to use the network as long as no other device is transmitting. The collision detection (CD) takes care of discarding data corrupted by a "collision" that occurs when two devices attempt to transmit at the same time, after which the devices arbitrate for transmission time.

BASIC OPERATION   All of the devices connected to an Ethernet network are independent and have equal opportunity to use the shared medium. When data is sent, all the devices look at the destination address in the data frame to determine if it belongs to them. If a device finds that the DA is not meant for it, then it shuts off the data frame. The device that matches the DA, accepts and processes the data frame. Once a frame has been sent,

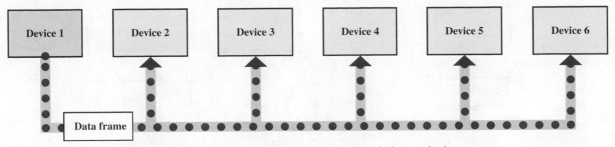

(a) A data frame is transmitted over the network from device 1 to device 4 and all the devices receive it.

(b) Each device "reads" the frame up to and including the destination address. All devices but device 4 determine that the DA is not their address and ignore the data frame.

(c) Device 4 sends an acknowledge receipt message back to device 1 and processes the data.

FIGURE 14–45   Simple illustration of a data transfer in a network.

received, and acknowledged, the medium is again available to all the devices. This operation is illustrated in Figure 14–45.

If a network has a bus topology, data collisions can occur when two devices send data at the same time. This is dealt with through the CSMA/CD protocol in Ethernet as illustrated in a very simple way in Figure 14–46.

ETHERNET STAR NETWORK   Using a star network topology reduces the risk of collision and helps avoid other system failures. Switches have largely replaced hubs and repeaters in many networks because switches route data from the transmitting device only to the device(s) for which the transmitted data are intended and not to all of the devices on the network, as with a hub. The switch reads the DA and routes the data frame to the appropriate device, as shown in Figure 14–47. None of the other devices receive the data from device 1. If at the same time, device 4 is sending data to device 2, the switches route the data through a separate connection and a collision is avoided, as shown. If two devices try to send data to the same receiving device at the same time, a collision will occur. However, the chance of a collision is greatly reduced by using the star topology with switches.

(a) Devices 2 and 5 try to send a data frame at the same time. Data frame collision creates corrupted data on the bus at time t_1. CSMA/CD detects collision, and bus arbitration begins between the two devices.

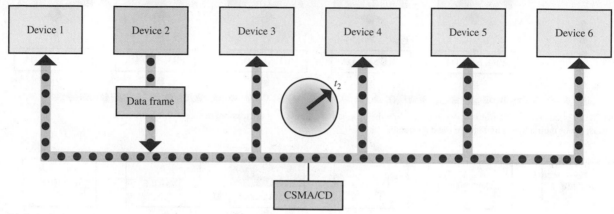

(b) Only device 2 is allowed to send data at time t_2.

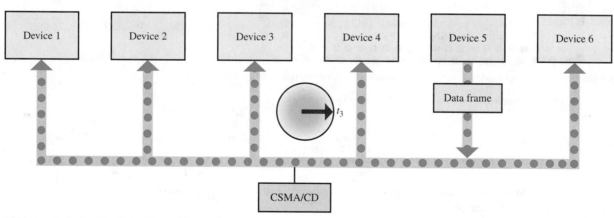

(c) Next, only device 5 is allowed to send data at time t_3.

FIGURE 14–46 **Simple illustration of bus network collision management by Ethernet CSMA/CD protocol.**

ETHERNET STRUCTURE   Two layers that are defined by the IEEE 802.3 standard are the physical layer and the data link layer. In the physical layer, Ethernet defines

- Interconnection methods
- Cables and connectors
- Data encoding (Manchester)
- Electrical properties

In the data link layer, Ethernet

- Creates a data frame for IP data (Internet protocol)
- Addresses devices on the network

FIGURE 14–47   **Simple illustration of switched star topology. Two devices are sending data simultaneously to different destinations.**

- Passes data within the LAN and to network interface controller (NIC) for IP data
- Provides CSMA/CD protocol
- Detects errors

ETHERNET CABLES AND CONNECTORS   Figure 14–48 shows an Ethernet connector and cable containing four twisted-pair conductors. The pin numbers and signals are shown as well as the color-coded twisted pairs.

FIGURE 14–48   **An Ethernet cable and connector.**

Four-twisted pair cable

87654321

Signal	Wire color
Transmit +	White/green
Transmit -	Green
Receive +	White/orange
Unused	Blue
Unused	White/blue
Receive -	Orange
Unused	White/brown
Unused	Brown

TCP/IP

TCP/IP stands for *Transmission Control Protocol/Internet Protocol* and is the communication protocol for the Internet. Actually, there are two protocol suites in TCP/IP. TCP provides for communication between applications; IP handles communication between computers and is responsible for sending data packets to the right destination across a network. This set of protocols enables different computers and other devices to communicate with each other in many ways including e-mail.

Each computer on the Internet has a unique address called the IP address. When data is sent or received, the message is divided into small packets (for example, up to 256 bytes) containing the IP address of the source and the IP address of the destination device. The IP packets are passed to a gateway or router for a local network. For example, an Ethernet LAN allows data packets of up to 1500 bytes; when it receives a group of smaller packets from the Internet, it reassembles them into the standard Ethernet format.

TCP/IP has four layers based on the OSI model; they are application, transport, Internet, and network interface. Programs communicate with the application layer where there are protocols such as SMTP for e-mail, FTP for file transfer, and HTTP for web browsing. After processing a program request, the application layer protocol communicates with another protocol on the transport layer. The transport layer divides the data into packets and sends them to the Internet layer. The Internet layer protocol (IP) receives the data from the transport layer and adds the source and destination addresses (IP addresses). The packets, called *datagrams*, are then sent to the network interface layer, which then passes them to the local network (Ethernet) where they are assembled into frames. The process reverses when data is being sent from the LAN. Figure 14–49 illustrates the basic process for a file transfer.

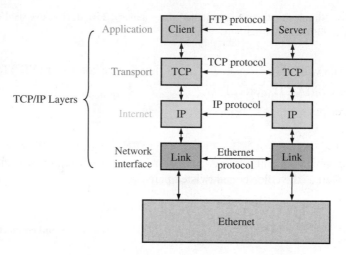

FIGURE 14–49   Diagram of network structure for Ethernet LAN running an FTP (file transfer protocol) on the Internet via TCP/IP.

A **client/server** model like shown in Figure 14–49 is used on the Internet. Client/server networking is an application architecture that handles tasks between service providers called *servers* and service requesters, call *clients*. Clients are usually PCs running network software applications that request information over the network. Servers store information, including files and databases, and provide information to clients upon request.

IP ADDRESSES   Internet protocol (IP) addresses have 32 bits expressed as four 8-bit numbers. Each of the four numbers are usually between 1 and 254 in the following binary format:

$$\text{wwwwwww . xxxxxxxx . yyyyyyyy . zzzzzzzz}$$

An example of an IP address in decimal form is **162.18.248.1**

An **IP address** is assigned to each device in a network that uses the Internet protocol and serves to identify the network and the location of the device within the network. A typical Ethernet/TCP/IP network is shown in Figure 14–50. The **router** routes the data packet to the correct destination based on the IP address. The **gateway** is a network point that provides access to another network. It can be thought of as an interface between two different networks such as, in this case, the Ethernet and the Internet.

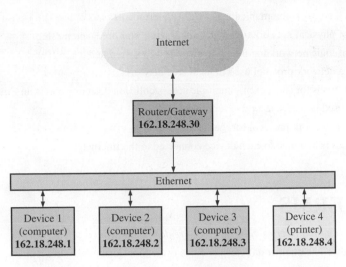

FIGURE 14–50   Example of an Ethernet/TCP/IP network with IP addresses assigned. (IP addresses are in decimal form.)

SECTION 14–7 CHECKUP

1. What is the IEEE 802.3 standard commonly called?

2. What does OSI stand for?

3. What does TCP/IP stand for?

4. Describe the structure of an IP address.

SUMMARY

- The physical properties of a bus include the number of conductors, the configuration and length of the conductors, and the types and configurations of the connectors.
- The electrical properties of a bus include signal format, signal voltage levels, clock frequency, data transfer speed, bandwidth, data frame format, data rate handshaking protocol, error detection, and impedances and line termination.
- A system bus, also known as a frontside bus, is a computer bus that is classified as a local bus. It connects the CPU, memories, and I/O and contains three component buses: data, address, and control.
- General bus characteristics are width, frequency, transfer speed, and bandwidth.
- A synchronous bus contains a clock, and an asynchronous bus is unclocked.
- Single-ended operation uses one wire for data and one wire for ground where the signal voltage on the wire is with respect to ground.
- Differential operation uses two wires for data and one wire for ground.
- Tristate devices are used to interface circuitry to a bus.
- A bridge is an interface between buses.
- The PCI (peripheral component interconnect) bus is an internal synchronous bus for interconnecting chips, expansion boards, and processor/memory subsystems.
- Both PCI and PCI-X use a shared bus configuration.
- PCI-Express uses a lane configuration via a switching device.
- In addition to PCI buses, other important parallel buses are AGP, ISA, IEEE-488, and SCSI.
- USB is a widely used standard serial bus for connecting peripherals to a computer.
- A USB packet can include the following fields: synchronization, packet identification, address, data, end point, CRC, and end of packet.
- Four types of USB packets are token, data, handshake, and start-of-frame.
- In addition to USB, some important serial buses are RS-232/422/423/485, SPI, I^{2}C, CAN, and Firewire.

- Devices on a network communicate using certain protocols (sets of rules).
- The types of physical network topology are bus, ring, star, tree, and mesh.
- The types of logic network topology are broadcast or shared and token-based.
- Ethernet is a network protocol used in local area networks.
- CSMA/CD stands for carrier sense multiple access/collision detection and is an Ethernet protocol.
- Ethernet is used in bus, star, or tree topologies.
- TCP/IP is the network protocol for the Internet.
- An IP address is assigned to each device connected to the Internet.

KEY TERMS

Key terms and other bold terms in the chapter are defined in the end-of-book glossary.

Bus bandwidth The number of bytes per clock cycle times the number of clock cycles per second.

Bus frequency The clock frequency at which the bus can operate.

Bus protocol A set of rules that allow two or more devices to communicate through a bus.

Bus transfer speed The number of bytes per clock cycle.

Bus width The number of bits that a bus can transmit at one time.

CAN Controller area network. A bus standard used in automotive and other applications.

Client/Server A networking application architecture that handles tasks between service providers called *servers* and service requesters called *clients*.

Ethernet A widely used standard local area network (LAN) technology or protocol; also known as IEEE-802.3.

Firewire A high-speed external serial bus standard developed by Apple Inc. and used in high-speed communications and real-time data transfer; also known as IEEE-1394.

GPIB General-purpose interface bus based on the IEEE-488 standard.

Handshake A routine by which two devices initiate and complete a bus transfer.

IP address Internet protocol address. An address assigned to each device in a network that uses the Internet protocol and serves to identify the network and the location of the device within the network.

ISA bus Industry standard architecture bus; also known at the AT bus.

I^2C Inter-integrated circuit bus. An internal serial bus primarily for connecting ICs on a PC board.

LAN Local area network.

Local bus The internal bus of a computer system which includes the system bus; the PCI bus, and the ISA bus, among others.

Network A set of computers and associated devices that are interconnected in a specified way in order to communicate to share information and resources.

Network topology The physical and logical arrangement of devices in a network.

OSI Open systems interconnection. A model in which all protocol operations and specifications for communicating on a network are broken down into seven parts called layers.

Parallel bus A bus that consists of multiple conductors and carries several data bits simultaneously, one on each conductor.

PCI Peripheral component interconnect. An internal synchronous bus for interconnecting chips, expansion boards, and processor/memory subsystems.

PCI-Express Also designated as PCIe or PCI-E. This bus differs from the PCI and PCI-X buses in that it does not use a shared bus.

PCI-X A high-performance enhancement of the PCI bus that is backward compatible with PCI.

RS-232 A bus standard, also known as EIA-232, used in industrial and telecommunication applications as well as scientific instrumentation, but largely replaced by USB in computer applications.

SCSI Small computer system interface bus; a parallel I/O bus.

Serial bus A bus that carries data bits sequentially one at a time on a single conductor.

SPI Serial-to-peripheral interface bus. A synchronous serial communications bus that uses four wires for communication between a "master" device and a "slave" device.

System bus   The data, address, and control buses in a computer.

TCP/IP   Transmission Control Protocol/Internet Protocol. The communication protocol for the Internet.

Tristate buffer   A circuit used to interface one device to another to prevent loading.

USB   Universal serial bus. A widely used standard serial bus for connecting peripherals to a computer.

TRUE/FALSE QUIZ

Answers are at the end of the chapter.

1. A bus connects two or more devices to allow them to communicate.

2. A bus is only defined by the wires and connectors.

3. The system bus in a computer is also known as the frontside bus.

4. A parallel bus is always faster than a serial bus.

5. Bus width is the width of each conductor in a parallel bus.

6. Handshaking is part of bus protocol.

7. A single-ended transmission is simpler and lower in cost compared to a differential transmission.

8. A tristate driver has a HIGH state, a LOW state, and a shorted state.

9. FSB stands for *frontside bus*.

10. PCI stands for *peripheral computer interface*.

11. Two types of PCI buses are PCI-X and PCI-E.

12. IEEE-488 is known as the GPIB.

13. SCSI stands for *serial computer system interface*.

14. USB is a widely used standard serial bus for connecting peripherals to a computer.

15. The SPI bus is typically used in embedded systems and on PC boards for communication between microprocessors or microcontrollers and peripheral IC chips or between two processors.

16. An internal serial bus primarily for connecting ICs on a PC board is the I^2C bus.

17. Most automotive systems use the Firewire bus.

18. CAN stands for *computer area node*.

19. The types of physical network topologies are bus, ring, star, tree, and mesh.

20. A hub is used in ring network topology.

21. OSI is a model for network protocols.

22. LAN is a local area network.

23. Ethernet is a WAN protocol.

24. CSMA/CD is part of the Ethernet protocol technology.

25. Data collisions are more common in star topology than in the bus topology.

26. TCP/IP is the Internet protocol.

27. An IP address is assigned to each device connected to the Internet.

28. A gateway is a network point that provides access to another network.

SELF-TEST

Answers are at the end of the chapter.

1. Properties that define a bus include
    - (a) type of connectors
    - (b) length and type of cable or connection
    - (c) data rate and encoding
    - (d) all of these

2. A typical local bus includes

   (a) PCI and USB

   (b) PCI, data bus, address bus, and control bus

   (c) data bus, address bus, control bus, and SPI

   (d) SPI, CAN, data bus, and power bus

3. AGP is the acronym for

   (a) advanced graphics protocol   **(b)** access granted peripheral

   (c) accelerated graphics port   **(d)** address group packet

4. The speed of a parallel bus can be limited by

   (a) cross talk   **(b)** EMI   **(c)** skew   **(d)** all of these

5. The method by which two devices initiate and complete a bus transfer is

   (a) handshaking   **(b)** saluting

   (c) give and take   **(d)** multiple access protocol

6. PCI is the acronym for

   (a) peripheral controller interface   **(b)** peripheral computer interface

   (c) protocol compatible interface   **(d)** peripheral component interconnect

7. Two types of bus bridges in a computer system are known as

   (a) northbridge and southbridge   **(b)** hook and ladder

   (c) eastbridge and westbridge   **(d)** local bridge and system bridge

8. In a PCI system, the individual paths from switch to peripherals are called

   (a) pipes   **(b)** lanes   **(c)** highways   **(d)** channels

9. The following is not a classification of USB:

   (a) low-speed   **(b)** full-speed

   (c) high-speed   **(d)** intermediate speed

10. A USB cable contains

   (a) two twisted pairs

   (b) one twisted pair

   (c) two twisted pairs and two straight wires

   (d) two straight wires

11. Four types of USB packets are

   (a) token, data, handshake, and start-of-frame

   (b) token, data, handshake, and control

   (c) identification, address, synchronization, and data

   (d) none of these

12. The RS-232 encoding method is

   (a) RZ   **(b)** Manchester   **(c)** biphase   **(d)** NRZ

13. The bus typically used to connect systems in an automobile is the

   (a) SPI   **(b)** CAN   **(c)** I^{2}C   **(d)** PCI

14. In a bus network topology, the bus must be terminated with a

   (a) node   **(b)** capacitor   **(c)** resistor   **(d)** open circuit

15. A daisy-chain network configuration is an example of

   (a) ring topology   **(b)** star topology

   (c) tree topology   **(d)** none of these

16. Two types of logical network topology are

   (a) duplex and biplex   **(b)** shared and token

   (c) unidirectional and bidirectional   **(d)** single level and multilevel

17. The IEEE-802.3 standard is commonly known as the

   (a) SCSI   **(b)** GPIB   **(c)** Firewire   **(d)** Ethernet

18. An Ethernet data collision is detected by the

   (a) CSMA/CD   **(b)** TCP/IP   **(c)** CRC   **(d)** SMTP

PROBLEMS

Answers to odd-numbered problems are at the end of the book.

SECTION 14–1   Bus Basics

1. List six physical characteristics and five electrical and performance characteristics of a bus.

2. Name the components of a computer system bus.

3. What is the purpose of the AGP bus?

4. List five devices that connect to a computer with an external bus.

5. Explain the difference between bus width and bus bandwidth.

6. A certain bus is specified with a width of 16 bits and a frequency of 100 MHz. Determine the bus bandwidth expressed as two different values.

7. Describe a simple example of a handshake.

8. State an advantage of a differential bus over a single-ended bus.

SECTION 14–2   Bus Interfacing

9. In a simple serial transfer of eight data bits from a source device to an acceptor device, the handshaking sequence in Figure 14–51 is observed on the four generic bus lines. By analyzing the time relationships, identify the function of each signal, and indicate if it originates at the source or at the receiver.

FIGURE 14–51

10. Determine the signal on the bus line in Figure 14–52 for the data-input and enable waveforms shown.

FIGURE 14–52

11. In Figure 14–53(a), data from the two sources are being placed on the data bus under control of the select line. The select waveform is shown in Figure 14–53(b). Determine the data-bus waveforms for the device output codes indicated.

FIGURE 14–53

SECTION 14–3 Parallel Buses

12. Describe the functions of the northbridge and the southbridge.

13. Explain the difference between PCI and PCI-X buses.

14. Explain how the PCI-Express differs from PCI and PCI-X.

15. What does x2 mean in a PCI-Express bus?

16. The terms *listener* and *talker* are associated with which bus standard?

17. Provide the description of each of the SCSI signals: BSY, SEL, RST, C/D, REQ, ACK, ATN, and MSG.

18. Consider the GPIB interface between a talker and a listener as shown in Figure 14–54(a). From the handshaking timing diagram in part (b), determine how many data bytes are actually transferred to the listening device.

(a) (b)

FIGURE 14–54

19. Describe the operations depicted in the GPIB timing diagram of Figure 14–55. Develop a basic block diagram of the system involved in this operation.

FIGURE 14–55

SECTION 14–4 The Universal Serial BUS (USB)

20. Identify each of the symbols in Figure 14–56.

21. List the four types of USB packets.

22. Describe each of the fields in the USB data packet in Figure 14–57.

	(a)	(b)	(c)

Sync 8/32	PID 8	Data 0-8192	CRC 16	EOP 3

FIGURE 14–56　　　　　　　　　　　　　　　　**FIGURE 14–57**

23. What type of data encoding is used in USB 3.0?

24. Determine the maximum number of bytes in a USB data field.

25. What is the maximum separation of two USB 2.0 devices?

SECTION 14–5   Other Serial Buses

26. Describe how RS-232 and RS-422 differ.

27. List the four signals in an SPI bus and describe each one.

28. Classify the two SPI configurations in Figure 14–58.

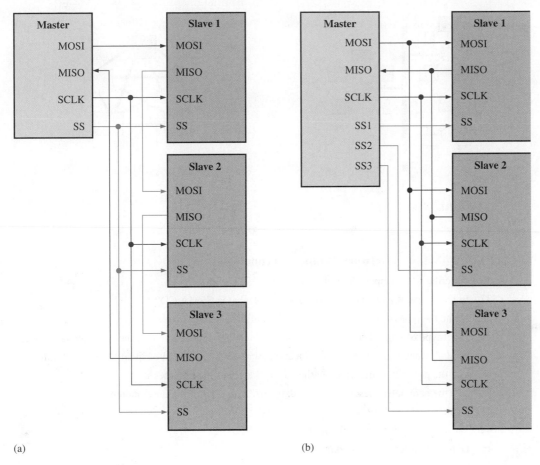

(a)   (b)

FIGURE 14–58

29. Describe the main use for the I^2C bus.

30. Fill in the field descriptions for the blank CAN bus data format in Figure 14–59.

FIGURE 14–59

31. Refer to Figure 14–37 and list additional units that may appear on a CAN automotive system.

32. What is the data rate for the IEEE-1394 bus standard in the S100 mode? What is the data rate in the S1600 mode?

SECTION 14–6   Network Topologies

33. Identify each network topology in Figure 14–60 on page 736.

34. Explain the difference between token-based and shared network logical topology.

35. Discuss the difference between physical and logical network topologies.

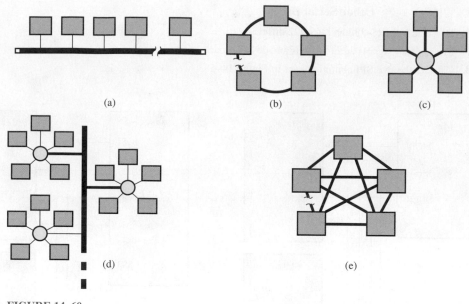

FIGURE 14–60

SECTION 14–7 Network Protocol Technologies

36. Describe what happens in each layer of the OSI network model.

37. In what type of network is Ethernet used and what is CSMA/CD?

38. Name the types of addresses contained in an Ethernet data frame.

39. What type of error check is done in the frame check sequence field of the Ethernet data frame?

40. Explain what causes a data collision and describe how it is handled in an Ethernet system.

41. List the parameters described in the physical layer of the Ethernet standard.

42. List the parameters described in the data link layer of the Ethernet standard.

43. Explain what TCP/IP is and where it is used.

44. Write the binary code for the IP address **162.18.248.1.**

45. Define client/server networking.

ANSWERS TO SECTION CHECKUPS

SECTION 14–1 Bus Basics

1. A bus bridge is a device that interfaces two buses, usually a faster bus with slower system bus.

2. Three internal computer buses are data, address, and control.

3. Speed of a parallel bus can be limited by cross talk, EMI, and clock skew.

4. An internal bus connects parts of a system that are within the computer. An external bus connects a computer or other device with peripheral devices.

5. Bus characteristics are width, frequency, transfer rate, and bandwidth.

6. Bus protocol is a set of rules used by two or more devices to establish and maintain communication.

7. A single-ended system uses one wire for data and one wire for ground, where the signal voltage on the wire is with respect to ground. In a differential system, two wires are used for data and one wire for ground. The data signal is sent on one wire and its complement (inversion) is sent on the other wire. The difference between the two data wires is the differential signal.

SECTION 14–2 Bus Interfacing

1. Tristate buffers allow devices to be completely disconnected from the bus when not in use, thus preventing interference with other devices.

2. A bus interconnects all the devices in a system and makes communication between devices possible.

SECTION 14–3   Parallel Buses

1. PCI is peripheral component interconnect.
2. PCI-Express is also designated PCIe and PCI-E.
3. A lane is a dedicated path to a single chip known as a switch.
4. GPIB is IEEE-488.
5. SCSI is small computer system interface.

SECTION 14–4   The Universal Serial Bus (USB)

1. USB is universal serial bus.
2. USB pins are D+, D−, +5 V, Ground.
3. The twisted pair reduces or eliminates noise.
4. USB 3.0 can run at higher speeds than USB 2.0. USB 3.0 has shorter cable lengths than USB 2.0.

SECTION 14–5   Other Serial Buses

1. RS-232, RS-422, RS-423, RS-485, SPI, I^2C, CAN, and Firewire (IEEE-1394)
2. SPI is serial-to-peripheral interface.
3. I^2C is inter-integrated circuit.
4. CAN is controller area network.
5. Firewire is IEEE-1394.

SECTION 14–6   Network Topologies

1. Physical network topology is the structure or the way in which devices are connected to the network with actual cables.
2. Logical network topology is the way in which data is moved through the network and is related to network protocol.
3. The major types of physical network topology are bus, ring, star, tree, and mesh.
4. Categories of logical network topology are *token-based* and *shared media*.

SECTION 14–7   Network Protocol Technologies

1. The IEEE 802.3 standard is the Ethernet.
2. OSI is open systems interconnection.
3. TCP/IP is Transmission Control Protocol/Internet Protocol.
4. An IP address contains four 8-bit numbers.

ANSWER TO RELATED PROBLEM FOR EXAMPLE

14–1   15.625 MHz

ANSWERS TO TRUE/FALSE QUIZ

1. T	**2.** F	**3.** T	**4.** F	**5.** F	**6.** T	**7.** T
8. F	**9.** T	**10.** F	**11.** T	**12.** T	**13.** F	**14.** T
15. T	**16.** T	**17.** F	**18.** F	**19.** T	**20.** F	**21.** T
22. T	**23.** F	**24.** T	**25.** F	**26.** T	**27.** T	**28.** T

ANSWERS TO SELF-TEST

1. (d)	**2.** (b)	**3.** (c)	**4.** (d)	**5.** (d)	**6.** (d)	**7.** (a)
8. (b)	**9.** (d)	**10.** (c)	**11.** (a)	**12.** (d)	**13.** (b)	**14.** (c)
15. (a)	**16.** (b)	**17.** (d)	**18.** (a)			

Conversions

DECIMAL	BCD(8421)	OCTAL	BINARY	DECIMAL	BCD(8421)	OCTAL	BINARY	DECIMAL	BCD(8421)	OCTAL	BINARY
0	0000	0	0	34	00110100	42	100010	68	01101000	104	1000100
1	0001	1	1	35	00110101	43	100011	69	01101001	105	1000101
2	0010	2	10	36	00110110	44	100100	70	01110000	106	1000110
3	0011	3	11	37	00110111	45	100101	71	01110001	107	1000111
4	0100	4	100	38	00111000	46	100110	72	01110010	110	1001000
5	0101	5	101	39	00111001	47	100111	73	01110011	111	1001001
6	0110	6	110	40	01000000	50	101000	74	01110100	112	1001010
7	0111	7	111	41	01000001	51	101001	75	01110101	113	1001011
8	1000	10	1000	42	01000010	52	101010	76	01110110	114	1001100
9	1001	11	1001	43	01000011	53	101011	77	01110111	115	1001101
10	00010000	12	1010	44	01000100	54	101100	78	01111000	116	1001110
11	00010001	13	1011	45	01000101	55	101101	79	01111001	117	1001111
12	00010010	14	1100	46	01000110	56	101110	80	10000000	120	1010000
13	00010011	15	1101	47	01000111	57	101111	81	10000001	121	1010001
14	00010100	16	1110	48	01001000	60	110000	82	10000010	122	1010010
15	00010101	17	1111	49	01001001	61	110001	83	10000011	123	1010011
16	00010110	20	10000	50	01010000	62	110010	84	10000100	124	1010100
17	00010111	21	10001	51	01010001	63	110011	85	10000101	125	1010101
18	00011000	22	10010	52	01010010	64	110100	86	10000110	126	1010110
19	00011001	23	10011	53	01010011	65	110101	87	10000111	127	1010111
20	00100000	24	10100	54	01010100	66	110110	88	10001000	130	1011000
21	00100001	25	10101	55	01010101	67	110111	89	10001001	131	1011001
22	00100010	26	10110	56	01010110	70	111000	90	10010000	132	1011010
23	00100011	27	10111	57	01010111	71	111001	91	10010001	133	1011011
24	00100100	30	11000	58	01011000	72	111010	92	10010010	134	1011100
25	00100101	31	11001	59	01011001	73	111011	93	10010011	135	1011101
26	00100110	32	11010	60	01100000	74	111100	94	10010100	136	1011110
27	00100111	33	11011	61	01100001	75	111101	95	10010101	137	1011111
28	00101000	34	11100	62	01100010	76	111110	96	10010110	140	1100000
29	00101001	35	11101	63	01100011	77	111111	97	10010111	141	1100001
30	00110000	36	11110	64	01100100	100	1000000	98	10011000	142	1100010
31	00110001	37	11111	65	01100101	101	1000001	99	10011001	143	1100011
32	00110010	40	100000	66	01100110	102	1000010				
33	00110011	41	100001	67	01100111	103	1000011				

Powers of Two

2^n	n	2^{-n}
1	0	1.0
2	1	0.5
4	2	0.25
8	3	0.125
16	4	0.062 5
32	5	0.031 25
64	6	0.015 625
128	7	0.007 812 5
256	8	0.003 906 25
512	9	0.001 953 125
1 024	10	0.000 976 562 5
2 048	11	0.000 488 281 25
4 096	12	0.000 244 140 625
8 192	13	0.000 122 070 312 5
16 384	14	0.000 061 035 156 25
32 768	15	0.000 030 517 578 125
65 536	16	0.000 015 258 789 062 5
131 072	17	0.000 007 629 394 531 25
262 144	18	0.000 003 814 697 265 625
524 288	19	0.000 001 907 348 632 812 5
1 048 576	20	0.000 000 953 674 316 406 25
2 097 152	21	0.000 000 476 837 158 203 125
4 194 304	22	0.000 000 238 418 579 101 562 5
8 388 608	23	0.000 000 119 209 289 550 781 25
16 777 216	24	0.000 000 059 604 644 775 390 625
33 554 432	25	0.000 000 029 802 322 387 695 312 5
67 108 864	26	0.000 000 014 901 161 193 847 656 25
134 217 728	27	0.000 000 007 450 580 596 923 828 125
268 435 456	28	0.000 000 003 725 290 298 461 914 062 5
536 870 912	29	0.000 000 001 862 645 149 230 957 031 25
1 073 741 824	30	0.000 000 000 931 322 574 615 478 515 625
2 147 483 648	31	0.000 000 000 465 661 287 307 739 257 812 5
4 294 967 296	32	0.000 000 000 232 830 643 653 869 628 906 25
8 589 934 592	33	0.000 000 000 116 415 321 826 934 814 453 125
17 179 869 184	34	0.000 000 000 058 207 660 913 467 407 226 562 5
34 359 738 368	35	0.000 000 000 029 103 830 456 733 703 613 281 25
68 719 476 736	36	0.000 000 000 014 551 915 228 366 851 806 640 625
137 438 953 472	37	0.000 000 000 007 275 957 614 183 425 903 320 312 5
274 877 906 944	38	0.000 000 000 003 637 978 807 091 712 951 660 156 25
549 755 813 888	39	0.000 000 000 001 818 989 403 545 856 475 830 078 125
1 099 511 627 776	40	0.000 000 000 000 909 494 701 772 928 237 915 039 062 5
2 199 023 255 552	41	0.000 000 000 000 454 747 350 886 464 118 957 519 531 25
4 398 046 511 104	42	0.000 000 000 000 227 373 675 443 232 059 478 759 765 625
8 796 093 022 208	43	0.000 000 000 000 113 686 837 721 616 029 739 379 882 812 5
17 592 186 044 416	44	0.000 000 000 000 056 843 418 860 808 014 869 689 941 406 25
35 184 372 088 832	45	0.000 000 000 000 028 421 709 430 404 007 434 844 970 703 125
70 368 744 177 664	46	0.000 000 000 000 014 210 854 715 202 003 717 422 485 351 562 5
140 737 488 355 328	47	0.000 000 000 000 007 105 427 357 601 001 858 711 242 675 781 25
281 474 976 710 656	48	0.000 000 000 000 003 552 713 678 800 500 929 355 621 337 890 625
562 949 953 421 312	49	0.000 000 000 000 001 776 356 839 400 250 464 677 810 668 945 312 5
1 125 899 906 842 624	50	0.000 000 000 000 000 888 178 419 700 125 232 338 905 334 472 656 25
2 251 799 813 685 248	51	0.000 000 000 000 000 444 089 209 850 062 616 169 452 667 236 328 125
4 503 599 627 370 496	52	0.000 000 000 000 000 222 044 604 925 031 308 084 726 333 618 164 062 5
9 007 199 254 740 992	53	0.000 000 000 000 000 111 022 302 462 515 654 042 363 166 809 082 031 25
18 014 398 509 481 984	54	0.000 000 000 000 000 055 511 151 231 257 827 021 181 583 404 541 015 625
36 028 797 018 963 968	55	0.000 000 000 000 000 027 755 575 615 628 913 510 590 791 702 270 507 812 5
72 057 594 037 927 936	56	0.000 000 000 000 000 013 877 787 807 814 456 755 295 395 851 135 253 906 25
144 115 188 075 855 872	57	0.000 000 000 000 000 006 938 893 903 907 228 377 647 697 925 567 626 953 125
288 230 376 151 711 744	58	0.000 000 000 000 000 003 469 446 951 953 614 188 823 848 962 783 813 476 562 5
576 460 752 303 423 488	59	0.000 000 000 000 000 001 734 723 475 976 807 094 411 924 481 391 906 738 281 25
1 152 921 504 606 846 976	60	0.000 000 000 000 000 000 867 361 737 988 403 547 205 962 240 695 953 369 140 625
2 305 843 009 213 693 952	61	0.000 000 000 000 000 000 433 680 868 994 201 773 602 981 120 347 976 684 570 312 5
4 611 686 018 427 387 904	62	0.000 000 000 000 000 000 216 840 434 497 100 886 801 490 560 173 988 342 285 156 25
9 223 372 036 854 775 808	63	0.000 000 000 000 000 000 108 420 217 248 550 443 400 745 280 086 994 171 142 578 125
18 446 744 073 709 551 616	64	0.000 000 000 000 000 000 054 210 108 624 275 221 700 372 640 043 497 085 571 289 062 5
36 893 488 147 419 103 232	65	0.000 000 000 000 000 000 027 105 054 312 137 610 850 186 320 021 748 542 785 644 531 25
73 786 976 294 838 206 464	66	0.000 000 000 000 000 000 013 552 527 156 088 805 425 093 160 010 874 271 392 822 265 625
147 573 952 589 676 412 928	67	0.000 000 000 000 000 000 006 776 263 578 034 402 712 546 580 005 437 135 696 411 132 812 5
295 147 905 179 352 825 856	68	0.000 000 000 000 000 000 003 388 131 789 017 201 356 273 290 002 718 567 848 205 566 406 25
590 295 810 358 705 651 712	69	0.000 000 000 000 000 000 001 694 065 894 508 600 678 136 645 001 359 283 924 102 783 203 125
1 180 591 620 717 411 303 424	70	0.000 000 000 000 000 000 000 847 032 947 254 300 339 068 322 500 679 641 962 051 391 601 562 5
2 361 183 241 434 822 606 848	71	0.000 000 000 000 000 000 000 423 516 473 627 150 169 534 161 250 339 820 981 025 695 800 781 25
4 722 366 482 869 645 213 696	72	0.000 000 000 000 000 000 000 211 758 236 813 575 084 767 080 625 169 910 490 512 847 900 390 625

Programs for Security System Components

BCDEncoder–VHDL

```vhdl
library ieee;
use ieee.std_logic_1164.all;

entity BCDEncoder is
port(D: in std_logic_vector(0 to 9);
     Q: out std_logic_vector(0 to 3));
end entity BCDEncoder;

architecture BCDBehavior of BCDEncoder is
begin
    Q(0) <= D(9) or D(7) or D(5) or D(3) or D(1);
    Q(1) <= D(7) or D(6) or D(3) or D(2);
    Q(2) <= D(7) or D(6) or D(5) or D(4);
    Q(3) <= D(9) or D(8);
end architecture BCDBehavior;
```

CodeSelection–VHDL

```vhdl
library ieee;
use ieee.std_logic_1164.all;

entity CodeSelection is
port(ShiftIn, Clk: in std_logic;
     Bout: out std_logic_vector(1 to 4));
end entity CodeSelection;

architecture CodeSelectionBehavior of CodeSelection is
component FourBitShiftReg is
port(R_L, Clk: in std_logic;
     D0, D1, D2, D3: in std_logic;
     Q0, Q1, Q2, Q3: buffer std_logic);
end component FourBitShiftReg;

signal A: std_logic_vector(0 to 3);
signal B: std_logic_vector(0 to 3);
signal C: std_logic_vector(0 to 3);
signal D: std_logic_vector(0 to 3);
signal Q: std_logic_vector(0 to 3);
signal Code1: std_logic_vector(0 to 3);
signal Code2: std_logic_vector(0 to 3);
signal Code3: std_logic_vector(0 to 3);
signal Code4: std_logic_vector(0 to 3);

begin
    Code1 <= "0001";
    Code2 <= "1001";
    Code3 <= "1001";
    Code4 <= "0001";
    A(0) <= Q(0) and Code1(0); B(0) <= Q(1) and Code2(0);
    C(0) <= Q(2) and Code3(0); D(0) <= Q(3) and Code4(0);
    A(1) <= Q(0) and Code1(1); B(1) <= Q(1) and Code2(1);
    C(1) <= Q(2) and Code3(1); D(1) <= Q(3) and Code4(1);
    A(2) <= Q(0) and Code1(2); B(2) <= Q(1) and Code2(2);
    C(2) <= Q(2) and Code3(2); D(2) <= Q(3) and Code4(2);
    A(3) <= Q(0) and Code1(3); B(3) <= Q(1) and Code2(3);
    C(3) <= Q(2) and Code3(3); D(3) <= Q(3) and Code4(3);
    Bout(1) <= D(0) or C(0) or B(0) or A(0);
    Bout(2) <= D(1) or C(1) or B(1) or A(1);
    Bout(3) <= D(2) or C(2) or B(2) or A(2);
    Bout(4) <= D(3) or C(3) or B(3) or A(3);

ShiftReg: FourBitShiftReg
port map(R_L=>ShiftIn, Clk=>Clk,D0=>'1',D1=>'0',D2=>'0',
         D3=>'0',Q0=>Q(0),Q1=>Q(1),Q2=>Q(2),
         Q3=>Q(3));
end architecture CodeSelectionBehavior;
```

EightBitShiftReg–VHDL

```vhdl
library ieee;
use ieee.std_logic_1164.all;

entity EightBitShiftReg is
port(R_L, Clk: in std_logic;
     D: in std_logic_vector(0 to 7);
     Q: buffer std_logic);
end entity EightBitShiftReg;

architecture EightBitBehavior of EightBitShiftReg is
component dff is
  port(D,Clk: in std_logic;
       Q: out std_logic);
end component dff;

signal D0, D1, D2, D3, D4, D5, D6, D7: std_logic;
signal Q0, Q1, Q2, Q3, Q4, Q5, Q6: std_logic;

begin
    D0 <= (not R_L and D(0));
    D1 <= (Q0 and R_L) or (not R_L and D(1));
    D2 <= (Q1 and R_L) or (not R_L and D(2));
    D3 <= (Q2 and R_L) or (not R_L and D(3));
    D4 <= (Q3 and R_L) or (not R_L and D(4));
    D5 <= (Q4 and R_L) or (not R_L and D(5));
    D6 <= (Q5 and R_L) or (not R_L and D(6));
    D7 <= (Q6 and R_L) or (not R_L and D(7));

DFF0: dff port map(D => D0, Clk => Clk, Q => Q0);
DFF1: dff port map(D => D1, Clk => Clk, Q => Q1);
DFF2: dff port map(D => D2, Clk => Clk, Q => Q2);
DFF3: dff port map(D => D3, Clk => Clk, Q => Q3);
DFF4: dff port map(D => D4, Clk => Clk, Q => Q4);
DFF5: dff port map(D => D5, Clk => Clk, Q => Q5);
DFF6: dff port map(D => D6, Clk => Clk, Q => Q6);
DFF7: dff port map(D => D7, Clk => Clk, Q => Q);
end architecture EightBitBehavior;
```

FourBitParSftReg–VHDL

```vhdl
library ieee;
use ieee.std_logic_1164.all;

entity FourBitParSftReg is
port(D: in std_logic_vector(0 to 3);
     Clk: in std_logic;
     Q: out std_logic_vector(0 to 3));
end entity FourBitParSftReg;

architecture FourBitBehavior of FourBitParSftReg is
component dff is
  port(D,Clk: in std_logic;
       Q: out std_logic);
end component dff;

begin
DFF0: dff port map(D => D(0), Clk => Clk, Q => Q(0));
DFF1: dff port map(D => D(1), Clk => Clk, Q => Q(1));
DFF2: dff port map(D => D(2), Clk => Clk, Q => Q(2));
DFF3: dff port map(D => D(3), Clk => Clk, Q => Q(3));
end architecture FourBitBehavior;
```

OneShot–VHDL

```vhdl
library ieee;
use ieee.std_logic_1164.all;

entity OneShot is
port(Enable, Clk: in std_logic;
     QOut: buffer std_logic);
end entity OneShot;

architecture OneShotBehavior of OneShot is
begin
 process(Enable, Clk)
 variable Flag    : boolean := true;
 variable Cnt     : integer range 0 to 255;
 variable SetCount : integer range 0 to 255;
 begin
    SetCount := 2;
    if (Clk'EVENT and Clk = '1') then
        if Enable = '0' then
           Flag := true;
        end if;

        if Enable = '1' and Flag then
           Cnt := 1;
           Flag :=False;
        end if;

        if cnt = SetCount then
           Qout <= '0';
           Cnt := 0;
           Flag := false;
        else
           if Cnt > 0 then
             Cnt := Cnt + 1;
             if Cnt > 1 then
                Qout <= '1';
             end if;
           end if;
        end if;
    end if;
 end process;
end architecture OneShotBehavior;
```

ComparatorFourBit–VHDL

```vhdl
library ieee;
use ieee.std_logic_1164.all;

entity ComparatorFourBit is
port(A, B: in std_logic_vector(0 to 3);
     EQ: out std_logic);
end entity ComparatorFourBit;

architecture ComparatorBehavior of
             ComparatorFourBit is

begin
  process (A,B)
  begin
     if (A = B) and not (A = "0000") then
        EQ <= '1';
     else
        EQ <= '0';
     end if;
  end process;
end architecture ComparatorBehavior;
```

BCDEncoder–Verilog

```verilog
module BCDEncoder(D, Q);
input  [9:0] D;
output [3:0] Q;

   assign Q[0] = D[9] || D[7] || D[5] || D[3] || D[1];
   assign Q[1] = D[7] || D[6] || D[3] || D[2];
   assign Q[2] = D[7] || D[6] || D[5] || D[4];
   assign Q[3] = D[9] || D[8];
endmodule
```

CodeSelection–Verilog

```verilog
module CodeSelection(ShiftIn, Clk, Bout);
input ShiftIn;
input Clk;
output [4:1]Bout;

  wire[3:0] A;
  wire[3:0] B;
  wire[3:0] C;
  wire[3:0] D;
  wire[3:0] Q;
  wire[3:0] Code1;
  wire[3:0] Code2;
  wire[3:0] Code3;
  wire[3:0] Code4;

  assign Code1 = 4'b1001;
  assign Code2 = 4'b1000;
  assign Code3 = 4'b1000;
  assign Code4 = 4'b1001;
  assign A[0]=Q[0] && Code1[0];assign B[0]=Q[1] && Code2[0]
  assign C[0]=Q[2] && Code3[0];assign D[0]=Q[3] && Code4[0];
  assign A[1]=Q[0] && Code1[1];assign B[1]=Q[1] && Code2[1];
  assign C[1]=Q[2] && Code3[1];assign D[1]=Q[3] && Code4[1];
  assign A[2]=Q[0] && Code1[2];assign B[2]=Q[1] && Code2[2];
  assign C[2]=Q[2] && Code3[2];assign D[2]=Q[3] && Code4[2];
  assign A[3]=Q[0] && Code1[3];assign B[3]=Q[1] && Code2[3];
  assign C[3]=Q[2] && Code3[3];assign D[3]=Q[3] && Code4[3];

  assign Bout[1] = D[0] || C[0] || B[0] || A[0];
  assign Bout[2] = D[1] || C[1] || B[1] || A[1];
  assign Bout[3] = D[2] || C[2] || B[2] || A[2];
  assign Bout[4] = D[3] || C[3] || B[3] || A[3];

FourBitShiftReg ShiftReg(.S_L(ShiftIn),
   .Clk(Clk),.D0(1),.D1(0),.D2(0),.D3(0),
   .Q0(Q[0]),.Q1(Q[1]),.Q2(Q[2]),.Q3(Q[3]));
endmodule;
```

EightBitShiftReg–Verilog

```verilog
module EightBitShiftReg(S_L, Clk, D, Q);
input  S_L;
input  Clk;
input  [7:0] D;
output Q;

wire D0, D1, D2, D3, D4, D5, D6, D7;
wire Q0, Q1, Q2, Q3, Q4, Q5, Q6;

   assign D0 = (!S_L && D[0]);
   assign D1 = (Q0 && S_L) || (!S_L && D[1]);
   assign D2 = (Q1 && S_L) || (!S_L && D[2]);
   assign D3 = (Q2 && S_L) || (!S_L && D[3]);
   assign D4 = (Q3 && S_L) || (!S_L && D[4]);
   assign D5 = (Q4 && S_L) || (!S_L && D[5]);
   assign D6 = (Q5 && S_L) || (!S_L && D[6]);
   assign D7 = (Q6 && S_L) || (!S_L && D[7]);

dff DFF0(.d(D0), .Clk(Clk), .q(Q0));
dff DFF1(.d(D1), .Clk(Clk), .q(Q1));
dff DFF2(.d(D2), .Clk(Clk), .q(Q2));
dff DFF3(.d(D3), .Clk(Clk), .q(Q3));
dff DFF4(.d(D4), .Clk(Clk), .q(Q4));
dff DFF5(.d(D5), .Clk(Clk), .q(Q5));
dff DFF6(.d(D6), .Clk(Clk), .q(Q6));
dff DFF7(.d(D7), .Clk(Clk), .q(Q));
endmodule
```

FourBitParSftReg–Verilog

```verilog
module FourBitParSftReg(D, Clk, Q);
input [3:0] D;
input Clk;
output [3:0]Q;

dff DFF0(.d(D[0]), .Clk(Clk), .Q(Q[0]));
dff DFF1(.d(D[1]), .Clk(Clk), .Q(Q[1]));
dff DFF2(.d(D[2]), .Clk(Clk), .Q(Q[2]));
dff DFF3(.d(D[3]), .Clk(Clk), .Q(Q[3]));
endmodule
```

OneShot–Verilog

```verilog
module OneShot(Enable, Clk, QOut);
input Enable;
input Clk;
output QOut;

integer  SetCount = 2;
integer  Cnt = 0           ;
reg[0:0] Flag;
reg QOut;

initial Flag = 1'b1;
always @(posedge Clk)
begin
  if (Enable == 0)
  begin
    Flag = 1;
  end
  else
  begin
    if ((Enable == 1) && (Flag == 1))
    begin
      Cnt = 1;
      Flag = 0;
    end    end
  if (Cnt == SetCount)
  begin
    QOut = 0;
    Cnt  = 0;
    Flag = 0;
  end
  else
    begin
    if (Cnt > 0)
    begin
      Cnt = Cnt + 1;
      if (Cnt > 1)
      begin
        QOut = 1;
      end
    end
  send
end
endmodule
```

ComparatorFourBit–Verilog

```verilog
module ComparatorFourBit(A, B, EQ);
input [3:0] A;
input [3:0] B;
output EQ;

reg Out;

    always @(A , B)
      if ((A == B) && !(A == 0))
      begin
        Out = 1'b1;
      end
      else
      begin
        Out = 1'b0;
      end
      assign EQ = Out;
endmodule
```

ANSWERS TO ODD-NUMBERED PROBLEMS

Chapter 1

1. Digital can be transmitted and stored more efficiently and reliably.

3. Clock

   Thermometer

   Speedometer

5. (a) 11010001   (b) 000101010

7. (a) 550 ns   (b) 600 ns   (c) 2.7 μs   (d) 10 V

9. 250 Hz

11. 50%

13. 8 μs; 1 μs

15. L_{on} = SW1 + SW2 + SW1 · SW2

17. AND gate

19. (a) adder   (b) multiplier

    (c) multiplexer   (d) comparator

21. 01010000

23. (a) Simple Programmable Logic Device

    (b) Complex Programmable Logic Device

    (c) Hardware Description Language

    (d) Field-Programmable Gate Array

    (e) Generic Array Logic

25. Place-and-route or fitting is the process where the logic structures described by the netlist are mapped into the actual structure of the specific target device. This results in an output called a bitstream.

27. DIP pins go through holes in a circuit board. SMT pins connect to surface pads.

29. The system can be implemented with a PLD, a microcontroller, or with fixed-function ICs.

31. Register A stores the maximum number of tablets/bottle. Register B stores the cumulative total of tablets bottled.

33. 6 V

Chapter 2

1. (a) 1   (b) 100   (c) 100,000

3. (a) 400; 70; 1   (b) 9000; 300; 50; 6

   (c) 100,000; 20,000; 5000; 0; 0; 0

5. (a) 3   (b) 4   (c) 7   (d) 8   (e) 9

   (f) 12   (g) 11   (h) 15

7. (a) 51.75   (b) 42.25   (c) 65.875

   (d) 120.625   (e) 92.65625   (f) 113.0625

   (g) 90.625   (h) 127.96875

9. (a) 5 bits   (b) 6 bits   (c) 6 bits   (d) 7 bits

   (e) 7 bits   (f) 7 bits   (g) 8 bits   (h) 8 bits

11. (a) 1010   (b) 10001   (c) 11000

    (d) 110000   (e) 111101   (f) 1011101

    (g) 1111101   (h) 10111010

13. (a) 1111   (b) 10101   (c) 11100

    (d) 100010   (e) 101000   (f) 111011

    (g) 1000001   (h) 1001001

15. (a) 100   (b) 100   (c) 1000

    (d) 1101   (e) 1110   (f) 11000

17. (a) 1001   (b) 1000   (c) 100011

    (d) 110110   (e) 10101001   (f) 10110110

19. all 0s or all 1s

21. (a) 010   (b) 001   (c) 0101

    (d) 00101000   (e) 0001010   (f) 11110

23. (a) 00011101   (b) 11010101

    (c) 01100100   (d) 11111011

25. (a) 00001100   (b) 10111100

    (c) 01100101   (d) 10000011

27. (a) −102   (b) +116   (c) −64

29. (a) 0 10001101 11110000101011000000000

    (b) 1 10001010 11000001100000000000000

31. (a) 00110000   (b) 00011101

    (c) 11101011   (d) 100111110

33. (a) 11000101   (b) 11000000

35. 100111001010

37. (a) 00111000   (b) 01011001

    (c) 101000010100   (d) 010111001000

    (e) 0100000100000000   (f) 1111101100010111

    (g) 1000101010011101

39. (a) 35   (b) 146   (c) 26   (d) 141

    (e) 243   (f) 235   (g) 1474   (h) 1792

41. (a) 60_{16} (b) $10B_{16}$ (c) $1BA_{16}$

43. (a) 10   (b) 23   (c) 46   (d) 52   (e) 67

    (f) 367   (g) 115   (h) 532   (i) 4085

45. (a) 001011   (b) 101111

    (c) 001000001   (d) 011010001

    (e) 101100000   (f) 100110101011

    (g) 001011010111001   (h) 100101110000000

    (i) 001000000010001011

47. (a) 00010000   (b) 00010011

     (c) 00011000   (d) 00100001

     (e) 00100101   (f) 00110110

     (g) 01000100   (h) 01010111

     (i) 01101001   (j) 10011000

     (k) 000100100101   (l) 000101010110

49. (a) 000100000100   (b) 000100101000

     (c) 000100110010   (d) 000101010000

     (e) 000110000110   (f) 001000010000

     (g) 001101011001   (h) 010101000111

     (i) 0001000001010001

51. (a) 80   (b) 237   (c) 346   (d) 421

     (e) 754   (f) 800   (g) 978   (h) 1683

     (i) 9018   (j) 6667

53. (a) 00010100   (b) 00010010

     (c) 00010111   (d) 00010110

     (e) 01010010   (f) 000100001001

     (g) 000110010101   (h) 0001001001101001

55. The Gray code makes only one bit change at a time when going from one number in the sequence to the next.

57. (a) 1100   (b) 00011   (c) 10000011110

59. (a) CAN   (b) J   (c) =

     (d) #   (e) >   (f) B

61. 48 65 6C 6C 6F 2E 20 48 6F 77 20 61 72 65 20 79 6F 75 3F

63. (b) is incorrect.

65. (a) 110100100   (b) 000001001   (c) 111111110

67. In each case, you get the other number.

69. The remainder is 10, indicating an error.

Chapter 3

1. $X = A + B + C + D$

3. $X = \overline{A} + \overline{B} + \overline{C}$

5. (a) $AB = 1$ when $A = 1, B = 1$

     (b) $A\overline{B}C = 1$ when $A = 1, B = 0, C = 1$

     (c) $A + B = 0$ when $A = 0, B = 0$

     (d) $\overline{A} + B + \overline{C} = 0$ when $A = 1, B = 0, C = 1$

     (e) $\overline{A} + \overline{B} + C = 0$ when $A = 1, B = 1, C = 0$

     (f) $\overline{A} + B = 0$ when $A = 1, B = 0$

     (g) $A\overline{B}\,\overline{C} = 1$ when $A = 1, B = 0, C = 0$

7. (a) Commutative   (b) Commutative   (c) Distributive

9. See Figure P–1.

FIGURE P–1

11. See Figure P–2.

FIGURE P–2

13. See Figure P–3.

FIGURE P–3

15. See Figure P–4.

FIGURE P–4

17. See Figure P–5.

FIGURE P–5

19. See Figure P–6.

FIGURE P–6

21. See Figure P–7.

FIGURE P–7

23. See Figure P–8.

FIGURE P–8

25. See Figure P–9.

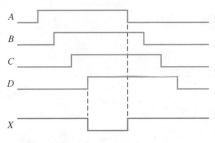

FIGURE P–9

27. See Figure P–10.

FIGURE P–10

29. See Figure P–11.

FIGURE P–11

31. $\text{XOR} = A\overline{B} + \overline{A}B$; $\text{OR} = A + B$

33. See Figure P–12.

FIGURE P–12

35. $t_{PLH} = 4.3$ ns; $t_{PHL} = 10.5$ ns

37. 20 mW

39. $X_1 = \overline{A}B, X_2 = \overline{A}\,\overline{B}, X_3 = A\overline{B}$.

41. **entity** ANDgate **is**

  port (A, B, C, D: **in** bit; X: **out** bit);

  end entity ANDgate;

  architecture ANDfunction **of** ANDgate **is**

  begin

    X <= A **and** B **and** C **and** D;

  end architecture ANDfunction;

43. **entity** NORgate **is**

  port (A, B, C, D, E: **in** bit; X: **out** bit);

  end entity NORgate;

  architecture NORfunction **of** NORgate **is**

  begin

    X <= **not**(A **or** B **or** C **or** D **or** E);

  end architecture NORfunction;

45. Troubleshooting is the process of recognizing, isolating, and correcting a fault or failure in a system.

47. In the signal-tracing method, a signal is tracked as it progresses through a system until a point is found where the signal disappears or is incorrect.

49. When a failure is reported, determine when and how it failed and what are the symptoms.

51. An incorrect output can be caused by an incorrect dc supply voltage, improper ground, incorrect component value, or a faulty component.

53. To isolate a fault in a system, apply half-splitting or signal tracing.

55. When a fault has been isolated to a particular circuit board, the options are to repair the board or replace the board with a known good board.

57. See Figure P–13.

FIGURE P–13

59. See Figure P–14.

FIGURE P–14

61. See Figure P–15.

FIGURE P–15

63. **Circuit fault:** Input B of AND gate U1 shorted to ground.

Predicted effect of fault: Output X is always LOW.

Observed effect of introduced fault: Output X is always LOW.

65. **Observed operation:** Output X is LOW whenever input B is HIGH.

Suspected fault: Input A of NOR gate shorted to ground.

Effect of introduced fault: Output X is LOW whenever B is HIGH.

Chapter 4

1. See Figure P–16.

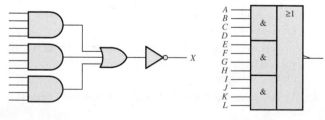

FIGURE P–16

3. (a) $X = ABB$

(b) $X = AB + B$

(c) $X = \overline{A} + B$

(d) $X = (A + B) + AB$

(e) $X = \overline{ABC}$

(f) $X = (A + B)(\overline{B} + C)$

5. (a) See Table P–1. $X = (A + B)(C + D)$

TABLE P–1				
A	**B**	**C**	**D**	**X**
0	0	0	0	0
0	0	0	1	0
0	0	1	0	0
0	0	1	1	0
0	1	0	0	0
0	1	0	1	1
0	1	1	0	1
0	1	1	1	1
1	0	0	0	0
1	0	0	1	1
1	0	1	0	1
1	0	1	1	1
1	1	0	0	0
1	1	0	1	1
1	1	1	0	1
1	1	1	1	1

(b) See Table P–2. $X = \overline{\overline{\overline{ABC} + \overline{CD}}}$

TABLE P–2				
A	**B**	**C**	**D**	**X**
0	0	0	0	0
0	0	0	1	0
0	0	1	0	0
0	0	1	1	1
0	1	0	0	0
0	1	0	1	0
0	1	1	0	0
0	1	1	1	1
1	0	0	0	0
1	0	0	1	0
1	0	1	0	0
1	0	1	1	1
1	1	0	0	0
1	1	0	1	0
1	1	1	0	0
1	1	1	1	0

5. (c) See Table P–3. $X = (AB + C)D + E$

(d) See Table P–4. $X = \overline{\overline{\overline{(A + B)(\overline{BC})}} + D}$

TABLE P–3

A	B	C	D	E	X
0	0	0	0	0	0
0	0	0	0	1	1
0	0	0	1	0	0
0	0	0	1	1	1
0	0	1	0	0	0
0	0	1	0	1	1
0	0	1	1	0	1
0	0	1	1	1	1
0	1	0	0	0	0
0	1	0	0	1	1
0	1	0	1	0	0
0	1	0	1	1	1
0	1	1	0	0	0
0	1	1	0	1	1
0	1	1	1	0	1
0	1	1	1	1	1
1	0	0	0	0	0
1	0	0	0	1	1
1	0	0	1	0	0
1	0	0	1	1	1
1	0	1	0	0	0
1	0	1	0	1	1
1	0	1	1	0	1
1	0	1	1	1	1
1	1	0	0	0	0
1	1	0	0	1	1
1	1	0	1	0	1
1	1	0	1	1	1
1	1	1	0	0	0
1	1	1	0	1	1
1	1	1	1	0	1
1	1	1	1	1	1

TABLE P–4

A	B	C	D	X
0	0	0	0	1
0	0	0	1	1
0	0	1	0	1
0	0	1	1	1
0	1	0	0	1
0	1	0	1	1
0	1	1	0	1
0	1	1	1	1
1	0	0	0	0
1	0	0	1	1
1	0	1	0	0
1	0	1	1	1
1	1	0	0	1
1	1	0	1	1
1	1	1	0	1
1	1	1	1	1

(e) See Table P–5. $X = \overline{\overline{\overline{(\overline{AB} + \overline{C})D}} + \overline{E}}$

TABLE P–5

A	B	C	D	E	X
0	0	0	0	0	1
0	0	0	0	1	0
0	0	0	1	0	1
0	0	0	1	1	1
0	0	1	0	0	1
0	0	1	0	1	0
0	0	1	1	0	1
0	0	1	1	1	0
0	1	0	0	0	1
0	1	0	0	1	0
0	1	0	1	0	1
0	1	0	1	1	1

TABLE P–5   (Continued)

A	B	C	D	E	X
0	1	1	0	0	1
0	1	1	0	1	0
0	1	1	1	0	1
0	1	1	1	1	0
1	0	0	0	0	1
1	0	0	0	1	0
1	0	0	1	0	1
1	0	0	1	1	1
1	0	1	0	0	1
1	0	1	0	1	0
1	0	1	1	0	1
1	0	1	1	1	0
1	1	0	0	0	1
1	1	0	0	1	0
1	1	0	1	0	1
1	1	0	1	1	1
1	1	1	0	0	1
1	1	1	0	1	0
1	1	1	1	0	1
1	1	1	1	1	1

5. (f) See Table P–6. $X = \overline{\overline{(\overline{AB} + \overline{CD})}\,\overline{(\overline{EF} + \overline{GH})}}$

TABLE P–6

A	B	C	D	E	F	G	H	X
0	X	0	X	X	X	X	X	1
X	0	0	X	X	X	X	X	1
0	X	X	0	X	X	X	X	1
X	0	X	0	0	X	X	X	1
X	X	X	X	0	X	0	X	1
X	X	X	X	X	0	0	X	1
X	X	X	X	0	X	X	0	1
X	X	X	X	X	0	X	0	1

For all other entries $X = 0$.
$X = $ *don't care*
An abbreviated table is shown because there are 256 combinations.

7. $\overline{ABCD + EFGH}$

9. See Figure P–17.

11. $X = \overline{A}\,\overline{B}\,\overline{C} + \overline{A}B\overline{C} + A\overline{B}\,\overline{C} + AB\overline{C} + ABC$
Five 3-input AND gates, one 5-input OR gate, and three inverters.

(a) $X = AB + \overline{B}C$

(b) $X = A(B + \overline{C})$

(c) $X = A\overline{B} + AB$

(d) $X = \overline{ABC} + B(EF + \overline{G})$

(e) $X = A[BC(A + B + C + D)]$

(f) $X = B(C\overline{D}E + \overline{E}FG)\,(\overline{AB} + C)$

FIGURE P–17

13. (a) See Table P–7.

   (b) See Table P–8.

TABLE P–7			
A	**B**	**C**	**X**
0	0	0	0
0	0	1	0
0	1	0	1
0	1	1	0
1	0	0	0
1	0	1	1
1	1	0	0
1	1	1	1

TABLE P–8			
X	**Y**	**Z**	**Q**
0	0	0	1
0	0	1	1
0	1	0	0
0	1	1	1
1	0	0	0
1	0	1	1
1	1	0	1
1	1	1	0

15. (a) $X = \overline{A}\,\overline{B}C + A\overline{B}\,\overline{C} + A\overline{B}C + ABC$

   $X = (A + B + C)(A + \overline{B} + C)(A + \overline{B} + \overline{C})(\overline{A} + \overline{B} + C)$

   (b) $X = AB\overline{C} + A\overline{B}C + ABC$

   $X = (A + B + C)(A + B + \overline{C})(A + \overline{B} + C)(A + \overline{B} + \overline{C})(\overline{A} + B + C)$

   (c) $X = \overline{A}\,\overline{B}\,\overline{C}D + \overline{A}\,\overline{B}\,CD + \overline{A}B\overline{C}D + \overline{A}BCD + \overline{A}BC\overline{D} + A\overline{B}\,\overline{C}D + AB\overline{C}D$

   $X = (A + B + \overline{C} + D)(A + \overline{B} + C + D)(A + \overline{B} + \overline{C} + \overline{D})(\overline{A} + B + C + D)(\overline{A} + B + \overline{C} + D)$
   $(\overline{A} + B + \overline{C} + \overline{D})(\overline{A} + \overline{B} + C + D)(\overline{A} + \overline{B} + \overline{C} + D)(\overline{A} + \overline{B} + \overline{C} + \overline{D})$

   (d) $X = \overline{A}\,\overline{B}C\overline{D} + \overline{A}B\overline{C}\,\overline{D} + \overline{A}BC\overline{D} + \overline{A}BCD + A\overline{B}CD + AB\overline{C}\,\overline{D} + ABCD$

   $X = (A + B + C + D)(A + B + C + \overline{D})(A + B + \overline{C} + \overline{D})(A + \overline{B} + \overline{C} + D)(\overline{A} + B + C + D)$
   $(\overline{A} + B + C + \overline{D})(\overline{A} + B + \overline{C} + D)(\overline{A} + \overline{B} + C + \overline{D})(\overline{A} + \overline{B} + \overline{C} + D)$

17. (a) $\overline{A} + B + \overline{C}D$

   (b) $\overline{A} + \overline{B} + (\overline{C} + D)(E + \overline{F})$

   (c) $\overline{A}B\overline{C}D + \overline{A} + \overline{B} + \overline{C} + D$

   (d) $\overline{A} + B + C + D + A\overline{B}\,\overline{C}D$

   (e) $AB + (\overline{C} + D)(E + \overline{F}) + ABCD$

19. See Figure P–18.

FIGURE P–18

(a)   (b)

21. See Figure P–19.

FIGURE P–19

(a)   (b)

23. $X = \overline{A} + \overline{B} + B = AB\overline{B} = 0$
The output X is always LOW.

25. X is HIGH when ABC are all HIGH or when A is HIGH and B is LOW and C is LOW or when A is HIGH and B is LOW and C is HIGH.

$X = ABC + A\overline{B}\overline{C} + A\overline{B}C$

See Figure P–20.

$X = ABC + A\overline{B}\overline{C} + A\overline{B}C$

FIGURE P–20

27. See Figure P–21.

FIGURE P–21

29. $X <= A$ **and** B **and** C; $X = A$ **&&** B **&&** C;

31. VHDL for 4–38(e):

entity Circuit 4_38e **is**
   port (A, B, C: **in** bit; X: **out** bit);
end entity Circuit4_38e;
architecture LogicFunction **of** Circuit4_38e **is**
begin
   X <= ((A **nand** B) **and** B) **nand** C;
end architecture LogicFunction;

Verilog for Figure 4–38(e):

module Circuit4_38e (A, B, C, X);
input (A, B, C);
output X;
assign X = !(((!(A **&&** B)) **&&** B) **&&** C);
endmodule

VHDL for 4–38(f):

entity Circuit4_38f **is**
   port (A, B, C: **in** bit; X: **out** bit);
end entity Circuit4_38f;
architecture LogicFunction **of** Circuit4_38f **is**
begin
   X <= (A **or** B) **and** (**not** B **or** C);
end architecture LogicFunction;

Verilog for Figure 4–38(f):

module Circuit4_38f (A, B, C, X);
input (A, B, C);
output X;
assign X = (A ‖ B) **&&** (! B ‖ C);
endmodule

33. Number gates from top to bottom and left to right G1, G2, G3, etc. Relabel inputs IN1, IN2, IN3, etc. and output OUT.

entity Circuit4_39f **is**
   port (IN1, IN2, IN3, IN4, IN5, IN6, IN7, IN8: **in** bit; OUT: **out** bit);
end entity Circuit4_39f;
architecture LogicFunction **of** Circuit4_39f **is**
component NAND_gate **is**
   port (A, B: **in** bit; X: **out** bit);
end component NAND_gate;
   signal G1OUT, G2OUT, G3OUT, G4OUT, G5OUT, G6OUT: bit;
begin
   G1: NAND_gate **port map** (A => IN1, B => IN2, X => G1OUT);
   G2: NAND_gate **port map** (A => IN3, B => IN4, X => G2OUT);
   G3: NAND_gate **port map** (A => IN5, B => IN6, X => G3OUT);
   G4: NAND_gate **port map** (A => IN7, B => IN8, X => G4OUT);
   G5: NAND_gate **port map** (A => G1OUT, B => G2OUT, X => G5OUT);
   G6: NAND_gate **port map** (A => G3OUT, B => G4OUT, X => G6OUT);
   G7: NAND_gate **port map** (A => G5OUT, B => G6OUT, X => OUT);
end architecture LogicFunction;

Verilog for 4–39(f):

module Circuit4_39f (A, B, C, D, E, F, G, H, X);
input (A, B, C, D, E, F, G, H);
output X;
assign X = !(((A **&&** B) ‖ (C **&&** D)) **&&** ((E **&&** F) ‖ (G **&&** H)));
endmodule

35. Data flow approach

entity Fig4_47a **is**
   port (A, B, C, D, E: **in** bit; X: **out** bit);
end entity Fig4_47a;
architecture DataFlow **of** Fig4_47a **is**
begin
   X <= (A **and** B **and** C) **or** (D **and not** E);
end architecture DataFlow;

Structural approach

entity Fig4_47a **is**
   port (IN1, IN2, IN3, IN4, IN5: **in** bit; OUT: **out** bit);
end entity Fig4_47a;
architecture Structure **of** Fig4_47a **is**
component AND_gate **is**
   port (A, B: **in** bit; X: **out** bit);
end component AND_gate;
component OR_gate **is**
   port (A, B: **in** bit; X: **out** bit);
end component OR_gate;

component Inverter **is**
   port (A: **in** bit; X: **out** bit);
end component Inverter;
   signal G1OUT, G2OUT, G3OUT, INVOUT: bit;
begin
   G1: AND_gate **port map** (A => IN1, B => IN2,
   X => G1OUT);
   G2: AND_gate **port map** (A => G1OUT, B => IN3,
   X => G2OUT);
   INV: Inverter **port map** (A => IN5, X => INVOUT);
   G3: AND_gate **port map** (A => IN4, B => INVOUT,
   X => G3OUT);
   G4: OR_gate **port map** (A => G2OUT, B => G3OUT,
   X => OUT);
end architecture Structure;

Verilog for 4–47(a):

module Circuit4_47a (A, B, C, D, E, X);
input (A, B, C, D, E);
output X;
assign X = ((A **&&** B) **&&** C) **||** (D **&&** !E)
endmodule

37. See Table P–9.

TABLE P–9

	INPUTS			OUTPUT
A	*B*	*C*	*D*	*X*
0	0	0	0	0
1	0	0	0	0
0	1	0	0	0
1	1	0	0	0
0	0	1	0	0
1	0	1	0	0
0	1	1	0	0
1	1	1	0	0
0	0	0	1	0
1	0	0	1	0
0	1	0	1	0
1	1	0	1	1
0	0	1	1	0
1	0	1	1	1
0	1	1	1	1
1	1	1	1	1

39. The AND gates are numbered top to bottom G1, G2, G3, G4. The OR gate is G5 and the inverters are, top to bottom,

G6 and G7. Change A_1, A_2, B_1, B_2 to IN1, IN2, IN3, IN4 respectively. Change X to OUT.
entity Circuit4_48 **is**
   port (IN1, IN2, IN3, IN4: **in** bit; OUT: **out** bit);
end entity Circuit4_48;
architecture Logic **of** Circuit4_48 **is**
component AND_gate **is**
   port (A, B: **in** bit; X: **out** bit);
end component AND_gate;
component OR_gate **is**
   port (A, B, C, D: **in** bit; X: **out** bit);
end component OR_gate;
component Inverter **is**
   port (A: **in** bit; X: **out** bit);
end component Inverter;
   signal G1OUT, G2OUT, G3OUT, G4OUT, G5OUT,
   G6OUT, G7OUT: bit;
begin
   G1: AND_gate **port map** (A => IN1, B => IN2,
   X => G1OUT);
   G2: AND_gate **port map** (A => IN2, B => G6OUT,
   X => G2OUT);
   G3: AND_gate **port map** (A => G6OUT,
   B => G7OUT, X => G3OUT);
   G4: AND_gate **port map** (A => G7OUT, B => IN1,
   X => G4OUT);
   G5: OR_gate **port map** (A => G1OUT, B => G2OUT,
   C => G3OUT, D => G4OUT, X => OUT);
   G6: Inverter **port map** (A => IN3, X => G6OUT);
   G7: Inverter **port map** (A => IN4, X => G7OUT);
end architecture Logic;

41. See Figure P–22.

FIGURE P–22

43. See Figure P–23.

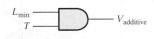

FIGURE P–23

45. $X = ABC + D\overline{E}$. Since X is the same as the G_3 output, either G_1 or G_2 has failed, with its output stuck LOW.

47. $X = \overline{\overline{AB} + \overline{CD} + \overline{EF}} = (\overline{\overline{AB}})(\overline{\overline{CD}})(\overline{\overline{EF}})$
   $= (A + B)(C + D)(E + F)$
Since X does not go HIGH when C or D is HIGH, the output of gate G_2 must be stuck LOW.

49. No, the output of the $\overline{CD}$ gate is stuck LOW.

51. X = lamp on, A = front door switch on, B = back door switch on. See Figure P–24.

FIGURE P–24

53. Circuit fault: Input C is shorted to ground.

Predicted effect of fault: X1 and X2 are always off. X3 is on only when both B and C are HIGH. X4 is on when either B or C is LOW.

Observed effect of introduced fault: X1 and X2 are always off. X3 is on only when both B and C are HIGH (switches B and C are both open).

55. Observed operation: X1 is on when either A or B is LOW. X2 is always off. X3 is on when both A and B are HIGH or when C is LOW. X4 is always ON. X5 is on when C is HIGH and either A or B is LOW.

Suspected fault: Output of U3 is shorted to ground.

Effect of introduced fault: X1 is on when either A or B is LOW. X2 is always off. X3 is on when both A and B are HIGH or when C is LOW. X4 is always ON. X5 is on when C is HIGH and either A or B is LOW.

Chapter 5

1. Register A: 000000110110; Register B: 10110100

3. 00100100

5. **(a)** $A \oplus B = 0, \Sigma = 1, (A \oplus B)C_{\text{in}} = 0, AB = 1,$ $C_{\text{out}} = 1$

   (b) $A \oplus B = 1, \Sigma = 0, (A \oplus B)C_{\text{in}} = 1, AB = 0,$ $C_{\text{out}} = 1$

   (c) $A \oplus B = 1, \Sigma = 1, (A \oplus B)C_{\text{in}} = 0, AB = 0,$ $C_{\text{out}} = 0$

7. **(a)** $\Sigma = 1, C_{\text{out}} = 0;$

   (b) $\Sigma = 1, C_{\text{out}} = 0;$

   (c) $\Sigma = 0, C_{\text{out}} = 1;$

   (d) $\Sigma = 1, C_{\text{out}} = 1$

9. 11100

11. $\Sigma_3\Sigma_2\Sigma_1\Sigma_0 = 1101$

13. $\Sigma_1 = 0110; \Sigma_2 = 1011; \Sigma_3 = 0110; \Sigma_4 = 0001;$ $\Sigma_5 = 1000$

15. 225 ns

17. $A = B$ is HIGH when $A_0 = B_0$ and $A_1 = B_1$; see Figure P–25.

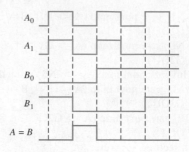

FIGURE P–25

19. **(a)** $A > B = 1; A = B = 0; A < B = 0$

    (b) $A < B = 1; A = B = 0; A > B = 0$

    (c) $A = B = 1; A < B = 0; A > B = 0$

21. See Figure P–26.

23. $X = \overline{A_3}\overline{A_2}\overline{A_1}A_0 + A_3\overline{A_2}A_1\overline{A_0} + A_3A_2\overline{A_1}\overline{A_0} + A_3\overline{A_2}A_1A_0$

25. See Figure P–27.

27. $A_3A_2A_1A_0 = 1011$, invalid BCD

29. **(a)** $2 = 0010_{\text{BCD}} = 0010_2$

    (b) $8 = 1000_{\text{BCD}} = 1000_2$

    (c) $13 = 00010011_{\text{BCD}} = 1101_2$

    (d) $26 = 00100110_{\text{BCD}} = 11010_2$

    (e) $33 = 00110011_{\text{BCD}} = 100001_2$

31. **(a)** 1010000000 Gray $\rightarrow$ 1100000000 binary

    (b) 0011001100 Gray $\rightarrow$ 0010001000 binary

    (c) 1111000111 Gray $\rightarrow$ 1010000101 binary

    (d) 0000000001 Gray $\rightarrow$ 0000000001 binary

    See Figure P–28.

(a)   (b)

(c)   (d)

(e)   (f)

(g)   (h)

FIGURE P–26

FIGURE P–27

FIGURE P–30

FIGURE P–28

33. See Figure P–29.

FIGURE P–29

35. See Figure P–30.

37. See Figure P–31.

39. **module** 8_Bit_Adder (A, B, C, Cin, SUM, Cout);
　　input [7:0] A;
　　input [7:0] B;
　　input Cin;
　　output [7:0] SUM;
　　output [7:0] Cout;
　　wire C1, C2, C3, C4, C5, C6, C7;
　　　　Full Adder Full Adder 1 (A[0], B[0], Cin, SUM[0], C1);
　　　　Full Adder Full Adder 2 (A[1], B[1], Cin, SUM[1], C2);
　　　　Full Adder Full Adder 3 (A[2], B[2], Cin, SUM[2], C3);
　　　　Full Adder Full Adder 4 (A[3], B[3], Cin, SUM[3], C4);
　　　　Full Adder Full Adder 5 (A[4], B[4], Cin, SUM[4], C5);
　　　　Full Adder Full Adder 6 (A[5], B[5], Cin, SUM[5], C6);
　　　　Full Adder Full Adder 7 (A[6], B[6], Cin, SUM[6], C7);
　　　　Full Adder Full Adder 8 (A[7], B[7], Cin, SUM[7], Cout);
　　endmodule

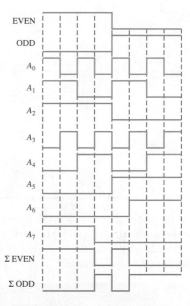

FIGURE P–31

41. **module** Encoder (D, X);
　　input [9:0] D; **output** [3:0] X;
　　X[0] = (D[1] ∥ D[3] ∥ D[5] ∥ D[7] ∥ D[9]) **&&** !(D[0] ∥
　　　　D[2] ∥ D[4] ∥ D[6] ∥ D[8]);
　　X[1] = (D[2] ∥ D[3] ∥ D[6] ∥ D[7]) **&&** !(D[0] ∥ D[1] ∥
　　　　D[4] ∥ D[5] ∥ D[8] ∥ D[9]);
　　X[2] = (D[4] ∥ D[5] ∥ D[6] ∥ D[7]) **&&** !(D[0] ∥ D[1]) ∥
　　　　D[2] ∥ D[(3] ∥ D[8] ∥ D[9]);
　　X[3] = (D[8] ∥ D[9]) **&&** !(D[0] ∥ D[1] ∥ D[2] ∥ D[3] ∥
　　　　D[4] ∥ D[5] ∥ D[6] ∥ D[7]);
　　endmodule

43. (a) OK

(b) segment g burned out; output G open

(c) Segment b output stuck LOW

45. (a) The A_1 input of the top adder is open: All binary values corresponding to a BCD number having a value of 0, 1, 4, 5, 8, or 9 will be off by 2. This will first be seen for a BCD value of 0000 0000.

(b) The carry out of the top adder is open: All values not normally involving an output carry will be off by 32. This will first be seen for a BCD value of 0000 0000.

(c) The Σ_4 output of the top adder is shorted to ground: Same binary values above 15 will be short by 16. The first BCD value to indicate this will be 0001 1000.

(d) The Σ_3 output of the bottom adder is shorted to ground: Every other set of 16 values starting with 16 will be short 16. The first BCD value to indicate this will be 0001 0110.

47. 1. Place a LOW on the Enable input.

2. Apply a HIGH to D_0 and a LOW to D_1 through D_7.

3. Go through the binary sequence on the select inputs and check Y and $\overline{Y}$ according to Table P–10.

TABLE P–10

S_2	S_1	S_0	Y	$\overline{Y}$
0	0	0	1	0
0	0	1	0	1
0	1	0	0	1
0	1	1	0	1
1	0	0	0	1
1	0	1	0	1
1	1	0	0	1
1	1	1	0	1

4. Repeat the binary sequence of select inputs for each set of data inputs listed in Table P–11. A HIGH on the Y output should occur only for the corresponding combinations of select inputs shown.

49. Apply a HIGH in turn to each Data input, D_0 through D_7 with LOWs on all the other inputs. For each HIGH applied to a data input, sequence through all eight binary combinations of select inputs ($S_2S_1S_0$) and check for HIGH on the corresponding data output and LOWs on all the other data outputs.

51. $\Sigma = \overline{A}\,\overline{B}C_{in} + \overline{A}B\overline{C}_{in} + A\overline{B}\,\overline{C}_{in} + ABC_{in}$
$C_{out} = \overline{A}BC_{in} + A\overline{B}C_{in} + AB\overline{C}_{in} + ABC_{in}$
See Figure P–32.

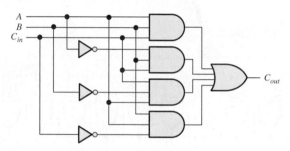

FIGURE P–32

53. See the block diagram in Figure P–33.

55. Circuit fault: Input B of 3-to-8 decoder is shorted to V_{CC}.

Predicted effect of fault: Inputs are decoded as follows: $0000 = 2, 0001 = 3, 0010 = 2, 0011 = 3, 0100 = 6, 0101 = 7, 0110 = 6, 0111 = 7, 1000$ and higher = no decoded output.

Observed effect of introduced fault: Inputs are decoded as follows: $0000 = 2, 0001 = 3, 0010 = 2, 0011 = 3, 0100 = 6, 0101 = 7, 0110 = 6, 0111 = 7, 1000$ and higher = no decoded output.

TABLE P–11

D_0	D_1	D_2	D_3	D_4	D_5	D_6	D_7	Y	$\overline{Y}$	S_2	S_1	S_0
L	H	L	L	L	L	L	L	1	0	0	0	1
L	L	H	L	L	L	L	L	1	0	0	1	0
L	L	L	H	L	L	L	L	1	0	0	1	1
L	L	L	L	H	L	L	L	1	0	1	0	0
L	L	L	L	L	H	L	L	1	0	1	0	1
L	L	L	L	L	L	H	L	1	0	1	1	0
L	L	L	L	L	L	L	H	1	0	1	1	1

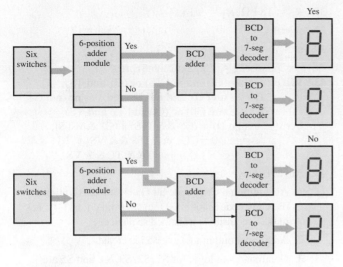

FIGURE P–33

57. Observed operation: Data output X is incorrect for binary inputs of 0011, 0111, 1011, and 1111.

Suspected fault: Line from input S2 to multiplexer input B is shorted to ground.

Effect of introduced fault: Data output X is incorrect for binary inputs of 0011, 0111, 1011, and 1111.

Chapter 6

1. The system remains in the first state for 25 s or as long as there is no vehicle on the side street.

3. $MR = G_1 G_0 + G_1 \overline{G_0}$ $MY = \overline{G_1} G_0$ $MG = \overline{G_1}\,\overline{G_0}$
   $SR = \overline{G_1}\,\overline{G_0} + \overline{G_1} G_0$ $SY = G_1 \overline{G_0}$ $SG = G_1 G_0$

5. See Figure P–34.

FIGURE P–34

7. See Figure P–35.

FIGURE P–35

9. See Figure P–36.

FIGURE P–36

11. See Figure P–37.

FIGURE P–37

13. See Figure P–38.

FIGURE P–38

15. See Figure P–39.

FIGURE P–39

17. See Figure P–40.

FIGURE P–40

19. See Figure P–41.

FIGURE P–41

21. See Figure P–42.

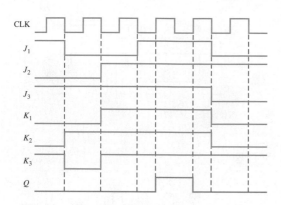

FIGURE P–42

23. divide-by-2; see Figure P–43.

FIGURE P–43

25. Direct current and dc supply voltage

27. 14.9 MHz

29. 150 mA, 750 mW

31. 7.26 μs

33. $C_1 = 1\,\mu\text{F}, R_1 = 227\,\text{k}\Omega$ (use 220 kΩ). See Figure P–44.

FIGURE P–44

35. $R_1 = 18\,\text{k}\Omega, R_2 = 9.1\,\text{k}\Omega$.

37. signal

39. assign

41. SetCountLong <= 25; SetCountShort <= 4;

43. VHDL: D1 <= (G0 **and not** TS) **or** (G1 **and** TS);
D0 <= (**not** G1 **and not** TL **and** VS) **or** (**not** G1 **and** G0) **or** (G0 **and** TL **and** VS);
Verilog: **assign** D1 = (G9 **&&** !TS) **||** (G1 **&&** TS);
assign D0 = (!G1 **&&** !TS **&&** VS) **||** (!G1 **&&** G0) **||** (G0 **&&** TL **&&** VS);

45. Answers may vary.

   1. An additional Gray code bit G3 is added. Additionally, the Gray code input must recycle from count 110 back to 000, requiring a truncated sequence.

   2. Additional output identifier 55 is created.

   3. Combinational logic for S1, S2, S3, S4, and S5 are modified to incorporate the following truncated 3-bit Gray code sequence:

      State 1 (S1): G2 = 0, G1 = 0, G0 = 0

      State 2 (S2): G2 = 0, G1 = 0, G0 = 1

      State 3 (S3): G2 = 0, G1 = 1, G0 = 1

      State 4 (S4): G2 = 0, G1 = 1, G0 = 0

      State 5 (S5): G2 = 1, G1 = 1, G0 = 0

47. Changes are

   1. The Walk portion of the sign is simply wired to a HIGH.

   2. Two new output variables are added: DontMain and DontSide are applied to the Don't input of the sign.

   3. Output DontMain is assigned the logic statement for SG. Output DontSide is assigned the logic statement for MG.

49. The wire from pin 6 to pin 10 and the ground wire are reversed on the protoboard.

51. $\overline{CLR}$ shorted to ground.

53. See Figure P–45. Delays not shown.

55. See Figure P–46.

57. Changes required for the system to incorporate a 15 s left turn signal on main:

   1. Change the 2-bit Gray code sequence to a 3-bit sequence.

   2. Add decoding logic to the State Decoder to decode the turn signal state.

   3. Change the Output Logic to incorporate the turn signal output.

   4. Change the Trigger Logic to incorporate a trigger output for the turn signal timer.

   5. Add a 15 second timer.

   See Figure P–47.

59. Circuit fault: Line to K input is shorted to V$_{CC}$.

Predicted effect of fault: If line to J input is LOW, Q output will go and remain LOW and notQ output will go and remain HIGH. If line to J input is HIGH, Q and notQ will toggle.

Observed effect of introduced fault: If line to J input is LOW, Q output will go and remain LOW and notQ output will go and remain HIGH. If line to J input is HIGH, Q and notQ will toggle.

(a)

(b) Same as (a)

(e)

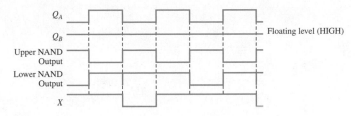

(c)

(d) $X = $ LOW if $Q_R = 1$; $X = \bar{Q}_A$ if $Q_R = 0$

FIGURE P–45

FIGURE P–46

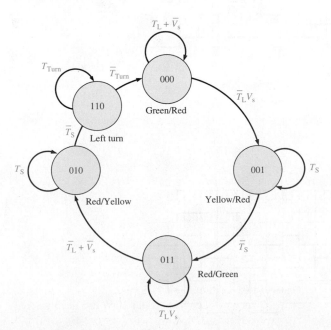

FIGURE P–47

61. Observed operation: Pulse width of one-shot is 690 μs rather than 6.9 μs.

Suspected fault: 1 kΩ resistor accidentally used in place of 10 kΩ timing resistor, or a 0.1 μF capacitor is used in place of the 1 μF timing capacitor.

Effect of introduced fault: Pulse width of one-shot is 690 μs rather than 6.9 μs.

Chapter 7

1. 1001001101110100

3. shift register A: 1001

shift register C: 00000100

5. Q0 = 1, Q1 = 0, Q2 = 0, Q3 = 0

7. The states of shift registers A and C after each key closure when entering 7645 are:

After key 7 is pressed:

Shift register A contains 0111

Shift register C contains 11000

After key 6 is pressed:

Shift register A contains 0110

Shift register C contains 11100

After key 4 is pressed:

Shift register A contains 0100

Shift register C contains 11110

After key 5 (an incorrect entry) is pressed:

Shift register A contains 0000

Shift register C contains 10000

9. Shift registers store binary data.

11. Shift data and store data.

13. See Figure P–48.

FIGURE P–48

15.

Initially:	101001111000
CLK1:	010100111100
CLK2:	001010011110
CLK3:	000101001111
CLK4:	000010100111
CLK5:	100001010011
CLK6:	110000101001
CLK7:	111000010100
CLK8:	011100001010
CLK9:	001110000101
CLK10:	000111000010
CLK11:	100011100001
CLK12:	110001110000

17. See Figure P–49.

FIGURE P–49

19. See Figure P–50.

FIGURE P–50

21. See Figure P–51.

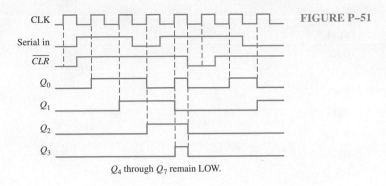

FIGURE P–51

Q_4 through Q_7 remain LOW.

23. See Figure P–52.

FIGURE P–52

25. See Figure P–53.

FIGURE P–53

27. See Figure P–54.

FIGURE P–54

29.

Initially (76):	01001100	
CLK1:	10011000	left
CLK2:	01001100	right
CLK3:	00100110	right
CLK4:	00010011	right
CLK5:	00100110	left
CLK6:	01001100	left
CLK7:	00100110	right
CLK8:	01001100	left
CLK9:	00100110	right
CLK10:	01001100	left
CLK11:	10011000	left

31. See Figure P–55.

FIGURE P–55

33. **(a)** 3   **(b)** 5   **(c)** 7   **(d)** 8

35. See Figure P–56.

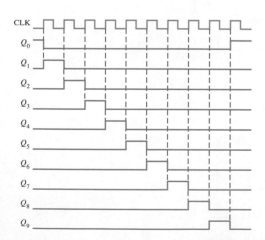

FIGURE P–56

37. The hard coded binary value stored in the 8-bit shift register C component will need to be changed from 00010000_2 to 00100000_2.

39. If two key presses were read by the code selection component before one shot A could timeout, we would skip a stored code value. The system would read the second value as an error and the system would not disarm.

41. D_3 input open

43. **(a)** No clock at switch closure because of faulty NAND (negative-OR) gate or one-shot; open clock (C) input to key code register; open $SH/\overline{LD}$ input to key code register

(b) Diode in third row open; Q_2 output of ring counter open

(c) The NAND (negative-OR) gate input connected to the first column is open or shorted.

(d) The "2" input to the column encoder is open.

45. **(a)** Contents of data output register remain constant.

(b) Contents of both registers do not change.

(c) Third stage output of data output register remains HIGH.

(d) Clock generator is disabled after each pulse by the flip-flop being continuously SET and then RESET.

47. See Figure P–57.

49. **Circuit fault:** Line to CLK input of U3 is open.

Predicted effect of fault: Q0 and Q1 will sequence normally. Q2 and Q3 will remain LOW.

Observed effect of introduced fault: Q0 and Q1 will sequence normally. Q2 and Q3 will remain LOW.

FIGURE P–57

51. Circuit fault: The QH output of the 74LS165 shift register is shorted to ground.

Predicted effect of fault: The QH output is always LOW but the notQH output behaves as expected.

Observed effect of introduced fault: The QH output is always LOW but the notQH output behaves as expected.

53. Observed operation: The ring counter can initiate circulation of 1s but not 0s.

Suspected fault: The notRESET line of the first stage is shorted to V_{CC}.

Effect of introduced fault: The ring counter can initiate circulation of 1s but not 0s.

Chapter 8

1. Hours tens: 0001

Hours units: 0010

Minutes tens: 0000

Minutes units: 0001

Seconds tens: 0000

Seconds units: 0010

3. This is a Moore machine. See Figure P–58.

FIGURE P–58

5. See Figure P–59.

FIGURE P–59

7. Worst-case delay is 24 ns; it occurs when all flip-flops change state from 011 to 100 or from 111 to 000.

9. 8 ns

11. Initially, each flip-flop is reset.

At CLK1:

$J_0 = K_0 = 1$ Therefore Q_0 goes to a 1.

$J_1 = K_1 = 0$ Therefore Q_1 remains a 0.

$J_2 = K_2 = 0$ Therefore Q_2 remains a 0.

$J_3 = K_3 = 0$ Therefore Q_3 remains a 0.

At CLK2:

$J_0 = K_0 = 1$ Therefore Q_0 goes to a 0.

$J_1 = K_1 = 1$ Therefore Q_1 goes to a 1.

$J_2 = K_2 = 0$ Therefore Q_2 remains a 0.

$J_3 = K_3 = 0$ Therefore Q_3 remains a 0.

At CLK3:

$J_0 = K_0 = 1$ Therefore Q_0 goes to a 1.

$J_1 = K_1 = 0$ Therefore Q_1 remains a 1.

$J_2 = K_2 = 0$ Therefore Q_2 remains a 0.

$J_3 = K_3 = 0$ Therefore Q_3 remains a 0.

A continuation of this procedure for the next seven clock pulses will show that the counter progresses through the BCD sequence.

13. See Figure P–60.

15. See Figure P–61.

FIGURE P–60

FIGURE P–61

17. See Figure P–62.

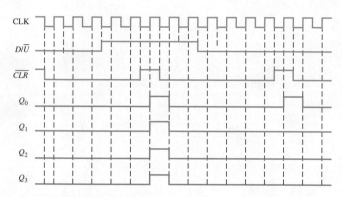

FIGURE P–62

19. See Figure P–63 for divide-by-10,000. Add one more DIV10 counter to create a divide-by-100,000.

FIGURE P–63

21. See Figure P–64.

FIGURE P–64

23. CLK2, output 0; CLK4, outputs 2, 0; CLK6, output 4; CLK8, outputs 6, 4, 0; CLK10, output 8; CLK12, outputs 10, 8; CLK14, output 12; CLK16, outputs 14, 12, 8

25. A glitch of the AND gate output occurs on the 111 to 000 transition. Eliminate by ANDing $\overline{CLK}$ with counter outputs (strobe) or use Gray code.

27. The addition of AND gate (G3) causes the additional fifth J-K flip-flop stage (FF4) to change when Q_0, Q_1, Q_2, and Q_3 are HIGH.

VHDL: G3 <= G2 **and** Q3;
Verilog: assign G3 = G2 **&&** Q3;

29. See Figure P–65.

FIGURE P–65

31. (a) Q_0 and Q_1 will not change from their initial state.

(b) normal operation except Q_0 floating

(c) Q_0 waveform is normal; Q_1 remains in initial state.

(d) normal operation

(e) The counter will not change from its initial state.

33. The K input of FF1 must be connected to ground rather than to the J input. Check for a wiring error.

35. Q_0 input to AND gate open and acting as a HIGH

37. See Table P–12.

39. The decode 6 gate interprets count 4 as a 6 (0110) and clears the counter back to 0 (actually 0010 since Q_1 is open). The apparent sequence of the tens portion of the counter is 0010, 0011, 0010, 0011, 0110.

41. 68

43. See Figure P–66.

STAGE	OPEN	LOADED COUNT	f_{OUT}
1	0	63C1	250.006 Hz
1	1	63C2	250.012 Hz
1	2	63C4	250.025 Hz
1	3	63C8	250.050 Hz
2	0	63D0	250.100 Hz
2	1	63E0	250.200 Hz
2	2	63C0	250 Hz
2	3	63C0	250 Hz
3	0	63C0	250 Hz
3	1	63C0	250 Hz
3	2	67C0	256.568 Hz
3	3	6BC0	263.491 Hz
4	0	73C0	278.520 Hz
4	1	63C0	250 Hz
4	2	63C0	250 Hz
4	3	E3C0	1.383 KHZ

TABLE P–12

FIGURE P–66

45. $65{,}536 - 50{,}000 = 15{,}536$

Preset the counter to 15,536 so that it counts from 15,536 up to 65,536 on each full cycle, thus producing a sequence of 50,000 states (modulus 50,000).
$15{,}536 = 11110010110000_2 = 3CB0_{16}$
See Figure P–67.

47. See Figure P–68.

49. Circuit fault: Line to CLK input of U3 is open.

Predicted effect of fault: Q0 and Q1 will sequence normally. Q2 and Q3 will remain LOW.

Observed effect of introduced fault: Q0 and Q1 will sequence normally. Q2 and Q3 will remain LOW.

FIGURE P–67

FIGURE P–68

51. Circuit fault: The Q output of U3 is shorted to ground.

Predicted effect of fault: The Q3Q2Q1Q0 count sequence is 0000, 0001, 0010, 0011, 0000, . . . rather than a decade up count.

Observed effect of introduced fault: The Q3Q2Q1Q0 count sequence is 0000, 0001, 0010, 0011, 0000, . . . rather than a decade up count.

53. Observed operation: The 74LS190 decade counter always functions as a down counter.

Suspected fault: Line to notU/D input of counter always HIGH.

Effect of introduced fault: The 74LS190 decade counter always functions as a down counter.

FIGURE P–69

Chapter 9

1. $X = \overline{A}\,\overline{B}\,\overline{C} + \overline{A}B\overline{C} + A\overline{B}C$

3. See Figure P–69.

5. A CPLD basically consists of multiple SPLDs that can be connected with a programmable interconnect array.

7. (a) $\overline{A}BC\overline{D}$

(b) $ABC(\overline{D} + \overline{E}) = ABC\overline{D} + ABC\overline{E}$

9. $X = A\overline{B} + \overline{A}B$

11. $X_1 = \overline{A}BC\overline{D} + \overline{A}BCD + ABC\overline{D};$
$X_2 = ABCD + AB\overline{C}D + \overline{A}B\overline{C}D + A\overline{B}CD$

13. (a) Combinational; 1   (b) Registered; 0

15. (a) Registered   (b) GCK1   (c) 0   (d) 0

17. SOP output $= \overline{A}\,\overline{B}\,\overline{C} + \overline{A}\,\overline{B}C + \overline{A}BC + A\overline{B}C + ABC$

19. See Figure P–70.

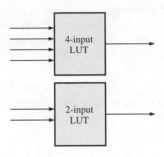

FIGURE P–70

21. See Figure P–71.

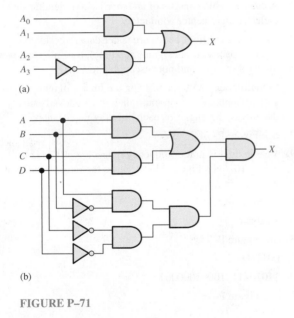

FIGURE P–71

23. See Figure P–72.

FIGURE P–72

25. Shift input $= 1$, data are applied to SDI, go through the MUX, and are clocked into Capture register A on the leading edge of the clock pulse. From the output of Capture register A, the data go through the upper MUX and are clocked into Capture register B on the trailing edge of the clock pulse.

27. PDI/O $= 0$ and OE $= 0$. The data are applied to the input pin and go through the selected MUX to the internal programmable logic.

29. 0000110010100011111011

0	0000110010100011111011
1	0000110010100011111011
3	0000110010100011111011
6	0000110010100011111011
12	0000110010100011111011
9	0000110010100011111011
2	0000110010100011111011
5	0000110010100011111011
10	0000110010100011111011
4	0000110010100011111011
8	0000110010100011111011
1	0000110010100011111011
3	0000110010100011111011
7	0000110010100011111011
15	0000110010100011111011
14	0000110010100011111011
13	0000110010100011111011
11	0000110010100011111011

Chapter 10

1. Cache is a small area of fast memory used by the central processing unit. The L1 cache is located in the processor, and the L2 cache is outside of the processor.

3. Hard disk has the highest latency.

5. Cache hit: 30 ns; cache miss: 105 ns.

7. (a) ROM

   (b) RAM

9. *Address bus* provides for transfer of address code to memory for accessing any memory location in any order for a read or write operation. *Data bus* provides for transfer of data between the microprocessor and the memory or I/O.

11.

	BIT 0	BIT 1	BIT 2	BIT 3
Row 0	1	0	0	0
Row 1	0	0	0	0
Row 2	0	0	1	0
Row 3	0	0	0	0

13. 512 rows $\times$ 128 8-bit columns

15. A SRAM stores bits in flip-flops indefinitely as long as power is applied. A DRAM stores bits in capacitors that must be refreshed periodically to retain the data.

17. See Table P–13.

TABLE P–13

INPUTS		OUTPUTS			
A_1	A_0	O_3	O_2	O_1	O_0
0	0	0	1	0	1
0	1	1	0	0	1
1	0	1	1	1	0
1	1	0	0	1	0

19. See Figure P–73.

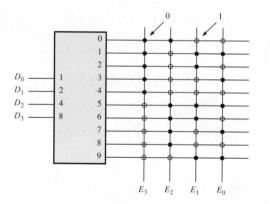

FIGURE P–73

21. Blown links: 1–17, 19–23, 25–31, 34, 37, 38, 40–47, 53, 55, 58, 61, 62, 63, 65, 67, 69

23. Use eight 16k $\times$ 4 DRAMs with sixteen address lines. Two of the address lines are decoded to enable the selected memory chips. Four data lines go to each chip.

25. 8 bits, 64k words; 4 bits, 256k words

27. lowest address: $FC0_{16}$

highest address: FFF_{16}

29. A hard disk is formatted into tracks and sectors. Each track is divided into a number of sectors with each sector of a track having a physical address. Hard disks typically have from a few hundred to a few thousand tracks.

31. Magnetic tape has a longer access time than disk because data must be accessed sequentially rather than randomly.

33. Checksum content is in error.

35. **(a)** ROM 2   **(b)** ROM 1   **(c)** All ROMs

Chapter 11

1. A coaxial cable consists of an outer jacket, metallic shield, dielectric, and center conductor.

3. Advantages of fiber optics are faster data rates, higher signal capacity (more signals at a time), transmission over longer distances, and not susceptible to EMI.

5. In multimode, the light entering the fiber will tend to propagate through the core in multiple rays (modes), basically due to varying angles as each light ray moves along.

7. See Figure P–74.

9. Visible light is in the frequency range 4×10^{14} Hz to 7.5×10^{14} Hz. Figure 11–10 shows the range in units of wavelength (nm).

11. 8 Mbps

13. 5 Mbaud

15. See Figure P–75.

17. 001110011

19. 11010111110001000001

21. See Figure P–76.

FIGURE P–74

FIGURE P–75

FIGURE P–76

23. See Figure P–77.

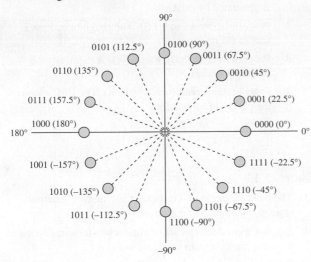

FIGURE P–77

25. The sawtooth intersects the sinusoidal modulating signal twice during each cycle. The sawtooth is either increasing above the sine wave or decreasing below the sine wave. When the sawtooth is increasing above the sine wave, a low level is generated, and when it is decreasing below the sine wave, a high level is generated. The resulting output is a series of pulses with widths proportional to the amplitude of the sine wave.

27. Data rate = 300 bps

29. Four bits

31. Bit interleaved: A single data bit from a source is transmitted on the channel, followed by a data bit from another source, and so on.

Byte interleaved: A byte of data from a source is transmitted on the channel, followed by a byte from another source, and so on.

33. In FDM, band-pass filters are used on the receiving end to separate the transmitted signals.

35. $A = 0.0209$; $V_{rec} = 2.94$ V

37. $A = 57.4$ dB

39. S/N $= -0.828$

Chapter 12

1. 0.25 V

3. 12 cycles

5. See Figure P–78.

7. 11, 11, 11, 11, 01, 11, 11, 11, 11

9. See Figure P–79.

FIGURE P–78

(a)

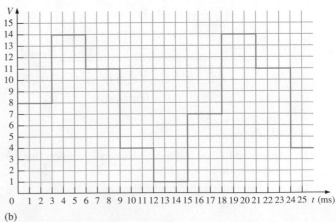

(b)

FIGURE P–79

11. 200

13. −21.4

15. 001, 010, 011, 101, 110, 111, 111, 111, 111, 110, 101, 101, 110, 110, 110, 101, 100, 011, 010, 001

17. 11, 11, 11

19. See Figure P–80.

(b) Output if 0111 code is missing

FIGURE P–80

21. See Figure P–81.

Output

FIGURE P–81

23. (a) 14.3%   (b) 0.098%   (c) 0.00038%

25. See Figure P–82.

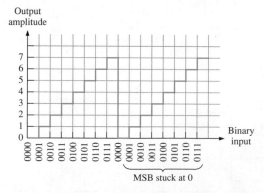

Output amplitude

Binary input

MSB stuck at 0

FIGURE P–82

27. An analog-to-digital converter converts an analog signal to a digital code.

29. A digital-to-analog converter changes a digital code to the corresponding analog signal.

31. $\dfrac{400 \text{ Mbps}}{32 \text{ bits/instruction}} = 12.5$ million instructions/s

33. **1.** Program address generate (PG). The program address is generated by the CPU.

    2. Program address send (PS). The program address is sent to the memory.

    3. Program access ready wait (PW). A memory read operation occurs.

    4. Program fetch packet receive (PR). The CPU receives the packet of instructions.

Chapter 13

1. The basic elements of a computer are the CPU, memory/storage, input/output, and buses.

3. A bus is a conductor or set of conductors for transferring data that meets certain specifications.

5. Tristate and open-collector outputs

7. (a) Nine loads   (b) Two loads

9. The wait state holds the state of the bus signals for one processor clock to allow the processor to complete an access operation.

11. The microprocessor controls system hardware, provides hardware support for the operating system, and executes application programs.

13. During fetch, an instruction is read from memory and decoded. During execute, the processor carries out the sequence of operations called for by the instruction.

15. (1) Address placed on address bus; (2) Data (instruction) placed on data bus and stored in data register; (3) Instruction decoded; (4) Instruction carried out.

17. *First fetch/execute cycle:* (1) Address of indexed op-code placed on address bus; (2) Indexed op-code placed on data bus and stored in data register; (3) Indexed instruction decoded; (4) Address of operand fetched.

Second fetch/execute cycle: (5) Offset address selected; (6) Offset address placed on data bus and stored in data register; (7) Offset address added to contents of index register to produce address of operand.

Third fetch/execute cycle: (8) Address of operand transferred to address register; (9) Address of operand placed on address bus; (10) Operand placed on data bus and stored in data register; (11) Operand sent to accumulator.

19. The interrupt vector table is used in autovectored interrupts to obtain the starting address for an interrupt service routine (ISR).

21. The sequence of events during a bus request operation is as follows.

    (1) The bus master requesting control of the system buses submits a request by asserting the processor's bus request (BR) line.

    (2) The processor tristates the system buses and signals that it has released control of the buses by asserting the bus grant (BG) line.

    (3) The requesting bus master uses the system address, data, and control lines to transfer data between system devices.

(4) After completing the data transfers, the requesting bus master tristates the system buses and signals the end of the bus request operation by asserting the bus grant acknowledge (BGACK) line.

23. The first group consists of application software, which includes word processors, spreadsheets, computer games, and other programs written to accomplish some specific task. The second group consists of system software, a major portion of which is the operating system. The operating system manages the system hardware, supervises the running of applications software, provides a standard operating environment for programs in which they can run, and interacts with the computer hardware.

25. Two ways in which computers execute more than one process are multitasking and multiprocessing.

27. MMUs handle memory accessing including memory protection, wait state generation, address translation for virtual memory, and cache control.

29. One possible flow chart is shown in Figure P–83.

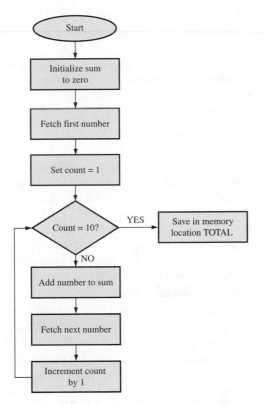

FIGURE P–83

31. Move contents of bx register into ax register.

33. A microcontroller is a device that combines a microprocessor with common peripheral units.

35. (a) GPIO: General-purpose input/output can be configured for serial data or parallel data input and output.

(b)(c) ADC and DAC: Analog-to-digital converter and digital-to-analog converter allow the digital system to interact with the analog world.

(d) PWM: Pulse width modulation varies pulse width proportional to an analog signal and is commonly used for motor control.

(e) Quadrature encoder is used to determine the speed, direction, and position of a moving object such as a mouse or stepper motor.

(f) Timer provides for frequency and timing requirements in a microcontroller.

(g) Communications controller allows microcontrollers to communicate with other devices using specified protocols.

37. The microcontroller includes certain peripherals that are external to a microprocessor.

Chapter 14

1. The physical characteristics: Number of conductors, length, configuration (serial or parallel), type of connector, number of connector pins, pin configuration.

Electrical/performance characteristics: Signal format, signal voltage, clock frequency, transfer speed, bandwidth, data frame format, handshaking protocol, error detection, impedances.

3. The AGP (accelerated graphics port) bus is an interface for a graphic card. The graphics controller can directly access the main memory through the AGP.

5. Bus width is the number of bits that a bus can transmit at one time; bus bandwidth, expressed in MBps, is the number of bytes per clock cycle times the number of clock cycles per second.

7. Simple handshake protocol: A requesting device sends a request for data to a responding device. The responding device acknowledges the request and sends the requested data.

9. See Figure P–84.

FIGURE P–84

11. See Figure P–85.

13. The PCI bus is an internal synchronous bus for interconnecting chips, expansion boards, and processor/memory subsystems. The original PCI bus has a width of 32 bits and a frequency of 33 MHz. Another version has a width of 64 bits and a frequency of 66 MHz. Still later versions enable 64-bit data transfers using up to a 133 MHz clock to enable bandwidths of up to 1066 MBps.

The PCI-X bus is a high-performance enhancement of the PCI and is backward compatible with PCI, although it is a faster bus and has some additional features. It is a 64-bit bus. PIC-X runs at a frequency of 133 MHz and the PCI-X 2.0 revision supports frequencies of 266 MHz and 533 MHz. Some of the additional features increase system reliability by minimizing errors at high transfer rates.

FIGURE P–85

15. × 2 indicates a two-lane bus.

17. BSY: Busy   MSG: Message

     SEL: Select   REQ: Request

     RST: Reset   ACK: Acknowledge a request

     C/D: Control/Data   ATN: Attention

19. A controller is sending data to two listeners. The first two bytes of data (3F and 41) go to the listener with address 001A. The second two bytes go to the listener with address 001B. The handshake signals (DAV, NRFD, and NDAC) indicate that the data transfer is successful. See Figure P–86.

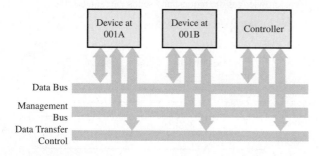

FIGURE P–86

21. Four types of USB packets: token, data, handshake, and start-of-frame.

23. NRZ

25. 82 feet

27. **1.** MOSI (master out slave in) is initiated by the master and received by the slave.

     2. MISO (master in slave out) is initiated by the slave and received by the master.

3. SCLK (serial clock) is generated by the master for synchronizing data transfers.

4. SS (slave select) is generated by the master to select an individual slave.

29. I^2C is an internal serial bus primarily for connecting ICs on a PC board.

31. Other possible units on an automotive CAN system include wiper control unit, parking control unit, entertainment system unit, tire pressure monitor, seat position unit, heads-up display unit.

33. **(a)** bus   **(b)** ring   **(c)** star

     (d) tree   **(e)** mesh

35. The physical structure or the way in which devices are connected to the network with actual cables is called physical topology. The way in which data are moved through the network is called logical topology and is related to network protocol.

37. Ethernet local area network (LAN) technology or protocol uses a process called CSMA/CD (carrier sense multiple access/collision detection) to access the network.

39. CRC (cyclic redundancy check)

41. The Ethernet physical layer describes interconnection methods, cables and connectors, data encoding (Manchester), and electrical properties.

43. TCP/IP stands for Transmission Control Protocol/Internet Protocol and is the communication protocol for the Internet. There are two protocol suites in TCP/IP. TCP provides for communication between applications; IP handles communication between computers and is responsible for sending data packets to the right destination across a network.

45. Client/server networking is an application architecture that handles tasks between service providers called *servers* and service requesters called *clients*.

GLOSSARY

acceptor A receiving device on a bus.

access time The time from the application of a valid memory address to the appearance of valid output data.

addend In addition, the number that is added to another number called the augend.

adder A logic circuit used to add two binary numbers.

address The location of a given storage cell or group of cells in a memory; a unique memory location containing one byte.

address bus A one-way group of conductors from the microprocessor to a memory, or other external device, on which the address code is sent.

adjacency Characteristic of cells in a Karnaugh map in which there is a single-variable change from one cell to another cell next to it on any of its four sides.

AGP Accelerated graphic port.

aliasing The effect created when a signal is sampled at less than twice the signal frequency. Aliasing creates unwanted frequencies that interfere with the signal frequency.

alphanumeric Consisting of numerals, letters, and other characters.

ALU Arithmetic logic unit; the key processing element of a microprocessor that performs arithmetic and logic operations.

amplitude In a pulse waveform, the height or maximum value of the pulse as measured from its low level.

analog Being continuous or having continuous values, as opposed to having a set of discrete values.

analog system A system that processes data in analog form only.

analog-to-digital (A/D) conversion The process of converting an analog signal to digital form.

analog-to-digital converter (ADC) A device used to convert an analog signal to a sequence of digital codes.

AND A basic logic operation in which a true (HIGH) output occurs only when all the input conditions are true (HIGH).

AND array An array of AND gates consisting of a matrix of programmable interconnections.

AND gate A logic gate that produces a HIGH output only when all of the inputs are HIGH.

ANSI American National Standards Institute.

antifuse A type of PLD nonvolatile programmable link that can be left open or can be shorted once as directed by the program.

architecture The VHDL unit that describes the internal operation of a logic function; the internal functional arrangement of the elements that give a device its particular operating characteristics.

array In a PLD, a matrix formed by rows of product-term lines and columns of input lines with a programmable cell at each junction. In VHDL, an array is an ordered set of individual items called elements with a single identifier name.

ASCII American Standard Code for Information Interchange; the most widely used alphanumeric code.

ASK Amplitude shift keying; a form of modulation in which a digital signal modulates the amplitude of a higher frequency sine wave.

assembler A program that converts English-like mnemonics into machine code.

assembly language A programming language that uses English-like words and has a one-to-one correspondence to machine language.

associative law In addition (ORing) and multiplication (ANDing) of three or more variables, the order in which the variables are grouped makes no difference.

astable Having no stable state. An astable multivibrator oscillates between two quasi-stable states.

asynchronous Having no fixed time relationship; a condition in which signals or systems are not aligned or synchronized in terms of timed events.

asynchronous counter A type of counter in which each stage is clocked from the output of the preceding stage.

attenuation The loss of energy as a signal propagates through a medium.

augend In addition, the number to which the addend is added.

backside bus A bus that connects the CPU to the cache to provide faster access to RAM.

bank A section of memory within a single memory array (chip).

base One of the three regions in a bipolar junction transistor.

base address The beginning address of a segment of memory.

baud The number of symbols per second in a data transmission.

BCD Binary coded decimal; a digital code in which each of the decimal digits, 0 through 9, is represented by a group of four bits.

BEDO DRAM Burst extended data output dynamic random-access memory.

bed-of-nails A method for the automated testing of printed circuit boards in which the board is mounted on a fixture that resembles a bed of nails that makes contact with test points.

BiCMOS A family of logic circuits that combines CMOS and bipolar logic.

bidirectional Having two directions. In a bidirectional shift register, the stored data can be shifted right or left.

binary Having two values or states; describes a number system that has a base of two and utilizes 1 and 0 as its digits.

BIOS Basic input/output system; a set of programs in ROM that interfaces the I/O devices in a computer system.

bipolar A class of integrated logic circuits implemented with bipolar transistors; also known as TTL.

bistable Having two stable states. Flip-flops and latches are bistable multivibrators.

bit A binary digit, which can be either a 1 or 0.

bit rate The number of bits per second in a data transmission.

bitstream A series of bits describing a final design that is sent to the target device during programming.

bit time The interval of time occupied by a single bit in a sequence of bits; the period of the clock.

BIU Bus interface unit; the portion of the CPU that interfaces with the system buses and fetches instructions, reads operands, and writes results.

BJT Bipolar junction transistor; a semiconductor device used for switching or amplification. A BJT has two junctions, the base-emitter junction and the base-collector junction.

Boolean addition In Boolean algebra, the OR operation.

Boolean algebra The mathematics of logic circuits.

Boolean expression An expression of variables and operators used to express the operation of a logic circuit.

Boolean multiplication In Boolean algebra, the AND operation.

boundary scan A method for internally testing a PLD based on the JTAG standard (IEEE Std. 1149.1).

break point A flag placed within a program source code to stop a program for investigation.

buffer A circuit that prevents loading of an input or output.

bus One or more interconnections that interface one or more devices based on a standardized specification.

bus arbitration The process that prevents two sources from using a bus at the same time.

bus bandwidth The number of bytes per clock cycle times the number of clock cycles per second.

bus bridge A device that interfaces two buses, usually a faster bus with a slower bus.

bus contention An adverse condition that could occur if two or more devices try to communicate at the same time on a bus.

bus frequency The clock frequency at which the bus can operate.

bus master Any device that can control and manage the system buses in a computer system.

bus network topology A type of physical network layout in which all devices are connected to a single common bus.

bus protocol A set of rules that allow two or more devices to communicate through a bus.

bus transfer speed The number of bytes per clock cycle.

bus width The number of bits that a bus can transmit at one time.

byte A group of eight bits.

cache memory A relatively small, high-speed memory that stores the most recently used instructions or data from the larger but slower main memory.

caching The process of copying frequently accessed program instructions from main memory into faster memory to increase processing speed.

CAN Controller area network; a bus standard used in automotive and other applications.

capacity The total number of data units (bits, nibbles, bytes, words) that a memory can store.

carry The digit generated when the sum of two binary digits exceeds 1.

carry generation The process of producing an output carry in a full-adder when both input bits are 1s.

carry propagation The process of rippling an input carry to become the output carry in a full-adder when either or both of the input bits are 1s and the input carry is a 1.

cascade To connect "end-to-end" as when several counters are connected from the terminal count output of one counter to the enable input of the next counter.

cascading Connecting two or more similar devices in a manner that expands the capability of one device.

CCD Charge-coupled device; a type of semiconductor memory that stores data in the form of charge packets and is serially accessed.

CD-R CD-Recordable; an optical disk storage device on which data can be stored once.

CD-ROM An optical disk storage device on which data are prestored and can only be read.

CD-RW CD-Rewritable; an optical disk storage on which data can be written and overwritten many times.

cell A single storage element in a memory; a fused cross point of a row and column in a PLD.

character A symbol, letter, or numeral.

circuit An arrangement of electrical and/or electronic components interconnected in such a way as to perform a specified function.

CLB Configurable logic block; a unit of logic in an FPGA that is made up of multiple smaller logic modules and a local programmable interconntect that is used to connect logic modules within the CLB.

clear An input used to reset a flip-flop (make the Q output 0); to place a register or counter in the state in which it contains all 0s.

client/server A networking application architecture that handles tasks between service providers called *servers* and service requesters called *clients*.

clock The basic timing signal in a digital system; a periodic waveform used to synchronize operation.

CMOS Complementary metal oxide semiconductor; a class of integrated logic circuits that is implemented with a type of field-effect transistor.

coaxial cable A type of data transmission media in which a shielded conductor is used to minimize EMI.

code A set of bits arranged in a unique pattern and used to represent such information as numbers, letters, and other symbols; in VHDL, program statements.

codec A combined coder and decoder.

collector One of the three regions in a bipolar transistor.

combinational logic A combination of logic gates interconnected to produce a specified Boolean function with no storage or memory capability; sometimes called *combinatorial logic*.

commutative law In addition (ORing) and multiplication (ANDing) of two variables, the order in which the variables are ORed or ANDed makes no difference.

comparator A digital circuit that compares the magnitudes of two quantities and produces an output indicating the relationship of the quantities.

compiler An application program in development software packages that controls the design flow process and translates source code into object code in a format that can be logically tested or downloaded to a target device.

complement The inverse or opposite of a number; in Boolean algebra, the inverse function, expressed with a bar over the variable. The complement of a 1 is a 0, and vice versa.

component A VHDL feature that can be used to predefine the logic function for multiple use throughout a program or programs.

contiguous Joined together.

control bus A set of conductive paths that connects the CPU to other parts of the computer to coordinate its operations and to communicate with external devices.

controller An instrument that can specify each of the other instruments on the bus as either a talker or a listener for the purpose of data transfer.

control unit The portion within the microprocessor that provides the timing and control signals for getting data into and out of the microprocessor and for synchronizing the execution of instructions.

counter A digital circuit capable of counting electronic events, such as pulses, by progressing through a sequence of binary states.

CPLD A complex programmable logic device that consists basically of multiple SPLD arrays with programmable interconnections.

CPU Central processing unit; the main part of a computer responsible for control and processing of data; the core of a DSP that processes the program instructions.

cross-assembler A program that translates an assembly language program for one type of microprocessor to an assembly language for another type of microprocessor.

current sinking The action of a circuit in which it accepts current into its output from a load.

current sourcing The action of a circuit in which it sends current out of its output and into a load.

cyclic redundancy check (CRC) A type of error detection code.

DAT Digital audio tape; a type of magnetic tape format.

data Information in numeric, alphabetic, or other form.

data bus A bidirectional set of conductive paths on which data or instruction codes are transferred into a microprocessor or on which the result of an operation is sent out from the microprocessor.

data selector A circuit that selects data from several inputs one at a time in a sequence and places them on the output; also called a multiplexer.

data sheet A document that specifies parameter values and operating conditions for an integrated circuit or other device.

DCE Data communications equipment.

DDR Double data rate.

decade Characterized by ten states or values.

decade counter A digital counter having ten states.

decimal Describes a number system with a base of ten.

decode A stage of the DSP pipeline operation in which instructions are assigned to functional units and are decoded.

decoder A digital circuit (device) that converts coded information into another (familiar) or noncoded form.

decrement To decrease the binary state of a counter by one.

delta modulation A method of analog-to-digital conversion using a 1-bit quantization process.

design flow The process or sequence of operations carried out to program a target device.

D flip-flop A type of bistable multivibrator in which the output assumes the state of the D input on the triggering edge of a clock pulse.

demultiplexer (demux) A circuit (digital device) that switches digital data from one input line to several output lines in a specified time sequence.

dependency notation A notational system for logic symbols that specifies input and output relationships, thus fully defining a given function; an integral part of ANSI/IEEE Std. 91-1984.

difference The result of a subtraction.

differential operation A bus operation that uses two wires for data (one for data and one for the complement of the data) and one wire for ground.

digit A symbol used to express a quantity.

digital Related to digits or discrete quantities; having a set of discrete values as opposed to continuous values.

digital system An arrangement of the individual logic functions connected to perform a specified operation or produce a defined output.

digital-to-analog (D/A) conversion The process of converting a sequence of digital codes to an analog form.

digital-to-analog converter (DAC) A device in which information in digital form is converted to analog form.

DIMM Dual in-line memory module.

diode A semiconductor device that conducts current in only one direction.

DIP Dual in-line package; a type of IC package whose leads must pass through holes to the other side of a PC board.

distortion The change in shape or other parameters of a signal waveform due to characteristics and imperfections in the transmission media.

distributive law The law that states that ORing several variables and then ANDing the result with a single variable is equivalent to ANDing the single variable with each of the several variables and then ORing the product.

dividend In a division operation, the quantity that is being divided.

divisor In a division operation, the quantity that is divided into the dividend.

DLT Digital linear tape; a type of magnetic tape format.

DMA Direct memory access; a method to directly interface a peripheral device to memory without using the CPU for control.

domain All of the variables in a Boolean expression.

"Don't care" A combination of input literals that cannot occur and can be used as a 1 or a 0 for simplification.

downloading A design flow process in which the logic design is transferred from software to hardware.

drain One of the terminals of a field-effect transistor.

DRAM Dynamic random-access memory; a type of semiconductor memory that uses capacitors as the storage elements and is a volatile, read/write memory.

DSP Digital signal processor; a special type of microprocessor that processes data in real time.

DSP core The central processing unit of a digital system processor.

DTE Data terminal equipment.

duty cycle The ratio of pulse width to period expressed as a percentage.

DVD-ROM Digital versatile disk-ROM; also known as digital video disk-ROM; a type of optical storage device on which data is prestored with a much higher capacity than a CD-ROM.

dynamic memory A type of semiconductor memory having capacitive storage cells that lose stored data over a period of time and, therefore, must be refreshed.

ECL Emitter-coupled logic; a class of integrated logic circuits that are implemented with nonsaturating bipolar junction transistors.

E²CMOS Electrically erasable CMOS (EECMOS); the circuit technology used for the reprogrammable cells in a PLD.

edge-triggered flip-flop A type of flip-flop in which the data are entered and appear on the output on the same clock edge.

EDIF Electronic design interchange format; a standard form of netlist.

EDO DRAM Extended data output dynamic random-access memory.

EEPROM Electrically erasable programmable read-only memory; a type of nonvolatile PLD reprogrammable link based on electrically-erasable programmable read-only memory cells and can be turned on or off repeatedly by programming.

EIA-232 Also known as RS-232.

8 mm A type of magnetic tape format.

electromagnetic waves Related to the electromagnetic spectrum, which includes radio waves, microwaves, infrared, visible, ultraviolet, X-rays, and gamma rays.

embedded system Generally, a single-purpose system, such as a processor, built into a larger system for the purpose of controlling the system.

EMI Electromagnetic interference.

emitter One of the three regions in a bipolar junction transistor.

encoder A digital circuit (device) that converts information to a coded form.

entity The VHDL unit that describes the inputs and outputs of a logic function.

EPROM Erasable programmable read-only memory; A type of PLD nonvolatile programmable link based on electrically programmable read-only memory cells and can be turned either on or off once with programming.

error detection The process of detecting bit errors in a digital code.

Ethernet A widely used standard local area network (LAN) technology or protocol; also known as IEEE-802.3.

EU Execution unit; the portion of a CPU that executes instructions; it contains the arithmetic logic unit (ALU), the general registers, and the flags.

even parity The condition of having an even number of 1s in every group of bits.

exception Any software event that requires special handling by the processor.

exclusive-NOR (XNOR)gate A logic gate that produces a LOW only when the two inputs are at opposite levels.

exclusive-OR (XOR) A basic logic operation in which a HIGH occurs when the two inputs are at opposite levels.

exclusive-OR (XOR) gate A logic gate that produces a HIGH only when the two inputs are at opposite levels.

execute A CPU process in which an instruction is carried out; a stage of the DSP pipeline operation in which the decoded instructions are carried out.

exponent The part of a floating-point number that represents the number of places that the decimal point (or binary point) is to be moved.

fall time The time interval between the 90% point and the 10% point on the negative-going edge of a pulse.

fan-out The number of equivalent gate inputs of the same family series that a logic gate can drive.

FDM Frequency division multiplexing; a broadband technique in which the total bandwidth available to a system is divided into frequency sub-bands and information is sent in analog form.

feedback The output voltage or a portion of it that is connected back to the input of a circuit.

FET Field-effect transistor.

fetch A CPU process in which an instruction is obtained from the memory; a stage of the DSP pipeline operation in which an instruction is obtained from the program memory.

FIFO First in–first out memory.

FireWire A high-speed external serial bus standard developed by Apple Inc. and used in high-speed communications and real-time data transfer, also known as IEEE-1394.

firmware Small fixed programs and/or data structures that internally control various electronic devices; usually stored in ROM.

fixed-function logic A category of digital integrated circuits having functions that cannot be altered.

flag A bit that indicates the result of an arithmetic or logic operation or is used to alter an operation.

flash A type of PLD nonvolatile reprogrammable link technology based on a single transistor cell.

flash ADC A simultaneous analog-to-digital converter.

flash memory A nonvolatile read/write random-access semiconductor memory in which data is stored as charge on the floating gate of a certain FET.

flip-flop A basic storage circuit that can store only one bit at a time; a synchronous bistable device.

floating-point number A number representation based on scientific notation in which the number consists of an exponent and a mantissa.

flying probe A method for the automated testing of printed circuit boards, in which a probe or probes move from place to place to contact test points.

forward bias A voltage polarity condition that allows a semiconductor *pn* junction in a transistor or diode to conduct current.

FPGA Field-programmable gate array; a programmable logic device that uses the LUT as the basic logic elements and generally employs either antifuse or SRAM-based process technology.

FPM DRAM Fast page mode dynamic random-access memory.

frequency (f) The number of pulses in one second for a periodic waveform. The unit of frequency is the hertz.

frontside bus A system bus that connects the major components of a computer system.

FSK Frequency shift keying; a form of modulation in which a digital signal modulates the frequency of a higher frequency sine wave.

full-adder A digital circuit that adds two bits and an input carry to produce a sum and an output carry.

full-duplex A connection in which the data flows both ways simultaneously in the same channel.

functional simulation A software process that tests the logical or functional operation of a design.

fuse A type of PLD nonvolatile programmable link that can be left shorted or can be opened once as directed by the program; also called a fusible link.

GAL Generic array logic; a reprogrammable type of SPLD that is similar to a PAL except that it uses a reprogrammable process technology, such as EEPROM (E^2 CMOS), instead of fuses.

gate A logic circuit that performs a basic logic operation, such as AND or OR; one of the three terminals of a field-effect transistor.

gateway A network point that provides access to another network.

glitch A voltage or current spike of short duration, usually unintentionally produced and unwanted.

graphic (schematic) entry A method of entering a logic design into software by graphically creating a logic diagram (schematic) on a design screen.

GPIB General-purpose interface bus based on the IEEE 488 standard.

Gray code An unweighted digital code characterized by a single bit change between adjacent code numbers in a sequence.

half-adder A digital circuit that adds two bits and produces a sum and an output carry. It cannot handle input carries.

half-duplex A connection in which the data flows both ways but not at the same time in the same channel.

handshake A routine by which two devices initiate and complete a bus transfer.

handshaking The process of signal interchange by which two digital devices or systems jointly establish communication.

hard core A fixed portion of logic in an FPGA that is put in by the manufacturer to provide a specific function.

hard disk A magnetic disk storage device; typically, a stack of two or more rigid disks enclosed in a sealed housing.

hardware The circuitry and physical components of a computer system (as opposed to the directions called software).

HDL Hardware description language; a language used for describing a logic design using software.

hexadecimal Describes a number system with a base of 16.

high-level language A type of computer language closest to human language that is a level above assembly language.

high-Z The high-impedance state of a tristate circuit in which the output is effectively disconnected from the rest of the circuit.

hit A successful attempt to read or write a block of data in a given level of memory.

hit rate The percentage of memory accesses that find the requested data in the given level of memory.

hold time The time interval required for the control levels to remain on the inputs to a flip-flop after the triggering edge of the clock in order to reliably activate the device.

HPIB Hewlett-Packard interface bus; same as GPIB (general-purpose interface bus).

hub A common connection point containing multiple ports for devices in a computer or network.

hysteresis A characteristic of a threshold-triggered circuit, such as the Schmitt trigger, where the device turns on and off at different input levels.

IEEE Institute of Electrical and Electronics Engineers.

IEEE 488 bus Same as GPIB (general-purpose interface bus); a standard parallel bus used widely for test and measurement interfacing.

IEEE 1394 A serial bus for high-speed data transfer; also known as FireWire.

I^2L Integrated injection logic; an IC technology.

implementation The software process where the logic structures described by the netlist are mapped into the structure of the target device.

increment To increase the binary state of a counter by one.

input The signal or line going into a circuit; a signal that controls the operation of a circuit.

input/output (I/O) A terminal of a device that can be used as either an input or as an output.

instruction One step in a computer program; a unit of information that tells the CPU what to do.

instruction pairing The process of combining certain independent instructions so that they can be executed simultaneously by two separate execution units.

In-system programming (ISP) A method for programming SPLDs after they are installed on a printed circuit board and operating in a system.

integer A whole number.

integrated circuit (IC) A type of circuit in which all of the components are integrated on a single chip of semiconductive material of very small size.

intellectual property (IP) Designs owned by the manufacturer of programmable logic devices.

interfacing The process of making two or more electronic devices or systems operationally compatible with each other so that they function properly together.

interrupt Any hardware event that requires special handling by the processor; an event that causes the current process to be temporarily stopped while a service routine is run.

inversion The conversion of a HIGH level to a LOW level or vice versa; also called complementation.

inverter A NOT circuit; a circuit that changes a HIGH to a LOW or vice versa.

I/O port Input/output port; the interface between an internal bus and a peripheral.

IP Instruction pointer; a special register within the CPU that holds the offset address of the next instruction to be executed.

IP address Internet protocol address; an address assigned to each device in a network that uses the Internet protocol and serves to identify the network and the location of the device within the network.

I²C Inter-integrated circuit bus; an internal serial bus primarily for connecting ICs on a PC board.

ISA bus Industry standard architecture bus; an internal parallel bus standard.

J-K flip-flop A type of flip-flop that can operate in the SET, RESET, no-change, and toggle modes.

Johnson counter A type of register in which a specific prestored pattern of 1s and 0s is shifted through the stages, creating a unique sequence of bit patterns.

JTAG Joint test action group; the IEEE Std. 1149.1 standard interface for in-system programming.

junction The boundary between an *n* region and a *p* region in a BJT.

Karnaugh map An arrangement of cells representing the combinations of literals in a Boolean expression and used for a systematic simplification of the expression.

LAB Logic array block; an SPLD array in a CPLD.

LAN Local area network.

latch A bistable digital circuit used for storing a bit.

latency period The time it takes for the desired sector to spin under the head once the head is positioned over the desired track of a magnetic hard disk.

LCC Leadless ceramic chip; an SMT package that has metallic contacts molded into its body.

LCD Liquid crystal display.

leading edge The first transition of a pulse.

least significant bit (LSB) Generally, the right-most bit in a binary whole number or code.

LED Light-emitting diode.

LIFO Last in–first out memory, memory stack.

listener An instrument capable of receiving data on a GPIB (general-purpose interface bus) when it is addressed by the computer.

literal A constant value assigned to a variable or the complement of a variable.

load To enter data into a shift register.

loading The effect of the multiple inputs degrading the voltage or timing specifications of an output.

local bus The internal bus of a computer system which includes the system bus, the PCI bus, and the ISA bus, among others.

local interconnect A set of lines that allows interconnections among the eight logic elements in a logic array block without using the row and column interconnects.

logic In digital electronics, the decision-making capability of gate circuits, in which a HIGH represents a true statement and a LOW represents a false one.

logical network topology The description of how devices in a network interact in order to send and receive data based on protocols.

logic array block (LAB) A group of macrocells that can be interconnected with other LABs or to other I/Os using a programmable interconnect array; also called a function block.

logic element The smallest section of logic in an FPGA that typically contains an LUT, associated logic, and a flip-flop.

look-ahead carry A method of binary addition whereby carries from preceding adder stages are anticipated, thus eliminating carry propagation delays.

LSI Large-scale integration; a level of fixed-function IC complexity in which there are from more than 100 to 10,000 equivalent gates per chip.

LUT Look-up table; a type of memory that can be programmed to produce SOP functions.

machine code The basic binary instructions understood by the processor.

machine language Computer instructions written in binary code that are understood by a computer; the lowest level of programming language.

macrocell An SOP logic array with combinational and registered outputs; part of a PAL or GAL that generally consists of one OR gate and some associated output logic. Multiple interconnected macrocells form a CPLD.

magneto-optical disk A storage device that uses electromagnetism and a laser beam to read and write data.

magnitude The size or value of a quantity.

main memory Memory used by computer systems to store the bulk of programs and associated data.

MAN Metropolitan area network.

Manchester encoding A method of encoding called biphase in which a 1 is represented by a positive-going transition and a 0 is represented by a negative-going transition.

mantissa The magnitude of a floating-point number.

Mealy state machine A state machine in which the outputs depend on both the internal present state and on the inputs.

memory The portion of a computer or other system that stores binary data.

memory array An array of memory cells arranged in rows and columns.

memory hierarchy The arrangement of various memory elements to maximize speed and minimize cost.

memory latency The time required to access a memory.

mesh network topology A type of physical network layout in which each device connects to all other devices providing multiple routes for data.

MFLOPS Million floating-point operations per second.

microcontroller A semiconductor device that combines a microprocessor, memory, and various hardware peripherals on a single IC.

microprocessor A large-scale digital integrated circuit that can be programmed to perform arithmetic, logic, or other operations; the CPU of a computer.

minuend The number from which another number is subtracted.

MIPS Million instructions per second.

miss A failed attempt by the processor to read or write a block or data in a given level of memory.

MMACS Million multiply/accumulates per second.

MMU Memory management unit; a device responsible for handling accesses to memory requested by the CPU.

mnemonic An English-like instruction that is converted by an assembler into a machine code for use by a processor.

modem A modulator/demodulator for interfacing digital devices to analog transmission systems such as telephone lines.

module A Verilog code block that defines inputs, outputs, and logic structure or function.

modulus The number of unique states through which a counter will sequence.

monostable Having only one stable state. A monostable multivibrator, commonly called a one-shot, produces a single pulse in response to a triggering input.

monotonic The characteristic of a DAC defined by the absence of any incorrect step reversals; one type of digital-to-analog linearity.

Moore state machine A state machine in which the outputs depend only on the internal present state.

MOS Metal-oxide semiconductor; a type of transistor technology.

MOSFET Metal-oxide semiconductor field-effect transistor.

most significant bit (MSB) The left-most bit in a binary whole number or code.

MSI Medium-scale integration; a level of fixed-function IC complexity in which there are from 10 to 100 equivalent gates per chip.

multicore processor A microprocessor chip with more than one processor.

multimode The characteristic of an optical fiber in which the light is propagated in multiple rays.

multiplexer (mux) A circuit (digital device) that switches digital data from several input lines onto a single output line in a specified time sequence.

multiplicand The number that is being multiplied by another number.

multiplier The number that multiplies the multiplicand.

multiprocessing A data-processing technique that uses more than one processor.

multitasking A technique by which a processor runs multiple programs concurrently.

multithreading The process of executing different parts of a program, called threads, simultaneously.

multivibrator A class of digital circuits in which the output is connected back to the input (an arrangement called feedback) to produce either two stable states, one stable state, or no stable states, depending on the configuration.

NAND gate A logic circuit in which a LOW output occurs only if all the inputs are HIGH.

negative-AND An equivalent NOR gate operation in which the HIGH is the active input when all inputs are LOW.

negative-OR An equivalent NAND gate operation in which the HIGH is the active input when one or more of the inputs are LOW.

netlist A detailed listing of information necessary to describe a circuit, such as types of elements, inputs, and outputs, and all interconnections.

network A set of computers and associated devices that are interconnected in a specified way in order to communicate to share information and resources.

network topology The physical and logical arrangement of devices in a network.

nibble A group of four bits.

NMOS An n-channel metal-oxide semiconductor.

node A common connection point in a circuit in which a gate output is connected to one or more gate inputs.

noise The unwanted electrical disturbance caused by both natural and man-made sources.

noise immunity The ability of a circuit to reject unwanted signals.

noise margin The amount by which the actual signal level exceeds the minimum acceptable level for an error-free transmission.

nonvolatile A term that describes a memory that can retain stored data when the power is removed.

NOR gate A logic gate in which the output is LOW when any or all of the inputs are HIGH.

Northbridge A bridge that serves as an interface generally to the AGP and RAM.

NOT A basic logic operation that performs inversions.

NRZ Nonreturn to zero; a type of data format in which the signal level does not return to zero during a bit time after a high-level (1) data bit occurs.

numeric Related to numbers.

Nyquist frequency The highest signal frequency that can be sampled at a specified sampling frequency; a frequency equal to or less than half the sampling frequency.

object program A machine language translation of a high-level source program.

octal Describes a number system with a base of eight.

odd parity The condition of having an odd number of 1s in every group of bits.

offset address The distance in number of bytes of a physical address from the base address.

OLMC Output logic macrocell; the part of a GAL that can be programmed for either combinational or registered outputs; a block of logic in a GAL that contains a fixed OR gate and other logic for handling inputs and/or outputs.

one-shot A monostable multivibrator.

op code Operation code; the code representing a particular microprocessor instruction; a mnemonic.

open-collector A type of output in a logic circuit in which the collector of the output transistor is left disconnected from any internal circuitry and is available for external connection; normally used for driving higher-current or higher-voltage loads.

operand The object to be manipulated by the instruction.

operating system The software that controls the computer system and oversees the execution of application software.

operational amplifier (op-amp) A device with two differential inputs that has very high gain, very high input impedance, and very low output impedance.

optical fiber A type of data transmission media used for transmitting light signals.

optical jukebox A type of auxiliary storage for very large amounts of data.

OR A basic logic operation in which a true (HIGH) output occurs when one or more of the input conditions are true (HIGH).

OR gate A logic gate that produces a HIGH output when one or more inputs are HIGH.

oscillator An electronic circuit that is based on the principle of regenerative feedback and produces a repetitive output waveform; a signal source.

OSI Open systems interconnection; a model in which all protocol operations and specifications for communicating on a network are broken down into seven parts called layers.

OTP One-time programmable.

output The signal or line coming out of a circuit.

overflow The condition that occurs when the number of bits in a sum exceeds the number of bits in each of the numbers added.

packet A formatted block of digital data.

PAL Programmable array logic; a type of one-programmable SPLD that consists of a programmable array of AND gates that connects to a fixed array of OR gates.

PAM Pulse amplitude modulation; a method of modulation in which the height or amplitude of the pulses are varied according to the modulating analog signal, and each pulse represents a value of amplitude of the analog signal.

parallel In digital systems, data occurring simultaneously on several lines; the transfer or processing of several bits simultaneously.

parallel bus A bus that consists of multiple conductors and carries several data bits simultaneously, one on each conductor.

parallel data Data that is represented by pulses sent simultaneously over multiple channels.

parity In relation to binary codes, the condition of evenness or oddness of the number of 1s in a code group.

parity bit A bit attached to each group of information bits to make the total number of 1s odd or even for every group of bits.

PCI Peripheral component interconnect.

PCI bus An internal synchronous bus for interconnecting chips, expansion boards, and processor/memory subsystems.

PCI-Express Also designated as PCIe or PCI-E. This bus differs from the PCI and PCI-X buses in that it does not use a shared bus.

PCI-X A high-performance enhancement of the PCI bus that is backward compatible with PCI.

PCM Pulse code modulation; A method of modulation that involves sampling of an analog signal amplitude at regular intervals and converting the sampled values to a digital code.

period (T) The time required for a periodic waveform to repeat itself.

periodic Describes a waveform that repeats itself at a fixed interval.

peripheral A device or instrument that provides communication with a computer or provides auxiliary services or functions for the computer.

physical address The actual location of a data unit in memory.

PIC Programable interrupt controller; handles the interrupts on a priority basis.

pipeline As applied to memories, an implementation that allows a read or write operation to be initiated before the previous operation is completed; part of the DSP architecture that allows multiple instructions to be processed simultaneously.

pipelining A technique where the processor begins executing the next instruction before the previous instruction has been completed.

PLA Programmable logic array; an SPLD with programmable AND and OR arrays.

platform FPGA An FPGA that contains either or both hard core and soft core embedded processors and other functions.

PLCC Plastic leaded chip carrier; an SMT package whose leads are turned up under its body in a J-type shape.

PLD Programmable logic device; an integrated circuit that can be programmed with any specified logic function.

PMOS A p-channel metal-oxide semiconductor.

pointer The contents of a register (or registers) that contain an address.

polling The process of checking a series of peripheral devices to determine if any require service from the CPU.

port A physical interface on a computer through which data are passed to or from peripherals.

POS Product-of-sums; a form of Boolean expression that is basically the ANDing of ORed terms.

positive logic The system of representing a binary 1 with a HIGH and a binary 0 with a LOW.

power dissipation The product of the dc supply voltage and the dc supply current in an electronic circuit; the amount of power required by a circuit.

PPM Pulse position modulation; a method of modulation in which the position of each pulse relative to a reference or timing signal is varied proportional to the amplitude of the modulating signal waveform.

prefetching The process of executing instructions at the same time as other instructions are "fetched," eliminating idle time; also called pipelining.

preset An asynchronous input used to set a flip-flop (make the Q output 1).

priority encoder An encoder in which only the highest value input digit is encoded and any other active input is ignored.

probe An accessory used to connect a voltage to the input of an oscilloscope or other instrument.

processes Instances of a computer program that are being executed.

product The result of a multiplication.

product term The Boolean product of two or more literals equivalent to an AND operation.

program A sequential set of computer instructions designed to accomplish a given task(s).

programmable interconnect array (PIA) An array consisting of conductors that run throughout the CPLD chip and to which connections from the macrocells in each LAB can be made.

programmable logic A category of digital integrated circuits capable of being programmed to perform specified functions.

programmable logic device (PLD) A type of integrated circuit (IC) that starts as a "blank slate" and into which a logic design is programmed.

PROM Programmable read-only semiconductor memory; an SPLD with a fixed AND array and programmable OR array; used as a memory device and normally not as a logic circuit device.

propagation delay time The time interval between the occurrence of an input transition and the occurrence of the corresponding output transition in a logic circuit.

pseudo-operation An instruction to the assembler (as opposed to a processor).

PSK Phase shift keying; a form of modulation in which a digital signal modulates the phase of a higher frequency sine wave.

pull-up resistor A resistor with one end connected to the dc supply voltage used to keep a given point in a circuit HIGH when in the inactive state.

pulse A sudden change from one level to another, followed after a time, called the pulse width, by a sudden change back to the original level.

pulse width (t_{PW}) The time interval between the 50% points of the leading and trailing edges of the pulse; the duration of the pulse.

PWM Pulse width modulation; a method of modulation in which the width or duration of the pulses and duty cycle are varied according to the modulating analog signal, and each pulse width represents an amplitude value of the analog signal.

QAM Quadrature amplitude modulation; a form of modulation that uses a combination of PSK and amplitude modulation to send information.

QIC Quarter-inch cassette; a type of magnetic tape.

quantization The process whereby a binary code is assigned to each sampled value during analog-to-digital conversion.

queue A high-speed memory that stores instructions or data.

quotient The result of a division.

race A condition in a logic network in which the difference in propagation times through two or more signal paths in the network can produce an erroneous output.

RAM Random-access memory; a volatile read/write semiconductor memory.

rank A group of chips that make up a memory module that stores data in units such as words or bytes.

read The process of retrieving data from a memory.

real mode Operation of an Intel processor in a manner to emulate the 8086's 1 MB of memory.

recycle To undergo transition (as in a counter) from the final or terminal state back to the initial state.

refresh To renew the contents of a dynamic memory by recharging the capacitor storage cells.

register A digital circuit capable of storing and shifting binary information; typically used as a temporary storage device.

register array A set of temporary storage locations within the microprocessor for keeping data and addresses that need to be accessed quickly by the program.

registered A CPLD macrocell output configuration where the output comes from a flip-flop.

relocatable code A program that can be moved anywhere within the memory space without changing the basic code.

remainder The amount left over after a division.

RESET The state of a flip-flop or latch when the output is 0; the action of producing a RESET state.

resolution The number of bits used in an ADC.

reverse bias A voltage polarity condition that prevents a *pn* junction of a transistor or diode from conducting current.

ring counter A register in which a certain pattern of 1s and 0s is continuously recirculated.

ring network topology A type of physical network layout in which the devices are daisy-chained and each device communicates only with its two neighboring devices.

ripple carry A method of binary addition in which the output carry from each adder becomes the input carry of the next higher-order adder.

ripple counter An asynchronous counter.

rise time The time required for the positive-going edge of a pulse to go from 10% of its full value to 90% of its full value.

ROM Read-only semiconductor memory, accessed randomly; also referred to as mask-ROM.

router A device that routes a data packet to the correct destination based on the IP address.

RS-232 A bus standard, also known as EIA-232, used in industrial and telecommunication applications as well as scientific instrumentation, but largely replaced by USB in computer applications.

RS-422 A bus standard for differential data transmission.

RS-423 A bus standard for single-ended data transmission.

RS-485 A bus standard for differential data transmission.

RZ Return to zero; a type of data format in which the signal level goes to or remains at zero after each data bit.

sampling The process of taking a sufficient number of discrete values at points on a waveform that will define the shape of the waveform.

schematic (graphic) entry A method of placing a logic design into software using schematic symbols.

Schottky A specific type of transistor-transistor logic circuit technology.

SCSI Small computer system interface; an external parallel bus standard.

SDRAM Synchronous dynamic random-access memory.

seek time The time for the read/write head in a hard drive to position itself over the desired track for a read operation.

segment A 64k block of memory.

serial Having one element following another, as in a serial transfer of bits; occurring, as pulses, in sequence rather than simultaneously.

serial bus A bus that carries data bits sequentially one at a time on a single conductor.

serial data Data that is represented by pulses sent one at a time over a single channel.

SET The state of a flip-flop or latch when the output is 1; the action of producing a SET state.

set-up time The time interval required for the control levels to be on the inputs to a digital circuit, such as a flip-flop, prior to the triggering edge of clock pulse.

shared bus A bus, such as PCI, that is shared by multiple devices.

shared media topology A type of logical network topology where all devices can access the physical network at any time as long as no other devices are attempting access.

signal A type of VHDL object that holds data.

signal-to-noise ratio (SNR) A measure of the signal strength relative to background noise, usually expressed in decibels (dB).

signal tracing A troubleshooting technique in which waveforms are observed in a step-by-step manner beginning at the input and working toward the output or vice versa. At each point the observed waveform is compared with the correct signal for that point.

sign bit The left-most bit of a binary number that designates whether the number is positive (0) or negative (1).

simplex A connection in which data flows in only one direction from the sender (transmitter) to the receiver.

single-ended operation A bus operation that uses one wire for data and one wire for ground.

single mode The characteristic of an optical fiber in which the light tends to propagate in a single beam or ray.

SMT Surface-mount technology; an IC package technique in which the packages are smaller than DIPs and are mounted on the printed surface of the PC board.

soft core A portion of logic in an FPGA; similar to hard core except it has some programmable features.

software Computer programs; programs that instruct a computer what to do in order to carry out a given set of tasks.

software interrupt An instruction that invokes an interrupt service routine.

SOIC Small-outline integrated circuit; an SMT package that resembles a small DIP but has its leads bent out in a "gull-wing" shape.

SOP Sum-of-products; a form of Boolean expression that is basically the ORing of ANDed terms.

source A sending device.

source program A program written in either assembly or high-level language.

Southbridge A bridge that handles all I/O functions.

speed-power product A performance parameter that is the product of the propagation delay time and the power dissipation in a digital circuit.

SPI Serial-to-peripheral interface bus; a synchronous serial communications bus that uses four wires for communication between a "master" device and a "slave" device.

SPLD Simple programmable logic device; an array of AND gates and OR gates that can be programmed to achieve specified logic functions. Four types are PROM, PLA, PAL, and GAL.

SRAM Static random-access memory; a type of PLD volatile reprogrammable link based on static random-access memory cells and can be turned on or off repeatedly with programming.

SSI Small-scale integration; a level of fixed-function IC complexity in which there are up to 10 equivalent gates per chip.

SSOP Shrink small-outline package.

stage One storage element (flip-flop) in a register.

star network topology A type of physical network layout containing a central connection point or hub, and all devices are connected through the hub with individual cables.

state diagram A graphic depiction of a sequence of states or values.

state machine A logic system exhibiting a sequence of states conditioned by internal logic and external inputs; any sequential circuit exhibiting a specified sequence of states.

static memory A volatile semiconductor memory that uses flip-flops as the storage cells and is capable of retaining data without refreshing.

storage The capability of a digital device to retain bits; the process of retaining digital data for later use.

STP Shielded twisted pair; a type of transmission media.

string A contiguous sequence of bytes or words.

strobing A process of using a pulse to sample the occurrence of an event at a specified time in relation to the event.

subroutine A series of instructions that can be assembled together and used repeatedly by a program but programmed only once.

subtracter A logic circuit used to subtract two binary numbers.

subtrahend The number that is being subtracted from the minuend.

sum The result when two or more numbers are added together.

sum term The Boolean sum of two or more literals equivalent to an OR operation.

synchronous A condition that describes signals or systems that are aligned or synchronized with each other in terms of timed events, two or more systems that have the same timing signal.

synchronous counter A type of counter in which each stage is clocked by the same pulse.

synthesis The software process where the design is translated into a netlist.

system bus The interconnecting paths in a computer system including the address bus, data bus and control bus.

talker A device capable of sending data over the GPIB.

tape library A type of auxiliary storage for very large amounts of data.

target device A PLD mounted on a programming fixture or development board into which a software logic design is to be downloaded; the programmable logic device that is being programmed.

TCP/IP Transmission Control Protocol/Internet Protocol; the communication protocol for the Internet.

TDM Time division multiplexing; a technique in which data from several sources are interleaved on a time basis and sent on a single communication channel or data link.

terminal count The final state in a counter's sequence.

terminated The condition of a bus connection in which the signal is prevented from reflecting back when it reaches the end of the bus or transmission line. The condition when a resistor is connected from the bus to ground.

text entry A method of entering a logic design into software using a hardware description language (HDL).

throughput The average speed with which a program is executed.

timer A circuit that can be used as a one-shot or as an oscillator; a circuit that produces a fixed time interval output.

timing diagram A graph of digital waveforms showing the proper time relationship of two or more waveforms and how each waveform changes in relation to the others.

timing simulation A software process that uses information on propagation delays and netlist data to test both the logical operation and the worst-case timing of a design.

toggle The action of a flip-flop when it changes state on each clock pulse.

token-based topology A type of logical network topology where a code, called a token, travels around the network allowing a device to send data only if it attached the data packet to a token.

totem-pole A type of output in TTL circuits.

trailing edge The second transition of a pulse.

transistor A semiconductor device exhibiting current and/or voltage gain. When used as a switching device, it approximates an open or closed switch.

tree topology A type of physical network layout which is a hybrid combining bus topology and star topology.

trigger A pulse used to initiate a change in the state of a logic circuit.

tristate A type of output in logic circuits that exhibits three states: HIGH, LOW, and high-Z; also known as 3-state.

tristate buffer A circuit used to interface one device to another to prevent loading.

troubleshooting The technique of systematically identifying, isolating, and correcting a fault in a circuit or system.

truth table A table showing the inputs and corresponding output level of a logic circuit.

TTL Transistor-transistor logic; a class of integrated logic circuit that uses bipolar junction transistors. Also called *bipolar*.

ULSI Ultra large-scale integration; a level of IC complexity in which there are more than 100,000 equivalent gates per chip.

unit load A measure of fan-out. One gate input represents a unit load to the output of a gate within the same IC family.

universal gate Either a NAND gate or a NOR gate. The term *universal* refers to the property of a gate that permits any logic function to be implemented by that gate or by a combination of gates of that kind.

up/down counter A counter that can progress in either direction through a certain sequence.

USB Universal serial bus; a widely used standard serial bus for connecting peripherals to a computer.

UTP Unshielded twisted pair; a type of transmission media.

UV EPROM Ultraviolet erasable programmable ROM.

variable symbol used to represent an action, a condition, or data that can have a value of 1 or 0, usually designated by an italic letter or word.

Verilog A standard hardware description language that uses a module structure to describe a function.

VHDL A standard hardware description language that describes a function with an entity/architecture structure; IEEE Std. 1076–1993.

VLSI Very large-scale integration; a level of IC complexity in which there are from more than 10,000 to 100,000 equivalent gates per chip.

volatile The characteristic of a programmable logic device that loses programmed data when power is turned off.

wait state A system bus delay equal to one processor clock cycle. Wait states are used to ensure that the system bus timing satisfies the address, data, and control timing specifications of a system.

WAN Wide area network.

weight The value of a digit in a number based on its position in the number.

word A group of bits or bytes that acts as a single entity that can be stored in one memory location; two bytes.

word capacity The number of words that a memory can store.

word length The number of bits in a word.

WORM Write once-read many; a type of optical storage device.

write The process of storing data in a memory.

zero suppression The process of blanking out leading or trailing zeros in a digital display.

INDEX